INTERMEDIATE
Algebra

FIFTH ▶ EDITION

Charles P. McKeague
Cuesta College

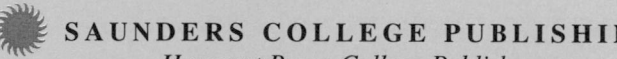

SAUNDERS COLLEGE PUBLISHING
Harcourt Brace College Publishers

Fort Worth   Philadelphia   San Diego   New York

Orlando   Austin   San Antonio   Toronto

Montreal   London   Sydney   Tokyo

Text Typeface: Times Roman
Compositor: Progressive Information Technologies
Acquisitions Editor: Deirdre Lynch
Developmental Editor: Marc Sherman
Managing Editor: Carol Field
Project Editor: Sarah Fitz-Hugh
Copy Editor: Martha Brown
Manager of Art and Design: Carol Bleistine
Art Directors: Christine Schueler, Carol Bleistine
Art and Design Coordinator: Sue Kinney
Text Designer: Gene Harris
Cover Designer: Louis Fuiano/Fuiano Art & Design
Text Artwork: Techsetters Inc./Rolin Graphics
Director of EDP: Tim Frelick
Production Manager: Charlene Squibb

Cover Credit: Shinobu HIRAI/Photonica

Printed in the United States of America

INTERMEDIATE ALGEBRA, Fifth Edition

ISBN: 0-03-097357-0 Student's Edition
       0-03-010859-4 Instructor's Edition

Library of Congress Catalog Card Number: 94-69874

78901234 032 10 9876543

CONTENTS

PREFACE TO THE INSTRUCTOR

This fifth edition of *Intermediate Algebra* retains the same basic format and style as the previous edition. The book is intended for a lecture-format class. Each section of the book can be discussed in a forty-five- to fifty-minute class session.

Features of the Book

Chapter Openings   New to this edition, each chapter begins with an opening that includes the following three elements:

1. *Introduction*   Each chapter opens with an introduction in which a real-world application, historical example, or a link between topics is used to stimulate interest in the chapter. Whenever possible, these introductions are expanded on later in the chapter and then carried through to topics found further on in the book.
2. *Overview*   A general overview of the chapter follows the chapter introduction. The overview lists the important topics that will be covered in the chapter, along with their connection to one another and to topics covered previously in the text. Most of the overviews end with a list of topics from previous chapters that students need to know in order to be successful in the current chapter.
3. *Study Skills*   Found in the first six chapter openings are a list of study skills intended to help students become organized and efficient with their time. The study skills point students in the direction of success. They are intended to benefit students in this course and throughout their college careers. These skills are more detailed than the general study skills listed in the Preface to the Student.

Organization of the Problem Sets   Six main ideas are incorporated into the problem sets.

1. *Drill*   There are enough problems in each set to ensure student proficiency in the material.

2. *Progressive difficulty*   The problems increase in difficulty as the problem set progresses.

3. *Odd–even similarities*   Each pair of consecutive problems is similar. Since the answers to the odd-numbered problems are listed in the back of the book, the similarity of the odd–even pairs of problems allows students to check their work on an odd-numbered problem and then to try the similar even-numbered problem.

4. *Application Problems*   Students are always curious about how the algebra they are learning can be applied, but at the same time many of them are apprehensive about attempting application problems. I have found that they are more likely to put some time and effort into trying application problems if they do not have to work an overwhelming number of them at one time, and if they work on them every day. For these reasons, I have placed a few application problems in almost every problem set.

5. *Review Problems*   As was the case in the fourth edition, each problem set, beginning with Chapter 2, contains a few review problems. Where appropriate, the review problems cover material that will be needed in the next section. Otherwise they cover material from the previous chapter. That is, the review problems in Chapter 5 reinforce the important points covered in Chapter 4. Likewise, the review problems in Chapter 6 review the important material from Chapter 5. If you give tests on two chapters at a time, you will find the reviews to be a time-saving feature. Your students will review one chapter as they study the next chapter.

6. *One Step Further problems*   Most of the problem sets end with a few problems under the heading ''One Step Further.'' These problems are more challenging than those in the problem set, or they are problems that extend some of the topics covered in the section.

Blueprint for Problem Solving   New to this edition, the Blueprint for Problem Solving is a detailed outline of the steps needed to successfully attempt application problems. Intended as a guide to problem solving in general, the Blueprint overlays the solution process to all the application problems in the first few chapters of the book. As students become more familiar with problem solving, the steps in the Blueprint are streamlined.

Using Technology   Scattered throughout the book is optional material that shows how graphing calculators, spreadsheet programs, and computer graphing programs may be used to enhance the topics being covered. The more common uses of technology are shown, with the emphasis on graphing calculators. The material on technology can be used effectively as a starting point for further use of technology, as material to be demonstrated by an instructor using an overhead graphing calculator or computer projection screen, or it can be treated as enrichment material for students who are interested in pursuing the topics further.

Research Projects   Scattered throughout this edition are problems for students to research and then report on. Although they appear at the end of a number of the problem sets, they are not intended to be part of the student's daily assignment. In my classes, I use these projects for extra credit. In most cases, I require students to type their reports, just as they would type an essay in their English classes. Do not be concerned if you are not familiar with the topics shown in the research projects. The idea behind these projects is to have your students do the research, and then tell you what they have learned.

Chapter Summaries   Each chapter summary lists the new properties and definitions found in the chapter. The margins in the chapter summaries contain examples that illustrate the topics being reviewed.

Chapter Reviews   Each chapter ends with a set of review problems that covers all the different types of problems found in the chapter. The chapter reviews are longer and more extensive than the chapter tests.

Chapter Tests   Each chapter test contains a representative sample of the problems covered in the chapter.

Changes in the Fifth Edition

In addition to the chapter openings, research projects, Blueprint for Problem Solving, and Using Technology material mentioned above, the following items are also new to this edition.

Increased Visualization of Topics   This edition contains many more diagrams, charts, and graphs than the previous edition. Their purpose is to provide additional information, in visual form, to help students understand the topics we cover.

Facts from Geometry   Many of the important facts from geometry are now listed under this heading. In most cases, an example or two accompanies each of the facts to give students a chance to see how topics from geometry are related to the algebra they are learning.

Conditional Statements   An introductory look at the type of reasoning that is the foundation of mathematics is contained in Section 1.5. This new section gives students experience with conditional statements and their associated forms: inverse, converse, and contrapositive.

Number Sequences and Inductive Reasoning   Section 1.6 contains an introductory coverage of number sequences and inductive reasoning. Chapter 2 expands upon this material, which is then carried through to other parts of the book. I find that there are many interesting topics I can cover if students have some experience with number sequences. Number sequences also provide the easiest way to demonstrate inductive reasoning.

Unit Analysis   Chapter 4 now contains problems requiring students to convert from one unit of measure to another. The method used to accomplish the conversions is the method students will use if they take a chemistry class. Since this method is similar to the method we use to multiply rational expressions, unit analysis is covered in Section 4.7 as an application of multiplication and division of rational expressions.

Challenging Problems and Application Problems   More challenging problems have been added to many of the problem sets. Some of the new problems are in the drill problem category, while many others are new, more realistic application problems.

Early Coverage of Functions and Graphs   The material on functions and graphs starts in Section 1.7 with coverage of paired data. Looking through that section you will see that the examples are chosen so that they can be referred to later on as a foundation on which to build the concept of a function: graphs are approached from an intuitive standpoint through the use of histograms and line graphs. Functions are introduced in Section 2.8, along with function notation.

Flexible Coverage of Functions and Graphs   The material on graphing equations in two variables (Chapter 5) can be covered anytime after Chapter 2. Cover graphing early or late as you wish. The reverse is true for functions. They are covered early in Sections 1.7 and 2.8. However, you can defer coverage of functions until Chapter 5 by skipping Sections 1.7 and 2.8.

The more specific changes for individual chapters are listed below.

- *Chapter 1*   Section 1.5, which covers conditional statements is new, as is Section 1.6, which covers number sequences and inductive reasoning. Section 1.7 on paired data sets the stage for the coverage of functions in Chapter 2.
- *Chapter 2*   New material showing the relationship between number line graphs and interval notation has been added to the sections on inequalities. The material on applications has been expanded and split into two sections. Section 2.8, which covers functions and function notation, is new.
- *Chapter 5*   New Section 5.5 extends the coverage of functions started in Chapter 2, while new Section 5.6 extends the material on function notation.
- *Chapter 9*   The material on functions has been condensed from four sections into three.

Supplements to the Textbook

This fifth edition of *Intermediate Algebra* is accompanied by a number of useful supplements.

For the Instructor

- **Instructor's Edition**   An instructor version of the text is available to facilitate accessible and effective teaching. It contains all answers to the problem sets.

- **Instructor's Solution Manual**   The manual supplies instructor-appropriate solutions for half the odd and all the even exercises in the problem sets. The Student's Solutions Manual comprises solutions for all other exercises.
- **Printed Test Bank and Prepared Tests**   The test bank consists of multiple-choice test items organized by chapter, section, and objective. The prepared tests comprise 11 sets of ready-to-copy tests: one set for each chapter and one set for the entire book. Each set comprises two multiple-choice and four show-your-work tests. Items for half of the tests are ordered according to the sequence of topics in the book; items for the other half of the tests are in mixed-up order. Answers for every test item are provided.
- ***ExaMaster +*™ Computerized Test Bank**   A flexible, powerful, computerized testing system, *ExaMaster +*™ offers teachers a wide range of integrated testing options and features. Available in either IBM, Macintosh, or Windows format, it offers teachers the ability to select, edit, or create not only test items but algorithms for test items as well. Teachers can tailor tests according to a variety of criteria, scramble the order of test items, and administer tests on-line. *ExaMaster +*™ also includes full-function gradebook and graphing features.

For the Student

- **Videotape Package**   Free to adopters, the videotape package consists of 11 VHS videotapes, one for each chapter of the book. Each chapter tape is an hour to an hour and a half in length and is divided into lessons that correspond to each section of the chapter.
- **Student's Solutions Manual**   This manual contains complete annotated solutions to every other odd problem in problem sets and all chapter review and chapter test exercises.
- **MathCue Tutorial**   This computer software package of tutorials has problems that correspond to every section in the series. The software presents problems to solve and tutors students by displaying annotated, step-by-step solutions. Students may view partial solutions to get started on a problem, see continuous record of progress, and back up to review missed problems. Student scores can also be printed. Available for IBM and Macintosh.
- **MathCue Solution Finder**   This software allows students to enter their own problems into the computer and get annotated, step-by-step solutions in return. This unique program simulates working with a tutor, tracks student progress, refers students to specific sections in the text when appropriate, and prints student scores. Available for IBM and Macintosh.
- **MathCue Practice**   This algorithm-based software allows students to generate large numbers of practice problems keyed to problem types from each section of the book. Practice scores students' performance, and saves students' scores session to session. Available for IBM and Macintosh.
- **MathCue F/C Graph: The Functions and Conics Grapher**   For use with INTERMEDIATE ALGEBRA, 5/E, this computer program allows students to graph and analyze any polynomial, logarithmic, exponential, or trigonometric

function or conic equation they choose. F/C Graph can zoom, trace, display function values or coordinates of selected points, graph up to four functions simultaneously, and save and retrieve setups. Students can use F/C Graph to relate algebraic and visual forms of functions and conic sections, and to explore how changing parameters affect the graph. A set of computer lab exercises accompanies the software to direct student investigations of a variety of topics. Available for IBM and Macintosh.

- **Intermediate Algebra and the Graphing Calculator: A Learning Resource**   This workbook helps both instructors and students understand how to use graphing calculators and presents exercises and exploratory investigations into graphing functions, linear equations and inequalities, polynomials and factoring, lines and inequalities, quadratic equations, and systems of linear equations.

Acknowledgments

A project of this size cannot be completed without help from many people. In particular, Deirdre Lynch, my editor at Saunders College Publishing, contributed a number of helpful suggestions on the content of this revision. She was encouraging throughout the process and has been a pleasure to work with. Kate Pawlik and Martha Brown coordinated the copy editing, accuracy checking, and proofreading for the book. Their attention to detail and ability to get work done on time are unmatched. Marc Sherman, the developmental editor on this project, has done an exceptional job of suggesting ways in which to incorporate new material into the books, especially the material on technology. He and project editor Sarah Fitz-Hugh held the project together so that the book was published on time. My son, Patrick, and my daughter, Amy, assisted me with this revision from the beginning to the end. I am pleased with the way the book has turned out, and much of what I like about it is due to their influence. My thanks to these seven people; this book would not have been possible without them.

Thanks also to Linda Kyle of Tarrant County Junior College, Nancy Olson of Johnson County Community College, and Chandra and Hali Hallock for their help with the proofreading and problem checking, to Christine Schueler for the design of the book and cover, to Stacey Lloyd for her word-processing skills, and to my wife, Diane, for continuing to encourage my writing endeavors.

Finally, I am grateful to the following instructors for their suggestions and comments on this revision. Some reviewed the entire manuscript, while others were asked to evaluate the development of specific topics or the overall sequence of topics. My thanks go to the people listed below:

Carole Bauer, Triton College
Deborah Bennett, Jersey City State College
Kim Castagna, West Hills College
Elaine Deutschman, Oregon Institute of Technology
Richard DiDio, LaSalle University
Rose Dubec, Purdue University—Calumet

Theresa Geiger, St. Petersburg Junior College
Jeanne Glover, Memphis State University
Marion Graziano, Montgomery County Community College
Vic Gummersheimer, Southeast Missouri State University
Michael Hamm, Brookhaven College
Kay Haralson, Austin Peay State University
Melissa Hardeman, University of Arkansas at Little Rock
Linda Horner, Broward Community College
Everett House, Nashville State Tech
David Jewkes, Wallawalla College
John Johnson, Purdue University—Calumet
Anita Johnston, Jackson Community College
Judy Matlock, Collin County Community College
Nancy Olson, Johnson County Community College
John Rayko, University of Texas at San Antonio
Donal B. Staake, Jr., Jackson Community College
Hazel Stewart, Charleston Southern University
Kathleen Conway Stiehl, Southeast Missouri State University
John Wisthoff, Anne Arundel Community College

Charles P. McKeague

PREFACE TO THE
STUDENT

Many of my algebra students are apprehensive at first because they are worried that they will not understand the topics we cover. When I present a new topic that they do not grasp completely, they think something is wrong with them for not understanding it.

On the other hand, some students are excited about the course from the beginning. They are not worried about understanding algebra and, in fact, *expect* to find some topics difficult.

What is the difference between these two types of students?

Those who are excited about the course know from experience (as you do) that a certain amount of confusion is associated with most new topics in mathematics. They don't worry about it because they also know that the confusion gives way to understanding in the process of reading the textbook, working the problems, and getting questions answered. If they find a topic they are having difficulty with, they work as many problems as necessary to grasp the subject. They don't wait for the understanding to come to them; they go out and get it by working lots of problems. In contrast, the students who lack confidence tend to give up when they become confused. Instead of working more problems, they sometimes stop working problems altogether—and that of course guarantees that they will remain confused.

If you are worried about this course because you lack confidence in your ability to understand algebra, and you want to change the way you feel about mathematics, then look forward to the first topic that causes you some confusion. As soon as that topic comes along, make it your goal to master it, in spite of your apprehension. You will see that each and every topic covered in this course is one you can eventually master, even if your initial encounter with it is accompanied by some confusion. As long as you have passed a college-level beginning algebra course (or its equivalent), you are ready to take this course.

If you have decided to do well in algebra, the following list will be important to you.

How to Be Successful in Algebra

1. **Attend all class sessions on time**   You cannot know exactly what goes on in class unless you are there. Missing class and then expecting to find out what went on from someone else is not the same as being there yourself.

2. **Read the book**   It is best to read the section that will be covered in class beforehand. Reading in advance, even if you do not understand everything you read, is still better than going to class with no idea of what will be discussed.

3. **Work problems every day and check your answers**   The key to success in mathematics is working problems. The more problems you work, the better you will become at working them. The answers to the odd-numbered problems are given in the back of the book. When you have finished an assignment, be sure to compare your answers with those in the book. If you have made a mistake, find out what it is and correct it.

4. **Do it on your own**   Don't be misled into thinking someone else's work is your own. Having someone else show you how to work a problem is not the same as working the same problem yourself. It is okay to get help when you are stuck. As a matter of fact, it is a good idea. Just be sure you do the work yourself.

5. **Review every day**   After you have finished the problems your instructor has assigned, take another fifteen minutes and review a section you have already completed. The more you review, the longer you will retain the material you have learned.

6. **Don't expect to understand every new topic the first time you see it**   Sometimes you will understand everything you are doing, and sometimes you won't. That's just the way things are in mathematics. Expecting to understand each new topic the first time you see it can lead to disappointment and frustration. The process of understanding algebra takes time. It requires that you read the book, work problems, and get your questions answered.

7. **Spend as much time as it takes for you to master the material**   No set formula exists for the exact amount of time you need to spend on algebra to master it. You will find out as you go along what is or isn't enough time for you. If you end up spending two or more hours on each section in order to master the material there, then that's how much time it takes; trying to get by with less will not work.

8. **Relax**   It's probably not as difficult as you think.

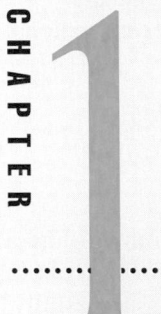

BASIC PROPERTIES AND DEFINITIONS

INTRODUCTION

The day after my daughter Amy turned 17, she and her friend Jenny had the following conversation with me.

AMY:	Dad, now that I'm 17, can my curfew be extended?
ME:	To how late?
AMY:	One o'clock.
ME:	No.
AMY:	Dad, everyone else gets to stay out as long as they want. It's not fair that I have to be in by midnight.
ME:	There is nothing to do here after midnight, except get in trouble.
AMY:	Okay then, can I spend the night at Jenny's?

There are messages within this conversation, some explicit and some implied, that mathematics, and the forms of reasoning on which mathematics is built, can clarify. For instance, one fact emerges clearly from this conversation:

If I don't give Amy permission to stay out late, then she will anyway.

Sections 1.5 and 1.6 of this chapter cover the kind of reasoning I would use to understand the conversation with my daughter. You may be surprised to find that it is the same kind of reasoning that accompanies most topics in mathematics.

OVERVIEW

The material in Chapter 1 is some of the most important material in the book. Be sure that you master it. Your success in the following chapters is directly related to how well you understand the material in Chapter 1.

Here is a list of the essential concepts from Chapter 1 that you will need to be successful in the succeeding chapters:

1. You must know how to add, subtract, multiply, and divide both positive and negative numbers.

1

2. You must understand and recognize the commutative, associative, and distributive properties—the three most important properties of real numbers.
3. You must know the difference between whole numbers, integers, rational numbers, and real numbers.

STUDY SKILLS

At the beginning of each of the first few chapters of this book you will find a section like this in which we list the skills that are necessary for success in algebra. If you have just completed an introductory algebra class successfully, you have acquired most of these skills. If it has been some time since you have taken a math class, you must pay attention to the sections on study skills.

Here is a list of things you can do to begin to develop effective study skills.

1. **Put Yourself on a Schedule**   The general rule is that you spend two hours on homework for every hour you are in class. Make a schedule for yourself in which you set aside two hours each day to work on algebra. Once you make the schedule, stick to it. Don't just complete your assignments and stop. Use all the time you have set aside. If you complete an assignment and have time left over, read the next section in the book, and work more problems. As the course progresses you may find that two hours a day is not enough time for you to master the material in this course. If it takes you longer than two hours a day to reach your goals for this course, then that's how much time it takes. Trying to get by with less will not work.
2. **Find Your Mistakes and Correct Them**   There is more to studying algebra than just working problems. You must always check your answers with the answers in the back of the book. When you have made a mistake, find out what it is, and correct it. Making mistakes is part of the process of learning mathematics. One of the number sequences we will study in this chapter is known as the *Fibonacci sequence* after the mathematician Fibonacci. Fibonacci published a book titled *The Book of Squares* in 1225. In the prologue he has this to say about his writing:

   > I have come to request indulgence if in any place it contains something more or less than right or necessary; for to remember everything and be mistaken in nothing is divine rather than human . . .

   Fibonacci knew, as you know, that human beings make mistakes. You cannot learn algebra without making mistakes. The key to discovering what you do not understand can be found by correcting your mistakes.
3. **Imitate Success**   Your work should look like the work you see in this book and the work your instructor shows. The steps shown in solving problems in this book were written by someone who has been successful in mathematics. The same is true of your instructor. Your work should imitate the work of people who have been successful in mathematics.

Basic Definitions

This section is, for the most part, simply a list of many of the basic symbols and definitions we will be using throughout the book.

We begin by reviewing the symbols we use to compare numbers and expressions; then we review the symbols associated with addition, subtraction, multiplication, and division.

Comparison Symbols

In Symbols	In Words
$a = b$	a is equal to b
$a \neq b$	a is not equal to b
$a < b$	a is less than b
$a \leq b$	a is less than or equal to b
$a \nless b$	a is not less than b
$a > b$	a is greater than b
$a \geq b$	a is greater than or equal to b
$a \ngtr b$	a is not greater than b
$a \Leftrightarrow b$	a is equivalent to b

Operation Symbols

Operation	In Symbols	In Words
Addition	$a + b$	The sum of a and b
Subtraction	$a - b$	The difference of a and b
Multiplication	ab, $a \cdot b$, $a(b)$, $(a)b$, or $(a)(b)$	The product of a and b
Division	$a \div b$, a/b, or $\dfrac{a}{b}$	The quotient of a and b

---

The key words are **sum, difference, product,** and **quotient.** They are used frequently in mathematics. For instance, we may say the product of 3 and 4 is 12. We mean both the statements $3 \cdot 4$ and 12 are called the product of 3 and 4. The important idea here is that the word *product* implies multiplication, regardless of whether it is written $3 \cdot 4$, 12, 3(4), or (3)4.

The following example shows how we translate sentences written in English into expressions written in symbols.

▶ **EXAMPLE 1**

In English	*In Symbols*
The sum of x and 5 is less than 2	$x + 5 < 2$
The product of 3 and x is 21	$3x = 21$
The quotient of y and 6 is 4	$\dfrac{y}{6} = 4$
Twice the difference of b and 7 is greater than 5	$2(b - 7) > 5$
The difference of twice b and 7 is greater than 5	$2b - 7 > 5$ ◀

Exponents

The next topic we will cover gives us a way to write repeated multiplications in a shorthand form.

Consider the expression 3^4. The 3 is called the **base** and the 4 is called the **exponent.** The exponent 4 tells us the number of times the base appears in the product. That is,

$$3^4 = 3 \cdot 3 \cdot 3 \cdot 3 = 81$$

The expression 3^4 is said to be in exponential form, while $3 \cdot 3 \cdot 3 \cdot 3$ is said to be in expanded form.

▶ **EXAMPLES** Expand and multiply.

2. $5^2 = 5 \cdot 5 = 25$ Base 5, exponent 2
3. $2^5 = 2 \cdot 2 \cdot 2 \cdot 2 \cdot 2 = 32$ Base 2, exponent 5
4. $4^3 = 4 \cdot 4 \cdot 4 = 64$ Base 4, exponent 3 ◀

Order of Operations

It is important when evaluating arithmetic expressions in mathematics that each expression have only one answer in reduced form. Consider the expression

$$3 \cdot 7 + 2$$

If we find the product of 3 and 7 first, then add 2, the answer is 23. On the other hand, if we first combine the 7 and 2, then multiply by 3, we have 27. The problem seems to have two distinct answers depending on whether we multiply first or add first. To avoid this situation, we will decide that multiplication in a situation like this will always be done before addition. In this case, only the first answer, 23, is correct.

Here is the complete set of rules for evaluating expressions. It is intended to avoid the type of confusion found in the preceding illustration.

> **Rule (Order of Operations)**
> When evaluating a mathematical expression, we will perform the operations in the following order, beginning with the expression in the innermost parentheses or brackets and working our way out.
>
> 1. Simplify all numbers with exponents, working from left to right if more than one of these expressions is present.
> 2. Then, do all multiplications and divisions left to right.
> 3. Perform all additions and subtractions left to right.

Here are some examples that illustrate the use of this rule.

▶ **EXAMPLES**   Simplify each expression using the rule for order of operations.

5. $5 + 3(2 + 4) = 5 + 3(6)$ Simplify inside parentheses
$= 5 + 18$ Then, multiply
$= 23$ Add

6. $5 \cdot 2^3 - 4 \cdot 3^2 = 5 \cdot 8 - 4 \cdot 9$ Simplify exponentials left to right
$= 40 - 36$ Multiply left to right
$= 4$ Subtract

7. $20 - (2 \cdot 5^2 - 30) = 20 - (2 \cdot 25 - 30)$ ⎫   Simplify inside
$= 20 - (50 - 30)$ ⎬   parentheses, evaluating
$= 20 - (20)$ ⎭   exponents first, then
$= 0$ multiplying, and
      finally subtracting

8. $40 - 20 \div 5 + 8 = 40 - 4 + 8$ Divide first
$= 36 + 8$ ⎫   Then, add and subtract left
$= 44$ ⎭   to right

9. $2 + 4[5 + (3 \cdot 2 - 2)] = 2 + 4[5 + (6 - 2)]$ ⎫   Simplify inside
$= 2 + 4(5 + 4)$ ⎬   innermost
$= 2 + 4(9)$ ⎭   parentheses
$= 2 + 36$ Then, multiply
$= 38$ Add   ◀

▶ **EXAMPLE 10**   Subtract 5 from the quotient of 24 and 3.

Solution   We begin by translating the sentence into symbols. Then, we simplify using the rule for order of operations.

$$24 \div 3 - 5 = 8 - 5 \qquad \text{Division before subtraction}$$
$$= 3 \qquad \text{Subtract} \qquad ◀$$

The next concept we will cover, that of a set, can be considered the starting point for all the branches of mathematics.

Sets

DEFINITION A **set** is a collection of objects or things. The objects in the set are called *elements* or *members* of the set.

Sets are usually denoted by capital letters and elements of sets by lowercase letters. We use braces, { }, to enclose the elements of a set.

To show that an element is contained in a set we use the symbol $\in$. That is,

$x \in A$ is read ''x is an element (member) of set A''
The symbol $\notin$ is read ''is *not* a member of''

For example, if A is the set $\{1, 2, 3\}$, then $2 \in A$. On the other hand, $5 \notin A$, meaning 5 is not an element of set A.

DEFINITION Set A is a **subset** of set B, written $A \subset B$, if every element in A is also an element of B. That is,

$A \subset B$ if and only if A is contained in B

Here are some examples of sets and subsets.

▶ **EXAMPLES**

11. The set of numbers used to count things is $\{1, 2, 3, \ . \ . \ .\}$. The dots mean the set continues indefinitely in the same manner. This is an example of an **infinite** set.
12. The set of all numbers represented by the dots on the faces of a regular die is $\{1, 2, 3, 4, 5, 6\}$. This set is a subset of the set in Example 11. It is an example of a **finite** set, since it has a limited number of elements. ◀

DEFINITION The set with no members is called the **empty** or **null set.** It is denoted by the symbol $\varnothing$. (*Note:* A mistake is sometimes made by trying to denote the empty set with the notation $\{\varnothing\}$. The set $\{\varnothing\}$ is not the empty set, since it contains one element, the empty set $\varnothing$.)

The empty set is considered a subset of every set.

Operations with Sets

There are two basic operations used to combine sets. The operations are union and intersection.

DEFINITION The **union** of two sets A and B, written $A \cup B$, is the set of all elements that are either in A or in B, or in both A and B. The key word here is *or*. For an element to be in $A \cup B$ it must be in A or B. In symbols, the definition looks like this:

$$x \in A \cup B \quad \text{if and only if} \quad x \in A \text{ or } x \in B$$

DEFINITION The **intersection** of two sets A and B, written $A \cap B$, is the set of elements in both A and B. The key word in this definition is the word *and*. For an element to be in $A \cap B$ it must be in both A and B, or

$$x \in A \cap B \quad \text{if and only if} \quad x \in A \text{ and } x \in B$$

▶ **EXAMPLES** Let $A = \{1, 3, 5\}$, $B = \{0, 2, 4\}$, and $C = \{1, 2, 3, \ldots\}$. Then

13. $A \cup B = \{0, 1, 2, 3, 4, 5\}$
14. $A \cap B = \varnothing$ (A and B have no elements in common)
15. $A \cap C = \{1, 3, 5\} = A$
16. $B \cup C = \{0, 1, 2, 3, \ldots\}$ ◀

Up to this point we have described the sets encountered by listing all the elements and then enclosing them with braces { }. There is another notation we can use to describe sets. It is called **set-builder notation.** Here is how we would write our definition for the union of two sets A and B using set-builder notation:

$$A \cup B = \{x \mid x \in A \text{ or } x \in B\}$$

The right side of this statement is read "the set of all x such that x is a member of A or x is a member of B." As you can see, the vertical line after the first x is read "such that."

▶ **EXAMPLE 17** If $A = \{1, 2, 3, 4, 5, 6\}$, find $C = \{x \mid x \in A \text{ and } x \geq 4\}$.

Solution We are looking for all the elements of A that are also greater than or equal to 4. They are 4, 5, and 6. Using set notation, we have

$$C = \{4, 5, 6\}$$ ◀

PROBLEM SET 1.1

Translate each of the following statements into symbols.

1. The sum of x and 5.
2. The sum of y and -3.
3. The difference of 6 and x.
4. The difference of x and 6.
5. The product of 2 and t is less than y.
6. The product of $5x$ and y is equal to z.
7. The quotient of $3x$ and $2y$ is greater than 6.
8. The quotient of $2y$ and $3x$ is not less than 7.
9. The sum of x and y is less than the difference of x and y.
10. Twice the sum of a and b is 15.
11. Three times the difference of x and 5 is more than y.
12. The product of x and y is greater than or equal to the quotient of x and y.
13. The difference of s and t is not equal to their sum.
14. The quotient of $2x$ and y is less than or equal to the sum of $2x$ and y.
15. Twice the sum of t and 3 is not greater than the difference of t and 6.
16. Three times the product of x and y is equal to the sum of $2y$ and $3z$.

Expand and multiply.

17. 6^2 **18.** 8^2
19. 10^2 **20.** 10^3
21. 2^3 **22.** 5^3
23. 2^4 **24.** 1^4
25. 10^4 **26.** 4^3
27. 11^2 **28.** 10^5

Simplify each expression using the rule for order of operations.

29. $3 \cdot 5 + 4$ **30.** $3 \cdot 7 - 6$
31. $3(5 + 4)$ **32.** $3(7 - 6)$
33. $2 + 8 \cdot 5$ **34.** $12 - 3 \cdot 3$
35. $(2 + 8)5$ **36.** $(12 - 3)3$
37. $6 + 3 \cdot 4 - 2$ **38.** $8 + 2 \cdot 7 - 3$
39. $6 + 3(4 - 2)$ **40.** $8 + 2(7 - 3)$
41. $(6 + 3)(4 - 2)$ **42.** $(8 + 2)(7 - 3)$

43. $4 \cdot 2^2 + 5 \cdot 2^3$ **44.** $3 \cdot 4^2 + 2 \cdot 4^3$
45. $(4 \cdot 2)^2 + (5 \cdot 2)^3$ **46.** $(3 \cdot 4)^2 + (2 \cdot 4)^3$
47. $(5 + 3)^2$ **48.** $(8 - 3)^2$
49. $5^2 + 3^2$ **50.** $8^2 - 3^2$
51. $5^2 + 2(5)(3) + 3^2$ **52.** $8^2 - 2(8)(3) + 3^2$
53. $(7 - 4)(7 + 4)$ **54.** $(8 - 5)(8 + 5)$
55. $7^2 - 4^2$ **56.** $8^2 - 5^2$
57. $2 + 3 \cdot 2^2 + 3^2$ **58.** $3 + 4 \cdot 4^2 + 5^2$
59. $2 + 3(2^2 + 3^2)$ **60.** $3 + 4(4^2 + 5^2)$
61. $(2 + 3)(2^2 + 3^2)$ **62.** $(3 + 4)(4^2 + 5^2)$
63. $40 - 10 \div 5 + 1$ **64.** $20 - 10 \div 2 + 3$
65. $(40 - 10) \div 5 + 1$ **66.** $(20 - 10) \div 2 + 3$
67. $(40 - 10) \div (5 + 1)$
68. $(20 - 10) \div (2 + 3)$
69. $24 \div 4 + 8 \div 2$ **70.** $36 \div 3 + 9 \div 3$
71. $24 \div (4 + 8) \div 2$ **72.** $36 \div (3 + 9) \div 3$
73. $5 \cdot 10^3 + 4 \cdot 10^2 + 3 \cdot 10 + 1$
74. $6 \cdot 10^3 + 5 \cdot 10^2 + 4 \cdot 10 + 3$
75. $2^3 + 3(8 + 12 \div 2)$
76. $3^2 + 2(10 + 15 \div 5)$
77. $16 - (4 \cdot 5^2 - 9 \cdot 10)$
78. $18 - (3 \cdot 4^3 - 19 \cdot 10)$
79. $40 - [10 - (4 - 2)]$
80. $50 - [17 - (8 - 3)]$
81. $40 - 10 - 4 - 2$ **82.** $50 - 17 - 8 - 3$
83. $3 + 2(2 \cdot 3^2 + 1)$ **84.** $4 + 5(3 \cdot 2^2 - 5)$
85. $(3 + 2)(2 \cdot 3^2 + 1)$
86. $(4 + 5)(3 \cdot 2^2 - 5)$
87. $3[2 + 4(5 + 2 \cdot 3)]$
88. $2[4 + 2(6 + 3 \cdot 5)]$
89. $6[3 + 2(5 \cdot 3 - 10)]$
90. $8[7 + 2(6 \cdot 9 - 14)]$
91. $5(7 \cdot 4 - 3 \cdot 4) + 8(5 \cdot 9 - 4 \cdot 9)$
92. $4(3 \cdot 9 - 2 \cdot 9) + 5(6 \cdot 8 - 5 \cdot 8)$
93. $5^3 + 4^3 \div 2^4 - 3^4$
94. $6^3 + 2^6 \div 4^2 - 3^4$

Translate each of the following into symbols and then simplify.

95. Five added to the product of 2 and 8.
96. Nine added to the quotient of 12 and 4.
97. Three subtracted from the quotient of 36 and 2.

98. Four subtracted from the product of 5 and 9.

99. Two added to the difference of 9 and 4.

100. Eight added to the difference of 10 and 7.

For the following problems, let $A = \{0, 2, 4, 6\}$, $B = \{1, 2, 3, 4, 5\}$, and $C = \{1, 3, 5, 7\}$.

101. $A \cup B$ **102.** $A \cup C$

103. $A \cap B$ **104.** $A \cap C$

105. $B \cap C$ **106.** $B \cup C$

107. $A \cup (B \cap C)$ **108.** $C \cup (A \cap B)$

109. $\{x \mid x \in A$ and $x < 4\}$

110. $\{x \mid x \in B$ and $x > 3\}$

111. $\{x \mid x \in A$ and $x \notin B\}$

112. $\{x \mid x \in B$ and $x \notin C\}$

113. $\{x \mid x \in A$ or $x \in C\}$

114. $\{x \mid x \in A$ or $x \in B\}$

115. $\{x \mid x \in B$ and $x \neq 3\}$

116. $\{x \mid x \in C$ and $x \neq 5\}$

One Step Further

Most of the problem sets in this book end with a few problems like the ones below. These problems challenge you to extend your knowledge of the material in the problem set. In most cases, there are no examples in the text similar to these problems. You should approach these problems with a positive point of view, because even though you may not complete them correctly, just the process of attempting them will increase your knowledge and ability in algebra. The notation $n(A)$ is used to denote the *number* of elements in set A. For example, if $A = \{a, b, c\}$, then $n(A) = 3$.

117. If A and B are sets such that $n(A) = 4$, $n(B) = 5$, and $A \cap B = \varnothing$, find $n(A \cup B)$. (How many elements are in the union of A and B?)

118. In general, if $A \cap B = \varnothing$, what is $n(A \cap B)$?

119. If $n(A) = 7$, $n(B) = 8$, and $n(A \cap B) = 2$, find $n(A \cup B)$.

120. If $n(A) = 4$, $n(B) = 10$, and $n(A \cap B) = 3$, find $n(A \cup B)$.

SECTION

1.2   # The Real Numbers

We begin this section by discussing the real numbers.

The Real Number Line

The real number line is constructed by drawing a straight line and labeling a convenient point with the number 0. Positive numbers are in increasing order to the right of 0; negative numbers are in decreasing order to the left of 0. The point on the line corresponding to 0 is called the **origin**.

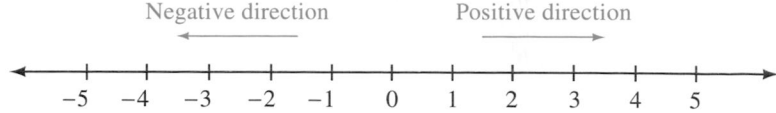

The numbers on the number line increase in size as we move to the right. When we compare the size of two numbers on the number line, the one on the left is always the smaller number.

The numbers associated with the points on the line are called **coordinates** of those points. Every point on the line has a number associated with it. The set of all these numbers makes up the set of real numbers.

DEFINITION   A **real number** is any number that is the coordinate of a point on the real number line.

▶ **EXAMPLE 1**   Locate the numbers -4.5, -0.75, $\frac{1}{2}$, $\sqrt{2}$, π, and 4.1 on the real number line.

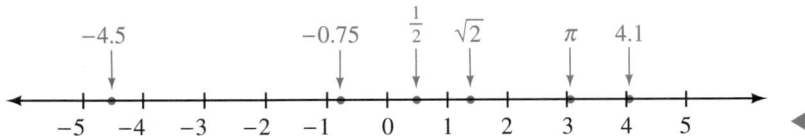

Note   In this book we will refer to real numbers as being on the real number line. Actually, real numbers are *not* on the line; only the points representing them are on the line. We can save some writing, however, if we simply refer to real numbers as being on the number line.

We can use the real number line to give a visual representation to inequality statements.

▶ **EXAMPLE 2**   Graph $\{x \,|\, x \leq 3\}$.

Solution   We want to graph all the real numbers less than or equal to 3—that is, all the real numbers below 3 and including 3. We label 0 on the number line for reference as well as 3 since the latter is what we call the endpoint. The graph is as follows.

We use a solid circle at 3 since 3 is included in the graph.   ◀

▶ **EXAMPLE 3**   Graph $\{x \,|\, x < 3\}$.

Solution   The graph will be identical to the graph in Example 2 except at the endpoint 3. In this case we will use an open circle since 3 is not included in the graph.

In Section 1.1 we defined the **union** of two sets A and B to be the set of all elements that are in either A or B. The word *or* is the key word in the definition. The **intersection** of two sets A and B is the set of all elements contained in both A and B, the key word here being *and*. We can put the words *and* and *or* together with our methods of graphing inequalities to graph some **compound inequalities.**

▶ **EXAMPLE 4**   Graph $\{x \,|\, x \leq -2 \text{ or } x > 3\}$.

Solution   The two inequalities connected by the word *or* are referred to as a **compound inequality.** We begin by graphing each inequality separately.

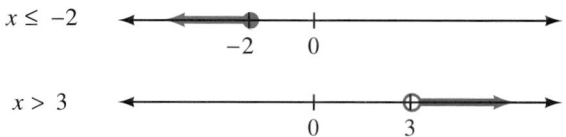

Since the two inequalities are connected by the word *or,* we graph their union. That is, we graph all points on either graph.

▶ **EXAMPLE 5**   Graph $\{x \,|\, x > -1 \text{ and } x < 2\}$.

Solution   We first graph each inequality separately.

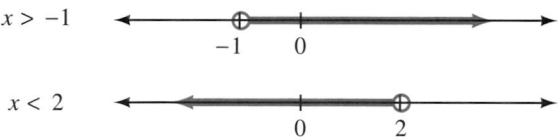

Since the two inequalities are connected by the word *and,* we graph their intersection—the part they have in common.

Notation   Sometimes compound inequalities that use the word *and* as the connecting word can be written in a shorter form. For example, the compound inequality $-3 \leq x$ and $x \leq 4$ can be written $-3 \leq x \leq 4$. The word *and* does not appear when an inequality is written in this form. It is implied. Inequalities of the form $-3 \leq x \leq 4$ are called **continued inequalities.** This new notation is useful because it takes fewer symbols to write it. The graph of $-3 \leq x \leq 4$ is

▶ **EXAMPLE 6** Graph $\{x \mid 1 \leq x < 2\}$.

Solution The word *and* is implied in the continued inequality $1 \leq x < 2$. That is, the continued inequality $1 \leq x < 2$ is equivalent to the compound inequality $1 \leq x$ and $x < 2$. Therefore, we graph all the numbers between 1 and 2 on the number line, including 1 but not including 2.

▶ **EXAMPLE 7** Graph $\{x \mid x < -2 \text{ or } 2 < x < 6\}$.

Solution Here we have a combination of compound and continued inequalities. We want to graph all real numbers that are either less than -2 or between 2 and 6.

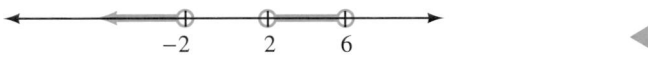

In addition to the phrases that translate directly into inequality statements, we have the following translations:

In Words	In Symbols
x is at least 40	$x \geq 40$
x is at most 30	$x \leq 30$
x is no more than 20	$x \leq 20$
x is no less than 10	$x \geq 10$
x is between 4 and 5	$4 < x < 5$

In the last case, we can include the endpoints 4 and 5 by saying ''x is between 4 and 5, inclusive,'' which translates to $4 \leq x \leq 5$.

▶ **EXAMPLE 8** Suppose you have a part-time job that requires that you work at least 10 hours but no more than 20 hours each week. Use the letter t to write an inequality that shows the number of hours you work per week.

Solution If t is at least 10 but no more than 20, then $10 \leq t$ and $t \leq 20$, or equivalently, $10 \leq t \leq 20$. Note that the word ''but,'' as used here, has the same meaning as the word ''and.'' ◀

▶ **EXAMPLE 9** If the highest temperature on Tuesday was 76° and the lowest temperature was 55°, write an inequality using the letter x that gives the range of temperatures on Tuesday.

Solution Since the smallest value of x is 55 and the largest value of x is 76, then $55 \leq x \leq 76$. We could say that the temperature on Tuesday was between 55° and 76°, inclusive. ◀

Subsets of the Real Numbers

Next, we consider some of the more important subsets of the real numbers. Each set listed here is a subset of the real numbers:

Counting (or natural) numbers $= \{1, 2, 3, \ldots\}$
Whole numbers $= \{0, 1, 2, 3, \ldots\}$
Integers $= \{\ldots, -3, -2, -1, 0, 1, 2, 3, \ldots\}$
Rational numbers $= \left\{\dfrac{a}{b} \,\middle|\, a \text{ and } b \text{ are integers}, b \neq 0\right\}$

Remember, the notation used to write the rational numbers is read "the set of numbers a/b, such that a and b are integers and b is not equal to 0." Any number that can be written in the form

$$\frac{\text{integer}}{\text{integer}}$$

is a rational number. Rational numbers are numbers that can be written as the ratio of two integers. Each of the following is a rational number:

$\dfrac{3}{4}$ Because it is the ratio of the integers 3 and 4

-8 Because it can be written as the ratio of -8 to 1

0.75 Because it is the ratio of 75 to 100 (or 3 to 4 if you reduce to lowest terms)

$0.333 \ldots$ Because it can be written as the ratio of 1 to 3

There are still other numbers on the number line that are not members of the subsets we have listed so far. They are real numbers, but they cannot be written as the ratio of two integers. That is, they are not rational numbers. For that reason, we call them irrational numbers.

Irrational numbers $= \{x \mid x \text{ is real, but not rational}\}$

The following are irrational numbers:

$$\sqrt{2} \qquad -\sqrt{3} \qquad 4 + 2\sqrt{3} \qquad \pi \qquad \pi + 5\sqrt{6}$$

▶ **EXAMPLE 10**   For the set $\{-5, -3.5, 0, \frac{3}{4}, \sqrt{3}, \sqrt{5}, 9\}$, list the numbers that are: (a) whole numbers, (b) integers, (c) rational numbers, (d) irrational numbers, and (e) real numbers.

Solution
(a) whole numbers: $0, 9$
(b) integers: $-5, 0, 9$
(c) rational numbers: $-5, -3.5, 0, \frac{3}{4}, 9$
(d) irrational numbers: $\sqrt{3}, \sqrt{5}$
(e) They are all real numbers.   ◀

Prime Numbers and Factoring

The following diagram shows the relationship between multiplication and factoring:

Multiplication

Factors $\rightarrow$ 3 $\cdot$ 4 = 12 $\leftarrow$ Product

Factoring

When we read the problem from left to right, we say the product of 3 and 4 is 12. Or we multiply 3 and 4 to get 12. When we read the problem in the other direction, from right to left, we say we have **factored** 12 into 3 times 4, or 3 and 4 are **factors** of 12.

The number 12 can be factored still further:

$$
\begin{aligned}
12 &= 4 \cdot 3 \\
&= 2 \cdot 2 \cdot 3 \\
&= 2^2 \cdot 3
\end{aligned}
$$

The numbers 2 and 3 are called **prime** factors of 12 because neither can be factored any further.

DEFINITION   If a and b represent integers, then a is said to be a **factor** (or divisor) of b if a divides b evenly—that is, if a divides b with no remainder.

DEFINITION   A **prime** number is any positive integer larger than 1 whose only positive factors (divisors) are itself and 1. An integer greater than 1 that is not prime is said to be **composite.**

Here is a list of the first few prime numbers:

Prime numbers = {2, 3, 5, 7, 11, 13, 17, 19, 23, 29, 31, 37, 41, . . .}

When a number is not prime, we can factor it into the product of prime numbers. To factor a number into the product of primes, we simply factor it until it cannot be factored further.

▶ **EXAMPLE 11**   Factor 525 into the product of primes.

Solution   Since 525 ends in 5, it is divisible by 5:

$$525 = 5 \cdot 105$$
$$= 5 \cdot 5 \cdot 21$$
$$= 5 \cdot 5 \cdot 3 \cdot 7$$
$$= 3 \cdot 5^2 \cdot 7 \qquad \blacktriangleleft$$

When we have factored a number into the product of its prime factors, not only do we know what prime numbers divide the original number, but we also know all of the other numbers that divide it. For instance, if we were to factor 210 into its prime factors, we would have $210 = 2 \cdot 3 \cdot 5 \cdot 7$, which means that 2, 3, 5, and 7 divide 210, as well as any combination of products of 2, 3, 5, and 7. That is, since 3 and 7 divide 210, then so does their product 21. Since 3, 5, and 7 each divide 210, then so does their product 105:

$$21 \text{ divides } 210$$
$$210 = 2 \cdot 3 \cdot 5 \cdot 7$$
$$105 \text{ divides } 210$$

Although there are many ways in which factoring is used in arithmetic and algebra, one simple application is in reducing fractions to lowest terms.

Recall that we reduce fractions to lowest terms by dividing the numerator and denominator by the same number. We can use the prime factorization of numbers to help us reduce fractions with large numerators and denominators.

▶ **EXAMPLE 12** Reduce $\dfrac{210}{231}$ to lowest terms.

Solution First we factor 210 and 231 into the product of prime factors. Then we reduce to lowest terms by dividing the numerator and denominator by any factors they have in common.

$$\frac{210}{231} = \frac{2 \cdot 3 \cdot 5 \cdot 7}{3 \cdot 7 \cdot 11} \qquad \text{Factor the numerator and denominator completely}$$

$$= \frac{2 \cdot \cancel{3} \cdot 5 \cdot \cancel{7}}{\cancel{3} \cdot \cancel{7} \cdot 11} \qquad \text{Divide the numerator and denominator by } 3 \cdot 7$$

$$= \frac{2 \cdot 5}{11}$$

$$= \frac{10}{11} \qquad \blacktriangleleft$$

The small lines we have drawn through the factors that are common to the numerator and denominator are used to indicate that we have divided the numerator and denominator by those factors.

If the original number is positive or 0, then its absolute value is the number itself. If the number is negative, its absolute value is its opposite (which must be positive). We see that $-x$, as written in this definition, is a positive quantity since x in this case is a negative number.

Note   It is important to recognize that if x is a real number, $-x$ is not necessarily negative. For example, if x is 5, then $-x$ is -5. On the other hand, if x were -5, then $-x$ would be $-(-5)$, which is 5.

▶ **EXAMPLES**   Write each expression without absolute value symbols.

9. $|5| = 5$ **12.** $-|-3| = -3$

10. $|-2| = 2$ **13.** $-|5| = -5$

11. $\left|-\dfrac{1}{2}\right| = \dfrac{1}{2}$ **14.** $-|-\sqrt{2}| = -\sqrt{2}$

◀

Properties of Real Numbers

The list of properties that follows is actually just an organized summary of the things we know from past experience to be true about numbers in general. For instance, we know that adding 3 and 7 gives the same answer as adding 7 and 3. The order of two numbers in an addition problem can be changed without changing the result. This fact about numbers and addition is called the **commutative property of addition.** We say addition is a commutative operation. Similarly, multiplication is a commutative operation.

We now state these properties formally. For all the properties listed in this section, a, b, and c represent real numbers.

Commutative Property of Addition

In symbols: $a + b = b + a$

In words: The *order* of the numbers in a sum does not affect the result.

Commutative Property of Multiplication

In symbols: $a \cdot b = b \cdot a$

In words: The *order* of the numbers in a product does not affect the result.

▶ **EXAMPLES**

15. The statement $3 + 7 = 7 + 3$ is an example of the commutative property of addition.

16. The statement $3 \cdot x = x \cdot 3$ is an example of the commutative property of multiplication.

◀

The other two basic operations (subtraction and division) are not commutative. If we change the order in which we are subtracting or dividing two numbers, we will change the result.

Another property of numbers you have used many times has to do with grouping. When adding $3 + 5 + 7$, we can add the 3 and 5 first and then the 7, or we can add the 5 and 7 first and then the 3. Mathematically, it looks like this: $(3 + 5) + 7 = 3 + (5 + 7)$. Operations that behave in this manner are called **associative** operations.

Associative Property of Addition

In symbols: $a + (b + c) = (a + b) + c$
In words: The *grouping* of the numbers in a sum does not affect the result.

Associative Property of Multiplication

In symbols: $a(bc) = (ab)c$
In words: The *grouping* of the numbers in a product does not affect the result.

The following examples illustrate how the associative properties can be used to simplify expressions that involve both numbers and variables.

▶ **EXAMPLES**   Simplify by using the associative property.

17. $2 + (3 + y) = (2 + 3) + y$ Associative property
$\qquad\qquad\quad = 5 + y$ Addition

18. $5(4x) = (5 \cdot 4)x$ Associative property
$\qquad\quad = 20x$ Multiplication

19. $\dfrac{1}{4}(4a) = \left(\dfrac{1}{4} \cdot 4\right)a$ Associative property

$\qquad\qquad = 1a$ Multiplication
$\qquad\qquad = a$ ◀

Our next property involves both addition and multiplication. It is called the **distributive property** and is stated as follows.

Distributive Property

In symbols: $a(b + c) = ab + ac$
In words: Multiplication *distributes* over addition.

You will see as we progress through the book that the distributive property is used very frequently in algebra. To see that the distributive property works, compare the following:

$$3(4 + 5) \qquad 3(4) + 3(5)$$
$$3(9) \qquad 12 + 15$$
$$27 \qquad 27$$

In both cases the result is 27. Since the results are the same, the original two expressions must be equal; or, $3(4 + 5) = 3(4) + 3(5)$.

▶ **EXAMPLES** Apply the distributive property to each expression and then simplify the result.

20. $5(4x + 3) = 5(4x) + 5(3)$ Distributive property
$\qquad\qquad\quad = 20x + 15$ Multiplication

21. $6(3x + 2y) = 6(3x) + 6(2y)$ Distributive property
$\qquad\qquad\quad\; = 18x + 12y$ Multiplication

22. $\dfrac{1}{2}(3x + 6) = \dfrac{1}{2}(3x) + \dfrac{1}{2}(6)$ Distributive property

$\qquad\qquad\quad = \dfrac{3}{2}x + 3$ Multiplication

23. $2(3y + 4) + 2 = 2(3y) + 2(4) + 2$ Distributive property
$\qquad\qquad\qquad\;\; = 6y + 8 + 2$ Multiplication
$\qquad\qquad\qquad\;\; = 6y + 10$ Addition   ◀

Note Although the properties we are listing are stated for only two or three real numbers, they hold for as many numbers as needed. For example, the distributive property holds for expressions like $3(x + y + z + 2)$. That is,

$$3(x + y + z + 2) = 3x + 3y + 3z + 6$$

With the distributive property, it is not important how many numbers are contained in the sum, only that it is a sum. Multiplication distributes over addition, whether there are two numbers in the sum or two hundred.

The distributive property can also be used to combine similar terms. (For now, a *term* is the product of a number with one or more variables. We will give a precise definition in Chapter 3.) Similar terms are terms with the same variable part. The terms $3x$ and $5x$ are similar, as are $2y$, $7y$, and $-3y$, because the variable parts are the same. To combine similar terms, we use the distributive property. The following examples illustrate this.

▶ **EXAMPLES**   Use the distributive property to combine similar terms.

24. $3x + 5x = (3 + 5)x$ \qquad Distributive property
$\qquad\qquad = 8x$ \qquad\qquad\quad Addition

25. $4a + 7a = (4 + 7)a$ \qquad Distributive property
$\qquad\qquad = 11a$ \qquad\qquad\quad Addition

26. $3y + y = (3 + 1)y$ \qquad Distributive property
$\qquad\qquad = 4y$ \qquad\qquad\quad Addition   ◀

Review of Addition with Fractions

The distributive property is also used to add fractions. For example, to add $\frac{3}{7}$ and $\frac{2}{7}$, we first write each as the product of a whole number and $\frac{1}{7}$. Then, we apply the distributive property as we did in the preceding examples:

$$\frac{3}{7} + \frac{2}{7} = 3 \cdot \frac{1}{7} + 2 \cdot \frac{1}{7}$$

$$= (3 + 2)\frac{1}{7} \qquad \text{Distributive property}$$

$$= 5 \cdot \frac{1}{7}$$

$$= \frac{5}{7}$$

To add fractions using the distributive property, each fraction must have the same denominator. Here is another example.

▶ **EXAMPLE 27**   Use the distributive property to add $\dfrac{4}{x} + \dfrac{5}{x} + \dfrac{6}{x}$.

Solution   Here are the steps.

$$\frac{4}{x} + \frac{5}{x} + \frac{6}{x} = 4 \cdot \frac{1}{x} + 5 \cdot \frac{1}{x} + 6 \cdot \frac{1}{x} \qquad \text{Distributive property}$$

$$= (4 + 5 + 6) \cdot \frac{1}{x}$$

$$= 15 \cdot \frac{1}{x}$$

$$= \frac{15}{x} \qquad\qquad\qquad ◀$$

Adding fractions is always accomplished by using the distributive property (even if it is not shown). Since the distributive property cannot be used unless the fractions have the same denominator, we begin all addition problems with fractions by making sure each fraction has the same denominator.

DEFINITION   The **least common denominator** (LCD) for a set of denominators is the smallest number divisible by *all* the denominators.

The first step in adding fractions is to find a common denominator for all the denominators. We then rewrite each fraction (if necessary) as an equivalent fraction with the common denominator. Finally, we add the numerators and reduce to lowest terms if necessary.

▶ **EXAMPLE 28**   Add $\dfrac{5}{12} + \dfrac{7}{18}$.

Solution   The least common denominator for the denominators 12 and 18 must be the smallest number divisible by both 12 and 18. We can factor 12 and 18 completely and then build the LCD from these factors.

$$\left.\begin{array}{l} 12 = 2 \cdot 2 \cdot 3 \\ 18 = 2 \cdot 3 \cdot 3 \end{array}\right\} \quad \text{LCD} = 2 \cdot 2 \cdot 3 \cdot 3 = 36$$

12 divides the LCD

18 divides the LCD

Next we rewrite our original fractions as equivalent fractions with denominators of 36. To do so, we multiply each original fraction by an appropriate form of the number 1:

$$\frac{5}{12} + \frac{7}{18} = \frac{5}{12} \cdot \frac{3}{3} + \frac{7}{18} \cdot \frac{2}{2}$$

$$= \frac{15}{36} + \frac{14}{36}$$

Finally, we add numerators and place the result over the common denominator, 36. (Remember, this is an application of the distributive property.)

$$\frac{15}{36} + \frac{14}{36} = \frac{15 + 14}{36} = \frac{29}{36}$$ ◀

Simplifying Expressions

We can use the commutative, associative, and distributive properties together to simplify expressions such as $3x + 4 + 5x + 8$. We begin by applying the commutative property to change the order of the terms and write:

$$3x + 5x + 4 + 8$$

Next, we use the associative property to group similar terms together:

$$(3x + 5x) + (4 + 8)$$

Applying the distributive property to the first two terms, we have

$$(3 + 5)x + (4 + 8)$$

Finally, we add 3 and 5, and 4 and 8 to get

$$8x + 12$$

Here are some additional examples.

▶ **EXAMPLE 29** Simplify $7x + 4 + 6x + 3$.

Solution We begin by applying the commutative and associative properties to group similar terms:

$$7x + 4 + 6x + 3$$
$$= (7x + 6x) + (4 + 3) \qquad \text{Commutative and associative properties}$$
$$= (7 + 6)x + (4 + 3) \qquad \text{Distributive property}$$
$$= 13x + 7 \qquad \text{Addition} \qquad ◀$$

▶ **EXAMPLE 30** Simplify $4 + 3(2y + 5) + 8y$.

Solution Since our rule for order of operations indicates that we are to multiply before adding, we must distribute the 3 across $2y + 5$ first:

$$4 + 3(2y + 5) + 8y$$
$$= 4 + 6y + 15 + 8y \qquad \text{Distributive property}$$
$$= (6y + 8y) + (4 + 15) \qquad \text{Commutative and associative properties}$$
$$= (6 + 8)y + (4 + 15) \qquad \text{Distributive property}$$
$$= 14y + 19 \qquad \text{Addition} \qquad ◀$$

The remaining properties of real numbers have to do with the numbers 0 and 1.

Additive Identity Property
There exists a unique number 0 such that
In symbols: $a + 0 = a$ and $0 + a = a$
In words: Zero preserves identities under addition. (The identity of the number is unchanged after addition with 0.)

Multiplicative Identity Property
There exists a unique number 1 such that
In symbols: $a(1) = a$ and $1(a) = a$
In words: The number 1 preserves identities under multiplication. (The identity of the number is unchanged after multiplication by 1.)

Note   0 and 1 are called the **additive identity** and **multiplicative identity,** respectively. Combining 0 with a number, under addition, does not change the identity of the number. Likewise, combining 1 with a number, under multiplication, does not alter the identity of the number. We see that 0 is to addition what 1 is to multiplication.

Additive Inverse Property

For each real number a, there exists a unique real number $-a$ such that

In symbols: $a + (-a) = 0$

In words: Opposites add to 0.

Multiplicative Inverse Property

For every real number a, except 0, there exists a unique real number $\dfrac{1}{a}$ such that

In symbols: $a \left(\dfrac{1}{a} \right) = 1$

In words: Reciprocals multiply to 1.

The following examples illustrate how we use the properties listed here. Each line contains an algebraic expression that has been changed in some way. The property that justifies the change is written to the right.

▶ **EXAMPLES**

31. $7(1) = 7$ Multiplicative identity property

32. $4 + (-4) = 0$ Additive inverse property

33. $6 \left(\dfrac{1}{6} \right) = 1$ Multiplicative inverse property

34. $(5 + 0) + 2 = 5 + 2$ Additive identity property   ◀

PROBLEM SET 1.3

Use the associative property to rewrite each of the following expressions and then simplify the result.

1. $4 + (2 + x)$ **2.** $6 + (5 + 3x)$

3. $(a + 3) + 5$ **4.** $(4a + 5) + 7$

5. $5(3y)$ **6.** $7(4y)$

7. $\frac{1}{3}(3x)$ **8.** $\frac{1}{5}(5x)$

9. $4\left(\frac{1}{4}a\right)$ **10.** $7\left(\frac{1}{7}a\right)$

11. $\frac{2}{3}\left(\frac{3}{2}x\right)$ **12.** $\frac{4}{3}\left(\frac{3}{4}x\right)$

Apply the distributive property to each expression. Simplify when possible.

13. $3(x + 6)$ **14.** $5(x + 9)$

15. $2(6x + 4)$ **16.** $3(7x + 8)$

17. $5(3a + 2b)$ **18.** $7(2a + 3b)$

19. $4(7 + 3y)$ **20.** $8(6 + 2y)$

21. $\frac{1}{3}(4x + 6)$ **22.** $\frac{1}{2}(3x + 8)$

23. $\frac{1}{2}(2a + 4)$ **24.** $\frac{1}{2}(4a + 2)$

25. $\frac{1}{5}(10 + 5y)$ **26.** $\frac{1}{6}(12 + 6y)$

Apply the distributive property to each expression. Simplify when possible.

27. $(5t + 1)8$

28. $(3t + 2)5$

29. $3(5x + 2) + 4$

30. $4(3x + 2) + 5$

31. $4(2y + 6) + 8$

32. $6(2y + 3) + 2$

33. $5(1 + 3t) + 4$

34. $2(1 + 5t) + 6$

35. $3 + (2 + 7x)4$

36. $4 + (1 + 3x)5$

37. $9(3x + 5y + 7) + 10$

38. $7(2x + 4y + 6) + 10$

Use the distributive property to add the following fractions.

39. $\frac{3}{7} + \frac{1}{7} + \frac{2}{7}$

40. $\frac{3}{8} + \frac{1}{8} + \frac{1}{8}$

41. $\dfrac{4}{\sqrt{3}} + \dfrac{5}{\sqrt{3}}$

42. $\dfrac{1}{\sqrt{5}} + \dfrac{8}{\sqrt{5}}$

43. $\dfrac{4}{x} + \dfrac{7}{x}$

44. $\dfrac{5}{y} + \dfrac{9}{y}$

Add the following fractions.

45. $\frac{2}{5} + \frac{1}{15}$

46. $\frac{5}{8} + \frac{1}{4}$

47. $\frac{5}{6} + \frac{7}{8}$

48. $\frac{3}{4} + \frac{2}{3}$

49. $\frac{9}{48} + \frac{3}{54}$

50. $\frac{6}{28} + \frac{5}{42}$

51. $\frac{3}{4} + \frac{1}{8} + \frac{2}{3}$

52. $\frac{1}{3} + \frac{5}{6} + \frac{5}{12}$

Use the commutative, associative, and distributive properties to simplify the following.

53. $3x + 5 + 4x + 2$

54. $5x + 1 + 7x + 8$

55. $x + 3 + 4x + 9$

56. $5x + 2 + x + 10$

57. $5a + 7 + 8a + a$

58. $6a + 4 + a + 4a$

59. $3y + y + 5 + 2y + 1$

60. $4y + 2y + 3 + y + 7$

61. $x + 1 + x + 2 + x + 3$

62. $5 + x + 6 + x + 7 + x$

63. $2(5x + 1) + 2x$

64. $3(4x + 1) + 9x$

65. $7 + 2(4y + 2)$

66. $6 + 3(5y + 2)$

67. $3 + 4(5a + 3) + 4a$

68. $8 + 2(4a + 2) + 5a$

69. $5x + 2(3x + 8) + 4$

70. $7x + 3(4x + 1) + 7$

71. $2t + 3(1 + 6t) + 2$

72. $3t + 2(4 + 2t) + 6$

Each of the following problems has a mistake in it. Correct the right-hand side.

73. $5(2x + 4) = 10x + 4$

74. $7(x + 8) = 7x + 15$

75. $3x + 4x = 7(2x)$

76. $3x + 4x = 7x^2$

77. $\frac{3}{5} + \frac{1}{5} = \frac{4}{10}$

78. $\frac{5}{9} + \frac{2}{9} = \frac{7}{18}$

Identify the property (or properties) of real numbers that justifies each of the following.

79. $3 + 2 = 2 + 3$

80. $3(ab) = (3a)b$

81. $5x = x5$

82. $2 + 0 = 2$

83. $4 + (-4) = 0$

84. $1(6) = 6$

85. $x + (y + 2) = (y + 2) + x$

86. $(a + 3) + 4 = a + (3 + 4)$

87. $4(5 \cdot 7) = 5(4 \cdot 7)$

88. $6(xy) = (xy)6$

89. $4 + (x + y) = (4 + y) + x$

90. $(r + 7) + s = (r + s) + 7$

91. $3(4x + 2) = 12x + 6$

92. $5(\frac{1}{5}) = 1$

Write each of the following without absolute value symbols.

93. $|-2|$

94. $|-7|$

95. $\left|-\frac{3}{4}\right|$

96. $\left|\frac{5}{6}\right|$

97. $|\pi|$

98. $|-\sqrt{2}|$

99. $-|4|$

100. $-|5|$

101. $-|-2|$

102. $-|-10|$

Multiply the following.

103. $\frac{3}{5} \cdot \frac{7}{8}$

104. $\frac{6}{7} \cdot \frac{9}{5}$

105. $\frac{1}{3} \cdot 6$

106. $\frac{1}{4} \cdot 8$

107. $(\frac{2}{3})^3$

108. $(\frac{4}{5})^2$

109. $(\frac{1}{10})^4$

110. $(\frac{1}{2})^5$

111. $\frac{3}{5} \cdot \frac{4}{7} \cdot \frac{6}{11}$

112. $\frac{4}{5} \cdot \frac{6}{7} \cdot \frac{3}{11}$

113. $9(\frac{1}{3})^2$

114. $25(\frac{1}{5})^2$

Applying the Concepts

115. Name two numbers that are their own reciprocals.

116. Give the number that has no reciprocal.

117. Name the number that is its own opposite.

118. The reciprocal of a negative number is negative—true or false?

119. Show that the statement $5x - 5 = x$ is not correct by replacing x with 4 and simplifying both sides.

120. Show that the statement $8x - x = 8$ is not correct by replacing x with 5 and simplifying both sides.

121. Simplify the expression $15 - (8 - 2)$ and $(15 - 8) - 2$ to show that subtraction is not an associative operation.

122. Simplify the expressions $(48 \div 6) \div 2$ and $48 \div (6 \div 2)$ to show that division is not an associative operation.

One Step Further

Let the operation $*$ be defined as $a * b = ab + a$. Use this definition to make the following calculations.

123. $5 * 3$

124. $3 * 5$

125. $6 * 2$

126. $2 * 6$

127. Is the operation $*$ a commutative operation?

128. Calculate $5 * (3 + 4)$ and $5 * 3 + 5 * 4$.

Let the operation ∇ be defined as $a \nabla b = aa + bb$. Use this definition to simplify each expression.

129. $3 \nabla 5$

130. $5 \nabla 3$

131. $4 \nabla 3$

132. $3 \nabla 4$

133. Is the operation ∇ a commutative operation?

SECTION

1.4   Arithmetic with Real Numbers

Adding Real Numbers

The purpose of this section is to review the rules for arithmetic with real numbers and the justification for those rules. We can justify the rules for addition of real numbers geometrically by use of the real number line. Consider the sum of -5 and 3:

$$-5 + 3$$

We can interpret this expression as meaning "start at the origin and move 5 units in the negative direction and then 3 units in the positive direction." With the aid of a number line we can visualize the process.

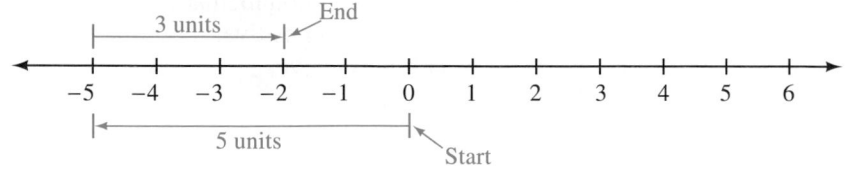

Since the process ends at -2, we say the sum of -5 and 3 is -2:

$$-5 + 3 = -2$$

We can use the real number line in this way to add any combination of positive and negative numbers.

The sum of -4 and -2, $-4 + (-2)$, can be interpreted as starting at the origin, moving 4 units in the negative direction, and then 2 more units in the negative direction:

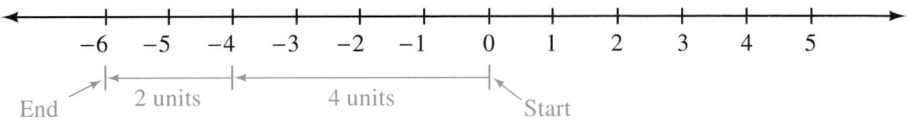

End   2 units   4 units   Start

Since the process ends at -6, we say the sum of -4 and -2 is -6:

$$-4 + (-2) = -6$$

We can eliminate actually drawing a number line by simply visualizing it mentally. The following example gives the results of all possible sums of positive and negative 5 and 7.

▶ **EXAMPLE 1**   Add all combinations of positive and negative 5 and 7.

Solution

$$5 + 7 = 12$$
$$-5 + 7 = 2$$
$$5 + (-7) = -2$$
$$-5 + (-7) = -12$$ ◀

Looking closely at the relationships in Example 1 (and trying other similar examples if necessary), we can arrive at the following rule for adding two real numbers.

Strategy for Adding Two Real Numbers

With the *same* sign:
    Step 1: Add their absolute values.
    Step 2: Attach their common sign. If both numbers are positive, their sum is positive; if both numbers are negative, their sum is negative.

With *opposite* signs:
    Step 1: Subtract the smaller absolute value from the larger.
    Step 2: Attach the sign of the number whose absolute value is larger.

Subtracting Real Numbers

In order to have as few rules as possible, we will not attempt to list new rules for the *difference* of two real numbers. We will instead define it in terms of addition and apply the rule for addition.

DEFINITION   If a and b are any two real numbers, then the **difference** of a and b, written $a - b$, is given by

$$\underbrace{a - b}\quad = \quad \underbrace{a + (-b)}$$

To subtract b,   add the opposite of b.

We define the process of subtracting b from a as being equivalent to adding the opposite of b to a. In short, we say, "subtraction is addition of the opposite." Here is how it works.

▶ **EXAMPLES**   Subtract.

2. $5 - 3 = 5 + (-3)$ Subtracting 3 is equivalent to adding -3
$\qquad\quad = 2$

3. $-7 - 6 = -7 + (-6)$ Subtracting 6 is equivalent to adding -6
$\qquad\qquad = -13$

4. $9 - (-2) = 9 + 2$ Subtracting -2 is equivalent to adding 2
$\qquad\qquad\; = 11$

5. $-6 - (-5) = -6 + 5$ Subtracting -5 is equivalent to adding 5
$\qquad\qquad\quad = -1$ ◀

▶ **EXAMPLE 6**   Subtract -3 from -9.

Solution   Since subtraction is not commutative, we must be sure to write the numbers in the correct order. Because we are subtracting -3, the problem looks like this when translated into symbols:

$$-9 - (-3) = -9 + 3 \qquad \text{Change to addition of the opposite}$$
$$\qquad\quad = -6 \qquad \text{Add}$$ ◀

▶ **EXAMPLE 7**   Add -4 to the difference of -2 and 5.

Solution   The difference of -2 and 5 is written $-2 - 5$. Adding -4 to that difference gives us

$$(-2 - 5) + (-4) = -7 + (-4) \qquad \text{Simplify inside parentheses}$$
$$\qquad\qquad = -11 \qquad\quad \text{Add}$$ ◀

Multiplying Real Numbers

Multiplication with whole numbers is simply a shorthand way of writing repeated addition.

Although this definition of multiplication does not hold for other kinds of numbers, such as fractions, it does, however, give us ways of interpreting products of positive and negative numbers. For example, $3(-2)$ can be evaluated as follows:

$$3(-2) = -2 + (-2) + (-2) = -6$$

We can evaluate the product $-3(2)$ in a similar manner if we first apply the commutative property of multiplication:

$$-3(2) = 2(-3) = -3 + (-3) = -6$$

From these results it seems reasonable to say that the product of a positive and a negative is a negative number.

The last case we must consider is the product of two negative numbers, such as $-3(-2)$. In order to evaluate this product we will look at the expression $-3[2 + (-2)]$ in two different ways. First, since $2 + (-2) = 0$, we know the expression $-3[2 + (-2)]$ is equal to 0. On the other hand, we can apply the distributive property to get

$$-3[2 + (-2)] = -3(2) + (-3)(-2) = -6 + ?$$

Since we know the expression is equal to 0, it must be true that our ? is 6, since 6 is the only number we can add to -6 to get 0. Therefore, we have

$$-3(-2) = 6$$

Here is a summary of what we have so far:

Original numbers have		The answer is
the same sign	$3(2) = 6$	positive
different signs	$3(-2) = -6$	negative
different signs	$-3(2) = -6$	negative
the same sign	$-3(-2) = 6$	positive

Strategy for Multiplying Two Real Numbers

To multiply two real numbers, simply
   Step 1: Multiply their absolute values.
   Step 2: If the two numbers have the *same* sign, the product is positive. If the two numbers have *opposite* signs, the product is negative.

▶ **EXAMPLE 8** Multiply all combinations of positive and negative 7 and 3.

Solution

$$7(3) = 21$$
$$7(-3) = -21$$
$$-7(3) = -21$$
$$-7(-3) = 21$$ ◀

Dividing Real Numbers

DEFINITION If a and b are any two real numbers, where $b \neq 0$, then the **quotient** of a and b, written $\dfrac{a}{b}$, is given by

$$\frac{a}{b} = a \cdot \left(\frac{1}{b}\right)$$

Dividing a by b is equivalent to multiplying a by the reciprocal of b. In short, we say, "division is multiplication by the reciprocal."

Since division is defined in terms of multiplication, the same rules hold for assigning the correct sign to a quotient as held for assigning the correct sign to a product. That is, *the quotient of two numbers with like signs is positive, while the quotient of two numbers with unlike signs is negative.*

▶ **EXAMPLES** Divide.

9. $\dfrac{6}{3} = 6 \cdot \left(\dfrac{1}{3}\right) = 2$

10. $\dfrac{6}{-3} = 6 \cdot \left(-\dfrac{1}{3}\right) = -2$

11. $\dfrac{-6}{3} = -6 \cdot \left(\dfrac{1}{3}\right) = -2$

12. $\dfrac{-6}{-3} = -6 \cdot \left(-\dfrac{1}{3}\right) = 2$

Notice these examples indicate that if a and b are positive real numbers then

$$\frac{-a}{b} = \frac{a}{-b} = -\frac{a}{b}$$

and

$$\frac{-a}{-b} = \frac{a}{b}.$$ ◀

The second step in the preceding examples is written only to show that each quotient can be written as a product. It is not actually necessary to show this step when working problems.

In the examples that follow, we find a combination of operations. In each case we use the rule for order of operations.

▶ **EXAMPLES**   Simplify each expression as much as possible.

13. $(-2 - 3)(5 - 9) = (-5)(-4)$ Simplify inside parentheses
$= 20$ Multiply

14. $2 - 5(7 - 4) - 6 = 2 - 5(3) - 6$ Simplify inside parentheses
$= 2 - 15 - 6$ Then, multiply
$= -19$ Finally, subtract, left to right

15. $4(-3)^2 - 5(-2)^3 = 4(9) - 5(-8)$ Exponents first
$= 36 - (-40)$ Multiply
$= 76$ Subtract

16. $2(4 - 7)^3 + 3(-2 - 3)^2 = 2(-3)^3 + 3(-5)^2$ Simplify inside parentheses
$= 2(-27) + 3(25)$ Evaluate numbers with exponents
$= -54 + 75$ Multiply
$= 21$ Add

17. $(3 - 7)[15 - 2(-4)] = (-4)(15 + 8)$
$= -4(23)$
$= -92$ ◀

▶ **EXAMPLES**   Simplify as much as possible.

18. $\dfrac{5(-3) - 10}{-4 - 1} = \dfrac{-15 - 10}{-4 - 1}$
$= \dfrac{-25}{-5}$
$= 5$

19. $\dfrac{-5(-4) + 2(-3)}{2(-1) - 5} = \dfrac{20 - 6}{-2 - 5}$
$= \dfrac{14}{-7}$
$= -2$

20. $\dfrac{2^3 + 3^3}{2^2 - 3^2} = \dfrac{8 + 27}{4 - 9}$
$= \dfrac{35}{-5}$
$= -7$ ◀

Remember, since subtraction is defined in terms of addition, we can restate the distributive property in terms of subtraction. That is, if a, b, and c are real numbers, then $a(b - c) = ab - ac$.

▶ **EXAMPLES** Apply the distributive property.

21. $3(x - 4) = 3x - 12$

22. $\frac{1}{2}(4y - 6) = \frac{1}{2}(4y) - \frac{1}{2}(6)$

$\qquad\qquad = 2y - 3$

23. $5(2a - 3) + 4 = 10a - 15 + 4$

$\qquad\qquad\qquad = 10a - 11$ ◀

▶ **EXAMPLE 24** Simplify $3(2y - 1) + y$.

Solution We begin by multiplying the 3 and $2y - 1$. Then, we combine similar terms:

$$3(2y - 1) + y = 6y - 3 + y \qquad \text{Distributive property}$$
$$= 7y - 3 \qquad \text{Combine similar terms} \quad ◀$$

▶ **EXAMPLE 25** Simplify $8 - 3(4x - 2) + 5x$.

Solution First we distribute the -3 across the $4x - 2$.

$$8 - 3(4x - 2) + 5x = 8 - 12x + 6 + 5x$$
$$= -7x + 14 \quad ◀$$

▶ **EXAMPLE 26** Simplify $5(2a + 3) - (6a - 4)$.

Solution We begin by applying the distributive property to remove the parentheses. The expression $-(6a - 4)$ can be thought of as $-1(6a - 4)$. Thinking of it in this way allows us to apply the distributive property.

$$-1(6a - 4) = -1(6a) - (-1)(4) = -6a + 4$$

Here is the complete problem:

$$5(2a + 3) - (6a - 4) = 10a + 15 - 6a + 4 \qquad \text{Distributive property}$$
$$= 4a + 19 \qquad \text{Combine similar}$$
$$\text{terms} \quad ◀$$

Dividing Fractions

We end this section by reviewing division with fractions and division with the number 0.

▶ **EXAMPLES**   Divide and reduce to lowest terms.

27. $\dfrac{3}{4} \div \dfrac{6}{11} = \dfrac{3}{4} \cdot \dfrac{11}{6}$ Definition of division

$\qquad\quad = \dfrac{33}{24}$ Multiply numerators, multiply denominators

$\qquad\quad = \dfrac{11}{8}$ Divide numerator and denominator by 3

28. $10 \div \dfrac{5}{6} = \dfrac{10}{1} \cdot \dfrac{6}{5}$ Definition of division

$\qquad\quad = \dfrac{60}{5}$ Multiply numerators, multiply denominators

$\qquad\quad = 12$ Divide

29. $-\dfrac{3}{8} \div 6 = -\dfrac{3}{8} \cdot \dfrac{1}{6}$ Definition of division

$\qquad\quad = -\dfrac{3}{48}$ Multiply numerators, multiply denominators

$\qquad\quad = -\dfrac{1}{16}$ Divide numerator and denominator by 3   ◀

Division with the Number 0

For every division problem there is an associated multiplication problem involving the same numbers. For example, the following two problems say the same thing about the numbers 2, 3, and 6:

$$\textit{Division} \qquad \textit{Multiplication}$$
$$\dfrac{6}{3} = 2 \qquad\quad 6 = 2(3)$$

We can use this relationship between division and multiplication to clarify division involving the number 0.

First of all, dividing 0 by a number other than 0 is allowed and always results in 0. To see this, consider dividing 0 by 5. We know the answer is 0 because of the relationship between multiplication and division. This is how we write it:

$$\dfrac{0}{5} = 0 \quad \text{because} \quad 0 = 0(5)$$

On the other hand, dividing a nonzero number by 0 is not allowed in the real numbers. Suppose we were attempting to divide 5 by 0. We don't know whether there is an answer to this problem, but if there is, let's say the answer is a number that we can represent with the letter n. If 5 divided by 0 is a number n, then

$$\frac{5}{0} = n \quad \text{and} \quad 5 = n(0)$$

But this is impossible, because no matter what number n is, when we multiply it by 0 the answer must be 0. It can never be 5. In algebra, we say expressions like $\frac{5}{0}$ are undefined, because there is no answer to them. That is, division by 0 is not allowed in the real numbers.

P R O B L E M   S E T   1 . 4

Find each of the following sums.

1. $6 + (-2)$ **2.** $11 + (-5)$
3. $-6 + 2$ **4.** $-11 + 5$
5. $-\frac{1}{2} + (-\frac{1}{6}) + (-\frac{1}{18})$
6. $-\frac{1}{2} + (-\frac{1}{4}) + (-\frac{1}{10})$

Find each of the following differences.

7. $-7 - 3$ **8.** $-6 - 9$
9. $-7 - (-3)$ **10.** $-6 - (-9)$
11. $\frac{3}{4} - (-\frac{5}{6})$ **12.** $\frac{2}{3} - (-\frac{7}{5})$
13. $\frac{11}{42} - \frac{17}{30}$ **14.** $\frac{13}{70} - \frac{19}{42}$

Simplify as much as possible.

15. $6 - (-2) + 11$ **16.** $8 - (-3) + 12$
17. $-\frac{4}{3} - (-\frac{1}{2}) - \frac{3}{2}$
18. $-\frac{1}{6} - (-\frac{1}{3}) - \frac{1}{2}$
19. $-5 - (2 - 6) - 3$ **20.** $-4 - (5 - 9) - 2$
21. $-(2 - 5) - (7 - 3)$
22. $-(8 - 10) - (6 - 1)$
23. Subtract 5 from -3.
24. Subtract -3 from 5.
25. Find the difference of -4 and 8.
26. Find the difference of 8 and -4.
27. Subtract $4x$ from $-3x$.
28. Subtract $-5x$ from $7x$.
29. What number do you subtract from 5 to get -8?
30. What number do you subtract from -3 to get 9?
31. Add -7 to the difference of 2 and 9.
32. Add -3 to the difference of 9 and 2.
33. Subtract $3a$ from the sum of $8a$ and a.
34. Subtract $-3a$ from the sum of $3a$ and $5a$.

Find the following products.

35. $3(-5)$ **36.** $-3(5)$
37. $-3(-5)$ **38.** $4(-6)$
39. $-8(3)$ **40.** $-7(-6)$
41. $-2(-1)(-6)$ **42.** $-3(-2)(5)$
43. $2(-3)(4)$ **44.** $-2(3)(-4)$
45. $-2(5x)$ **46.** $-5(4x)$
47. $-\frac{1}{3}(-3x)$ **48.** $-\frac{1}{6}(-6x)$
49. $-\frac{2}{3}(-\frac{3}{2}y)$ **50.** $-\frac{2}{5}(-\frac{5}{2}y)$
51. $-2(4x - 3)$ **52.** $-6(2x - 1)$
53. $-4(-3t + 7)$ **54.** $-2(-5t + 6)$
55. $-\frac{1}{2}(6a - 8)$ **56.** $-\frac{1}{3}(6a - 9)$
57. $-\frac{1}{2}(-3x - 4)$ **58.** $-\frac{1}{2}(-5x - 8)$

Simplify each expression as much as possible.

59. $3(-4) - 2$ **60.** $-3(-4) - 2$
61. $4(-3) - 6(-5)$ **62.** $-6(-3) - 5(-7)$
63. $2 - 5(-4) - 6$ **64.** $3 - 8(-1) - 7$
65. $2 - 5(-4 - 6)$ **66.** $3 - 8(-1 - 7)$
67. $(2 - 5)(-4 - 6)$ **68.** $(3 - 8)(-1 - 7)$
69. $4 - 3(7 - 1) - 5$ **70.** $8 - 5(6 - 3) - 7$
71. $2(-3)^2 - 4(-2)^3$ **72.** $5(-2)^2 - 2(-3)^3$
73. $(2 - 8)^2 - (3 - 7)^2$
74. $(5 - 8)^2 - (4 - 8)^2$
75. $7(3 - 5)^3 - 2(4 - 7)^3$
76. $3(-7 + 9)^3 - 5(-2 + 4)^3$
77. $-3(2 - 9) - 4(6 - 1)$
78. $-5(5 - 6) - 7(2 - 8)$
79. $-5(-8 - 2) - 3(-2 - 8)$
80. $-3(-5 - 15) - 4(-12 - 8)$
81. $2 - 4[3 - 5(-1)]$ **82.** $6 - 5[2 - 4(-8)]$

83. $(8 - 7)[4 - 7(-2)]$
84. $(6 - 9)[15 - 3(-4)]$
85. $-3 + 4[6 - 8(-3 - 5)]$
86. $-2 + 7[2 - 6(-3 - 4)]$
87. $5 - 6[-3(2 - 9) - 4(8 - 6)]$
88. $9 - 4[-2(4 - 8) - 5(3 - 1)]$

Simplify each expression.

89. $3(5x + 4) - x$
90. $4(7x + 3) - x$
91. $6 - 7(3 - m)$
92. $3 - 5(5 - m)$
93. $7 - 2(3x - 1) + 4x$
94. $8 - 5(2x - 3) + 4x$
95. $5(3y + 1) - (8y - 5)$
96. $4(6y + 3) - (6y - 6)$
97. $4(2 - 6x) - (3 - 4x)$
98. $7(1 - 2x) - (4 - 10x)$
99. $10 - 4(2x + 1) - (3x - 4)$
100. $7 - 2(3x + 5) - (2x - 3)$

Use the definition of division to write each division problem as a multiplication problem, then simplify.

101. $\dfrac{8}{-4}$
102. $\dfrac{-8}{4}$
103. $\dfrac{-8}{-4}$
104. $\dfrac{-12}{-4}$
105. $\dfrac{4}{0}$
106. $\dfrac{-7}{0}$

107. $\dfrac{0}{-3}$
108. $\dfrac{0}{5}$
109. $-\frac{3}{4} \div \frac{9}{8}$
110. $-\frac{2}{3} \div \frac{4}{9}$
111. $-8 \div (-\frac{1}{4})$
112. $-12 \div (-\frac{2}{3})$
113. $-40 \div (-\frac{5}{8})$
114. $-30 \div (-\frac{5}{6})$
115. $\frac{4}{9} \div (-8)$
116. $\frac{3}{7} \div (-6)$

Simplify as much as possible.

117. $\dfrac{6(-2) - 8}{-15 - (-10)}$
118. $\dfrac{8(-3) - 6}{-7 - (-2)}$
119. $\dfrac{3(-1) - 4(-2)}{8 - 5}$
120. $\dfrac{6(-4) - 5(-2)}{7 - 6}$
121. $8 - (-6)\left[\dfrac{2(-3) - 5(4)}{-8(6) - 4}\right]$
122. $-9 - 5\left[\dfrac{11(-1) - 9}{4(-3) + 2(5)}\right]$
123. $6 - (-3)\left[\dfrac{2 - 4(3 - 8)}{1 - 5(1 - 3)}\right]$
124. $8 - (-7)\left[\dfrac{6 - 1(6 - 10)}{4 - 3(5 - 7)}\right]$

125. Subtract -5 from the product of 12 and $-\frac{2}{3}$.
126. Subtract -3 from the product of -12 and $\frac{3}{4}$.
127. Add -5 to the quotient of -3 and $\frac{1}{2}$.
128. Add -7 to the quotient of 6 and $-\frac{1}{2}$.
129. Add $8x$ to the product of -2 and $3x$.
130. Add $7x$ to the product of -5 and $-2x$.

SECTION

1.5 Conditional Statements

Consider the two statements below:

Statement 1: If I study, then I will get good grades.
Statement 2: If x is a negative number, then $-x$ is a positive number.

In Statement 1, if we let A represent the phrase "I study" and B represent the phrase "I will get good grades," then Statement 1 has the form

If A, then B

Likewise, in Statement 2, if A is the phrase "x is a negative number" and B is "$-x$ is a positive number," then Statement 2 has the form

$$\text{If } A, \text{ then } B$$

Each statement has the same form: If A, then B. We call this the "if/then" form, and any statement that has this form is called a **conditional statement.** For every conditional statement, the first phrase, A, is called the **hypothesis,** and the second phrase, B, is called the **conclusion.** All conditional statements can be written in the form

$$\text{If } \textit{hypothesis, } \text{then } \textit{conclusion}$$

Notation   A shorthand way to write an "if/then" statement is with the implies symbol

$$A \Rightarrow B$$

This statement is read "A implies B." It is equivalent to saying "If A, then B."

▶ **EXAMPLE 1**   Identify the hypothesis and conclusion in each statement.
(a) If a and b are positive numbers, then $-a(-b) = ab$.
(b) If it is raining, then the streets are wet.
(c) $C = 90° \Rightarrow c^2 = a^2 + b^2$.

Solution
(a) *Hypothesis: a and b are positive numbers*
    Conclusion: $-a(-b) = ab$
(b) *Hypothesis:* It is raining
    Conclusion: The streets are wet
(c) *Hypothesis: $C = 90°$*
    Conclusion: $c^2 = a^2 + b^2$ ◀

For each conditional statement $A \Rightarrow B$, we can find three related statements that may or may not be true depending on whether or not the original conditional statement is true.

DEFINITIONS   For every conditional statement $A \Rightarrow B$ there exist the following associated statements:

The **converse:**	$B \Rightarrow A$	If B, then A
The **inverse:**	not $A \Rightarrow$ not B	If not A, then not B
The **contrapositive:**	not $B \Rightarrow$ not A	If not B, then not A

▶ **EXAMPLE 2**   For the statement below, write the converse, the inverse, and the contrapositive.

If it is raining, then the streets are wet

Solution   It is sometimes easier to work a problem like this if we write out the phrases *A*, *B*, not *A*, and not *B*.

Let *A* = it is raining, then not *A* = it is not raining
Let *B* = the streets are wet, then not *B* = the streets are not wet

Here are the three associated statements:

The *converse:* (If *B*, then *A*.) If the streets are wet, then it is raining.
The *inverse:* (If not *A*, then not *B*.) If it is not raining, then the streets are not wet.
The *contrapositive:* (If not *B*, then not *A*.) If the streets are not wet, then it is not raining.   ◀

True or False?

Next, we want to answer this question: "If a conditional statement is true, which, if any, of the associated statements are true also?" Consider the statement below.

If it is a square, then it has four sides

We know from our experience with squares that this is a true statement. Now, is the converse necessarily true? Here is the converse:

If it has four sides, then it is a square

Obviously the converse is *not* true, because there are many four-sided figures that are not squares—rectangles, parallelograms, and trapezoids, to mention a few. So, the converse of a true conditional statement is not necessarily true itself.
     Next, we consider the inverse of our original statement.

If it is not a square, then it does not have four sides

Again, the inverse is not true since there are many nonsquare figures that do have four sides. For example, a 3-inch by 5-inch rectangle fits that description.
     Finally, we consider the contrapositive:

If it does not have four sides, then it is not a square

As you can see, the contrapositive is true. That is, if something doesn't have four sides, it can't possibly be a square.
     The discussion above leads us to the following theorem.

> **Theorem**
> If a conditional statement is true, then so is its contrapositive. That is, the two statements
>
> $$\text{If } A, \text{ then } B \quad \text{and} \quad \text{If not } B, \text{ then not } A$$
>
> are equivalent; one can't be true without the other being true also. That is, they are either both true or both false.

The theorem doesn't mention the inverse and the converse because they are true or false independent of the original statement. That is, knowing that a conditional statement is true tells us that the contrapositive is also true—but the truth of the inverse and the converse does not follow from the truth of the original statement.

The next two examples are intended to clarify the discussion above and our theorem. As you read through them, be careful not to let your intuition, experience, or opinion get in the way.

▶ **EXAMPLE 3**   If the statement "If you are guilty, then you will be convicted" is true, give another statement that must also be true.

Solution   From our theorem, and the discussion preceding it, we know that the contrapositive of a true conditional statement is also true. Here is the contrapositive of our original statement:

If you are not convicted, then you are not guilty

Remember, we are not asking for your opinion; we are simply asking for another conditional statement that must be true if the original statement is true. The answer is *always* the contrapositive. Now, you may be wondering about the converse:

If you are convicted, then you must be guilty

It may be that the converse is actually true. But if it is, it is not because of the original conditional statement. That is, the truth of the converse *does not follow* from the truth of the original statement.   ◀

▶ **EXAMPLE 4**   If the statement below is true, what other conditional statement must also be true?

$$\text{If } a = b, \text{ then } a^2 = b^2$$

Solution   Again, every true conditional statement has a true contrapositive. Therefore, the statement below is also true:

$$\text{If } a^2 \neq b^2, \text{ then } a \neq b$$

In this case, we know from experience that the original statement is true; that is, if two numbers are equal, then so are their squares. We also know from experience that the contrapositive is true; if the squares of two numbers are not equal, then the numbers themselves can't be equal. Do you think the inverse and converse are true also? Here is the converse:

$$\text{If } a^2 = b^2, \text{ then } a = b$$

The converse is not true. If a is -3 and b is 3, then a^2 and b^2 are equal, but a and b are not. This same kind of reasoning will show that the inverse is not necessarily true. This example, then, gives further evidence that our theorem is true: A true conditional statement has a true contrapositive. No conclusion can be drawn about the inverse or the converse.   ◄

Everyday Language

In everyday life, we don't always use the ''if/then'' form exactly as we have illustrated it here. Many times we use shortened, reversed, or otherwise altered forms of ''if/then'' statements. For instance, each of the following statements is a variation of the ''if/then'' form, and each carries the same meaning.

> If it is raining, then the streets are wet
> If it is raining, the streets are wet
> When it rains, the streets get wet
> The streets are wet if it is raining
> The streets are wet because it is raining
> Rain will make the streets wet

▶ **EXAMPLE 5**   Write the following statement in ''if/then'' form.

> Romeo loves Juliet

Solution   We must be careful that we do not change the meaning of the statement when we write it in ''if/then'' form. Here is an ''if/then'' form that has the same meaning as the original statement:

> If he is Romeo, then he loves Juliet

We can see that it would be incorrect to rewrite the original statement as

> If she is Juliet, then she loves Romeo

because the original statement is Romeo loves Juliet, not Juliet loves Romeo. It would also be incorrect to rewrite our statement as either

> If he is not Romeo, then he does not love Juliet
>
> or
>
> If he loves Juliet, then he is Romeo

because people other than Romeo may also love Juliet. (The statements above are actually the inverse and converse, respectively, of the original statement.) Finally, another statement that has the same meaning as our original statement is

If he does not love Juliet, then he is not Romeo

because this is the contrapositive of our original statement and we know that the contrapositive is always true when the original statement is true. ◀

PROBLEM SET 1.5

For each conditional statement below, state the hypothesis and the conclusion.

1. If you argue for your limitations, then they are yours.
2. If you think you can, then you can.
3. If x is an even number, then x is divisible by 2.
4. If x is an odd number, then x is not divisible by 2.
5. If a triangle is equilateral, then all of its angles are equal.
6. If a triangle is isosceles, then two of its angles are equal.
7. If $x + 5 = -2$, then $x = -7$.
8. If $x - 5 = -2$, then $x = 3$.

For each conditional statement below, give the converse, the inverse, and the contrapositive.

9. If $a = 8$, then $a^2 = 64$.
10. If $x = y$, then $x^2 = y^2$.
11. If $\frac{a}{b} = 1$, then $a = b$.
12. If $a + b = 0$, then $a = -b$.
13. If it is a square, then it is a rectangle.
14. If you live in a glass house, then you shouldn't throw stones.
15. If better is possible, then good is not enough.
16. If a and b are positive, then ab is positive.

For each statement below, write an equivalent statement in "if/then" form.

17. $E \Rightarrow F$
18. $a^3 = b^3 \Rightarrow a = b$
19. Misery loves company.
20. Rollerblading is not a crime.
21. The squeaky wheel gets the grease.

22. The girl who can't dance says the band can't play.

In each problem below, a conditional statement is given. If the given conditional statement is true, which of the three statements that follow it must also be true?

23. If you heard it first, then you heard it on Eyewitness News.
    (a) If you heard it on Eyewitness News, then you heard it first.
    (b) If you didn't hear it first, then you didn't hear it on Eyewitness News.
    (c) If you didn't hear it on Eyewitness News, then you didn't hear it first.
24. If it is raining, then the streets are wet.
    (a) If the streets are wet, then it is raining.
    (b) If the streets are not wet, then it is not raining.
    (c) If it is not raining, then the streets are not wet.
25. If $C = 90°$, then $c^2 = a^2 + b^2$.
    (a) If $c^2 \neq a^2 + b^2$, then $C \neq 90°$.
    (b) If $c^2 = a^2 + b^2$, then $C = 90°$.
    (c) If $C \neq 90°$, then $c^2 \neq a^2 + b^2$.
26. If you graduate from college, then you will get a good job.
    (a) If you get a good job, then you graduated from college.
    (b) If you do not get a good job, then you did not graduate from college.
    (c) If you do not graduate from college, then you will not get a good job.

27. If you get a B average, then your car insurance will cost less.
   (a) If your car insurance costs less, then you got a B average.
   (b) If you do not get a B average, then your car insurance will not cost less.
   (c) If your car insurance does not cost less, then you did not get a B average.
28. If you go out without a sweater, then you will get sick.
   (a) If you do not get sick, then you went out with a sweater.
   (b) If you go out with a sweater, then you will not get sick.
   (c) If you get sick, then you went out without a sweater.
29. If a and b are negative, then $a + b$ is negative.
   (a) If a and b are not both negative, then $a + b$ is not negative.
   (b) If $a + b$ is not negative, then a and b are not both negative.
   (c) If $a + b$ is negative, then a and b are negative.
30. If it is a square, then all its sides have the same length.
   (a) If all its sides are not the same length, then it is not a square.
   (b) If all its sides have the same length, then it is a square.
   (c) If it is not a square, then its sides are not all the same length.

Research Project 1

To see the connection between the topics we studied in this section and mathematics in general, here is a quote taken from the beginning of the first sentence in the book *Principles of Mathematics* by the British philosopher and mathematician Bertrand Russell.

> Pure Mathematics is the class of all propositions of the form ''p implies q,'' where p and q are propositions containing one or more variables . . .

Russell is using the phrase ''p implies q'' in the same way we use the phrase ''If A, then B.'' Our work with conditional statements is actually an introduction to the foundations upon which all of mathematics is built.

Write an essay on the life of Bertrand Russell. In the essay, indicate what purpose he had for writing and publishing his book *Principles of Mathematics*. Write in complete sentences and organize your work just as you would if you were writing a paper for an English class.

SECTION 1.6   Number Sequences and Inductive Reasoning

Much of what we do in mathematics is concerned with recognizing patterns and classifying together groups of numbers that share a common characteristic. For instance, suppose you were asked to give the next number in this sequence:

$$3, 5, 7, \ldots$$

Looking for a pattern, you may observe that each number is 2 more than the number preceding it. That being the case, the next number in the sequence will be 9 because 9 is 2 more than 7. Reasoning in this manner is called **inductive reasoning.** In mathematics, we use inductive reasoning when we notice a pattern to a sequence of numbers and then extend the sequence using the pattern.

(a) 2, 10, 50, . . . : Starting with 2, each number is obtained from the previous number by multiplying by 5 each time. The next number will be $50 \cdot 5 = 250$.

(b) $3, -15, 75, \ldots$: The sequence starts with 3. After that, each number is obtained by multiplying by -5 each time. The next number will be $75(-5) = -375$.

(c) $\frac{1}{8}, \frac{1}{4}, \frac{1}{2}, \ldots$: This sequence starts with $\frac{1}{8}$. Multiplying each number in the sequence by 2 produces the next number in the sequence. To extend the sequence we multiply $\frac{1}{2}$ by 2:

$$\frac{1}{2} \cdot 2 = 1$$

The next number in the sequence is 1.   ◄

The Fibonacci Sequence

In the introduction to this chapter we quoted the mathematician Fibonacci. There is a special sequence in mathematics named for Fibonacci.

Fibonacci sequence: 1, 1, 2, 3, 5, 8, . . .

To construct the Fibonacci sequence we start with two 1's. The rest of the numbers in the sequence are found by adding the two previous terms. Adding the first two terms, 1 and 1, we have 2. Then, adding 1 and 2 we have 3. In general, adding any two consecutive terms of the Fibonacci sequence gives us the next term.

A Mathematical Model

One of the reasons we study number sequences is because they can be used to model some of the patterns and events we see in the world around us. The discussion that follows shows how the Fibonacci sequence can be used to predict the number of bees in each generation of the family tree of a male honeybee. It is based on an example from Chapter 2 of the book *Mathematics: A Human Endeavor,* by Harold Jacobs. If you find that you enjoy discovering patterns in mathematics, Mr. Jacobs' book has many interesting examples and problems involving patterns in mathematics.

A male honeybee has one parent, its mother, while a female honeybee has two parents, a mother and a father. (A male honeybee comes from an unfertilized egg; a female honeybee comes from a fertilized egg.) Using these facts, we construct the family tree of a male honeybee using M to represent a male honeybee and F to represent a female honeybee.

Generation   Number of Bees

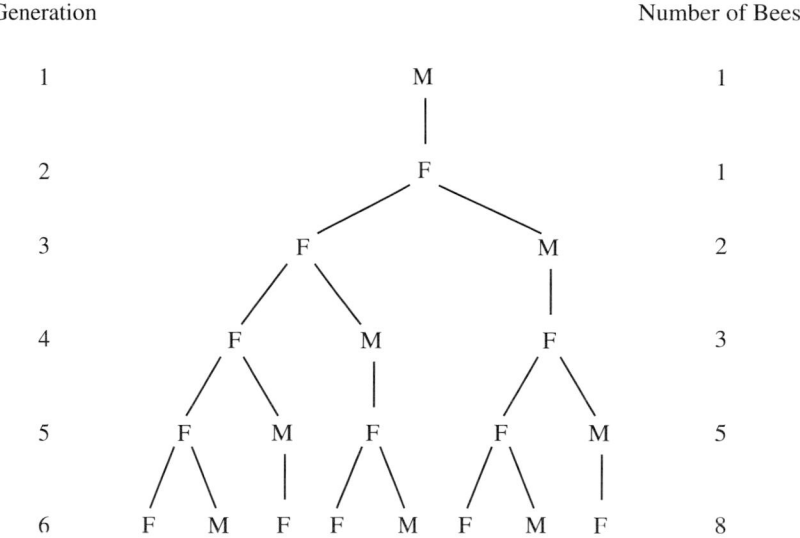

Looking at the numbers in the right column in our diagram, the sequence that gives us the number of bees in each generation of the family tree of a male honeybee is

$$1 \quad 1 \quad 2 \quad 3 \quad 5 \quad 8$$

As you can see, this is the Fibonacci sequence. We have taken our original diagram (the family tree of the male honeybee) and reduced it to a mathematical model (the Fibonacci sequence). The model can be used in place of the diagram to find the number of bees in any generation back from our first bee.

▶ **EXAMPLE 4**   Find the number of bees in the tenth generation of the family tree of a male honeybee.

Solution   We can continue the diagram above, and simply count the number of bees in the tenth generation, or we can use inductive reasoning to conclude that the number of bees in the tenth generation will be the tenth term of the Fibonacci sequence. Let's make it easy on ourselves and find the first ten terms of the Fibonacci sequence.

Generation:	1	2	3	4	5	6	7	8	9	10
Number of bees:	1	1	2	3	5	8	13	21	34	55

As you can see, the number of bees in the tenth generation is 55.   ◄

Note   It is not unusual for two students to see different patterns in the same number sequence. For example, what is the next number in this sequence:

$$3, 6, 9, 12, \ldots$$

▶ **EXAMPLE 1**   Table 2 gives the net price of a textbook on January 1 of each year in which a new edition was published. Use the information in the table to construct a histogram and a line graph.

TABLE 2   Price of Textbook

Edition	Year Published	Price ($)
First	1979	14.95
Second	1982	21.50
Third	1986	27.50
Fourth	1990	36.25

Solution   We can pair the edition number with price or we can pair the year published with price. Note that the number of years from one edition to the next varies; the first edition was in print for 3 years, while the second and third editions were in print for four years. When we label an axis with numbers, we should try to keep the distances between them consistent. The histogram is shown in Figure 4; it pairs the edition number with the price. The line graph is shown in Figure 5; it pairs the year published with the price. Note how we have labeled the horizontal axis in Figure 5 so that the distance from one number to the next is consistent.

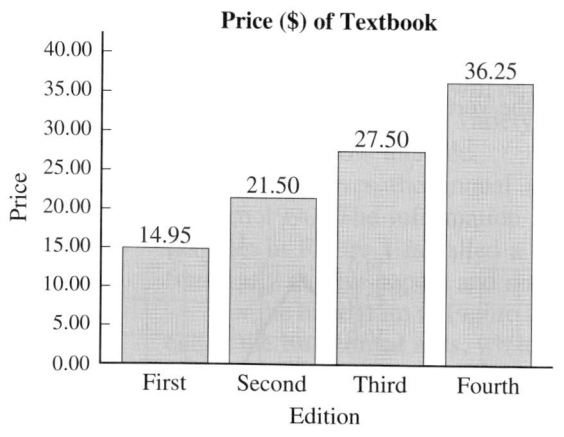

Figure 4

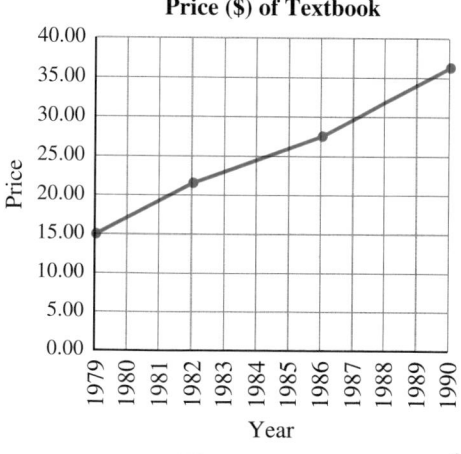

Figure 5

▶ **EXAMPLE 2**   Use the line graph in Figure 5 to estimate the price of the textbook in 1983 and in 1993.

Solution   Figure 6 shows the line graph from Figure 5 with two additional lines drawn. The vertical line that starts at 1983 is used to estimate the price of the textbook that year. If we follow the dotted line from 1983 up to the line graph itself and then over to the vertical axis (follow the arrows), we see that a reasonable estimate of the textbook price in 1983 is $23.00.

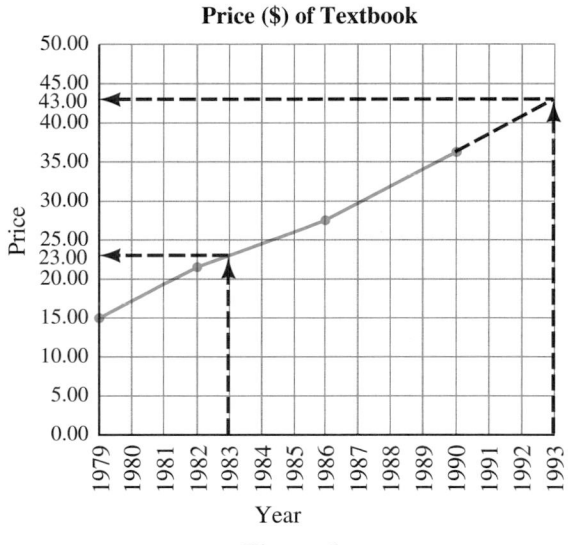

Figure 6

To estimate the price of the textbook in 1993, we must extend our line graph, which requires that we extend both axes (as shown in Figure 6). To extend our line graph, we assume that it continues from 1990 to 1993 in the same manner as from 1986 to 1990. As you can see, a reasonable estimate of the price of this book in 1993 is $43.00. ◄

When we estimate a pair of numbers that falls within the data we are given, we are **interpolating** to obtain our estimate. When we estimate a pair of numbers that falls outside the data we are given, we are **extrapolating** to obtain our estimate. Estimating that the price of this textbook was $23.00 in 1983 is interpolation. The estimation of $43.00 for the 1993 price of the textbook was found by extrapolation.

We can incorporate negative numbers into our tables and diagrams as the next example illustrates.

▶ **EXAMPLE 3**   Table 3 gives an indication of how one category of air pollutant changed during the 1980s. It gives the percent change from the previous year for the total amount of nitrogen oxides in the air. Use the information in the table to construct a histogram.

TABLE 3   Nitrogen Oxides Emissions

Year	Percent Change from Previous Year
1980	16.3
1981	−9.3
1982	−4.7
1983	−2.9
1984	1.5
1985	−3.5
1986	−3.1
1987	1.1
1988	3.7
1989	−2.0

Solution   Figure 7 is a histogram constructed from the information in Table 3. Note that we aligned the horizontal axis with 0 on the vertical axis, and that the rectangles associated with negative numbers are drawn below the horizontal axis.

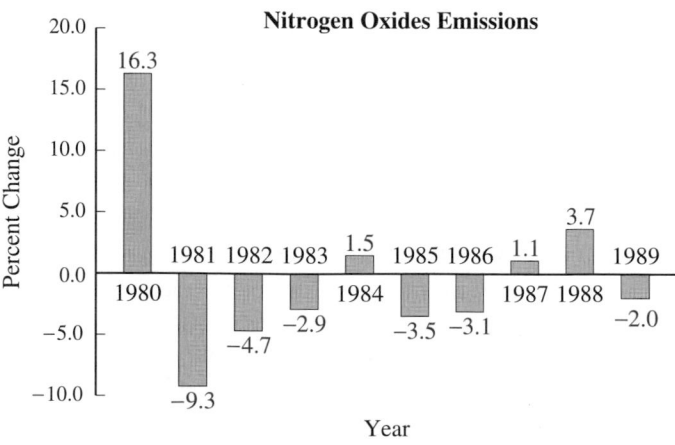

Figure 7

The number sequences we worked with in the previous section can also be written as paired data by associating each number in the sequence with its position in the sequence. For instance, in the sequence of even numbers

$$2, 4, 6, 8, 10, \ldots$$

the number 8 is the fourth number in the sequence. Its position is 4 and its value is 8. Here is the sequence of even numbers written so that the position of each term is noted.

Position: 1, 2, 3, 4, 5, . . .
Value: 2, 4, 6, 8, 10, . . .

▶ **EXAMPLE 4**   Tables 4 and 5 give the first five terms in each of two sequences. The arithmetic sequence is formed by starting with 3 and adding 2 each time. The geometric sequence is formed by starting with 3 and multiplying by 2 each time. In each case, use the information to construct a scatter diagram.

TABLE 4 An Arithmetic Sequence		TABLE 5 A Geometric Sequence	
Position	**Value**	**Position**	**Value**
1	3	1	3
2	5	2	6
3	7	3	12
4	9	4	24
5	11	5	48

Solution   The two scatter diagrams are shown in Figures 8 and 9. Notice how the dots in Figure 8 seem to line up in a straight line, while the dots in Figure 9 give the impression of a curve. We say that the points in Figure 8 suggest a *linear* relationship between the two sets of data, while the points in Figure 9 suggest a *nonlinear* relationship.

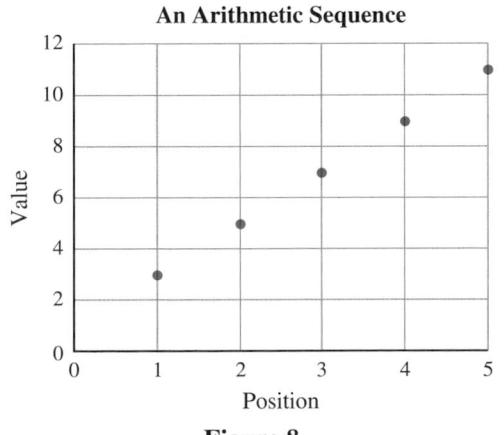

Figure 8

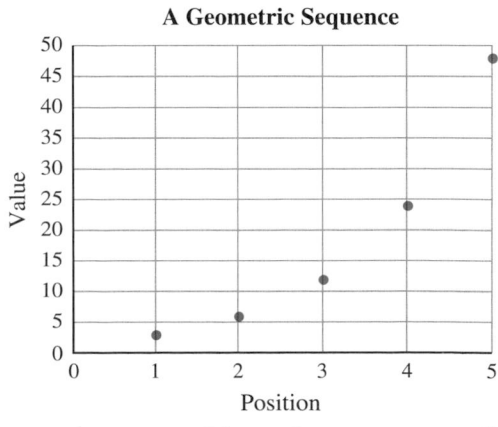

Figure 9   ◀

In 1993 I bought a new Ford Mustang LX with a 5.0 liter V-8 engine. A year later I decided to check the newspaper for prices of used Mustangs. Table 6 gives the result.

TABLE 6   Used Mustang Prices

Year	Age from 1994	Price ($)
1993	1	13,925
1993	1	11,850
1993	1	9,995
1992	2	10,200
1992	2	9.600
1991	3	9,525
1990	4	8,675
1990	4	7,900
1989	5	6,975

Figure 10 shows a scatter diagram for the information in Table 6. Note that this data is different from all the previous data in that some numbers appearing on the horizontal axis are paired with more than one number on the vertical axis. This fact will become more important when we discuss functions in the next chapter. For now, it is enough to simply notice that this is true.

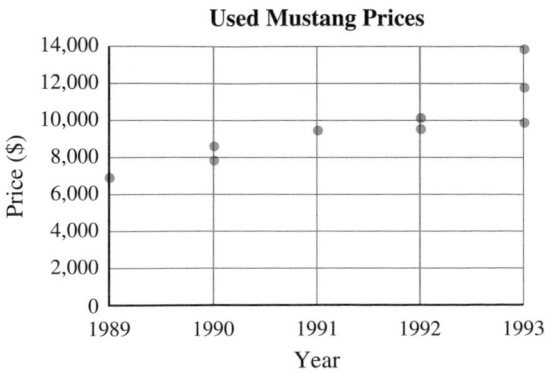

Figure 10

USING TECHNOLOGY: Spreadsheet Programs
....

From time to time we will point out how a particular topic can be enhanced with the use of technology.

When I put together the manuscript for this book, I used a spreadsheet program to draw the histograms, scatter diagrams, and line graphs that you see in this section. Every current spreadsheet program has graphing capabilities. If you don't own a computer with a spreadsheet program, you can often rent time on one at a copy center.

Figure 11 shows what the screen of my computer looked like when I was preparing the histogram for Example 1 of this section. The histogram was drawn by the computer from the data in the table. A histogram was just one of many choices for ways in which to display the data. You should consider producing the histograms and other graphics asked for in Problem Set 1.7 using a spreadsheet program, instead of doing them by hand.

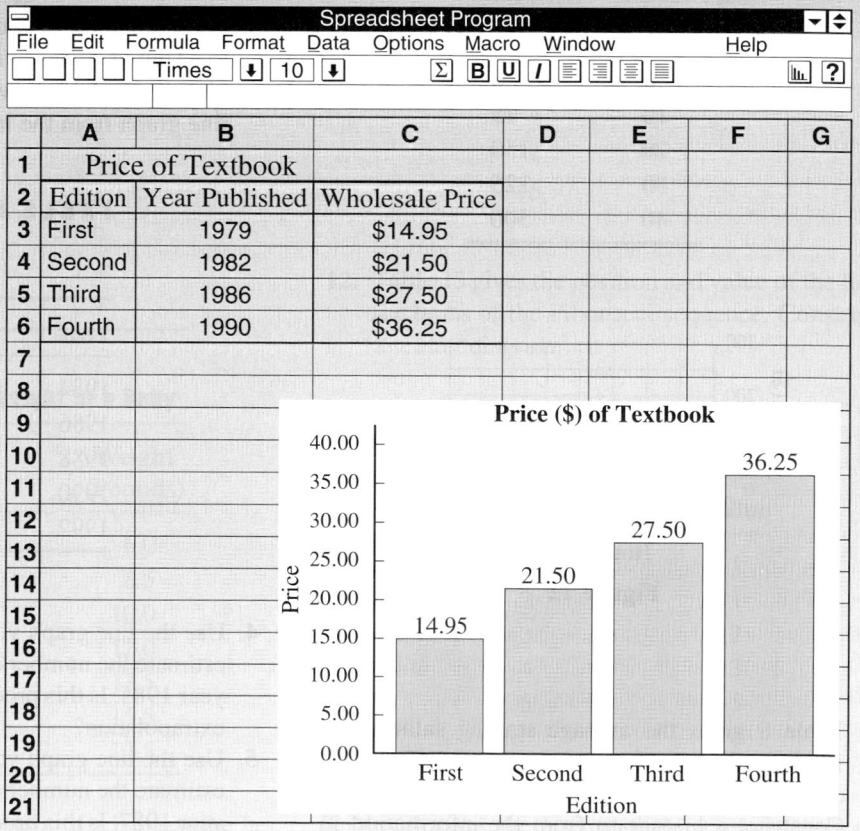

Figure 11

2. $10 + (2 \cdot 3^2 - 4 \cdot 2)$

$= 10 + (2 \cdot 9 - 4 \cdot 2)$

$= 10 + (18 - 8)$

$= 10 + 10$

$= 20$

Order of Operations [1.1]

When evaluating a mathematical expression, we will perform the operations in the following order, beginning with the expression in the innermost parentheses or brackets and working our way out.

1. Simplify all numbers with exponents, working from left to right if more than one of these numbers is present.
2. Then, do all multiplications and divisions left to right.
3. Finally, perform all additions and subtractions left to right.

3. If $A = \{0, 1, 2\}$ and $B = \{2, 3\}$, then $A \cup B = \{0, 1, 2, 3\}$ and $A \cap B = \{2\}$.

Sets [1.1]

A *set* is a collection of objects or things.

The *union* of two sets A and B, written $A \cup B$, is all the elements that are in A *or* are in B.

The *intersection* of two sets A and B, written $A \cap B$, is the set consisting of all elements common to both A *and* B.

Set A is a *subset* of set B, written $A \subset B$, if all elements in set A are also in set B.

4. 5 is a counting number, a whole number, an integer, a rational number, and a real number.

$\frac{3}{4}$ is a rational number and a real number.

$\sqrt{2}$ is an irrational number and a real number.

Special Sets [1.2]

Counting numbers $= \{1, 2, 3, \ldots\}$

Whole numbers $= \{0, 1, 2, 3, \ldots\}$

Integers $= \{\ldots, -3, -2, -1, 0, 1, 2, 3, \ldots\}$

Rational numbers $= \left\{ \dfrac{a}{b} \,\middle|\, a \text{ and } b \text{ are integers, } b \neq 0 \right\}$

Irrational numbers $= \{x \,|\, x \text{ is real, but not rational}\}$

Real numbers $= \{x \,|\, x \text{ is rational or } x \text{ is irrational}\}$

Prime numbers $= \{2, 3, 5, 7, 11, \ldots\} = \{x \,|\, x$ is a positive integer greater than 1 whose only positive divisors are itself and 1$\}$

5. Graph each inequality mentioned at right.

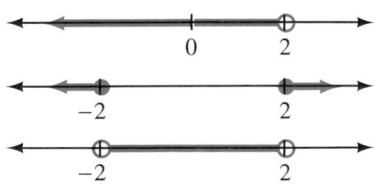

Inequalities [1.2]

The set $\{x \,|\, x < 2\}$ is the set of all real numbers that are less than 2. To graph this set we place an open circle at 2 on the real number line and then draw an arrow that starts at 2 and points to the left.

The set $\{x \,|\, x \leq -2 \text{ or } x \geq 2\}$ is the set of all real numbers that are either less than or equal to -2 or greater than or equal to 2.

The set $\{x \,|\, -2 < x < 2\}$ is the set of all real numbers that are between -2 and 2, that is, the real numbers that are greater than -2 and less than 2.

6. The numbers 5 and -5 are opposites; their sum is 0.

$$5 + (-5) = 0$$

Opposites [1.3]

Any two real numbers the same distance from 0 on the number line, but in opposite directions from 0, are called *opposites* or *additive inverses*. Opposites always add to 0.

7. The numbers 3 and $\frac{1}{3}$ are reciprocals; their product is 1.

$$3\left(\frac{1}{3}\right) = 1$$

Reciprocals [1.3]
Any two real numbers whose product is 1 are called *reciprocals*. Every real number has a reciprocal except 0.

8. $|5| = 5$

$|-5| = 5$

Absolute Value [1.3]
The *absolute value* of a real number is its distance from 0 on the number line. If $|x|$ represents the absolute value of x, then

$$|x| = \begin{cases} x & \text{if} \quad x \geq 0 \\ -x & \text{if} \quad x < 0 \end{cases}$$

The absolute value of a real number is never negative.

Properties of Real Numbers [1.3]

	For Addition	*For Multiplication*
Commutative	$a + b = b + a$	$ab = ba$
Associative	$a + (b + c) = (a + b) + c$	$a(bc) = (ab)c$
Identity	$a + 0 = a$	$a \cdot 1 = a$
Inverse	$a + (-a) = 0$	$a\left(\dfrac{1}{a}\right) = 1$
Distributive	$a(b + c) = ab + ac$	

9. $5 + 3 = 8$

$5 + (-3) = 2$

$-5 + 3 = -2$

$-5 + (-3) = -8$

Addition [1.4]
To add two real numbers with

1. *the same sign:* simply add absolute values and use the common sign.
2. *different signs:* subtract the smaller absolute value from the larger absolute value. The answer has the same sign as the number with the larger absolute value.

10. $6 - 2 = 6 + (-2) = 4$

$6 - (-2) = 6 + 2 = 8$

Subtraction [1.4]
If a and b are real numbers,

$$a - b = a + (-b)$$

To subtract b, add the opposite of b.

11. $5(4) = 20$

$5(-4) = -20$

$-5(4) = -20$

$-5(-4) = 20$

Multiplication [1.4]
To multiply two real numbers, simply multiply their absolute values. Like signs give a positive answer. Unlike signs give a negative answer.

12. $\dfrac{12}{-3} = -4$

$\dfrac{-12}{-3} = 4$

13. The statement below is a conditional statement.

$$\text{If } x = 3, \text{ then } x^2 = 9$$

The hypothesis is $x = 3$, while the conclusion is $x^2 = 9$.

For the conditional statement above, we have the following three associated statements:

The *converse:*
$\qquad$ If $x^2 = 9$, then $x = 3$
The *inverse:*
$\qquad$ If $x \neq 3$, then $x^2 \neq 9$
The *contrapositive:*
$\qquad$ If $x^2 \neq 9$, then $x \neq 3$

14. We use inductive reasoning when we conclude that the next number in the sequence below is 25.

$$1, 4, 9, 16, \ldots$$

15. The sequence below is an arithmetic sequence since each term is obtained from the preceding term by adding 3 each time.

$$4, 7, 10, 13, \ldots$$

16. The sequence below is a geometric sequence since each term is obtained from the previous term by multiplying by 3 each time.

$$4, 12, 36, 108, \ldots$$

Division [1.4]

If a and b are real numbers and $b \neq 0$, then

$$\frac{a}{b} = a \cdot \left(\frac{1}{b} \right)$$

To divide by b, multiply by the reciprocal of b.

Conditional Statements [1.5]

A *conditional statement* is any statement that can be written in the form "If A, then B." The first phrase, A, is called the *hypothesis,* and the second phrase, B, is called the *conclusion.* A shorthand way to write an "if/then" statement is with the implies symbol:

$$A \Rightarrow B$$

For every conditional statement "If A, then B" there exist the following associated statements:

$\qquad$ The *converse:* If B, then A
$\qquad$ The *inverse:* If not A, then not B
$\qquad$ The *contrapositive:* If not B, then not A

If the original conditional statement is true, then the contrapositive is also true.

Inductive Reasoning [1.6]

Inductive reasoning is reasoning in which a conclusion is drawn based on evidence and observations that support that conclusion. In mathematics this usually involves noticing that a few items in a group have a trait or characteristic in common, and then concluding that all items in the group have that same trait.

Arithmetic Sequence [1.6]

An *arithmetic sequence* is a sequence of numbers in which each number (after the first number) comes from adding the same amount to the number before it. The number we add to each term to obtain the next term is called the *common difference.*

Geometric Sequence [1.6]

A *geometric sequence* is a sequence of numbers in which each number (after the first number) comes from the number before it by multiplying by the same amount each time. The amount by which we multiply each term to obtain the next term is called the *common ratio.*

COMMON MISTAKES

1. Interpreting absolute value symbols to mean that the sign (whether positive or negative) of the number inside the absolute value symbols changes. That is, $|-5| = +5$, $|+5| = -5$. To avoid this mistake, remember, absolute value is defined as a distance and distance is always measured in positive units.

2. Confusing $-(-5)$ with $-|-5|$. The first answer is $+5$, while the second answer is -5.

CHAPTER 1 REVIEW

The numbers in brackets refer to the section(s) in the text where similar problems can be found.

Translate each expression into symbols. [1.1]

1. The sum of x and 2.
2. The difference of x and 2.
3. The quotient of x and 2.
4. The product of 2 and x.
5. Twice the sum of x and y.
6. The sum of twice x and y.

Expand and multiply. [1.1]

7. 3^3 **8.** 5^3
9. 8^2 **10.** 1^8
11. 2^5 **12.** 3^4

Simplify each expression. [1.1]

13. $2 + 3 \cdot 5$ **14.** $10 - 2 \cdot 3$
15. $20 \div 2 + 3$ **16.** $30 \div 6 + 4 \div 2$
17. $3 + 2(5 - 2)$ **18.** $(10 - 2)(7 - 3)$
19. $3 \cdot 4^2 - 2 \cdot 3^2$ **20.** $3 + 5(2 \cdot 3^2 - 10)$

Find each of the following if $A = \{1, 3, 5\}$, $B = \{2, 4, 6\}$, and $C = \{0, 1, 2, 3, 4$: [1.1]

21. $A \cup B$ **22.** $A \cap C$
23. $\{x \mid x \in A \text{ and } x \notin C\}$
24. $\{x \mid x \in B \text{ and } x > 4\}$
25. Locate the numbers $-4, -2.5, -1, 0, 1.5, 3.1,$ 4.75 on the number line. [1.2]
26. Locate the numbers $-4.75, -3, -1, 0, 1.5, 2.3,$ 3 on the number line. [1.2]

Give the opposite and reciprocal of each number. [1.3]

27. 2   **28.** $-\frac{2}{5}$

Write each expression without absolute value symbols, then simplify. [1.3]

29. $|-3|$ **30.** $-|-5|$
31. $|-7| - |-3|$ **32.** $|-3| - |-7|$

For the set $\{-7, -4.2, -\sqrt{3}, 0, \frac{3}{4}, \pi, 5\}$ list all the elements that are in the following sets: [1.2]

33. Integers   **34.** Rational numbers
35. Irrational numbers
36. Factor 4,356 into the product of prime factors. [1.2]
37. Reduce $\dfrac{4,356}{5,148}$ to lowest terms. [1.2]

Multiply. [1.3]

38. $\frac{3}{4} \cdot \frac{8}{5} \cdot \frac{5}{6}$ **39.** $(\frac{3}{4})^3$
40. $\frac{1}{4} \cdot 8$

Graph each inequality. [1.2]

41. $\{x \mid x < -2 \text{ or } x > 3\}$
42. $\{x \mid x > 2 \text{ and } x < 5\}$
43. $\{x \mid -3 \leq x \leq 4\}$
44. $\{x \mid 0 \leq x \leq 5 \text{ or } x > 10\}$

Translate each statement into an equivalent inequality. [1.2]

45. x is at least 4
46. x is no more than 5
47. x is between 0 and 8
48. x is between 0 and 8, inclusive

Combine similar terms. [1.3, 1.4]

49. $-2y + 4y$ **50.** $-3x - x + 7x$
51. $3x - 2 + 5x + 7$ **52.** $2y + 4 - y - 2$

Match each expression on the left with the letter of the appropriate property (or properties) on the right. [1.3]

53. $x + 3 = 3 + x$

54. $(x + 2) + 3 = x + (2 + 3)$

55. $3(x + 4) = 3(4 + x)$

56. $(5x)y = x(5y)$

57. $(x + 2) + y = (x + y) + 2$

58. $3(1) = 3$

59. $5 + 0 = 5$

60. $5 + (-5) = 0$

    a. Commutative property of addition

    b. Commutative property of multiplication

    c. Associative property of addition

    d. Associative property of multiplication

    e. Additive identity

    f. Multiplicative identity

    g. Additive inverse

    h. Multiplicative inverse

Find the following sums and differences. [1.4]

61. $5 - 3$

62. $-5 - (-3)$

63. $7 + (-2) - 4$

64. $6 - (-3) + 8$

65. $|-4| - |-3| + |-2|$

66. $(7 - 9) - (-3 - 5)$

67. $6 - (-3) - 2 - 5$

68. $2 \cdot 3^2 - 4 \cdot 2^3 + 5 \cdot 4^2$

69. $-\frac{1}{12} - \frac{1}{6} - \frac{1}{4} - \frac{1}{3}$

70. $-\frac{1}{3} - \frac{1}{4} - \frac{1}{6} - \frac{1}{12}$

Find the following products. [1.4]

71. $6(-7)$

72. $-3(5)(-2)$

73. $7(3x)$

74. $-3(2x)$

Apply the distributive property. [1.4]

75. $-2(3x - 5)$

76. $-3(2x - 7)$

77. $-\frac{1}{2}(2x - 6)$

78. $-3(5x - 1)$

Divide. [1.4]

79. $-\frac{5}{8} \div \frac{3}{4}$

80. $-12 \div \frac{1}{3}$

81. $\frac{3}{5} \div 6$

82. $\frac{4}{7} \div (-2)$

Simplify each expression as much as possible. [1.4]

83. $2(-5) - 3$

84. $3(-4) - 5$

85. $6 + 3(-2)$

86. $7 + 2(-4)$

87. $-3(2) - 5(6)$

88. $-4(3)^2 - 2(-1)^3$

89. $8 - 2(6 - 10)$

90. $(8 - 2)(6 - 10)$

91. $\dfrac{3(-4) - 8}{-5 - 5}$

92. $\dfrac{9(-1)^3 - 3(-6)^2}{6 - 9}$

93. $4 - (-2)\left[\dfrac{6 - 3(-4)}{1 + 5(-2)}\right]$

Simplify. [1.3, 1.4]

94. $7 - 2(3y - 1) + 4y$

95. $4(3x - 1) - 5(6x + 2)$

96. $4(2a - 5) - (3a + 2)$

For each statement below, write the converse, the inverse, and the contrapositive. [1.5]

97. If $x = -7$, then $|x| = 7$.

98. If Therese lives in Amarillo, then she lives in Texas.

For each conditional statement below, assume that the statement is true and then write another conditional statement that must also be true. [1.5]

99. If $x = 3$, then $x^4 = 81$.

100. If Maria goes to the beach, then she does not go to school.

Find the next number in each sequence. Identify any sequences that are arithmetic and any that are geometric. [1.6]

101. 11, 8, 5, 2, . . .

102. 10, −30, 90, −270, . . .

103. 1, 1, 2, 3, 5, . . .

104. 1, 16, 81, 256, . . .

105. $1, \frac{1}{2}, 0, -\frac{1}{2}, \ldots$

106. $1, -\frac{1}{2}, \frac{1}{4}, -\frac{1}{8}, \ldots$

107. Table 1 gives the position and value of the first five terms in an arithmetic sequence (start with 5, add 3 each time). Construct a histogram and a line graph from the information in the table. [1.7]

TABLE 1 Arithmetic Sequence

Position	Value
1	5
2	8
3	11
4	14
5	17

108. Table 2 gives the position and value of the first four terms of a geometric sequence (start with 5, multiply by 3 each time). Construct a histogram and a scatter diagram from the information in the table. [1.7]

TABLE 2 Geometric Sequence

Position	Value
1	5
2	15
3	45
4	135

Table 3 gives the interest earned on $500 if it is invested for 1 year at a variety of simple interest rates. [1.7]

109. Construct a scatter diagram from the information in the table.
110. Construct a histogram from the information in the table.

TABLE 3 Interest Earned in 1 Year

Interest Rate (%)	Interest Earned ($)
2.0	10.00
2.5	12.50
3.0	15.00
3.5	17.50
4.0	20.00

111. Does your scatter diagram suggest that the relationship between interest rate and interest earned is a linear one?
112. Use your scatter diagram to predict the amount of interest that is earned on $500 if it is invested at 4.5% for 1 year. Was this estimate found by interpolation or extrapolation?

CHAPTER 1 TEST

The numbers in brackets indicate the section(s) to which the problems correspond.

Write each of the following in symbols. [1.1]

1. Twice the sum of $3x$ and $4y$.
2. The difference of $2a$ and $3b$ is less than their sum.

Simplify each expression using the rule for order of operations. [1.1]

3. $3 \cdot 2^2 + 5 \cdot 3^2$ **4.** $6 + 2(4 \cdot 3 - 10)$
5. $12 - 8 \div 4 + 2 \cdot 3$
6. $20 - 4[3^2 - 2(2^3 - 6)]$

Find the following if $A = \{1, 2, 3, 4\}$, $B = \{2, 4, 6\}$, and $C = \{1, 3, 5\}$: [1.1]

7. $A \cap B$
8. $\{x \mid x \in B \text{ and } x \in C\}$

Give the opposite and reciprocal of each of the following. [1.3]

9. -3 **10.** $\frac{4}{3}$

Simplify each of the following. [1.3]

11. $-(-3)$ **12.** $-|-2|$

Indicate which elements from the set
$\{-5, -4.1, -3.75, -\frac{5}{6}, -\sqrt{2}, 0, \sqrt{3}, 1, 1.8, 4\}$,
belong in each of the following. [1.2]

13. Integers   **14.** Rational numbers
15. Irrational numbers
16. Factor 585 into the product of prime factors. [1.2]

Graph each of the following. [1.2]

17. $\{x \mid x \le -1 \text{ or } x > 5\}$
18. $\{x \mid -2 \le x \le 4\}$

State the property or properties that justify each of the following. [1.3]

19. $4 + x = x + 4$ **20.** $5(1) = 5$
21. $3(x \cdot y) = (3y) \cdot x$
22. $(a + 1) + b = (a + b) + 1$

Simplify each of the following as much as possible. [1.4]

23. $5(-4) + 1$ **24.** $-4(-3) + 2$
25. $3(2 - 4)^3 - 5(2 - 7)^2$
26. $-2 + 5[7 - 3(-4 - 8)]$
27. $\dfrac{-4(-1) - (-10)}{5 - (-2)}$
28. $3 - 2\left[\dfrac{8(-1) - 7}{-3(2) - 4}\right]$
29. $-\frac{3}{8} + \frac{5}{12} - (-\frac{7}{9})$
30. $-\frac{1}{2}(8x)$ **31.** $-4(3x + 2) + 7x$
32. $5(2y - 3) - (6y - 5)$
33. $3 + 4(2x - 5) - 5x$
34. $2 + 5a + 3(2a - 4)$
35. Add $-\frac{2}{3}$ to the product of -2 and $\frac{5}{6}$.
36. Subtract $\frac{3}{4}$ from the product of -4 and $\frac{7}{16}$.
37. For the statement below, write the converse, the inverse, and the contrapositive. [1.5]

If Emily goes out at night,
then she does not study

38. Assume the conditional statement below is true, then write another conditional statement that must also be true. [1.5]

If $x = -3$, then $|x| = 3$

Find the next number in each sequence. Identify any sequences that are arithmetic and any that are geometric. [1.6]

39. $5, -20, 80, \ldots$ **40.** $1, 0, -1, \ldots$
41. $7, 8, 10, 13, \ldots$ **42.** $1, \frac{1}{5}, \frac{1}{25}, \ldots$

Table 1 gives the relationship between the number of 12-ounce bottles of beer an average size person (154 pounds) consumes, and their approximate blood alcohol level. The table is taken from the 1987 edition of the *American Medical Association Family Medical Guide*. [1.7]

T A B L E 1   Blood Alcohol Levels

Bottles of Beer Consumed	Blood Alcohol Level
2	0.05
3	0.08
5	0.13
10	0.26
12	0.32

43. Construct a histogram from the information in the table.
44. Construct a line graph from the information in the table.
45. Estimate the blood alcohol level after drinking 4 bottles of beer. Was this estimate found by interpolation or extrapolation?
46. Estimate the blood alcohol level after drinking 13 bottles of beer. Was this estimate found by interpolation or extrapolation?

2 LINEAR EQUATIONS AND INEQUALITIES IN ONE VARIABLE

INTRODUCTION

The topics we cover in this course are interdependent. Each topic is linked to other topics, which are linked to still more topics. The links form chains that hold all of mathematics together. The diagram below shows the relationships that exist among some of the major topics we will cover. Some topics are linked to other topics, and many topics are linked to the real world by applications.

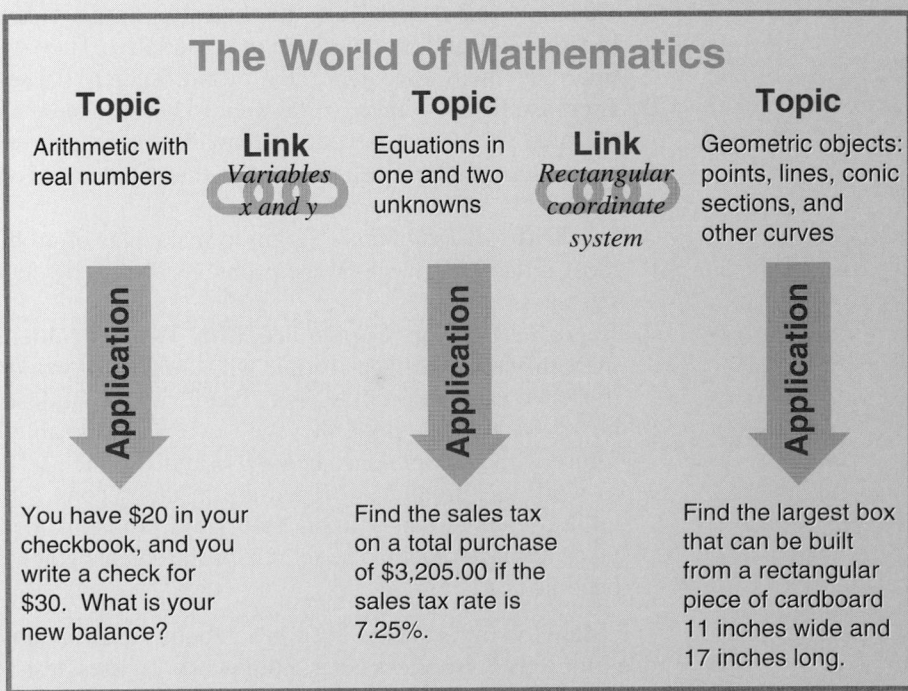

The World of Mathematics

Topic	**Link**	**Topic**	**Link**	**Topic**
Arithmetic with real numbers	*Variables x and y*	Equations in one and two unknowns	*Rectangular coordinate system*	Geometric objects: points, lines, conic sections, and other curves

Application

You have $20 in your checkbook, and you write a check for $30. What is your new balance?

Application

Find the sales tax on a total purchase of $3,205.00 if the sales tax rate is 7.25%.

Application

Find the largest box that can be built from a rectangular piece of cardboard 11 inches wide and 17 inches long.

OVERVIEW

In this chapter you will learn the basic steps used to solve linear equations and inequalities in one variable. The methods we develop here will be used again and again throughout the rest of the book. Here is a list of the more important concepts needed to begin this chapter:

1. You must know how to add, subtract, multiply, and divide positive and negative numbers.
2. You should know the commutative, associative, and distributive properties.
3. You should understand that opposites add to 0 and reciprocals multiply to 1.
4. You must know the definition of absolute value.
5. Finally, you should know that the least common denominator for a set of fractions is divisible by each of the denominators used to find it.

STUDY SKILLS

If you have successfully completed Chapter 1, then you have made a good start at developing the study skills necessary to succeed in all math classes. Some of the study skills for this chapter are a continuation of the skills from Chapter 1, while others are new to this chapter.

1. **Continue to Set and Keep a Schedule**   Sometimes I find students do well in Chapter 1 and then become overconfident. They will begin to put in less time with their homework. Don't do it. Keep to the same schedule.
2. **Increase Effectiveness**   You want to become more and more effective with the time you spend on your homework. Increase those activities that are the most beneficial and decrease those that have not given you the results you want.
3. **List Difficult Problems**   Begin to make lists of problems that give you the most difficulty. These are the problems in which you are repeatedly making mistakes.
4. **Begin to Develop Confidence with Word Problems**   It seems that the main difference between people who are good at working word problems and those who are not is confidence. People with confidence know that no matter how long it takes them, they will eventually be able to solve the problem. Those without confidence begin by saying to themselves, ''I'll never be able to work this problem.'' If you are in this second category, then instead of telling yourself that you can't do word problems, decide to do whatever it takes to master them. The more word problems you work the better you will become at it.

   Many of my students keep a notebook that contains everything that they need for the course: class notes, homework, quizzes, tests, and research projects.

A three-ring binder with tabs is ideal. Organize your notebook so that you can easily get to any item you wish to look at.

SECTION
2.1

Linear Equations in One Variable

A **linear equation in one variable** is any equation that can be put in the form

$$ax + b = c$$

where a, b, and c are constants and $a \neq 0$. For example, each of the equations

$$5x + 3 = 2 \qquad 2x = 7 \qquad 2x + 5 = 0$$

are linear because they can be put in the form $ax + b = c$. In the first equation, $5x$, 3, and 2 are called **terms** of the equation—$5x$ is a variable term; 3 and 2 are constant terms.

DEFINITION   The **solution set** for an equation is the set of all numbers that, when used in place of the variable, make the equation a true statement.

▶ **EXAMPLE 1**   The solution set for $2x - 3 = 9$ is $\{6\}$, since replacing x with 6 makes the equation a true statement.

$$
\begin{aligned}
\text{If} \qquad\qquad\qquad x &= 6 \\
\text{then} \qquad\quad 2x - 3 &= 9 \\
\text{becomes} \qquad 2(6) - 3 &= 9 \\
12 - 3 &= 9 \\
9 &= 9 \qquad \text{A true statement} \quad ◀
\end{aligned}
$$

DEFINITION   Two or more equations with the same solution set are called **equivalent equations.**

▶ **EXAMPLE 2**   The equations $2x - 5 = 9$, $x - 1 = 6$, and $x = 7$ are all equivalent equations since the solution set for each is $\{7\}$.   ◀

In addition to the properties from Chapter 1, we need two new properties—one for addition or subtraction and one for multiplication or division—to assist us in solving linear equations.

Properties of Equality

The first property states that adding the same quantity to both sides of an equation preserves equality. Or, more importantly, adding the same amount to both sides of an equation *never changes* the solution set. This property is called the **addition property of equality** and is stated in symbols as follows.

Addition Property of Equality

For any three algebraic expressions A, B, and C,

$$\text{if} \qquad A = B$$

$$\text{then} \qquad A + C = B + C$$

In words: Adding the same quantity to both sides of an equation will not change the solution.

Our second new property is called the **multiplication property of equality** and is stated like this.

Multiplication Property of Equality

For any three algebraic expressions A, B, and C, where $C \neq 0$,

$$\text{if} \qquad A = B$$

$$\text{then} \qquad AC = BC$$

In words: Multiplying both sides of an equation by the same nonzero quantity will not change the solution.

Note   Since subtraction is defined in terms of addition, and division is defined in terms of multiplication, we do not need to introduce separate properties for subtraction and division. The solution set for an equation will never be changed by subtracting the same amount from both sides or by dividing both sides by the same nonzero quantity.

The following examples illustrate how we use the properties from Chapter 1 along with the addition property of equality and the multiplication property of equality to solve linear equations.

▶ **EXAMPLE 3** Solve for x: $2x - 3 = 9$.

Solution We begin by using the addition property of equality to add $+3$, the opposite of -3, to both sides of the equation:

$$2x - 3 + 3 = 9 + 3$$
$$2x = 12$$

To get x alone on the left side, we use the multiplication property of equality and multiply both sides by $\frac{1}{2}$, the reciprocal of 2:

$$\frac{1}{2}(2x) = \frac{1}{2}(12)$$
$$x = 6$$

Note Even though we have not shown it, we are using the associative property when we say that $\frac{1}{2}$ times $2x$ is x. Remember, it is the associative property that allows us to change the grouping in a product. If we were to show the associative property, it would look like this:

$$\frac{1}{2}(2x) = \left(\frac{1}{2} \cdot 2\right)x = 1x = x$$

Since the addition and multiplication properties of equality always produce equations equivalent to the original equations, our last equation, $x = 6$, is equivalent to our first equation, $2x - 3 = 9$. The solution set is, therefore, $\{6\}$. ◀

▶ **EXAMPLE 4** Solve $\frac{3}{4}x + 5 = -4$.

Solution We begin by adding -5 to both sides of the equation. Once this has been done, we multiply both sides by the reciprocal of $\frac{3}{4}$, which is $\frac{4}{3}$.

$$\frac{3}{4}x + 5 = -4$$

$$\frac{3}{4}x + 5 + (-5) = -4 + (-5) \quad \text{Add } -5 \text{ to both sides}$$

$$\frac{3}{4}x = -9$$

$$\frac{4}{3}\left(\frac{3}{4}x\right) = \frac{4}{3}(-9) \qquad \text{Multiply both sides by } \frac{4}{3}$$

$$x = -12 \qquad \frac{4}{3}(-9) = \frac{4}{3}\left(-\frac{9}{1}\right) = -\frac{36}{3} = -12 \quad ◀$$

Our next example involves solving an equation that has variable terms on both sides of the equal sign.

▶ **EXAMPLE 5**   Find the solution set for $3a - 5 = -6a + 1$.

Solution   To solve for a we must isolate it on one side of the equation. Let's decide to isolate a on the left side by adding $6a$ to both sides of the equation.

$$3a - 5 = -6a + 1$$

$$3a + \mathbf{6a} - 5 = -6a + \mathbf{6a} + 1 \qquad \text{Add } \mathbf{6a} \text{ to both sides}$$

$$9a - 5 = 1$$

$$9a - 5 + \mathbf{5} = 1 + \mathbf{5} \qquad \text{Add } \mathbf{5} \text{ to both sides}$$

$$9a = 6$$

$$\frac{\mathbf{1}}{\mathbf{9}}(9a) = \frac{\mathbf{1}}{\mathbf{9}}(6) \qquad \text{Multiply both sides by } \frac{\mathbf{1}}{\mathbf{9}}$$

$$a = \frac{2}{3} \qquad\qquad \frac{1}{9}(6) = \frac{6}{9} = \frac{2}{3} \qquad ◀$$

Note   From Chapter 1 we know that multiplication by a number and division by its reciprocal always produce the same result. Because of this fact, instead of multiplying each side of our equation by $\frac{1}{9}$, we could just as easily divide each side by 9. If we did so, the last two lines in our solution would look like this:

$$\frac{9a}{9} = \frac{6}{9}$$

$$a = \frac{2}{3}$$

We can check our solution in Example 5 by replacing a in the original equation with $\frac{2}{3}$.

When $\qquad\qquad\qquad a = \dfrac{2}{3}$

the equation $\qquad\qquad 3a - 5 = -6a + 1$

becomes $\qquad\qquad 3\left(\dfrac{2}{3}\right) - 5 \stackrel{?}{=} -6\left(\dfrac{2}{3}\right) + 1$

$$2 - 5 \stackrel{?}{=} -4 + 1$$

$$-3 = -3 \qquad \text{A true statement}$$

There will be times when we solve equations and end up with a negative sign in front of the variable. The next example shows how to handle this situation.

▶ **EXAMPLE 6**   Solve each equation.
(a) $-x = 4$ (b) $-y = -8$

Solution   Neither equation can be considered solved because of the negative sign in front of the variable. To eliminate the negative signs we simply multiply each side of the equations by -1.

(a) $-x = 4$ (b) $-y = -8$

$-1(-x) = -1(4)$ $-1(-y) = -1(-8)$ Multiply each side by -1 ◀

$x = -4$ $y = 8$

The next example involves fractions. The least common denominator, which is the smallest expression that is divisible by each of the denominators, can be used with the multiplication property of equality to simplify equations containing fractions.

▶ **EXAMPLE 7**   Solve $\dfrac{2}{3}x + \dfrac{1}{2} = -\dfrac{3}{8}$.

Solution   We can solve this equation by applying our properties and working with fractions, or we can begin by eliminating the fractions. Let's use both methods.

Method 1   Working with the fractions.

$$\frac{2}{3}x + \frac{1}{2} + \left(-\frac{1}{2}\right) = -\frac{3}{8} + \left(-\frac{1}{2}\right) \qquad \text{Add } -\frac{1}{2} \text{ to each side}$$

$$\frac{2}{3}x = -\frac{7}{8} \qquad\qquad -\frac{3}{8} + \left(-\frac{1}{2}\right) = -\frac{3}{8} + \left(-\frac{4}{8}\right)$$

$$\frac{3}{2}\left(\frac{2}{3}x\right) = \frac{3}{2}\left(-\frac{7}{8}\right) \qquad\qquad \text{Multiply each side by } \frac{3}{2}$$

$$x = -\frac{21}{16}$$

Method 2   Eliminating the fractions in the beginning.

Our original equation has denominators of 3, 2, and 8. The least common denominator, abbreviated LCD, for these three denominators is 24, and it has the property that all three denominators will divide it evenly. Therefore, if we multiply both sides of our equation by 24, each denominator will divide into 24, and we will be left with an equation that does not contain any denominators other than 1.

$$24\left(\frac{2}{3}x + \frac{1}{2}\right) = 24\left(-\frac{3}{8}\right)$$ Multiply each side
by the LCD **24**

$$24\left(\frac{2}{3}x\right) + 24\left(\frac{1}{2}\right) = 24\left(-\frac{3}{8}\right)$$ Distributive property
on the left side

$$16x + 12 = -9$$ Multiply

$$16x = -21$$ Add -12 to each side

$$x = -\frac{21}{16}$$ Multiply each side by $\frac{1}{16}$

As the third line above indicated, multiplying each side of the equation by the LCD eliminates all the fractions from the equation. Both methods yield the same solution. To check our solution, we substitute $x = -21/16$ back into our original equation to obtain

$$\frac{2}{3}\left(-\frac{21}{16}\right) + \frac{1}{2} \overset{?}{=} -\frac{3}{8}$$

$$-\frac{7}{8} + \frac{1}{2} \overset{?}{=} -\frac{3}{8}$$

$$-\frac{7}{8} + \frac{4}{8} \overset{?}{=} -\frac{3}{8}$$

$$-\frac{3}{8} = -\frac{3}{8}$$

As we can see, our solution checks.

Note   We are placing question marks over the equal signs because we don't know yet if the expressions on the left will be equal to the expressions on the right.   ◀

▶ **EXAMPLE 8**   Solve the equation $0.06x + 0.05(10,000 - x) = 560$.

Solution   We can solve the equation in its original form by working with the decimals, or we can eliminate the decimals first by using the multiplication property of equality and solve the resulting equation. Here are both methods.

Method 1   Working with the decimals.

$$0.06x + 0.05(10,000 - x) = 560$$ Original equation

$$0.06x + 0.05(10,000) - 0.05x = 560$$ Distributive property

$$0.01x + 500 = 560$$ Simplify the left side

$$0.01x + 500 + (-500) = 560 + (-500)$$ Add **−500** to each side

$$0.01x = 60$$

$$\frac{0.01x}{\mathbf{0.01}} = \frac{60}{\mathbf{0.01}} \qquad \text{Divide each side by } \mathbf{0.01}$$

$$x = 6{,}000$$

Method 2 Eliminating the decimals in the beginning.

To move the decimal point two places to the right in $0.06x$ and 0.05, we multiply each side of the equation by 100.

$0.06x + 0.05(10{,}000 - x) = 560$	Original equation
$0.06x + 500 - 0.05x = 560$	Distributive property
$\mathbf{100}(0.06x) + \mathbf{100}(500) - \mathbf{100}(0.05x) = \mathbf{100}(560)$	Multiply each side by $\mathbf{100}$
$6x + 50{,}000 - 5x = 56{,}000$	Multiply
$x + 50{,}000 = 56{,}000$	Simplify the left side
$x = 6{,}000$	Add $-50{,}000$ to each side

Using either method, the solution to our equation is 6,000. We check our work (to be sure we have not made a mistake in applying the properties or an arithmetic mistake) by substituting 6,000 into our original equation and simplifying each side of the result separately.

Check: Substituting 6,000 for x in the original equation, we have

$$0.06(6{,}000) + 0.05(10{,}000 - 6{,}000) \overset{?}{=} 560$$

$$0.06(6{,}000) + 0.05(4{,}000) \overset{?}{=} 560$$

$$360 + 200 \overset{?}{=} 560$$

$$560 = 560 \qquad \text{A true statement} \quad \blacktriangleleft$$

Here is a list of steps to use as a guideline for solving linear equations in one variable.

Strategy for Solving Linear Equations in One Variable

Step 1a: Use the distributive property to separate terms, if necessary.

    1b: If fractions are present, consider multiplying both sides by the LCD to eliminate the fractions. If decimals are present, consider multiplying both sides by a power of 10 to clear the equation of decimals.

    1c: Combine similar terms on each side of the equation.

> *Step 2:* Use the addition property of equality to get all variable terms on one side of the equation and all constant terms on the other side. A variable term is a term that contains the variable (for example, $5x$). A constant term is a term that does not contain the variable (the number 3, for example).
>
> *Step 3:* Use the multiplication property of equality to get x (that is, $1x$) by itself on one side of the equation.
>
> *Step 4:* Check your solution in the original equation to be sure that you have not made a mistake in the solution process.

As you will see as you work through the problems in the problem set, it is not always necessary to use all four steps when solving equations. The number of steps used depends upon the equation. In Example 9 below, there are no fractions or decimals in the original equation, so step 1b will not be used.

▶ **EXAMPLE 9** Solve $3(2y - 1) + y = 5y + 3$.

Solution Applying the steps outlined above we have

$$
\textit{Step 1a:} \begin{cases} 3(2y - 1) + y = 5y + 3 & \text{Distributive} \\ \qquad \downarrow \quad \downarrow & \text{Property} \\ 6y - 3 + y = 5y + 3 \end{cases}
$$

Step 1c: $\qquad 7y - 3 = 5y + 3$ Simplify

$$
\textit{Step 2:} \begin{cases} 7y + (\mathbf{-5y}) - 3 = 5y + (\mathbf{-5y}) + 3 & \text{Add } \mathbf{-5y} \text{ to both} \\ 2y - 3 = 3 & \text{sides} \\ 2y - 3 + \mathbf{3} = 3 + \mathbf{3} & \text{Add } \mathbf{+3} \text{ to both} \\ 2y = 6 & \text{sides} \end{cases}
$$

$$
\textit{Step 3:} \begin{cases} \dfrac{1}{2}(2y) = \dfrac{1}{2}(6) & \text{Multiply by } \dfrac{1}{2} \\ y = 3 \end{cases}
$$

Check:
$$
\textit{Step 4:} \begin{cases} \text{When} & y = 3 \\ \text{the equation} & 3(2y - 1) + y = 5y + 3 \\ \text{becomes} & 3(2 \cdot 3 - 1) + 3 \stackrel{?}{=} 5 \cdot 3 + 3 \\ & 3(5) + 3 \stackrel{?}{=} 15 + 3 \\ & 18 = 18 \qquad \text{A true statement} \end{cases}
$$

Our solution checks in the original equation. ◀

▶ **EXAMPLE 10**   Solve the equation $8 - 3(4x - 2) + 5x = 35$.

Solution   We must begin by distributing the -3 across the quantity $4x - 2$. (It would be a mistake to subtract 3 from 8 first, since the rule for order of operations indicates we are to do multiplication before subtraction.) After we have simplified the left side of our equation, we apply the addition property and the multiplication property. In this example, we will show only the result:

Step 1a:
$$\begin{cases} 8 - 3(4x - 2) + 5x = 35 & \text{Original equation} \\ \quad\quad\downarrow\quad\quad\downarrow \\ 8 - 12x + 6 + 5x = 35 & \text{Distributive property} \end{cases}$$

Step 1c: $-7x + 14 = 35$ Simplify

Step 2: $-7x = 21$ Add -14 to each side

Step 3: $x = -3$ Multiply by $-\dfrac{1}{7}$

Step 4:   When x is replaced by -3 in the original equation, a true statement results. Therefore, -3 is the solution to our equation.   ◀

Identities and Equations with No Solution

There are two special cases associated with solving linear equations in one variable, each of which is illustrated below.

▶ **EXAMPLE 11**   Solve for x: $2(3x - 4) = 3 + 6x$.

Solution   Applying the distributive property to the left side gives us

$$6x - 8 = 3 + 6x \quad\quad \text{Distributive property}$$

Now, if we add $-6x$ to each side, we are left with

$$-8 = 3$$

which is a false statement. This means that there is no solution to our equation. Any number we substitute for x in the original equation will lead to a similar false statement.   ◀

▶ **EXAMPLE 12**   Solve for x: $-15 + 3x = 3(x - 5)$.

Solution   We start by applying the distributive property to the right side.

$$-15 + 3x = 3x - 15 \quad\quad \text{Distributive property}$$

If we add $-3x$ to each side, we are left with the true statement

$$-15 = -15$$

$$\boxed{\pi}\boxed{\times}\boxed{3}\boxed{\wedge}\boxed{2}\boxed{\times}\boxed{4}$$

Pressing the ENTER key will give the first volume as 113 to the nearest whole number. To find the volume for the second set of numbers, recall the original formula and simply change the numbers. That is, use the recall feature of your calculator to bring back the formula for the first volume. Once you have it, move the cursor to the radius and replace it with a new radius. Do the same for the height. Pressing the ENTER key gives the new volume. Here are the volumes for all four sets of numbers.

$r = 3, h = 4$	formula: $\pi * 3 \wedge 2 * 4$	$V = 113$
$r = 4, h = 3$	formula: $\pi * 4 \wedge 2 * 3$	$V = 151$
$r = 10, h = 100$	formula: $\pi * 10 \wedge 2 * 100$	$V = 31,416$
$r = 100, h = 10$	formula: $\pi * 100 \wedge 2 * 10$	$V = 314,159$

PROBLEM SET 2.2

Use the formula $3x - 4y = 12$ to find y if

1. x is 0
2. x is -2
3. x is 4
4. x is -4

Use the formula $y = 2x - 3$ to find x when

5. y is 0
6. y is -3
7. y is 5
8. y is -5

A company that manufactures typewriter ribbons finds that they can sell x ribbons each week at a price of p dollars each, according to the formula $x = 1,300 - 100p$. What price should they charge for each ribbon if they want to sell

9. 800 ribbons each week
10. 400 ribbons each week
11. 300 ribbons each week
12. 900 ribbons each week

The volume of a cylinder is given by the formula $V = \pi r^2 h$, where r is the radius and h is the height. Find the height if

13. the volume is 308 cubic centimeters and the radius is 7 centimeters. (Use $\frac{22}{7}$ for π.)

14. the volume is 308 cubic centimeters and the radius is $\frac{7}{2}$ centimeters. (Use $\frac{22}{7}$ for π.)
15. the volume is 628 cubic inches and the radius is 10 inches. (Use 3.14 for π.)
16. the volume is 12.56 cubic inches and the radius is 5 inches. (Use 3.14 for π.)

The surface area of a cylinder that is closed at the top and bottom is given by the formula $S = 2\pi r^2 + 2\pi rh$. Find the height if

17. the surface area is 942 square feet and the radius is 10 feet. (Use 3.14 for π.)
18. the surface area is 471 square feet and the radius is 5 feet. (Use 3.14 for π.)

Solve each of the following formulas for the indicated variable.

19. $A = lw$ for l
20. $A = \frac{1}{2} bh$ for b
21. $I = prt$ for t
22. $I = prt$ for r
23. $PV = nRT$ for T
24. $PV = nRT$ for R
25. $y = mx + b$ for b
26. $y = mx + b$ for x
27. $s = \frac{1}{2}(a + b + c)$ for c
28. $s = \frac{1}{2}(a + b + c)$ for b

29. $A = P + Prt$ for r **30.** $A = P + Prt$ for t

31. $C = \frac{5}{9}(F - 32)$ for F

32. $F = \frac{9}{5}C + 32$ for C

33. $h = vt + 16t^2$ for v

34. $h = vt - 16t^2$ for v

35. $A = a + (n - 1)d$ for d

36. $A = a + (n - 1)d$ for n

37. $2x + 3y = 6$ for y

38. $2x - 3y = 6$ for y

39. $-3x + 5y = 15$ for y

40. $-2x - 7y = 14$ for y

41. $9x - 3y = 6$ for y

42. $9x + 3y = 15$ for y

43. $2x - 6y + 12 = 0$ for y

44. $7x - 2y - 6 = 0$ for y

45. $8x - 10y - 16 = 0$ for x

46. $6x - 3y + 12 = 0$ for x

47. $z = \dfrac{x - \mu}{s}$ for x

48. $z = \dfrac{x - \mu}{s}$ for μ

49. $ax + 4 = bx + 9$ for x

50. $ax - 5 = cx - 2$ for x

51. $A = P + Prt$ for P

52. $S = 2\pi r + \pi rh$ for π

53. $ax + b = cx + d$ for x

54. $4x + 2y = 3x + 5y$ for y

Solve for y.

55. $\dfrac{x}{8} + \dfrac{y}{2} = 1$ **56.** $\dfrac{x}{7} + \dfrac{y}{9} = 1$

57. $\dfrac{x}{5} + \dfrac{y}{-3} = 1$ **58.** $\dfrac{x}{16} + \dfrac{y}{-2} = 1$

Translate each of the following into a linear equation and then solve the equation.

59. What number is 54% of 38?

60. What number is 11% of 67?

61. What number is 5% of 10,000?

62. What number is 6% of 6,000?

63. What percent of 36 is 9?

64. What percent of 50 is 5?

65. 37 is 4% of what number?

66. 8 is 2% of what number?

Write the first five terms of the sequences with the following general terms.

67. $a_n = 3n + 1$ **68.** $a_n = 4n - 1$

69. $a_n = n^2 + 3$ **70.** $a_n = n^3 + 1$

71. $a_n = \dfrac{n}{n + 3}$ **72.** $a_n = \dfrac{n}{n + 2}$

73. $a_n = \dfrac{1}{n^2}$ **74.** $a_n = \dfrac{1}{n^3}$

75. $a_n = 2^n$ **76.** $a_n = 3^n$

77. $a_n = 1 + \dfrac{1}{n}$ **78.** $a_n = 1 - \dfrac{1}{n}$

79. Find the seventh term of the sequence given by $a_n = \dfrac{(-1)^n}{n^2}$.

80. Find the tenth term of the sequence given by $a_n = \dfrac{(-1)^n}{n^2}$.

81. Find a_4 for the sequence given by $a_n = \dfrac{(-1)^{n+1}}{n^2 + 1}$.

82. Find a_7 for the sequence given by $a_n = \dfrac{(-1)^{n+1}}{n^2 + 1}$.

83. Find the first four terms of the sequence given by $a_n = \dfrac{(-1)^n}{n^3}$.

84. Find the first four terms of the sequence given by $a_n = \dfrac{(-1)^{n+1}}{n^3}$.

Review Problems

The following problems review some of the material we covered in Section 1.1. Reviewing these problems will help you with the next section.

Translate each of the following into symbols.

85. Twice the sum of x and 3.

86. The sum of twice x and 3.

87. Twice the sum of x and 3 is 16.

88. The sum of twice x and 3 is 16.

89. Five times the difference of x and 3.

90. Five times the difference of x and 3 is 10.

91. The sum of $3x$ and 2 is equal to the difference of x and 4.

92. The sum of x and $x + 2$ is 12 more than their difference.

One Step Further

93. Solve for x: $\dfrac{x}{a} + \dfrac{y}{b} = 1$.

94. Solve for y: $\dfrac{x}{a} + \dfrac{y}{b} = 1$.

95. Solve for a: $\dfrac{1}{a} + \dfrac{1}{b} = \dfrac{1}{c}$.

96. Solve for b: $\dfrac{1}{a} + \dfrac{1}{b} = \dfrac{1}{c}$.

97. Solve for R: $\dfrac{1}{R} = \dfrac{1}{a} + \dfrac{1}{b} + \dfrac{1}{c}$.

98. Solve for a: $\dfrac{1}{R} = \dfrac{1}{a} + \dfrac{1}{b} + \dfrac{1}{c}$.

USING TECHNOLOGY

The formula below gives the amount of money A in an account in which P dollars has been invested for t years at an interest rate of r compounded n times a year.

$$A = P\left(1 + \frac{r}{n}\right)^{nt}$$

Use the recall formula function on your calculator to answer the following questions.

99. How much money is in an account in which $500 has been invested at 5%, compounded quarterly, for 6 years? (That is, find A when $P = 500$, $r = 0.05$, $n = 4$, and $t = 6$.)

100. How much money is in an account in which $500 has been invested at 5%, compounded monthly, for 6 years?

101. How much money is in an account in which $500 has been invested at 5%, compounded daily, for 6 years?

102. How much money is in an account in which $500 has been invested at 6%, compounded monthly, for 6 years?

103. How much money is in an account in which $500 has been invested at 7%, compounded monthly, for 6 years?

104. In general, is it more desirable to increase the interest rate or the number of compounding periods to increase the amount of money in the account?

Research Project 3

The triangular array of numbers shown below is known as Pascal's triangle, after the French philosopher Blaise Pascal (1623–1662). Part of the solution to Research Project 4 that you will see in the next chapter depends on Pascal's triangle.

```
            1
          1   1
        1   2   1
      1   3   3   1
    1   4   6   4   1
  1   5  10  10   5   1
```

Part 1 To begin your research into Pascal's triangle, you need to discover how the numbers in each row of the triangle are obtained from the numbers in the row above it. Once you have discovered how to extend the triangle, write the next two rows.

Part 2 Pascal's triangle can be linked to the Fibonacci sequence by rewriting Pascal's triangle so that the 1's on the left side of the triangle line up under one another, and the other columns are equally spaced to the right of the first column.

Rewrite Pascal's triangle as indicated and then look up along the upward diagonals of the new array until you discover how the Fibonacci sequence can be obtained from it.

Summarize your results in an essay that starts with a description of how Pascal's triangle is formed and then goes on to explain how it is related to the Fibonacci sequence.

SECTION

2.3   # Applications

In this section we usc the skills we have developed for solving equations to solve problems written in words. You may find that some of the examples and problems are more realistic than others. Since we are just beginning our work with application problems, even the ones that seem unrealistic are good practice. What is important in this section is the *method* we use to solve application problems, not the applications themselves. The method, or strategy, that we use to solve application problems is called the *Blueprint for Problem Solving.* It is an outline that will overlay the solution process we use on all application problems.

Blueprint for Problem Solving

Step 1: ***Read*** the problem and then mentally ***list*** the items that are known and the items that are unknown.

Step 2: ***Assign a variable*** to one of the unknown items. (In most cases this will amount to letting $x =$ the item that is asked for in the problem.) Then ***translate*** the other ***information*** in the problem to expressions involving the variable.

Step 3: ***Reread*** the problem, and then ***write an equation,*** using the items and variables listed in steps 1 and 2, that describes the situation.

Step 4: ***Solve the equation*** found in step 3.

Step 5: ***Write*** your ***answer*** using a complete sentence.

Step 6: ***Reread*** the problem and ***check*** your solution with the original words in the problem.

There are a number of substeps within each of the steps in our blueprint. For instance, with steps 1 and 2 it is always a good idea to draw a diagram or picture, if it helps visualize the relationship between the items in the problem.

▶ **EXAMPLE 1**   Twice the sum of a number and 3 is 16. Find the number.

Solution   Using the Blueprint for Problem Solving as an outline, we solve the problem as follows.

Step 1: **Read** the problem and then mentally **list** the items that are known and the items that are unknown.

Known items: The numbers 3 and 16

Unknown item: The number in question

Step 2: **Assign a variable** to one of the unknown items. Then **translate** the other **information** in the problem to expressions involving the variable.

Let x = the number asked for in the problem.

Then "Twice the sum of a number and 3" translates to $2(x + 3)$.

Step 3: **Reread** the problem, and then **write an equation,** using the items and variables listed in steps 1 and 2, that describes the situation.

With all word problems, the word "is" translates to =.

Twice the sum of x and 3 is 16

$$2(x + 3) \qquad = 16$$

Step 4: **Solve the equation** found in step 3.

$$2(x + 3) = 16$$
$$2x + 6 = 16$$
$$2x = 10$$
$$x = 5$$

Step 5: **Write** your **answer** using a complete sentence.

The number is 5.

Step 6: **Reread** the problem and **check** your solution with the original words in the problem.

Twice the sum of 5 and 3 is twice 8, which is 16; a true statement. ◄

To help with other problems of the type shown in Example 1, here are some common English words and phrases and their translation into algebraic expressions.

English Phrase	Algebraic Expression
The sum of a and b	$a + b$
The difference of a and b	$a - b$
The product of a and b	$a \cdot b$
The quotient of a and b	$\dfrac{a}{b}$
4 more than x	$4 + x$
Twice the sum of a and 5	$2(a + 5)$
The sum of twice a and 5	$2a + 5$
8 decreased by y	$8 - y$
3 less than m	$m - 3$

In our next example, we abbreviate the steps in the blueprint.

▶ **EXAMPLE 2**   The sum of two consecutive even integers is 3 times the smaller integer. Find the two integers.

Solution   When we ask for consecutive even integers, we mean even integers that are next to each other on the number line, such as 6 and 8, 12 and 14, and 20 and 22. Consecutive means following one another in uninterrupted order. To move from one even integer to the next consecutive even integer, we must add 2.

Step 1: *Read and list*
>*Known items:* Integers are even and consecutive. Their sum is three times the smaller.
>*Unknown items:* The two integers

Step 2: *Assign a variable and translate information*
>Let $x =$ the smaller integer, then $x + 2$ is the next consecutive even integer after x.

Step 3: *Reread and write an equation*
>The sum of the two integers is 3 times the smaller

$$x + (x + 2) \qquad = \qquad 3x$$

Step 4: *Solve the equation*

$$x + (x + 2) = 3x$$
$$2x + 2 = 3x$$
$$2 = x$$

Step 5: *Write answer*
>The two integers are $x = 2$ and $x + 2 = 4$.

Step 6: Reread and check

The numbers 2 and 4 are consecutive even integers. Their sum is 6, which is 3 times the smaller of the two. ◄

Remember as you read through the steps in the solution to each example in this section that step 1 is done mentally. Read the problem and then *mentally* list the items that you know and the items that you don't know. The purpose of step 1 is to give you direction as you begin to work application problems. Finding the solution to an application problem is a process; it doesn't happen all at once. The first step is to read the problem with a purpose in mind. That purpose is to mentally note the items that are known and the items that are unknown.

▶ **EXAMPLE 3**   The length of a rectangle is 3 inches less than twice the width. The perimeter is 45 inches. Find the length and width.

Solution   When working problems that involve geometric figures, a sketch of the figure helps organize and visualize the problem.

Step 1: Read and list

Known items: The figure is a rectangle. The length is 3 inches less than twice the width. The perimeter is 45 inches.
Unknown items: The length and the width

Step 2: Assign a variable and translate information

Since the length is given in terms of the width (the length is 3 less than twice the width), we let x = the width of the rectangle. The length is 3 less than twice the width, so it must be $2x - 3$. The diagram in Figure 1 is a visual description of the relationships we have listed so far.

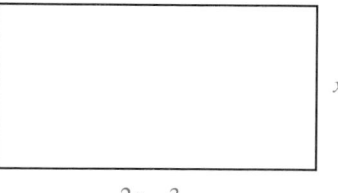

$2x - 3$

Figure 1

Step 3: Reread and write an equation

The equation that describes the situation is

Twice the length + twice the width is the perimeter
$$2(2x - 3) \quad + \quad 2x \quad = \quad 45$$

Step 4: Solve the equation

$$2(2x - 3) + 2x = 45$$
$$4x - 6 + 2x = 45$$
$$6x - 6 = 45$$
$$6x = 51$$
$$x = 8.5$$

Step 5: Write answer
The width is 8.5 inches. The length is
$2x - 3 = 2(8.5) - 3 = 14$ inches.

Step 6: Reread and check
If the length is 14 inches and the width is 8.5 inches, then the perimeter must be $2(14) + 2(8.5) = 28 + 17 = 45$ inches. Also, the length, 14, is 3 less than twice the width. ◄

▶ **EXAMPLE 4** In April 1993, Pat bought a new Ford Mustang with a 5.0 liter engine. The total price, which includes the price of the car plus sales tax, was $17,481.75. If the sales tax rate is 7.25%, what was the price of the car?

Solution

Step 1: Read and list
Known items: The total price is $17,481.75. The sales tax rate is 7.25%, which is 0.0725 in decimal form.
Unknown item: The price of the car

Step 2: Assign a variable and translate information
If we let x = the price of the car, then to calculate the sales tax, we multiply the price of the car x by the sales tax rate:

$$\text{Sales tax} = (\text{price of the car})(\text{sales tax rate})$$
$$= x \cdot 0.0725$$
$$= 0.0725x$$

Step 3: Reread and write an equation

$$\text{Car price} + \text{sales tax} = \text{total price}$$
$$x + 0.0725x = 17,481.75$$

Step 4: Solve the equation

$$x + 0.0725x = 17,481.75$$
$$1.0725x = 17,481.75$$
$$x = \frac{17,481.75}{1.0725}$$
$$= 16,300.00$$

Step 5: *Write answer*

The price of the car is $16,300.00.

Step 6: *Reread and check*

The price of the car is $16,300.00. The tax is 0.0725(16,300) = $1,181.75. Adding the retail price and the sales tax we have a total bill of $17,481.75. ◀

Facts from Geometry: Angles in General

An angle is formed by two rays with the same endpoint. The common endpoint is called the **vertex** of the angle, and the rays are called the **sides** of the angle.

In Figure 2, angle θ (theta) is formed by the two rays *OA* and *OB*. The vertex of θ is *O*. Angle θ is also denoted as angle *AOB*, where the letter associated with the vertex is always the middle letter in the three letters used to denote the angle.

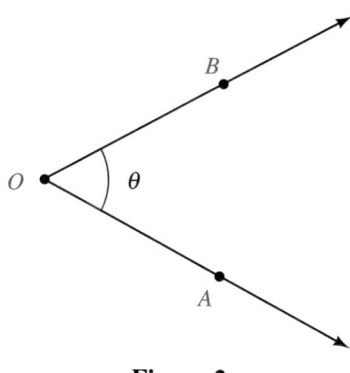

Figure 2

Degree Measure

The angle formed by rotating a ray through one complete revolution about its endpoint (Figure 3) has a measure of 360 degrees, which we write as 360°.

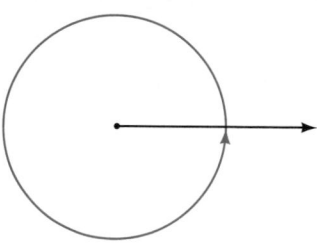

One complete revolution = 360°

Figure 3

One degree of angle measure, written 1°, is 1/360 of a complete rotation of a ray about its endpoint; there are 360° in one full rotation. (The number 360 was decided upon by early civilizations because it was believed that the earth was at the center of the universe and the sun would rotate once around the earth every 360 days.) Similarly, 180° is half of a complete rotation, and 90° is a quarter of a full rotation. Angles that measure 90° are called **right angles,** while angles that measure 180° are called **straight angles.** If an angle measures between 0° and 90° it is called an **acute angle,** while an angle that measures between 90° and 180° is an **obtuse angle.** Figure 4 illustrates.

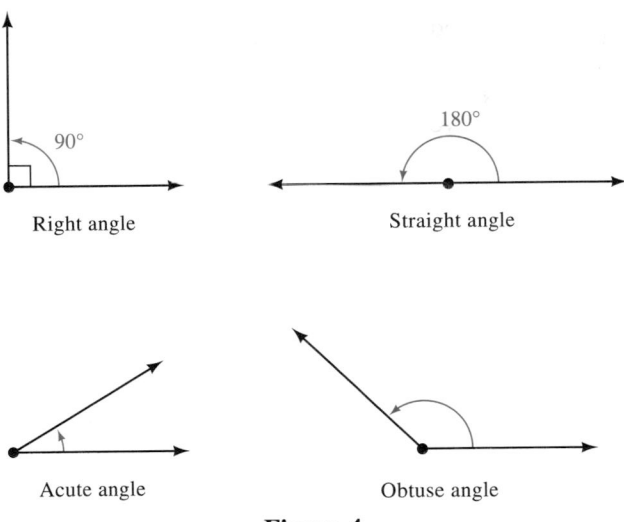

Figure 4

Complementary Angles and Supplementary Angles

If two angles add up to 90°, then we call them **complementary angles,** and each is called the **complement** of the other. If two angles have a sum of 180°, then we call them **supplementary angles,** and each is called the **supplement** of the other. Figure 5 illustrates the relationship between angles that are complementary and angles that are supplementary.

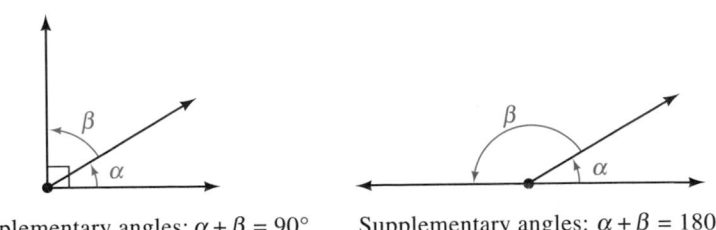

Complementary angles: $\alpha + \beta = 90°$ Supplementary angles: $\alpha + \beta = 180°$

Figure 5

▶ **EXAMPLE 5**   Two complementary angles are such that one is twice as large as the other. Find the two angles.

Solution   Applying the Blueprint for Problem Solving we have

Step 1: **Read and list**
Known items: Two complementary angles. One is twice as large as the other.
Unknown items: The sizes of the angles

Step 2: **Assign a variable and translate information**
Let x = the smaller angle. The larger angle is twice the smaller so we represent the larger angle with $2x$.

Step 3: **Reread and write an equation**
Since the two angles are complementary, their sum is 90. Therefore,

$$x + 2x = 90$$

Step 4: **Solve the equation**

$$x + 2x = 90$$
$$3x = 90$$
$$x = 30$$

Step 5: **Write answer**
The smaller angle is 30° and the larger angle is $2 \cdot 30 = 60°$.

Step 6: **Reread and check**
The larger angle is twice the smaller angle and their sum is 90°.

◀

When you begin working the problems in the problem set that follows, remember that word problems are not always solved correctly the first time you try them. Sometimes it takes a number of attempts, and some wrong answers, before you can set up and solve a problem correctly.

P R O B L E M   S E T   2 . 3

Solve each of the following applications. Be sure to show the equation used in each case.

Number Problems

1. Three times the sum of a number and 4 is 3. Find the number.
2. Five times the difference of a number and 3 is 10. Find the number.
3. Twice the sum of 2 times a number and 1 is the same as 3 times the difference of the number and 5. Find the number.
4. The sum of 3 times a number and 2 is the same as the difference of the number and 4. Find the number.
5. If 5 times a number is increased by 2, the result is 8 more than 3 times the number. Find the number.

6. If 6 times a number is decreased by 5, the result is 7 more than 4 times the number decreased by 5. Find the number.

7. The sum of two consecutive even integers is 18. Find the two integers.

8. The sum of two consecutive odd integers is 16. Find the two integers.

9. The sum of two consecutive integers is 1 less than 3 times the smaller. Find the two integers.

10. The sum of two consecutive integers is 5 less than 3 times the larger. Find the integers.

11. If twice the smaller of two consecutive integers is added to the larger, the result is 7. Find the smaller one.

12. If twice the larger of two consecutive integers is added to the smaller, the result is 23. Find the smaller one.

13. If the larger of two consecutive odd integers is subtracted from twice the smaller, the result is 5. Find the two integers.

14. If the smaller of two consecutive even integers is subtracted from twice the larger, the result is 12. Find the two integers.

Geometry Problems

15. A rectangle is twice as long as it is wide. The perimeter is 60 feet. Find the dimensions.

16. The length of a rectangle is 5 times the width. The perimeter is 48 inches. Find the dimensions.

17. A square has a perimeter of 28 feet. Find the length of each side.

18. A square has a perimeter of 36 centimeters. Find the length of each side.

19. A triangle has a perimeter of 23 inches. The medium side is 3 inches more than the shortest side, and the longest side is twice the shortest side. Find the shortest side.

20. The longest side of a triangle is 2 times the shortest side, while the medium side is three meters more than the shortest side. The perimeter is 27 meters. Find the dimensions.

21. The length of a rectangle is 3 meters less than twice the width. The perimeter is 18 meters. Find the width.

22. The length of a rectangle is one foot more than twice the width. The perimeter is 20 feet. Find the dimensions.

23. A livestock pen is built in the shape of a rectangle that is twice as long as it is wide. The perimeter is 48 feet. If the material used to build the pen is $1.75 per foot for the longer sides, and $2.25 per foot for the shorter sides (the shorter sides have gates, which increase the cost per foot), find the cost to build the pen.

24. A garden is in the shape of a square with a perimeter of 42 feet. The garden is surrounded by two fences. One fence is around the perimeter of the garden, while the second fence is 3 feet from the first fence, as Figure 6 indicates. If the material used to build the two fences is $1.28 per foot, what was the total cost of the fences?

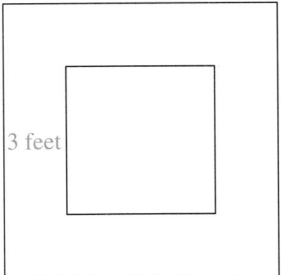

Figure 6

Percent Problems

25. Suppose the total price, including sales tax, of a new Ford F-150 pickup truck is $10,039.43. If the sales tax rate is 7.5%, what was the price of the truck? Round your answer to the nearest cent.

26. Shane returned from a trip to Las Vegas with $300.00, which was 50% more money than she had at the beginning of the trip. How much money did Shane have at the beginning of her trip?

27. Every item in the Just a Dollar store is priced at $1.00. When Mary Jo opens the store, there is $125.50 in the cash register. When she counts

the money in the cash register at the end of the day, the total is $1,058.60. If the sales tax rate is 8.5%, how many items were sold that day?

28. During one ratings period in 1993, it was estimated that 5.4 million people watched *The Tonight Show,* with Jay Leno. If the number of people watching Jay Leno was 1.8% more than the number of viewers watching *Nightline* with Ted Koppel, approximately how many viewers did Ted Koppel have? Round your answer to the nearest tenth of a million viewers.

29. On Monday, August 16, 1993, Continental Airlines announced it was laying off 2,500 workers, or 6% of its workforce, by December 31 of that year. How many workers did Continental Airlines have on Friday, August 13, 1993?

30. Suppose a college bookstore buys a textbook from a publishing company and then marks up the price they paid for the book 33% and sells it to a student at the marked up price. If the student pays $45.00 for the textbook, what did the bookstore pay for it? Round your answer to the nearest cent.

31. An accountant earns $3,440 per month after receiving a 5.5% raise. What was the accountant's monthly income before the raise? Round your answer to the nearest cent.

32. A sheet metal worker earns $26.80 per hour after receiving a 4.5% raise. What was the sheet metal worker's hourly pay before the raise? Round your answer to the nearest cent.

More Geometry Problems

33. Two supplementary angles are such that one is eight times as large as the other. Find the two angles.

34. Two complementary angles are such that one is five times as large as the other. Find the two angles.

35. One angle is 12 degrees less than four times another. Find the measure of each angle if
    (a) they are complements of each other.
    (b) they are supplements of each other.

36. One angle is 4 degrees more than three times another. Find the measure of each angle if
    (a) they are complements of each other.
    (b) they are supplements of each other.

Review Problems

The following problems review material we covered in Section 1.2.

For the set $\{-4, -\pi, -\frac{2}{5}, 0, 2, \sqrt{5}, 3\}$, name all the

37. integers   38. irrational numbers

Reduce each fraction to lowest terms.

39. $\frac{10}{12}$ 40. $\frac{65}{39}$

41. $\frac{364}{490}$ 42. $\frac{495}{975}$

SECTION

2.4   Additional Applications

In this section we continue our work with application problems, using our Blueprint for Problem Solving to organize our work. You have probably noticed that step 3, in which we write an equation that describes the situation, is the key step. Anyone with experience solving application problems will tell you that there will be times when your first attempt at step 3 results in the wrong equation. Remember, mistakes are part of the process of learning to do things correctly. Many times the correct equation will become obvious after you have written an equation that is partially wrong. In any case, it is better to write an equation that is partially wrong and be actively involved with the problem than to write nothing at all. Application problems, like other problems in algebra, are not always solved correctly the first time.

Suppose we know that the sum of two numbers is 50. If we let x represent one of the two numbers, how can we represent the other? Let's suppose for a moment that x turns out to be 30. Then the other number will be 20, because their sum is 50. That is, if two numbers add up to 50, and one of them is 30, then the other must be $50 - 30 = 20$. Generalizing this to any number x, we see that if two numbers have a sum of 50, and one of the numbers is x, then the other must be $50 - x$. The table that follows shows some additional examples.

If two numbers have a sum of	and one of them is	then the other must be
50	x	$50 - x$
10	y	$10 - y$
12	n	$12 - n$

Number Problem

▶ **EXAMPLE 1**   The sum of two numbers is 12. If one of the numbers is twice as large as the other, find the two numbers.

Solution

Step 1: ***Read and list*** (Remember, this step is done mentally.)
Known items: Two numbers that add to 12. One is twice as large as the other.
Unknown items: The two numbers

Step 2: ***Assign a variable and translate information***
If we let $x =$ one of the numbers, then the other number must be $12 - x$, because their sum is 12.

Step 3: ***Reread and write an equation***
Since one of the numbers is twice the other, the equation that describes the situation is

$$x = 2(12 - x)$$

Step 4: ***Solve the equation***

$$x = 24 - 2x \qquad \text{Multiply out the right side}$$

$$3x = 24 \qquad \text{Add } 2x \text{ to each side}$$

$$x = 8 \qquad \text{Multiply each side by } \tfrac{1}{3}$$

Step 5: ***Write answer***
One number is $x = 8$, so the other is $12 - x = 12 - 8 = 4$.

Step 6: Reread and check

The two solutions, 4 and 8, check in the original problem since their sum is $4 + 8 = 12$ and 8 is twice 4. ◄

Coin Problem

▶ **EXAMPLE 2** Suppose Bob has a collection of dimes and nickels that totals $3.50. If he has a total of 50 coins, how many of each type does he have?

Solution

Step 1: Read and list

Known items: The coins are dimes and nickels. There are 50 coins with a total value of $3.50.

Unknown items: The number of dimes and the number of nickels

Step 2: Assign a variable and translate information

If we let x = the number of dimes, then $50 - x$ is the number of nickels. (The number of coins is 50, so if he has x of one kind, he must have $50 - x$ of the other.) Since each dime is worth 10 cents, the value of the x dimes is $10x$. Likewise, since each nickel is worth 5 cents, the value of the $50 - x$ nickels is $5(50 - x)$. Here is all the information we have summarized in a table.

	Dimes	**Nickels**	**Total**
Number of	x	$50 - x$	50
Value of	$10x$	$5(50 - x)$	350

Step 3: Reread and write an equation

The second line in our table gives us the equation we need to solve the problem. Since the amount of money he has in dimes plus the amount of money he has in nickels must total $3.50, we have

$$10x + 5(50 - x) = 350$$

Note that we have written this equation in terms of cents. Next, we solve the equation.

Step 4: Solve the equation

$$10x + 5(50 - x) = 350$$
$$10x + 250 - 5x = 350$$
$$5x + 250 = 350$$

$$5x = 100$$
$$x = 20$$

Step 5: *Write answer*
The number of dimes is $x = 20$. The number of nickels is $50 - x = 50 - 20 = 30$.

Step 6: *Reread and check*
We check our results as follows:

20 dimes are worth $10(20) = 200$ cents
30 nickels are worth $5(30) = 150$ cents

The total value is   350 cents or $3.50   ◀

Interest Problem

▶ **EXAMPLE 3** Suppose a person invests a total of $10,000 in two accounts. One account earns 5% annually and the other earns 6% annually. If the total interest earned from both accounts in a year is $560, how much is invested in each account?

Solution

Step 1: *Read and list*
Known items: Two accounts. One pays interest of 5% and the other pays 6%. The total dollars invested is $10,000. The total interest earned is $560.
Unknown items: The number of dollars invested in each individual account

Step 2: *Assign a variable and translate information*
The form of the solution to this problem is very similar to that of Example 2. If we let x equal the amount invested at 6%, then $10,000 - x$ is the amount invested at 5%. The total interest earned from both accounts is $560. The amount of interest earned on x dollars at 6% is $0.06x$, while the amount of interest earned on $10,000 - x$ dollars at 5% is $0.05(10,000 - x)$.

	Dollars at 6%	Dollars at 5%	Total
Number of	x	$10,000 - x$	10,000
Interest on	$0.06x$	$0.05(10,000 - x)$	560

Step 3: *Reread and write an equation*
Again, the last line gives us the equation we are after:

$$0.06x + 0.05(10,000 - x) = 560$$

Table Building

We can use our knowledge of formulas from Section 2.2 to build tables of paired data. In this section we will begin to investigate equations in two variables. As you will see, equations, or formulas, that contain exactly two variables produce pairs of numbers that can be used to construct tables similar to the ones we encountered in Section 1.7. The histograms, scatter diagrams, and line graphs that are produced from the tables of paired data are a visual representation of the relationship between the variables given in the original equation that was used to produce the table in the first place.

▶ **EXAMPLE 5**   A 12-inch-long piece of string is to be formed into a rectangle. Build a table that gives the length of the rectangle if the width is 1, 2, 3, 4, or 5 inches. Then find the area of each of the rectangles formed. Construct a histogram that shows the relationship between the width and length of each rectangle. Then construct a histogram that shows the relationship between the width and the area of each rectangle.

Solution   Since the formula for the perimeter of a rectangle is $P = 2l + 2w$, and our piece of string is 12 inches long, then the formula we will use to find the lengths for the given widths is $12 = 2l + 2w$. To solve this formula for l, we divide each side by 2 and then subtract w. The result is $l = 6 - w$. The table below organizes our work so that the formula we use to find l for a given value of w is showing, and we have added a last column to give us the area of the rectangles formed. The units for the first three columns are inches, while the units for the numbers in column 4 are square inches.

TABLE 1   Length, Width, and Area

w	$l = 6 - w$	l	$A = lw$
1	$l = 6 - 1$	5	5
2	$l = 6 - 2$	4	8
3	$l = 6 - 3$	3	9
4	$l = 6 - 4$	2	8
5	$l = 6 - 5$	1	5

Figures 3 and 4 show two histograms that are constructed using the information in Table 1.

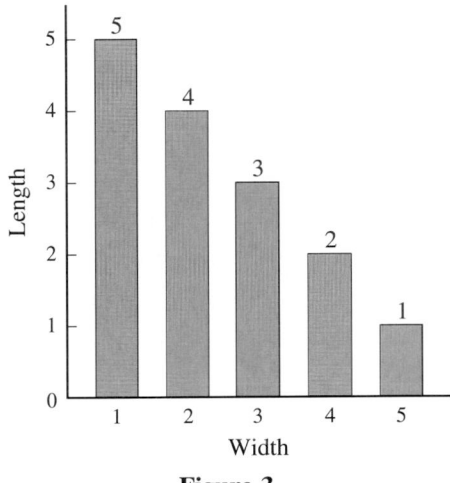

Figure 3

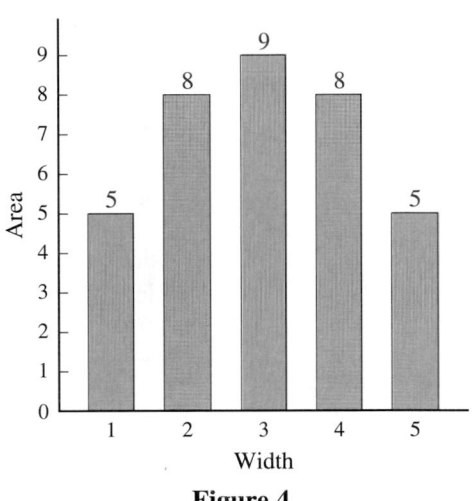

Figure 4

Notice that Figure 3 seems to indicate that the relationship between w and l is linear, while Figure 4 shows that the relationship between w and A is nonlinear.

◀

USING TECHNOLOGY: Graphing Calculators and Spreadsheet Programs

Graphing Calculators

A number of graphing calculators have table-building capabilities. We can let the calculator variable X represent the width of the rectangles in Example 5. To find the length, we set variable Y_1 equal to $6 - X$. The area of each rectangle can be found by setting variable Y_2 equal to $X * Y_1$. To have the calculator produce the table automatically, we use a table minimum of 0 and a table increment of 1. Here is a summary of how the graphing calculator is set up:

Table Setup	*Y Variables Setup*
Table minimum = 0	$Y_1 = 6 - X$
Table increment = 1	$Y_2 = X * Y_1$
Independent variable: Auto	
Dependent variable: Auto	

On the next page is what one such table will look like.

X	Y_1	Y_2
0	6	0
1	5	5
2	4	8
3	3	9
4	2	8
5	1	5
6	0	0

Once the table has been constructed, the graphing calculator can be used to draw a histogram, scatter diagram, or line graph.

Spreadsheet Programs

A table similar to the one above can be created with almost any spreadsheet program. Using formulas to produce the numbers in the cells allows the spreadsheet to construct the table automatically. The advantage to using a spreadsheet program is that the program can create, label, and print a variety of graphics, including histograms, from the data in the table. Figure 5 shows the formulas in each cell of a spreadsheet program that will produce the numbers in Table 1 in Example 5.

	Spreadsheet Program					▼|♦
File Edit Formula Format Data Options Macro Window Help						
☐☐☐☐ Times ⬇ 10 ⬇ Σ **B** U *I* ▤ ▤ ▤ ▤ ▥ ?						

	A	B	C	D	E	F	G
1	Width	Length	Area				
2	0	=6-A2	=A2*B2				
3	1	=6-A3	=A3*B3				
4	2	=6-A4	=A4*B4				
5	3	=6-A5	=A5*B5				
6	4	=6-A6	=A6*B6				
7	5	=6-A7	=A7*B7				
8	6	=6-A8	=A8*B8				
9							
10							
11							
12							
13							
14							
15							

Figure 5

P R O B L E M   S E T   2 . 4

Number Problems

1. The sum of two numbers is 24. If one of the numbers is twice as large as the other, find the two numbers.
2. The sum of two numbers is 16. If one of the numbers is three times as large as the other, find the two numbers.
3. The sum of two numbers is 16. One of the numbers is 2 less than twice the other. Find the two numbers.
4. The sum of two numbers is 21. One of the numbers is 3 more than twice the other. Find the two numbers.

Coin Problems

5. Sylvia has a collection of dimes and nickels that has a total value of $2.80. If she has a total of 36 coins, how many of each type does she have?

	Dimes	Nickels	Total
Number of			
Value of			

6. Sharon has a collection of dimes and quarters that has a total value of $2.00. If she has a total of 14 coins, how many of each type does she have?

	Dimes	Quarters	Total
Number of			
Value of			

7. A coin collection consists of nickels and quarters. If there are 26 coins in the collection with a total value of $2.50, how many of each coin are there?
8. A coin collection consists of nickels and quarters. If there are 24 coins in the collection with a total value of $3.00, how many of each coin are there?

Interest Problems

9. A woman has a total of $9,000 to invest. She invests part of the money in an account that pays 8% per year and the rest in an account that pays 9% per year. If the interest earned in the first year is $750, how much did she invest in each account?

	Dollars at 8%	Dollars at 9%	Total
Number of			
Interest on			

10. A man invests $12,000 in two accounts. If one account pays 10% per year, and the other pays 7% per year, how much was invested in each account if the total interest earned in the first year was $960?

	Dollars at 10%	Dollars at 7%	Total
Number of			
Interest on			

11. A total of $15,000 is invested in two accounts. One of the accounts earns 12% per year, while the other earns 10% per year. If the total interest earned in the first year is $1,600, how much was invested in each account?
12. A total of $11,000 is invested in two accounts. One of the two accounts pays 9% per year, and the other account pays 11% per year. If the total interest paid in the first year is $1,150, how much was invested in each account?
13. Stacey has a total of $6,000 in two accounts. The total amount of interest she earns from both accounts in the first year is $500. If one of the accounts earns 8% interest per year and the other earns 9% interest per year, how much did she invest in each account?

14. Travis has a total of $6,000 invested in two accounts. The total amount of interest he earns from the accounts in the first year is $410. If one account pays 6% per year and the other pays 8% per year, how much did he invest in each account?

Geometry Problems

15. A triangle is such that the largest angle is three times the smallest angle. The third angle is 9° less than the largest angle. Find the measure of each angle.

16. The smallest angle in a triangle is half of the largest angle. The third angle is 15° less than the largest angle. Find the measure of all three angles.

17. The smallest angle in a triangle is one third of the largest angle. The third angle is 10° more than the smallest angle. Find the measure of all three angles.

18. The third angle in an isosceles triangle is half as large as each of the two base angles. Find the measure of each angle.

19. The third angle in an isosceles triangle is 8° more than twice as large as each of the two base angles. Find the measure of each angle.

20. The third angle in an isosceles triangle is 4° more than one fifth of each of the two base angles. Find the measure of each angle.

Miscellaneous Problems

21. Tickets for the father and son breakfast were $2.00 for fathers and $1.50 for sons. If a total of 75 tickets were sold for $127.50, how many fathers and how many sons attended the breakfast?

22. A Girl Scout troop sells 62 tickets to their mother and daughter dinner for a total of $216. If the tickets cost $4.00 for mothers and $3.00 for daughters, how many of each ticket did they sell?

23. A woman owns a small, cash-only business in a state that requires her to charge a 6% sales tax on each item she sells. At the beginning of the day she has $250 in the cash register.

At the end of the day she has $1,204 in the register. How much money should she send to the state government for the sales tax she collected?

24. A store is located in a state that requires a 6% tax on all items sold. If the store brings in a total of $3,392 in one day, how much of that total was sales tax?

25. Patrick goes away to college. The first week he is away from home he calls his girlfriend, using his parents' telephone credit card, and talks for a long time. The telephone company charges 40 cents for the first minute and 30 cents for each additional minute, and then adds on a 50-cent service charge for using the credit card. If his parents receive a bill for $13.80 for Patrick's call, how long did he talk?

26. A person makes a long distance person-to-person call to Santa Barbara, California. The telephone company charges 41 cents for the first minute and 32 cents for each additional minute. Because the call is person-to-person, there is also a service charge of $3.00. If the cost of the call is $6.29, how many minutes did the person talk?

Problems 27–34 may be solved using a graphing calculator.

Table Building

27. A farmer buys 48 feet of fencing material to build a rectangular livestock pen. Build a table that gives the length of the pen if the width is 2, 4, 6, 8, 10, or 12 feet. Then find the area of each of the pens formed. Construct a histogram that shows the relationship between the width and length of each rectangle. Then construct a histogram that shows the relationship between the width and the area of each rectangle.

28. A store selling art supplies finds that they can sell x sketch pads each week at a price of p dollars each according to the formula $x = 900 - 300p$. Use this formula to build a table that gives the number of pads sold each week if the price per pad is $0.50, $1.00, $1.50,

$2.00, or $2.50. Construct a histogram from the information in the table.

29. A ball is thrown straight up into the air with a velocity of 128 feet per second. The formula that gives the height h of the ball t seconds after it is tossed is

$$h = -16t^2 + 128t$$

Use this formula to find the height of the ball after 1, 2, 3, 4, 5, and 6 seconds. Construct a histogram from the table.

If you were to buy a portable telephone in California and access it through GTE Mobilnet, you would have a choice of rates. Here is a problem that involves the cost of using a portable telephone.

30. One of the rates listed on the GTE Mobilnet rate card in 1993 was a flat rate of $15 per month plus $1.50 for each minute you use the phone. Write an equation in two variables that will let you calculate the monthly charge for talking x minutes. Then build a table that shows the cost for talking 10, 15, 20, 25, or 30 minutes in one month. Construct a histogram from the information in the table.

For each sequence below, use the formula to construct a table that gives the first six terms in the sequence by substituting 1, 2, 3, 4, 5, and 6 for n in the formula. (Label the first column in the table n, and the second column a_n.) Then use the paired data from the table to construct a scatter diagram.

31. $a_n = 2n - 3$
32. $a_n = n^2 - 9$
33. $a_n = 2^n$
34. $a_n = \dfrac{12}{n}$

Review Problems

The problems below review material we covered in Section 1.2. Reviewing these problems will help you with the next section.

Graph each inequality.

35. $\{x \mid x > -5\}$
36. $\{x \mid x \le 4\}$
37. $\{x \mid x \le -2 \text{ or } x > 5\}$
38. $\{x \mid x < 3 \text{ or } x \ge 5\}$
39. $\{x \mid x > -4 \text{ and } x < 0\}$
40. $\{x \mid x \ge 0 \text{ and } x \le 2\}$
41. $\{x \mid 1 \le x \le 4\}$
42. $\{x \mid -4 < x < -2\}$

USING TECHNOLOGY

43. Use a graphing calculator or a spreadsheet program to reproduce Table 1 from this section (page 108). This time, start with a width of 0 and increase the width by 0.5 inch each time. What is the largest area produced by this table?

44. Use a graphing calculator or a spreadsheet program to reproduce Table 1 from this section (page 108). This time, start with a width of 0 and increase the width by 0.25 inch each time. What is the largest area produced by this table?

45. If you invest $100 in an account that earns 6% annual interest, compounded monthly, the amount of money in that account t months later is given by the formula $A =$ $100(1.005)^t$. Use a graphing calculator or a spreadsheet program to produce a table that gives the amount of money in the account each month for the first two years it is on deposit. (On your graphing calculator, let X represent t, and Y_1 represent A. The minimum X is 0 and the table increment is 1.)

46. A spherical balloon is being inflated at a constant rate so that the radius increases by 1 centimeter every second. Since the balloon is a sphere, its volume when the radius is r is given by the formula $V = \frac{4}{3}\pi r^3$. Construct a table that gives the volume of the balloon every second, starting when the radius is 0 centimeters and ending when the radius is 10 centimeters. (Round to the nearest whole number.)

Linear Inequalities in One Variable

A linear inequality in one variable is any inequality that can be put in the form

$$ax + b < c \qquad (a, b, \text{ and } c \text{ constants}, a \neq 0)$$

where the inequality symbol ($<$) can be replaced with any of the other three inequality symbols ($\leq$, $>$, or $\geq$).

Some examples of linear inequalities are

$$3x - 2 \geq 7 \qquad -5y < 25 \qquad 3(x - 4) > 2x$$

Our first property for inequalities is similar to the addition property we used when solving equations.

Addition Property for Inequalities

For any algebraic expressions A, B, and C,

$$\text{if} \qquad A < B$$

$$\text{then} \qquad A + C < B + C$$

In words: Adding the same quantity to both sides of an inequality will not change the solution set.

Note Since subtraction is defined as addition of the opposite, our new property holds for subtraction as well as addition. That is, we can subtract the same quantity from each side of an inequality and always be sure that we have not changed the solution.

▶ **EXAMPLE 1** Solve $3x + 3 < 2x - 1$ and graph the solution.

Solution We use the addition property for inequalities to write all the variable terms on one side and all constant terms on the other side:

$$3x + 3 < 2x - 1$$
$$3x + (-2x) + 3 < 2x + (-2x) - 1 \qquad \text{Add } -2x \text{ to each side}$$
$$x + 3 < -1$$
$$x + 3 + (-3) < -1 + (-3) \qquad \text{Add } -3 \text{ to each side}$$
$$x < -4$$

The solution set is all real numbers that are less than -4. To show this we can use set notation and write

$$\{x \mid x < -4\}$$

Or we can graph the solution set on the number line using an open circle at -4 to show that -4 is not part of the solution set:

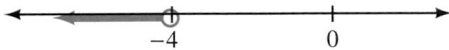

This graph gives rise to the following notation, called **interval notation,** that is an alternate way to write the solution set.

$$(-\infty, -4)$$

The expression above indicates that the solution set is all real numbers from negative infinity up to, but not including, -4.

We have three equivalent representations for the solution set to our original inequality. Here are all three together.

Set Notation	Line Graph	Interval Notation

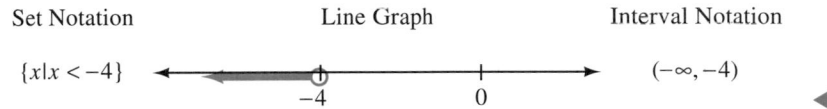

◀

Before we state the multiplication property for inequalities, we will take a look at what happens to an inequality statement when we multiply both sides by a positive number and what happens when we multiply by a negative number.

We begin by writing three true inequality statements:

$$3 < 5 \qquad -3 < 5 \qquad -5 < -3$$

We multiply both sides of each inequality by a positive number—say, 4:

$$4(3) < 4(5) \qquad 4(-3) < 4(5) \qquad 4(-5) < 4(-3)$$
$$12 < 20 \qquad\quad -12 < 20 \qquad\quad -20 < -12$$

Notice in each case that the resulting inequality symbol points in the same direction as the original inequality symbol. Multiplying both sides of an inequality by a positive number preserves the *sense* of the inequality.

Let's take the same three original inequalities and multiply both sides by -4:

$$3 < 5 \qquad\qquad -3 < 5 \qquad\qquad -5 < -3$$

$$-4(3) > -4(5) \qquad -4(-3) > -4(5) \qquad -4(-5) > -4(-3)$$
$$-12 > -20 \qquad\quad 12 > -20 \qquad\qquad 20 > 12$$

Notice in this case that the resulting inequality symbol always points in the opposite direction from the original one. Multiplying both sides of an inequality by a negative number *reverses* the sense of the inequality. Keeping this in mind, we will now state the multiplication property for inequalities.

Multiplication Property for Inequalities

Let A, B, and C represent algebraic expressions.

If $\qquad A < B$

then $\qquad AC < BC$ if C is positive $(C > 0)$

or $\qquad AC > BC$ if C is negative $(C < 0)$

In words: Multiplying both sides of an inequality by a positive number always produces an equivalent inequality. Multiplying both sides of an inequality by a negative number reverses the sense of the inequality.

Note Since division is defined as multiplication by the reciprocal, we can apply our new property to division as well as to multiplication. We can divide both sides of an inequality by any nonzero number as long as we reverse the direction of the inequality symbol when the number we are dividing by is a negative number.

The multiplication property for inequalities does not limit what we can do with inequalities. We are still free to multiply both sides of an inequality by any nonzero number we choose. If the number we multiply by happens to be *negative,* then we *must* also *reverse* the direction of the inequality symbol.

▶ **EXAMPLE 2** Solve $3x - 5 \leq 7$.

Solution We apply the addition property first, then the multiplication property:

$$3x - 5 \leq 7$$
$$3x - 5 + \mathbf{5} \leq 7 + \mathbf{5} \qquad \text{Add } \mathbf{5} \text{ to both sides}$$
$$3x \leq 12$$
$$\frac{1}{3}(3x) \leq \frac{1}{3}(12) \qquad \text{Multiply by } \frac{1}{3}$$
$$x \leq 4$$

The solution set consists of all real numbers that are less than, or equal to, 4. Writing the solution set with set notation and interval notation, we have

$$\{x \mid x \leq 4\} = (-\infty, 4]$$

Notice how a bracket is used with interval notation to show that 4 is part of the solution set.

The graph of the solution set will contain a solid circle at 4, since 4 is included in the solution set.

Here is the graph:

▶ **EXAMPLE 3**   Find the solution set for $-2y - 3 < 7$.

Solution   We begin by adding 3 to each side of the inequality:

$-2y - 3 < 7$

$-2y < 10$ Add $+3$ to both sides

$-\dfrac{1}{2}(-2y) > -\dfrac{1}{2}(10)$ Multiply by $-\frac{1}{2}$ and reverse the
                                 direction of the inequality symbol

$y > -5$

The solution set is all real numbers that are greater than -5. Below are three equivalent ways to represent this solution set.

Set Notation	Line Graph	Interval Notation
$\{y \mid y > -5\}$		$(-5, \infty)$

When our inequalities become more complicated, we use the same basic steps we used in Section 2.1 when we were solving equations. That is, we simplify each side of the inequality before we apply the addition property or multiplication property. When we have solved the inequality, we graph the solution on a number line.

▶ **EXAMPLE 4**   Solve $3(2x - 4) - 7x \le -3x$.

Solution   We begin by using the distributive property to separate terms. Next, simplify both sides:

$3(2x - 4) - 7x \le -3x$ Original inequality

$6x - 12 - 7x \le -3x$ Distributive property

$-x - 12 \le -3x$ $6x - 7x = (6 - 7)x = -x$

$-12 \le -2x$ Add x to both sides

$-\dfrac{1}{2}(-12) \ge -\dfrac{1}{2}(-2x)$ Multiply both sides by $-\frac{1}{2}$ and
                                reverse the direction of the

$6 \ge x$ inequality symbol

This last line is equivalent to $x \leq 6$. The solution set can be represented with any of the three items below.

Set Notation	Line Graph	Interval Notation

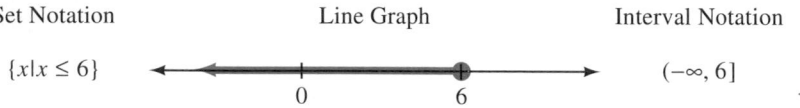

$\{x \mid x \leq 6\}$ $(-\infty, 6]$

Note In Examples 3 and 4, notice that each time we multiplied both sides of the inequality by a negative number we also reversed the direction of the inequality symbol. Failure to do so would cause our graph to lie on the wrong side of the endpoint.

Before we solve more complicated inequalities, let's look at the relationship between interval notation and the graphs of compound inequalities.

In Section 1.2 we found the graph of the set $\{x \mid x \leq -2 \text{ or } x > 3\}$ by using the fact that the word *or* indicates the union of two sets. Here is the graph:

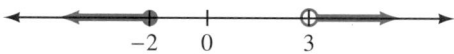

To represent this set of numbers with interval notation we use two intervals connected with the symbol for the union of two sets. Here is the equivalent set of numbers described with interval notation:

$$(-\infty, -2] \cup (3, \infty)$$

Further, since the word *and* is associated with the intersection of two sets, the graph of $\{x \mid x > -1 \text{ and } x < 2\}$ is the part they have in common:

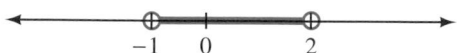

This graph corresponds to the interval $(-1, 2)$, which is called an **open interval** since neither endpoint is included in the interval.

Likewise, using our shorthand notation for inequalities connected by the word *and*, we know the graph of $\{x \mid -3 \leq x \leq 4\}$ is

The corresponding interval is $[-3, 4]$, which is called a **closed interval** since both endpoints are included in the interval.

▶ **EXAMPLE 5** Solve and graph $-3 \leq 2x - 5 \leq 3$.

Solution We can extend our properties for addition and multiplication to cover this situation. If we add a number to the middle expression, we must add the same number to the outside expressions. If we multiply the center expression by a number, we must do the same to the outside expressions, remembering to reverse the direction of the inequality symbols if we multiply by a negative number. We begin by adding 5 to all three parts of the inequality:

$$-3 \leq 2x - 5 \leq 3$$

$$2 \leq \quad 2x \quad \leq 8 \qquad \text{Add 5 to all three members}$$

$$1 \leq \quad x \quad \leq 4 \qquad \text{Multiply through by } \frac{1}{2}$$

Here are three ways to write this solution set:

Set Notation	Line Graph	Interval Notation

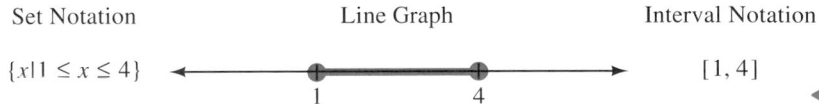

$\{x \mid 1 \leq x \leq 4\}$ ⟵ ①———① (1 ... 4) $[1, 4]$

▶ **EXAMPLE 6** Solve the compound inequality

$$3t + 7 \leq -4 \quad \text{or} \quad 3t + 7 \geq 4$$

Solution We solve each half of the compound inequality separately, then we graph the solution set:

$$3t + 7 \leq -4 \quad \text{or} \quad 3t + 7 \geq 4$$

$$3t \leq -11 \quad \text{or} \qquad 3t \geq -3 \qquad \text{Add } -7$$

$$t \leq -\frac{11}{3} \quad \text{or} \qquad t \geq -1 \qquad \text{Multiply by } \frac{1}{3}$$

The solution set can be written in any of the following ways:

Set Notation	Line Graph	Interval Notation

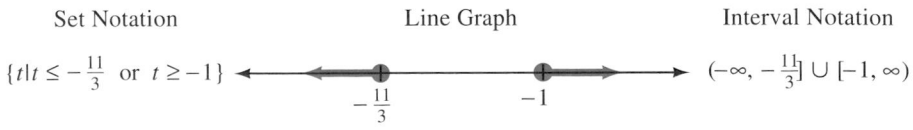

$\{t \mid t \leq -\frac{11}{3} \text{ or } t \geq -1\}$ ⟵ $-\frac{11}{3}$ -1 $(-\infty, -\frac{11}{3}] \cup [-1, \infty)$

▶ **EXAMPLE 7** A company that manufactures typewriter ribbons finds that they can sell x ribbons each week at a price of p dollars each, according to the formula $x = 1{,}300 - 100p$. What price should they charge for each ribbon if they want to sell at least 300 ribbons a week?

Solution Since x is the number of ribbons they sell each week, an inequality that corresponds to selling at least 300 ribbons a week is

$$x \geq 300$$

Solution   The quantities $3a + 2$ and $2a + 3$ have equal absolute values. They are, therefore, the same distance from 0 on the number line. They must be equals or opposites:

$$|3a + 2| = |2a + 3|$$

Equals		*Opposites*
$3a + 2 = 2a + 3$	or	$3a + 2 = -(2a + 3)$
$a + 2 = 3$		$3a + 2 = -2a - 3$
$a = 1$		$5a + 2 = -3$
		$5a = -5$
		$a = -1$

The solution set is $\{1, -1\}$.

It makes no difference in the outcome of the problem if we take the opposite of the first or second expression. It is very important, once we have decided which one to take the opposite of, that we take the opposite of both its terms and not just the first term. That is, the opposite of $2a + 3$ is $-(2a + 3)$, which we can think of as $-1(2a + 3)$. Distributing the -1 across *both* terms, we have

$$-1(2a + 3) = -2a - 3$$ ◀

▶ **EXAMPLE 6**   Solve $|x - 5| = |x - 7|$.

Solution   As was the case in Example 5, the quantities $x - 5$ and $x - 7$ must be equal or they must be opposites, because their absolute values are equal:

Equals		*Opposites*
$x - 5 = x - 7$	or	$x - 5 = -(x - 7)$
$-5 = -7$		$x - 5 = -x + 7$
No solution here		$2x - 5 = 7$
		$2x = 12$
		$x = 6$

Since the first equation leads to a false statement, it will not give us a solution. (If either of the two equations were to reduce to a true statement, it would mean all real numbers would satisfy the original equation.) In this case, our only solution is $x = 6$.   ◀

PROBLEM SET 2.6

Use the definition of absolute value to solve each of the following problems.

1. $|x| = 4$ **2.** $|x| = 7$ **5.** $|x| = -3$ **6.** $|x| = -4$

3. $2 = |a|$ **4.** $5 = |a|$ **7.** $|a| + 2 = 3$ **8.** $|a| - 5 = 2$

9. $|y| + 4 = 3$

10. $|y| + 3 = 1$

11. $4 = |x| - 2$

12. $3 = |x| - 5$

13. $|x - 2| = 5$

14. $|x + 1| = 2$

15. $|a - 4| = \frac{5}{3}$

16. $|a + 2| = \frac{7}{5}$

17. $1 = |3 - x|$

18. $2 = |4 - x|$

19. $\left|\frac{3}{5}a + \frac{1}{2}\right| = 1$

20. $\left|\frac{2}{7}a + \frac{3}{4}\right| = 1$

21. $60 = |20x - 40|$

22. $800 = |400x - 200|$

23. $|2x + 1| = -3$

24. $|2x - 5| = -7$

25. $\left|\frac{3}{4}x - 6\right| = 9$

26. $\left|\frac{4}{5}x - 5\right| = 15$

27. $\left|1 - \frac{1}{2}a\right| = 3$

28. $\left|2 - \frac{1}{3}a\right| = 10$

Solve each equation.

29. $|3x + 4| + 1 = 7$

30. $|5x - 3| - 4 = 3$

31. $|3 - 2y| + 4 = 3$

32. $|8 - 7y| + 9 = 1$

33. $3 + |4t - 1| = 8$

34. $2 + |2t - 6| = 10$

35. $\left|9 - \frac{3}{5}x\right| + 6 = 12$

36. $\left|4 - \frac{2}{7}x\right| + 2 = 14$

37. $5 = \left|\frac{2x}{7} + \frac{4}{7}\right| - 3$

38. $7 = \left|\frac{3x}{5} + \frac{1}{5}\right| + 2$

39. $2 = -8 + \left|4 - \frac{1}{2}y\right|$

40. $1 = -3 + \left|2 - \frac{1}{4}y\right|$

41. $|3(x + 5) + 2| = 1$

42. $|2(x - 4) + 3| = 7$

43. $|1 + 2(x - 1)| = 7$

44. $|3 + 4(x + 2)| = 5$

45. $1 = |2(k + 4) - 3|$

46. $4 = |3(k - 2) + 1|$

Solve the following equations.

47. $|3a + 1| = |2a - 4|$

48. $|5a + 2| = |4a + 7|$

49. $\left|x - \frac{1}{3}\right| = \left|\frac{1}{2}x + \frac{1}{6}\right|$

50. $\left|\frac{1}{10}x - \frac{1}{2}\right| = \left|\frac{1}{5}x + \frac{1}{10}\right|$

51. $|y - 2| = |y + 3|$

52. $|y - 5| = |y - 4|$

53. $|3x - 1| = |3x + 1|$

54. $|5x - 8| = |5x + 8|$

Solve the following equations.

55. $|3 - m| = |m + 4|$

56. $|5 - m| = |m + 8|$

57. $|0.03 - 0.01x| = |0.04 + 0.05x|$

58. $|0.07 - 0.01x| = |0.08 - 0.02x|$

59. $|x - 2| = |2 - x|$

60. $|x - 4| = |4 - x|$

61. $\left|\frac{x}{5} - 1\right| = \left|1 - \frac{x}{5}\right|$

62. $\left|\frac{x}{3} - 1\right| = \left|1 - \frac{x}{3}\right|$

63. Each of the equations in Problems 59 through 62 has the form $|a - b| = |b - a|$. The solution set for each of these equations is all real numbers. This means that the statement $|a - b| = |b - a|$ must be true no matter what numbers a and b are. The statement itself is a property of absolute value. Show that the statement is true when $a = 4$ and $b = -7$, as well as when $a = -5$ and $b = -8$.

64. Show that the statement $|ab| = |a||b|$ is true when $a = 3$ and $b = -6$, and when $a = -8$ and $b = -2$.

65. Name all the numbers in the set $\{-4, -3, -2, -1, 0, 1, 2, 3, 4\}$ that are solutions to the equation $|x + 2| = x + 2$.

66. Name all the numbers in the set $\{-4, -3, -2, -1, 0, 1, 2, 3, 4\}$ that are solutions to the equation $|x - 2| = x - 2$.

67. The equation $|a| = a$ is true only if $a \geq 0$. We can use this fact to actually solve equations like the one in Problem 65. The only way the equation $|x + 2| = x + 2$ can be true is if $x + 2 \geq 0$. Adding -2 to both sides of this last inequality we see that $x \geq -2$. Use the same kind of reasoning to solve the equation $|x - 2| = x - 2$.

68. Solve the equation $|x + 3| = x + 3$.

Problems 69–72 may be solved using a graphing calculator.

Table Building

To obtain a visual representation for expressions that contain absolute value, we can use number sequences. For each sequence below, use the formula to construct a table that gives the first five terms in the sequence by substituting 1, 2, 3, 4, and 5 for n in the formula. (Label the first column in the table n, and the second column a_n.) Then use the paired data from the table to construct a scatter diagram.

69. $a_n = |n - 3|$

70. $a_n = |3 - n|$

71. $a_n = |2n - 6|$

72. $a_n = |6 - 2n|$

Review Problems

The problems below review material we covered in Sections 1.2 and 2.5. Reviewing these problems will help you with the next section.

Graph each inequality. [1.2]

73. $x < -2$ or $x > 8$ **74.** $1 < x < 4$

75. $-2 \le x \le 1$ **76.** $x \le -\frac{3}{2}$ or $x \ge 3$

Solve each inequality. [2.5]

77. $4t - 3 \le -9$ **78.** $-3 < 2a - 5 < 3$

79. $-3x > 15$ **80.** $-2x \le 10$

81. $\frac{1}{2} < \frac{3}{4}a < \frac{3}{5}$ **82.** $\frac{3}{7}a + 2 < \frac{1}{4}$

One Step Further

Solve each formula for x. (Assume a, b, and c are positive.)

83. $|x - a| = b$ **84.** $|x + a| - b = 0$

85. $|ax + b| = c$ **86.** $|ax - b| - c = 0$

87. $\left| \dfrac{x}{a} + \dfrac{y}{b} \right| = 1$ **88.** $\left| \dfrac{x}{a} + \dfrac{y}{b} \right| = c$

SECTION

2.7   # Inequalities Involving Absolute Value

In this section we will again apply the definition of absolute value to solve inequalities involving absolute value. Again, the absolute value of x, which is $|x|$, represents the distance that x is from 0 on the number line. We will begin by considering three absolute value expressions and their English translations:

Expression	*In Words*
$|x| = 7$	x is exactly 7 units from 0 on the number line
$|a| < 5$	a is less than 5 units from 0 on the number line
$|y| \ge 4$	y is greater than or equal to 4 units from 0 on the number line

Once we have translated the expression into words, we can use the translation to graph the original equation or inequality. The graph is then used to write a final equation or inequality that does not involve absolute value.

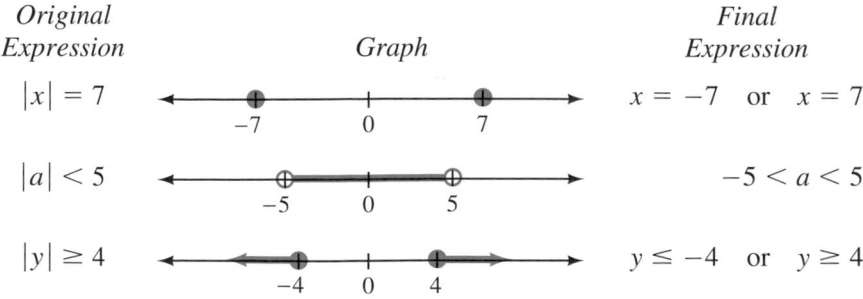

Original Expression	*Graph*	*Final Expression*
$|x| = 7$		$x = -7$ or $x = 7$
$|a| < 5$		$-5 < a < 5$
$|y| \ge 4$		$y \le -4$ or $y \ge 4$

Although we will not always write out the English translation of an absolute value inequality, it is important that we understand the translation. Our second expression, $|a| < 5$, means a is within 5 units of 0 on the number line. The graph of this relationship is which can be written with the following continued inequality:

$$-5 < a < 5$$

We can follow this same kind of reasoning to solve more complicated absolute value inequalities.

▶ **EXAMPLE 1**   Graph the solution set: $|2x - 5| < 3$.

Solution   The absolute value of $2x - 5$ is the distance that $2x - 5$ is from 0 on the number line. We can translate the inequality as, "$2x - 5$ is less than 3 units from 0 on the number line." That is, $2x - 5$ must appear between -3 and 3 on the number line.

A picture of this relationship is

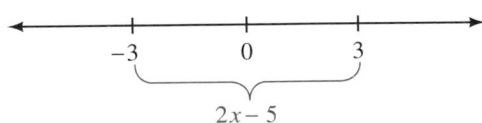

Using the picture, we can write an inequality without absolute value that describes the situation:

$$-3 < 2x - 5 < 3$$

Next, we solve the continued inequality by first adding $+5$ to all three members and then multiplying all three by $\frac{1}{2}$:

$$-3 < 2x - 5 < 3$$
$$2 < \quad 2x \quad < 8 \qquad \text{Add } +5 \text{ to all three expressions}$$
$$1 < \quad x \quad < 4 \qquad \text{Multiply each expression by } \frac{1}{2}$$

The graph of the solution set is

◀

We can see from the solution that in order for the absolute value of $2x - 5$ to be within 3 units of 0 on the number line, x must be between 1 and 4.

▶ **EXAMPLE 2**   Solve and graph $|3a + 7| \leq 4$.

Solution   We can read the inequality as, "The distance between $3a + 7$ and 0 is less than or equal to 4." Or, "$3a + 7$ is within 4 units of 0 on the number line.'' This relationship can be written without absolute value as

$$-4 \leq 3a + 7 \leq 4$$

Solving as usual, we have

$$-4 \leq 3a + 7 \leq 4$$

$$-11 \leq \quad 3a \quad \leq -3 \qquad \text{Add } -7 \text{ to all three members}$$

$$-\frac{11}{3} \leq \quad a \quad \leq -1 \qquad \text{Multiply each expression by } \frac{1}{3}$$

We can see from Examples 1 and 2 that in order to solve an inequality involving absolute value, we must be able to write an equivalent expression that does not involve absolute value.

▶ **EXAMPLE 3** Solve $|x - 3| > 5$ and graph the solution.

Solution We interpret the absolute value inequality to mean that $x - 3$ is more than 5 units from 0 on the number line. The quantity $x - 3$ must be either above $+5$ or below -5. Here is a picture of the relationship:

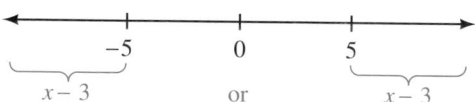

An inequality without absolute value that also describes this situation is

$$x - 3 < -5 \quad \text{or} \quad x - 3 > 5$$

Adding $+3$ to both sides of each inequality we have

$$x < -2 \quad \text{or} \quad x > 8$$

the graph of which is

▶ **EXAMPLE 4** Graph the solution set: $|4t - 3| \geq 9$.

Solution The quantity $4t - 3$ is greater than or equal to 9 units from 0. It must be either above $+9$ or below -9.

$$4t - 3 \leq -9 \quad \text{or} \quad 4t - 3 \geq 9$$

$$4t \leq -6 \quad \text{or} \qquad 4t \geq 12 \qquad \text{Add } +3$$

$$t \le -\frac{6}{4} \quad \text{or} \quad t \ge \frac{12}{4} \quad \text{Multiply by } \frac{1}{4}$$

$$t \le -\frac{3}{2} \quad \text{or} \quad t \ge 3$$

We can use the results of our first few examples and the material in the previous section to summarize the information we have related to absolute value equations and inequalities.

Rewriting Absolute Value Equations and Inequalities

If c is a positive real number, then each statement on the left is equivalent to the corresponding statement on the right.

With Absolute Value	*Without Absolute Value*
$\lvert x \rvert = c$	$x = -c \quad \text{or} \quad x = c$
$\lvert x \rvert < c$	$-c < x < c$
$\lvert x \rvert > c$	$x < -c \quad \text{or} \quad x > c$
$\lvert ax + b \rvert = c$	$ax + b = -c \quad \text{or} \quad ax + b = c$
$\lvert ax + b \rvert < c$	$-c < ax + b < c$
$\lvert ax + b \rvert > c$	$ax + b < -c \quad \text{or} \quad ax + b > c$

▶ **EXAMPLE 5** Solve and graph $\lvert 2x + 3 \rvert + 4 < 9$.

Solution Before we can apply the method of solution we used in the previous examples, we must isolate the absolute value on one side of the inequality. To do so, we add -4 to each side.

$$\lvert 2x + 3 \rvert + 4 < 9$$
$$\lvert 2x + 3 \rvert + 4 + (-4) < 9 + (-4)$$
$$\lvert 2x + 3 \rvert < 5$$

From this last line we know that $2x + 3$ must be between -5 and $+5$.

$$-5 < 2x + 3 < 5$$
$$-8 < \quad 2x \quad < 2 \qquad \text{Add } -3 \text{ to each expression}$$
$$-4 < \quad x \quad < 1 \qquad \text{Multiply each expression by } \frac{1}{2}$$

The graph is

▶ **EXAMPLE 6**   Solve and graph $|4 - 2t| > 2$.

Solution   The inequality indicates that $4 - 2t$ is less than -2 or greater than $+2$. Writing this without absolute value symbols we have

$$4 - 2t < -2 \quad \text{or} \quad 4 - 2t > 2$$

To solve these inequalities we begin by adding -4 to each side.

$$4 + (-4) - 2t < -2 + (-4) \quad \text{or} \quad 4 + (-4) - 2t > 2 + (-4)$$

$$-2t < -6 \qquad\qquad \text{or} \qquad\qquad -2t > -2$$

Next we must multiply both sides of each inequality by $-\frac{1}{2}$. When we do so, we must also reverse the direction of each inequality symbol.

$$-2t < -6 \qquad\qquad \text{or} \qquad\qquad -2t > -2$$

$$-\frac{1}{2}(-2t) > -\frac{1}{2}(-6) \quad \text{or} \quad -\frac{1}{2}(-2t) < -\frac{1}{2}(-2)$$

$$t > 3 \qquad\qquad \text{or} \qquad\qquad t < 1$$

Although in situations like this we are used to seeing the "less than" symbol written first, the meaning of the solution is clear. We want to graph all real numbers that are either greater than 3 or less than 1. Here is the graph.

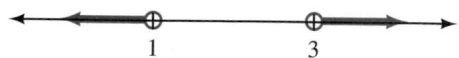

Since absolute value always results in a nonnegative quantity, we sometimes come across special solution sets when a negative number appears on the right side of an absolute value inequality.

▶ **EXAMPLE 7**   Solve $|7y - 1| < -2$.

Solution   The *left* side is never negative, because it is an absolute value. The *right* side is negative. We have a positive quantity less than a negative quantity, which is impossible. The solution set is the empty set, $\varnothing$. There is no real number to substitute for y to make the above inequality a true statement. ◀

▶ **EXAMPLE 8**   Solve $|6x + 2| > -5$.

Solution   This is the opposite case from that in Example 7. No matter what real number we use for x on the *left* side, the result will always be positive, or zero. The *right* side is negative. We have a positive quantity greater than a negative quantity. Every real number we choose for x gives us a true statement. The solution set is the set of all real numbers. ◄

PROBLEM SET 2.7

Solve each of the following inequalities using the definition of absolute value. Graph the solution set in each case.

1. $|x| < 3$

2. $|x| \leq 7$

3. $|x| \geq 2$

4. $|x| > 4$

5. $|x| + 2 < 5$

6. $|x| - 3 < -1$

7. $|t| - 3 > 4$

8. $|t| + 5 > 8$

9. $|y| < -5$

10. $|y| > -3$

11. $|x| \geq -2$

12. $|x| \leq -4$

13. $|x - 3| < 7$

14. $|x + 4| < 2$

15. $|a + 5| \geq 4$

16. $|a - 6| \geq 3$

Solve each inequality and graph the solution set.

17. $|a - 1| < -3$

18. $|a + 2| \geq -5$

19. $|2x - 4| < 6$

20. $|2x + 6| < 2$

21. $|3y + 9| \geq 6$

22. $|5y - 1| \geq 4$

23. $|2k + 3| \geq 7$

24. $|2k - 5| \geq 3$

25. $|x - 3| + 2 < 6$

26. $|x + 4| - 3 < -1$

27. $|2a + 1| + 4 \geq 7$

28. $|2a - 6| - 1 \geq 2$

29. $|3x + 5| - 8 < 5$

30. $|6x - 1| - 4 \leq 2$

Solve each inequality and graph the solution set. Keep in mind that if you multiply or divide both sides of an inequality by a negative number you must reverse the sense of the inequality.

31. $|5 - x| > 3$

32. $|7 - x| > 2$

33. $\left|3 - \frac{2}{3}x\right| \geq 5$

34. $\left|3 - \frac{3}{4}x\right| \geq 9$

35. $\left|2 - \frac{1}{2}x\right| > 1$

36. $\left|3 - \frac{1}{3}x\right| > 1$

Solve each inequality.

37. $|x - 1| < 0.01$

38. $|x + 1| < 0.01$

39. $|2x + 1| \geq \frac{1}{5}$

40. $|2x - 1| \geq \frac{1}{8}$

41. $\left|\dfrac{3x - 2}{5}\right| \leq \dfrac{1}{2}$

42. $\left|\dfrac{4x - 3}{2}\right| \leq \dfrac{1}{3}$

43. $\left|2x - \frac{1}{5}\right| < 0.3$

44. $\left|3x - \frac{3}{5}\right| < 0.2$

45. Write the continued inequality $-4 \leq x \leq 4$ as a single inequality involving absolute value.

46. Write the continued inequality $-8 \leq x \leq 8$ as a single inequality involving absolute value.

47. Write $-1 \leq x - 5 \leq 1$ as a single inequality involving absolute value.

48. Write $-3 \leq x + 2 \leq 3$ as a single inequality involving absolute value.

Review Problems

The problems that follow review material we covered in Section 1.4. Simplify each expression as much as possible.

49. $-9 \div \frac{3}{2}$

50. $-\frac{4}{5} \div (-4)$

51. $3 - 7(-6 - 3)$

52. $(3 - 7)(-6 - 3)$

53. $-4(-2)^3 - 5(-3)^2$

54. $4(2 - 5)^3 - 3(4 - 5)^5$

55. $-2(-3 + 8) - 7(-9 + 6)$

56. $-3 - 6[5 - 2(-3 - 1)]$

57. $\dfrac{2(-3) - 5(-6)}{-1 - 2 - 3}$

58. $\dfrac{4 - 8(3 - 5)}{2 - 4(3 - 5)}$

One Step Further

Assume a, b, and c are positive and solve each formula for x.

59. $|x - a| < b$

60. $|x - a| > b$

61. $|ax - b| > c$

62. $|ax - b| < c$

63. $\left|\dfrac{x}{a} + \dfrac{y}{b}\right| < 1$

64. $\left|\dfrac{x}{a} + \dfrac{y}{b}\right| > 1$

65. $\left|\dfrac{x}{a} + \dfrac{y}{b}\right| < c$

66. $\left|\dfrac{x}{a} + \dfrac{y}{b}\right| > c$

Introduction to Functions

We begin this section with a discussion that is intended to give you an intuitive introduction to the idea of a function. Following this discussion, we will give the formal definition of a function. The ideas presented first are ideas that are familiar to you. What is of interest is how we describe these ideas mathematically—by an equation, a diagram, a table, a graph, or a combination thereof.

To begin, suppose you have a job that pays $7.50 per hour and that you work anywhere from 0 to 40 hours per week. The amount of money you make in one week depends on the number of hours you work that week. In mathematics we say that your weekly earnings are a **function** of the number of hours you work. If we let the variable x represent the number of hours you work in one week and the variable y represent the amount of money you make that week, then the relationship between x and y can be written as

$$y = 7.5x \quad \text{for} \quad 0 \le x \le 40$$

Table 1 is similar to the tables we encountered in Section 2.4. It gives some of the paired data that satisfy the equation $y = 7.5x$. Figure 1 is the line graph constructed using the information from Table 1.

TABLE 1 Weekly Wages

Hours Worked x	Rule $y = 7.5x$	Income y
0	$y = 7.5(0)$	0
10	$y = 7.5(10)$	75
20	$y = 7.5(20)$	150
30	$y = 7.5(30)$	225
40	$y = 7.5(40)$	300

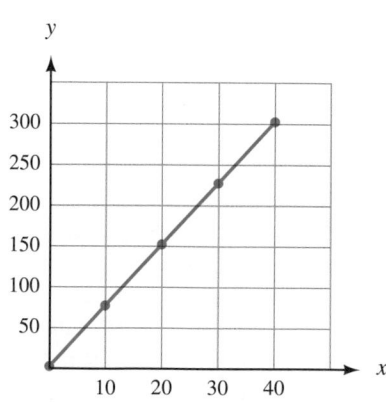

Figure 1

We began this discussion by saying that the number of hours worked during the week was from 0 to 40, so these are the values that x can assume. From the line graph in Figure 1, we see that the values of y range from 0 to 300. We call the complete set of values that x can assume the **domain** of the function, and the values that are assigned to y are called the **range** of the function.

Another way to visualize the relationship between x and y is with the following diagram, which we call a **function map**.

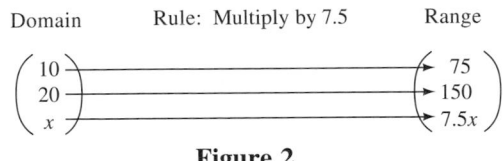

Figure 2

Although the diagram in Figure 2 does not show all the values that x and y can assume, it does give us a visual description of how x and y are related. It shows that values of y in the range come from values of x in the domain according to a specific rule (multiply by 7.5 each time).

What is apparent from the discussion above is that we are working with paired data: the solutions to the equation $y = 7.5x$ are pairs of numbers; the points on the line graph in Figure 1 come from paired data; and the diagram in Figure 2 pairs numbers in the domain with numbers in the range. We have worked with paired data in these first two chapters, so the only thing new in our discussion are the words *function, domain,* and *range.* We continue our discussion with the formal definition for a function.

DEFINITION A **function** is a rule that pairs each element in one set, called the domain, with exactly one element from a second set, called the range.

In other words, a function is a rule that pairs numbers in the domain with numbers in the range, so that each number in the domain is paired with exactly one number in the range. Further, sometimes we know the rule and sometimes we don't. It can be given explicitly, as was the case in the discussion above, or it can be implied by paired data.

You may be wondering if any of the sets of paired data we have worked with so far fail to qualify as functions. The answer is yes. Specifically, the set of data we found on the price of used Ford Mustangs cannot be considered a function. Table 2 and the scatter diagram in Figure 3 were considered previously in Section 1.7.

TABLE 2   Used Mustang Prices

Year	Age from 1994	Price ($)
1993	1	13,925
1993	1	11,850
1993	1	9,995
1992	2	10,200
1992	2	9,600
1991	3	9,525
1990	4	8,675
1990	4	7,900
1989	5	6,975

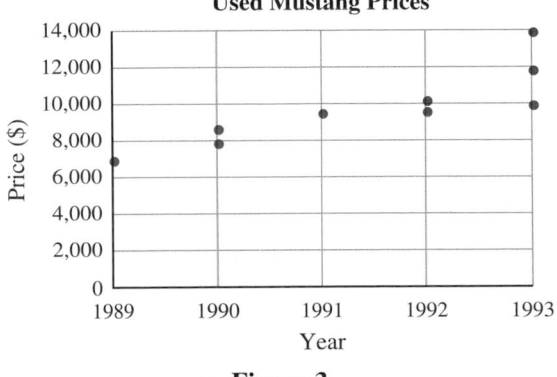

Figure 3

In Table 2, if we pair the numbers in column 2 with the numbers in column 3, we see that the number 1 is paired with three different prices, \$13,925, \$11,850, and \$9,995. That is enough to disqualify the data from belonging to a function. For a set of paired data to be considered a function, each number in the domain must be paired with exactly one number in the range.

We will return to this discussion of what does, and does not, qualify as a function when we look at functions in more detail in Chapter 5.

Function Notation

Let's return to the discussion that introduced us to functions at the beginning of this section. If a job pays \$7.50 per hour for working from 0 to 40 hours a week, then the amount of money earned in one week, y, is a function of the number of hours worked, x. The exact relationship between x and y is written

$$y = 7.5x \quad \text{for} \quad 0 \le x \le 40$$

Since the amount of money earned, y, depends on the number of hours worked, x, we call y the **dependent variable** and x the **independent variable.** Furthermore, if we let f represent all the pairs of numbers produced by the equation, then we can write

$f = \{$all pairs of numbers x and y where $y = 7.5x$ and $0 \le x \le 40\}$

Once we have named a function with a letter, as we have done in the paragraph above, we can use an alternate notation to represent the dependent variable y. The alternate notation for y is $f(x)$. It is read "f of x" and can be used instead of the variable y when working with functions. The notation y and the notation $f(x)$ are equivalent. That is,

$$y = 7.5x \quad \Leftrightarrow \quad f(x) = 7.5x$$

When we use the notation $f(x)$ we are using **function notation.** The benefit of using function notation is that we can write more information with fewer symbols than we can by using just the variable y. For example, asking how much money a person will make for working 20 hours is simply a matter of asking for $f(20)$. Without function notation, we would have to say "find the value of y that corresponds to a value of x of 20." To illustrate further, using the variable y, we can say "y is 150 when x is 20." Using the notation $f(x)$ we simply say "$f(20) = 150$." Each expression indicates that you will earn \$150 for working 20 hours.

▶ **EXAMPLE 1**   If $f(x) = 7.5x$, find $f(0)$, $f(10)$, and $f(20)$.

Solution   To find $f(0)$ we substitute 0 for x in the expression $7.5x$ and simplify. We find $f(10)$ and $f(20)$ in a similar manner—by substitution.

If $\qquad f(x) = 7.5x$

then $\qquad f(0) = 7.5(0) = 0$

$$f(10) = 7.5(10) = 75$$

$$f(20) = 7.5(20) = 150$$

Here are the same three problems again, this time using y instead of $f(x)$.

If $y = 7.5x$, then

when $x = 0$, $y = 7.5(0) = 0$

when $x = 10$, $y = 7.5(10) = 75$

when $x = 20$, $y = 7.5(20) = 150$ ◀

The letters we usually use to represent functions are lowercase letters from the middle of the alphabet; the most common are f, g, and h. If we changed the example in the discussion that opened this section so that the hourly wage was $6.50 per hour, we would have a new equation to work with, namely,

$$y = 6.5x \quad \text{for} \quad 0 \leq x \leq 40$$

Suppose we name this new function with the letter g. Then,

$g = \{$all the pairs of numbers x and y where $y = 6.5x$ and $0 \leq x \leq 40\}$

and

$$g(x) = 6.5x$$

If we want to talk about both functions in the same discussion, having two different letters, f and g, makes it easy to distinguish between them. For example, since $f(x) = 7.5x$ and $g(x) = 6.5x$, asking how much money a person makes for working 20 hours is simply a matter of asking for $f(20)$ or $g(20)$, avoiding any confusion over which hourly wage we are talking about.

The diagrams shown in Figure 4 further illustrate the similarities and differences between the two functions we have been discussing.

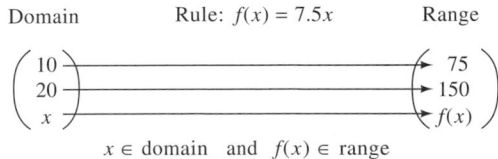

$x \in$ domain   and $f(x) \in$ range

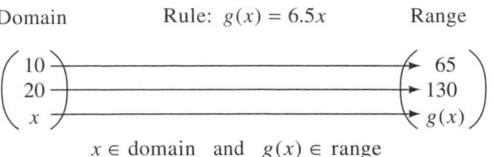

$x \in$ domain   and $g(x) \in$ range

Figure 4

▶ **EXAMPLE 2** If $f(x) = 3x^2 + 2x - 1$, find $f(0)$, $f(3)$, and $f(-2)$.

Solution Since $f(x) = 3x^2 + 2x - 1$, we have

$$f(0) = 3(0)^2 + 2(0) - 1 = 0 + 0 - 1 = -1$$

$$f(3) = 3(3)^2 + 2(3) - 1 = 27 + 6 - 1 = 32$$

$$f(-2) = 3(-2)^2 + 2(-2) - 1 = 12 - 4 - 1 = 7 \qquad ◀$$

In Example 2, the function f is defined by the equation $f(x) = 3x^2 + 2x - 1$. We could just as easily have said $y = 3x^2 + 2x - 1$. That is, $y = f(x)$. Saying $f(-2) = 7$ is exactly the same as saying y is 7 when x is -2.

▶ **EXAMPLE 3** If $f(x) = 4x - 1$ and $g(x) = x^2 + 2$, then

$$f(5) = 4(5) - 1 = 19 \quad \text{and} \quad g(5) = 5^2 + 2 = 27$$

$$f(-2) = 4(-2) - 1 = -9 \quad \text{and} \quad g(-2) = (-2)^2 + 2 = 6$$

$$f(0) = 4(0) - 1 = -1 \quad \text{and} \quad g(0) = 0^2 + 2 = 2$$

$$f(z) = 4z - 1 \quad \text{and} \quad g(z) = z^2 + 2$$

$$f(a) = 4a - 1 \quad \text{and} \quad g(a) = a^2 + 2 \qquad ◀$$

P R O B L E M   S E T   2 . 8

Let $f(x) = 2x - 5$ and $g(x) = x^2 + 3x + 4$. Evaluate the following.

1. $f(2)$

2. $f(3)$

3. $f(-3)$

4. $g(-2)$

5. $g(-1)$

6. $f(-4)$

7. $g(-3)$

8. $g(2)$

9. $g(4) + f(4)$

10. $f(2) - g(3)$

11. $f(3) - g(2)$

12. $g(-1) + f(-1)$

Let $f(x) = 3x^2 - 4x + 1$ and $g(x) = 2x - 1$. Evaluate the following.

13. $f(0)$

14. $g(0)$

15. $g(-4)$

16. $f(1)$

17. $f(-1)$

18. $g(-1)$

19. $g(10)$

20. $f(10)$

21. $f(3)$

22. $g(3)$

23. $g(\frac{1}{2})$

24. $g(\frac{1}{4})$

25. $f(a)$

26. $g(b)$

Suppose that you have a job that pays $8.50 per hour and that you work anywhere from 0 to 40 hours per week.

27. Write an equation, with a restriction on the variable x, that gives the amount of money, $f(x)$, you will earn for working x hours in one week.

28. What does $f(20)$ represent?

A salesperson has a base salary of $200 per week and earns a commission of 6% of the amount he sells each week.

29. How much will the salesperson earn if he sells $500 worth of merchandise in one week?

30. If x is the amount of the salesperson's weekly sales and $f(x)$ is the amount of money earned in one week, write a formula that shows the relationship between x and $f(x)$. (Be sure to note any restrictions on the variable x.)

31. The length of a rectangle is 3 inches more than twice the width. Let x represent the width of the rectangle and $P(x)$ represent the perimeter of the rectangle. Use function notation to write the relationship between x and $P(x)$, noting any restrictions on the variable x.

32. The length of a rectangle is 3 inches more than twice the width. Let x represent the width of the rectangle and $A(x)$ represent the area of the rectangle. Use function notation to write the relationship between x and $A(x)$, noting any restrictions on the variable x.

The formula for the area, A, of a circle with radius r can be written with function notation as $A(r) = \pi r^2$, where $A(r)$ stands for the area.

33. Find $A(2)$, $A(5)$, and $A(10)$.

34. Why doesn't it make sense to ask for $A(-10)$?

35. Suppose a phone company charges 33¢ for the first minute and 24¢ for each additional minute to place a long-distance call out of state between 5 P.M. and 11 P.M. If x is the number of additional minutes and $f(x)$ is the cost of the call, then $f(x) = 24x + 33$.

(a) How much does it cost to talk for 10 minutes?

(b) What does $f(5)$ represent in this problem?

(c) If a call costs $1.29, how long was it?

36. The same phone company mentioned in Problem 35 charges 52¢ for the first minute and 36¢ for each additional minute to place an out-of-state call between 8 A.M. and 5 P.M.

(a) Let $g(x)$ be the total cost of an out-of-state call between 8 A.M. and 5 P.M. and write an equation for $g(x)$.

(b) Find $g(5)$.

(c) Find the difference in price between a 10-minute call made between 8 A.M. and 5 P.M. and the same call made between 5 P.M. and 11 P.M.

Review Problems

The problems that follow review material we covered in Section 2.7. Solve each inequality.

37. $|x| < 3$ **38.** $|x| > 3$

39. $|2y + 1| > -5$ **40.** $|2y + 1| < -5$

41. $|1 - 3t| \geq 5$ **42.** $|3t - 1| \leq 5$

CHAPTER 2 SUMMARY

Examples

1. We can solve $x + 3 = 5$ by adding -3 to both sides:

$$x + 3 + (-3) = 5 + (-3)$$

$$x = 2$$

Addition Property of Equality [2.1]

For algebraic expressions A, B, and C,

$$\text{if} \qquad A = B$$

$$\text{then} \qquad A + C = B + C$$

This property states that we can add the same quantity to both sides of an equation without changing the solution set.

2. We can solve $3x = 12$ by multiplying both sides by $\frac{1}{3}$:

$$3x = 12$$

$$\frac{1}{3}(3x) = \frac{1}{3}(12)$$

$$x = 4$$

Multiplication Property of Equality [2.1]

For algebraic expressions A, B, and C,

$$\text{if} \qquad A = B$$

$$\text{then} \qquad AC = BC, C \neq 0$$

Multiplying both sides of an equation by the same nonzero quantity never changes the solution set.

3. Solve $3(2x - 1) = 9$.

$$3(2x - 1) = 9$$
$$6x - 3 = 9$$
$$6x - 3 + 3 = 9 + 3$$
$$6x = 12$$
$$\frac{1}{6}(6x) = \frac{1}{6}(12)$$
$$x = 2$$

Strategy for Solving Linear Equations in One Variable [2.1]

Step 1a: Use the distributive property to separate terms, if necessary.

1b: If fractions are present, consider multiplying both sides by the LCD to eliminate the fractions. If decimals are present, consider multiplying both sides by a power of 10 to clear the equation of decimals.

1c: Combine similar terms on each side of the equation.

Step 2:   Use the addition property of equality to get all variable terms on one side of the equation and all constant terms on the other side. A variable term is a term that contains the variable (for example, $5x$). A constant term is a term that does not contain the variable (the number 3, for example).

Step 3:   Use the multiplication property of equality to get x (that is, $1x$) by itself on one side of the equation.

Step 4:   Check your solution in the original equation to be sure that you have not made a mistake in the solution process.

4. Solve for w:

$$P = 2l + 2w$$
$$P - 2l = 2w$$
$$\frac{P - 2l}{2} = w$$

Formulas [2.2]

A *formula* in algebra is an equation involving more than one variable. To solve a formula for one of its variables, simply isolate that variable on one side of the equation.

5. The perimeter of a rectangle is 32 inches. If the length is 3 times the width, find the dimensions.

Step 1:   This step is done mentally.

Step 2:   Let x = the width. Then the length is $3x$.

Step 3:   The perimeter is 32; therefore

$$2x + 2(3x) = 32$$

Step 4: $8x = 32$
         $x = 4$

Step 5:   The width is 4 inches. The length is $3(4) = 12$ inches.

Step 6:   The perimeter is $2(4) + 2(12)$ which is 32. The length is 3 times the width.

Blueprint for Problem Solving [2.3, 2.4]

Step 1:   **Read** the problem and then mentally **list** the items that are known and the items that are unknown.

Step 2:   **Assign a variable** to one of the unknown items. (In most cases this will amount to letting x = the item that is asked for in the problem.) Then **translate** the other **information** in the problem to expressions involving the variable.

Step 3:   **Reread** the problem, and then **write an equation,** using the items and variables listed in steps 1 and 2, that describes the situation.

Step 4:   **Solve the equation** found in step 3.

Step 5:   **Write** your **answer** using a complete sentence.

Step 6:   **Reread** the problem and **check** your solution with the original words in the problem.

6. Adding **5** to both sides of the inequality $x - 5 < -2$ gives

$$x - 5 + 5 < -2 + 5$$

$$x < 3$$

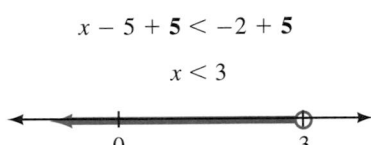

Addition Property for Inequalities [2.5]

For expressions A, B, and C,

if $A < B$

then $A + C < B + C$

Adding the same quantity to both sides of an inequality never changes the solution set.

7. Multiplying both sides of $-2x \geq 6$ by $-\frac{1}{2}$ gives

$$-2x \geq 6$$

$$-\frac{1}{2}(-2x) \leq -\frac{1}{2}(6)$$

$$x \leq -3$$

Multiplication Property for Inequalities [2.5]

For expressions A, B, and C,

if $A < B$

then $AC < BC$ if $C > 0$ (C is positive)

or $AC > BC$ if $C < 0$ (C is negative)

We can multiply both sides of an inequality by the same nonzero number without changing the solution set as long as each time we multiply by a negative number we also reverse the direction of the inequality symbol.

8. To solve $|2x - 1| + 2 = 7$, we first isolate the absolute value on the left side by adding -2 to each side to obtain

$$|2x - 1| = 5$$

$$2x - 1 = 5 \quad \text{or} \quad 2x - 1 = -5$$

$$2x = 6 \quad \text{or} \quad 2x = -4$$

$$x = 3 \quad \text{or} \quad x = -2$$

Absolute Value Equations [2.6]

To solve an equation that involves absolute value, we isolate the absolute value on one side of the equation and then rewrite the absolute value equation as two separate equations that do not involve absolute value. In general, if b is a positive real number, then

$$|a| = b \text{ is equivalent to } a = b \text{ or } a = -b$$

9. To solve $|x - 3| + 2 < 6$, we first add -2 to both sides to obtain

$$|x - 3| < 4$$

which is equivalent to

$$-4 < x - 3 < 4$$

$$-1 < \quad x \quad < 7$$

Absolute Value Inequalities [2.7]

To solve an inequality that involves absolute value, we first isolate the absolute value on the left side of the inequality symbol. Then, we rewrite the absolute value inequality as an equivalent continued or compound inequality that does not contain absolute value symbols. In general, if b is a positive real number, then

$$|a| < b \text{ is equivalent to } -b < a < b$$

and $|a| > b$ is equivalent to $a < -b$ or $a > b$

Functions [2.8]

A *function* is a rule that pairs each element in one set, called the *domain,* with exactly one element from a second set, called the *range.*

10. If $f(x) = 5x - 3$ then

$$f(0) = 5(0) - 3 = -3$$

$$f(1) = 5(1) - 3 = 2$$

$$f(-2) = 5(-2) - 3 = -13$$

$$f(a) = 5a - 3$$

Function Notation [2.8]

The alternate notation for y is $f(x)$. It is read "f of x," and can be used instead of the variable y when working with functions. The notation y and the notation $f(x)$ are equivalent. That is, $y = f(x)$.

COMMON MISTAKES

A very common mistake in solving inequalities is to forget to reverse the direction of the inequality symbol when multiplying both sides by a negative number. When this mistake occurs, the graph of the solution set is always on the wrong side of the endpoint.

CHAPTER 2 REVIEW

Solve each equation. [2.1]

1. $x - 3 = 7$

2. $x + 5 = 4$

3. $5x - 2 = 8$

4. $3x - 4 = 5$

5. $400 - 100a = 200$

6. $300 - 100a = -500$

7. $5 - \frac{2}{3}a = 7$

8. $3 - \frac{4}{5}a = -5$

9. $4x - 2 = 7x + 7$

10. $8x - 3 = 2x - 6$

11. $\frac{3}{2}x - \frac{1}{6} = -\frac{7}{6}x - \frac{1}{6}$

12. $\frac{1}{2}x + \frac{3}{8} = -\frac{3}{4}x + \frac{3}{8}$

13. $7y - 5 - 2y = 2y - 3$

14. $8y - 4 - 6y = 5y - 1$

15. $\frac{3y}{4} - \frac{1}{2} + \frac{3y}{2} = 2 - y$

16. $\frac{y}{4} - 1 + \frac{3y}{8} = \frac{3}{4} - y$

17. $3(2x + 1) = 18$

18. $6(4x - 2) = 12$

19. $-\frac{1}{2}(4x - 2) = -x$

20. $-\frac{1}{2}(4x - 3) = -x$

21. $8 - 3(2t + 1) = 5(t + 2)$

22. $7 - 2(8t - 3) = 4(t - 2)$

23. $8 + 4(1 - 3t) = -3(t - 4) + 2$

24. $4 + 5(4 - 2t) = -2(t - 1) - 4$

25. $0.08x + 0.07(900 - x) = 67$

26. $0.07x + 0.06(1,400 - x) = 90$

Substitute the given values in each formula and then solve for the variable that does not have a numerical replacement. [2.2]

27. $P = 2b + 2h$: $P = 40$, $b = 3$
28. $P = 2b + 2h$: $P = 100$, $h = 10$
29. $A = P + Prt$: $A = 2,000$, $P = 1,000$, $r = 0.05$
30. $A = P + Prt$: $A = 1,000$, $P = 500$, $r = 0.1$
31. $A = a + (n - 1)d$: $A = 40$, $a = 4$, $d = 9$
32. $A = a + (n - 1)d$: $A = 32$, $a = 2$, $d = 10$

Solve each formula for the indicated variable. [2.2]

33. $I = prt$ for p
34. $I = prt$ for t
35. $y = mx + b$ for x
36. $y = mx + b$ for m
37. $4x - 3y = 12$ for y
38. $4x - 3y = 12$ for x
39. $d = vt + 16t^2$ for v
40. $d = vt - 16t^2$ for v
41. $C = \frac{5}{9}(F - 32)$ for F
42. $F = \frac{9}{5}C + 32$ for C

Solve each application. In each case, be sure to show the equation that describes the situation. [2.3, 2.4]

43. A number increased by 8 is 2 less than twice the number. Find the number.
44. Five more than 3 times a number is 6 less than twice the number. What is the number?
45. Four times the sum of two consecutive integers equals 2 more than 14 times the larger integer. Find the integers.
46. If 3 times the smaller of two consecutive even integers is added to the larger, the result is 4 more than 3 times the larger. Find the two integers.
47. The length of a rectangle is 3 times the width. The perimeter is 32 feet. Find the length and width.
48. The length of a rectangle is 4 less than 3 times the width. The perimeter is 32 inches. Find the dimensions.

49. The three sides of a triangle are given by three consecutive integers. If the perimeter is 12 meters, find the length of each side.
50. The three sides of a triangle are given by three consecutive even integers. If the perimeter is 24 yards, find the length of each side.
51. A triangle is such that the largest angle is five times the smallest angle. The third angle is 7° less than the largest angle. Find the measure of each angle.
52. The third angle in an isosceles triangle is one fourth as large as each of the two base angles. Find the measure of each angle.
53. The sum of two numbers is 16. If one of the numbers is 3 times the other, find the two numbers.
54. The sum of two numbers is 25. If one of the numbers is 4 times the other, find the two numbers.
55. A man has a collection of 23 coins with a total value of $1.90. If the coins are all dimes and nickels, how many of each type does he have?
56. A collection of dimes and quarters has a total value of $2.75. If there are 23 coins total, how many of each type are there?
57. A woman invests a total of $900 in two accounts. One of the accounts pays 8% in annual interest, while the other pays 7% in annual interest. If the total interest for the first year is $67, how much did she invest in each account?
58. A man invests a total of $1,400 in two accounts. One of the accounts pays 7% in annual interest, while the other pays 6% in annual interest. If his total interest for the first year is $90, how much did he invest in each account?

Solve each inequality. Write your answer using interval notation. [2.5]

59. $-8a > -4$ **60.** $-9a > -3$
61. $6 - a \geq -2$ **62.** $7 - a \geq -6$
63. $\frac{3}{4}x + 1 \leq 10$ **64.** $\frac{2}{5}x - 1 \leq 9$
65. $800 - 200x < 1000$
66. $600 - 300x < 900$
67. $\frac{1}{3} \leq \frac{1}{6}x \leq 1$

68. $-\frac{1}{2} \le \frac{1}{6}x \le \frac{1}{3}$
69. $-0.01 \le 0.02x - 0.01 \le 0.01$
70. $-0.01 < 0.01x - 0.02 < 0.01$
71. $5t + 1 \le 3t - 2$ or $-7t \le -21$
72. $6t - 3 \le t + 1$ or $-8t \le -16$
73. $3(x + 1) < 2(x + 2)$ or $2(x - 1) \ge x + 2$
74. $5(x + 3) < 3(x - 5)$ or $4(x - 6) \ge x + 3$

Solve each equation. [2.6]

75. $|x| = 2$ **76.** $|x| = 3$

77. $|a| - 3 = 1$ **78.** $|a| - 2 = 3$

79. $|x - 3| = 1$ **80.** $|x - 2| = 3$

81. $|2y - 3| = 5$ **82.** $|3y - 2| = 7$

83. $|4x - 3| + 2 = 11$ **84.** $|6x - 2| + 4 = 16$

85. $\left| \dfrac{7}{3} - \dfrac{x}{3} \right| + \dfrac{4}{3} = 2$

86. $\left| 4 - \frac{1}{2}x \right| - 1 = \frac{3}{2}$
87. $|5t - 3| = |3t - 5|$
88. $|6t - 2| = |4t + 8|$
89. $\left| \frac{1}{2} - x \right| = \left| x + \frac{1}{2} \right|$
90. $\left| 1 - \frac{2}{3}x \right| = \left| \frac{2}{3}x + 1 \right|$

Solve each inequality and graph the solution set. [2.7]

91. $|x| < 5$ **92.** $|a| > 2$

93. $|0.01a| \ge 5$ **94.** $|0.01a| \ge 2$
95. $|x| \le 0$ **96.** $|y - 2| < 3$
97. $|y + 5| \ge 0.02$ **98.** $|5x - 1| > 3$
99. $|2t + 1| - 3 < 2$ **100.** $|2t + 1| - 1 < 5$

Solve each equation or inequality, if possible. [2.1, 2.5, 2.6, 2.7]

101. $2x - 3 = 2(x - 3)$
102. $4(x + 1) = 4x + 1$
103. $3(5x - \frac{1}{2}) = 15x + 2$
104. $4(\frac{1}{2}x - 1) = 2x - 4$
105. $|4y + 8| = -1$ **106.** $|5y - 2| = -3$
107. $|x| > 0$ **108.** $|x| \ge 0$
109. $|5 - 8t| + 4 \le 1$ **110.** $|7 - 9t| + 5 \le 3$
111. $|2x + 1| \ge -4$ **112.** $|5x - 8| \ge -2$

Let $f(x) = 3x - 5$ and $g(x) = 2x^2 + 1$ and find the following. [2.8]

113. $f(0)$ **114.** $g(0)$
115. $g(3)$ **116.** $f(3)$
117. $f(-1)$ **118.** $g(-1)$
119. $f(\frac{1}{3})$ **120.** $g(\frac{1}{2})$

CHAPTER 2 TEST

Solve the following equations. [2.1]

1. $x - 5 = 7$ **2.** $3y = -4$
3. $5 - \frac{4}{7}a = -11$
4. $\frac{1}{5}x - \frac{1}{2} - \frac{1}{10}x + \frac{2}{5} = \frac{3}{10}x + \frac{1}{2}$
5. $5(x - 1) - 2(2x + 3) = 5x - 4$
6. $0.07 - 0.02(3x + 1) = -0.04x + 0.01$

Solve for the indicated variable. [2.2]

7. $P = 2l + 2w$ for w
8. $A = \frac{1}{2}h(b + B)$ for B

Solve each of the following. [2.3, 2.4]

9. Find two consecutive even integers whose sum is 18. (Be sure to write an equation that describes the situation.)
10. A rectangle is twice as long as it is wide. The perimeter is 36 inches. Find the dimensions.
11. The third angle in an isosceles triangle is 8° less than twice as large as each of the two base angles. Find the measure of each angle.
12. The sum of two numbers is 19. If one of the numbers is 4 more than twice the other, find the numbers.
13. Diane has a collection of dimes and nickels

worth $1.10. If the total number of coins is 14, how many of each kind does she have?

14. Howard invests a total of $8,000 in two accounts. One account earns 8% annually, and the other earns 10% annually. If the total interest earned from both accounts in a year is $680, how much is invested in each account?

Solve the following inequalities. Write the solution set using interval notation, then graph the solution set. [2.5]

15. $-5t \le 30$ **16.** $5 - \frac{3}{2}x > -1$
17. $1.6x - 2 < 0.8x + 2.8$
18. $3(2y + 4) \ge 5(y - 8)$

Solve the following equations. [2.6]

19. $\left|\frac{1}{4}x - 1\right| = \frac{1}{2}$ **20.** $\left|\frac{2}{3}a + 4\right| = 6$
21. $|3 - 2x| + 5 = 2$ **22.** $5 = |3y + 6| - 4$

Solve the following inequalities and graph the solutions. [2.7]

23. $|6x - 1| > 7$ **24.** $|3x - 5| - 4 \le 3$
25. $|5 - 4x| \ge -7$ **26.** $|4t - 1| < -3$

Let $f(x) = x^2 - 4$ and $g(x) = 2x + 3$ and find the following. [2.8]

27. $f(2)$ **28.** $f(-2)$
29. $g\left(\frac{1}{2}\right)$ **30.** $g\left(-\frac{1}{2}\right)$

3

EXPONENTS AND
POLYNOMIALS

INTRODUCTION

As we mentioned in the introduction to Chapter 2, many of the topics we cover
are linked to the ''real world'' by applications. In this chapter we find a number
of topics that can be used to solve application problems in the business world.
The links to the business world are formed by some well-known formulas. The
most basic of the business formulas is the formula for profit:

$$\text{Profit} = \text{Revenue} - \text{Cost}$$

where revenue is the total amount of money a company brings in by selling their
product and cost is the total cost to produce the product they sell. If we let the
variable P represent profit, R represent revenue, and C represent cost, then our
formula becomes

$$P = R - C$$

Revenue itself can be broken down further by another formula common in the
business world. Suppose a store sells x items at a price of p dollars per item. The
revenue they obtain from selling all x items is the product of the number of items
sold and the price per item. That is,

$$\text{Revenue} = (\text{number of items sold})(\text{price of each item})$$

For example, if 100 items are sold for \$9 each, the revenue is $100(9) = \$900$.
Likewise, if 500 items are sold for \$11 each, then the revenue is $500(11) =$
\$5,500. In general, if x is the number of items sold and p is the selling price of
each item, then we can write

$$R = xp$$

In this chapter, the profit formula and the revenue equation serve as links that
connect the topics in the chapter to the business world. If you go on to take
classes in business, you will see these equations many times.

OVERVIEW

We begin this chapter by expanding our knowledge of exponents to include a list of properties that describe how exponents behave in certain situations. Our expanded study of exponents leads us to polynomials, which we study in detail. The chapter ends with a look at some equations that can be solved by a method that depends on factoring polynomials. Your success in this chapter hinges on your ability to understand and apply the distributive property: The distributive property is the foundation upon which algebra with polynomials is built.

STUDY SKILLS

The study skills for this chapter have to do with the way you approach new situations in mathematics. The first study skill is a point of view you hold about your natural instincts for what does and doesn't work in mathematics. The second study skill gives you a way of testing your instincts.

1. **Don't Let Your Intuition Fool You**   As you become more experienced and more successful in mathematics you will be able to trust your mathematical intuition. For now, though, it can get in the way of your success. For example, if you ask some students to ''subtract 3 from -5'' many will answer -2 or 2. Both answers are incorrect, even though they may seem intuitively true. Likewise, some students will expand $(a + b)^2$ and arrive at $a^2 + b^2$, which is incorrect. In both cases, intuition leads directly to the wrong answer.
2. **Test Properties of Which You Are Unsure**   From time to time you will be in a situation where you would like to apply a property or rule, but you are not sure it is true. You can always test a property or statement by substituting numbers for variables. For instance, I always have students that rewrite $(x + 3)^2$ as $x^2 + 9$, thinking that the two expressions are equivalent. The fact that the two expressions are not equivalent becomes obvious when we substitute 10 for x in each one.

    When $x = 10$, the expression $(x + 3)^2$ is $(10 + 3)^2 = 13^2 = 169$

    When $x = 10$, the expression $x^2 + 9$ is $10^2 + 9 = 100 + 9 = 109$

    When you test the equivalence of expressions by substituting numbers for the variable, make it easy on yourself by choosing numbers that are easy to work with, such as 10. Don't try to verify the equivalence of expressions by substituting 0, 1, or 2 for the variable, as these numbers will occasionally give you false results.

It is not good practice to trust your intuition or instincts in every new situation in algebra. If you have any doubt about the generalizations you are making, test them by replacing variables with numbers and simplifying.

SECTION

3.1   # Properties of Exponents I

In Chapter 1, we defined positive integer exponents in terms of repeated multiplication. That is, $3^4 = 3 \cdot 3 \cdot 3 \cdot 3 = 81$. Also, recall that for the expression 3^4, the base is 3 and the exponent is 4. The expression 3^4 is written in *exponential form,* while the expression $3 \cdot 3 \cdot 3 \cdot 3$ is in *expanded form.*

In this section we will be concerned with the simplification of products that involve more than one base and more than one exponent. We begin by making some generalizations about exponents.

▶ **EXAMPLE 1**   Write the product $x^3 \cdot x^4$ with a single exponent.

Solution

$$
\begin{aligned}
x^3 \cdot x^4 &= (x \cdot x \cdot x)(x \cdot x \cdot x \cdot x) \\
&= (x \cdot x \cdot x \cdot x \cdot x \cdot x \cdot x) \\
&= x^7 \qquad \textit{Notice}: 3 + 4 = 7 \qquad ◀
\end{aligned}
$$

We can generalize this result into the first property of exponents.

Property 1 for Exponents
If a is a real number and r and s are integers, then

$$a^r \cdot a^s = a^{r+s}$$

Note   We are stating our properties of exponents for integer exponents instead of just positive integer exponents. As you will see, the definition for negative integer exponents is stated in such a way that we can change any expression with a negative exponent to an equivalent expression with a positive exponent. This allows us to state our properties for all integers, not just for positive ones.

▶ **EXAMPLE 2**   Write $(5^3)^2$ with a single exponent.

Solution

$$
\begin{aligned}
(5^3)^2 &= 5^3 \cdot 5^3 \\
&= 5^6 \qquad \textit{Notice}: 3 \cdot 2 = 6 \qquad ◀
\end{aligned}
$$

Generalizing this result we have a second property of exponents.

> **Property 2 for Exponents**
> If a is a real number and r and s are integers, then
> $$(a^r)^s = a^{r \cdot s}$$

A third property of exponents arises when we have the product of two or more numbers raised to an integer power.

▶ **EXAMPLE 3** Expand $(3x)^4$ and then multiply.

Solution

$$(3x)^4 = (3x)(3x)(3x)(3x)$$
$$= (3 \cdot 3 \cdot 3 \cdot 3)(x \cdot x \cdot x \cdot x)$$
$$= 3^4 \cdot x^4 \qquad \textit{Notice}: \text{The exponent 4 distributes over the product } 3x.$$
$$= 81x^4 \qquad ◀$$

Generalizing Example 3 we have Property 3 for exponents.

> **Property 3 for Exponents**
> If a and b are any two real numbers and r is an integer, then
> $$(ab)^r = a^r \cdot b^r$$

Here are some examples that use combinations of the first three properties of exponents to simplify expressions involving exponents.

▶ **EXAMPLES** Simplify each expression using the properties of exponents.

4. $(-3x^2)(5x^4) = -3(5)(x^2 \cdot x^4)$ Commutative, associative properties
$$= -15x^6 \qquad\qquad \text{Property 1 for exponents}$$

5. $(-2x^2)^3(4x^5) = (-2)^3(x^2)^3(4x^5)$ Property 3
$$= -8x^6 \cdot (4x^5) \qquad \text{Property 2}$$
$$= (-8 \cdot 4)(x^6 \cdot x^5) \quad \text{Commutative and associative}$$
$$= -32x^{11} \qquad\qquad \text{Property 1}$$

6. $(x^2)^4(x^2y^3)^2(y^4)^3 = x^8 \cdot x^4 \cdot y^6 \cdot y^{12}$ Properties 2 and 3
$$= x^{12}y^{18} \qquad\qquad \text{Property 1} \qquad ◀$$

The last property of exponents for this section deals with negative integer exponents.

Property 4 for Exponents

If a is any nonzero real number and r is a positive integer, then

$$a^{-r} = \frac{1}{a^r}$$

Note   This property is actually a definition. That is, we are defining negative integer exponents as indicating reciprocals. Doing so gives us a way to write an expression with a negative exponent as an equivalent expression with a positive exponent.

▶ **EXAMPLES**   Write with positive exponents, then simplify.

7. $5^{-2} = \dfrac{1}{5^2} = \dfrac{1}{25}$

8. $(-2)^{-3} = \dfrac{1}{(-2)^3} = \dfrac{1}{-8} = -\dfrac{1}{8}$

9. $\left(\dfrac{3}{4}\right)^{-2} = \dfrac{1}{(\frac{3}{4})^2} = \dfrac{1}{\frac{9}{16}} = \dfrac{16}{9}$ ◀

If we generalize the result in Example 9, we have the following extension of Property 4,

$$\left(\frac{a}{b}\right)^{-r} = \left(\frac{b}{a}\right)^{r}$$

which indicates that raising a fraction to a negative power is equivalent to raising the reciprocal of the fraction to the positive power.

The next example shows how we simplify expressions that contain both positive and negative exponents.

▶ **EXAMPLES**   Simplify and write your answers with positive exponents only. (Assume all variables are nonzero.)

10. $(2x^{-3})^4 = 2^4(x^{-3})^4$ Property 3

$\phantom{(2x^{-3})^4} = 16x^{-12}$ Property 2

$\phantom{(2x^{-3})^4} = 16 \cdot \dfrac{1}{x^{12}}$ Property 4

$\phantom{(2x^{-3})^4} = \dfrac{16}{x^{12}}$ Multiplication of fractions

11. $(5y^{-4})^2(2y^5) = 25y^{-8}(2y^5)$ Properties 2 and 3

$= (25 \cdot 2)(y^{-8}y^5)$ Commutative and associative

$= 50y^{-3}$ Property 1

$= 50 \cdot \dfrac{1}{y^3}$ Property 4

$= \dfrac{50}{y^3}$ Multiplication of fractions

12. $(5x^{-2}y^3)^2(2x^4y^{-5})^{-3} = (25x^{-4}y^6)\left(\dfrac{1}{8}x^{-12}y^{15}\right)$ Properties 2, 3, and 4

$= \left(25 \cdot \dfrac{1}{8}\right)(x^{-4}x^{-12})(y^6y^{15})$ Commutative and associative

$= \dfrac{25}{8}x^{-16}y^{21}$ Property 1

$= \dfrac{25}{8} \cdot \dfrac{1}{x^{16}} \cdot \dfrac{y^{21}}{1}$ Property 4

$= \dfrac{25y^{21}}{8x^{16}}$ ◀

Scientific Notation

Scientific notation is a way in which to write very large or very small numbers in a more manageable form. Here is the definition.

DEFINITION   A number is written in **scientific notation** if it is written as the product of a number between 1 and 10 and an integer power of 10. A number written in scientific notation has the form

$$n \times 10^r$$

where $1 \leq n < 10$ and $r =$ an integer.

▶ **EXAMPLE 13**   Write 376,000 in scientific notation.

Solution   We must rewrite 376,000 as the product of a number between 1 and 10 and a power of 10. To do so we move the decimal point 5 places to the left so that it appears between the 3 and the 7. Then we multiply this number by 10^5. The number that results has the same value as our original number and is written in scientific notation.

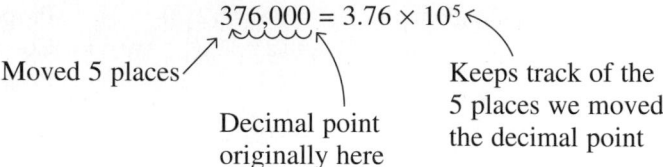

Moved 5 places

Decimal point
originally here

Keeps track of the
5 places we moved
the decimal point

◀

If a number written in expanded form is greater than or equal to 10, then when the number is written in scientific notation the exponent on 10 will be positive. A number that is less than 1 will have a negative exponent when written in scientific notation.

▶ **EXAMPLE 14** Write 4.52×10^3 in expanded form.

Solution Since 10^3 is 1,000, we can think of this as simply a multiplication problem. That is

$$4.52 \times 10^3 = 4.52 \times 1,000 = 4,520$$

On the other hand, we can think of the exponent 3 as indicating the number of places we need to move the decimal point in order to write our number in expanded form. Since our exponent is positive 3, we move the decimal point three places to the right.

$$4.52 \times 10^3 = 4,520$$ ◀

The table that follows lists some additional examples of numbers written in expanded form and in scientific notation. In each case, note the relationship between the number of places the decimal point is moved and the exponent on 10.

Number Written in Expanded Form		Number Written Again in Scientific Notation
376,000	=	3.76×10^5
49,500	=	4.95×10^4
3,200	=	3.2×10^3
591	=	5.91×10^2
46	=	4.6×10^1
8	=	8×10^0
0.47	=	4.7×10^{-1}
0.093	=	9.3×10^{-2}
0.00688	=	6.88×10^{-3}
0.0002	=	2×10^{-4}
0.000098	=	9.8×10^{-5}

Calculator Note   Some scientific calculators have a key that allows you to enter numbers in scientific notation. The key is labeled

$$\boxed{\text{EXP}} \text{ or } \boxed{\text{EE}} \text{ or } \boxed{\text{SCI}}$$

To enter the number 3.45×10^6 you would first enter the decimal number, then press the scientific notation key, and finally, enter the exponent.

$$3.45 \boxed{\text{EXP}} 6$$

To enter 6.2×10^{-27} you would use the following sequence:

$$6.2 \boxed{\text{EXP}} 27 \boxed{+/-}$$

PROBLEM SET 3.1

Evaluate each of the following.

1. 4^2

2. $(-4)^2$

3. -4^2

4. $-(-4)^2$

5. -0.3^3

6. $(-0.3)^3$

7. 2^5

8. 2^4

9. $(\frac{1}{2})^3$

10. $(\frac{3}{4})^2$

11. $(-\frac{5}{6})^2$

12. $(-\frac{7}{8})^2$

Use the properties of exponents to simplify each of the following as much as possible.

13. $x^5 \cdot x^4$

14. $x^6 \cdot x^3$

15. $(2^3)^2$

16. $(3^2)^2$

17. $(-\frac{2}{3}x^2)^3$

18. $(-\frac{3}{5}x^4)^3$

19. $-3a^2(2a^4)$

20. $5a^7(-4a^6)$

21. $6x^2(-3x^4)(2x^5)$

22. $(5x^3)(-7x^4)(-2x^6)$

23. $(-\frac{1}{3}n)^4 (2n^3)^2 (\frac{3}{2}n^6)^4$

24. $(\frac{1}{8}n^6)^2(-2n^3)^2(-3n^7)^2$

Write each of the following with positive exponents. Then simplify as much as possible.

25. 3^{-2}

26. $(-5)^{-2}$

27. $(-2)^{-5}$

28. 2^{-5}

29. $(-3)^{-2}$

30. $(-7)^{-2}$

31. $(\frac{3}{4})^{-2}$

32. $(\frac{3}{5})^{-2}$

33. $(\frac{1}{3})^{-2} + (\frac{1}{2})^{-3}$

34. $(\frac{1}{2})^{-2} + (\frac{1}{3})^{-3}$

35. $(\frac{2}{3})^{-2} - (\frac{2}{5})^{-2}$

36. $(\frac{3}{2})^{-2} - (\frac{3}{4})^{-2}$

Simplify each expression. Write all answers with positive exponents only. (Assume all variables are nonzero.)

37. $x^{-4}x^7$

38. $x^{-3}x^8$

39. $(a^2b^{-5})^3$

40. $(a^4b^{-3})^3$

41. $(\frac{1}{2}x^{-3})^3(6x^4)$

42. $(4x^{-4})^3(\frac{1}{8}x^8)$

43. $(5y^4)^{-3}(2y^{-2})^3$

44. $(3y^5)^{-2}(2y^{-4})^3$

45. $(\frac{1}{2}x^3)(\frac{2}{3}x^4)(\frac{3}{5}x^{-7})$

46. $(\frac{1}{7}x^{-3})(\frac{7}{8}x^{-5})(\frac{8}{9}x^8)$

47. $(0.2a^3)^2(5a^{-4})^2(4a^3)$

48. $(0.3a^{-4})^2(3a^4)^{-2}(\frac{1}{10}a^{16})$

49. $(2x^4y^{-3})(7x^{-8}y^5)$

50. $(4x^7y^{-2})(8x^3y^{-4})$

51. $(3x^2y^5z^{-3})(5x^7y^{-2}z^5)$

52. $(9x^{-3}y^4z^{-2})(7x^5y^{-1}z^4)$

53. $(4a^5b^2)(2b^{-5}c^2)(3a^7c^4)$

54. $(3a^{-2}c^3)(5b^{-6}c^5)(4a^6b^{-2})$

55. $(2x^2y^{-5})^3(3x^{-4}y^2)^{-4}$

56. $(4x^{-4}y^9)^{-2}(5x^4y^{-3})^2$

57. $(3r^2s^{-1})^4(9r^{-6}s^4)^{-3}$

58. $(8r^{-2}s^4)^2(4r^{-3}s^2)^{-2}$

Write each number in scientific notation.

59. 378,000

60. 3,780,000

61. 4,900

62. 490

63. 0.00037

64. 0.000037

65. 0.00495

66. 0.0495

67. 0.562

68. 0.0562

Write each number in expanded form.

69. 5.34×10^3

70. 5.34×10^2

71. 7.8×10^6

72. 7.8×10^4

73. 3.44×10^{-3}

74. 3.44×10^{-5}

75. 4.9×10^{-1}

76. 4.9×10^{-2}

Property 4 for exponents states that $a^{-r} = \dfrac{1}{a^r}$. We could also have stated it this way: $\dfrac{1}{a^{-r}} = a^r$ because

$$\frac{1}{a^{-r}} = \frac{1}{\dfrac{1}{a^r}} = 1 \cdot \frac{a^r}{1} = a^r$$

Use this idea to simplify the following.

77. $\dfrac{1}{2^{-3}}$ **78.** $\dfrac{1}{3^{-2}}$

79. $\dfrac{1}{5^{-2}}$ **80.** $\dfrac{1}{3^{-4}}$

81. Which of these two expressions is larger, $(2^2)^3$ or 2^{2^3}?
82. Which of these two expressions is larger, $(3^4)^2$ or 3^{4^2}?
83. Let $d = 174$, $v = 10$, and $t = 3$ in the formula $d = vt + \frac{1}{2}gt^2$, and then solve for g.
84. Let $d = 36$, $v = 60$, and $t = 3$ in the formula $d = vt - \frac{1}{2}gt^2$, and then solve for g.
85. The statement $(a + b)^{-1} = a^{-1} + b^{-1}$ is false for all pairs of real numbers a and b. Show that it is false when $a = 2$ and $b = 4$.
86. Is the statement $(a + b)^2 = a^2 + b^2$ true in general? Try it with $a = 2$ and $b = 3$ and see.

Applying the Concepts

87. The number 237×10^4 is not written in scientific notation because 237 is larger than 10. Write 237×10^4 in scientific notation.
88. Write 46.2×10^{-3} in scientific notation.
89. If you are 20 years old, you have been alive for more than 630,000,000 seconds. Write this last number in scientific notation.

90. Use the information from Problem 89 to give the approximate number of seconds you have lived if you are 40 years old. Write your answer in scientific notation.
91. The mass of the earth is approximately 5.98×10^{24} kilograms. If this number were written in expanded form, how many zeros would it contain?
92. The mass of a single hydrogen atom is approximately 1.67×10^{-27} kilograms. If this number were written in expanded form, how many digits would there be to the right of the decimal point?

Review Problems

The following problems review material we covered in Section 1.4.

Simplify each expression.

93. $6 - (-8)$ **94.** $-6 - (-8)$
95. $8 - (-6)$ **96.** $-8 - (-6)$
97. $-4 - (-3)$ **98.** $4 - (-3)$
99. $-4 - (-9)$ **100.** $4 - (-9)$

One Step Further

Assume m and n are positive integers and simplify each expression.

101. $x^{m+2} \cdot x^{-2m}x^{m-5}$
102. $x^{m-4}x^{m+9}x^{-2m}$
103. $(y^m)^2(y^{-3})^m(y^{m+3})$
104. $(y^m)^{-4}(y^3)^m(y^{m+6})$
105. $(2x^my^n)^2(3x^{-m}y^{-n})^2$
106. $(4x^{3m}y^{2n})^{-4}(2x^{-6m}y^{-4n})^2$

Properties of Exponents II

We begin this section by stating two properties of exponents that show how exponents affect division. The first property shows that exponents distribute over quotients. This is Property 5 in our list of properties.

> ## Property 5 for Exponents
> If a and b are any two real numbers with $b \neq 0$ and r is an integer, then
>
> $$\left(\frac{a}{b}\right)^r = \frac{a^r}{b^r}$$

PROOF OF PROPERTY 5

$$\left(\frac{a}{b}\right)^r = \underbrace{\left(\frac{a}{b}\right)\left(\frac{a}{b}\right)\left(\frac{a}{b}\right) \cdots \left(\frac{a}{b}\right)}_{r \text{ factors}}$$

$$= \frac{a \cdot a \cdot a \cdots a \leftarrow r \text{ factors}}{b \cdot b \cdot b \cdots b \leftarrow r \text{ factors}}$$

$$= \frac{a^r}{b^r}$$

Since we will be working with quotients in this section, let's assume all our variables represent nonzero numbers.

▶ **EXAMPLES** Simplify each expression and write all answers with positive exponents only.

1. $\left(\dfrac{x^2}{y^3}\right)^4 = \dfrac{(x^2)^4}{(y^3)^4}$ Property 5

$= \dfrac{x^8}{y^{12}}$ Property 2

2. $\left(\dfrac{2x^{-2}}{y^{-1}}\right)^3 = \dfrac{(2x^{-2})^3}{(y^{-1})^3}$ Property 5

$= \dfrac{8x^{-6}}{y^{-3}}$ Properties 2 and 3

$= \dfrac{\dfrac{8}{x^6}}{\dfrac{1}{y^3}}$ Definition of negative exponents (Property 4)

$= \dfrac{8}{x^6} \cdot \dfrac{y^3}{1}$ Division of fractions

$= \dfrac{8y^3}{x^6}$ ◀

Since multiplication with the same base resulted in addition of exponents, it seems reasonable to expect division with the same base to result in subtraction of exponents.

Let's begin with an example in which the exponent in the numerator is larger than the exponent in the denominator:

$$\frac{6^5}{6^3} = \frac{6 \cdot 6 \cdot 6 \cdot 6 \cdot 6}{6 \cdot 6 \cdot 6}$$

Dividing out the 6's common to the numerator and denominator, we have

$$\frac{6^5}{6^3} = \frac{\cancel{6} \cdot \cancel{6} \cdot \cancel{6} \cdot 6 \cdot 6}{\cancel{6} \cdot \cancel{6} \cdot \cancel{6}}$$

$$= 6 \cdot 6$$

$$= 6^2 \qquad \textit{Notice:} \quad 5 - 3 = 2$$

If we were simply to subtract the exponent in the denominator from the exponent in the numerator, we would obtain the correct result, which is 6^2.

Let's consider an example where the exponent in the denominator is larger than the exponent in the numerator:

$$\frac{5^3}{5^7} = \frac{\cancel{5} \cdot \cancel{5} \cdot \cancel{5}}{\cancel{5} \cdot \cancel{5} \cdot \cancel{5} \cdot 5 \cdot 5 \cdot 5 \cdot 5}$$

$$= \frac{1}{5 \cdot 5 \cdot 5 \cdot 5}$$

$$= \frac{1}{5^4}$$

$$= 5^{-4} \qquad \textit{Notice:} \quad 3 - 7 = -4$$

Here again, simply subtracting the exponent in the denominator from the exponent in the numerator would give the correct result.

We summarize this discussion with Property 6 for exponents.

Property 6 for Exponents

If a is any nonzero real number and r and s are any two integers, then

$$\frac{a^r}{a^s} = a^{r-s}$$

Notice again we have specified r and s to be any integers. Our definition of negative exponents is such that the properties of exponents hold for all integer exponents, whether positive integers or negative.

▶ **EXAMPLES** Apply Property 6 to each expression and then simplify the result. All answers that contain exponents should contain positive exponents only.

3. $\dfrac{2^8}{2^3} = 2^{8-3} = 2^5 = 32$

4. $\dfrac{x^2}{x^{18}} = x^{2-18} = x^{-16} = \dfrac{1}{x^{16}}$

5. $\dfrac{a^6}{a^{-8}} = a^{6-(-8)} = a^{14}$

6. $\dfrac{m^{-5}}{m^{-7}} = m^{-5-(-7)} = m^2$ ◀

In the next example we use more than one property of exponents to simplify an expression.

▶ **EXAMPLE 7**   Simplify $\left(\dfrac{x^{-3}}{x^5}\right)^{-2}$.

Solution   Let's begin by applying Property 6 to simplify inside the parentheses.

$$\left(\frac{x^{-3}}{x^5}\right)^{-2} = (x^{-3-5})^{-2} \qquad \text{Property 6}$$

$$= (x^{-8})^{-2}$$

$$= x^{16} \qquad\qquad \text{Property 2} \quad ◀$$

Let's complete our list of properties by looking at how the numbers 0 and 1 behave when used as exponents.

We can use the original definition for exponents when the number 1 is used as an exponent.

$$a^1 = \underbrace{a}_{1 \text{ factor}}$$

For 0 as an exponent, consider the expression $\dfrac{3^4}{3^4}$. Since $3^4 = 81$, we have

$$\frac{3^4}{3^4} = \frac{81}{81} = 1$$

On the other hand, since we have the quotient of two expressions with the same base, we can subtract exponents.

$$\frac{3^4}{3^4} = 3^{4-4} = 3^0$$

Hence, 3^0 must be the same as 1.

Summarizing these results, we have our last property for exponents.

Property 7 for Exponents

If a is any real number, then

$$a^1 = a$$

and

$$a^0 = 1 \quad \text{(as long as } a \neq 0\text{)}$$

▶ **EXAMPLES** Simplify.

8. $(2x^2y^4)^0 = 1$

9. $(2x^2y^4)^1 = 2x^2y^4$ ◀

Here are some examples that use many of the properties of exponents. There are a number of different ways to proceed on problems like these. You should use the method that works best for you.

▶ **EXAMPLES** Simplify.

10. $\dfrac{(x^3)^{-2}(x^4)^5}{(x^{-2})^7} = \dfrac{x^{-6}x^{20}}{x^{-14}}$ Property 2

$\qquad = \dfrac{x^{14}}{x^{-14}}$ Property 1

$\qquad = x^{28}$ Property 6: $x^{14-(-14)} = x^{28}$

11. $\dfrac{6a^5b^{-6}}{12a^3b^{-9}} = \dfrac{6}{12} \cdot \dfrac{a^5}{a^3} \cdot \dfrac{b^{-6}}{b^{-9}}$ Write as separate fractions

$\qquad = \dfrac{1}{2}a^2b^3$ Property 6

Note This last answer can also be written as $\dfrac{a^2b^3}{2}$. Either answer is correct.

12. $\dfrac{(4x^{-5}y^3)^2}{(x^4y^{-6})^{-3}} = \dfrac{16x^{-10}y^6}{x^{-12}y^{18}}$ Properties 2 and 3

$\qquad = 16x^2y^{-12}$ Property 6

$\qquad = 16x^2 \cdot \dfrac{1}{y^{12}}$ Property 4

$\qquad = \dfrac{16x^2}{y^{12}}$ Multiplication ◀

We can use our properties of exponents to do arithmetic with numbers written in scientific notation. Here are some examples.

▶ **EXAMPLES**   Simplify each expression and write all answers in scientific notation.

13. $(2 \times 10^8)(3 \times 10^{-3}) = (2)(3) \times (10^8)(10^{-3})$
$$= 6 \times 10^5$$

14. $\dfrac{4.8 \times 10^9}{2.4 \times 10^{-3}} = \dfrac{4.8}{2.4} \times \dfrac{10^9}{10^{-3}}$
$$= 2 \times 10^{9-(-3)}$$
$$= 2 \times 10^{12}$$

15. $\dfrac{(6.8 \times 10^5)(3.9 \times 10^{-7})}{7.8 \times 10^{-4}} = \dfrac{(6.8)(3.9)}{7.8} \times \dfrac{(10^5)(10^{-7})}{10^{-4}}$
$$= 3.4 \times 10^2 \qquad \blacktriangleleft$$

Calculator Note   If you have a scientific calculator with a scientific notation key, then the sequence of keys you would use to do Example 14 would look like this:

$$4.8 \;\boxed{\text{EXP}}\; 9 \;\boxed{\div}\; 2.4 \;\boxed{\text{EXP}}\; 3 \;\boxed{+/-}\;\boxed{=}$$

P R O B L E M   S E T   3 . 2

Use the properties of exponents to simplify each expression. All answers should contain positive exponents only. Assume all variables are nonzero.

1. $\left(\dfrac{x^3}{y^2}\right)^2$

2. $\left(\dfrac{x^5}{y^2}\right)^3$

3. $\left(\dfrac{2a^{-2}}{b^{-1}}\right)^2$

4. $\left(\dfrac{2a^{-3}}{b^{-2}}\right)^3$

5. $\dfrac{3^4}{3^6}$

6. $\dfrac{3^6}{3^4}$

7. $\dfrac{2^{-2}}{2^{-5}}$

8. $\dfrac{2^{-5}}{2^{-2}}$

9. $\dfrac{x^{-1}}{x^9}$

10. $\dfrac{x^{-3}}{x^5}$

11. $\dfrac{a^4}{a^{-6}}$

12. $\dfrac{a^5}{a^{-2}}$

13. $\dfrac{t^{-10}}{t^{-4}}$

14. $\dfrac{t^{-8}}{t^{-5}}$

15. $\left(\dfrac{x^5}{x^3}\right)^6$

16. $\left(\dfrac{x^7}{x^4}\right)^5$

17. $\dfrac{(a^3)^4}{a^7}$

18. $\dfrac{(a^5)^3}{a^{10}}$

19. $\dfrac{a^7}{(a^3)^4}$

20. $\dfrac{a^{10}}{(a^5)^3}$

21. $\dfrac{(x^5)^6}{(x^3)^4}$

22. $\dfrac{(x^7)^3}{(x^4)^5}$

23. $\dfrac{x^5 x^6}{x^3}$

24. $\dfrac{x^7 x^8}{x^4}$

25. $\dfrac{a^3}{a^5 a^6}$

26. $\dfrac{a^4}{a^7 a^8}$

27. $\dfrac{(m^3)^2 m^5}{(m^4)^3}$

28. $\dfrac{(m^6)^2 m^4}{(m^5)^8}$

29. $\dfrac{(x^{-2})^3 (x^3)^{-2}}{x^{10}}$

30. $\dfrac{(x^{-4})^3 (x^3)^{-4}}{x^{10}}$

31. $\dfrac{15 x^6 y^7}{5 x^4 y^3}$

32. $\dfrac{24 x^8 y^6}{8 x^5 y^2}$

33. $\dfrac{5 a^8 b^3}{20 a^5 b^{-4}}$

34. $\dfrac{7 a^6 b^{-2}}{21 a^2 b^{-5}}$

35. $\dfrac{27x^3y^{-4}z}{9x^7y^{-6}z^4}$

36. $\dfrac{28x^5y^{-2}z}{14x^8y^{-5}z^6}$

37. $\dfrac{12r^{-6}s^0t^{-3}}{3r^{-4}s^{-3}t^{-5}}$

38. $\dfrac{18r^{-8}s^{-2}t^0}{6r^{-10}s^{-4}t^{-3}}$

39. $\dfrac{(2x^3y^4)^5}{(x^2y^3)^3}$

40. $\dfrac{(5x^8y^{10})^2}{(x^3y^4)^3}$

41. $\dfrac{(3x^{-2}y^8)^4}{(9x^4y^{-3})^2}$

42. $\dfrac{(6x^{-3}y^{-5})^2}{(3x^{-4}y^{-3})^4}$

43. $\dfrac{(2a^3b^{-2}c)^5}{(a^{-2}b^4c^{-3})^{-2}}$

44. $\dfrac{(4a^{-2}bc^4)^3}{(a^5b^{-2}c^{-7})^{-3}}$

45. $\left(\dfrac{8x^2y}{4x^4y^{-3}}\right)^4$

46. $\left(\dfrac{5x^4y^5}{10xy^{-2}}\right)^3$

47. $\left(\dfrac{x^{-5}y^2}{x^{-3}y^5}\right)^{-2}$

48. $\left(\dfrac{x^{-8}y^{-3}}{x^{-5}y^6}\right)^{-1}$

49. $\left(\dfrac{2x^{-3}y^0}{4x^6y^{-5}}\right)^{-2}$

50. $\left(\dfrac{2x^6y^4z^0}{8x^{-3}y^0z^{-5}}\right)^{-1}$

51. $\left(\dfrac{ab^{-3}c^{-2}}{a^{-3}b^0c^{-5}}\right)^{-1}$

52. $\left(\dfrac{a^3b^2c^1}{a^{-1}b^{-2}c^{-3}}\right)^{-2}$

53. $\left(\dfrac{x^{-3}y^2}{x^4y^{-5}}\right)^{-2}\left(\dfrac{x^{-4}y}{x^0y^2}\right)$

54. $\left(\dfrac{x^{-1}y^4}{x^{-5}y^0}\right)^{-1}\left(\dfrac{x^3y^{-1}}{xy^{-3}}\right)$

55. $2x^2y^3\left(\dfrac{7xy^4}{14x^3y^6}\right)^{-2}$

56. $3xy^5\left(\dfrac{2x^4y}{6x^5y^3}\right)^{-2}$

57. $7x^{-3}y^4\left(\dfrac{3x^{-1}y^5}{9x^3y^{-2}}\right)^{-3}$

58. $8x^4y^{-3}\left(\dfrac{12x^{-3}y^{-2}}{24x^4y^{-5}}\right)^{-3}$

Use the properties of exponents to simplify each of the following expressions. Write all answers in scientific notation.

59. $(4.5 \times 10^6)(2 \times 10^4)$

60. $(4.3 \times 10^8)(2 \times 10^5)$

61. $\dfrac{6.8 \times 10^6}{3.4 \times 10^{10}}$

62. $\dfrac{9.6 \times 10^{11}}{4.8 \times 10^{15}}$

63. $\dfrac{(5 \times 10^6)(4 \times 10^{-8})}{8 \times 10^4}$

64. $\dfrac{(6 \times 10^{-7})(3 \times 10^9)}{5 \times 10^6}$

65. $\dfrac{(2.4 \times 10^{-3})(3.6 \times 10^{-7})}{(4.8 \times 10^6)(1 \times 10^{-9})}$

66. $\dfrac{(7.5 \times 10^{-6})(1.5 \times 10^9)}{(1.8 \times 10^4)(2.5 \times 10^{-2})}$

Convert each number to scientific notation and then simplify. Write all answers in scientific notation.

67. $(2,000,000)(0.0000249)$

68. $(30,000)(0.000192)$

69. $\dfrac{69,800}{0.000349}$

70. $\dfrac{0.000545}{1,090,000}$

71. $\dfrac{(40,000)(0.0007)}{0.0014}$

72. $\dfrac{(800,000)(0.00002)}{4,000}$

For each expression that follows, find a value of x that makes it a true statement.

73. $4^3 \cdot 4^x = 4^9$

74. $(4^3)^x = 4^9$

75. $\dfrac{3^x}{3^5} = 3^4$

76. $\dfrac{3^x}{3^5} = 3^{-4}$

77. $2^x \cdot 2^3 = \dfrac{1}{16}$

78. $(2^x)^3 = \dfrac{1}{8}$

Applying the Concepts

79. A light-year, the distance light travels in one year, is approximately 5.9×10^{12} miles. The Andromeda galaxy is approximately 1.7×10^6 light-years from our galaxy. Find the distance in miles between our galaxy and the Andromeda galaxy.

80. The distance from the earth to the sun is approximately 9.3×10^7 miles. If light travels 1.2×10^7 miles in one minute, how many minutes does it take the light from the sun to reach the earth?

Review Problems

The following problems review some of the material we covered in Sections 1.3 and 1.4.

Simplify each expression.

81. $6 + 2(x + 3)$ **82.** $8 + 3(x + 4)$
83. $3(2a + 1) - 6a$ **84.** $5(3a + 3) - 8a$
85. $\frac{1}{2}(4y - 2) - \frac{1}{3}(6y - 3)$
86. $\frac{1}{5}(5y - 10) - \frac{1}{3}(3y + 6)$
87. $3 - 7(x - 5) + 3x$ **88.** $4 - 9(x - 3) + 5x$

Combine the following fractions.

89. $-\frac{1}{3} + \frac{7}{12} - \frac{1}{4}$ **90.** $\frac{1}{2} - \frac{1}{4} + \frac{1}{3}$
91. $\frac{5}{12} + \frac{9}{14} - \frac{1}{7} + \frac{1}{2}$ **92.** $\frac{4}{15} + \frac{3}{10} - \frac{1}{5} - \frac{1}{2}$

One Step Further

If you have a calculator that has a key for entering numbers in scientific notation, then use it to simplify the following expressions.

93. $\dfrac{(1.98 \times 10^{25})(3.85 \times 10^{35})}{6.93 \times 10^{40}}$

94. $\dfrac{(1.43 \times 10^{33})(1.87 \times 10^{47})}{2.21 \times 10^{50}}$

95. $\dfrac{(1.05 \times 10^{-18})(5.25 \times 10^{23})}{3.15 \times 10^{-37}}$

96. $\dfrac{(3.51 \times 10^{-51})(5.13 \times 10^{37})}{2.47 \times 10^{-18}}$

Assume all variable exponents represent positive integers and simplify each expression.

97. $\dfrac{x^{n+2}}{x^{n-3}}$ **98.** $\dfrac{x^{n-3}}{x^{n-7}}$

99. $\dfrac{a^{3m}a^{m+1}}{a^{4m}}$ **100.** $\dfrac{a^{2m}a^{m-5}}{a^{3m-7}}$

101. $\dfrac{(y^r)^{-2}}{y^{-2r}}$ **102.** $\dfrac{(y^r)^2}{y^{2r-1}}$

103. $\dfrac{(x^r y^s)^{-2}(8x^2 y^s)^2}{16x^{-4r+2}y^{s+5}}$

104. $\dfrac{(3x^{-2}y^s)^{-3}(2x^r y^4)^3}{6x^{3r-2}y^{-4s+1}}$

3.3 Polynomials, Sums, and Differences

We begin this section with the definition around which polynomials are defined. Once we have listed all the terminology associated with polynomials, we will show how the distributive property is used to find sums and differences of polynomials.

Polynomials in General

DEFINITION A **term** or **monomial** is a constant or the product of a constant and one or more variables raised to whole-number exponents.

The following are monomials or terms:

$$-16 \qquad 3x^2y \qquad -\frac{2}{5}a^3b^2c \qquad xy^2z$$

The numerical part of each monomial is called the **numerical coefficient,** or just **coefficient** for short. For the preceding terms, the coefficients are -16, 3, $-\frac{2}{5}$, and 1. Notice that the coefficient for xy^2z is understood to be 1.

DEFINITION A **polynomial** is any finite sum of terms. Since subtraction can be written in terms of addition, finite differences are also included in this definition.

The following are polynomials:

$$2x^2 - 6x + 3 \qquad -5x^2y + 2xy^2 \qquad 4a - 5b + 6c + 7d$$

Polynomials can be classified further according to the number of terms present. If a polynomial consists of two terms, it is said to be a **binomial.** If it has three terms, it is called a **trinomial.** And, as stated above, a polynomial with only one term is said to be a **monomial.**

DEFINITION The **degree** of a polynomial with one variable is the highest power to which the variable is raised in any one term.

▶ **EXAMPLES**

1. $6x^2 + 2x - 1$ A trinomial of degree 2
2. $5x - 3$ A binomial of degree 1
3. $7x^6 - 5x^3 + 2x - 4$ A polynomial of degree 6
4. $-7x^4$ A monomial of degree 4
5. 15 A monomial of degree 0   ◀

Polynomials in one variable are usually written in decreasing powers of the variable. When this is the case, the coefficient of the first term is called the **leading coefficient.** In Example 1, the leading coefficient is 6. In Example 2, it is 5. The leading coefficient in Example 3 is 7.

DEFINITION Two or more terms that differ only in their numerical coefficients are called **similar** or **like** terms. Since similar terms differ only in their coefficients, they have identical variable parts—that is, the same variables raised to the same power. For example, $3x^2$ and $-5x^2$ are similar terms. So are $15x^2y^3z$, $-27x^2y^3z$, and $\frac{3}{4}x^2y^3z$.

We can use the distributive property to combine the similar terms $6x^2$ and $9x^2$ as follows:

$$6x^2 + 9x^2 = (6 + 9)x^2 \qquad \text{Distributive property}$$
$$= 15x^2 \qquad \text{The sum of 6 and 9 is 15}$$

The distributive property can also be used to combine more than two like terms.

▶ **EXAMPLE 6** Combine $7x^2y + 4x^2y - 10x^2y + 2x^2y$.

Solution

$$7x^2y + 4x^2y - 10x^2y + 2x^2y = (7 + 4 - 10 + 2)x^2y \quad \text{Distributive}$$
$$\text{property}$$
$$= 3x^2y \qquad\qquad\qquad \text{Addition} \qquad ◀$$

Addition and Subtraction of Polynomials

To add two polynomials, we simply apply the commutative and associative properties to group similar terms together and then use the distributive property as we have in the preceding example.

▶ **EXAMPLE 7** Add $5x^2 - 4x + 2$ and $3x^2 + 9x - 6$.

Solution

$$(5x^2 - 4x + 2) + (3x^2 + 9x - 6)$$
$$= (5x^2 + 3x^2) + (-4x + 9x) + (2 - 6) \qquad \text{Commutative and}$$
$$\text{associative properties}$$
$$= (5 + 3)x^2 + (-4 + 9)x + (2 - 6) \qquad \text{Distributive property}$$
$$= 8x^2 + 5x + (-4)$$
$$= 8x^2 + 5x - 4 \qquad\qquad\qquad\qquad ◀$$

▶ **EXAMPLE 8** Find the sum of $-8x^3 + 7x^2 - 6x + 5$ and $10x^3 + 3x^2 - 2x - 6$.

Solution We can add the two polynomials using the method of Example 7, or we can arrange similar terms in columns and add vertically. Using the column method, we have

$$
\begin{array}{r}
-8x^3 + 7x^2 - 6x + 5 \\
10x^3 + 3x^2 - 2x - 6 \\
\hline
2x^3 + 10x^2 - 8x - 1
\end{array}
$$

◀

$$R(p) = (900 - 300p)p$$
$$R(p) = 900p - 300p^2$$

This last formula gives the revenue in terms of the price p. If $p = 1.60$, then

$$R(1.60) = 900(1.60) - 300(1.60)^2$$
$$= 900(1.60) - 300(2.56)$$
$$= 1{,}440 - 768$$
$$= 672$$

The weekly revenue will be \$672 if they charge \$1.60 for each sketch pad. ◀

▶ **EXAMPLE 3**   Multiply $2x - 3$ and $x + 5$.

Solution   Distributing the $2x - 3$ across the sum $x + 5$ gives us

$(2x - 3)(x + 5)$

$= (2x - 3)x + (2x - 3)5$	Distributive property
$= 2x(x) + (-3)x + 2x(5) + (-3)5$	Distributive property
$= 2x^2 - 3x + 10x - 15$	
$= 2x^2 + 7x - 15$	Combine like terms

Notice the third line in this example. It consists of all possible products of terms in the first binomial and those of the second binomial. We can generalize this into a rule for multiplying two polynomials. ◀

Rule

To multiply two polynomials, multiply each term in the first polynomial by each term in the second polynomial.

Multiplying polynomials can be accomplished by a method that looks very similar to long multiplication with whole numbers. We line up the polynomials vertically and then apply our rule for multiplication of polynomials.

▶ **EXAMPLE 4**   Multiply $(2x - 3y)$ and $(3x^2 - xy + 4y^2)$ vertically.

Solution

$$
\begin{array}{r}
3x^2 - xy + 4y^2 \\
2x - 3y \\
\hline
6x^3 - 2x^2y + 8xy^2 \\
- 9x^2y + 3xy^2 - 12y^3 \\
\hline
6x^3 - 11x^2y + 11xy^2 - 12y^3
\end{array}
$$

Multiply $(3x^2 - xy + 4y^2)$ by $2x$
Multiply $(3x^2 - xy + 4y^2)$ by $-3y$
Add similar terms ◀

Multiplying Binomials — The FOIL Method

Consider the product of $(2x - 5)$ and $(3x - 2)$. Distributing $(3x - 2)$ over $2x$ and -5, we have

$$
\begin{aligned}
(2x - 5)(3x - 2) &= (2x)(3x - 2) + (-5)(3x - 2) \\
&= (2x)(3x) + (2x)(-2) + (-5)(3x) + (-5)(-2) \\
&= 6x^2 - 4x - 15x + 10 \\
&= 6x^2 - 19x + 10
\end{aligned}
$$

Looking closely at the second and third lines, we notice the following:

1. $6x^2$ comes from multiplying the *first* terms in each binomial:

$$
(2x - 5)(3x - 2) \qquad 2x(3x) = 6x^2 \qquad \textit{First} \text{ terms}
$$

2. $-4x$ comes from multiplying the *outside* terms in the product.

$$
(2x - 5)(3x - 2) \qquad 2x(-2) = -4x \qquad \textit{Outside} \text{ terms}
$$

3. $-15x$ comes from multiplying the *inside* terms in the product:

$$
(2x - 5)(3x - 2) \qquad -5(3x) = -15x \qquad \textit{Inside} \text{ terms}
$$

4. 10 comes from multiplying the *last* two terms in the product:

$$
(2x - 5)(3x - 2) \qquad -5(-2) = 10 \qquad \textit{Last} \text{ terms}
$$

Once we know where the terms in the answer come from, we can reduce the number of steps used in finding the product:

$$
(2x - 5)(3x - 2) = \underset{\text{First}}{6x^2} - \underset{\text{Outside}}{4x} - \underset{\text{Inside}}{15x} + \underset{\text{Last}}{10} = 6x^2 - 19x + 10
$$

▶ **EXAMPLES** Multiply using the FOIL method.

5. $(4a - 5b)(3a + 2b) = \underset{\text{F}}{12a^2} + \underset{\text{O}}{8ab} - \underset{\text{I}}{15ab} - \underset{\text{L}}{10b^2}$

$$
= 12a^2 - 7ab - 10b^2
$$

6. $(3 - 2t)(4 + 7t) = \underset{\text{F}}{12} + \underset{\text{O}}{21t} - \underset{\text{I}}{8t} - \underset{\text{L}}{14t^2}$

$$
= 12 + 13t - 14t^2
$$

7. $\left(2x + \dfrac{1}{2}\right)\left(4x - \dfrac{1}{2}\right) = 8x^2 - x + 2x - \dfrac{1}{4} = 8x^2 + x - \dfrac{1}{4}$

$$
\qquad\qquad\qquad \text{F} \quad \text{O} \quad \text{I} \quad \text{L}
$$

8. $(a^5 + 3)(a^5 - 7) = a^{10} - 7a^5 + 3a^5 - 21$
$$\qquad\qquad\qquad\quad\; \text{F}\quad\; \text{O}\quad\; \text{I}\quad\; \text{L}$$
$$= a^{10} - 4a^5 - 21$$

9. $(2x + 3)(5y - 4) = 10xy - 8x + 15y - 12$
$$\qquad\qquad\qquad\quad\; \text{F}\quad\;\; \text{O}\quad\;\; \text{I}\quad\;\; \text{L}$$ ◀

▶ **EXAMPLE 10**   The lengths of the three sides of a rectangular box that is closed at the top are given by three consecutive even integers. Write a formula that will give the surface area of the box.

Solution   If we let $x =$ the first of the even integers, then $x + 2$ is the next consecutive even integer, and $x + 4$ is the one after that. A diagram of the box looks like this:

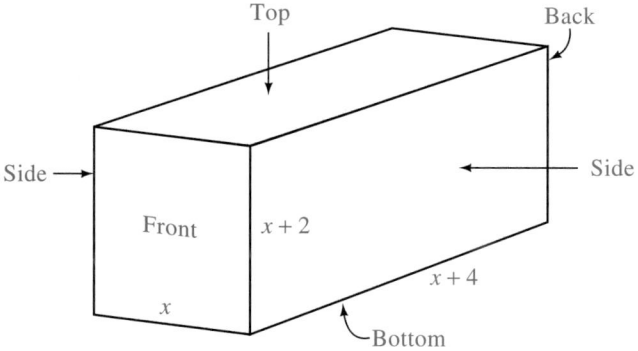

To find the surface area, we add the areas of the two sides, the top and the bottom, and the front and back. Since the surface area depends on the variable x, we use function notation to denote it as $S(x)$.

Total surface area	Area of the two sides	Area of the top and bottom	Area of the front and back

$$S(x) \;=\; 2(x + 2)(x + 4) \;+\; 2x(x + 4) \;+\; 2x(x + 2)$$
$$=\; 2x^2 + 12x + 16 + 2x^2 + 8x \;+\; 2x^2 + 4x$$
$$=\; 6x^2 + 24x + 16$$

If the shortest side x were 4 inches, the total surface area would be

$$S(4) = 6(4)^2 + 24(4) + 16 = 208 \text{ square inches}$$ ◀

The Square of a Binomial

▶ **EXAMPLE 11**   Find $(4x - 6)^2$.

Solution   Applying the definition of exponents and then the FOIL method, we have

$$(4x - 6)^2 = (4x - 6)(4x - 6)$$
$$= 16x^2 - 24x - 24x + 36$$
$$ \text{F} \quad \text{O} \quad \text{I} \quad \text{L}$$
$$= 16x^2 - 48x + 36 \qquad \blacktriangleleft$$

This example is the square of a binomial. This type of product occurs frequently enough in algebra that we have special formulas for it. Here are the formulas for binomial squares:

$$(a + b)^2 = (a + b)(a + b) = a^2 + ab + ab + b^2 = a^2 + 2ab + b^2$$
$$(a - b)^2 = (a - b)(a - b) = a^2 - ab - ab + b^2 = a^2 - 2ab + b^2$$

Observing the results in both cases, we have the following rule.

Rule

The square of a binomial is the sum of the square of the first term, twice the product of the two terms, and the square of the last term. Or

$$(a + b)^2 = \quad a^2 \quad + \quad 2ab \quad + \quad b^2$$

	Square of first term	Twice the product of the two terms	Square of last term

$$(a - b)^2 = \quad a^2 \quad - \quad 2ab \quad + \quad b^2$$

▶ **EXAMPLES** Use the preceding formulas to expand each binomial square.

12. $(x + 7)^2 = x^2 + 2(x)(7) + 7^2 = x^2 + 14x + 49$

13. $(3t - 5)^2 = (3t)^2 - 2(3t)(5) + 5^2 = 9t^2 - 30t + 25$

14. $(4x + 2y)^2 = (4x)^2 + 2(4x)(2y) + (2y)^2 = 16x^2 + 16xy + 4y^2$

15. $(5 - a^3)^2 = 5^2 - 2(5)(a^3) + (a^3)^2 = 25 - 10a^3 + a^6$

16. $\left(x + \dfrac{1}{3}\right)^2 = x^2 + 2(x)\left(\dfrac{1}{3}\right) + \left(\dfrac{1}{3}\right)^2 = x^2 + \dfrac{2}{3}x + \dfrac{1}{9}$ $\qquad \blacktriangleleft$

Products Resulting in the Difference of Two Squares

Another frequently occurring kind of product is found when multiplying two binomials that differ only in the sign between their terms.

▶ **EXAMPLE 17** Multiply $(3x - 5)$ and $(3x + 5)$.

Solution Applying the FOIL method, we have

$$(3x - 5)(3x + 5) = 9x^2 + 15x - 15x - 25 \qquad \text{Two middle terms}$$
$$\qquad\qquad\qquad\qquad \text{F} \quad \text{O} \quad \text{I} \quad \text{L} \qquad \text{add to 0}$$
$$= 9x^2 - 25 \qquad\qquad\qquad\qquad \blacktriangleleft$$

The outside and inside products in Example 17 are opposites and therefore add to 0. Here it is in general:

$$(a - b)(a + b) = a^2 + ab - ab - b^2 \qquad \text{Two middle terms add to 0}$$
$$= a^2 - b^2$$

> **Rule**
>
> To multiply two binomials that differ only in the sign between their two terms, simply subtract the square of the second term from the square of the first term:
>
> $$(a + b)(a - b) = a^2 - b^2$$

The expression $a^2 - b^2$ is called the **difference of two squares.**

▶ **EXAMPLES**   Find the following products.

18. $(x - 5)(x + 5) = x^2 - 25$
19. $(2a - 3)(2a + 3) = 4a^2 - 9$
20. $(x^2 + 4)(x^2 - 4) = x^4 - 16$
21. $(x^3 - 2a)(x^3 + 2a) = x^6 - 4a^2$ ◀

P R O B L E M   S E T   3 . 4

Multiply the following by applying the distributive property.

1. $2x(6x^2 - 5x + 4)$
2. $-3x(5x^2 - 6x - 4)$
3. $-3a^2(a^3 - 6a^2 + 7)$
4. $4a^3(3a^2 - a + 1)$
5. $2a^2b(a^3 - ab + b^3)$
6. $5a^2b^2(8a^2 - 2ab + b^2)$
7. $-4x^2y^3(7x^2 - 3xy + 6y^2)$
8. $-3x^3y^2(6x^2 - 3xy + 4y^2)$
9. $3r^3s^2(r^3 - 2r^2s + 3rs^2 + s^3)$
10. $5r^2s^3(2r^3 + 3r^2s - 4rs^2 + 5s^3)$

Multiply the following vertically.

11. $(x - 5)(x + 3)$ **12.** $(x + 4)(x + 6)$

13. $(2x^2 - 3)(3x^2 - 5)$ **14.** $(3x^2 + 4)(2x^2 - 5)$
15. $(x + 3)(x^2 + 6x + 5)$
16. $(x - 2)(x^2 - 5x + 7)$
17. $(3a + 5)(2a^3 - 3a^2 + a)$
18. $(2a - 3)(3a^2 - 5a + 1)$
19. $(a - b)(a^2 + ab + b^2)$
20. $(a + b)(a^2 - ab + b^2)$
21. $(2x + y)(4x^2 - 2xy + y^2)$
22. $(x - 3y)(x^2 + 3xy + 9y^2)$
23. $(2a - 3b)(a^2 + ab + b^2)$
24. $(5a - 2b)(a^2 - ab - b^2)$
25. $(3x - 4y)(6x^2 + 3xy + 4y^2)$
26. $(2x - 6y)(7x^2 - 6xy + 3y^2)$
27. $2x^2(x - 5)(3x - 7)$
28. $-5x^3(3x - 2)(x + 4)$

29. $(x - 2)(2x + 3)(3x - 4)$
30. $(x + 5)(2x - 6)(x - 3)$

Multiply the following using the FOIL method.

31. $(x - 2)(x + 3)$ 　　　**32.** $(x + 2)(x - 3)$
33. $(x^2 - 2)(x^2 + 3)$ 　　**34.** $(x^2 + 2)(x^2 - 3)$
35. $(2a + 3)(3a + 2)$ 　　**36.** $(5a - 4)(2a + 1)$
37. $(5 - 3t)(4 + 2t)$ 　　**38.** $(7 - t)(6 - 3t)$
39. $(x^3 + 3)(x^3 - 5)$ 　　**40.** $(x^3 + 4)(x^3 - 7)$
41. $(4a + 1)(5a + 1)$ 　　**42.** $(3a - 1)(2a - 1)$
43. $(5x - 6y)(4x + 3y)$ 　**44.** $(6x - 5y)(2x - 3y)$
45. $(3t + \frac{1}{3})(6t - \frac{2}{3})$ 　　**46.** $(5t - \frac{1}{5})(10t + \frac{3}{5})$
47. $(b - 4a^2)(b + 3a^2)$ 　**48.** $(b + 5a^2)(b - 2a^2)$

Find the following special products.

49. $(5x + 2y)^2$ 　　　**50.** $(3x - 4y)^2$
51. $(5 - 3t^3)^2$ 　　　**52.** $(7 - 2t^4)^2$
53. $(2a + 3b)(2a - 3b)$ 　**54.** $(6a - 1)(6a + 1)$
55. $(3r^2 + 7s)(3r^2 - 7s)$
56. $(5r^2 - 2s)(5r^2 + 2s)$
57. $(\frac{1}{3}x - \frac{2}{5})(\frac{1}{3}x + \frac{2}{5})$ 　**58.** $(\frac{3}{4}x - \frac{1}{7})(\frac{3}{4}x + \frac{1}{7})$

Find the following products.

59. $(x - 2)^3$ 　　　　**60.** $(x + 4)^3$
61. $(2x - 1)^3$ 　　　　**62.** $(4x + 1)^3$
63. $(x - \frac{1}{2})^3$ 　　　**64.** $(x + \frac{1}{4})^3$
65. $3(x - 1)(x - 2)(x - 3)$
66. $2(x + 1)(x + 2)(x + 3)$
67. $(b^2 + 8)(a^2 + 1)$ 　**68.** $(b^2 + 1)(a^4 - 5)$
69. $(x - 2)(3y^2 + 4)$ 　**70.** $(x - 4)(2y^3 + 1)$

Simplify each expression.

71. $(x + 1)^2 + (x + 2)^2 + (x + 3)^2$
72. $(x - 1)^2 + (x - 2)^2 + (x - 3)^2$
73. $(2x + 3)^2 - (2x - 3)^2$
74. $(5x - 4)^2 - (5x + 4)^2$
75. $(x - 1)^3 - (x + 1)^3$ 　**76.** $(x - 3)^3 - (x + 3)^3$
77. Multiply $(x + y - 4)(x + y + 5)$ by first writing it like this:

$$[(x + y) - 4][(x + y) + 5]$$

and then applying the FOIL method.
78. Multiply $(x - 5 - y)(x - 5 + y)$ by first writing it like this:

$$[(x - 5) - y][(x - 5) + y]$$

79. Expand and multiply $(x + y + z)^2$ by first grouping the first two terms within the parentheses together.

$$[(x + y) + z]^2$$

80. Repeat Problem 79, but this time group the last two terms together.

$$[x + (y + z)]^2$$

81. The flower color (red, pink, or white) of a certain species of sweet pea plant is due to a single gene. If p is the proportion of the dominant form of the gene in a population, and q is the proportion of the recessive form of this gene in the same population, then the proportion of plants in the next generation that have pink flowers is given by the middle term in the expansion of $(p + q)^2$. If $p = \frac{1}{4}$ and $q = \frac{3}{4}$, find the proportion of the next generation that will have pink flowers.
82. Repeat Problem 81 with $p = \frac{1}{2}$ and $q = \frac{1}{2}$.
83. Let $a = 2$ and $b = 3$ and evaluate each of the following expressions.

$$a^4 - b^4 \quad (a - b)^4 \quad (a^2 + b^2)(a + b)(a - b)$$

84. Let $a = 2$ and $b = 3$ and evaluate each of the following expressions.

$$a^3 + b^3 \quad (a + b)^3 \quad a^3 + 3a^2b + 3ab^2 + b^3$$

Applying the Concepts

Problems 85–88 may be solved using a graphing calculator.

85. A company manufacturing prerecorded videotapes finds that it can sell x tapes per day at p dollars per tape, according to the formula $x = 230 - 20p$. Write a formula for the daily revenue and use it to find the revenue obtained by selling the tapes for $6.50 each.

86. A company selling diskettes for home computers finds that it can sell x diskettes per day at p dollars per diskette, according to the formula $x = 800 - 100p$. Write a formula for the daily revenue and use it to find the revenue obtained by selling the diskettes for $3.80 each.

87. A company sells an inexpensive accounting program for home computers. If it can sell x programs per week at p dollars per program, according to the formula $x = 350 - 10p$, find an equation for the weekly revenue. How much will the weekly revenue be if it charges $28.50 for the program?

88. A company sells boxes of greeting cards through the mail. It finds that it can sell x boxes of cards each week at p dollars per box, according to the formula $x = 1,475 - 250p$. Find the equation for the weekly revenue. What revenue will it bring in each week if the price of each box is $5.10?

89. The lengths of the three sides of a rectangular box that is closed at the top are given by three consecutive integers. Write a formula for the total surface area of the box.

90. A rectangular box is closed at the top. The lengths of the three sides are given by three consecutive odd integers. Write a formula for the total surface area of the box. Compare your answer with the answer to Example 10 in this section.

91. If you deposit $100 in an account with an interest rate r that is compounded annually, then the amount of money in that account at the end of four years is given by the formula $A = 100(1 + r)^4$. Expand the right side of this formula.

92. If you deposit P dollars in an account with an annual interest rate r that is compounded twice a year, then at the end of a year the amount of money in that account is given by the formula

$$A = P\left(1 + \frac{r}{2}\right)^2$$

Expand the right side of this formula.

Review Problems

The problems that follow review material we covered in Section 2.6.

Solve each equation.

93. $|3x - 5| = 7$

94. $|0.04 - 0.03x| = 0.02$

95. $|4y + 2| - 8 = -2$

96. $4 = |3 - 2y| - 5$

97. $5 + |6t + 2| = 3$

98. $7 + |3 - \frac{3}{4}t| = 10$

99. $|\frac{1}{10}x - \frac{1}{5}| = |\frac{1}{2} - \frac{1}{10}x|$

100. $|3x + 4| = |2 - 5x|$

One Step Further

Assume n is a positive integer and multiply.

101. $(x^n - 2)(x^n - 3)$ **102.** $(x^n + 4)(x^n - 1)$

103. $(x^{2n} + 3)(x^{2n} - 3)$

104. $(x^{3n} + 4)(x^{3n} - 4)$

105. $(2x^n + 3)(5x^n - 1)$

106. $(4x^n - 3)(7x^n + 2)$

107. $(x^n + 5)^2$ **108.** $(x^n - 3)^2$

109. $(x^n + 1)^3$ **110.** $(x^n - 1)^3$

Research Project 4

The title on the diagram below is *Binomial Expansions* because each line gives the expansion of the binomial $x + y$ raised to a whole-number power.

Binomial Expansions

$$
\begin{aligned}
(x + y)^0 &= 1 \\
(x + y)^1 &= x + y \\
(x + y)^2 &= x^2 + 2xy + y^2 \\
(x + y)^3 &= x^3 + 3x^2y + 3xy^2 + y^3 \\
(x + y)^4 &= \\
(x + y)^5 &=
\end{aligned}
$$

Justify the fourth row in the diagram by expanding $(x + y)^3$ using the methods developed in this section. Then complete the diagram by expanding the binomials $(x + y)^4$ and $(x + y)^5$ using the multiplication procedures you have learned in this section.

Finally, study the completed diagram until you see patterns there that will allow you to continue the diagram one more row without using multiplication. (One pattern you will see is Pascal's triangle, which we mentioned in Research Project 3 in Chapter 2. You will see it if you rewrite the triangular array with just the coefficients of each term in the array.) When you are finished write an essay in which you describe what you have done and the results you have obtained.

SECTION

3.5

The Greatest Common Factor and Factoring by Grouping

In general, factoring is the reverse of multiplication. The following diagram illustrates the relationship between factoring and multiplication:

Multiplication

Factors $\rightarrow 3 \cdot 7 = 21 \leftarrow$ Product

Factoring

Reading from left to right, we say the product of 3 and 7 is 21. Reading in the other direction, from right to left, we say 21 factors into 3 times 7. Or, 3 and 7 are factors of 21.

> **DEFINITION** The **greatest common factor** for a polynomial is the largest monomial that divides (is a factor of) each term of the polynomial.

Note The term *largest monomial,* as used here, refers to the monomial with the largest integer exponents whose coefficient has the greatest absolute value.

The greatest common factor for the polynomial $25x^5 + 20x^4 - 30x^3$ is $5x^3$ since it is the largest monomial that is a factor of each term. We can apply the distributive property and write

$$25x^5 + 20x^4 - 30x^3 = 5x^3(5x^2) + 5x^3(4x) - 5x^3(6)$$
$$= 5x^3(5x^2 + 4x - 6)$$

The last line is written in factored form.

▶ **EXAMPLE 1** Factor the greatest common factor from

$$8a^3 - 8a^2 - 48a$$

Solution The greatest common factor is $8a$. It is the largest monomial that divides each term of our polynomial. We can write each term in our polynomial as the product of $8a$ and another monomial. Then, we apply the distributive property to factor $8a$ from each term:

$$8a^3 - 8a^2 - 48a = 8a(a^2) - 8a(a) - 8a(6)$$
$$= 8a(a^2 - a - 6) \qquad \blacktriangleleft$$

▶ **EXAMPLE 2**   Factor the greatest common factor from

$$16a^5b^4 - 24a^2b^5 - 8a^3b^3$$

Solution   The largest monomial that divides each term is $8a^2b^3$. We write each term of the original polynomial in terms of $8a^2b^3$ and apply the distributive property to write the polynomial in factored form:

$$16a^5b^4 - 24a^2b^5 - 8a^3b^3 = 8a^2b^3(2a^3b) - 8a^2b^3(3b^2) - 8a^2b^3(a)$$
$$= 8a^2b^3(2a^3b - 3b^2 - a) \qquad \blacktriangleleft$$

▶ **EXAMPLE 3**   Factor the greatest common factor from

$$5x^2(a + b) - 6x(a + b) - 7(a + b)$$

Solution   The greatest common factor is $a + b$. Factoring it from each term, we have

$$5x^2(a + b) - 6x(a + b) - 7(a + b) = (a + b)(5x^2 - 6x - 7) \qquad \blacktriangleleft$$

▶ **EXAMPLE 4**   A company manufacturing prerecorded videotapes finds that the total daily revenue for selling x tapes is given by

$$R(x) = 11.5x - 0.05x^2$$

Factor x from each term on the right side of the equation to find the formula that gives the price p in terms of x. Then, use it to find the price they should charge if they want to sell 120 videotapes per day.

Solution   We begin by factoring x from the right side of the equation:

$$\text{If} \qquad R(x) = 11.5x - 0.05x^2$$
$$\text{then} \qquad R(x) = x(11.5 - 0.05x)$$

Because R is always xp, the quantity in parentheses must be p. Therefore, the price they should charge if they want to sell x items per day is

$$p = 11.5 - 0.05x$$

To find the price at which they will sell 120 videotapes per day, we let $x = 120$ in our last equation:

$$p = 11.5 - 0.05(120)$$
$$= 11.5 - 6$$
$$= 5.5$$

They should charge \$5.50 for each videotape.

We should also note here that if they sell $x = 120$ tapes at $p = \$5.50$ per tape, their daily revenue will be $R = xp = 120(5.50) = \$660$. This is the same amount we would obtain by substituting $x = 120$ into the original revenue equation $R(x) = 11.5x - 0.05x^2$. ◄

Factoring by Grouping

Many polynomials have no greatest common factor other than the number 1. Some of these can be factored using the distributive property if those terms with a common factor are grouped together.

For example, the polynomial $5x + 5y + x^2 + xy$ can be factored by noticing that the first two terms have a 5 in common, whereas the last two have an x in common.

Applying the distributive property, we have

$$5x + 5y + x^2 + xy = 5(x + y) + x(x + y)$$

This last expression can be thought of as having two terms, $5(x + y)$ and $x(x + y)$, each of which has a common factor $(x + y)$. We apply the distributive property again to factor $(x + y)$ from each term:

$$5(x + y) + x(x + y)$$
$$= (x + y)(5 + x)$$

▶ **EXAMPLE 5**   Factor $a^2b^2 + b^2 + 8a^2 + 8$.

Solution   The first two terms have b^2 in common; the last two have 8 in common:

$$a^2b^2 + b^2 + 8a^2 + 8 = b^2(a^2 + 1) + 8(a^2 + 1)$$
$$= (a^2 + 1)(b^2 + 8)$$ ◄

▶ **EXAMPLE 6**   Factor $15 - 5y^4 - 3x^3 + x^3y^4$.

Solution   Let's try factoring a 5 from the first two terms and an x^3 from the last two terms:

$$15 - 5y^4 - 3x^3 + x^3y^4 = 5(3 - y^4) + x^3(-3 + y^4)$$

Now, $3 - y^4$ and $-3 + y^4$ are not equal and we cannot factor further. Notice, however, that we can factor $-x^3$ instead of x^3 from the last two terms and obtain the desired result:

$$15 - 5y^4 - 3x^3 + x^3y^4 = 5(3 - y^4) - x^3(3 - y^4)$$
$$= (3 - y^4)(5 - x^3)$$ ◄

▶ **EXAMPLE 7**   Factor by grouping $x^3 + 2x^2 + 9x + 18$.

Solution   We begin by factoring x^2 from the first two terms and 9 from the second two terms:

$$x^3 + 2x^2 + 9x + 18 = x^2(x + 2) + 9(x + 2)$$
$$= (x + 2)(x^2 + 9) \quad \blacktriangleleft$$

P R O B L E M   S E T   3 . 5

Factor the greatest common factor from each of the following. (The answers in the back of the book all show greatest common factors whose coefficients are positive.)

1. $10x^3 - 15x^2$
2. $12x^5 + 18x^7$
3. $9y^6 + 18y^3$
4. $24y^4 - 8y^2$
5. $9a^2b - 6ab^2$
6. $30a^3b^4 + 20a^4b^3$
7. $21xy^4 + 7x^2y^2$
8. $14x^6y^3 - 6x^2y^4$
9. $3a^2 - 21a + 30$
10. $3a^2 - 3a - 6$
11. $4x^3 - 16x^2 - 20x$
12. $2x^3 - 14x^2 + 20x$
13. $10x^4y^2 + 20x^3y^3 - 30x^2y^4$
14. $6x^4y^2 + 18x^3y^3 - 24x^2y^4$
15. $-x^2y + xy^2 - x^2y^2$
16. $-x^3y^2 - x^2y^3 - x^2y^2$
17. $4x^3y^2z - 8x^2y^2z^2 + 6xy^2z^3$
18. $7x^4y^3z^2 - 21x^2y^2z^2 - 14x^2y^3z^4$
19. $20a^2b^2c^2 - 30ab^2c + 25a^2bc^2$
20. $8a^3bc^5 - 48a^2b^4c + 16ab^3c^5$
21. $5x(a - 2b) - 3y(a - 2b)$
22. $3a(x - y) - 7b(x - y)$
23. $3x^2(x + y)^2 - 6y^2(x + y)^2$
24. $10x^3(2x - 3y) - 15x^2(2x - 3y)$
25. $2x^2(x + 5) + 7x(x + 5) + 6(x + 5)$
26. $2x^2(x + 2) + 13x(x + 2) + 15(x + 2)$

Factor each of the following by grouping.

27. $3xy + 3y + 2ax + 2a$
28. $5xy^2 + 5y^2 + 3ax + 3a$
29. $x^2y + x + 3xy + 3$
30. $x^3y^3 + 2x^3 + 5x^2y^3 + 10x^2$
31. $3xy^2 - 6y^2 + 4x - 8$
32. $8x^2y - 4x^2 + 6y - 3$
33. $x^2 - ax - bx + ab$
34. $ax - x^2 - bx + ab$
35. $ab + 5a - b - 5$
36. $x^2 - xy - ax + ay$

37. $a^4b^2 + a^4 - 5b^2 - 5$
38. $2a^2 - a^2b - bc^2 + 2c^2$
39. $x^3 + 3x^2 - 4x - 12$
40. $x^3 + 5x^2 - 4x - 20$
41. $x^3 + 2x^2 - 25x - 50$
42. $x^3 + 4x^2 - 9x - 36$
43. $2x^3 + 3x^2 - 8x - 12$
44. $3x^3 + 2x^2 - 27x - 18$
45. $4x^3 + 12x^2 - 9x - 27$
46. $9x^3 + 18x^2 - 4x - 8$
47. The greatest common factor of the binomial $3x - 9$ is 3. The greatest common factor of the binomial $6x - 2$ is 2. What is the greatest common factor of their product, $(3x - 9)(6x - 2)$, when it has been multiplied out?
48. The greatest common factors of the binomials $5x - 10$ and $2x + 4$ are 5 and 2, respectively. What is the greatest common factor of their product, $(5x - 10)(2x + 4)$, when it has been multiplied out?

Applying the Concepts

49. If P dollars are placed in a savings account in which the rate of interest r is compounded yearly, then at the end of 1 year the amount of money in the account can be written as $P + Pr$. At the end of two years the amount of money in the account is

$$P + Pr + (P + Pr)r$$

Use factoring by grouping to show that this last expression can be written as $P(1 + r)^2$.

50. At the end of three years, the amount of money in the savings account in Problem 49 will be

$$P(1 + r)^2 + P(1 + r)^2r$$

Use factoring to show that this last expression can be written as $P(1 + r)^3$.

Use Example 4 as a guide in solving the next four problems.

51. A company manufacturing prerecorded videotapes finds that the total daily revenue R for selling x tapes at p dollars per tape is given by

$$R(x) = 11.5x - 0.05x^2$$

Factor x from each term on the right side of the equation to find the formula that gives the price p in terms of x. Then, use it to find the price they should charge if they want to sell 125 videotapes per day.

52. A company producing diskettes for home computers finds that the total daily revenue for selling x diskettes at p dollars per diskette is given by

$$R(x) = 8x - 0.01x^2$$

Use the fact that $R = xp$ with your knowledge of factoring to find a formula that gives the price p in terms of x. Then, use it to find the price they should charge if they want to sell 420 diskettes per day.

53. The weekly revenue equation for a company selling an inexpensive accounting program for home computers is given by the equation

$$R(x) = 35x - 0.1x^2$$

where x is the number of programs they sell per week. What price p should they charge if they want to sell 65 programs per week?

54. The weekly revenue equation for a small mail-order company selling boxes of greeting cards is

$$R(x) = 5.9x - 0.004x^2$$

where x is the number of boxes they sell per week. What price p should they charge if they want to sell 200 boxes each week?

Review Problems

The problems that follow review material we covered in Sections 2.5 and 3.4. Reviewing the problems from Section 3.4 will help you understand the next section.

Multiply using the FOIL method. [3.4]

55. $(x + 2)(x + 3)$ **56.** $(x - 2)(x - 3)$
57. $(2y + 5)(3y - 7)$ **58.** $(2y - 5)(3y + 7)$
59. $(4 - 3a)(5 - a)$ **60.** $(4 - 3a)(5 + a)$

Solve each of the following inequalities. [2.5]

61. $-5x < 30$ **62.** $-\frac{2}{5}x > 12$

63. $\frac{1}{3} + \frac{y}{5} \leq \frac{26}{15}$ **64.** $-\frac{1}{3} \geq \frac{1}{6} - \frac{y}{2}$

65. $5t - 4 > 3t - 8$
66. $-3(t - 2) < 6 - 5(t + 1)$
67. $-9 < -4 + 5t < 6$
68. $-3 < 2t + 1 < 3$

One Step Further

Assume n is a positive integer and factor each expression by factoring out the greatest common factor.

69. $x^{n+2} + x^{n+1} + x^n$
70. $x^{n+4} + x^{n+2} + x^n$
71. $x^{n+5} + x^{n+4} + x^{n+3}$
72. $x^{n+3} + x^{n+2} + x^{n+1}$

SECTION

3.6   **Factoring Trinomials**

Factoring trinomials is probably the most common type of factoring found in algebra. We begin this section by considering trinomials that have a leading coefficient of 1.

RATIONAL EXPRESSIONS

INTRODUCTION

As we mentioned in the introduction to Chapter 2, many of the topics we cover are linked to one another. One of the most satisfying experiences in studying mathematics is discovering links between topics that do not seem to be related. Here are some items that may initially seem unrelated, but which are in fact very closely related to one another.

A Continued Fraction   *The Fibonacci Sequence*   *The Golden Rectangle*

$$1 + \cfrac{1}{1 + \cfrac{1}{1 + \cfrac{1}{1 + \cdots}}}$$

$1, 1, 2, 3, 5, \ldots$

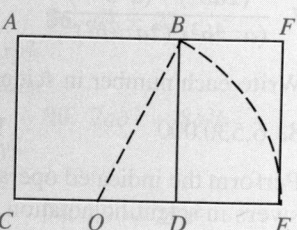

Each of these items was discovered at a different time in the history of mathematics. For instance, the golden rectangle was discovered over 2,000 years ago, while the Fibonacci sequence was discovered more than 1,000 years later. As you progress through the book, the relationships that exist between and among these items will begin to appear. In this chapter we will introduce the continued fraction shown above, and show how it is related to the Fibonacci sequence you studied in Chapter 1. In Chapter 6 we construct the golden rectangle and discover how it is related to both the continued fraction and the Fibonacci sequence.

OVERVIEW

This chapter is mostly concerned with simplifying a certain kind of algebraic expression. The expressions are called rational expressions because they are to algebra what rational numbers are to arithmetic. Most of the work we will do with rational expressions parallels the work you have done in previous math classes with fractions. Once we have learned to add, subtract, multiply, and divide rational expressions, we will turn our attention to equations involving rational expressions. The single most important tool needed for success in this chapter is factoring. Almost every problem encountered in this chapter involves factoring at one point or another. You may be able to understand all the theory and steps involved in solving the problems, but unless you can factor the polynomials in the problems, you will be unable to work any of them.

STUDY SKILLS

The study skills for this chapter are concerned with getting ready to take an exam.

1. **Getting Ready to Take an Exam**   Try to arrange your daily study habits so that you have very little studying to do the night before your next exam. The next two goals will help you achieve goal number 1.
2. **Review with the Exam in Mind**   You should review material that will be covered on the next exam every day. Your review should consist of working problems. Preferably, the problems you work should be problems from your list of difficult problems.
3. **Continue to List Difficult Problems**   This study skill was started in Chapter 2. You should continue to list and rework the problems that give you the most difficulty. It is this list that you will use to study for the next exam. Your goal is to go into the next exam knowing that you can successfully work any problem from your list of hard problems.
4. **Pay Attention to Instructions**   Taking a test is different from doing homework. When you take a test, the problems will be mixed up. When you do your homework, you usually work a number of similar problems. I sometimes have students who do very well on their homework become confused when they see the same problems on a test, because they have not paid attention to the instructions on their homework. For example, look at the two problems below:

$$\textit{Problem 1:} \quad \frac{5}{x^2 - 3x + 2} - \frac{1}{x - 2}$$

$$\textit{Problem 2:} \quad \frac{5}{x^2 - 3x + 2} - \frac{1}{x - 2} = \frac{1}{3x - 3}$$

Without instructions to accompany them, we don't know what to do with the problems. Here are two valid instructions that could accompany either problem:

List any restrictions on the variable.

Find the least common denominator.

On the other hand, "Subtract" is an instruction that could accompany Problem 1, but not Problem 2. Likewise, "Solve for *x*" could accompany Problem 2, but not Problem 1.

Train yourself to pay attention to the instructions that accompany the problems you work on your assignments, so that you will know what to do with those same problems when you see them on your next test.

Basic Properties and Reducing to Lowest Terms

We will begin this section with the definition of a rational expression. We will then state the two basic properties associated with rational expressions, and go on to apply one of the properties to reduce rational expressions to lowest terms.

Recall from Chapter 1 that a *rational number* is any number that can be expressed as the ratio of two integers:

$$\text{Rational numbers} = \left\{ \frac{a}{b} \,\middle|\, a \text{ and } b \text{ are integers, } b \neq 0 \right\}$$

A rational expression is defined similarly as any expression that can be written as the ratio of two polynomials:

$$\text{Rational expressions} = \left\{ \frac{P}{Q} \,\middle|\, P \text{ and } Q \text{ are polynomials, } Q \neq 0 \right\}$$

Some examples of rational expressions are

$$\frac{2x - 3}{x + 5} \qquad \frac{x^2 - 5x - 6}{x^2 - 1} \qquad \frac{a - b}{b - a}$$

Basic Properties

For rational expressions, multiplying the numerator and denominator by the same nonzero expression may change the form of the rational expression, but it will always produce an expression equivalent to the original one. The same is true when dividing the numerator and denominator by the same nonzero quantity.

Properties of Rational Expressions

If P, Q, and K are polynomials with $Q \neq 0$ and $K \neq 0$, then

$$\frac{P}{Q} = \frac{PK}{QK} \qquad \text{and} \qquad \frac{P}{Q} = \frac{P/K}{Q/K}$$

Reducing to Lowest Terms

The fraction $\frac{6}{8}$ can be written in lowest terms as $\frac{3}{4}$. The process is shown here:

$$\frac{6}{8} = \frac{3 \cdot \overset{1}{\cancel{2}}}{4 \cdot \underset{1}{\cancel{2}}} = \frac{3}{4}$$

Reducing $\frac{6}{8}$ to $\frac{3}{4}$ involves dividing the numerator and denominator by 2, the factor they have in common. Before dividing out the common factor 2, we must notice that the common factor *is* 2! (This may not be obvious since we are very familiar with the numbers 6 and 8 and therefore do not have to put much thought into finding what number divides both of them.)

We reduce rational expressions to lowest terms by first factoring the numerator and denominator and then dividing both numerator and denominator by any factors they have in common.

▶ **EXAMPLE 1**   Reduce $\dfrac{x^2 - 9}{x - 3}$ to lowest terms.

Solution   Factoring, we have

$$\frac{x^2 - 9}{x - 3} = \frac{(x + 3)(x - 3)}{x - 3}$$

The numerator and denominator have the factor $x - 3$ in common. Dividing the numerator and denominator by $x - 3$, we have

$$\frac{(x + 3)\overset{1}{\cancel{(x - 3)}}}{\underset{1}{\cancel{x - 3}}} = \frac{x + 3}{1} = x + 3$$

Note that the lines drawn through the $x - 3$ in the numerator and denominator indicate that we have divided through by $x - 3$. As the problems become more involved, these lines will help keep track of which factors have been divided out and which have not.   ◀

Note   For the problem in Example 1, there is an implied restriction on the variable x; it cannot be 3. If x were 3, the expression $(x^2 - 9)/(x - 3)$ would become 0/0, an expression that we cannot associate with a real number. For all problems involving rational expressions, we restrict the variable to only those values that result in a nonzero denominator. When we state the relationship

$$\frac{x^2 - 9}{x - 3} = x + 3$$

we are assuming that it is true for all values of x except $x = 3$.

Here are some other examples of reducing rational expressions to lowest terms.

▶ **EXAMPLES** Reduce to lowest terms.

2.
$$\frac{y^2 - 5y - 6}{y^2 - 1} = \frac{(y - 6)(y + 1)}{(y - 1)(y + 1)}$$
Factor numerator and denominator

$$= \frac{y - 6}{y - 1}$$
Divide out common factor $y + 1$

3.
$$\frac{2a^3 - 16}{4a^2 - 12a + 8} = \frac{2(a^3 - 8)}{4(a^2 - 3a + 2)}$$
$$= \frac{2(a - 2)(a^2 + 2a + 4)}{4(a - 2)(a - 1)}$$
Factor numerator and denominator

$$= \frac{a^2 + 2a + 4}{2(a - 1)}$$
Divide out common factor $2(a - 2)$

4.
$$\frac{x^2 - 3x + ax - 3a}{x^2 - ax - 3x + 3a} = \frac{x(x - 3) + a(x - 3)}{x(x - a) - 3(x - a)}$$
$$= \frac{(x - 3)(x + a)}{(x - a)(x - 3)}$$
Factor numerator and denominator

$$= \frac{x + a}{x - a}$$
Divide out common factor $x - 3$ ◀

The answer to Example 4 is $(x + a)/(x - a)$. The problem cannot be reduced further. It is a fairly common mistake to attempt to divide out an x or an a in this last expression. Remember, we can divide out only the factors common to the numerator and denominator of a rational expression. For the last expression in Example 4, neither the numerator nor the denominator can be factored further; x is not a factor of the numerator or the denominator and neither is a. The expression is in lowest terms.

The next example involves what we call a trick. The trick is to reverse the order of the terms in a difference by factoring -1 from each term. The next examples illustrate how this is done.

▶ **EXAMPLE 5** Reduce to lowest terms: $\dfrac{a - b}{b - a}$.

Solution The relationship between $a - b$ and $b - a$ is that they are opposites. We can show this fact by factoring -1 from each term in the numerator:

$$\frac{a - b}{b - a} = \frac{-1(-a + b)}{b - a}$$
Factor -1 from each term in the numerator

$$= \frac{-1(b - a)}{b - a}$$
Reverse the order of the terms in the numerator

$$= -1$$
Divide out common factor $b - a$ ◀

▶ **EXAMPLE 6** Reduce to lowest terms: $\dfrac{x^2 - 25}{5 - x}$.

Solution We begin by factoring the numerator:

$$\frac{x^2 - 25}{5 - x} = \frac{(x - 5)(x + 5)}{5 - x}$$

The factors $x - 5$ and $5 - x$ are similar but are not exactly the same. We can reverse the order of either by factoring -1 from it. That is, $5 - x = -1(-5 + x) = -1(x - 5)$.

$$\frac{(x - 5)(x + 5)}{5 - x} = \frac{(x - 5)(x + 5)}{-1(x - 5)}$$
$$= \frac{x + 5}{-1}$$
$$= -(x + 5) \qquad \blacktriangleleft$$

Applications

You may have noticed that light bulbs often are rated in lumens as well as in watts. For example, a 100-watt light bulb may be rated at 1,500 lumens. A lumen is a measure of the amount of light given off by the bulb. To find the intensity of light E that falls on a surface d feet away from a bulb that gives off L lumens, you use the formula

$$E = \frac{7L}{88d^2}$$

To have enough light to read comfortably, the value of E should be around 10 lumens per square foot, on the surface on which you are reading.

▶ **EXAMPLE 7** The lumen rating on a certain brand of 100-watt light bulbs is given as 1,500. Make a table that gives the intensity of light that falls on a surface that is 1, 2, 3, 4, 5, or 6 feet away from this bulb.

Solution We substitute 1,500 for L in the formula above. Then we calculate E for each of the given values of d. Here is the calculation for $d = 4$:

$$E = \frac{7(1,500)}{88(4^2)} = \frac{10,500}{1,408} = 7.5 \text{ to the nearest tenth}$$

Note that we did not reduce 10,500/1,408 to lowest terms. Instead, we simply divided 10,500 by 1,408 to obtain 7.5. A similar calculation for the other values of d yields the numbers shown in Table 1 on the next page.

TABLE 1 Intensity of Light from a 100-Watt Bulb

Distance above Surface (in feet)	Intensity in Lumens/Square Foot
1	119.3
2	29.8
3	13.3
4	7.5
5	4.8
6	3.3

As the table indicates, the light should be placed between 3 and 4 feet above the surface in order to light the surface for comfortable reading. ◀

Ratios   You may recall from previous math classes that the ratio of a to b is the same as the fraction $\frac{a}{b}$. To illustrate, if the ratio of men to women in a math class is 3 to 2, then $\frac{3}{5}$ of the class are men and $\frac{2}{5}$ of the class are women. Here are two ratios that are used frequently in mathematics:

1. The number π is defined as the ratio of the circumference of a circle to the diameter of a circle. That is

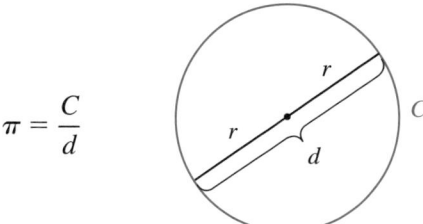

$$\pi = \frac{C}{d}$$

Multiplying both sides of this formula by d, we have the more common form $C = \pi d$.

Note   Since the diameter of a circle is twice the radius, the formula for circumference is sometimes written $C = 2\pi r$. Neither formula should be confused with the formula for the area of a circle, which is $A = \pi r^2$.

2. The *average speed* of a moving object is defined to be the ratio of distance to time. If you drive your car for 5 hours and travel a distance of 200 miles, then your average rate of speed is

$$\text{Average speed} = \frac{200 \text{ miles}}{5 \text{ hours}} = 40 \text{ miles/hour}$$

The formula we use for the relationship between average speed r, distance d, and time t is

$$r = \frac{d}{t}$$

The formula is sometimes called the **rate equation.** Multiplying both sides by t, we have the equivalent form of the rate equation, $d = rt$.

Our next example involves both the formula for the circumference of a circle and the rate equation.

▶ **EXAMPLE 8**   The first Ferris wheel was designed and built by George Ferris in 1893. The diameter of the wheel was 250 feet. It had 40 carriages, equally spaced around the wheel, each of which held a maximum of 40 people. One trip around the wheel took 20 minutes. Find the average speed of a rider on the first Ferris wheel. (Use 3.14 as an approximation for π.)

Solution   The distance traveled is the circumference of the wheel, which is

$$C = 250\pi = 250(3.14) = 785 \text{ feet}$$

To find the average speed, we divide the distance traveled by the amount of time it took to go once around the wheel.

$$r = \frac{d}{t} = \frac{785 \text{ feet}}{20 \text{ minutes}} = 39.3 \text{ feet/minute to the nearest tenth}$$

Later in this chapter we will convert this ratio into an equivalent ratio that gives the speed of the rider in miles per hour.   ◀

P R O B L E M   S E T   4 . 1

Reduce each fraction to lowest terms.

1. $-\dfrac{12}{36}$

2. $-\dfrac{45}{60}$

3. $\dfrac{2a^2b^3}{4a^2}$

4. $\dfrac{3a^3b^2}{6b^2}$

5. $-\dfrac{24x^3y^5}{16x^4y^2}$

6. $-\dfrac{36x^6y^8}{24x^3y^9}$

7. $\dfrac{144a^2b^3c^4}{56a^4b^3c^2}$

8. $\dfrac{108a^5b^2c^5}{27a^2b^5c^2}$

Reduce each rational expression to lowest terms.

9. $\dfrac{x^2 - 16}{6x + 24}$

10. $\dfrac{5x + 25}{x^2 - 25}$

11. $\dfrac{12x - 9y}{3x^2 + 3xy}$

12. $\dfrac{x^3 - xy^2}{4x + 4y}$

13. $\dfrac{a^4 - 81}{a - 3}$

14. $\dfrac{a + 4}{a^2 - 16}$

15. $\dfrac{a^2 - 4a - 12}{a^2 + 8a + 12}$

16. $\dfrac{a^2 - 7a + 12}{a^2 - 9a + 20}$

17. $\dfrac{20y^2 - 45}{10y^2 - 5y - 15}$

18. $\dfrac{54y^2 - 6}{18y^2 - 60y + 18}$

19. $\dfrac{20x^2 - 93x + 34}{4x^2 - 9x - 34}$

20. $\dfrac{15x^2 - 59x + 52}{5x^2 - 33x + 52}$

21. $\dfrac{12y - 2xy - 2x^2y}{6y - 4xy - 2x^2y}$

22. $\dfrac{250a + 100ax + 10ax^2}{50a - 2ax^2}$

23. $\dfrac{(x - 3)^2(x + 2)}{(x + 2)^2(x - 3)}$ **24.** $\dfrac{(x - 4)^3(x + 3)}{(x + 3)^2(x - 4)}$

25. $\dfrac{a^3 + b^3}{a^2 - b^2}$ **26.** $\dfrac{a^2 - b^2}{a^3 - b^3}$

27. $\dfrac{8x^4 - 8x}{4x^4 + 4x^3 + 4x^2}$

28. $\dfrac{6x^5 - 48x^2}{12x^3 + 24x^2 + 48x}$

29. $\dfrac{6x^2 + 7xy - 3y^2}{6x^2 + xy - y^2}$

30. $\dfrac{4x^2 - y^2}{4x^2 - 8xy - 5y^2}$

31. $\dfrac{ax + 2x + 3a + 6}{ay + 2y - 4a - 8}$

32. $\dfrac{ax - x - 5a + 5}{ax + x - 5a - 5}$

33. $\dfrac{x^2 + bx - 3x - 3b}{x^2 - 2bx - 3x + 6b}$

34. $\dfrac{x^2 - 3ax - 2x + 6a}{x^2 - 3ax + 2x - 6a}$

35. $\dfrac{x^3 + 3x^2 - 4x - 12}{x^2 + x - 6}$

36. $\dfrac{x^3 + 5x^2 - 4x - 20}{x^2 + 7x + 10}$

37. $\dfrac{3x^3 + 21x^2 + 36x}{3x^4 + 12x^3 - 27x^2 - 108x}$

38. $\dfrac{2x^4 + 14x^3 + 20x^2}{2x^5 + 4x^4 - 50x^3 - 100x^2}$

39. $\dfrac{4x^4 - 25}{6x^3 - 4x^2 + 15x - 10}$

40. $\dfrac{16x^4 - 49}{8x^3 - 12x^2 + 14x - 21}$

Refer to Examples 5 and 6 in this section and reduce the following to lowest terms.

41. $\dfrac{x - 4}{4 - x}$ **42.** $\dfrac{6 - x}{x - 6}$

43. $\dfrac{y^2 - 36}{6 - y}$ **44.** $\dfrac{1 - y}{y^2 - 1}$

45. $\dfrac{1 - 9a^2}{9a^2 - 6a + 1}$

46. $\dfrac{1 - a^2}{a^2 - 2a + 1}$

Reduce each rational expression to lowest terms, then subtract.

47. $\dfrac{28x^2 - 41x + 15}{7x - 5} - \dfrac{12x^2 - 41x + 24}{3x - 8}$

48. $\dfrac{42x^2 + 47x - 55}{7x - 5} - \dfrac{18x^2 - 15x - 88}{3x - 8}$

49. $\dfrac{x^3 - 8}{x - 2} - \dfrac{x^3 + 8}{x + 2}$

50. $\dfrac{x^4 - 16}{x + 2} - \dfrac{x^4 - 16}{x - 2}$

Explain the mistake made in each of the following problems:

51. $\dfrac{\dfrac{x}{\cancel{x}^2 - \cancel{9}^3}}{\cancel{x} - \cancel{3}} = x - 3$

52. $\dfrac{\cancel{x}^2 - 6\cancel{x} + \cancel{9}^3}{\cancel{x}^2 + \cancel{x} - \cancel{6}_2} = \dfrac{3}{2}$

53. $\dfrac{\cancel{x} + y}{\cancel{x}} = y$

54. $\dfrac{x + \cancel{3}}{\cancel{3}} = x$

55. Replace x with 3 in the expression $\dfrac{x^3 - 1}{x - 1}$, and then simplify. The result should be the same as what you would get if you replaced x with 3 in the expression $x^2 + x + 1$.

56. Replace x with 7 in the expression $\dfrac{x - 4}{4 - x}$ and simplify. Now, replace x with 10 and simplify. The result in both cases should be the same. Can you think of a number to replace x with that will not give the same result?

Applying the Concepts

For Problems 57–66, round all answers to the nearest tenth.

Problems 57–60 may be solved using a graphing calculator.

57. A 150-watt light bulb is rated at 2,250 lumens. Use the formula from Example 7 in this section to make a table that gives the intensity of light that falls on a surface that is 1, 2, 3, 4, 5, or 6 feet away from this bulb.

58. A 90-watt light bulb is rated at 1,540 lumens. Use the formula from Example 7 in this section to make a table that gives the intensity of light that falls on a surface that is 1, 2, 3, 4, 5, or 6 feet away from this bulb.

59. A 52-watt light bulb is rated at 780 lumens. Use the formula from Example 7 in this section to make a table that gives the intensity of light that falls on a surface that is 1, 1.5, 2, 2.5, 3, or 3.5 feet away from this bulb.

60. A 25-watt light bulb is rated at 480 lumens. Use the formula from Example 7 in this section to make a table that gives the intensity of light that falls on a surface that is 1, 1.5, 2, 2.5, 3, or 3.5 feet away from this bulb.

61. A jogger covers 3.5 miles in 29.75 minutes. Find the average speed of the jogger in miles per minute.

62. A bullet fired from a gun travels a distance of 4,750 feet in 3.2 seconds. Find the average speed of the bullet in feet per second.

63. A pickup truck travels 175.8 miles on 16.3 gallons of gas. Give the average rate of fuel consumption of the truck in miles per gallon.

64. A luxury car travels 200 miles on 16.5 gallons of gas. Give the average fuel consumption of the car in miles per gallon.

For Problems 65–68, use 3.14 as an approximation for π.

65. A person riding a Ferris wheel with a diameter of 65 feet travels once around the wheel in 30 seconds. What is the average speed of the rider in feet per second?

66. A person riding a Ferris wheel with a diameter of 102 feet travels once around the wheel in 3.5 minutes. What is the average speed of the rider in feet per minute?

The abbreviation "rpm" stands for revolutions per minute. If a point on a circle rotates at 300 rpm, then it rotates through one complete revolution 300 times every minute. The length of time it takes to rotate once around the circle is 1/300 minute.

67. A $3\frac{1}{2}$-inch diskette, when placed in the disk drive of a computer, rotates at 300 rpm (meaning one revolution takes 1/300 minute). Find the average speed of a point 2 inches from the center of the diskette. Then find the average speed of a point 1.5 inches from the center of the diskette.

68. A 5-inch fixed disk in a computer rotates at 3,600 rpm. Find the average speed of a point 2 inches from the center of the disk. Then find the average speed of a point 1.5 inches from the center.

If two resistors R_1 and R_2 in an electrical circuit are connected in parallel, the resistance they give to the circuit is equivalent to one resistor with a resistance R of

$$\frac{1}{R} = \frac{R_1 + R_2}{R_1 R_2}$$

Find R in the formula above if

69. $R_1 = 10$ ohms and $R_2 = 5$ ohms
70. $R_1 = 6$ ohms and $R_2 = 18$ ohms

Let $f(x) = \dfrac{x^2 - 4}{x - 2}$ and $g(x) = x + 2$, and evaluate the following expressions if possible.

71. $f(0)$ and $g(0)$ **72.** $f(1)$ and $g(1)$
73. $f(2)$ and $g(2)$ **74.** $f(3)$ and $g(3)$

Let $f(x) = \dfrac{x^2 - 1}{x - 1}$ and $g(x) = x + 1$, and evaluate the following expressions if possible.

75. $f(0)$ and $g(0)$ **76.** $f(1)$ and $g(1)$
77. $f(2)$ and $g(2)$ **78.** $f(3)$ and $g(3)$

Review Problems

The problems below review material we covered in Section 3.3. Reviewing these problems will help you with the next section.

Subtract as indicated.

79. Subtract $x^2 + 2x + 1$ from $4x^2 - 5x + 5$.
80. Subtract $3x^2 - 5x + 2$ from $7x^2 + 6x + 4$.
81. Subtract $10x - 20$ from $10x - 11$.
82. Subtract $-6x - 18$ from $-6x + 5$.
83. Subtract $4x^3 - 8x^2$ from $4x^3$.
84. Subtract $2x^2 + 6x$ from $2x^2$.

One Step Further

As you know, multiplying the numerator and denominator of a rational expression by the same non-zero number will not change its value. Use this fact to clear each expression of fractions, then reduce.

85. $\dfrac{\frac{1}{2}x^2 + \frac{5}{12}x - \frac{1}{2}}{\frac{1}{6}x^2 + \frac{1}{2}x + \frac{3}{8}}$

86. $\dfrac{\frac{1}{3}x^2 + \frac{5}{12}x - \frac{1}{2}}{\frac{2}{3}x^2 - x + \frac{3}{8}}$

Assume n is a positive integer and reduce each expression to lowest terms.

87. $\dfrac{6x^{2n} - 5x^n - 21}{9x^{2n} - 49}$

88. $\dfrac{30x^{2n} - 19x^n - 4}{36x^{2n} - 1}$

89. $\dfrac{x^{3n} + 2x^{2n} - 9x^n - 18}{x^{2n} - x^n - 6}$

90. $\dfrac{x^{3n} + 5x^{2n} - 4x^n - 20}{x^{2n} + 10x^n + 25}$

Simplify each expression.

91. $\dfrac{(4x^2 + 3)(x - 3) + (29x + 4)(x - 3)}{4x^2 - 11x - 3}$

92. $\dfrac{(5x^2 + 21)(x - 4) + (43x + 3)(x - 4)}{5x^2 - 17x - 12}$

Research Project 5

In 1897, a large Ferris wheel was built in Vienna that is still in operation today. Known as the Reisenrad, or *Great Wheel*, it has a diameter of 197 feet and can carry a total of 800 people. A brochure that gives some statistics associated with the Reisenrad indicates that passengers riding it travel at 2 feet 6 inches per second. You can check the accuracy of this number by watching the movie *The Third Man*. In the movie, Orson Welles rides the Reisenrad through one complete revolution. Play *The Third Man* on a VCR, so you can view the Reisenrad in operation. Use the pause button and the timer on the VCR to time how long it takes Orson Welles to ride once around the wheel. Then calculate his average speed during the ride. Use your results to either prove or disprove the claim that passengers travel at 2 feet 6 inches per second on the Reisenrad. When you have finished, write your procedures and results in essay form.

SECTION

4.2 Division of Polynomials

We begin this section by considering division of a polynomial by a monomial. This is the simplest kind of polynomial division. The rest of the section is devoted to division of a polynomial by a polynomial. This kind of division is similar to long division with whole numbers.

Dividing a Polynomial by a Monomial

To divide a polynomial by a monomial, we use the definition of division and apply the distributive property. The following example illustrates the procedure.

▶ **EXAMPLE 1** Divide $\dfrac{10x^5 - 15x^4 + 20x^3}{5x^2}$.

Solution Writing division as multiplication by the reciprocal we have

$= (10x^5 - 15x^4 + 20x^3) \cdot \dfrac{1}{5x^2}$ Dividing by $5x^2$ is the same as multiplying by $\dfrac{1}{5x^2}$

$= 10x^5 \cdot \dfrac{1}{5x^2} - 15x^4 \cdot \dfrac{1}{5x^2} + 20x^3 \cdot \dfrac{1}{5x^2}$ Distributive property

$= \dfrac{10x^5}{5x^2} - \dfrac{15x^4}{5x^2} + \dfrac{20x^3}{5x^2}$ Multiplying by $\dfrac{1}{5x^2}$ is the same as dividing by $5x^2$

$= 2x^3 - 3x^2 + 4x$ Divide coefficients, subtract exponents ◀

Notice that division of a polynomial by a monomial is accomplished by dividing each term of the polynomial by the monomial. The first two steps are usually not shown in a problem like this. They are part of Example 1 to justify distributing $5x^2$ under all three terms of the polynomial $10x^5 - 15x^4 + 20x^3$.

Here are some more examples of this kind of division.

▶ **EXAMPLES** Divide. Write all results with positive exponents.

2. $\dfrac{8x^3y^5 - 16x^2y^2 + 4x^4y^3}{-2x^2y} = \dfrac{8x^3y^5}{-2x^2y} + \dfrac{-16x^2y^2}{-2x^2y} + \dfrac{4x^4y^3}{-2x^2y}$

$\qquad\qquad\qquad\qquad\qquad = -4xy^4 + 8y - 2x^2y^2$

3. $\dfrac{10a^4b^2 + 8ab^3 - 12a^3b + 6ab}{4a^2b^2}$

$\qquad\qquad\qquad = \dfrac{10a^4b^2}{4a^2b^2} + \dfrac{8ab^3}{4a^2b^2} - \dfrac{12a^3b}{4a^2b^2} + \dfrac{6ab}{4a^2b^2}$

$\qquad\qquad\qquad = \dfrac{5a^2}{2} + \dfrac{2b}{a} - \dfrac{3a}{b} + \dfrac{3}{2ab}$ ◀

Notice in Example 3 that the result is not a polynomial because of the last three terms. If we were to write each as a product, some of the variables would have negative exponents. For example, the second term would be

$$\frac{2b}{a} = 2a^{-1}b$$

The divisor in each of the examples above was a monomial. We now want to turn our attention to division of polynomials in which the divisor has two or more terms.

Dividing a Polynomial by a Polynomial

▶ **EXAMPLE 4**   Divide $\dfrac{x^2 - 6xy - 7y^2}{x + y}$.

Solution   In this case, we can factor the numerator and perform our division by simply dividing out common factors, just like we did in the previous section:

$$\frac{x^2 - 6xy - 7y^2}{x + y} = \frac{\cancel{(x + y)}(x - 7y)}{\cancel{x + y}}$$
$$= x - 7y$$ ◀

For the type of division shown in Example 4, the denominator must be a factor of the numerator. When the denominator is not a factor of the numerator, or in the case where we can't factor the numerator, the method used in Example 4 won't work. We need to develop a new method for these cases. Since this new method is very similar to long division with whole numbers, we will review it here.

▶ **EXAMPLE 5**   Divide $25\overline{)4{,}628}$.

Solution

$$
\begin{array}{r}
1 \\
25\overline{)4{,}628} \\
2\,5 \\
\hline
2\,1
\end{array}
$$
← Estimate 25 into 46
← Multiply $1 \times 25 = 25$
← Subtract $46 - 25 = 21$

$$
\begin{array}{r}
1 \\
25\overline{)4{,}628} \\
2\,5\downarrow \\
\hline
2\,12
\end{array}
$$
← Bring down the 2

These are the four basic steps in long division: estimate, multiply, subtract, and bring down the next term. To complete the problem, we simply perform the same four steps

$$
\begin{array}{r}
18 \\
25\overline{)4{,}628} \\
2\,5 \\
\hline
2\,12 \\
2\,00\downarrow \\
\hline
128
\end{array}
$$
← 8 is the estimate
← Multiply to get 200
← Subtract to get 12, then bring down the 8

One more time:

$$\begin{array}{r} 185 \leftarrow 5 \text{ is the estimate} \\ 25\overline{)4{,}628} \\ \underline{2\,5} \\ 2\,12 \\ \underline{2\,00}\downarrow \\ 128 \\ \underline{125} \leftarrow \text{Multiply to get 125} \\ 3 \leftarrow \text{Subtract to get 3} \end{array}$$

Since 3 is less than 25 and we have no more terms to bring down, we have our answer:

$$\frac{4{,}628}{25} = 185 + \frac{3}{25}$$

To check our answer, we multiply 185 by 25 and then add 3 to the result:

$$25(185) + 3 = 4{,}625 + 3 = 4{,}628 \qquad \blacktriangleleft$$

Long division with polynomials is very similar to long division with whole numbers. Both use the same four basic steps: estimate, multiply, subtract, and bring down the next term. We use long division with polynomials when the denominator has two or more terms and is not a factor of the numerator. Here is an example.

▶ **EXAMPLE 6** Divide $\dfrac{2x^2 - 7x + 9}{x - 2}$.

Solution

$$\begin{array}{r} 2x \qquad\qquad \leftarrow \text{Estimate } 2x^2 \div x = 2x \\ x - 2\overline{)\,2x^2 - 7x + 9} \\ \underset{+}{\overset{-}{\cancel{2x^2}}} \; \cancel{4x} \leftarrow \text{Multiply } 2x(x - 2) = 2x^2 - 4x \\ - 3x \leftarrow \text{Subtract } (2x^2 - 7x) - (2x^2 - 4x) = -3x \end{array}$$

$$\begin{array}{r} 2x \qquad\qquad \\ x - 2\overline{)\,2x^2 - 7x + 9} \\ \underset{+}{\overset{-}{\cancel{2x^2}}} \; \cancel{4x} \quad \downarrow \\ - 3x + 9 \leftarrow \text{Bring down the 9} \end{array}$$

Notice we change the signs on $2x^2 - 4x$ and add in the subtraction step. Subtracting a polynomial is equivalent to adding its opposite.

We repeat the four steps.

$$\begin{array}{r} 2x\ -\ 3 \\ x-2\overline{)\ 2x^2-7x+9} \end{array}$$ ← −3 is the estimate: $-3x \div x = -3$

$$\cancel{+}2x^2 \cancel{+} 4x$$

$$-3x+9$$

$$\cancel{+}3x\cancel{+}6$$ ← Multiply $-3(x-2) = -3x+6$

$$3$$ ← Subtract

$$(-3x+9)-(-3x+6)=3$$

Since we have no other term to bring down, we have our answer:

$$\frac{2x^2-7x+9}{x-2} = 2x-3+\frac{3}{x-2}$$

To check, we multiply $(2x-3)(x-2)$ to get $2x^2-7x+6$; then, adding the remainder 3 to this result, we have $2x^2-7x+9$. ◄

In setting up a division problem involving two polynomials, we must remember two things: (1) both polynomials should be in decreasing powers of the variable, and (2) neither should skip any powers from the highest power down to the constant term. If there are any missing terms, they can be filled in using a coefficient of 0.

▶ **EXAMPLE 7**   Divide $2x-4\overline{)4x^3-6x-11}$.

Solution   Since the trinomial is missing a term in x^2, we can fill it in with $0x^2$:

$$4x^3-6x-11 = 4x^3+0x^2-6x-11$$

Adding $0x^2$ does not change our original problem.

$$\begin{array}{r} 2x^2+4x\ +5 \\ 2x-4\overline{)\ 4x^3+0x^2-\ 6x-11} \end{array}$$

$$\cancel{+}4x^3\cancel{+}8x^2$$

$$+8x^2-6x$$

$$\cancel{+}8x^2\cancel{+}16x$$

$$+10x-11$$

$$\cancel{+}10x\cancel{+}20$$

$$+9$$

Notice: Adding the $0x^2$ term gives us a column in which to write $+8x^2$.

$$\frac{4x^3-6x-11}{2x-4} = 2x^2+4x+5+\frac{9}{2x-4}$$

To check this result, we multiply $2x - 4$ and $2x^2 + 4x + 5$

$$
\begin{array}{r}
2x^2 + 4x + 5 \\
2x - 4 \\
\hline
4x^3 + 8x^2 + 10x \\
-8x^2 - 16x - 20 \\
\hline
4x^3 \qquad - 6x - 20
\end{array}
$$

Adding 9 (the remainder) to this result gives us the polynomial $4x^3 - 6x - 11$. Our answer checks. ◄

For our next example in this section, let's do Example 4 again, but this time use long division.

▶ **EXAMPLE 8**   Divide $\dfrac{x^2 - 6xy - 7y^2}{x + y}$.

Solution

$$
\begin{array}{r}
x - 7y \\
x + y \overline{)\ x^2 - 6xy - 7y^2} \\
\underline{\ \cancel{-}x^2 \cancel{-}\ xy \quad \downarrow} \\
-7xy - 7y^2 \\
\underline{+ \qquad +} \\
\cancel{+}\ 7xy \cancel{+} 7y^2 \\
\hline
0
\end{array}
$$

In this case, the remainder is 0 and we have

$$
\frac{x^2 - 6xy - 7y^2}{x + y} = x - 7y
$$

which is easy to check since

$$
(x + y)(x - 7y) = x^2 - 6xy - 7y^2 \qquad ◄
$$

▶ **EXAMPLE 9**   Factor $x^3 + 9x^2 + 26x + 24$ completely if $x + 2$ is one of its factors.

Solution   Since $x + 2$ is one of the factors of the polynomial we are trying to factor, it must divide that polynomial evenly; that is, without a remainder.

Therefore, we begin by dividing the polynomial by $x + 2$:

$$
\begin{array}{r}
x^2 + 7x\ \ + 12 \\
x + 2\overline{)\ \ x^3 + 9x^2 + 26x + 24} \\
\end{array}
$$

$$
\require{cancel}
\begin{aligned}
&\cancel{x^3}\ \cancel{2x^2} \\
&\hphantom{xx} + 7x^2 + 26x \\
&\hphantom{xxxx} \cancel{7x^2}\ \cancel{14x} \\
&\hphantom{xxxxxx} + 12x + 24 \\
&\hphantom{xxxxxxxx} \cancel{12x}\ \cancel{24} \\
&\hphantom{xxxxxxxxxx} 0
\end{aligned}
$$

From the division problem above, we know that the polynomial we are trying to factor is equal to the product of $x + 2$ and $x^2 + 7x + 12$. To factor completely, we simply factor $x^2 + 7x + 12$:

$$
\begin{aligned}
x^3 + 9x^2 + 26x + 24 &= (x + 2)(x^2 + 7x + 12) \\
&= (x + 2)(x + 3)(x + 4) \qquad \blacktriangleleft
\end{aligned}
$$

P R O B L E M   S E T   4 . 2

Find the following quotients.

1. $\dfrac{4x^3 - 8x^2 + 6x}{2x}$

2. $\dfrac{6x^3 + 12x^2 - 9x}{3x}$

3. $\dfrac{10x^4 + 15x^3 - 20x^2}{-5x^2}$

4. $\dfrac{12x^5 - 18x^4 - 6x^3}{6x^3}$

5. $\dfrac{8y^5 + 10y^3 - 6y}{4y^3}$

6. $\dfrac{6y^4 - 3y^3 + 18y^2}{9y^2}$

7. $\dfrac{5x^3 - 8x^2 - 6x}{-2x^2}$

8. $\dfrac{-9x^5 + 10x^3 - 12x}{-6x^4}$

9. $\dfrac{28a^3b^5 + 42a^4b^3}{7a^2b^2}$

10. $\dfrac{a^2b + ab^2}{ab}$

11. $\dfrac{10x^3y^2 - 20x^2y^3 - 30x^3y^3}{-10x^2y}$

12. $\dfrac{9x^4y^4 + 18x^3y^4 - 27x^2y^4}{-9xy^3}$

Divide by factoring numerators and then dividing out common factors.

13. $\dfrac{x^2 - x - 6}{x - 3}$

14. $\dfrac{x^2 - x - 6}{x + 2}$

15. $\dfrac{2a^2 - 3a - 9}{2a + 3}$

16. $\dfrac{2a^2 + 3a - 9}{2a - 3}$

17. $\dfrac{5x^2 - 14xy - 24y^2}{x - 4y}$

18. $\dfrac{5x^2 - 26xy - 24y^2}{5x + 4y}$

19. $\dfrac{x^3 - y^3}{x - y}$

20. $\dfrac{x^3 + 8}{x + 2}$

21. $\dfrac{y^4 - 16}{y - 2}$

22. $\dfrac{y^4 - 81}{y - 3}$

23. $\dfrac{x^3 + 2x^2 - 25x - 50}{x - 5}$

24. $\dfrac{x^3 + 2x^2 - 25x - 50}{x + 5}$

25. $\dfrac{4x^3 + 12x^2 - 9x - 27}{x + 3}$

26. $\dfrac{9x^3 + 18x^2 - 4x - 8}{x + 2}$

Divide using the long division method.

27. $\dfrac{x^2 - 5x - 7}{x + 2}$ **28.** $\dfrac{x^2 + 4x - 8}{x - 3}$

29. $\dfrac{6x^2 + 7x - 18}{3x - 4}$ **30.** $\dfrac{8x^2 - 26x - 9}{2x - 7}$

31. $\dfrac{2x^3 - 3x^2 - 4x + 5}{x + 1}$

32. $\dfrac{3x^3 - 5x^2 + 2x - 1}{x - 2}$

33. $\dfrac{2y^3 - 9y^2 - 17y + 39}{2y - 3}$

34. $\dfrac{3y^3 - 19y^2 + 17y + 4}{3y - 4}$

35. $\dfrac{2x^3 - 9x^2 + 11x - 6}{2x^2 - 3x + 2}$

36. $\dfrac{6x^3 + 7x^2 - x + 3}{3x^2 - x + 1}$

37. $\dfrac{6y^3 - 8y + 5}{2y - 4}$ **38.** $\dfrac{9y^3 - 6y^2 + 8}{3y - 3}$

39. $\dfrac{a^4 - 2a + 5}{a - 2}$ **40.** $\dfrac{a^4 + a^3 - 1}{a + 2}$

41. $\dfrac{y^4 - 16}{y - 2}$ **42.** $\dfrac{y^4 - 81}{y - 3}$

43. $\dfrac{x^4 + x^3 - 3x^2 - x + 2}{x^2 + 3x + 2}$

44. $\dfrac{2x^4 + x^3 + 4x - 3}{2x^2 - x + 3}$

45. Factor $x^3 + 6x^2 + 11x + 6$ completely if one of its factors is $x + 3$.
46. Factor $x^3 + 10x^2 + 29x + 20$ completely if one of its factors is $x + 4$.
47. Factor $x^3 + 5x^2 - 2x - 24$ completely if one of its factors is $x + 3$.
48. Factor $x^3 + 3x^2 - 10x - 24$ completely if one of its factors is $x + 2$.

49. Problems 21 and 41 are the same problem. Are the two answers you obtained equivalent?
50. Problems 22 and 42 are the same problem. Are the two answers you obtained equivalent?
51. Find the value of the polynomial $x^2 - 5x - 7$ when x is -2. Compare it with the remainder in Problem 27.
52. Find the value of the polynomial $x^2 + 4x - 8$ when x is 3. Compare it with the remainder in Problem 28.

Review Problems

The problems that follow review material we covered in Sections 3.1 and 1.4. Reviewing the problems from Section 1.4 will help you with the next section.

Divide. [1.4]

53. $\dfrac{3}{5} \div \dfrac{2}{7}$ **54.** $\dfrac{2}{7} \div \dfrac{3}{5}$

55. $\dfrac{3}{4} \div \dfrac{6}{11}$ **56.** $\dfrac{6}{8} \div \dfrac{3}{5}$

57. $\dfrac{4}{9} \div 8$ **58.** $\dfrac{3}{7} \div 6$

59. $8 \div \dfrac{1}{4}$ **60.** $12 \div \dfrac{2}{3}$

Write each expression with positive exponents and simplify as much as possible. [3.1]

61. $\left(\dfrac{1}{3}\right)^{-2} + \left(\dfrac{1}{2}\right)^{-3}$ **62.** $\left(\dfrac{1}{2}\right)^{-3} - \left(\dfrac{1}{3}\right)^{-3}$

Simplify and write your answers with positive exponents only. [3.1]

63. $(9x^{-4}y^9)^{-2}(3x^2y^{-1})^4$
64. $(4x^4y^{-3})^2(2x^{-6}y^4)^{-3}$

One Step Further

Divide.

65. $\dfrac{4x^5 - x^4 - 20x^3 + 8x^2 - 15}{x^2 - 5}$

66. $\dfrac{4x^5 + 2x^4 + x^3 - 20x^2 - 10x - 5}{x^3 - 5}$

67. $\dfrac{0.5x^3 - 0.3x^2 + 0.22x + 0.06}{x + 0.2}$

68. $\dfrac{0.6x^3 - 1.1x^2 - 0.1x + 0.6}{0.3x + 0.2}$

71. $\dfrac{2x^2 + \frac{1}{3}x + \frac{5}{3}}{3x - 1}$

72. $\dfrac{x^2 + \frac{3}{5}x + \frac{8}{5}}{5x - 2}$

69. $\dfrac{3x^2 + x - 9}{2x + 4}$

70. $\dfrac{2x^2 - x - 9}{3x + 6}$

SECTION

4.3

Multiplication and Division of Rational Expressions

In Section 4.1 we found the process of reducing rational expressions to lowest terms to be the same process used in reducing fractions to lowest terms. The similarity also holds for the process of multiplication or division of rational expressions.

Multiplication with fractions is the simplest of the four basic operations. To multiply two fractions we simply multiply numerators and multiply denominators. That is, if a, b, c, and d are real numbers, with $b \neq 0$ and $d \neq 0$, then

$$\frac{a}{b} \cdot \frac{c}{d} = \frac{ac}{bd}$$

▶ **EXAMPLE 1**   Multiply $\dfrac{6}{7} \cdot \dfrac{14}{18}$.

Solution

$$\frac{6}{7} \cdot \frac{14}{18} = \frac{6(14)}{7(18)} \qquad \text{Multiply numerators and denominators}$$

$$= \frac{\cancel{2} \cdot \cancel{3}(2 \cdot \cancel{7})}{\cancel{7}(\cancel{2} \cdot \cancel{3} \cdot 3)} \qquad \text{Factor}$$

$$= \frac{2}{3} \qquad \text{Divide out common factors} \qquad ◀$$

Our next example is similar to some of the problems we worked in Chapter 3. We multiply fractions whose numerators and denominators are monomials by multiplying numerators and multiplying denominators and then reducing to lowest terms. Here is how it looks.

▶ **EXAMPLE 2**   Multiply $\dfrac{8x^3}{27y^8} \cdot \dfrac{9y^3}{12x^2}$.

Solution   We multiply numerators and denominators without actually carrying out the multiplication:

$$\frac{8x^3}{27y^8} \cdot \frac{9y^3}{12x^2} = \frac{8 \cdot 9x^3 y^3}{27 \cdot 12x^2 y^8} \qquad \text{Multiply numerators}$$
$$\text{Multiply denominators}$$

$$= \frac{\cancel{4} \cdot 2 \cdot \cancel{9}x^3 y^3}{\cancel{9} \cdot 3 \cdot \cancel{4} \cdot 3x^2 y^8} \qquad \text{Factor coefficients}$$

$$= \frac{2x}{9y^5} \qquad \text{Divide out common factors} \quad \blacktriangleleft$$

The product of two rational expressions is the product of their numerators over the product of their denominators.

Once again, we should mention that the little slashes we have drawn through the factors are simply used to denote the factors we have divided out of the numerator and denominator.

▶ **EXAMPLE 3**   Multiply $\dfrac{x - 3}{x^2 - 4} \cdot \dfrac{x + 2}{x^2 - 6x + 9}$.

Solution   We begin by multiplying numerators and denominators. We then factor all polynomials and divide out factors common to the numerator and denominator:

$$\frac{x - 3}{x^2 - 4} \cdot \frac{x + 2}{x^2 - 6x + 9}$$

$$= \frac{(x - 3)(x + 2)}{(x^2 - 4)(x^2 - 6x + 9)} \qquad \text{Multiply}$$

$$= \frac{\cancel{(x - 3)}\cancel{(x + 2)}}{\cancel{(x + 2)}(x - 2)\cancel{(x - 3)}(x - 3)} \qquad \text{Factor}$$

$$= \frac{1}{(x - 2)(x - 3)} \qquad \begin{array}{l} \text{Divide out} \\ \text{common factors} \end{array} \quad \blacktriangleleft$$

The first two steps can be combined to save time. We can perform the multiplication and factoring steps together.

▶ **EXAMPLE 4**   Multiply $\dfrac{2y^2 - 4y}{2y^2 - 2} \cdot \dfrac{y^2 - 2y - 3}{y^2 - 5y + 6}$.

Solution

$$\frac{2y^2 - 4y}{2y^2 - 2} \cdot \frac{y^2 - 2y - 3}{y^2 - 5y + 6} = \frac{\cancel{2}y\cancel{(y - 2)}\cancel{(y - 3)}(y + 1)}{\cancel{2}\cancel{(y + 1)}(y - 1)\cancel{(y - 3)}\cancel{(y - 2)}}$$

$$= \frac{y}{y - 1} \quad \blacktriangleleft$$

Notice in both of the preceding examples that we did not actually multiply the polynomials as we did in Chapter 3. It would be senseless to do that since we would then have to factor each of the resulting products to reduce them to lowest terms.

The quotient of two rational expressions is the product of the first and the reciprocal of the second. That is, we find the quotient of two rational expressions the same way we find the quotient of two fractions. Here is an example that reviews division with fractions.

▶ **EXAMPLE 5** Divide $\dfrac{6}{8} \div \dfrac{3}{5}$.

Solution

$$\dfrac{6}{8} \div \dfrac{3}{5} = \dfrac{6}{8} \cdot \dfrac{5}{3} \qquad \text{Write division in terms of multiplication}$$

$$= \dfrac{6(5)}{8(3)} \qquad \text{Multiply numerators and denominators}$$

$$= \dfrac{\cancel{2} \cdot \cancel{3}(5)}{\cancel{2} \cdot 2 \cdot 2(\cancel{3})} \qquad \text{Factor}$$

$$= \dfrac{5}{4} \qquad \text{Divide out common factors} \qquad ◀$$

To divide one rational expression by another, we use the definition of division to multiply by the reciprocal of the expression that follows the division symbol.

▶ **EXAMPLE 6** Divide $\dfrac{8x^3}{5y^2} \div \dfrac{4x^2}{10y^6}$.

Solution First we rewrite the problem in terms of multiplication. Then we multiply.

$$\dfrac{8x^3}{5y^2} \div \dfrac{4x^2}{10y^6} = \dfrac{8x^3}{5y^2} \cdot \dfrac{10y^6}{4x^2}$$

$$= \dfrac{\overset{2}{\cancel{8}} \cdot \overset{2}{\cancel{10}}x^3 y^6}{\cancel{4} \cdot \cancel{5}x^2 y^2}$$

$$= 4xy^4 \qquad ◀$$

▶ **EXAMPLE 7** Divide $\dfrac{x^2 - y^2}{x^2 - 2xy + y^2} \div \dfrac{x^3 + y^3}{x^3 - x^2y}$.

Solution We begin by writing the problem as the product of the first and the reciprocal of the second and then proceed as in the previous two examples:

$$\frac{x^2 - y^2}{x^2 - 2xy + y^2} \div \frac{x^3 + y^3}{x^3 - x^2 y}$$

$$= \frac{x^2 - y^2}{x^2 - 2xy + y^2} \cdot \frac{x^3 - x^2 y}{x^3 + y^3}$$ Multiply by the reciprocal of the divisor

$$= \frac{(x - y)(x + y)(x^2)(x - y)}{(x - y)(x - y)(x + y)(x^2 - xy + y^2)}$$ Factor and multiply

$$= \frac{x^2}{x^2 - xy + y^2}$$ Divide out common factors   ◄

Here are some more examples of multiplication and division with rational expressions.

▶ **EXAMPLE 8** Perform the indicated operations.

$$\frac{a^2 - 8a + 15}{a + 4} \cdot \frac{a + 2}{a^2 - 5a + 6} \div \frac{a^2 - 3a - 10}{a^2 + 2a - 8}$$

Solution First we rewrite the division as multiplication by the reciprocal. Then we proceed as usual.

$$\frac{a^2 - 8a + 15}{a + 4} \cdot \frac{a + 2}{a^2 - 5a + 6} \div \frac{a^2 - 3a - 10}{a^2 + 2a - 8}$$

$$= \frac{(a^2 - 8a + 15)(a + 2)(a^2 + 2a - 8)}{(a + 4)(a^2 - 5a + 6)(a^2 - 3a - 10)}$$ Change division to multiplication by the reciprocal

$$= \frac{(a - 5)(a - 3)(a + 2)(a + 4)(a - 2)}{(a + 4)(a - 3)(a - 2)(a - 5)(a + 2)}$$ Factor

$$= 1$$ Divide out common factors   ◄

Our next example involves factoring by grouping. As you may have noticed, working the problems in this chapter gives you a very detailed review of factoring.

▶ **EXAMPLE 9** Multiply $\dfrac{xa + xb + ya + yb}{xa - xb - ya + yb} \cdot \dfrac{xa + xb - ya - yb}{xa - xb + ya - yb}.$

Solution We will factor each polynomial by grouping, which takes two steps.

$$\frac{xa + xb + ya + yb}{xa - xb - ya + yb} \cdot \frac{xa + xb - ya - yb}{xa - xb + ya - yb}$$

$$= \frac{x(a + b) + y(a + b)}{x(a - b) - y(a - b)} \cdot \frac{x(a + b) - y(a + b)}{x(a - b) + y(a - b)} \Bigg\} \quad \text{Factor by grouping}$$

$$= \frac{(a + b)(x + y)(a + b)(x - y)}{(a - b)(x - y)(a - b)(x + y)}$$

$$= \frac{(a + b)^2}{(a - b)^2}$$ ◀

▶ **EXAMPLE 10**   Multiply $(4x^2 - 36) \cdot \dfrac{12}{4x + 12}$.

Solution   We can think of $4x^2 - 36$ as having a denominator of 1. Thinking of it in this way allows us to proceed as we did in the previous examples.

$$(4x^2 - 36) \cdot \frac{12}{4x + 12}$$

$$= \frac{4x^2 - 36}{1} \cdot \frac{12}{4x + 12} \qquad \text{Write } 4x^2 - 36 \text{ with denominator 1}$$

$$= \frac{4(x - 3)(x + 3)12}{4(x + 3)} \qquad \text{Factor}$$

$$= 12(x - 3) \qquad\qquad \text{Divide out common factors}$$ ◀

▶ **EXAMPLE 11**   Multiply $3(x - 2)(x - 1) \cdot \dfrac{5}{x^2 - 3x + 2}$.

Solution   This problem is very similar to the problem in Example 10. Writing the first rational expression with a denominator of 1 we have

$$\frac{3(x - 2)(x - 1)}{1} \cdot \frac{5}{x^2 - 3x + 2} = \frac{3(x - 2)(x - 1)5}{(x - 2)(x - 1)}$$

$$= 3 \cdot 5$$

$$= 15$$ ◀

P R O B L E M   S E T   4 . 3

Perform the indicated operations involving fractions.

1. $\frac{2}{9} \cdot \frac{3}{4}$ **2.** $\frac{5}{6} \cdot \frac{7}{8}$

3. $\frac{3}{4} \div \frac{1}{3}$ **4.** $\frac{3}{8} \div \frac{5}{4}$

5. $\frac{3}{7} \cdot \frac{14}{24} \div \frac{1}{2}$ **6.** $\frac{6}{5} \cdot \frac{10}{36} \div \frac{3}{4}$

7. $\dfrac{10x^2}{5y^2} \cdot \dfrac{15y^3}{2x^4}$ **8.** $\dfrac{8x^3}{7y^4} \cdot \dfrac{14y^6}{16x^2}$

9. $\dfrac{11a^2b}{5ab^2} \div \dfrac{22a^3b^2}{10ab^4}$ **10.** $\dfrac{8ab^3}{9a^2b} \div \dfrac{16a^2b^2}{18ab^3}$

11. $\dfrac{6x^2}{5y^3} \cdot \dfrac{11z^2}{2x^2} \div \dfrac{33z^5}{10y^8}$ **12.** $\dfrac{4x^3}{7y^2} \cdot \dfrac{6z^5}{5x^6} \div \dfrac{24z^2}{35x^6}$

Perform the indicated operations. Be sure to write all answers in lowest terms.

13. $\dfrac{x^2 - 9}{x^2 - 4} \cdot \dfrac{x - 2}{x - 3}$

14. $\dfrac{x^2 - 16}{x^2 - 25} \cdot \dfrac{x - 5}{x - 4}$

15. $\dfrac{y^2 - 1}{y + 2} \cdot \dfrac{y^2 + 5y + 6}{y^2 + 2y - 3}$

16. $\dfrac{y - 1}{y^2 - y - 6} \cdot \dfrac{y^2 + 5y + 6}{y^2 - 1}$

17. $\dfrac{3x - 12}{x^2 - 4} \cdot \dfrac{x^2 + 6x + 8}{x - 4}$

18. $\dfrac{x^2 + 5x + 1}{4x - 4} \cdot \dfrac{x - 1}{x^2 + 5x + 1}$

19. $\dfrac{5x + 2y}{25x^2 - 5xy - 6y^2} \cdot \dfrac{20x^2 - 7xy - 3y^2}{4x + y}$

20. $\dfrac{7x + 3y}{42x^2 - 17xy - 15y^2} \cdot \dfrac{12x^2 - 4xy - 5y^2}{2x + y}$

21. $\dfrac{a^2 - 5a + 6}{a^2 - 2a - 3} \div \dfrac{a - 5}{a^2 + 3a + 2}$

22. $\dfrac{a^2 + 7a + 12}{a - 5} \div \dfrac{a^2 + 9a + 18}{a^2 - 7a + 10}$

23. $\dfrac{4t^2 - 1}{6t^2 + t - 2} \div \dfrac{8t^3 + 1}{27t^3 + 8}$

24. $\dfrac{9t^2 - 1}{6t^2 + 7t - 3} \div \dfrac{27t^3 + 1}{8t^3 + 27}$

25. $\dfrac{2x^2 - 5x - 12}{4x^2 + 8x + 3} \div \dfrac{x^2 - 16}{2x^2 + 7x + 3}$

26. $\dfrac{x^2 - 2x + 1}{3x^2 + 7x - 20} \div \dfrac{x^2 + 3x - 4}{3x^2 - 2x - 5}$

27. $\dfrac{6a^2b + 2ab^2 - 20b^3}{4a^2b - 16b^3} \cdot \dfrac{10a^2 - 22ab + 4b^2}{27a^3 - 125b^3}$

28. $\dfrac{12a^2b - 3ab^2 - 42b^3}{9a^2 - 36b^2} \cdot \dfrac{6a^2 - 15ab + 6b^2}{8a^3b - b^4}$

29. $\dfrac{360x^3 - 490x}{36x^2 + 84x + 49} \cdot \dfrac{30x^2 + 83x + 56}{150x^3 + 65x^2 - 280x}$

30. $\dfrac{490x^2 - 640}{49x^2 - 112x + 64} \cdot \dfrac{28x^2 - 95x + 72}{56x^3 - 62x^2 - 144x}$

31. $\dfrac{x^5 - x^2}{5x^5 - 5x} \cdot \dfrac{10x^4 - 10x^2}{2x^4 + 2x^3 + 2x^2}$

32. $\dfrac{2x^4 - 16x}{3x^6 - 48x^2} \cdot \dfrac{6x^5 + 24x^3}{4x^4 + 8x^3 + 16x^2}$

33. $\dfrac{a^2 - 16b^2}{a^2 - 8ab + 16b^2} \cdot \dfrac{a^2 - 9ab + 20b^2}{a^2 - 7ab + 12b^2} \div \dfrac{a^2 - 25b^2}{a^2 - 6ab + 9b^2}$

34. $\dfrac{a^2 - 6ab + 9b^2}{a^2 - 4b^2} \cdot \dfrac{a^2 - 5ab + 6b^2}{(a - 3b)^2} \div \dfrac{a^2 - 9b^2}{a^2 - ab - 6b^2}$

35. $\dfrac{2y^2 - 7y - 15}{42y^2 - 29y - 5} \cdot \dfrac{12y^2 - 16y + 5}{7y^2 - 36y + 5} \div \dfrac{4y^2 - 9}{49y^2 - 1}$

36. $\dfrac{8y^2 + 18y - 5}{21y^2 - 16y + 3} \cdot \dfrac{35y^2 - 22y + 3}{6y^2 + 17y + 5} \div \dfrac{16y^2 - 1}{9y^2 - 1}$

37. $\dfrac{xy - 2x + 3y - 6}{xy + 2x - 4y - 8} \cdot \dfrac{xy + x - 4y - 4}{xy - x + 3y - 3}$

38. $\dfrac{ax + bx + 2a + 2b}{ax - 3a + bx - 3b} \cdot \dfrac{ax - bx - 3a + 3b}{ax - bx - 2a + 2b}$

39. $\dfrac{xy^2 - y^2 + 4xy - 4y}{xy - 3y + 4x - 12} \div \dfrac{xy^3 + 2xy^2 + y^3 + 2y^2}{xy^2 - 3y^2 + 2xy - 6y}$

40. $\dfrac{4xb - 8b + 12x - 24}{xb^2 + 3b^2 + 3xb + 9b} \div \dfrac{4xb - 8b - 8x + 16}{xb^2 + 3b^2 - 2xb - 6b}$

41. $\dfrac{2x^3 + 10x^2 - 8x - 40}{x^3 + 4x^2 - 9x - 36} \cdot \dfrac{x^2 + x - 12}{2x^2 + 14x + 20}$

42. $\dfrac{x^3 + 2x^2 - 9x - 18}{x^4 + 3x^3 - 4x^2 - 12x} \cdot \dfrac{x^3 + 5x^2 + 6x}{x^2 - x - 6}$

Use the method shown in Examples 10 and 11 to find the following products.

43. $(3x - 6) \cdot \dfrac{x}{x - 2}$

44. $(4x + 8) \cdot \dfrac{x}{x + 2}$

45. $(x^2 - 25) \cdot \dfrac{2}{x - 5}$

46. $(x^2 - 49) \cdot \dfrac{5}{x + 7}$

47. $(x^2 - 3x + 2) \cdot \dfrac{3}{3x - 3}$

48. $(x^2 - 3x + 2) \cdot \dfrac{-1}{x - 2}$

49. $(y - 3)(y - 4)(y + 3) \cdot \dfrac{-1}{y^2 - 9}$

50. $(y + 1)(y + 4)(y - 1) \cdot \dfrac{3}{y^2 - 1}$

51. $a(a + 5)(a - 5) \cdot \dfrac{a + 1}{a^2 + 5a}$

52. $a(a + 3)(a - 3) \cdot \dfrac{a - 1}{a^2 - 3a}$

Review Problems

The problems below review material we covered in Sections 3.4 and 1.3.

Multiply. [3.4]

53. $2x^2(5x^3 + 4x - 3)$
54. $3x^3(7x^2 - 4x - 8)$
55. $(3a - 1)(4a + 5)$
56. $(6a - 3)(2a + 1)$
57. $(x + 3)(x - 3)(x^2 + 9)$
58. $(2x - 3)(4x^2 + 6x + 9)$
59. $(4y - 5)^2$
60. $(2y - \frac{1}{2})^2$

61. $(3x + 7)(4y - 2)$
62. $(x + 2a)(2 - 3b)$
63. $(3 - t^2)^2$
64. $(2 - t^3)^2$
65. $3(x + 1)(x + 2)(x + 3)$
66. $4(x - 1)(x - 2)(x - 3)$

Combine. [1.3]

67. $\frac{3}{14} + \frac{7}{30}$
68. $\frac{5}{12} + \frac{7}{18}$
69. $\frac{4}{15} - \frac{5}{21}$
70. $\frac{3}{14} - \frac{5}{22}$

One Step Further

Simplify each expression.

71. $(1 + 2^{-1})(1 + 3^{-1})(1 + 4^{-1})(1 + 5^{-1})$
72. $(1 - 2^{-1})(1 - 3^{-1})(1 - 4^{-1})(1 - 5^{-1})$

The dots in the problems below represent factors not written that are in the same pattern as the surrounding factors. Simplify.

73. $(1 + 2^{-1})(1 + 3^{-1})(1 + 4^{-1}) \cdots (1 + 99^{-1})(1 + 100^{-1})$
74. $(1 - 2^{-1})(1 - 3^{-1})(1 - 4^{-1}) \cdots (1 - 99^{-1})(1 - 100^{-1})$

SECTION

4.4   Addition and Subtraction of Rational Expressions

This section is concerned with addition and subtraction of rational expressions. In the first part of this section we will look at addition of expressions that have the same denominator. In the second part of this section we will look at addition of expressions that have different denominators.

Addition and Subtraction with the Same Denominator

To add two expressions that have the same denominator, we simply add numerators and put the sum over the common denominator. Since the process we use to add and subtract rational expressions is the same process used to add and subtract fractions, we will begin with an example involving fractions.

▶ **EXAMPLE 1**   Add $\dfrac{4}{9} + \dfrac{2}{9}$.

Solution   Recall from Chapter 1 that we add fractions with the same denominator by using the distributive property. Here is a detailed look at the steps involved.

$$\frac{4}{9} + \frac{2}{9} = 4\left(\frac{1}{9}\right) + 2\left(\frac{1}{9}\right)$$

$$= (4 + 2)\left(\frac{1}{9}\right) \qquad \text{Distributive property}$$

$$= 6\left(\frac{1}{9}\right)$$

$$= \frac{6}{9}$$

$$= \frac{2}{3} \qquad \begin{array}{l}\text{Divide numerator and denominator}\\ \text{by common factor 3}\end{array}$$

Note that the important thing about the fractions in this example is that they each have a denominator of 9. If they did not have the same denominator, we could not have written them as two terms with a factor of $\frac{1}{9}$ in common. Without the $\frac{1}{9}$ common to each term, we couldn't apply the distributive property. And without the distributive property, we would not have been able to add the two fractions.   ◀

     In the examples that follow, we will not show all the steps we have shown in Example 1. The steps are shown in Example 1 so that you will see why both fractions must have the same denominator before we can add them. In actual practice we simply add numerators and place the result over the common denominator.

     We add and subtract rational expressions with the same denominator by combining numerators and writing the result over the common denominator. Then we reduce the result to lowest terms if possible. Example 2 shows this process in detail. If you see the similarities between operations on rational numbers and operations on rational expressions, this chapter will look like an extension of rational numbers rather than a completely new set of topics.

▶ **EXAMPLE 2**   Add $\dfrac{x}{x^2 - 1} + \dfrac{1}{x^2 - 1}$.

Solution   Since the denominators are the same, we simply add numerators:

$$\frac{x}{x^2 - 1} + \frac{1}{x^2 - 1} = \frac{x + 1}{x^2 - 1} \qquad \text{Add numerators}$$

$$= \frac{\cancel{x + 1}}{(x - 1)\cancel{(x + 1)}} \qquad \text{Factor denominator}$$

$$= \frac{1}{x - 1} \qquad \begin{array}{l}\text{Divide out common} \\ \text{factor } x + 1\end{array} \qquad \blacktriangleleft$$

Our next example involves subtraction of rational expressions. Pay careful attention to what happens to the signs of the terms in the numerator of the second expression when we subtract it from the first expression.

▶ **EXAMPLE 3** Subtract $\dfrac{2x - 5}{x - 2} - \dfrac{x - 3}{x - 2}$.

Solution Since each expression has the same denominator, we simply subtract the numerator in the second expression from the numerator in the first expression and write the difference over the common denominator $x - 2$. We must be careful, however, that we subtract both terms in the second numerator. To ensure that we do, we will enclose that numerator in parentheses.

$$\frac{2x - 5}{x - 2} - \frac{x - 3}{x - 2} = \frac{2x - 5 - (x - 3)}{x - 2} \qquad \text{Subtract numerators}$$

$$= \frac{2x - 5 - x + 3}{x - 2} \qquad \text{Remove parentheses}$$

$$= \frac{\cancel{x - 2}}{\cancel{x - 2}} \qquad \begin{array}{l}\text{Combine similar terms} \\ \text{in the numerator}\end{array}$$

$$= 1 \qquad \text{Reduce (or divide)}$$

Note the $+3$ in the numerator of the second step. It is a very common mistake to write this as -3, by forgetting to subtract both terms in the numerator of the second expression. Whenever the expression we are subtracting has two or more terms in its numerator, we have to watch for this mistake. ◀

Next we consider addition and subtraction of fractions and rational expressions that have different denominators.

Addition and Subtraction with Different Denominators

Before we look at an example of addition of fractions with different denominators, we need to review the definition for the least common denominator.

DEFINITION The **least common denominator,** abbreviated LCD, for a set of denominators is the smallest expression that is divisible by each of the denominators.

The first step in combining two fractions is to find the LCD. Once we have the common denominator, we rewrite each fraction as an equivalent fraction with the common denominator. After that, we simply add or subtract as we did in our first three examples.

Example 4 is a review of the step-by-step procedure used to add two fractions with different denominators.

▶ **EXAMPLE 4** Add $\dfrac{3}{14} + \dfrac{7}{30}$.

Solution

Step 1: Find the LCD.

To do this we first factor both denominators into prime factors.

$$\text{Factor 14:} \quad 14 = 2 \cdot 7$$
$$\text{Factor 30:} \quad 30 = 2 \cdot 3 \cdot 5$$

Since the LCD must be divisible by 14, it must have factors of $2 \cdot 7$. It must also be divisible by 30 and, therefore, have factors of $2 \cdot 3 \cdot 5$. We do not need to repeat the 2 that appears in both the factors of 14 and those of 30. Therefore,

$$\text{LCD} = 2 \cdot 3 \cdot 5 \cdot 7 = 210$$

Step 2: Change to equivalent fractions.

Since we want each fraction to have a denominator of 210 and at the same time keep its original value, we multiply each by 1 in the appropriate form.
Change $\frac{3}{14}$ to a fraction with denominator 210:

$$\frac{3}{14} \cdot \frac{\mathbf{15}}{\mathbf{15}} = \frac{45}{210}$$

Change $\frac{7}{30}$ to a fraction with denominator 210:

$$\frac{7}{30} \cdot \frac{\mathbf{7}}{\mathbf{7}} = \frac{49}{210}$$

Step 3: Add numerators of equivalent fractions found in step 2:

$$\frac{45}{210} + \frac{49}{210} = \frac{94}{210}$$

Step 4: Reduce to lowest terms if necessary:

$$\frac{94}{210} = \frac{47}{105}$$ ◄

The main idea in adding fractions is to write each fraction again with the LCD for a denominator. In doing so, we must be sure not to change the value of either of the original fractions.

► **EXAMPLE 5** Add $\dfrac{-2}{x^2 - 2x - 3} + \dfrac{3}{x^2 - 9}$.

Solution

Step 1: Factor each denominator and build the LCD from the factors:

$$\left.\begin{array}{l} x^2 - 2x - 3 = (x-3)(x+1) \\ x^2 - 9 \quad\quad = (x-3)(x+3) \end{array}\right\} \text{LCD} = (x-3)(x+3)(x+1)$$

Step 2: Change each rational expression to an equivalent expression that has the LCD for a denominator:

$$\frac{-2}{x^2 - 2x - 3} = \frac{-2}{(x-3)(x+1)} \cdot \frac{(x+3)}{(x+3)} = \frac{-2x - 6}{(x-3)(x+3)(x+1)}$$

$$\frac{3}{x^2 - 9} = \frac{3}{(x-3)(x+3)} \cdot \frac{(x+1)}{(x+1)} = \frac{3x+3}{(x-3)(x+3)(x+1)}$$

Step 3: Add numerators of the rational expressions found in step 2:

$$\frac{-2x - 6}{(x-3)(x+3)(x+1)} + \frac{3x+3}{(x-3)(x+3)(x+1)} = \frac{x-3}{(x-3)(x+3)(x+1)}$$

Step 4: Reduce to lowest terms by dividing out the common factor $x - 3$:

$$= \frac{1}{(x+3)(x+1)}$$ ◄

► **EXAMPLE 6** Subtract $\dfrac{x+4}{2x+10} - \dfrac{5}{x^2 - 25}$.

Solution We begin by factoring each denominator:

$$\frac{x+4}{2x+10} - \frac{5}{x^2 - 25} = \frac{x+4}{2(x+5)} - \frac{5}{(x+5)(x-5)}$$

The LCD is $2(x + 5)(x - 5)$. Completing the problem we have

$$= \frac{x + 4}{2(x + 5)} \cdot \frac{(x - 5)}{(x - 5)} - \frac{5}{(x + 5)(x - 5)} \cdot \frac{2}{2}$$

$$= \frac{x^2 - x - 20}{2(x + 5)(x - 5)} - \frac{10}{2(x + 5)(x - 5)}$$

$$= \frac{x^2 - x - 30}{2(x + 5)(x - 5)}$$

To see if this expression will reduce, we factor the numerator into $(x - 6)(x + 5)$.

$$= \frac{(x - 6)\cancel{(x + 5)}}{2\cancel{(x + 5)}(x - 5)}$$

$$= \frac{x - 6}{2(x - 5)} \qquad \blacktriangleleft$$

▶ **EXAMPLE 7** Subtract $\dfrac{2x - 2}{x^2 + 4x + 3} - \dfrac{x - 1}{x^2 + 5x + 6}$.

Solution We factor each denominator and build the LCD from those factors:

$$\frac{2x - 2}{x^2 + 4x + 3} - \frac{x - 1}{x^2 + 5x + 6}$$

$$= \frac{2x - 2}{(x + 3)(x + 1)} - \frac{x - 1}{(x + 3)(x + 2)}$$

$$= \frac{2x - 2}{(x + 3)(x + 1)} \cdot \frac{(x + 2)}{(x + 2)} - \frac{x - 1}{(x + 3)(x + 2)} \cdot \frac{(x + 1)}{(x + 1)} \qquad \begin{array}{l} \text{The LCD is} \\ (x + 1)(x + 2)(x + 3) \end{array}$$

$$= \frac{2x^2 + 2x - 4}{(x + 1)(x + 2)(x + 3)} - \frac{x^2 - 1}{(x + 1)(x + 2)(x + 3)} \qquad \begin{array}{l} \text{Multiply out each} \\ \text{numerator} \end{array}$$

$$\left.\begin{array}{l} = \dfrac{(2x^2 + 2x - 4) - (x^2 - 1)}{(x + 1)(x + 2)(x + 3)} \\[3mm] = \dfrac{x^2 + 2x - 3}{(x + 1)(x + 2)(x + 3)} \end{array}\right\} \qquad \begin{array}{l} \text{Subtract} \\ \text{numerators} \end{array}$$

$$= \frac{\cancel{(x + 3)}(x - 1)}{(x + 1)(x + 2)\cancel{(x + 3)}} \qquad \begin{array}{l} \text{Factor numerator} \\ \text{to see if we} \\ \text{can reduce} \end{array}$$

$$= \frac{x - 1}{(x + 1)(x + 2)} \qquad \text{Reduce} \qquad \blacktriangleleft$$

▶ **EXAMPLE 8** Add $\dfrac{x^2}{x - 7} + \dfrac{6x + 7}{7 - x}$.

Solution   In Section 4.1 we were able to reverse the terms in a factor such as $7 - x$ by factoring -1 from each term. In a problem like this, the same result can be obtained by multiplying the numerator and denominator by -1:

$$\frac{x^2}{x-7} + \frac{6x+7}{7-x} \cdot \frac{-1}{-1} = \frac{x^2}{x-7} + \frac{-6x-7}{x-7}$$

$$= \frac{x^2 - 6x - 7}{x - 7} \qquad \text{Add numerators}$$

$$= \frac{(x-7)(x+1)}{(x-7)} \qquad \text{Factor numerator}$$

$$= x + 1 \qquad\qquad \text{Divide out } x - 7 \quad \blacktriangleleft$$

For our next example we will look at a problem in which we combine a whole number and a rational expression.

▶ **EXAMPLE 9**   Subtract $2 - \dfrac{9}{3x+1}$.

Solution   To subtract these two expressions, we think of 2 as a rational expression with a denominator of 1.

$$2 - \frac{9}{3x+1} = \frac{2}{1} - \frac{9}{3x+1}$$

The LCD is $3x + 1$. Multiplying the numerator and denominator of the first expression by $3x + 1$ gives us a rational expression equivalent to 2, but with a denominator of $3x + 1$.

$$\frac{2}{1} \cdot \frac{(3x+1)}{(3x+1)} - \frac{9}{3x+1} = \frac{6x + 2 - 9}{3x+1}$$

$$= \frac{6x - 7}{3x + 1}$$

The numerator and denominator of this last expression do not have any factors in common other than 1, so the expression is in lowest terms. $\blacktriangleleft$

▶ **EXAMPLE 10**   Write an expression for the sum of a number and twice its reciprocal. Then, simplify that expression.

Solution   If x is the number, then its reciprocal is $\frac{1}{x}$. Twice its reciprocal is $\frac{2}{x}$. The sum of the number and twice its reciprocal is

$$x + \frac{2}{x}$$

To combine these two expressions, we think of the first term x as a rational expression with a denominator of 1. The least common denominator is x:

$$x + \frac{2}{x} = \frac{x}{1} + \frac{2}{x}$$

$$= \frac{x}{1} \cdot \frac{x}{x} + \frac{2}{x}$$

$$= \frac{x^2 + 2}{x}$$

◀

PROBLEM SET 4.4

Combine the following fractions.

1. $\frac{3}{4} + \frac{1}{2}$

2. $\frac{5}{6} + \frac{1}{3}$

3. $\frac{2}{5} - \frac{1}{15}$

4. $\frac{5}{8} - \frac{1}{4}$

5. $\frac{5}{6} + \frac{7}{8}$

6. $\frac{3}{4} + \frac{2}{3}$

7. $\frac{9}{48} - \frac{3}{54}$

8. $\frac{6}{28} - \frac{5}{42}$

9. $\frac{3}{4} - \frac{1}{8} + \frac{2}{3}$

10. $\frac{1}{3} - \frac{5}{6} + \frac{5}{12}$

Combine the following rational expressions. Reduce all answers to lowest terms.

11. $\frac{x}{x+3} + \frac{3}{x+3}$

12. $\frac{5x}{5x+2} + \frac{2}{5x+2}$

13. $\frac{4}{y-4} - \frac{y}{y-4}$

14. $\frac{8}{y+8} + \frac{y}{y+8}$

15. $\frac{x}{x^2-y^2} - \frac{y}{x^2-y^2}$

16. $\frac{x}{x^2-y^2} + \frac{y}{x^2-y^2}$

17. $\frac{2x-3}{x-2} - \frac{x-1}{x-2}$

18. $\frac{2x-4}{x+2} - \frac{x-6}{x+2}$

19. $\frac{1}{a} + \frac{2}{a^2} - \frac{3}{a^3}$

20. $\frac{3}{a} + \frac{2}{a^2} - \frac{1}{a^3}$

21. $\frac{7x-2}{2x+1} - \frac{5x-3}{2x+1}$

22. $\frac{7x-1}{3x+2} - \frac{4x-3}{3x+2}$

23. $\frac{2}{t^2} - \frac{3}{2t}$

24. $\frac{5}{3t} - \frac{4}{t^2}$

25. $\frac{3x+1}{2x-6} - \frac{x+2}{x-3}$

26. $\frac{x+1}{x-2} - \frac{4x+7}{5x-10}$

27. $\frac{6x+5}{5x-25} - \frac{x+2}{x-5}$

28. $\frac{4x+2}{3x+12} - \frac{x-2}{x+4}$

29. $\frac{x+1}{2x-2} - \frac{2}{x^2-1}$

30. $\frac{x+7}{2x+12} + \frac{6}{x^2-36}$

31. $\frac{1}{a-b} - \frac{3ab}{a^3-b^3}$

32. $\frac{1}{a+b} + \frac{3ab}{a^3+b^3}$

33. $\frac{1}{2y-3} - \frac{18y}{8y^3-27}$

34. $\frac{1}{3y-2} - \frac{18y}{27y^3-8}$

35. $\frac{x}{x^2-5x+6} - \frac{3}{3-x}$

36. $\frac{x}{x^2+4x+4} - \frac{2}{2+x}$

37. $\frac{2}{4t-5} + \frac{9}{8t^2-38t+35}$

38. $\frac{3}{2t-5} + \frac{21}{8t^2-14t-15}$

39. $\frac{1}{a^2-5a+6} + \frac{3}{a^2-a-2}$

40. $\frac{-3}{a^2+a-2} + \frac{5}{a^2-a-6}$

41. $\frac{1}{8x^3-1} - \frac{1}{4x^2-1}$

42. $\dfrac{1}{27x^3 - 1} - \dfrac{1}{9x^2 - 1}$

43. $\dfrac{4}{4x^2 - 9} - \dfrac{6}{8x^2 - 6x - 9}$

44. $\dfrac{9}{9x^2 + 6x - 8} - \dfrac{6}{9x^2 - 4}$

45. $\dfrac{4a}{a^2 + 6a + 5} - \dfrac{3a}{a^2 + 5a + 4}$

46. $\dfrac{3a}{a^2 + 7a + 10} - \dfrac{2a}{a^2 + 6a + 8}$

47. $\dfrac{2x - 1}{x^2 + x - 6} - \dfrac{x + 2}{x^2 + 5x + 6}$

48. $\dfrac{4x + 1}{x^2 + 5x + 4} - \dfrac{x + 3}{x^2 + 4x + 3}$

49. $\dfrac{2x - 8}{3x^2 + 8x + 4} + \dfrac{x + 3}{3x^2 + 5x + 2}$

50. $\dfrac{5x + 3}{2x^2 + 5x + 3} - \dfrac{3x + 9}{2x^2 + 7x + 6}$

51. $\dfrac{2}{x^2 + 5x + 6} - \dfrac{4}{x^2 + 4x + 3} + \dfrac{3}{x^2 + 3x + 2}$

52. $\dfrac{-5}{x^2 + 3x - 4} + \dfrac{5}{x^2 + 2x - 3} + \dfrac{1}{x^2 + 7x + 12}$

53. $\dfrac{2x + 8}{x^2 + 5x + 6} - \dfrac{x + 5}{x^2 + 4x + 3} - \dfrac{x - 1}{x^2 + 3x + 2}$

54. $\dfrac{2x + 11}{x^2 + 9x + 20} - \dfrac{x + 1}{x^2 + 7x + 12} - \dfrac{x + 6}{x^2 + 8x + 15}$

55. $2 + \dfrac{3}{2x + 1}$ **56.** $3 - \dfrac{2}{2x + 3}$

57. $5 + \dfrac{2}{4 - t}$ **58.** $7 + \dfrac{3}{5 - t}$

59. $x - \dfrac{4}{2x + 3}$

60. $x - \dfrac{5}{3x + 4} + 1$

61. $\dfrac{x}{x + 2} + \dfrac{1}{2x + 4} - \dfrac{3}{x^2 + 2x}$

62. $\dfrac{x}{x + 3} + \dfrac{7}{3x + 9} - \dfrac{2}{x^2 + 3x}$

63. $\dfrac{1}{x} + \dfrac{x}{2x + 4} - \dfrac{2}{x^2 + 2x}$

64. $\dfrac{1}{x} + \dfrac{x}{3x + 9} - \dfrac{3}{x^2 + 3x}$

65. The formula $P = \frac{1}{a} + \frac{1}{b}$ is used by optometrists to help determine how strong to make the lenses for a pair of eyeglasses. If a is 10 and b is 0.2, find the corresponding value of P.

66. Show that the formula in Problem 65 can be written $P = \dfrac{a + b}{ab}$, and then let $a = 10$ and $b = 0.2$ in this new form of the formula to find P.

67. Show that the expressions $(x + y)^{-1}$ and $x^{-1} + y^{-1}$ are not equal when $x = 3$ and $y = 4$.

68. Show that the expressions $(x + y)^{-1}$ and $x^{-1} + y^{-1}$ are not equal. (Begin by writing each with positive exponents only.)

Simplify each of the following expressions. (Change to positive exponents first.)

69. $(1 - 3^{-2}) \div (1 - 3^{-1})$
70. $(1 - 5^{-2}) \div (1 - 5^{-1})$
71. $(1 - x^{-2}) \div (1 - x^{-1})$
72. $(1 - x^{-3}) \div (1 - x^{-2})$
73. Write an expression for the sum of a number and four times its reciprocal. Then, simplify that expression.
74. Write an expression for the sum of a number and three times its reciprocal. Then, simplify that expression.
75. Write an expression for the sum of the reciprocals of two consecutive integers. Then, simplify that expression.
76. Write an expression for the sum of the reciprocals of two consecutive even integers. Then, simplify that expression.

Review Problems

The problems that follow review material we covered in Sections 3.1 and 3.2 on scientific notation.

Write each number in scientific notation.

77. 54,000   **78.** 768,000

79. 0.00034   **80.** 0.0359

Write each number in expanded form.

81. 6.44×10^3 **82.** 2.5×10^2

83. 6.44×10^{-3} **84.** 2.5×10^{-2}

Simplify each expression as much as possible. Write all answers in scientific notation.

85. $(3 \times 10^8)(4 \times 10^{-5})$

86. $(6 \times 10^3)(3 \times 10^{-7})$

87. $\dfrac{8 \times 10^{-3}}{4 \times 10^{-6}}$ **88.** $\dfrac{8 \times 10^5}{2 \times 10^{-8}}$

One Step Further

Simplify.

89. $\left(1 - \dfrac{1}{x}\right)\left(1 - \dfrac{1}{x+1}\right)\left(1 - \dfrac{1}{x+2}\right)\left(1 - \dfrac{1}{x+3}\right)$

90. $\left(1 + \dfrac{1}{x}\right)\left(1 + \dfrac{1}{x+1}\right)\left(1 + \dfrac{1}{x+2}\right)\left(1 + \dfrac{1}{x+3}\right)$

91. $\left(1 - \dfrac{1}{x}\right)\left(1 - \dfrac{1}{x+1}\right)\left(1 - \dfrac{1}{x+2}\right)\cdots\left(1 - \dfrac{1}{x+49}\right)\left(1 - \dfrac{1}{x+50}\right)$

92. $\left(1 + \dfrac{1}{x}\right)\left(1 + \dfrac{1}{x+1}\right)\left(1 + \dfrac{1}{x+2}\right)\cdots\left(1 + \dfrac{1}{x+98}\right)\left(1 + \dfrac{1}{x+99}\right)$

93. $\left(1 + \dfrac{1}{x}\right)\left(1 + \dfrac{1}{x-1}\right)\left(1 + \dfrac{1}{x-2}\right)\cdots\left(1 + \dfrac{1}{x-99}\right)\left(1 + \dfrac{1}{x-100}\right)$

94. $\left(1 - \dfrac{1}{x-1}\right)\left(1 - \dfrac{1}{x-2}\right)\left(1 - \dfrac{1}{x-3}\right)\cdots\left(1 - \dfrac{1}{x-48}\right)\left(1 - \dfrac{1}{x-49}\right)$

SECTION

....

4.5   ## Complex Fractions

The quotient of two fractions or two rational expressions is called a **complex fraction.** This section is concerned with the simplification of complex fractions.

▶ **EXAMPLE 1** Simplify $\dfrac{\frac{3}{4}}{\frac{5}{8}}$.

Solution There are generally two methods that can be used to simplify complex fractions.

Method 1 We can multiply the numerator and denominator of the complex fraction by the LCD for both of the fractions, which in this case is 8.

$$\frac{\dfrac{3}{4}}{\dfrac{5}{8}} = \frac{\dfrac{3}{4} \cdot 8}{\dfrac{5}{8} \cdot 8} = \frac{6}{5}$$

Method 2 Instead of dividing by $\frac{5}{8}$ we can multiply by $\frac{8}{5}$.

$$\frac{\dfrac{3}{4}}{\dfrac{5}{8}} = \frac{3}{4} \cdot \frac{8}{5} = \frac{24}{20} = \frac{6}{5}$$

◀

Here are some examples of complex fractions involving rational expressions. Most can be solved using either of the two methods shown in Example 1.

▶ **EXAMPLE 2** Simplify $\dfrac{\dfrac{1}{x} + \dfrac{1}{y}}{\dfrac{1}{x} - \dfrac{1}{y}}$.

Solution This problem is most easily solved using Method 1. We begin by multiplying both the numerator and denominator by the quantity xy, which is the LCD for all the fractions:

$$\frac{\dfrac{1}{x} + \dfrac{1}{y}}{\dfrac{1}{x} - \dfrac{1}{y}} = \frac{\left(\dfrac{1}{x} + \dfrac{1}{y}\right) \cdot xy}{\left(\dfrac{1}{x} - \dfrac{1}{y}\right) \cdot xy}$$

$$= \frac{\dfrac{1}{x}(xy) + \dfrac{1}{y}(xy)}{\dfrac{1}{x}(xy) - \dfrac{1}{y}(xy)} \qquad \begin{array}{l}\text{Apply the distributive property to}\\ \text{distribute } xy \text{ over both terms in}\\ \text{the numerator and denominator}\end{array}$$

$$= \frac{y + x}{y - x}$$

◀

▶ **EXAMPLE 3** Simplify $\dfrac{\dfrac{x - 2}{x^2 - 9}}{\dfrac{x^2 - 4}{x + 3}}$.

Solution Applying Method 2, we have

$$\frac{\dfrac{x-2}{x^2-9}}{\dfrac{x^2-4}{x+3}} = \frac{x-2}{x^2-9} \cdot \frac{x+3}{x^2-4}$$

$$= \frac{(x-2)(x+3)}{(x+3)(x-3)(x+2)(x-2)}$$

$$= \frac{1}{(x-3)(x+2)} \qquad \blacktriangleleft$$

▶ **EXAMPLE 4** Simplify $\dfrac{1-\dfrac{4}{x^2}}{1-\dfrac{1}{x}-\dfrac{6}{x^2}}$.

Solution The simplest way to simplify this complex fraction is to multiply the numerator and denominator by the LCD, x^2:

$$\frac{1-\dfrac{4}{x^2}}{1-\dfrac{1}{x}-\dfrac{6}{x^2}} = \frac{\boldsymbol{x^2}\left(1-\dfrac{4}{x^2}\right)}{\boldsymbol{x^2}\left(1-\dfrac{1}{x}-\dfrac{6}{x^2}\right)} \qquad \begin{array}{l}\text{Multiply numerator and}\\ \text{denominator by } \boldsymbol{x^2}\end{array}$$

$$= \frac{x^2 \cdot 1 - x^2 \cdot \dfrac{4}{x^2}}{x^2 \cdot 1 - x^2 \cdot \dfrac{1}{x} - x^2 \cdot \dfrac{6}{x^2}} \qquad \text{Distributive property}$$

$$= \frac{x^2-4}{x^2-x-6} \qquad \text{Simplify}$$

$$= \frac{(x-2)(x+2)}{(x-3)(x+2)} \qquad \text{Factor}$$

$$= \frac{x-2}{x-3} \qquad \text{Reduce} \qquad \blacktriangleleft$$

▶ **EXAMPLE 5** Simplify $2 - \dfrac{3}{x+\dfrac{1}{3}}$.

Solution First we simplify the expression that follows the subtraction sign.

$$2 - \frac{3}{x+\dfrac{1}{3}} = 2 - \frac{\boldsymbol{3\cdot 3}}{\boldsymbol{3}\left(x+\dfrac{1}{3}\right)} = 2 - \frac{9}{3x+1}$$

Now we subtract by rewriting the first term, 2, with the LCD, $3x + 1$.

$$2 - \frac{9}{3x + 1} = \frac{2}{1} \cdot \frac{3x + 1}{3x + 1} - \frac{9}{3x + 1}$$

$$= \frac{6x + 2 - 9}{3x + 1} = \frac{6x - 7}{3x + 1} \quad \blacktriangleleft$$

P R O B L E M   S E T   4 . 5

Simplify each of the following as much as possible.

1. $\dfrac{\frac{3}{4}}{\frac{2}{3}}$

2. $\dfrac{\frac{5}{9}}{\frac{7}{12}}$

3. $\dfrac{\frac{1}{3} - \frac{1}{4}}{\frac{1}{2} + \frac{1}{8}}$

4. $\dfrac{\frac{1}{6} - \frac{1}{3}}{\frac{1}{4} - \frac{1}{8}}$

5. $\dfrac{3 + \frac{2}{5}}{1 - \frac{3}{7}}$

6. $\dfrac{2 + \frac{5}{6}}{1 - \frac{7}{8}}$

7. $\dfrac{\frac{1}{x}}{1 + \frac{1}{x}}$

8. $\dfrac{1 - \frac{1}{x}}{\frac{1}{x}}$

9. $\dfrac{1 + \frac{1}{a}}{1 - \frac{1}{a}}$

10. $\dfrac{1 - \frac{2}{a}}{1 - \frac{3}{a}}$

11. $\dfrac{\frac{1}{x} - \frac{1}{y}}{\frac{1}{x} + \frac{1}{y}}$

12. $\dfrac{\frac{1}{x} + \frac{2}{y}}{\frac{2}{x} + \frac{1}{y}}$

13. $\dfrac{\frac{x - 5}{x^2 - 4}}{\frac{x^2 - 25}{x + 2}}$

14. $\dfrac{\frac{3x + 1}{x^2 - 49}}{\frac{9x^2 - 1}{x - 7}}$

15. $\dfrac{\frac{4a}{2a^3 + 2}}{\frac{8a}{4a + 4}}$

16. $\dfrac{\frac{2a}{3a^3 - 3}}{\frac{4a}{6a - 6}}$

17. $\dfrac{1 - \frac{9}{x^2}}{1 - \frac{1}{x} - \frac{6}{x^2}}$

18. $\dfrac{4 - \frac{1}{x^2}}{4 + \frac{4}{x} + \frac{1}{x^2}}$

19. $\dfrac{2 + \frac{5}{a} - \frac{3}{a^2}}{2 - \frac{5}{a} + \frac{2}{a^2}}$

20. $\dfrac{3 + \frac{5}{a} - \frac{2}{a^2}}{3 - \frac{10}{a} + \frac{3}{a^2}}$

21. $\dfrac{2 + \frac{3}{x} - \frac{18}{x^2} - \frac{27}{x^3}}{2 + \frac{9}{x} + \frac{9}{x^2}}$

22. $\dfrac{3 + \frac{5}{x} - \frac{12}{x^2} - \frac{20}{x^3}}{3 + \frac{11}{x} + \frac{10}{x^2}}$

23. $\dfrac{1 + \frac{1}{x + 3}}{1 - \frac{1}{x + 3}}$

24. $\dfrac{1 + \frac{1}{x - 2}}{1 - \frac{1}{x - 2}}$

25. $\dfrac{1 + \frac{1}{x + 3}}{1 + \frac{7}{x - 3}}$

26. $\dfrac{1 + \frac{1}{x - 2}}{1 - \frac{3}{x + 2}}$

27. $\dfrac{1 - \frac{1}{a + 1}}{1 + \frac{1}{a - 1}}$

28. $\dfrac{\frac{1}{a - 1} + 1}{\frac{1}{a + 1} - 1}$

29. $\dfrac{\frac{1}{x + 3} + \frac{1}{x - 3}}{\frac{1}{x + 3} - \frac{1}{x - 3}}$

30. $\dfrac{\frac{1}{x + a} + \frac{1}{x - a}}{\frac{1}{x + a} - \frac{1}{x - a}}$

31. $\dfrac{\frac{y + 1}{y - 1} + \frac{y - 1}{y + 1}}{\frac{y + 1}{y - 1} - \frac{y - 1}{y + 1}}$

32. $\dfrac{\frac{y - 1}{y + 1} - \frac{y + 1}{y - 1}}{\frac{y - 1}{y + 1} + \frac{y + 1}{y - 1}}$

33. $1 - \dfrac{x}{1 - \frac{1}{x}}$

34. $x - \dfrac{1}{x - \frac{1}{2}}$

35. $1 + \dfrac{1}{1 + \dfrac{1}{1 + 1}}$

36. $1 - \dfrac{1}{1 - \dfrac{1}{1 - \frac{1}{2}}}$

37. $\dfrac{1 - \dfrac{1}{x + \frac{1}{2}}}{1 + \dfrac{1}{x + \frac{1}{2}}}$

38. $\dfrac{2 + \dfrac{1}{x - \frac{1}{3}}}{2 - \dfrac{1}{x - \frac{1}{3}}}$

39. The formula $f = \dfrac{ab}{a + b}$ is used in optics to find the focal length of a lens. Show that the formula $f = (a^{-1} + b^{-1})^{-1}$ is equivalent to the preceding formula by rewriting it without the negative exponents and then simplifying the results.

40. Show that the expression $(a^{-1} - b^{-1})^{-1}$ can be simplified to $\dfrac{ab}{b - a}$ by first writing it without the negative exponents and then simplifying the result.

41. Show that the expression $\dfrac{1 - x^{-1}}{1 + x^{-1}}$ can be written as $\dfrac{x - 1}{x + 1}$.

42. Show that the expression $(1 + x^{-1})^{-1}$ can be written as $\dfrac{x}{x + 1}$.

Applying the Concepts

Below is the continued fraction that we mentioned in the introduction to this chapter.

$$1 + \dfrac{1}{1 + \dfrac{1}{1 + \dfrac{1}{1 + \cdots}}}$$

Although it looks similar to some of the complex fractions we studied in this section, we can't simplify it directly because there is no end to it. We can, however, use our work with complex fractions and number sequences to become more familiar with this continued fraction.

43. Simplify the first three terms in the sequence below, then use the results, along with your knowledge of the Fibonacci sequence, to predict the next three terms in the sequence.

$$1 + \dfrac{1}{1 + 1}, 1 + \dfrac{1}{1 + \dfrac{1}{1 + 1}}, 1 + \dfrac{1}{1 + \dfrac{1}{1 + \dfrac{1}{1 + 1}}}, \cdots$$

44. Use a calculator to find a decimal representation for each term in the sequence in Problem 43. Round your answers to the nearest thousandth. You will see these decimals again in Chapter 6 when we discuss the golden rectangle.

45. Simplify the first three terms in the sequence below.

$$1 + \dfrac{1}{2 + 1}, 1 + \dfrac{1}{2 + \dfrac{1}{2 + 1}}, 1 + \dfrac{1}{2 + \dfrac{1}{2 + \dfrac{1}{2 + 1}}}, \cdots$$

46. Use a calculator to find a decimal representation for each term in the sequence in Problem 45. Round your answers to the nearest thousandth. You will see these decimals again in Chapter 6 as approximations to $\sqrt{2}$.

The diagrams on page 248 show part of two electric circuits. The jagged lines represent resistors, which are named that way because they resist the flow of electricity. The resistors that are connected one after the other are said to be connected "in series," while the other resistors are connected "in parallel." For each of the diagrams, R_T is the total resistance between points A and B. As you can see, R_T for the two resistors connected in series is the sum of the individual resistances, while R_T for the resistors connected in parallel is given by a complex fraction.

Resistors Connected in Series

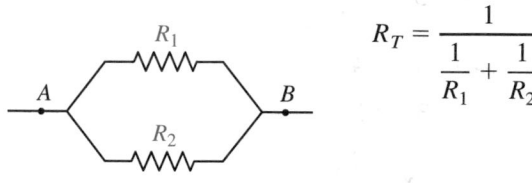

$$R_T = R_1 + R_2$$

Resistors Connected in Parallel

$$R_T = \cfrac{1}{\cfrac{1}{R_1} + \cfrac{1}{R_2}}$$

Suppose that the resistance R_1 is twice resistance R_2.

47. Show that R_T for the resistors connected in series is $3R_2$.

48. Show that R_T for the resistors connected in parallel is $\frac{2}{3}R_2$.

Suppose that the resistance R_1 is three times resistance R_2.

49. Show that R_T for the resistors connected in series is $4R_2$.

50. Show that R_T for the resistors connected in parallel is $\frac{3}{4}R_2$.

51. If $R_1 = kR_2$, where k is a positive integer, show that R_T for the resistors connected in series is $(k + 1)R_2$.

52. If $R_i = kR_2$, where k is a positive integer, show that R_T for the resistors connected in parallel is $\dfrac{k}{k + 1} R_2$.

Review Problems

The following problems review material we covered in Sections 2.1 and 3.9. Reviewing these problems will help you with the next section.

Solve each equation.

53. $3x + 60 = 15$ **54.** $3x - 18 = 4$
55. $3(y - 3) = 2(y - 2)$
56. $5(y + 2) = 4(y + 1)$
57. $10 - 2(x + 3) = x + 1$
58. $15 - 3(x - 1) = x - 2$
59. $x^2 - x - 12 = 0$ **60.** $x^2 + x - 12 = 0$
61. $3x^2 + x - 10 = 0$ **62.** $10x^2 - x - 3 = 0$
63. $(x + 1)(x - 6) = -12$
64. $(x + 1)(x - 4) = -6$

One Step Further

Try both methods shown in this section on each of the following problems. (One of the two methods is much easier than the other.)

65. $\dfrac{1 + \dfrac{1}{x - 3}}{x - \dfrac{10}{x + 3}}$ **66.** $\dfrac{1 + \dfrac{1}{x + 2}}{x - \dfrac{10}{x - 2}}$

67. $\dfrac{1 + \dfrac{1}{x + a}}{1 + \dfrac{2a + 1}{x - a}}$ **68.** $\dfrac{1 + \dfrac{1}{x - a}}{1 - \dfrac{2a - 1}{x + a}}$

Equations Involving Rational Expressions

The first step in solving an equation that contains one or more rational expressions is to find the LCD for all denominators in the equation. We then multiply both sides of the equation by the LCD to clear the equation of all fractions. That is, after we have multiplied through by the LCD, each term in the resulting equation will have a denominator of 1.

▶ **EXAMPLE 1** Solve $\dfrac{x}{2} - 3 = \dfrac{2}{3}$.

Solution The LCD for 2 and 3 is 6. Multiplying both sides by 6, we have

$$6\left(\frac{x}{2} - 3\right) = 6\left(\frac{2}{3}\right)$$

$$6\left(\frac{x}{2}\right) - 6(3) = 6\left(\frac{2}{3}\right)$$

$$3x - 18 = 4$$

$$3x = 22$$

$$x = \frac{22}{3} \qquad \blacktriangleleft$$

Multiplying both sides of an equation by the LCD clears the equation of fractions because the LCD has the property that all the denominators divide it evenly.

▶ **EXAMPLE 2** Solve $\dfrac{6}{a - 4} = \dfrac{3}{8}$.

Solution The LCD for $a - 4$ and 8 is $8(a - 4)$. Multiplying both sides by this quantity yields

$$8(a - 4) \cdot \frac{6}{a - 4} = 8(a - 4) \cdot \frac{3}{8}$$

$$48 = (a - 4) \cdot 3$$

$$48 = 3a - 12$$

$$60 = 3a$$

$$20 = a$$

The solution set is $\{20\}$, which checks in the original equation. ◀

When we multiply both sides of an equation by an expression containing the variable, we must be sure to check our solutions. The multiplication property of equality does not allow multiplication by 0. If the expression we multiply by contains the variable, then it has the possibility of being 0. In the last example we multiplied both sides by $8(a - 4)$. This gives a restriction $a \neq 4$ for any solution we come up with.

▶ **EXAMPLE 3** Solve $\dfrac{x}{x - 2} + \dfrac{2}{3} = \dfrac{2}{x - 2}$.

Solution   The LCD is $3(x - 2)$. We are assuming $x \neq 2$ when we multiply both sides of the equation by $3(x - 2)$:

$$3(x - 2) \cdot \left[\frac{x}{x - 2} + \frac{2}{3} \right] = 3(x - 2) \cdot \frac{2}{x - 2}$$

$$3x + (x - 2) \cdot 2 = 3 \cdot 2$$

$$3x + 2x - 4 = 6$$

$$5x - 4 = 6$$

$$5x = 10$$

$$x = 2$$

The only possible solution is $x = 2$. Checking this value back in the original equation gives

$$\frac{2}{2 - 2} + \frac{2}{3} \stackrel{?}{=} \frac{2}{2 - 2}$$

$$\frac{2}{0} + \frac{2}{3} \stackrel{?}{=} \frac{2}{0}$$

The first and last terms are undefined. The proposed solution, $x = 2$, does not check in the original equation. The solution set is the empty set. There is no solution to the original equation.   ◀

When the proposed solution to an equation is not actually a solution, it is called an **extraneous solution.** In the last example, $x = 2$ is an extraneous solution.

▶ **EXAMPLE 4**   Solve $\dfrac{5}{x^2 - 3x + 2} - \dfrac{1}{x - 2} = \dfrac{1}{3x - 3}$.

Solution   Writing the equation again with the denominators in factored form, we have

$$\frac{5}{(x - 2)(x - 1)} - \frac{1}{x - 2} = \frac{1}{3(x - 1)}$$

The LCD is $3(x - 2)(x - 1)$. Multiplying through by the LCD, we have

$$3(x - 2)(x - 1) \cdot \frac{5}{(x - 2)(x - 1)} - 3(x - 2)(x - 1) \cdot \frac{1}{(x - 2)}$$

$$= 3(x - 2)(x - 1) \cdot \frac{1}{3(x - 1)}$$

$$3 \cdot 5 - 3(x - 1) \cdot 1 = (x - 2) \cdot 1$$

$$15 - 3x + 3 = x - 2$$

$$-3x + 18 = x - 2$$

$$-4x + 18 = -2$$

$$-4x = -20$$

$$x = 5$$

Checking the proposed solution $x = 5$ in the original equation yields a true statement. Try it and see. ◄

▶ **EXAMPLE 5** Solve $3 + \dfrac{1}{x} = \dfrac{10}{x^2}$.

Solution To clear the equation of denominators, we multiply both sides by x^2:

$$x^2\left(3 + \frac{1}{x}\right) = x^2\left(\frac{10}{x^2}\right)$$

$$3(x^2) + \left(\frac{1}{x}\right)(x^2) = \left(\frac{10}{x^2}\right)(x^2)$$

$$3x^2 + x = 10$$

Rewrite in standard form and solve:

$$3x^2 + x - 10 = 0$$

$$(3x - 5)(x + 2) = 0$$

$$3x - 5 = 0 \quad \text{or} \quad x + 2 = 0$$

$$x = \frac{5}{3} \quad \text{or} \quad x = -2$$

The solution set is $\{-2, \frac{5}{3}\}$. Both solutions check in the original equation. Remember: We have to check *all solutions* any time we multiply both sides of the equation by an expression that contains the variable, just to be sure we haven't multiplied by 0. ◄

▶ **EXAMPLE 6** Solve $\dfrac{y - 4}{y^2 - 5y} = \dfrac{2}{y^2 - 25}$.

Solution   Factoring each denominator, we find the LCD is $y(y - 5)(y + 5)$. Multiplying each side of the equation by the LCD clears the equation of denominators and leads us to our possible solutions:

$$y(y - 5)(y + 5) \cdot \frac{y - 4}{y(y - 5)} = \frac{2}{(y - 5)(y + 5)} \cdot y(y - 5)(y + 5)$$

$$(y + 5)(y - 4) = 2y$$

$$y^2 + y - 20 = 2y \qquad \text{Multiply out the left side}$$

$$y^2 - y - 20 = 0 \qquad \text{Add } -2y \text{ to each side}$$

$$(y - 5)(y + 4) = 0$$

$$y - 5 = 0 \quad \text{or} \quad y + 4 = 0$$

$$y = 5 \quad \text{or} \qquad y = -4$$

The two possible solutions are 5 and -4. If we substitute -4 for y in the original equation, we find that it leads to a true statement. It is therefore a solution. On the other hand, if we substitute 5 for y in the original equation, we find that both sides of the equation are undefined. The only solution to our original equation is $y = -4$. The other possible solution $y = 5$ is extraneous.

◀

▶ **EXAMPLE 7**   Solve for y: $x = \dfrac{y - 4}{y - 2}$.

Solution   To solve for y we first multiply each side by $y - 2$ to obtain

$$x(y - 2) = y - 4$$

$$xy - 2x = y - 4 \qquad \text{Distributive property}$$

$$xy - y = 2x - 4 \qquad \text{Collect all terms containing } y \text{ on the left side}$$

$$y(x - 1) = 2x - 4 \qquad \text{Factor } y \text{ from each term on the left side}$$

$$y = \frac{2x - 4}{x - 1} \qquad \text{Divide each side by } x - 1$$

◀

▶ **EXAMPLE 8**   Solve the formula $\dfrac{1}{x} = \dfrac{1}{b} + \dfrac{1}{a}$ for x.

Solution   We begin by multiplying both sides by the least common denominator xab. As you can see from our previous examples, multiplying both sides of an equation by the LCD is equivalent to multiplying each term of both sides by the LCD:

$$xab \cdot \frac{1}{x} = \frac{1}{b} \cdot xab + \frac{1}{a} \cdot xab$$

$$ab = xa + xb$$

$$ab = (a + b)x \qquad \text{Factor } x \text{ from the right side}$$

$$\frac{ab}{a + b} = x$$

We know we are finished because the variable we were solving for is alone on one side of the equation and does not appear on the other side. ◀

PROBLEM SET 4.6

Solve each of the following equations.

1. $\dfrac{x}{5} + 4 = \dfrac{5}{3}$

2. $\dfrac{x}{5} = \dfrac{x}{2} - 9$

3. $\dfrac{a}{3} + 2 = \dfrac{4}{5}$

4. $\dfrac{a}{4} + \dfrac{1}{2} = \dfrac{2}{3}$

5. $\dfrac{y}{2} + \dfrac{y}{4} + \dfrac{y}{6} = 3$

6. $\dfrac{y}{3} - \dfrac{y}{6} + \dfrac{y}{2} = 1$

7. $\dfrac{5}{2x} = \dfrac{1}{x} + \dfrac{3}{4}$

8. $\dfrac{1}{2a} = \dfrac{2}{a} - \dfrac{3}{8}$

9. $\dfrac{1}{x} = \dfrac{1}{3} - \dfrac{2}{3x}$

10. $\dfrac{5}{2x} = \dfrac{2}{x} - \dfrac{1}{12}$

11. $\dfrac{2x}{x - 3} + 2 = \dfrac{2}{x - 3}$

12. $\dfrac{2}{x + 5} = \dfrac{2}{5} - \dfrac{x}{x + 5}$

13. $1 - \dfrac{1}{x} = \dfrac{12}{x^2}$

14. $2 + \dfrac{5}{x} = \dfrac{3}{x^2}$

15. $y - \dfrac{4}{3y} = -\dfrac{1}{3}$

16. $\dfrac{y}{2} - \dfrac{4}{y} = -\dfrac{7}{2}$

17. $\dfrac{x + 2}{x + 1} = \dfrac{1}{x + 1} + 2$

18. $\dfrac{x + 6}{x + 3} = \dfrac{3}{x + 3} + 2$

19. $\dfrac{3}{a - 2} = \dfrac{2}{a - 3}$

20. $\dfrac{5}{a + 1} = \dfrac{4}{a + 2}$

21. $6 - \dfrac{5}{x^2} = \dfrac{7}{x}$

22. $10 - \dfrac{3}{x^2} = -\dfrac{1}{x}$

23. $\dfrac{1}{x - 1} - \dfrac{1}{x + 1} = \dfrac{3x}{x^2 - 1}$

24. $\dfrac{5}{x - 1} + \dfrac{2}{x - 1} = \dfrac{4}{x + 1}$

25. $\dfrac{2}{x - 3} + \dfrac{x}{x^2 - 9} = \dfrac{4}{x + 3}$

26. $\dfrac{2}{x + 5} + \dfrac{3}{x + 4} = \dfrac{2x}{x^2 + 9x + 20}$

27. $\dfrac{3}{2} - \dfrac{1}{x - 4} = \dfrac{-2}{2x - 8}$

28. $\dfrac{2}{x} - \dfrac{1}{x + 1} = \dfrac{-2}{5x + 5}$

29. $\dfrac{t - 4}{t^2 - 3t} = \dfrac{-2}{t^2 - 9}$

30. $\dfrac{t + 3}{t^2 - 2t} = \dfrac{10}{t^2 - 4}$

31. $\dfrac{3}{y - 4} - \dfrac{2}{y + 1} = \dfrac{5}{y^2 - 3y - 4}$

32. $\dfrac{1}{y + 2} - \dfrac{2}{y - 3} = \dfrac{-2y}{y^2 - y - 6}$

33. $\dfrac{2}{1 + a} = \dfrac{3}{1 - a} + \dfrac{5}{a}$

34. $\dfrac{1}{a + 3} - \dfrac{a}{a^2 - 9} = \dfrac{2}{3 - a}$

35. $\dfrac{3}{2x - 6} - \dfrac{x + 1}{4x - 12} = 4$

36. $\dfrac{2x - 3}{5x + 10} + \dfrac{3x - 2}{4x + 8} = 1$

37. $\dfrac{y + 2}{y^2 - y} - \dfrac{6}{y^2 - 1} = 0$

38. $\dfrac{y + 3}{y^2 - y} - \dfrac{8}{y^2 - 1} = 0$

39. $\dfrac{4}{2x - 6} - \dfrac{12}{4x + 12} = \dfrac{12}{x^2 - 9}$

40. $\dfrac{1}{x + 2} + \dfrac{1}{x - 2} = \dfrac{4}{x^2 - 4}$

41. $\dfrac{2}{y^2 - 7y + 12} - \dfrac{1}{y^2 - 9} = \dfrac{4}{y^2 - y - 12}$

42. $\dfrac{1}{y^2 + 5y + 4} + \dfrac{3}{y^2 - 1} = \dfrac{-1}{y^2 + 3y - 4}$

43. Solve the equation $6x^{-1} + 4 = 7$ by multiplying both sides by x. (Remember, $x^{-1} \cdot x = x^{-1} \cdot x^1 = x^0 = 1$.)

44. Solve the equation $3x^{-1} - 5 = 2x^{-1} - 3$ by multiplying both sides by x.

45. Solve the equation $1 + 5x^{-2} = 6x^{-1}$ by multiplying both sides by x^2.

46. Solve the equation $1 + 3x^{-2} = 4x^{-1}$ by multiplying both sides by x^2.

47. Solve the formula $\dfrac{1}{x} = \dfrac{1}{b} - \dfrac{1}{a}$ for x.

48. Solve $\dfrac{1}{x} = \dfrac{1}{a} - \dfrac{1}{b}$ for x.

49. Solve for R in the formula $\dfrac{1}{R} = \dfrac{1}{R_1} + \dfrac{1}{R_2}$.

50. Solve for R in the formula
$$\dfrac{1}{R} = \dfrac{1}{R_1} + \dfrac{1}{R_2} + \dfrac{1}{R_3}.$$

Solve for y.

51. $x = \dfrac{y - 3}{y - 1}$

52. $x = \dfrac{y - 2}{y - 3}$

53. $x = \dfrac{2y + 1}{3y + 1}$

54. $x = \dfrac{3y + 2}{5y + 1}$

Review Problems

The following problems review material we covered in Sections 2.3, 2.4, and 3.9. Reviewing these problems will get you ready for the next section. In each case, be sure to show the equation used.

55. Twice the sum of a number and 3 is 16. Find the number.

56. The sum of two consecutive odd integers is 48. Find the two integers.

57. The length of a rectangle is 3 less than twice the width. The perimeter is 42 meters. Find the length and width.

58. The smallest angle in a triangle is one fourth as large as the largest angle. The third angle is 9° more than the smallest angle. Find the measure of all three angles.

59. The sum of the squares of two consecutive integers is 61. Find the integers.

60. The square of the sum of two consecutive integers is 121. Find the two integers.

61. The lengths of the sides of a right triangle are given by three consecutive integers. Find the lengths of the three sides.

62. The longest side of a right triangle is 8 inches more than the shortest side. The other side is 7 inches more than the shortest side. Find the lengths of the three sides.

One Step Further

Solve each equation.

63. $\dfrac{12}{x} + \dfrac{8}{x^2} - \dfrac{75}{x^3} - \dfrac{50}{x^4} = 0$

64. $\dfrac{45}{x} + \dfrac{18}{x^2} - \dfrac{80}{x^3} - \dfrac{32}{x^4} = 0$

65. $\dfrac{1}{x^3} - \dfrac{1}{3x^2} - \dfrac{1}{4x} + \dfrac{1}{12} = 0$

66. $\dfrac{1}{x^3} - \dfrac{1}{2x^2} - \dfrac{1}{9x} + \dfrac{1}{18} = 0$

Applications

We begin this section with some application problems, the solutions to which involve equations that contain rational expressions. As you will see, the solutions to the examples show only the essential steps from our Blueprint for Problem Solving. Recall that step 1 was done mentally; we read the problem and mentally list the items that are known and the items that are unknown. This is an essential part of problem solving. However, now that you have had experience with application problems, you are doing step 1 automatically.

Also in this section we will look at a method of solving conversion problems that is called **unit analysis.** With unit analysis, we can convert expressions with units of feet per minute to equivalent expressions in miles per hour. This method of converting between different units of measure is used often in chemistry, physics, and engineering classes.

▶ **EXAMPLE 1**   One number is twice another. The sum of their reciprocals is 2. Find the numbers.

Solution   Let x = the smaller number. The larger number is $2x$. Their reciprocals are $\frac{1}{x}$ and $\frac{1}{2x}$. The equation that describes the situation is

$$\frac{1}{x} + \frac{1}{2x} = 2$$

Multiplying both sides by the LCD $2x$, we have

$$2x \cdot \frac{1}{x} + 2x \cdot \frac{1}{2x} = 2x(2)$$

$$2 + 1 = 4x$$

$$3 = 4x$$

$$x = \frac{3}{4}$$

The smaller number is $\frac{3}{4}$. The larger is $2(\frac{3}{4}) = \frac{6}{4} = \frac{3}{2}$. Adding their reciprocals, we have

$$\frac{4}{3} + \frac{2}{3} = \frac{6}{3} = 2$$

The sum of the reciprocals of $\frac{3}{4}$ and $\frac{3}{2}$ is 2.   ◀

▶ **EXAMPLE 2**   The speed of a boat in still water is 20 miles/hour. It takes the same amount of time for the boat to travel 3 miles downstream (with the current) as it does to travel 2 miles upstream (against the current). Find the speed of the current.

We treat the units common to the numerator and denominator in the same way we treat factors common to the numerator and denominator: we divide out common units, just as we divide out common factors. In the expression above, we have feet common to the numerator and denominator. Dividing them out leaves us with miles only. Here is the complete problem.

$$65{,}000 \text{ feet} = \frac{65{,}000 \text{ feet}}{1} \cdot \frac{1 \text{ mile}}{5{,}280 \text{ feet}}$$

$$= \frac{65{,}000}{5{,}280} \text{ mile}$$

$$= 12.3 \text{ miles to the nearest tenth of a mile}$$

The key to solving a problem like this one lies in choosing the appropriate conversion factor. The fact that 1 mile = 5,280 feet yields two conversion factors, each of which is equal to the number 1. They are

$$\frac{1 \text{ mile}}{5{,}280 \text{ feet}} \quad \text{and} \quad \frac{5{,}280 \text{ feet}}{1 \text{ mile}}$$

The conversion factor we choose will depend on the units we are given and the units with which we want to end up. Multiplying any expression by either of the two conversion factors will leave the value of the original expression unchanged because each of the conversion factors is simply the number 1.

▶ **EXAMPLE 6**   In Example 8 of Section 4.1 we found a rider on the first Ferris wheel was traveling at approximately 39.3 feet per minute. Convert 39.3 feet per minute to miles per hour.

Solution   We know that 5,280 feet = 1 mile and 60 minutes = 1 hour. Therefore, we have the following conversion factors, each of which is equal to 1.

$$\frac{5{,}280 \text{ feet}}{1 \text{ mile}} \qquad \frac{1 \text{ mile}}{5{,}280 \text{ feet}} \qquad \frac{60 \text{ minutes}}{1 \text{ hour}} \qquad \frac{1 \text{ hour}}{60 \text{ minutes}}$$

The conversion factors we choose to multiply by are the ones that will allow us to divide out the units we are converting from and leave us with the units we are converting to. Specifically, we want to get rid of feet and be left with miles. Likewise, we want to get rid of minutes and be left with hours. Here is the conversion process that will accomplish these goals:

$$39.3 \text{ feet/minute} = \frac{39.3 \text{ feet}}{1 \text{ minute}} \cdot \frac{1 \text{ mile}}{5{,}280 \text{ feet}} \cdot \frac{60 \text{ minutes}}{1 \text{ hour}}$$

$$= \frac{39.3 \cdot 60 \text{ miles}}{5{,}280 \text{ hours}}$$

$$= 0.45 \text{ mile/hour} \quad \text{to the nearest hundredth} \qquad ◀$$

▶ **EXAMPLE 7** In 1993 a ski resort in Vermont advertised their new high-speed chair lift as "the world's fastest chair lift with a speed of 1,100 feet per second." Show why the speed cannot be correct.

Solution To solve this problem, we can convert feet per second into miles per hour, a unit of measure we are more familiar with on an intuitive level.

$$1{,}100 \text{ feet/second} = \frac{1{,}100 \text{ feet}}{1 \text{ second}} \cdot \frac{1 \text{ mile}}{5{,}280 \text{ feet}} \cdot \frac{60 \text{ seconds}}{1 \text{ minute}} \cdot \frac{60 \text{ minutes}}{1 \text{ hour}}$$

$$= \frac{1{,}100 \cdot 60 \cdot 60 \text{ miles}}{5{,}280 \text{ hours}}$$

$$= 750 \text{ miles/hour}$$

Obviously there is a mistake in the advertisement. ◀

PROBLEM SET 4.7

Solve each of the following word problems. Be sure to show the equation in each case.

Number Problems

1. One number is 3 times another. The sum of their reciprocals is $\frac{20}{3}$. Find the numbers.
2. One number is 3 times another. The sum of their reciprocals is $\frac{4}{9}$. Find the numbers.
3. The sum of a number and its reciprocal is $\frac{10}{3}$. Find the number.
4. The sum of a number and twice its reciprocal is $\frac{27}{5}$. Find the number.
5. The sum of the reciprocals of two consecutive integers is $\frac{7}{12}$. Find the two integers.
6. Find two consecutive even integers, the sum of whose reciprocals is $\frac{3}{4}$.
7. If a certain number is added to the numerator and denominator of $\frac{7}{9}$, the result is $\frac{5}{6}$. Find the number.
8. Find the number you would add to both the numerator and denominator of $\frac{8}{11}$ so the result would be $\frac{6}{7}$.

Rate Problems

9. The speed of a boat in still water is 5 miles per hour. If the boat travels 3 miles downstream in the same amount of time it takes to travel 1.5 miles upstream, what is the speed of the current?
10. A boat, which moves at 18 miles/hour in still water, travels 14 miles downstream in the same amount of time it takes to travel 10 miles upstream. Find the speed of the current.
11. The current of a river is 2 miles/hour. A boat travels to a point 8 miles upstream and back again in 3 hours. What is the speed of the boat in still water?
12. A motorboat travels at 4 miles/hour in still water. It goes 12 miles upstream and 12 miles back again in a total of 8 hours. Find the speed of the current of the river.
13. Train A has a speed 15 miles/hour greater than that of train B. If train A travels 150 miles in the same time train B travels 120 miles, what are the speeds of the two trains?
14. A train travels 30 miles/hour faster than a car. If the train covers 120 miles in the same time the car covers 80 miles, what is the speed of each of them?
15. A small airplane flies 810 miles from Los Angeles to Portland, OR, with an average speed of 270 miles/hour. An hour and a half after the plane leaves, a Boeing 747 leaves Los Angeles

for Portland. Both planes arrive in Portland at the same time. What was the average speed of the 747?

16. Lou leaves for a cross country excursion on a bicycle traveling at 20 miles/hour. His friends are driving the trip and will meet him at several rest stops along the way. The first stop is scheduled 30 miles from the original starting point. If the people driving leave 15 minutes after Lou from the same place, how fast will they have to drive to reach the first rest stop at the same time as Lou?

17. A tour bus leaves Sacramento every Friday evening at 5:00 for a 270-mile trip to Las Vegas. This week, however, the bus leaves at 5:30 P.M. In order to arrive in Las Vegas on time, the driver drives 6 miles/hour faster than usual. What is the bus's usual speed?

18. A bakery delivery truck leaves the bakery at 5:00 A.M. each morning on its 140-mile route. One day the driver gets a late start and does not leave the bakery until 5:30 A.M. In order to finish her route on time the driver drives 5 miles per hour faster than usual. At what speed does she usually drive?

Work Problems

19. If Sam can do a certain job in 3 days, while it takes Fred 6 days to do the same job, how long will it take them, working together, to complete the job?

20. Tim can finish a certain job in 10 hours. It takes his wife JoAnn only 8 hours to do the same job. If they work together, how long will it take them to complete the job?

21. Two people working together can complete a job in 6 hours. If one of them works twice as fast as the other, how long would it take the faster person, working alone, to do the job?

22. If two people working together can do a job in 3 hours, how long will it take the slower person to do the same job if one of them is 3 times as fast as the other?

23. A water tank can be filled by an inlet pipe in 8 hours. It takes twice that long for the outlet pipe

to empty the tank. How long will it take to fill the tank if both pipes are open?

24. A sink can be filled from the faucet in 5 minutes. It takes only 3 minutes to empty the sink when the drain is open. If the sink is full and both the faucet and the drain are open, how long will it take to empty the sink?

25. It takes 10 hours to fill a pool with the inlet pipe. It can be emptied in 15 hours with the outlet pipe. If the pool is half full to begin with, how long will it take to fill it from there if both pipes are open?

26. A sink is $\frac{1}{4}$ full when both the faucet and the drain are opened. The faucet alone can fill the sink in 6 minutes, while it takes 8 minutes to empty it with the drain. How long will it take to fill the remaining $\frac{3}{4}$ of the sink?

27. A sink has two faucets, one for hot water and one for cold water. The sink can be filled by the cold-water faucet in 3.5 minutes. If both faucets are open, the sink is filled in 2.1 minutes. How long does it take to fill the sink with just the hot-water faucet open?

28. A water tank is being filled by two inlet pipes. Pipe A can fill the tank in $4\frac{1}{2}$ hours, while both pipes together can fill the tank in 2 hours. How long does it take to fill the tank using only pipe B?

Unit Analysis Problems

Give your answers to the following problems to the nearest tenth.

29. The South Coast Shopping Mall in Costa Mesa, California, covers an area of 2,224,750 square feet. If 1 acre = 43,560 square feet, how many acres does the South Coast Shopping Mall cover?

30. The relationship between liters and cubic inches, both of which are measures of volume, is 0.0164 liters = 1 cubic inch. If a Ford Mustang has a motor with a displacement of 4.9 liters, what is the displacement in cubic inches?

31. The Forest chair lift at the Northstar ski resort in Lake Tahoe is 5,750 feet long. If a ride on this chair lift takes 11 minutes, what is the average speed of the lift in miles per hour?

32. The Bear Paw chair lift at the Northstar ski resort in Lake Tahoe is 790 feet long. If a ride on this chair lift takes 2.2 minutes, what is the average speed of the lift in miles per hour?

33. A sprinter runs 100 meters in 10.8 seconds. What is the sprinter's average speed in miles per hour? (1 meter = 3.28 feet)

34. A runner covers 400 meters in 49.8 seconds. What is the average speed of the runner in miles per hour?

35. A person riding a Ferris wheel with a diameter of 65 feet travels once around the wheel in 30 seconds. What is the average speed of the rider in miles per hour?

36. A person riding a Ferris wheel with a diameter of 102 feet travels once around the wheel in 3.5 minutes. What is the average speed of the rider in miles per hour?

37. A $3\frac{1}{2}$-inch diskette, when placed in the disk drive of a computer, rotates at 300 rpm (meaning one revolution takes 1/300 minute). Find the average speed of a point 2 inches from the center of the diskette in miles per hour.

38. A 5-inch fixed disk in a computer rotates at 3,600 rpm. Find the average speed of a point 2 inches from the center of the disk in miles per hour.

Review Problems

The problems that follow review material we covered in Sections 4.3, 4.4, 4.5, and 4.6. Reviewing these problems will help clarify the different methods we have used in this chapter.

Perform the indicated operations. [4.3, 4.4]

39. $\dfrac{2a + 10}{a^3} \cdot \dfrac{a^2}{3a + 15}$

40. $\dfrac{4a + 8}{a^2 - a - 6} \div \dfrac{a^2 + 7a + 12}{a^2 - 9}$

41. $(x^2 - 9)\left(\dfrac{x + 2}{x + 3}\right)$

42. $\dfrac{1}{x + 4} + \dfrac{8}{x^2 - 16}$

43. $\dfrac{2x - 7}{x - 2} - \dfrac{x - 5}{x - 2}$ **44.** $2 + \dfrac{25}{5x - 1}$

Simplify each expression. [4.5]

45. $\dfrac{\dfrac{1}{x} - \dfrac{1}{3}}{\dfrac{1}{x} + \dfrac{1}{3}}$

46. $\dfrac{1 - \dfrac{9}{x^2}}{1 - \dfrac{1}{x} - \dfrac{6}{x^2}}$

Solve each equation. [4.6]

47. $\dfrac{x}{x - 3} + \dfrac{3}{2} = \dfrac{3}{x - 3}$

48. $1 - \dfrac{3}{x} = \dfrac{-2}{x^2}$

CHAPTER 4 SUMMARY

Examples

1. $\frac{3}{4}$ is a rational number.

$\dfrac{x - 3}{x^2 - 9}$ is a rational expression.

Rational Numbers and Expressions [4.1]

A *rational number* is any number that can be expressed as the ratio of two integers:

$$\text{Rational numbers} = \left\{\dfrac{a}{b} \;\middle|\; a \text{ and } b \text{ are integers}, b \neq 0\right\}$$

This makes no sense at all. The numerator and denominator must be factored completely before any factors they have in common can be recognized:

$$\frac{x^2 - 9x + 20}{x^2 - 3x - 10} = \frac{(x - 5)(x - 4)}{(x - 5)(x + 2)}$$

$$= \frac{x - 4}{x + 2}$$

2. Forgetting to check solutions to equations involving rational expressions. When we multiply both sides of an equation by a quantity containing the variable, we must be sure to check for extraneous solutions (see Section 4.6).

CHAPTER 4 REVIEW

Reduce to lowest terms. [4.1]

1. $\dfrac{125x^4yz^3}{35x^2y^4z^3}$

2. $\dfrac{28xy^3z^2}{14x^2y^3z}$

3. $\dfrac{a^3 - ab^2}{4a + 4b}$

4. $\dfrac{12a - 9b}{16a^2 - 9b^2}$

5. $\dfrac{x^2 - 25}{x^2 + 10x + 25}$

6. $\dfrac{x^2 - 9x + 20}{x^2 - 7x + 12}$

7. $\dfrac{ax + x - 5a - 5}{ax - x - 5a + 5}$

8. $\dfrac{x^3 + 2x^2 - 9x - 18}{x^2 - x - 6}$

9. $\dfrac{6 - x - x^2}{x^2 - 5x + 6}$

10. $\dfrac{x^2 + 5x - 6}{6 - 5x - x^2}$

Divide. If the denominator is a factor of the numerator, as in Problem 17, you may want to factor the numerator and divide out the common factor. [4.2]

11. $\dfrac{12x^3 + 8x^2 + 16x}{4x^2}$

12. $\dfrac{10x^4 - 5x^3 + 15x^2}{5x^2}$

13. $\dfrac{27a^2b^3 - 15a^3b^2 + 21a^4b^4}{-3a^2b^2}$

14. $\dfrac{18a^4b^2 - 9a^2b^2 + 27a^2b^4}{-9a^2b^2}$

15. $\dfrac{x^{6n} - x^{5n}}{x^{3n}}$

16. $\dfrac{10x^{3n} - 15x^{4n}}{5x^n}$

17. $\dfrac{x^2 - x - 6}{x - 3}$

18. $\dfrac{x^2 - x - 6}{x + 2}$

19. $\dfrac{5x^2 - 14xy - 24y^2}{x - 4y}$

20. $\dfrac{5x^2 - 26xy - 24y^2}{5x + 4y}$

21. $\dfrac{y^4 - 16}{y - 2}$

22. $\dfrac{y^4 - 81}{y - 3}$

23. $\dfrac{8x^2 - 26x - 9}{2x - 7}$

24. $\dfrac{9x^2 + 9x - 18}{3x - 4}$

25. $\dfrac{2y^3 - 9y^2 - 17y + 39}{2y - 3}$

26. $\dfrac{3y^3 - 19y^2 + 17y + 4}{3y - 4}$

27. $\dfrac{a^4 - 2a + 5}{a - 2}$

28. $\dfrac{a^4 + a^3 - 1}{a + 2}$

Multiply and divide as indicated. [4.3]

29. $\frac{3}{4} \cdot \frac{12}{15} \div \frac{1}{3}$

30. $\frac{2}{3} \cdot \frac{9}{8} \div \frac{1}{4}$

31. $\dfrac{15x^2y}{8xy^2} \div \dfrac{10xy}{4x}$

32. $\dfrac{27x^3y^2}{13x^2y^4} \div \dfrac{9xy}{26y}$

33. $\dfrac{x^3 - 1}{x^4 - 1} \cdot \dfrac{x^2 - 1}{x^2 + x + 1}$

34. $\dfrac{x^4 - 16}{x^3 - 8} \cdot \dfrac{x^2 + 2x + 4}{x^2 + 4}$

35. $\dfrac{a^2 + 5a + 6}{a + 1} \cdot \dfrac{a + 5}{a^2 + 2a - 3} \div \dfrac{a^2 + 7a + 10}{a^2 - 1}$

36. $\dfrac{2a^2 - a - 1}{a^2 - 2a - 15} \cdot \dfrac{a^2 - 9}{a - 1} \div \dfrac{4a^2 - 1}{a - 5}$

37. $\dfrac{ax + bx + 2a + 2b}{ax - 3a + bx - 3b} \div \dfrac{ax - bx - 2a + 2b}{ax - bx - 3a + 3b}$

38. $\dfrac{xy + 2x + y + 2}{xy - 3y + 2x - 6} \cdot \dfrac{xy - 3y + 4x - 12}{xy - y + 4x - 4}$

39. $(4x^2 - 9) \cdot \dfrac{x + 3}{2x + 3}$

40. $(9x^2 - 25) \cdot \dfrac{x + 5}{3x - 5}$

41. $\dfrac{2x^3 + 3x^2 - 18x - 27}{10x^2 + 13x - 3} \cdot \dfrac{25x^2 - 1}{5x^2 - 14x - 3}$

42. $\dfrac{3x^3 + 2x^2 - 12x - 8}{12x^2 + 5x - 2} \cdot \dfrac{16x^2 - 1}{4x^2 + 9x + 2}$

Add and subtract as indicated. [4.4]

43. $\frac{3}{5} - \frac{1}{10} + \frac{8}{15}$

44. $\frac{3}{4} - \frac{1}{8} + \frac{3}{2}$

45. $\dfrac{5}{x - 5} - \dfrac{x}{x - 5}$

46. $\dfrac{y}{x^2 - y^2} - \dfrac{x}{x^2 - y^2}$

47. $\dfrac{1}{x} + \dfrac{1}{x^2} + \dfrac{1}{x^3}$

48. $\dfrac{1}{x} - \dfrac{1}{x^2} - \dfrac{1}{x^3}$

49. $\dfrac{8}{y^2 - 16} - \dfrac{7}{y^2 - y - 12}$

50. $\dfrac{6}{y^2 - 9} - \dfrac{5}{y^2 - y - 6}$

51. $\dfrac{x - 2}{x^2 + 5x + 4} - \dfrac{x - 4}{2x^2 + 12x + 16}$

52. $\dfrac{x - 1}{x^2 + 4x + 3} - \dfrac{x - 1}{2x^2 + 10x + 12}$

53. $3 + \dfrac{4}{5x - 2}$

54. $7 - \dfrac{2}{7x - 3}$

55. $\dfrac{-4}{x^2 + 5x + 6} - \dfrac{5}{x^2 + 7x + 12} + \dfrac{10}{x^2 + 6x + 8}$

56. $\dfrac{3}{x^2 + 8x + 15} - \dfrac{1}{x^2 + 7x + 12} - \dfrac{1}{x^2 + 9x + 20}$

57. $\dfrac{5}{3x^2 + x - 2} - \dfrac{1}{3x^2 + 5x + 2}$

58. $\dfrac{7}{4x^2 - x - 3} - \dfrac{1}{4x^2 - 7x + 3}$

Simplify each complex fraction. [4.5]

59. $\dfrac{1 + \frac{2}{3}}{1 - \frac{2}{3}}$

60. $\dfrac{1 - \frac{3}{4}}{1 + \frac{3}{4}}$

61. $\dfrac{\frac{4a}{2a^3 + 2}}{\frac{8a}{4a + 4}}$

62. $\dfrac{\frac{2a}{3a^3 - 3}}{\frac{4a}{6a - 6}}$

63. $1 + \dfrac{1}{x + \frac{1}{x}}$

64. $1 + \dfrac{x}{1 + \frac{1}{x}}$

65. $\dfrac{1 - \frac{9}{x^2}}{1 - \frac{1}{x} - \frac{6}{x^2}}$

66. $\dfrac{4 - \frac{1}{x^2}}{4 + \frac{4}{x} + \frac{1}{x^2}}$

Solve each equation. [4.6]

67. $\dfrac{3}{x - 1} = \dfrac{3}{5}$

68. $\dfrac{2}{x + 1} = \dfrac{4}{5}$

69. $\dfrac{x + 1}{3} + \dfrac{x - 3}{4} = \dfrac{1}{6}$

70. $\dfrac{x + 2}{3} + \dfrac{x - 5}{5} = -\dfrac{3}{5}$

71. $\dfrac{5}{y+1} = \dfrac{4}{y+2}$ **72.** $\dfrac{3}{y-2} = \dfrac{2}{y-3}$

73. $\dfrac{x+6}{x+3} - 2 = \dfrac{3}{x+3}$

74. $\dfrac{x+2}{x+1} - 2 = \dfrac{1}{x+1}$

75. $\dfrac{4}{x^2-x-12} + \dfrac{1}{x^2-9} = \dfrac{2}{x^2-7x+12}$

76. $\dfrac{1}{x^2+3x-4} + \dfrac{3}{x^2-1} = \dfrac{-1}{x^2+5x+4}$

77. $\dfrac{a+4}{a^2+5a} = \dfrac{-2}{a^2-25}$

78. $\dfrac{a}{2} + \dfrac{3}{a-3} = \dfrac{a}{a-3}$

79. $\dfrac{3x}{x-5} - \dfrac{2x}{x+1} = \dfrac{-42}{x^2-4x-5}$

80. $\dfrac{2x}{x+2} = \dfrac{x}{x+3} - \dfrac{3}{x^2+5x+6}$

81. $1 - \dfrac{1}{x} = \dfrac{6}{x^2}$ **82.** $2 - \dfrac{11}{x} = \dfrac{-12}{x^2}$

Solve each application. [4.7]

83. One number is twice another. The sum of their reciprocals is $\frac{1}{2}$. Find the two numbers.

84. One number is 4 times another. The sum of their reciprocals is $\frac{5}{8}$. Find the numbers.

85. A bathtub can be filled by the cold water faucet in 10 minutes and by the hot water faucet in 12 minutes. How long does it take to fill the tub if both faucets are left open?

86. A water faucet can fill a sink in 6 minutes, while the drain can empty it in 4 minutes. If the sink is full, how long will it take to empty if both the water faucet and the drain are open?

87. The sum of a number and twice its reciprocal is $\frac{9}{2}$. Find the number.

88. The sum of a number and three times its reciprocal is 4. Find the number.

89. A boat takes 1.5 hours to travel 6 miles downstream and the same distance back. If the boat travels at 9 mph, what is the speed of the current?

90. A car makes a 120-mile trip 10 miles/hour faster than a truck. The truck takes 2 hours longer to make the trip. What are the speeds of the car and the truck?

91. A jogger covers 3.5 miles in 28 minutes. Find the average speed of the jogger in miles per hour.

92. The speed of sound is 1,088 feet per second. Convert the speed of sound to miles per hour. Round your answer to the nearest whole number.

CHAPTER 4 TEST

Reduce to lowest terms. [4.1]

1. $\dfrac{x^2 - y^2}{x - y}$

2. $\dfrac{2x^2 - 5x + 3}{2x^2 - x - 3}$

Divide. [4.2]

3. $\dfrac{24x^3y + 12x^2y^2 - 16xy^3}{4xy}$

4. $\dfrac{2x^3 - 9x^2 + 10}{2x - 1}$

Multiply and divide as indicated. [4.3]

5. $\dfrac{a^2 - 16}{5a - 15} \cdot \dfrac{10(a-3)^2}{a^2 - 7a + 12}$

6. $\dfrac{a^4 - 81}{a^2 + 9} \div \dfrac{a^2 - 8a + 15}{4a - 20}$

7. $\dfrac{x^3 - 8}{2x^2 - 9x + 10} \div \dfrac{x^2 + 2x + 4}{2x^2 + x - 15}$

Add and subtract as indicated. [4.4]

8. $\frac{4}{21} + \frac{6}{35}$ **9.** $\frac{3}{4} - \frac{1}{2} + \frac{5}{8}$

10. $\dfrac{a}{a^2 - 9} + \dfrac{3}{a^2 - 9}$ **11.** $\dfrac{1}{x} + \dfrac{2}{x - 3}$

12. $\dfrac{4x}{x^2 + 6x + 5} - \dfrac{3x}{x^2 + 5x + 4}$

13. $\dfrac{2x + 8}{x^2 + 4x + 3} - \dfrac{x + 4}{x^2 + 5x + 6}$

Simplify each complex fraction. [4.5]

14. $\dfrac{3 - \dfrac{1}{a + 3}}{3 + \dfrac{1}{a + 3}}$ **15.** $\dfrac{1 - \dfrac{9}{x^2}}{1 + \dfrac{1}{x} - \dfrac{6}{x^2}}$

Solve each of the following equations. [4.6]

16. $\dfrac{1}{x} + 3 = \dfrac{4}{3}$

17. $\dfrac{x}{x - 3} + 3 = \dfrac{3}{x - 3}$

18. $\dfrac{y + 3}{2y} + \dfrac{5}{y - 1} = \dfrac{1}{2}$

19. $1 - \dfrac{1}{x} = \dfrac{6}{x^2}$

Solve the following applications. Be sure to show the equation in each case. [4.7]

20. What number must be subtracted from the denominator of $\frac{10}{23}$ to make the result $\frac{1}{3}$?

21. The current of a river is 2 miles/hour. It takes a motorboat a total of 3 hours to travel 8 miles upstream and return 8 miles downstream. What is the speed of the boat in still water?

22. An inlet pipe can fill a pool in 10 hours while an outlet pipe can empty it in 15 hours. If the pool is half full and both pipes are left open, how long will it take to fill the pool the rest of the way?

23. The top of Mount Whitney, the highest point in California, is 14,494 feet above sea level. Give this height in miles to the nearest tenth of a mile.

24. A bullet fired from a gun travels a distance of 4,750 feet in 3.2 seconds. Find the average speed of the bullet in miles per hour. Round to the nearest whole number.

CHAPTER 5

EQUATIONS AND INEQUALITIES IN TWO VARIABLES

INTRODUCTION

Below are three problems that we have encountered in previous chapters. The solution to each problem is connected to an equation in two variables. The figures below each of the problems give a visual representation of the information contained in each equation.

From Section 2.2 An art supply store finds that they can sell x sketch pads each week at p dollars each, according to the equation $x = 900 - 300p$.

From Section 3.4 The revenue R obtained by selling x sketch pads in one week at p dollars each is given by the equation $R = (900 - 300p)p$.

From Section 4.1 A person riding on a Ferris wheel with a diameter of 250 feet will have an average speed r if one trip around the wheel takes t minutes according to the equation $r = \dfrac{785}{t}$.

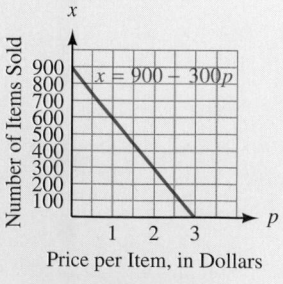

Figure 1

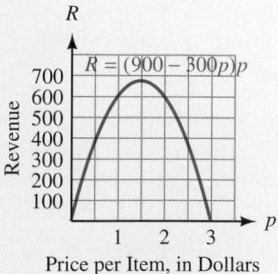

Figure 2

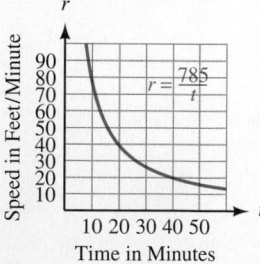

Figure 3

Each of the curves shown in Figures 1, 2, and 3 is drawn on a **rectangular coordinate system.** A rectangular coordinate system allows us to connect algebra and geometry by associating geometric shapes (the curves shown in the diagrams) with algebraic equations. The French philosopher and mathematician

270

René Descartes (1596–1650) is the person usually credited with the invention of the rectangular coordinate system. As a philosopher, Descartes is responsible for the statement "I think, therefore I am." Until Descartes invented his coordinate system in 1637, algebra and geometry were treated as separate subjects.

Overview

Most of this chapter is concerned with the graphs of linear equations in two variables. Equations of this type have graphs that are straight lines. The graph shown in Figure 1 is an example. We begin by finding the graph of a line from its equation. Then we define some important characteristics of lines: their slope and intercepts. We conclude our work with straight lines by finding the equation of a line in a variety of situations. Later in the chapter we look at graphs of inequalities in two variables, as well as functions, function notation, and inverse and direct variation. The background material needed for this chapter is in Chapter 2. You need to know how to solve a linear equation in one variable.

Study Skills

The study skills for this chapter are about attitude. They are points of view that point toward success.

1. **Be Focused, Not Distracted**   I have students who begin their assignments by asking themselves "Why am I taking this class?" If you are asking yourself similar questions, you are distracting yourself away from doing the things that will produce the results you want in this course. Don't dwell on questions and evaluations of the class that can be used as excuses for not doing well. If you want to succeed in this course, focus your energy and efforts toward success, rather than distracting yourself away from your goals.
2. **Be Resilient**   Don't let setbacks keep you from your goals. You want to put yourself on the road to becoming a person who can succeed in this class, or any class in college. Failing a test or quiz, or having a difficult time on some topics, is normal. No one goes through college without some setbacks. Don't let a temporary disappointment keep you from succeeding in this course. A low grade on a test or quiz is simply a signal that some reevaluation of your study habits needs to take place.
3. **Intend to Succeed**   I have a few students who simply go through the motions of studying without intending on mastering the material. It is more important to them to look like they are studying than to actually study. You need to study with the intention of being successful in the course. Intend to master the material, no matter what it takes.

Graphing in Two Dimensions

In Section 2.4 we found a number of values for the length and width of a rectangle made from a 12-inch piece of string. If we let x represent the width of the rectangle and y represent the length, then the equation that gives the length for any width is $y = 6 - x$. Here is a table similar to the one shown in Section 2.4.

TABLE 1   Length and Width of a Rectangle

Width x	Rule $y = 6 - x$	Length y
0	$y = 6 - 0$	6
1	$y = 6 - 1$	5
2	$y = 6 - 2$	4
3	$y = 6 - 3$	3
4	$y = 6 - 4$	2
5	$y = 6 - 5$	1
6	$y = 6 - 6$	0

The third row of the table indicates that $x = 2$ and $y = 4$ are a pair of numbers that satisfy the equation $y = 6 - x$.

In order to discuss solutions to any equation that contains two variables, we need to expand the concept of paired data. We do so with the following definition.

DEFINITION   A pair of numbers enclosed in parentheses and separated by a comma, such as $(-2, 1)$, is called an **ordered pair** of numbers. The first number in the pair is called the **x-coordinate** of the ordered pair, while the second number is called the **y-coordinate.** For the ordered pair $(-2, 1)$, the x-coordinate is -2 and the y-coordinate is 1.

Applying our definition of an ordered pair to the numbers in Table 1, we find that the table gives us the ordered pairs $(0, 6)$, $(1, 5)$, $(2, 4)$, $(3, 3)$, $(4, 2)$, $(5, 1)$, and $(6, 0)$ as solutions to the equation $y = 6 - x$.

Up to this point, we have used histograms, scatter diagrams, and line graphs to give a visual representation to paired data. Our next step is to standardize the way in which we display paired data visually. We do so with the **rectangular coordinate system,** which we discuss next.

A rectangular coordinate system is made by drawing two real number lines at right angles to each other. The two number lines, called **axes,** cross each other at 0. This point is called the **origin.** Positive directions are to the right and up. Negative directions are down and to the left. The rectangular coordinate system is shown in Figure 1.

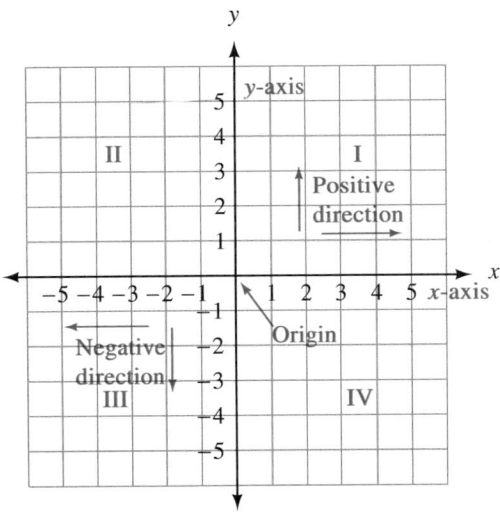

Figure 1

The horizontal number line is called the ***x*-axis** and the vertical number line is called the ***y*-axis.** The two number lines divide the coordinate system into four **quadrants,** which we number I through IV in a counterclockwise direction. Points on the axes are not considered as being in any quadrant.

To graph the ordered pair (a, b) on a rectangular coordinate system we start at the origin and move a units right or left (right if a is positive, left if a is negative). Then we move b units up or down (up if b is positive and down if b is negative). The point where we end up is the graph of the ordered pair (a, b).

▶ **EXAMPLE 1** Plot (graph) the ordered pairs $(2, 5)$, $(-2, 5)$, $(-2, -5)$, and $(2, -5)$.

Solution To graph the ordered pair $(2, 5)$, we start at the origin and move 2 units to the right, then 5 units up. We are now at the point whose coordinates are $(2, 5)$. We graph the other three ordered pairs in a similar manner (see Figure 2 on the next page).

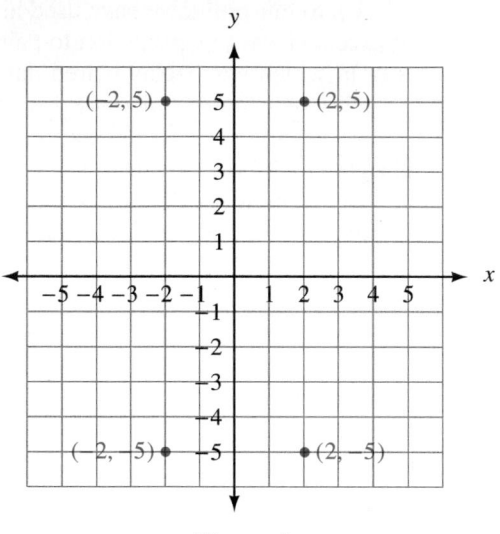

Figure 2   ◀

Note   From Example 1 we see that any point in quadrant I has both its *x*- and *y*-coordinates positive (+, +). Points in quadrant II have negative *x*-coordinates and positive *y*-coordinates (−, +). In quadrant III both coordinates are negative (−, −). In quadrant IV the form is (+, −).

▶ **EXAMPLE 2**   Graph the ordered pairs $(1, -3)$, $(\frac{1}{2}, 2)$, $(3, 0)$, $(0, -2)$, $(-1, 0)$, and $(0, 5)$.

Solution

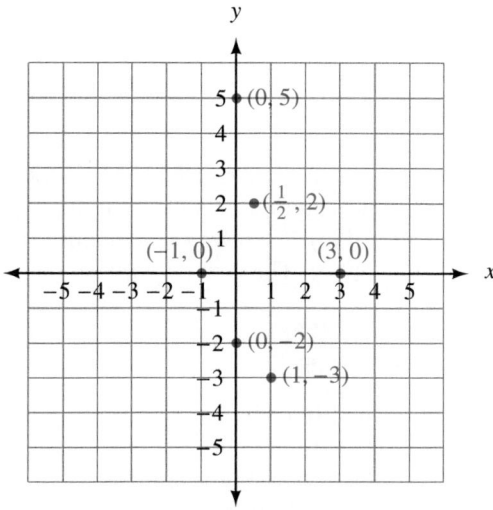

Figure 3   ◀

From Example 2 we see that any point on the x-axis has a y-coordinate of 0 (it has no vertical displacement), and any point on the y-axis has an x-coordinate of 0 (no horizontal displacement).

DEFINITION Any equation that can be put in the form $ax + by = c$, where a, b, and c are real numbers and a and b are not both 0, is called a **linear equation** in two variables. The graph of any equation of this form is a straight line (that is why these equations are called ''linear''). The form $ax + by = c$ is called *standard form.*

To graph a linear equation in two variables, we simply graph its solution set. That is, we draw a line through all the points whose coordinates satisfy the equation. For example, to graph the relationship between the variables x and y in the equation $y = 6 - x$ that we discussed at the beginning of this section, we simply plot each of the ordered pairs obtained in Table 1 and draw a line through them, as shown in Figure 4.

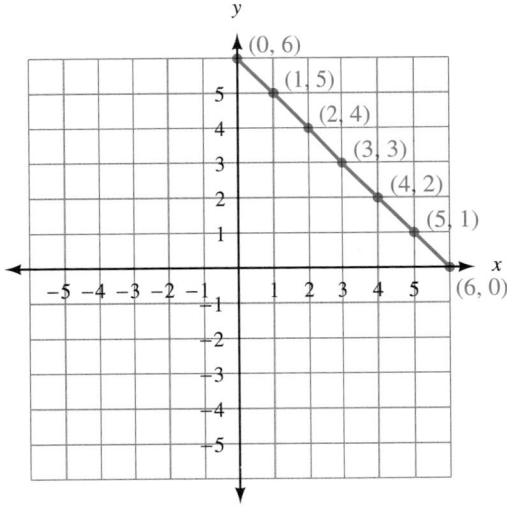

Figure 4

Note that, in this case, we have restricted our graph to the first quadrant because x and y represent the width and length of a rectangle and those quantities cannot be negative. When we graph linear equations in two variables that are not associated with objects, we are free to use negative numbers for the variables. Our next example illustrates.

▶ **EXAMPLE 3** Graph $y = 2x - 3$.

Solution Since $y = 2x - 3$ can be put in the form $ax + by = c$, it is a linear equation in two variables. Hence, the graph of its solution set is a straight line.

We can find some specific solutions by substituting numbers for x and then solving for the corresponding values of y. We are free to choose any convenient numbers for x, so let's use the numbers -1, 0, and 2:

When	$x = -1$	In table form
the equation	$y = 2x - 3$	
becomes	$y = 2(-1) - 3$	
	$y = -5$	

In table form

x	y
-1	-5
0	-3
2	1

The ordered pair $(-1, -5)$ is a solution.

When	$x = 0$
we have	$y = 2(0) - 3$
	$y = -3$

The ordered pair $(0, -3)$ is also a solution.

Using	$x = 2$
we have	$y = 2(2) - 3$
	$y = 1$

The ordered pair $(2, 1)$ is another solution.

Note   It actually takes only two points to determine a straight line. We have included a third point for "insurance." If all three points do not line up in a straight line, we have made a mistake.

Graphing these three ordered pairs and drawing a line through them, we have the graph of $y = 2x - 3$:

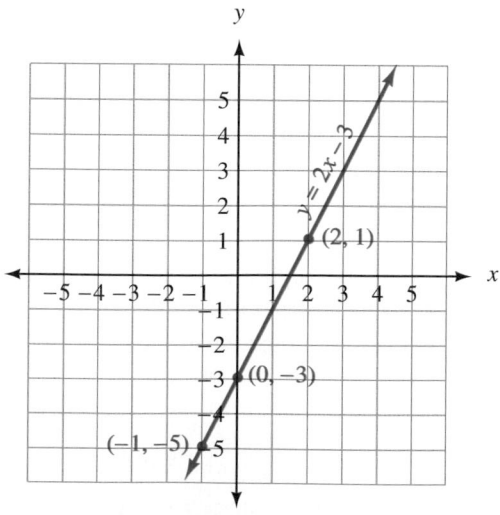

Figure 5

Example 3 illustrates again the connection between algebra and geometry that we mentioned in the introduction to this chapter. Descartes' rectangular coordinate system allows us to associate the equation $y = 2x - 3$ (an algebraic concept) with a specific straight line (a geometric concept). The study of the relationship between equations in algebra and their associated geometric figures is called *analytic geometry*. The rectangular coordinate system is often referred to as the *Cartesian coordinate system* in honor of Descartes.

▶ **EXAMPLE 4**   Graph the equation $y = -\dfrac{1}{3}x + 2$.

Solution   We need to find three ordered pairs that satisfy the equation. To do so, we can let x equal any numbers we choose and find corresponding values of y. But, since every value of x we substitute into the equation is going to be multiplied by $-\frac{1}{3}$, let's use numbers for x that are divisible by 3, like -3, 0, and 3. That way, when we multiply them by $-\frac{1}{3}$, the result will be an integer.

Let $x = -3$; $y = -\dfrac{1}{3}(-3) + 2$

$$y = 1 + 2$$

$$y = 3$$

The ordered pair $(-3, 3)$ is one solution.

In table form

x	y
-3	3
0	2
3	1

Let $x = 0$; $y = -\dfrac{1}{3}(0) + 2$

$$y = 0 + 2$$

$$y = 2$$

The ordered pair $(0, 2)$ is a second solution.

Let $x = 3$; $y = -\dfrac{1}{3}(3) + 2$

$$y = -1 + 2$$

$$y = 1$$

The ordered pair $(3, 1)$ is a third solution.

Plotting the ordered pairs $(-3, 3)$, $(0, 2)$, and $(3, 1)$ and drawing a straight line through their graphs, we have the graph of the equation $y = -\frac{1}{3}x + 2$, as shown in Figure 6 on the next page.

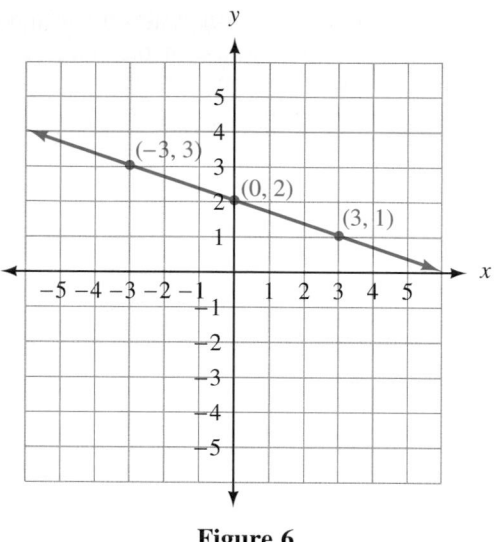

Figure 6

Note In Example 4 the values of x we used, -3, 0, and 3, are referred to as convenient values of x because they are easier to work with than some other numbers. For instance, if we let $x = 2$ in our original equation, we would have to add $-\frac{2}{3}$ and 2 to find the corresponding value of y. Not only would the arithmetic be more difficult, but the ordered pair we would obtain would have a fraction for its y-coordinate, making it more difficult to graph accurately.

Intercepts

Two important points on the graph of a straight line, if they exist, are the points where the graph crosses the axes.

> **DEFINITION** The **x-intercept** of the graph of an equation is the x-coordinate of the point where the graph crosses the x-axis. The **y-intercept** is defined similarly.

Since any point on the x-axis has a y-coordinate of 0, we can find the x-intercept by letting $y = 0$ and solving the equation for x. We find the y-intercept by letting $x = 0$ and solving for y.

▶ **EXAMPLE 5** Find the x- and y-intercepts for $2x + 3y = 6$; then graph the solution set.

Solution   To find the *y*-intercept we let $x = 0$.

$$\text{When} \qquad\qquad x = 0$$

$$\text{we have} \qquad 2(0) + 3y = 6$$

$$3y = 6$$

$$y = 2$$

The *y*-intercept is 2, and the graph crosses the *y*-axis at the point (0, 2).

$$\text{When} \qquad\qquad y = 0$$

$$\text{we have} \qquad 2x + 3(0) = 6$$

$$2x = 6$$

$$x = 3$$

The *x*-intercept is 3, so the graph crosses the *x*-axis at the point (3, 0). We use these results to graph the solution set for $2x + 3y = 6$. The graph is shown in Figure 7.

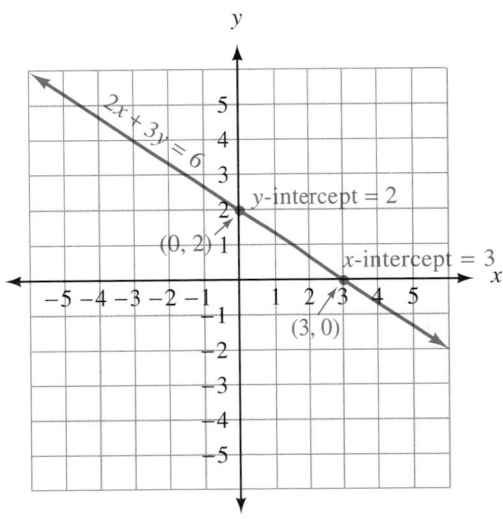

Figure 7

Graphing straight lines by finding the intercepts works best when the coefficients of *x* and *y* are factors of the constant term.   ◀

▶ **EXAMPLE 6** Graph the line $x = 3$ and the line $y = -2$.

Solution The line $x = 3$ is the set of all points whose x-coordinate is 3. The variable y does not appear in the equation, so the y-coordinate can be any number.

The line $y = -2$ is the set of all points whose y-coordinate is -2. The x-coordinate can be any number. Here are the graphs:

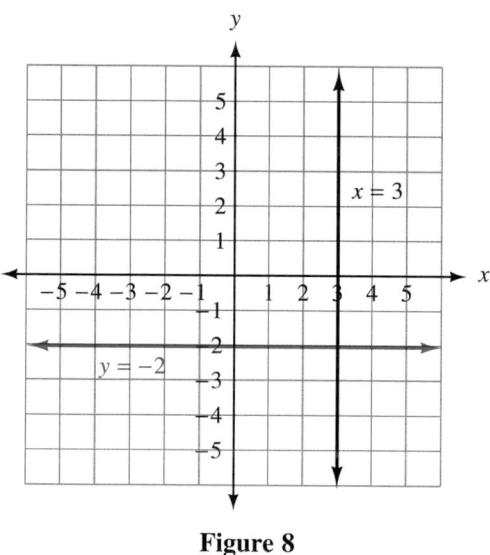

Figure 8 ◀

Our next example involves a problem encountered in two previous chapters.

▶ **EXAMPLE 7** An art supply store finds that they can sell x sketch pads each week at p dollars each, according to the equation $x = 900 - 300p$. Note any restrictions on the variable, then graph the relationship between x and p, with the values of p along the horizontal axis and the values of x on the vertical axis.

Solution The reason we look for restrictions on the variables is that this is an applied problem and, as such, there may be values of x or p that are not appropriate. Since x is the number of items sold, we know that $x \geq 0$, since we cannot sell a negative number of items. Likewise, $p \geq 0$, since the price cannot be a negative number. This tells us that our graph will appear in the first quadrant only. Below is a table of values for p and x, along with the graph of $x = 900 - 300p$.

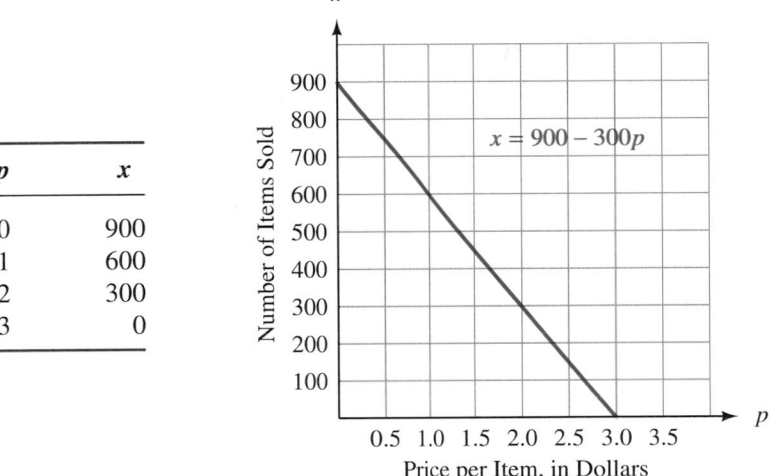

p	x
0	900
1	600
2	300
3	0

Figure 9

Note that the values of p range from 0 to 3, meaning that the price can be set from \$0 to \$3 for each sketch pad. The values of x range from 900 to 0, meaning that they will sell from 0 to 900 sketch pads each week. The first point on the graph, $p = 0$, $x = 900$, can be interpreted this way: If they give away the sketch pads, they will give every one they produce, which is 900 sketch pads. ◀

USING TECHNOLOGY: Graphing Calculators and Computer Graphing Programs

There are a variety of computer programs and graphing calculators currently available to help us graph equations and then obtain information from those graphs much faster than we can do with paper and pencil. Our examples will be aimed toward graphing calculators, but most of the skills we will learn translate to the popular computer graphing programs. The F/C Graph software that accompanies this book is one such program. Ask your instructor about the F/C Graph. As we mentioned earlier, we will not give the instructions for all the calculators available. The instructions here are generic in form. You will have to use the manual that came with your calculator to find the instructions that are appropriate to your calculator.

Graphing with Trace and Zoom
All graphing calculators have the ability to graph a function and then trace over the points on the graph, giving their coordinates as it goes. Further, all graphing calculators can zoom in and out on a graph that has been drawn. To graph a linear equation on a graphing calculator, we first set the graph

window. Most calculators call the smallest value of x Xmin and the largest value of x Xmax. The counterpart values of y are Ymin and Ymax. We will use the notation

> Window: X from -5 to 4, Y from -3 to 2

to stand for a window in which

$$\text{Xmin} = -5$$
$$\text{Xmax} = 4$$
$$\text{Ymin} = -3$$
$$\text{Ymax} = 2$$

Set your calculator with the following window:

> Window: X from -10 to 10, Y from -10 to 10

Graph the equation $Y = -X + 8$. The graph will be similar to the one shown in Figure 10.

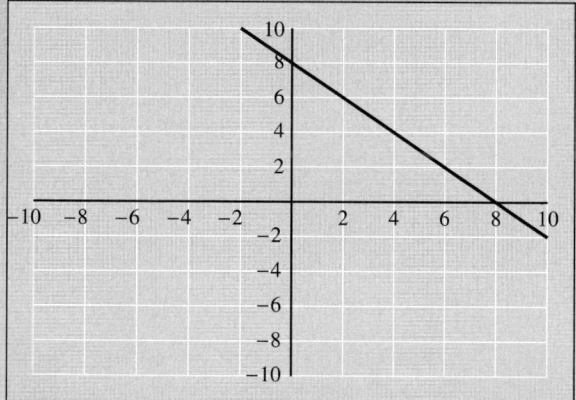

Figure 10

Use the Trace feature of your calculator to name three points on the graph. Next, use the Zoom feature of your calculator to zoom out so your window is twice as large.

Working with Variables Other than X and Y

To graph the equation $x = 900 - 300p$ shown in Example 7, you must let Y take the place of x in the equation and X take the place of p. Set your graph window as follows:

> Window: X from 0 to 4, Y from 0 to 1000

Graph $Y = 900 - 300X$. Compare your graph with the graph shown in Figure 9. Use the Trace feature on your calculator to confirm the values in the table next to Figure 9.

PROBLEM SET 5.1

Graph each of the following ordered pairs on a rectangular coordinate system.

1. $(1, 2)$ **2.** $(-1, 2)$
3. $(-1, -2)$ **4.** $(1, -2)$
5. $(3, 4)$ **6.** $(-3, -4)$
7. $(5, 0)$ **8.** $(0, -3)$
9. $(0, 2)$ **10.** $(4, 0)$
11. $(-5, -5)$ **12.** $(-4, -1)$
13. $(\frac{1}{2}, 2)$ **14.** $(3, \frac{1}{4})$
15. $(5, -2)$ **16.** $(0, 4)$

Give the coordinates of each of the following points.

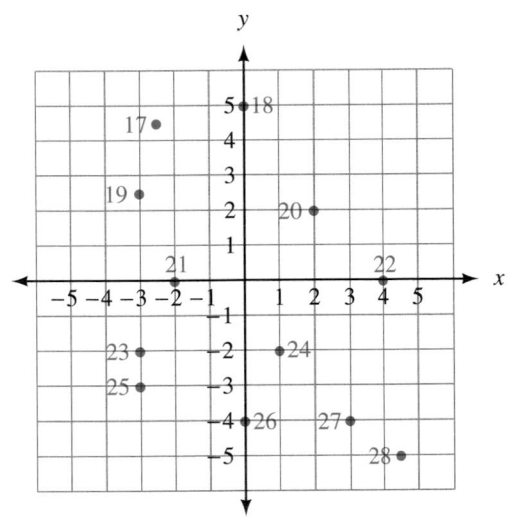

Graph each of the following linear equations by first finding the intercepts.

29. $2x - 3y = 6$
30. $3x - 2y = 6$
31. $y + 2x = 4$
32. $y - 2x = 4$
33. $4x - 5y = 20$
34. $4x + 5y = 20$
35. $3x + 5y - 15 = 0$
36. $5x - 3y - 15 = 0$
37. $y = 2x + 3$
38. $y = 3x - 2$

Graph each of the following straight lines.

39. $2y = 4x - 8$ **40.** $2y = 4x + 8$
41. $y = \frac{1}{3}x$ **42.** $y = \frac{1}{2}x$
43. $y = \frac{1}{2}x + 3$ **44.** $y = \frac{1}{2}x - 3$
45. $-2x + y = -3$ **46.** $-3x + y = -2$
47. $y = -\frac{2}{3}x + 1$ **48.** $y = -\frac{2}{3}x - 1$
49. $y = \frac{1}{2}x + 1$ **50.** $y = \frac{1}{3}x + 1$

51. $\dfrac{x}{3} + \dfrac{y}{4} = 1$ **52.** $\dfrac{x}{-2} + \dfrac{y}{3} = 1$

53. $0.02x + 0.03y = 0.06$
54. $0.05x - 0.03y = 0.15$
55. Graph the lines $x = -3$ and $y = 5$ on the same coordinate system.
56. Graph the lines $x = 4$ and $y = -3$ on the same coordinate system.
57. Graph the lines $y = x + 1$ and $y = x - 3$ on the same coordinate system. Can you tell from looking at the two graphs where the graph of $y = x + 3$ will be?
58. Graph the lines $y = 2x + 2$ and $y = 2x - 1$ on the same coordinate system. Use the similarities between the two graphs in order to graph the line $y = 2x - 4$.
59. The ordered pairs that satisfy the equation $y = 3x$ all have the form $(x, 3x)$, because y is always 3 times x. Graph all ordered pairs of the form $(x, 3x)$.
60. Graph all ordered pairs of the form $(x, -3x)$.
61. Graph all ordered pairs of the form (x, x).
62. Graph all ordered pairs of the form $(x, -x)$.

Applying the Concepts

63. If the cost of a long-distance phone call is 50¢ for the first minute and 25¢ for each additional minute, then the total cost y (in cents) of a call that goes x minutes past the first minute is $y = 25x + 50$. Let 1 unit on the x-axis equal 1 minute, and 1 unit on the y-axis equal 25¢, and graph this equation.
64. If the cost of a taxi ride in Las Vegas is $1.50 for the first mile and $0.50 for each $\frac{1}{10}$ of a mile after the first mile, then the total cost y (in

cents) of a ride that goes x tenths of a mile after the first mile is $y = 50x + 150$. Let each unit on the x-axis equal $\frac{1}{10}$ of a mile and each unit on the y-axis equal 50¢ and graph this equation.

65. A company that manufactures typewriter ribbons knows that the number of ribbons they can sell each week x is related to the price of each ribbon p by the equation $x = 1,200 - 100p$. Note any restrictions on the variables, then graph the relationship between x and p, with the values of p along the horizontal axis and the values of x on the vertical axis.

66. A company that manufactures diskettes for home computers finds that they can sell x diskettes each day at p dollars per diskette according to the equation $x = 800 - 100p$. Note any restrictions on the variables, then graph the relationship between x and p, with the values of p along the horizontal axis and the values of x on the vertical axis.

67. The relationship between the number of calculators a company sells each day x and the price of each calculator p is given by the equation $x = 1,700 - 100p$. Note any restrictions on the variables, then graph the relationship between x and p, with the values of p along the horizontal axis and the values of x on the vertical axis.

68. The relationship between the number of pencil sharpeners a company sells each week and the price of each sharpener is given by the equation $x = 1,800 - 100p$. Note any restrictions on the variables, then graph the relationship between x and p, with the values of p along the horizontal axis and the values of x on the vertical axis.

Review Problems

The problems that follow review material we covered in Sections 4.1 and 4.2.

Reduce to lowest terms. [4.1]

69. $\dfrac{x^2 - 9}{x^4 - 81}$

70. $\dfrac{6 - a - a^2}{3 - 2a - a^2}$

Divide. [4.2]

71. $\dfrac{15x^2y - 20x^4y^2}{5xy}$

72. $\dfrac{12x^3y^2 - 24x^2y^3}{6xy}$

Divide using long division. [4.2]

73. $\dfrac{10x^2 + 7x - 12}{2x + 3}$

74. $\dfrac{6x^2 - x - 35}{2x - 5}$

75. $\dfrac{x^3 - 125}{x - 5}$

76. $\dfrac{x^3 + 64}{x + 4}$

One Step Further

Find the x- and y-intercepts for each equation. Your answers will contain the constants a, b, and c.

77. $ax + by = c$

78. $ax - by = c$

79. $\dfrac{x}{a} + \dfrac{y}{b} = 1$

80. $y = ax + b$

81. Complete the following ordered pairs so they are solutions to $y = |x + 2|$. Then graph the equation by connecting the points in the way that makes the most sense to you.

$(-5, \ )\ (-4, \ )\ (-3, \ )\ (-2, \ )$

$(-1, \ )\ (0, \ )\ (1, \ )$

82. Complete the following ordered pairs so they are solutions to $y = |x - 2|$. Then use these points to graph $y = |x - 2|$.

$(-1, \ )\ (0, \ )\ (1, \ )\ (2, \ )\ (3, \ )\ (4, \ )\ (5, \ )$

83. Graph each equation.
    (a) $y = x - 3$
    (b) $y = -x + 3$
    (c) $y = |x - 3|$

84. Graph each equation.
    (a) $y = x + 3$
    (b) $y = -x - 3$
    (c) $y = |x + 3|$

USING TECHNOLOGY

Use your graphing calculator to graph each equation in the given window. Use the Trace feature to name the x-intercept, y-intercept, and one other point on the line. Then use the Zoom feature to view the graph in the window twice as large as the original.

85. $y = 9 - x$ Window: X from -10 to 10, Y from -10 to 10

86. $y = x - 9$ Window: X from -12 to 12, Y from -12 to 12

87. $y = 3 - \frac{1}{2}x$ Window: X from -5 to 5, Y from -5 to 5

88. $y = \frac{1}{2}x - 3$ Window: X from -5 to 5, Y from -5 to 5

89. $y = 300 - 30x$ Window: X from -8 to 20, Y from -100 to 400

90. $y = 30x - 300$ Window: X from -8 to 20, Y from -400 to 100

91. Graph the equation given in Problem 65. Name the intercepts and two other points in the first quadrant.

92. Graph the equation given in Problem 66. Name the intercepts and two other points in the first quadrant.

SECTION

5.2 The Slope of a Line

In defining the slope of a straight line, we are looking for a number to associate with a straight line that does two things. First of all, we want the slope of a line to measure the "steepness" of the line. That is, in comparing two lines, the slope of the steeper line should have the larger numerical value. Secondly, we want a line that *rises* going from left to right to have a **positive slope.** We want a line that *falls* going from left to right to have a **negative slope.** (A line that neither rises nor falls going from left to right must, therefore, have 0 slope.)

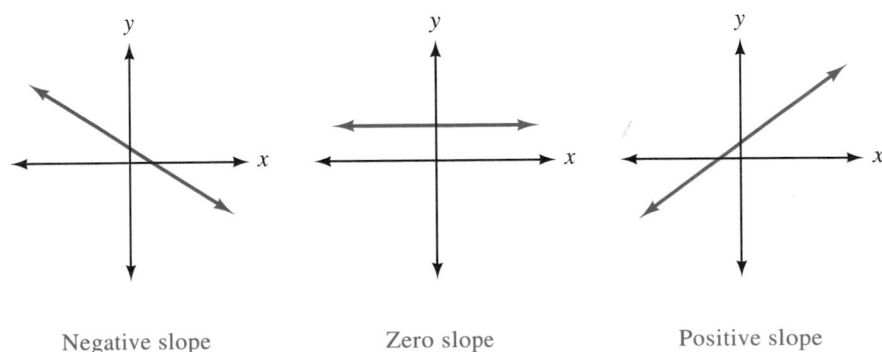

Negative slope   Zero slope   Positive slope

Geometrically, we can define the **slope** of a line as the ratio of the vertical change to the horizontal change encountered when moving from one point to another on the line. The vertical change is sometimes called the **rise.** The horizontal change is called the **run.**

▶ **EXAMPLE 1**   Find the slope of the line $y = 2x - 3$.

Solution   In order to make use of our geometric definition, we first graph $y = 2x - 3$. We then pick any two convenient points and find the ratio of rise to run. By convenient points we mean points with integer coordinates. If we let $x = 2$ in the equation, then $y = 1$. Likewise if we let $x = 4$, then y is 5.

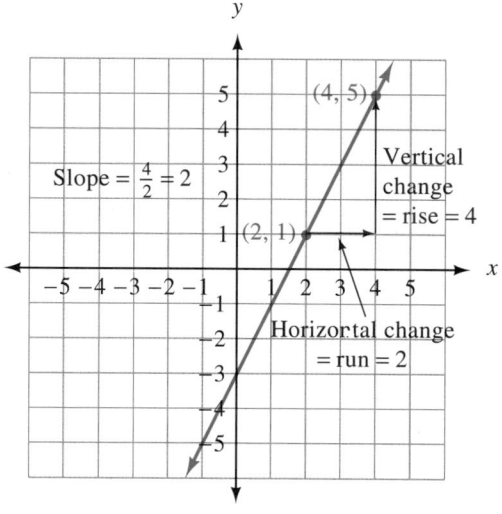

Figure 1

Our line has a slope of 2.   ◀

Notice that we can measure the vertical change by subtracting the y-coordinates of the two points shown: $5 - 1 = 4$. The horizontal change is the difference of the x-coordinates: $4 - 2 = 2$. This gives us a second way of defining the slope of a line. Algebraically, we say the slope of a line between two points whose coordinates are given is the ratio of the difference in the y-coordinates to the difference in the x-coordinates. We can summarize the preceding discussion by formalizing our definition for slope.

DEFINITION   The **slope** of the line between two points (x_1, y_1) and (x_2, y_2) is given by

$$\text{Slope} = m = \frac{\text{rise}}{\text{run}} = \frac{y_2 - y_1}{x_2 - x_1}$$

The letter m is usually used to designate slope. Our definition includes both the geometric form (rise/run) and the algebraic form $(y_2 - y_1)/(x_2 - x_1)$.

▶ **EXAMPLE 2** Find the slope of the line through $(-2, -3)$ and $(-5, 1)$.

Solution

$$m = \frac{y_2 - y_1}{x_2 - x_1} = \frac{1 - (-3)}{-5 - (-2)} = \frac{4}{-3} = -\frac{4}{3}$$

Looking at the graph of the line between the two points, we can see our geometric approach does not conflict with our algebraic approach.

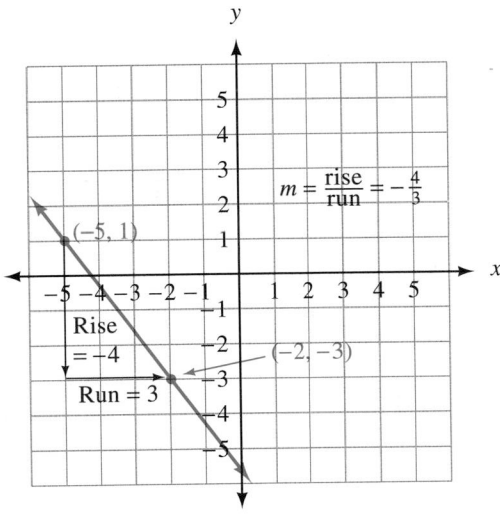

Figure 2

We should note here that it does not matter which ordered pair we call (x_1, y_1) and which we call (x_2, y_2). If we were to reverse the order of subtraction of both the x- and y-coordinates in the preceding example, we would have

$$m = \frac{-3 - 1}{-2 - (-5)} = \frac{-4}{3} = -\frac{4}{3}$$

which is the same as our previous result.   ◀

Note The two most common mistakes students make when first working with the formula for the slope of a line are

1. Putting the difference of the x-coordinates over the difference of the y-coordinates.
2. Subtracting in one order in the numerator and then subtracting in the opposite order in the denominator. You would make this mistake in Example 2 if you wrote $1 - (-3)$ in the numerator and then $-2 - (-5)$ in the denominator.

▶ **EXAMPLE 3**   Find the slope of the line containing (3, −1) and (3, 4).

Solution   Using the definition for slope, we have

$$m = \frac{-1 - 4}{3 - 3} = \frac{-5}{0}$$

The expression $\dfrac{-5}{0}$ is undefined. That is, there is no real number to associate with it. In this case, we say the line has *no slope.*

The graph of our line is as follows:

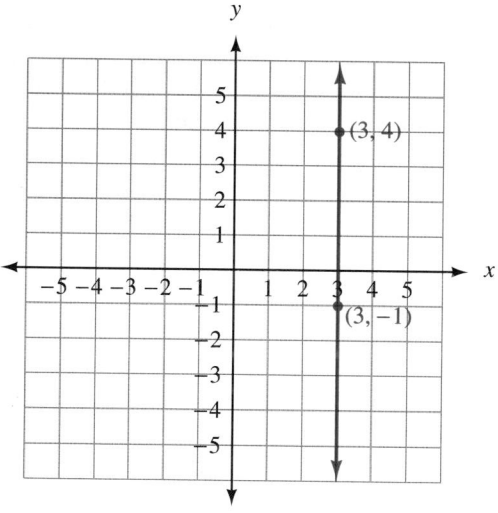

Figure 3

Our line with no slope is a vertical line. All vertical lines have no slope. (And all horizontal lines, as we mentioned earlier, have 0 slope.)   ◀

Slope of Parallel and Perpendicular Lines

In geometry we call lines in the same plane that never intersect **parallel.** In order for two lines to be nonintersecting, they must rise or fall at the same rate. That is, the ratio of vertical change to horizontal change must be the same for each line. In other words, two lines are *parallel* if and only if they have the *same slope.*

Although it is not as obvious, it is also true that two nonvertical lines are **perpendicular** if and only if the *product of their slopes is* −1. This is the same as saying their slopes are negative reciprocals.

We can state these facts with symbols as follows.

If line l_1 has slope m_1, and line l_2 has slope m_2, then

$$l_1 \text{ and } l_2 \text{ are parallel} \Leftrightarrow m_1 = m_2$$

and

$$l_1 \text{ and } l_2 \text{ are perpendicular} \Leftrightarrow m_1 \cdot m_2 = -1$$

$$\left(\text{or } m_1 = \frac{-1}{m_2}\right)$$

To clarify this, if a line has a slope of $\frac{2}{3}$, then any line parallel to it has a slope of $\frac{2}{3}$. Any line perpendicular to it has a slope of $-\frac{3}{2}$ (the negative reciprocal of $\frac{2}{3}$).

Although we cannot give a formal proof of the relationship between the slopes of perpendicular lines at this level of mathematics, we can offer some justification for the relationship. Figure 4 shows the graphs of two lines. One of the lines has a slope of $\frac{2}{3}$, while the other line has a slope of $-\frac{3}{2}$. As you can see, the lines are perpendicular. If you need more justification for the relationship, then you should draw some other pairs of lines with slopes that are negative reciprocals. For instance, graph a line with a slope of 2 and another line with a slope of $-\frac{1}{2}$ on the same coordinate system.

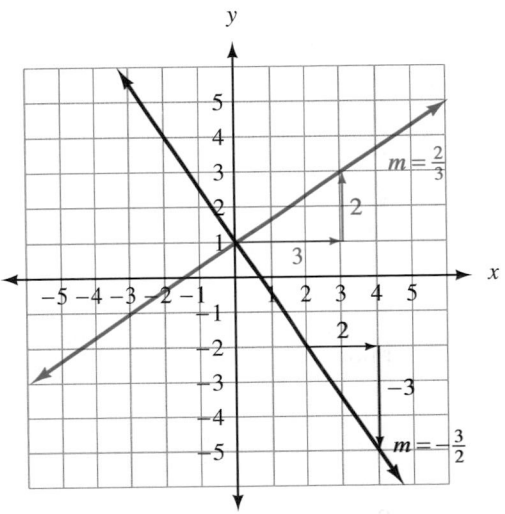

Figure 4

▶ **EXAMPLE 4** Find a if the line through $(3, a)$ and $(-2, -8)$ is perpendicular to a line with slope $-\frac{4}{5}$.

Solution   The slope of the line through the two points is

$$m = \frac{a - (-8)}{3 - (-2)} = \frac{a + 8}{5}$$

Since the line through the two points is perpendicular to a line with slope $-\frac{4}{5}$, we can also write its slope as $\frac{5}{4}$.

$$\frac{a + 8}{5} = \frac{5}{4}$$

Multiplying both sides by 20, we have

$$4(a + 8) = 5 \cdot 5$$

$$4a + 32 = 25$$

$$4a = -7$$

$$a = -\frac{7}{4}$$

◀

USING TECHNOLOGY: Graphing Calculators

We can use a graphing calculator or a computer graphing program to investigate the effect of the numbers a and b on the graph of $y = ax + b$. To see how the number b affects the graph, we can hold a constant and let b vary. Doing so will give us a *family* of curves. Suppose we set $a = 1$ and then let b take on integer values from -3 to 3. The equations we obtain are

$$y = x - 3$$
$$y = x - 2$$
$$y = x - 1$$
$$y = x$$
$$y = x + 1$$
$$y = x + 2$$
$$y = x + 3$$

There are two ways to graph this set of equations on a graphing calculator. The first is to use the list of Y variables and the second is to write a program that substitutes the given values of b into $y = x + b$ and then graph the result. To use the Y variables list, enter each equation at one of the Y variables, set the graph Window, then graph. The calculator will graph the equations in order, starting with Y_1 and ending with Y_7. Following is the Y variables list, an appropriate Window, and a sample of the type of graph obtained.

$Y_1 = X - 3$

$Y_2 = X - 2$

$Y_3 = X - 1$

$Y_4 = X$

$Y_5 = X + 1$

$Y_6 = X + 2$

$Y_7 = X + 3$

Window:
  X from -4 to 4,
  Y from -4 to 4

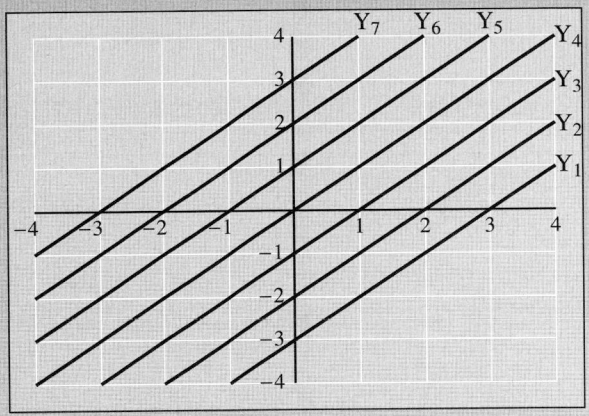

Figure 5

The same result can be obtained by programming your calculator to graph $y = x + b$ for $b = -3, -2, -1, 0, 1, 2,$ and 3. Here is an outline of a program that will do this. Check the manual that came with your calculator to find the commands appropriate to your calculator.

Step 1: Clear screen

Step 2: Set Window for X from -4 to 4 and Y from -4 to 4

Step 3: $-3 \rightarrow B$

Step 4: Label 1

Step 5: Graph $Y = X + B$

Step 6: $B + 1 \rightarrow B$

Step 7: If $B < 4$, Goto 1

Step 8: End

Running this program will produce graphs similar to those in Figure 5.

P R O B L E M   S E T   5 . 2

Find the slope of each of the following lines from
the given graph.

1.

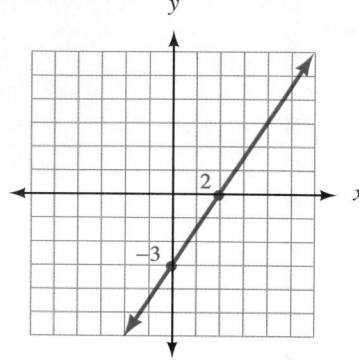

4.

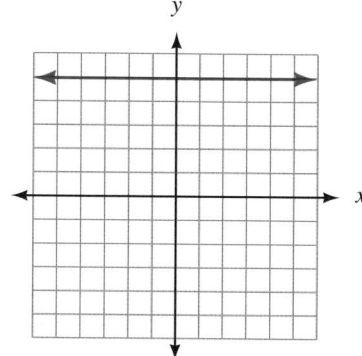

2.

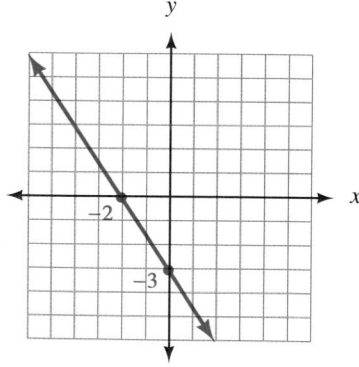

5.

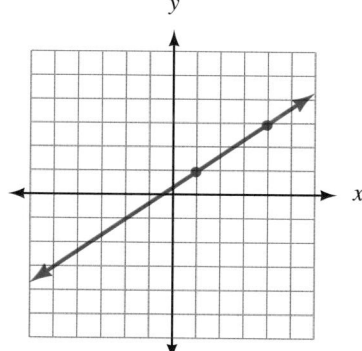

3.

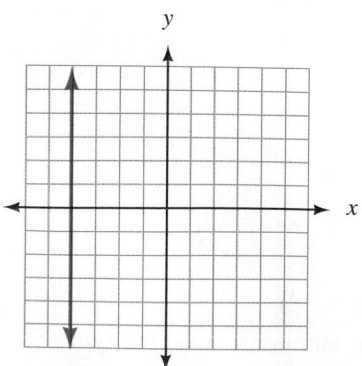

6.

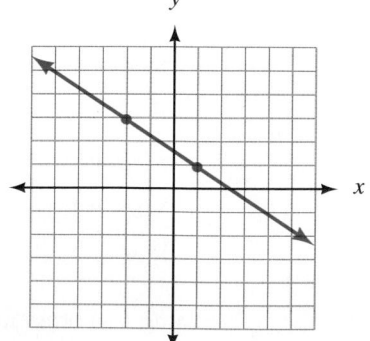

Find the slope of the line through the following pairs of points. Then, plot each pair of points, draw a line through them, and indicate the rise and run in the graph in the same manner shown in Example 2.

7. $(2, 1)$, $(4, 4)$ **8.** $(3, 1)$, $(5, 4)$
9. $(1, 4)$, $(5, 2)$ **10.** $(1, 3)$, $(5, 2)$
11. $(1, -3)$, $(4, 2)$ **12.** $(2, -3)$, $(5, 2)$
13. $(-3, -2)$, $(1, 3)$ **14.** $(-3, -1)$, $(1, 4)$
15. $(-3, 2)$, $(3, -2)$ **16.** $(-3, 3)$, $(3, -1)$
17. $(2, -5)$, $(3, -2)$ **18.** $(2, -4)$, $(3, -1)$

Solve for the indicated variable if the line through the two given points has the given slope.

19. $(5, a)$, $(4, 2)$; $m = 3$
20. $(3, a)$, $(1, 5)$; $m = -4$
21. $(2, 6)$, $(3, y)$; $m = -7$
22. $(-4, 9)$, $(-5, y)$; $m = 3$
23. $(7, y^2)$, $(5, y)$; $m = 3$
24. $(4, y^2)$, $(8, y)$; $m = -5$
25. $(3, b^2)$, $(4, 3b)$; $m = 2$
26. $(5, b^2)$, $(2, 4b)$; $m = -1$
27. Graph the line that has an x-intercept of 3 and a y-intercept of -2. What is the slope of this line?
28. Graph the line that has an x-intercept of 2 and a y-intercept of -3. What is the slope of this line?
29. Graph the line with x-intercept 4 and y-intercept 2. What is the slope of this line?
30. Graph the line with x-intercept -4 and y-intercept -2. What is the slope of this line?
31. Find the slope of any line parallel to the line through $(2, 3)$ and $(-8, 1)$.
32. Find the slope of any line parallel to the line through $(2, 5)$ and $(5, -3)$.
33. Line l contains the points $(5, -6)$ and $(5, 2)$. Give the slope of any line perpendicular to l.
34. Line l contains points $(3, 4)$ and $(-3, 1)$. Give the slope of any line perpendicular to l.
35. Line l has a slope of $\frac{2}{3}$. A horizontal change of 12 will always be accompanied by how much of a vertical change?
36. For any line with slope $\frac{4}{5}$, a vertical change of 8 is always accompanied by how much of a horizontal change?

37. The line through $(2, y^2)$ and $(1, y)$ is perpendicular to a line with slope $-\frac{1}{6}$. What are the possible values for y?
38. The line through $(7, y^2)$ and $(3, 6y)$ is parallel to a line with slope -2. What are the possible values for y?
39. A pile of sand at a construction site is in the shape of a cone. If the slope of the side of the pile is $\frac{2}{3}$ and the pile is 8 feet high, how wide is the diameter of the base of the pile?
40. The slope of the sides of one of the Great Pyramids in Egypt is $\frac{13}{10}$. If the base of the pyramid is 750 feet, how tall is the pyramid?

Review Problems

The problems below review material we covered in Sections 2.2 and 4.3. Reviewing the problems from Section 2.2 will help you with some of the next section.

Solve each equation for y. [2.2]

41. $2x - 3y = 6$ **42.** $3x + 2y = 6$
43. $2x - 3y = 5$ **44.** $3x - 2y = 5$

Perform the indicated operations. [4.3]

45. $\dfrac{8xy^3}{9x^2y} \div \dfrac{16x^2y^2}{18xy^3}$ **46.** $\dfrac{25x^2}{5y^4} \cdot \dfrac{30y^3}{2x^5}$

47. $\dfrac{12a^2 - 4a - 5}{2a + 1} \cdot \dfrac{7a + 3}{42a^2 - 17a - 15}$

48. $\dfrac{20a^2 - 7a - 3}{4a + 1} \cdot \dfrac{25a^2 - 5a - 6}{5a + 2}$

49. $\dfrac{8x^3 + 27}{27x^3 + 1} \div \dfrac{6x^2 + 7x - 3}{9x^2 - 1}$

50. $\dfrac{27x^3 + 8}{8x^3 + 1} \div \dfrac{6x^2 + x - 2}{4x^2 - 1}$

One Step Further

Find the slope of the line through each pair of points.

51. $\left(\frac{2}{3}, \frac{1}{8}\right)$ and $\left(\frac{5}{6}, -\frac{3}{2}\right)$
52. $\left(-\frac{5}{7}, \frac{4}{3}\right)$ and $\left(-\frac{1}{3}, \frac{3}{7}\right)$
53. $(0.04, 0.18)$ and $(0.16, -0.02)$

54. $(1.23, -3.21)$ and $(4.56, 1.23)$
55. (a^2, a) and $(b^2, -b)$
56. $(2a, 8a^3)$ and (b, b^3)
57. Graph each line on the same coordinate system, then use the graphs to find the slope of each line.

    (a) $y = 2x + 3$ (b) $y = -\frac{1}{2}x + 3$
58. Graph each line on the same coordinate system, then use the graphs to find the slope of each line.
    (a) $y = \frac{2}{3}x + 1$ (b) $y = -\frac{3}{2}x + 1$

USING TECHNOLOGY

59. Use your Y variables list, or write a program, to graph the family of curves $Y = 2X + B$ for $B = -3, -2, -1, 0, 1, 2,$ and 3.
60. Use your Y variables list, or write a program, to graph the family of curves $Y = -2X + B$ for $B = -3, -2, -1, 0, 1, 2,$ and 3.
61. Use your Y variables list, or write a program, to graph the family of curves $Y = AX$ for $A = -3, -2, -1, 0, 1, 2,$ and 3.
62. Use your Y variables list, or write a program, to graph the family of curves $Y = AX + 2$ for $A = -3, -2, -1, 0, 1, 2,$ and 3.
63. Use your Y variables list, or write a program, to graph the family of curves $Y = AX$ for $A = \frac{1}{4}, \frac{1}{3}, \frac{1}{2}, 1, 2,$ and 3.
64. Use your Y variables list, or write a program, to graph the family of curves $Y = AX - 2$ for $A = \frac{1}{4}, \frac{1}{3}, \frac{1}{2}, 1, 2,$ and 3.
65. Summarize the results of Problems 59–64 by giving a written description of the effect of a and b on the graph of $y = ax + b$.

SECTION

5.3 The Equation of a Straight Line

In the first section of this chapter we defined the y-intercept of a line to be the y-coordinate of the point where the graph crosses the y-axis. We can use this definition, along with the definition of slope from the preceding section, to derive the slope-intercept form of the equation of a straight line.

Suppose line l has slope m and y-intercept b. What is the equation of l?

Since the y-intercept is b, we know the point $(0, b)$ is on the line. If (x, y) is any other point on l, then using the definition for slope, we have

$$\frac{y - b}{x - 0} = m \qquad \text{Definition of slope}$$

$$y - b = mx \qquad \text{Multiply both sides by } x$$

$$y = mx + b \qquad \text{Add } b \text{ to both sides}$$

This last equation is known as the **slope-intercept form** of the equation of a straight line.

Slope-Intercept Form of the Equation of a Line

The equation of any line with slope m and y-intercept b is given by

$$y = mx + b$$

Slope y-intercept

When the equation is in this form, the *slope* of the line is always the *coefficient of x,* and the *y-intercept* is always the *constant term.*

▶ **EXAMPLE 1**   Find the equation of the line with slope $-\frac{4}{3}$ and y-intercept 5. Then, graph the line.

Solution   Substituting $m = -\frac{4}{3}$ and $b = 5$ into $y = mx + b$, we have

$$y = -\frac{4}{3}x + 5$$

Finding the equation from the slope and y-intercept is just that easy. If the slope is m and the y-intercept is b, then the equation is always $y = mx + b$. Now, let's graph the line.

Since the y-intercept is 5, the graph goes through the point (0, 5). To find a second point on the graph, we start at (0, 5) and move 4 units down (that's a rise of -4) and 3 units to the right (a run of 3). The point we end up at is (3, 1). Drawing a line that passes through (0, 5) and (3, 1), we have the graph of our equation.

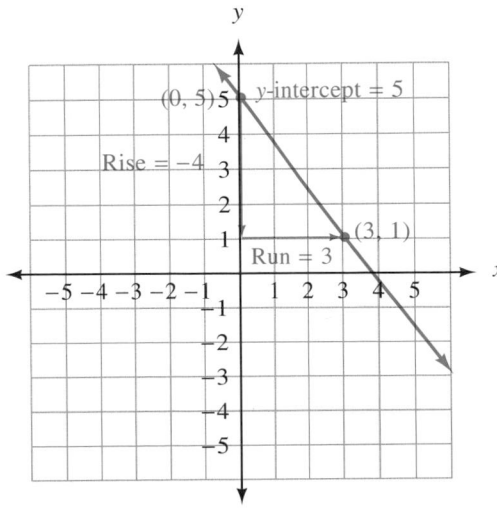

Figure 1

▶ **EXAMPLE 2**   Give the slope and y-intercept for the line $2x - 3y = 5$.

Solution   To use the slope-intercept form we must solve the equation for y in terms of x:

$$2x - 3y = 5$$

$$-3y = -2x + 5 \qquad \text{Add } -2x \text{ to both sides}$$

$$y = \frac{2}{3}x - \frac{5}{3} \qquad \text{Divide by } -3$$

The last equation has the form $y = mx + b$. The slope must be $m = \frac{2}{3}$, and the y-intercept is $b = -\frac{5}{3}$.   ◀

▶ **EXAMPLE 3**   Graph the equation $2x + 3y = 6$ using the slope and y-intercept.

Solution   Although we could graph this equation using the methods developed in Section 5.1 (by finding ordered pairs that are solutions to the equation and drawing a line through their graphs), it is sometimes easier to graph a line using the slope-intercept form of the equation.

Solving the equation for y, we have

$$2x + 3y = 6$$

$$3y = -2x + 6 \qquad \text{Add } -2x \text{ to both sides}$$

$$y = -\frac{2}{3}x + 2 \qquad \text{Divide by } 3$$

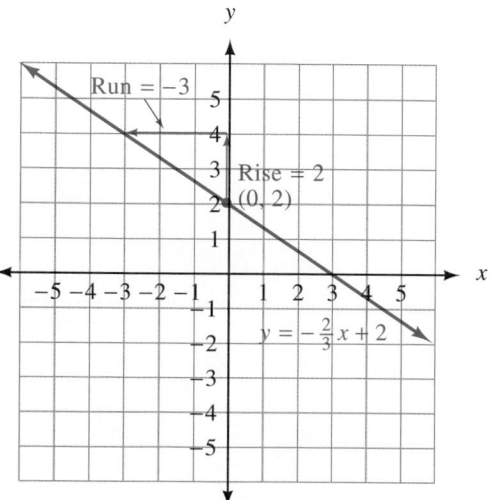

Figure 2

The slope is $m = -\frac{2}{3}$ and the y-intercept is $b = 2$. Therefore, the point $(0, 2)$ is on the graph and the ratio rise/run going from $(0, 2)$ to any other point on the line is $-\frac{2}{3}$. If we start at $(0, 2)$ and move 2 units up (that's a rise of 2) and 3 units to the left (a run of -3), we will be at another point on the graph. (We could also go down 2 units and right 3 units and still be assured of ending up at another point on the line, since $\frac{2}{-3}$ is the same as $\frac{-2}{3}$.)   ◀

As we mentioned in the introduction to this chapter, the rectangular coordinate system is the tool we use to connect algebra and geometry. Example 3 illustrates this connection, as do the many other examples in this chapter. In Example 3, Descartes' rectangular coordinate system allows us to associate the equation $2x + 3y = 6$ (an algebraic concept) with the straight line (a geometric concept) shown in Figure 2.

A second useful form of the equation of a straight line is the point-slope form.

Let line l contain the point (x_1, y_1) and have slope m. If (x, y) is any other point on l, then by the definition of slope we have

$$\frac{y - y_1}{x - x_1} = m$$

Multiplying both sides by $(x - x_1)$ gives us

$$(x - x_1) \cdot \frac{y - y_1}{x - x_1} = m(x - x_1)$$

$$y - y_1 = m(x - x_1)$$

This last equation is known as the **point-slope form** of the equation of a straight line.

Point-Slope Form of the Equation of a Line

The equation of the line through (x_1, y_1) with slope m is given by

$$y - y_1 = m(x - x_1)$$

This form of the equation of a straight line is used to find the equation of a line, either given one point on the line and the slope, or given two points on the line.

▶ **EXAMPLE 4** Find the equation of the line with slope -2 that contains the point $(-4, 3)$. Write the answer in slope-intercept form.

Solution

Using $(x_1, y_1) = (-4, 3)$ and $m = -2$

in $\quad\quad y - y_1 = m(x - x_1)$ $\quad\quad$ Point-slope form

gives us $\quad\quad y - 3 = -2(x + 4)$ $\quad\quad$ *Note:* $x - (-4) = x + 4$

$\quad\quad\quad\quad\quad y - 3 = -2x - 8$ $\quad\quad$ Multiply out right side

$\quad\quad\quad\quad\quad\quad\quad y = -2x - 5$ $\quad\quad$ Add 3 to each side

Figure 3 is the graph of the line that contains $(-4, 3)$ and has a slope of -2. Notice that the y-intercept on the graph matches that of the equation we found.

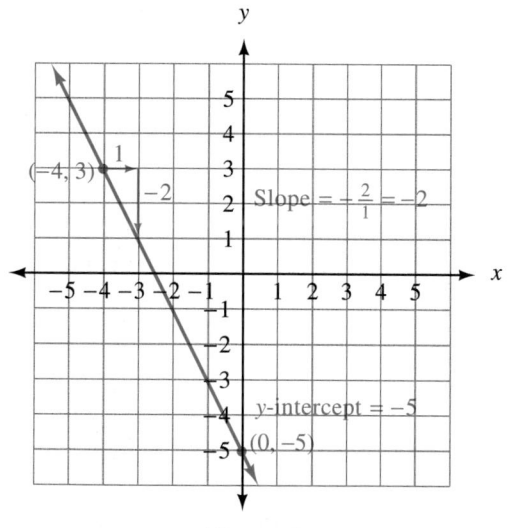

Figure 3

▶ **EXAMPLE 5** Find the equation of the line that passes through the points $(-3, 3)$ and $(3, -1)$.

Solution We begin by finding the slope of the line:

$$m = \frac{3 - (-1)}{-3 - 3} = \frac{4}{-6} = -\frac{2}{3}$$

Using $(x_1, y_1) = (3, -1)$ and $m = -\frac{2}{3}$ in $y - y_1 = m(x - x_1)$ yields

$$y + 1 = -\frac{2}{3}(x - 3)$$

$$y + 1 = -\frac{2}{3}x + 2 \qquad \text{Multiply out right side}$$

$$y = -\frac{2}{3}x + 1 \qquad \text{Add } -1 \text{ to each side}$$

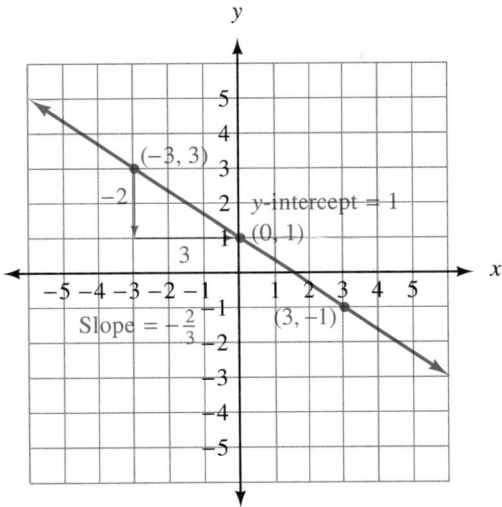

Figure 4

Figure 4 shows the graph of the line that passes through the points $(-3, 3)$ and $(3, -1)$. As you can see, the slope and y-intercept are $-\frac{2}{3}$ and 1, respectively.   ◀

Note   In Example 5 we could have used the point $(-3, 3)$ instead of $(3, -1)$ and obtained the same equation. That is, using $(x_1, y_1) = (-3, 3)$ and $m = -\frac{2}{3}$ in $y - y_1 = m(x - x_1)$ gives us

$$y - 3 = -\frac{2}{3}(x + 3)$$

$$y - 3 = -\frac{2}{3}x - 2$$

$$y = -\frac{2}{3}x + 1$$

which is the same result we obtained using $(3, -1)$.

The last form of the equation of a line that we will consider in this section is called standard form. It is used mainly to write equations in a form that is free of fractions and is easy to compare with other equations.

Standard Form for the Equation of a Line

If a, b, and c are integers, then the equation of a line is in standard form when it has the form

$$ax + by = c$$

If we were to write the equation

$$y = -\frac{2}{3}x + 1$$

in standard form, we would first multiply both sides by 3 to obtain

$$3y = -2x + 3$$

Then we would add $2x$ to each side yielding

$$2x + 3y = 3$$

which is a linear equation in standard form.

▶ **EXAMPLE 6**   Give the equation of the line through $(-1, 4)$ whose graph is perpendicular to the graph of $2x - y = -3$. Write the answer in standard form.

Solution   To find the slope of $2x - y = -3$, we solve for y:

$$2x - y = -3$$

$$y = 2x + 3$$

The slope of this line is 2. The line we are interested in is perpendicular to the line with slope 2 and must, therefore, have a slope of $-\frac{1}{2}$.
Using $(x_1, y_1) = (-1, 4)$ and $m = -\frac{1}{2}$, we have

$$y - y_1 = m(x - x_1)$$

$$y - 4 = -\frac{1}{2}(x + 1)$$

Since we want our answer in standard form we multiply each side by 2.

$$2y - 8 = -1(x + 1)$$

$$2y - 8 = -x - 1$$

$$x + 2y - 8 = -1$$
$$x + 2y = 7$$

The last equation is in standard form.   ◀

As a final note, we should mention again that all horizontal lines have equations of the form $y = b$ and slopes of 0. Vertical lines have no slope and have equations of the form $x = a$. These two special cases do not lend themselves well to either the slope-intercept form or the point-slope form of the equation of a line.

USING TECHNOLOGY: Graphing Calculators

One advantage to using a graphing calculator to graph lines is that a calculator does not care whether the equation has been simplified or not. To illustrate, in Example 5 we found that the equation of the line with slope $-\frac{2}{3}$ that passes through the point $(3, -1)$ is

$$y + 1 = -\frac{2}{3}(x - 3)$$

Normally, to graph this equation we would simplify it first. With a graphing calculator we add -1 to each side and enter the equation this way:

$$Y_1 = -(2/3)(X - 3) - 1$$

No simplification is necessary. We can graph the equation in this form and the graph will be the same as the simplified form of the equation, which is $y = -\frac{2}{3}x + 1$. To convince yourself that this is true, graph both the simplified form for the equation and the unsimplified form in the same window. As you will see, the two graphs coincide.

PROBLEM SET 5.3

Give the equation of the line with the following slope and y-intercept.

1. $m = 2, b = 3$ **2.** $m = -4, b = 2$
3. $m = 1, b = -5$ **4.** $m = -5, b = -3$
5. $m = \frac{1}{2}, b = \frac{3}{2}$ **6.** $m = \frac{2}{3}, b = \frac{5}{6}$
7. $m = 0, b = 4$ **8.** $m = 0, b = -2$

Give the slope and y-intercept for each of the following equations. Sketch the graph using the slope and y-intercept. Give the slope of any line perpendicular to the given line.

9. $y = 3x - 2$ **10.** $y = 2x + 3$
11. $2x - 3y = 12$ **12.** $3x - 2y = 12$
13. $4x + 5y = 20$
14. $5x - 4y = 20$

Write each equation in slope-intercept form. Then, graph each line using the slope and y-intercept.

15. $-2x + y = 4$ **16.** $-2x + y = 2$
17. $3x + y = 3$ **18.** $3x + y = 6$
19. $-2x - 5y = 10$
20. $-4x + 5y = 20$

For each problem below, the slope and one point on a line are given. In each case, find the equation of that line. (Write the equation for each line in slope-intercept form.)

21. $(-2, -5)$, $m = 2$ **22.** $(-1, -5)$, $m = 2$
23. $(-4, 1)$, $m = -\frac{1}{2}$ **24.** $(-2, 1)$, $m = -\frac{1}{2}$
25. $(2, -3)$, $m = \frac{3}{2}$ **26.** $(3, -4)$, $m = \frac{4}{3}$
27. $(-\frac{1}{3}, 2)$, $m = -3$ **28.** $(-\frac{2}{3}, 5)$, $m = -3$
29. $(\frac{2}{5}, \frac{3}{2})$, $m = 1$ **30.** $(\frac{3}{4}, \frac{1}{7})$, $m = -1$

Find the equation of the line that passes through each pair of points. Write your answers in standard form.

31. $(-2, -4)$, $(1, -1)$ **32.** $(2, 4)$, $(-3, -1)$
33. $(-1, -5)$, $(2, 1)$ **34.** $(-1, 6)$, $(1, 2)$
35. $(-3, -2)$, $(3, 6)$ **36.** $(-3, 6)$, $(3, -2)$
37. $(\frac{1}{3}, -\frac{1}{5})$, $(-\frac{1}{3}, -1)$ **38.** $(-\frac{1}{2}, -\frac{1}{5})$, $(\frac{1}{2}, \frac{1}{10})$
39. $(-\frac{2}{3}, \frac{1}{8})$, $(\frac{1}{3}, -\frac{1}{4})$
40. $(\frac{1}{4}, -\frac{1}{3})$, $(\frac{1}{2}, \frac{1}{6})$
41. Give the slope and y-intercept, and sketch the graph, of $y = -2$.
42. For the line $x = -3$ sketch the graph, give the slope, and name any intercepts.
43. Find the equation of the line parallel to the graph of $3x - y = 5$ that contains the point $(-1, 4)$.
44. Find the equation of the line parallel to the graph of $2x - 4y = 5$ that contains the point $(0, 3)$.
45. Line l is perpendicular to the graph of the equation $2x - 5y = 10$ and contains the point $(-4, -3)$. Find the equation for l.
46. Line l is perpendicular to the graph of the equation $-3x - 5y = 2$ and contains the point $(2, -6)$. Find the equation for l.
47. Give the equation of the line perpendicular to the graph of $y = -4x + 2$ that has an x-intercept of -1.
48. Write the equation of the line parallel to the graph of $7x - 2y = 14$ that has an x-intercept of 5.
49. Find the equation of the line with x-intercept 3 and y-intercept 2.
50. Find the equation of the line with x-intercept 2 and y-intercept 3.

51. Find the equation of the line with x-intercept $\frac{1}{2}$ and y-intercept $-\frac{1}{4}$.
52. Find the equation of the line with x-intercept $-\frac{1}{3}$ and y-intercept $\frac{1}{6}$.

Review Problems

The problems that follow review material we covered in Section 4.4. Add and subtract as indicated.

53. $\dfrac{2a - 4}{a + 2} - \dfrac{a - 6}{a + 2}$ **54.** $\dfrac{2a - 3}{a - 2} - \dfrac{a - 1}{a - 2}$

55. $3 + \dfrac{4}{3 - t}$ **56.** $6 + \dfrac{2}{5 - t}$

57. $\dfrac{3}{2x - 5} - \dfrac{39}{8x^2 - 14x - 15}$

58. $\dfrac{2}{4x - 5} + \dfrac{9}{8x^2 - 38x + 35}$

59. $\dfrac{1}{x - y} - \dfrac{3xy}{x^3 - y^3}$

60. $\dfrac{1}{x + y} + \dfrac{3xy}{x^3 + y^3}$

One Step Further

61. Label the units on your graph paper in multiples of 10 and graph the line $2x + 5y = 100$.
62. Label the units on your graph paper in multiples of 20 and graph the line $-4x + 10y = 100$.
63. Label your y-axis in multiples of 10 and your x-axis in multiples of 1 and graph the equation $y = 20x - 50$.
64. Label your y-axis in multiples of 10 and your x-axis in multiples of 1 and graph the equation $y = -20x + 30$.

Write each equation in slope-intercept form and then name the slope, the y-intercept, and the x-intercept.

65. $\dfrac{x}{2} + \dfrac{y}{3} = 1$

66. $\dfrac{x}{5} + \dfrac{y}{4} = 1$

67. $\dfrac{x}{-2} + \dfrac{y}{3} = 1$ **68.** $\dfrac{x}{2} + \dfrac{y}{-3} = 1$

69. When a linear equation is written in the form

$$\frac{x}{a} + \frac{y}{b} = 1$$

it is said to be in *two-intercept form*. Find the *x*-intercept, the *y*-intercept, and the slope of this line.

Research Project 6

In the introduction to this chapter we mentioned that René Descartes, the inventor of the rectangular coordinate system, is the person who made the statement "I think, therefore I am." Blaise Pascal, who we have mentioned previously, is responsible for the statement, "The heart has its reasons which reason does not know." Although Pascal and Descartes were contemporaries, the philosophies of the two men differed greatly. Research the philosophy of both Descartes and Pascal and then write an essay that gives the main points of each man's philosophy. In the essay, show how the quotations given above fit in with the philosophy of the man responsible for the quotation.

SECTION

5.4   # Linear Inequalities in Two Variables

A **linear inequality in two variables** is any expression that can be put in the form

$$ax + by < c$$

where a, b, and c are real numbers (a and b not both 0). The inequality symbol can be any one of the following four: $<, \leq, >, \geq$.

Some examples of linear inequalities are

$$2x + 3y < 6 \qquad y \geq 2x + 1 \qquad x - y \leq 0$$

Although not all of these examples have the form $ax + by < c$, each one can be put in that form.

The solution set for a linear inequality is a section of the coordinate plane. The boundary for the section is found by replacing the inequality symbol with an equal sign and graphing the resulting equation. The boundary is included in the solution set (and represented with a solid line) if the inequality symbol used originally is $\leq$ or $\geq$. The boundary is not included (and is represented with a broken line) if the original symbol is $<$ or $>$.

Let's look at some examples.

▶ **EXAMPLE 1**   Graph the solution set for $x + y \leq 4$.

Solution   The boundary for the graph is the graph of $x + y = 4$; the *x*- and *y*-intercepts are both 4. (Remember, the *x*-intercept is found by letting $y = 0$, and the *y*-intercept by letting $x = 0$.) The boundary is included in the solution set because the inequality symbol is $\leq$.

Here is the graph of the boundary:

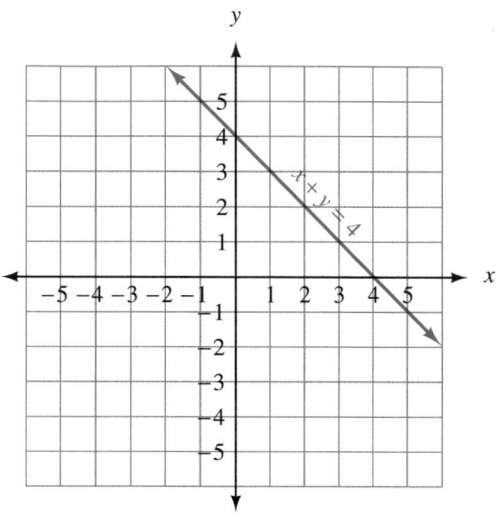

Figure 1

The boundary separates the coordinate plane into two sections or regions—the region above the boundary and the region below the boundary. The solution set for $x + y \leq 4$ is one of these two regions along with the boundary. To find the correct region, we simply choose any convenient point that is *not* on the boundary. We then substitute the coordinates of the point into the original inequality $x + y \leq 4$. If the point we choose satisfies the inequality, then it is a member of the solution set, and we can assume that all points on the same side of the boundary as the chosen point are also in the solution set. If the coordinates of our point do not satisfy the original inequality, then the solution set lies on the other side of the boundary.

In this example, a convenient point that is not on the boundary is the origin.

Substituting	$(0, 0)$
into	$x + y \leq 4$
gives us	$0 + 0 \leq 4$
	$0 \leq 4$ A true statement

Since the origin is a solution to the inequality $x + y \leq 4$, and the origin is below the boundary, all other points below the boundary are also solutions.

Here is the graph of $x + y \leq 4$:

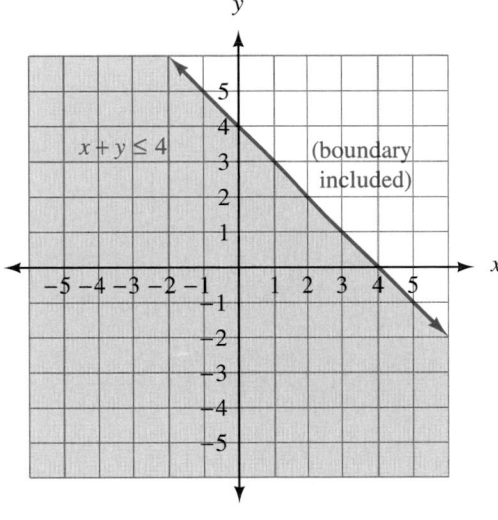

Figure 2

The region above the boundary is described by the inequality $x + y > 4$.
◀

Here is a list of steps to follow when graphing the solution set for linear inequalities in two variables.

Strategy for Graphing a Linear Inequality in Two Variables

Step 1: Replace the inequality symbol with an equal sign. The resulting equation represents the boundary for the solution set.

Step 2: Graph the boundary found in step 1 using a *solid line* if the boundary is included in the solution set (that is, if the original inequality symbol was either $\leq$ or $\geq$). Use a *broken line* to graph the boundary if it is *not* included in the solution set. (It is not included if the original inequality was either $<$ or $>$.)

Step 3: Choose any convenient point not on the boundary and substitute the coordinates into the *original* inequality. If the resulting statement is *true,* the graph lies on the *same* side of the boundary as the chosen point. If the resulting statement is *false,* the solution set lies on the *opposite* side of the boundary.

▶ **EXAMPLE 2**   Graph the solution set for $y < 2x - 3$.

Solution   The boundary is the graph of $y = 2x - 3$: a line with slope 2 and y-intercept -3. The boundary is not included since the original inequality symbol is $<$. Therefore, we use a broken line to represent the boundary.

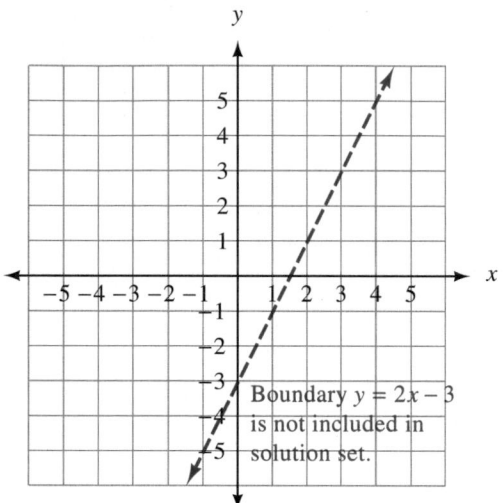

Boundary $y = 2x - 3$ is not included in solution set.

Figure 3

A convenient test point is again the origin:

Using $(0, 0)$ in $y < 2x - 3$

we have $0 < 2(0) - 3$

                      $0 < -3$ A false statement

Since our test point gives us a false statement and it lies above the boundary, the solution set must lie on the other side of the boundary.

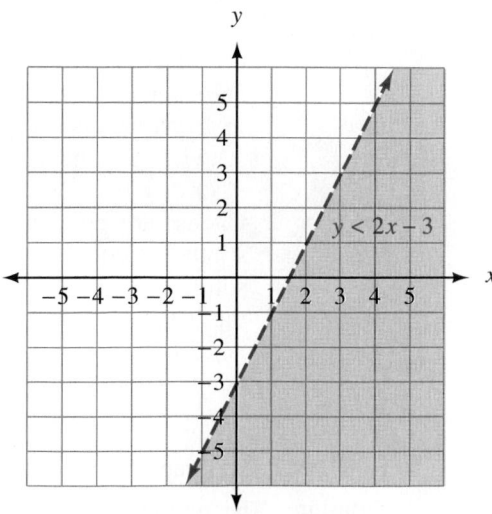

$y < 2x - 3$

Figure 4

▶ **EXAMPLE 3**   Graph the solution set for $x \le 5$.

Solution   The boundary is $x = 5$, which is a vertical line. All points to the left have x-coordinates less than 5 and all points to the right have x-coordinates greater than 5.

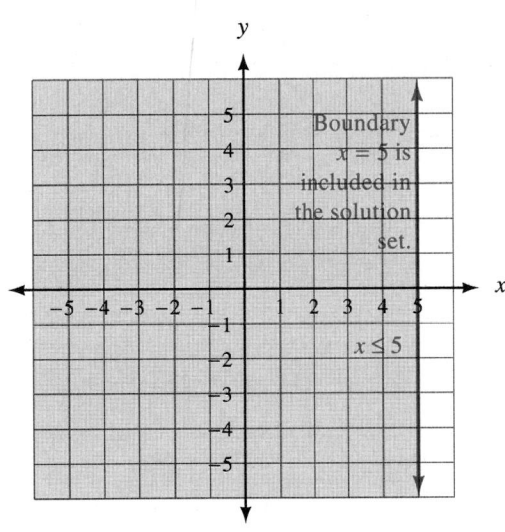

Figure 5   ◀

PROBLEM SET 5.4

Graph the solution set for each of the following.

1. $x + y < 5$

2. $x + y \le 5$

3. $x - y \ge -3$

4. $x - y > -3$

5. $2x + 3y < 6$

6. $2x - 3y > -6$

7. $-x + 2y > -4$

8. $-x - 2y < 4$

9. $2x + y < 5$

10. $2x + y < -5$

11. $3x + 5y > 7$

12. $3x - 5y > 7$

13. $y < 2x - 1$

14. $y \ge 2x - 1$

15. $y \ge -3x - 4$

16. $y < -3x + 4$

17. $x \ge 3$

18. $x < -2$

19. $y \le 4$

20. $y > -5$

21. $y < 2x$

22. $y > -3x$

23. $y \ge \frac{1}{2}x$

24. $y \le \frac{1}{3}x$

25. $y \ge \frac{3}{4}x - 2$

26. $y > -\frac{2}{3}x + 3$

27. $0.03x + 0.02y - 0.06 < 0$

28. $0.05x - 0.03y - 0.15 > 0$

29. $\dfrac{x}{3} + \dfrac{y}{2} > 1$

30. $\dfrac{x}{5} + \dfrac{y}{4} < 1$

31. $\dfrac{x}{-3} + \dfrac{y}{5} \le 1$

32. $\dfrac{x}{2} + \dfrac{y}{-4} \ge 1$

Review Problems

The problems that follow review material we covered in Section 4.5.

Simplify each complex fraction.

33. $\dfrac{\frac{1}{4} - \frac{1}{3}}{\frac{1}{2} + \frac{1}{6}}$

34. $\dfrac{\frac{1}{8} - \frac{1}{3}}{\frac{1}{4} - \frac{1}{3}}$

35. $\dfrac{1 - \dfrac{2}{y}}{1 + \dfrac{2}{y}}$

36. $\dfrac{1 + \dfrac{3}{y}}{1 - \dfrac{3}{y}}$

37. $\dfrac{4 + \dfrac{4}{x} + \dfrac{1}{x^2}}{4 - \dfrac{1}{x^2}}$

38. $\dfrac{1 - \dfrac{1}{x} - \dfrac{6}{x^2}}{1 - \dfrac{9}{x^2}}$

One Step Further

Graph each pair of inequalities on the same coordinate system. Then shade the region where their solution sets intersect.

39. $y \geq 2x - 1$ and $y \leq -\frac{1}{2}x + 4$
40. $y \geq \frac{2}{3}x - 4$ and $y \geq -\frac{3}{2}x + 4$
41. $x + y \leq 3$ and $x - y \leq 3$
42. $-2x + y \leq 4$ and $2x + y \leq 4$

Graph each inequality.

43. $y < |x + 2|$ **44.** $y > |x - 2|$
45. $y > |x - 3|$ **46.** $y < |x + 3|$

SECTION

5.5

Relations, Functions, and Graphs

In Section 2.8 we introduced the concept of a function from an intuitive point of view by considering the amount of money you would earn by working from 0 to 40 hours per week on a job that paid $7.50 per hour. The equation and its graph are shown below. Note that the values that x can assume are restricted to the number of hours per week that can be worked.

$$y = 7.5x \quad \text{for} \quad 0 \leq x \leq 40$$

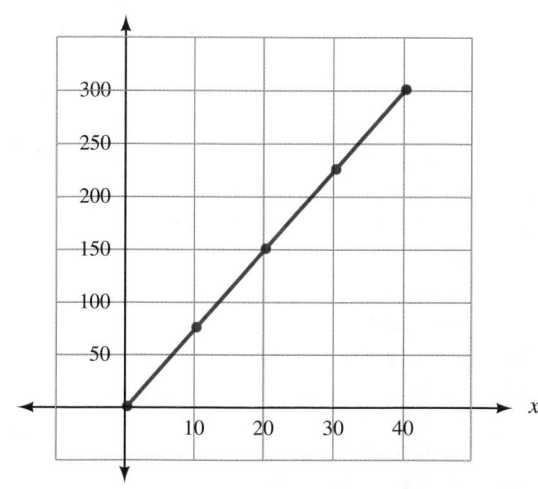

Figure 1

Now that we have found the connection between paired data and ordered pairs of numbers, we can consider the concept of a function in terms of ordered pairs of numbers. We begin with two definitions; first the definition of a **relation,** which is a new definition. then the definition of a **function,** which we have encountered before.

Relations

> **DEFINITION**   A **relation** is any set of ordered pairs. The set of all first coordinates is called the *domain* of the relation, and the set of all second coordinates is the *range* of the relation.

There are two ways to specify the ordered pairs in a relation. One method is simply to list them. The other method is to give the rule (equation) for obtaining them.

▶ **EXAMPLE 1**   The set $\{(1, 2), (-3, \frac{1}{2}), (\pi, -4), (0, 1)\}$ is a relation. The domain for this relation is $\{1, -3, \pi, 0\}$ and the range is $\{2, \frac{1}{2}, -4, 1\}$. ◀

▶ **EXAMPLE 2**   The set of ordered pairs given by

$$\{(x, y) \mid y = 7.5x, 0 \le x \le 40\}$$

is an example of a relation. In this case, the relation is written in terms of the equation that will produce the ordered pairs. As we indicated earlier, the graph in Figure 1 shows the domain to be the set

$$\{x \mid 0 \le x \le 40\}$$

Figure 1 also shows the range to be the set

$$\{y \mid 0 \le y \le 300\}$$ ◀

▶ **EXAMPLE 3**   The set of ordered pairs given by $\{(x, y) \mid x + y = 5\}$ is a relation. This relation is the set of all ordered pairs whose coordinates have a sum of 5. Note that there is no restriction on the domain. That is, unlike the relation in Example 2, we have not indicated explicitly what the values of x are. When there is no restriction on the domain, we assume that the domain contains all real numbers. (There are some exceptions to this, which we will discover later in the book.) ◀

Functions

A **function** is a relation in which no two different ordered pairs have the same first coordinates. The **domain** and **range** of a function are the sets of first and second coordinates, respectively. Here is the precise definition:

DEFINITION A **function** is a rule that pairs each element in one set, called the domain, with exactly one element from a second set, called the range.

As you can see from this definition, a function is a set of ordered pairs. In each ordered pair, the first number comes from the domain and the second number comes from the range. Since each element in the domain is paired with exactly one element in the range, no two ordered pairs will have the same first coordinates. A function is simply a relation that does not repeat any first coordinates.

▶ **EXAMPLE 4** The relation $\{(2, 3), (5, 2), (6, 3)\}$ is also a function, since no two ordered pairs have the same first coordinates. The relation $\{(1, 7), (3, 7), (1, 5)\}$ is not a function since two of its ordered pairs, $(1, 7)$ and $(1, 5)$, have the same first coordinates. ◀

Graphing Relations and Functions

Up to this point we have only graphed relations and functions whose graphs are straight lines. To give us a better perspective on functions, we need to look at some graphs that are not straight lines. The graphing we will do here will be accomplished by making tables that give us ordered pairs that satisfy the equations we are trying to graph.

▶ **EXAMPLE 5** Sketch the graph of $y = x^2$. Use the graph to specify the domain and range.

Solution Since the equation $y = x^2$ is not a linear equation, its graph will not be a straight line. We can get an idea of the shape of the graph by finding some ordered pairs that satisfy the equation, and then graphing them. We can let x take on values of -3, -2, -1, 0, 1, 2, and 3, and then find the corresponding values of y. Here is a table showing how that is done:

x	$y = x^2$	y	**Solution**
-3	$y = (-3)^2 = 9$	9	$(-3, 9)$
-2	$y = (-2)^2 = 4$	4	$(-2, 4)$
-1	$y = (-1)^2 = 1$	1	$(-1, 1)$
0	$y = 0^2 = 0$	0	$(0, 0)$
1	$y = 1^2 = 1$	1	$(1, 1)$
2	$y = 2^2 = 4$	4	$(2, 4)$
3	$y = 3^2 = 9$	9	$(3, 9)$

Graphing the seven solutions we found in the table, and then drawing a smooth curve through them, we have the graph shown in Figure 2.

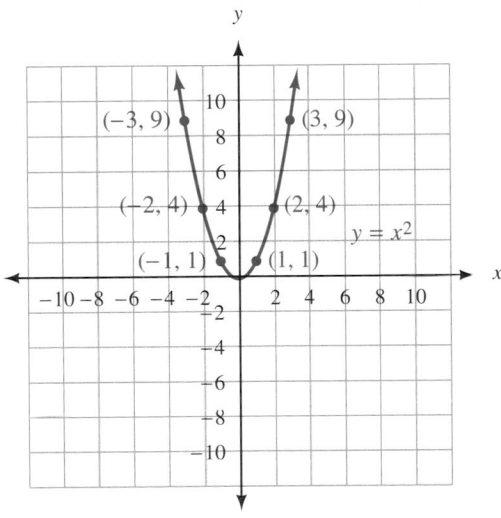

Figure 2

Since there is no restriction on the domain, we assume that it is the set of real numbers. To find the range, we look at Figure 2 and note that the graph appears only above (or on) the x-axis, which means that all the points on the graph have y-coordinates that are positive (or 0). Therefore, the range is the set $\{y \mid y \geq 0\}$. ◄

The graph shown in Figure 2 is the graph of a relation. But is it also the graph of a function? The answer is yes, because for every value of x there is one, and only one, value of y. That is, no two ordered pairs have the same first coordinate. But how can we see this by looking at the graph? The answer will become apparent when we compare the graph in Figure 2 with the graph of a relation that is not also a function. Our next example illustrates.

▶ **EXAMPLE 6**   Sketch the graph of $x = y^2$.

Solution   As you can see, the roles of x and y in the equation $x = y^2$ are reversed from the equation $y = x^2$ discussed in Example 5. Without going into much detail, we graph the equation $x = y^2$ by finding a number of ordered pairs that satisfy the equation, plotting these points, then drawing a smooth curve that connects them. A table of values for x and y that satisfy the equation is shown below, along with the graph of $x = y^2$.

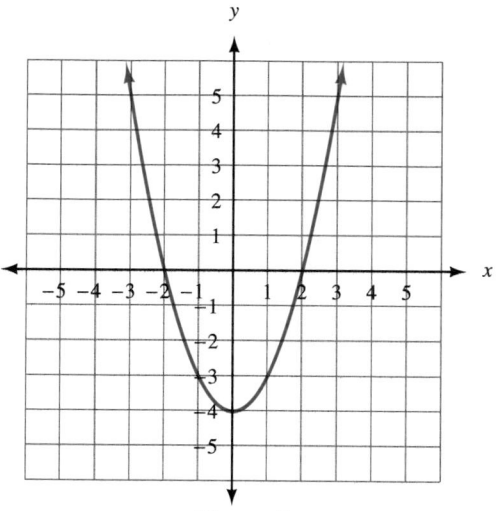

Figure 5

By the vertical line test, the graph is the graph of a function. (No vertical line can be found that will cross the graph in more than one place.) Since there is no restriction on the domain, we assume it is all real numbers. The range is $\{y \mid y \geq -4\}$ because the graph opens up vertically from -4 on the y-axis.

◀

▶ **EXAMPLE 8** Graph $y = |x|$. Use the graph to determine if we have the graph of a function. State the domain and range.

Solution We let x take on values of $-4, -3, -2, -1, 0, 1, 2, 3,$ and 4. The corresponding values of y are shown in the table.

x	y
-4	4
-3	3
-2	2
-1	1
0	0
1	1
2	2
3	3
4	4

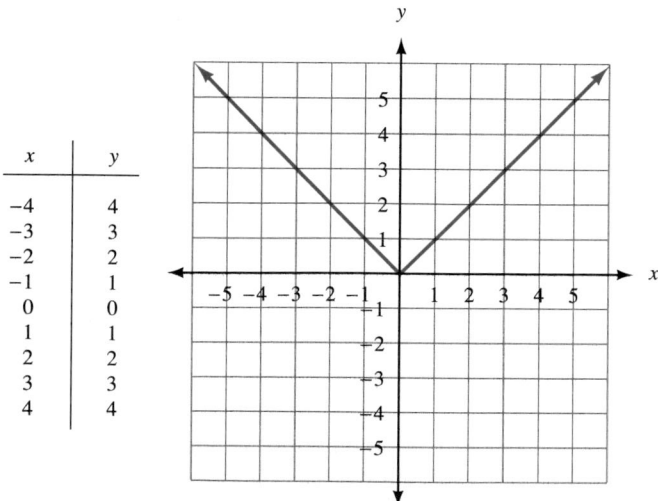

Figure 6

Since no vertical line can be found that crosses the graph in more than one place, $y = |x|$ is a function. The domain is all real numbers. The range is $\{y|y \geq 0\}$. ◀

USING TECHNOLOGY: Graphing Calculators

We can use function notation to write the formulas for the circumference and area of a circle with radius r.

$$C(r) = 2\pi r \quad \text{and} \quad A(r) = \pi r^2$$

To graph these two functions on a graphing calculator we set

$$Y_1 = 2\pi X \quad \text{and} \quad Y_2 = \pi X^2$$

Since the radius, circumference, and area of a circle are always positive quantities, we set our Window so that X and Y take on only positive numbers. Graph each of the functions above in the following windows:

Window 1: X from 0 to 3, Y from 0 to 10

Window 2: X from 0 to 3, Y from 0 to 20

Figure 7 shows graphs similar to the ones you will obtain using Window 2. Next, use the Trace feature to find the coordinates of the point at which the two curves intersect. This is the point at which the numerical value of the circumference and area of a circle are equal.

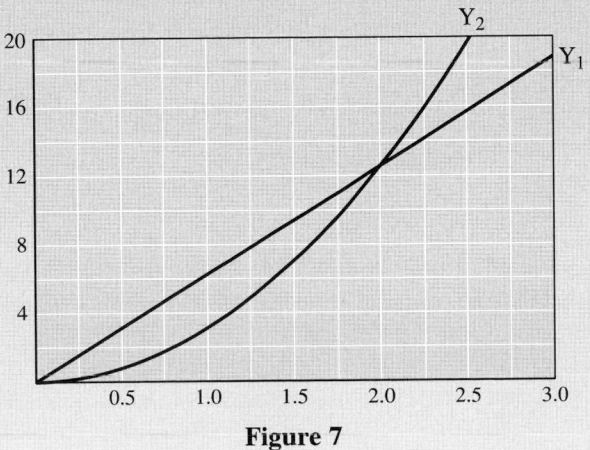

Figure 7

P R O B L E M   S E T   5 . 5

For each of the following relations, give the domain and range and indicate which are also functions.

1. {(1, 3), (2, 5), (4, 1)}
2. {(3, 1), (5, 7), (2, 3)}
3. {(−1, 3), (1, 3), (2, −5)}
4. {(3, −4), (−1, 5), (3, 2)}
5. {(7, −1), (3, −1), (7, 4)}
6. {(5, −2), (3, −2), (5, −1)}
7. {(4, 3), (3, 4), (3, 5)}
8. {(4, 1), (1, 4), (−1, −4)}
9. {(5, −3), (−3, 2), (2, −3)}
10. {(2, 4), (3, 4), (4, 4)}

Which of the following graphs represent functions?

11.

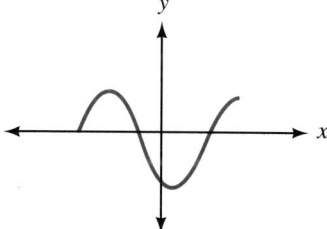

12.

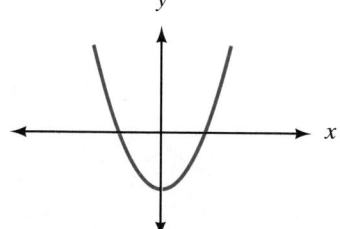

13.

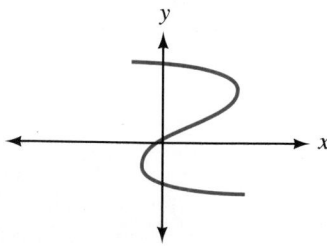

14.

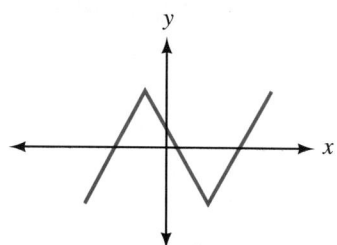

15.

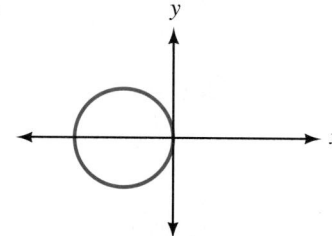

16.

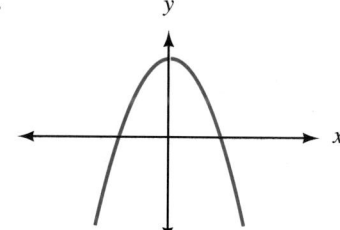

17.

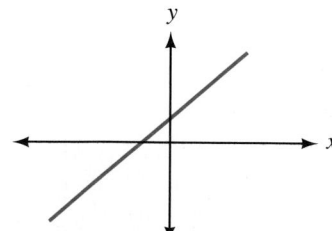

18.

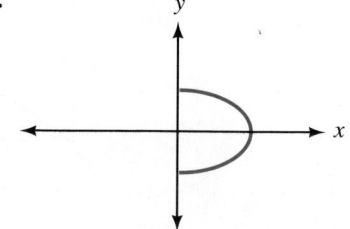

19.

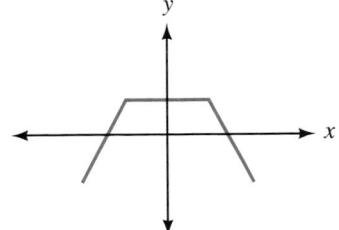

20.

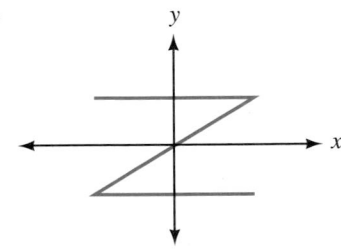

Graph each of the following relations. In each case, use the graph to find the domain and range. Then, indicate whether the graph is the graph of a function.

21. $y = x^2 - 1$

22. $y = x^2 + 1$

23. $y = x^2 + 4$

24. $y = x^2 - 9$

25. $x = y^2 - 1$

26. $x = y^2 + 1$

27. $x = y^2 + 4$

28. $x = y^2 - 9$

29. $y = |x - 2|$

30. $y = |x + 2|$

31. $y = |x| - 2$

32. $y = |x| + 2$

33. $x = |y|$

34. $x = |y| + 2$

Applying the Concepts

Suppose that you have a job that pays $7.50 per hour and that you work from 10 to 30 hours per week.

35. Write an equation, with a restriction on the variable x, that gives the amount of money, y, you will earn for working x hours in one week.

36. Graph the equation you found in Problem 35.

37. Is the equation you found in Problem 35 a function? Specify the domain and range.

38. What is the minimum amount you can earn in one week under the conditions given above? What is the maximum amount you can earn?

Suppose that you have a job that pays $8.50 per hour and that you work from 0 to 40 hours per week.

39. Write an equation, with a restriction on the variable x, that gives the amount of money, y, you will earn for working x hours in one week.

40. Graph the equation you found in Problem 39.

41. Is the equation you found in Problem 39 a function? Specify the domain and range.

42. What is the minimum amount you can earn in one week under the conditions given above? What is the maximum amount you can earn?

The formula for the area, A, of a circle with radius r is given by $A = \pi r^2$. The formula shows that A is a function of r.

43. Graph the function $A = \pi r^2$ for $0 \leq r \leq 3$. (On the graph, let the horizontal axis be the r-axis, and let the vertical axis be the A-axis.)

44. Give the domain and range of the function $A = \pi r^2$.

A rectangle is 2 inches longer than it is wide. Let x = the width, P = the perimeter, and A = the area of the rectangle.

45. Write an equation that will give the perimeter, P, in terms of the width, x, of the rectangle. Are there any restrictions on the values that x can assume?

46. Graph the relationship between P and x.

47. Write an equation that will give the area, A, in terms of the width, x, of the rectangle. Are there any restrictions on the values that x can assume?

48. Graph the relationship between A and x.

49. A ball is thrown straight up into the air from ground level. The relationship between the height, h, of the ball at any time, t, is illustrated by the following graph:

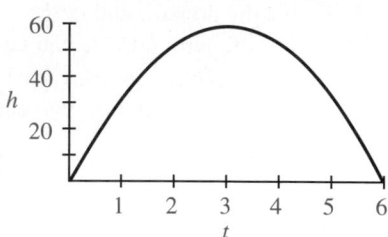

The horizontal axis represents time, t, and the vertical axis represents height, h.
(a) Is this graph the graph of a function?
(b) Identify the domain and range.
(c) At what time does the ball reach its maximum height?
(d) What is the maximum height of the ball?
(e) At what time does the ball hit the ground?

50. The following graph shows the relationship between a company's profits, P, and the number of items it sells, x. (P is in dollars.)

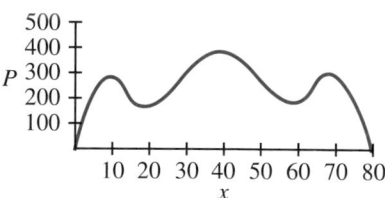

(a) Is this graph the graph of a function?
(b) Identify the domain and range.
(c) How many items must the company sell to make their maximum profit?
(d) What is their maximum profit?

Review Problems

The problems that follow review material we covered in Section 2.2.
For the equation $y = 3x - 2$,

51. find y if x is 4.
52. find y if x is 0.
53. find y if x is -4.

54. find y if x is -2.

For the equation $y = x^2 - 3$,

55. find y if x is 2.
56. find y if x is -2.
57. find y if x is 0.
58. find y if x is -4.

A company manufactures and sells prerecorded videotapes. They find that they can sell x videotapes each day at p dollars per tape, according to the equation $x = 230 - 20p$.

59. How many videotapes will they sell if $p = \$5$?
60. Use the revenue equation $R = xp$ to find the revenue they will obtain from selling the tapes for $p = \$5$ each.

One Step Further

Look back to the graph shown in Figure 5, and note that it crosses the y-axis at -4, and the x-axis at -2 and 2. As you know, the y-intercept, -4, can be found by letting $x = 0$ in the equation and solving for y. Similarly, the x-intercepts can be found by letting $y = 0$ and solving for x. For the graph shown in Figure 5, finding the x-intercepts would look like this:

$0 = x^2 - 4$	Let $y = 0$ in the equation
$0 = (x + 2)(x - 2)$	Factor right side
$x + 2 = 0$ or $x - 2 = 0$	Set factors to 0
$x = -2$ or $x = 2$	Solve for x-intercepts

Find the x- and y-intercepts for the following functions. Do not graph.

61. $y = x^2 - 9$
62. $y = x^2 - 16$
63. $y = x^2 - x - 6$
64. $y = x^2 + x - 6$
65. $y = x^2 + 10x + 25$
66. $y = x^2 - 10x + 25$

USING TECHNOLOGY

67. Using a graphing calculator, graph the functions for the perimeter and area that you found in Problems 45 and 47. Set the Y variables this way:

$$Y_1 = 4X+4 \qquad \text{This is the perimeter } P(x)$$

$$Y_2 = X(X+2) \qquad \text{This is the area } A(x)$$

Window: X from 0 to 4, Y from 0 to 20

Find the coordinates of the point where the two graphs intersect.

68. Use your Y variables list, or write a program, to graph the family of curves $Y = |X + B|$ for B = −3, −2, −1, 0, 1, 2, and 3.

69. Use your Y variables list, or write a program, to graph the family of curves $Y = |X| + B$ for B = −3, −2, −1, 0, 1, 2, and 3.

70. Use your Y variables list, or write a program, to graph the family of curves $Y = X^2 + B$ for B = −3, −2, −1, 0, 1, 2, and 3.

71. Use your Y variables list, or write a program, to graph the family of curves $Y = (X + B)^2$ for B = −3, −2, −1, 0, 1, 2, and 3.

SECTION

5.6 More about Function Notation

We introduced function notation in Section 2.8. To review, the notation $f(x)$ can be used in place of the variable y when working with functions. That is, if $f(x) = 3x - 5$, then we find $f(2)$ by replacing x with 2 in the expression $3x - 5$ and simplifying.

$$\text{If } f(x) = 3x - 5, \text{ then } f(2) = 3(2) - 5 = 1$$

In this case, we say the value of the function f at 2 is 1. That is, $f(2) = 1$. If we have a second function $g(x) = x^2 - 4$, then $g(2) = 2^2 - 4 = 0$. Again, we say the value of the function g at 2 is 0, or $g(2) = 0$. Here are some additional function values:

If $f(x) = 3x - 5$ and $g(x) = x^2 - 4$, then

$$f(0) = 3(0) - 5 = -5 \qquad\qquad g(0) = 0^2 - 4 = -4$$

$$f(4) = 3(4) - 5 = 7 \qquad\qquad g(4) = 4^2 - 4 = 12$$

$$f(-2) = 3(-2) - 5 = -11 \qquad g(-2) = (-2)^2 - 4 = 0$$

$$f\left(\frac{1}{3}\right) = 3\left(\frac{1}{3}\right) - 5 = -4 \qquad g\left(\frac{1}{3}\right) = \left(\frac{1}{3}\right)^2 - 4 = -\frac{35}{9}$$

$$f(a) = 3a - 5 \qquad\qquad g(a) = a^2 - 4$$

Let's return to the wage functions we discussed in Section 2.8 to see if we can find a visual representation for function notation. Here are the two functions from Section 2.8:

$$f(x) = 7.5x, \ 0 \le x \le 40 \quad \text{and} \quad g(x) = 6.5x, \ 0 \le x \le 40$$

where $f(x)$ is the amount of money earned by working x hours per week at $7.50 per hour. Likewise, $g(x)$ is the amount of money earned by working x hours per week at $6.50 per hour.

We can visualize the relationship between x and $f(x)$ or $g(x)$ on the graphs of the two functions. Figure 1 shows the graph of $f(x) = 7.5x$ along with two additional line segments. The horizontal line segment corresponds to $x = 20$, while the vertical line segment corresponds to $f(20)$. Figure 2 shows the graph of $g(x) = 6.5x$ along with the horizontal line segment that corresponds to $x = 20$, and the vertical line segment that corresponds to $g(20)$. (Note that the domain in each case is restricted to $0 \le x \le 40$.)

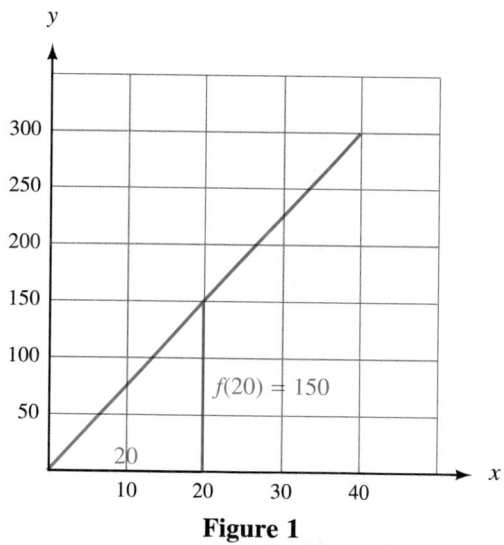

Figure 1

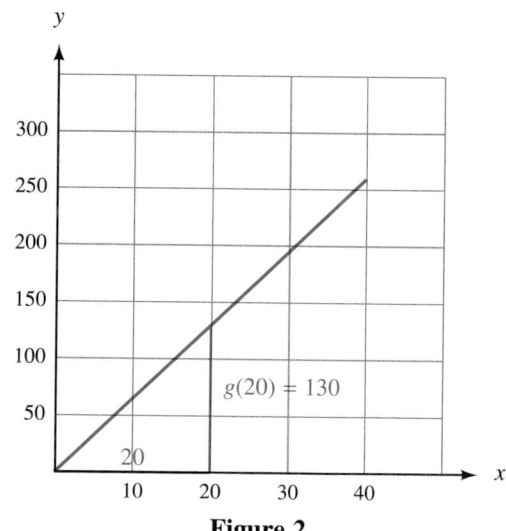

Figure 2

The remaining examples in this section will reinforce the ideas we have presented here concerning function notation.

▶ **EXAMPLE 1** A balloon has the shape of a sphere with a radius of 3 inches. Use the formulas below to find the volume and the surface area of the balloon.

$$V(r) = \frac{4}{3} \pi r^3 \qquad S(r) = 4 \pi r^2$$

Solution As you can see, we have used function notation to write the two formulas for volume and surface area, because each quantity is a function of the radius. To find these quantities when the radius is 3 inches, we evaluate $V(3)$ and $S(3)$:

$$V(3) = \frac{4}{3}\pi 3^3 = \frac{4}{3}\pi 27 = 36\pi \text{ inches}^3 \quad \text{or} \quad 113 \text{ inches}^3 \text{ to the nearest} \\ \text{whole number}$$

$$S(3) = 4\pi 3^2 = 36\pi \text{ inches}^2 \quad \text{or} \quad 113 \text{ inches}^2 \text{ to the nearest whole number}$$

The fact that $V(3) = 36\pi$ means that the ordered pair $(3, 36\pi)$ belongs to the function V. Likewise, the fact that $S(3) = 36\pi$ tells us that the ordered pair $(3, 36\pi)$ is a member of function S. ◀

We can generalize the discussion at the end of Example 1 this way:

$$(a, b) \in f \quad \text{if and only if} \quad f(a) = b$$

where $\in$ is read "belongs to."

▶ **EXAMPLE 2**   If the function f is given by

$$f = \{(-2, 0), (3, -1), (2, 4), (7, 5)\}$$

then $f(-2) = 0, f(3) = -1, f(2) = 4,$ and $f(7) = 5$. ◀

▶ **EXAMPLE 3**   If $f(x) = x^2 - x + 1$, find $f(3), f(a),$ and $f(a + 3)$. Use the results to show that $f(a + 3) \neq f(a) + f(3)$.

Solution   Replacing x with 3, a, and $a + 3$ we have

$$f(3) = 3^2 - 3 + 1 = 9 - 3 + 1 = 7$$

$$f(a) = a^2 - a + 1$$

$$f(a + 3) = (a + 3)^2 - (a + 3) + 1$$
$$= (a^2 + 6a + 9) - (a + 3) + 1$$
$$= a^2 + 5a + 7$$

To show that this last expression is not equal to $f(a) + f(3)$ we simply note that

$$f(a) + f(3) = (a^2 - a + 1) + 7 = a^2 - a + 8 \neq f(a + 3) \qquad ◀$$

▶ **EXAMPLE 4**   If $f(x) = 2x^2$ and $g(x) = 3x - 1$,   find   (a) $f[g(2)]$; (b) $g[f(2)]$.

Solutions   The expression $f[g(2)]$ is read "f of g of 2."
(a) Since $g(2) = 3(2) - 1 = 5$,
$$f[g(2)] = f(5) = 2(5)^2 = 50$$
(b) Since $f(2) = 2(2)^2 = 8$,
$$g[f(2)] = g(8) = 3(8) - 1 = 23 \qquad ◀$$

Many of the equations and formulas we have worked with previously can be written in terms of function notation. For example, if a company sells x items at a price of p dollars per item, then, in function notation

$R(x)$ is the revenue function that gives the revenue R in terms of the number of items x.

$R(p)$ is the revenue function that gives the revenue R in terms of the price per item p.

With function notation we can see exactly which variables we want our formulas written in terms of.

In the next two examples, we will use function notation to combine a number of problems we have worked previously.

▶ **EXAMPLE 5**   A company manufactures and sells prerecorded videotapes. They find that they can sell x videotapes each day at p dollars per tape, according to the equation $x = 230 - 20p$. Find $R(x)$ and $R(p)$.

Solution   The notation $R(p)$ tells us we are to write the revenue equation in terms of the variable p. To do so, we use the formula $R(p) = xp$ and substitute $230 - 20p$ for x to obtain

$$R(p) = xp = (230 - 20p)p = 230p - 20p^2$$

The notation $R(x)$ indicates that we are to write the revenue equation in terms of the variable x. We need to solve the equation $x = 230 - 20p$ for p. Let's begin by interchanging the two sides of the equation:

$$230 - 20p = x$$

$$-20p = -230 + x \qquad \text{Add } -230 \text{ to each side}$$

$$p = \frac{-230 + x}{-20} \qquad \text{Divide each side by } -20$$

$$p = 11.5 - 0.05x \qquad \frac{230}{20} = 11.5 \text{ and } \frac{1}{20} = 0.05$$

Now we can find $R(x)$ by substituting $11.5 - 0.05x$ for p in the formula $R(x) = xp$:

$$R(x) = xp = x(11.5 - 0.05x) = 11.5x - 0.05x^2$$

Our two revenue functions are actually equivalent. To offer some justification for this, suppose that the company decides to sell each tape for $5. The equation $x = 230 - 20p$ indicates that, at $5 per tape, they will sell $x = 230 - 20(5) = 230 - 100 = 130$ tapes per day. To find the revenue from selling the tapes for $5 each, we use $R(p)$ with $p = 5$:

$$\text{If} \qquad p = 5$$

$$\text{then} \qquad R(p) = R(5)$$

$$= 230(5) - 20(5)^2$$
$$= 1{,}150 - 500$$
$$= \$650$$

On the other hand, to find the revenue from selling 130 tapes, we use $R(x)$ with $x = 130$:

$$\text{If} \qquad x = 130$$
$$\text{then} \quad R(x) = R(130)$$
$$= 11.5(130) - 0.05(130)^2$$
$$= 1{,}495 - 845$$
$$= \$650 \qquad \blacktriangleleft$$

▶ **EXAMPLE 6** Suppose the daily cost function for the videotapes in Example 5 is $C(x) = 200 + 2x$. Find the profit function $P(x)$ and then find $P(130)$.

Solution Since profit is equal to the difference of the revenue and the cost, we have

$$P(x) = R(x) - C(x)$$
$$= 11.5x - 0.05x^2 - (200 + 2x)$$
$$= -0.05x^2 + 9.5x - 200$$

Notice that we used the formula for $R(x)$ from Example 5 instead of the formula for $R(p)$. We did so because we were asked to find $P(x)$, meaning we want the profit P only in terms of the variable x.

Next, we use the formula we just obtained to find $P(130)$:

$$P(130) = -0.05(130)^2 + 9.5(130) - 200$$
$$= -0.05(16{,}900) + 9.5(130) - 200$$
$$= -845 + 1{,}235 - 200$$
$$= \$190$$

Since $P(130) = \$190$, the company will make a profit of $190 per day by selling 130 tapes per day. $\blacktriangleleft$

USING TECHNOLOGY: Graphing Calculators

More about Example 1 If we look back to Example 1 of this section (page 320) we see that when the radius of a sphere is 3, the numerical values of the volume and surface area are equal. How unusual is this? Are there other values of r for which $V(r)$ and $S(r)$ are equal? We can answer this question by looking at the graphs of both V and S.

To graph the function $V(r) = \frac{4}{3}\pi r^3$, set $Y_1 = 4\pi X^3/3$. To graph $S(r) = 4\pi r^2$, set $Y_2 = 4\pi X^2$. Graph the two functions in each of the following windows:

Window 1: X from -4 to 4, Y from -2 to 10

Window 2: X from 0 to 4, Y from 0 to 50

Window 3: X from 0 to 4, Y from 0 to 100

Then use the Trace and Zoom features of your calculator to locate the point in the first quadrant where the two graphs intersect. How do the coordinates of this point compare with the results in Example 1?

More about Example 6   We can visualize the three functions given in Example 6 (page 323) if we set up the functions list and graphing window on our calculator this way:

$Y_1 = 11.5X - 0.05X^2$ This gives the graph of $R(x)$

$Y_2 = 200 + 2X$ This gives the graph of $C(x)$

$Y_3 = Y_1 - Y_2$ This gives the graph of $P(x)$

Window: X from 0 to 250, Y from 0 to 750

The graphs in Figure 3 are similar to what you will obtain using the functions list and window above. Next, find the value of $P(x)$ when $R(x)$ and $C(x)$ intersect.

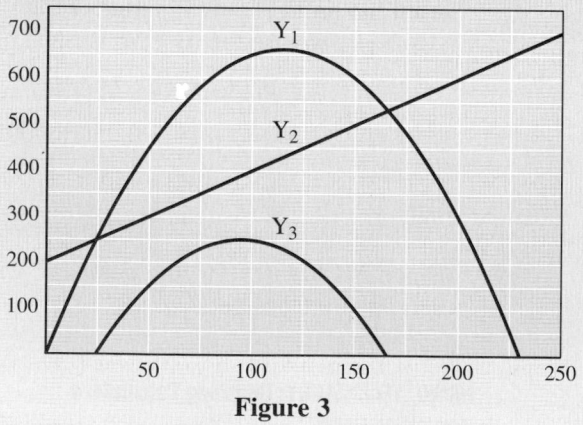

Figure 3

PROBLEM SET 5.6

Let $f(x) = 5x - 3$ and $g(x) = x^2 - 9$. Evaluate the following.

1. $f(2)$ **2.** $f(3)$

3. $f(-3)$ **4.** $g(-2)$

5. $g(-1)$

6. $f(-4)$

7. $g(-3)$

8. $g(2)$

9. $g(4) + f(4)$

10. $f(2) - g(3)$

11. $f(3) - g(2)$

12. $g(-1) + f(-1)$

If $f = \{(1, 4),\ (-2, 0),\ (3, \frac{1}{2}),\ (\pi, 0)\}$ and $g = \{(1, 1),\ (-2, 2),\ (\frac{1}{2}, 0)\}$, find each of the following values of f and g.

13. $f(1)$

14. $g(1)$

15. $g(\frac{1}{2})$

16. $f(3)$

17. $g(-2)$

18. $f(\pi)$

19. $f(-2) + g(-2)$

20. $g(1) + f(1)$

Let $f(x) = 2x^2 - 8$ and $g(x) = \frac{1}{2}x + 1$. Evaluate each of the following.

21. $f(0)$

22. $g(0)$

23. $g(-4)$

24. $f(1)$

25. $f(a)$

26. $g(z)$

27. $f(a + 3)$

28. $g(a - 2)$

29. $f[g(2)]$

30. $g[f(2)]$

31. $g[f(-1)]$

32. $f[g(-2)]$

33. $g[f(0)]$

34. $f[g(0)]$

35. Graph the function $f(x) = \frac{1}{2}x + 2$, then draw and label the line segments that represent $x = 4$ and $f(4)$.

36. Graph the function $f(x) = -\frac{1}{2}x + 6$, then draw and label the line segments that represent $x = 4$ and $f(4)$.

37. For the function $f(x) = \frac{1}{2}x + 2$, find the value of x for which $f(x) = x$.

38. For the function $f(x) = -\frac{1}{2}x + 6$, find the value of x for which $f(x) = x$.

39. Graph the function $f(x) = x^2$, then draw and label the line segments that represent $x = 1$ and $f(1)$, $x = 2$ and $f(2)$, and finally, $x = 3$ and $f(3)$.

40. Graph the function $f(x) = x^2 - 2$, then draw and label the line segments that represent $x = 2$ and $f(2)$, and the line segments corresponding to $x = 3$ and $f(3)$.

41. For the function $f(x) = x^2$, find the values of x for which $f(x) = x$.

42. For the function $f(x) = x^2 - 2$, find the values of x for which $f(x) = x$.

Applying the Concepts

Suppose you have a job that pays $7.25 per hour and that you work anywhere from 0 to 40 hours per week.

43. Write an equation, with a restriction on the variable x, that gives the amount of money, $f(x)$, you will earn for working x hours in one week.

44. What does $f(20)$ represent?

45. Graph the equation you found in Problem 43, and then show the relationship between $x = 20$, $f(20)$, and the graph.

46. Why doesn't it make sense to ask for $f(50)$?

A salesperson has a base salary of $200 per week and earns a commission of 5.5% of the amount he sells each week.

47. How much will the salesperson earn if he sells $500 worth of merchandise in one week?

48. How much will the salesperson earn if he sells $1,000 worth of merchandise in one week?

49. If x is the amount of the salesperson's weekly sales and $f(x)$ is the amount of money earned in one week, write a formula that shows the relationship between x and $f(x)$. (Be sure to note any restrictions on the variable x.)

50. Graph the function you wrote in Problem 49 and draw a vertical line segment, starting at 200 on the x-axis, that represents $f(200)$.

51. A company selling diskettes for home computers finds it can sell x diskettes per day at p dollars per diskette, according to the formula $x = 800 - 100p$. Find $R(p)$ and $R(x)$.

52. A company sells an inexpensive accounting program for home computers. If they can sell x programs per week at p dollars per program, according to the formula $x = 350 - 10p$, find formulas for $R(p)$ and $R(x)$.

53. If the cost to produce the x diskettes in Problem 51 is $C(x) = 2x + 200$, find $P(x)$ and $P(40)$.

54. If the cost to produce the x programs in Problem 52 is $C(x) = 5x + 500$, find $P(x)$ and $P(60)$.

55. Find the volume and surface area of a sphere with radius 5 inches.

56. For the sphere in Problem 55, find the increase in the volume and surface area if the radius is increased to 6 inches.

57. If $f(x) = x^2 - 9$, show that
$f(x + 3) \neq f(x) + f(3)$.

58. If $g(x) = 4x^2 - 1$, show that
$g(x + 2) \neq g(x) + g(2)$.

59. If $f(x) = 3x - 5$, show that
$f(a + b) \neq f(a) + f(b)$.

60. If $g(x) = 2x - 1$, show that
$g(a + b) \neq g(a) + g(b)$.

Review Problems

The problems that follow review material we covered in Section 2.6. Solve each equation.

61. $|x| = 3$ **62.** $|x| = -3$

63. $|2y + 1| = -5$ **64.** $|2y + 1| = 5$

65. $|1 - 3t| = 5$ **66.** $|3t - 1| = 5$

One Step Further

A function is called an **even function** if $f(-x) = f(x)$ for every value of x. Show that each of the following functions is an even function.

67. $f(x) = x^2 - 4$ **68.** $f(x) = x^4 + x^2$

69. $f(x) = x^{-2}$ **70.** $f(x) = x^2 - x^{-2}$

A function is called an **odd function** if $f(-x) = -f(x)$ for every value of x. Show that each of the following functions is an odd function.

71. $f(x) = 3x$ **72.** $f(x) = x^3$

73. $f(x) = x^3 - x$ **74.** $f(x) = \dfrac{5}{x}$

USING TECHNOLOGY

75. Using a graphing calculator, we can visualize the relationship between the demand equation $x = 800 - 100p$ and the revenue function $R(p) = (800 - 100p)p$ as given in Problem 51. Set up the functions this way:

$$Y_1 = 800 - 100X$$

$$Y_2 = Y_1 * X$$

Graph the two functions in the following windows:

Window 1: X from 0 to 8, Y from 0 to 800

Window 2: X from 0 to 8, Y from 0 to 1750

Use the Trace feature to find the largest value $R(p)$ can assume. What situation concerning the sale of diskettes does this number correspond to?

76. Convert the function $C(x) = 2x + 200$ to a function of p and add its graph to the two graphs in Window 2 of Problem 75. Use the Trace feature to find the points where the graphs of C and R intersect. These are called the breakeven points.

SECTION

5.7   Variation

There are two main types of variation—direct variation and inverse variation. Variation problems are most common in the sciences, particularly in chemistry and physics.

Direct Variation

When we say the variable y **varies directly** with the variable x, we mean that the relationship can be written in symbols as $y = Kx$ where K is a nonzero constant called the **constant of variation** (proportionality constant).

Another way of saying y varies directly with x is to say y is **directly proportional** to x.

Study the following list. It gives the mathematical equivalent of some direct-variation statements.

English Phrase	Algebraic Equation
y varies directly with x	$y = Kx$
s varies directly with the square of t	$s = Kt^2$
y is directly proportional to the cube of z	$y = Kz^3$
u is directly proportional to the square root of v	$u = K\sqrt{v}$

▶ **EXAMPLE 1** y varies directly with x. If y is 15 when x is 5, find y when x is 7.

Solution   The first sentence gives us the general relationship between x and y. The equation equivalent to the statement "y varies directly with x" is

$$y = Kx$$

The first part of the second sentence in our example gives us the information necessary to evaluate the constant K:

When $\qquad y = 15$

and $\qquad x = 5$

the equation $\qquad y = Kx$

becomes $\qquad 15 = K \cdot 5$

or $\qquad K = 3$

The equation can now be written specifically as

$$y = 3x$$

Letting $x = 7$, we have

$$y = 3 \cdot 7$$

$$y = 21$$

◀

▶ **EXAMPLE 2**   The distance a body falls from rest toward the earth is directly proportional to the square of the time it has been falling. If a body falls 64 feet in 2 seconds, how far will it fall in 3.5 seconds?

Solution   We will let d = distance and t = time. Since distance is directly proportional to the square of time, we have

$$d = Kt^2$$

Next we evaluate the constant K:

When	$t = 2$
and	$d = 64$
the equation	$d = Kt^2$
becomes	$64 = K(2)^2$
or	$64 = 4K$
and	$K = 16$

Specifically, then, the relationship between d and t is

$$d = 16t^2$$

Finally, we find d when $t = 3.5$:

$$d = 16(3.5)^2$$
$$d = 16(12.25)$$
$$d = 196 \text{ feet}$$ ◀

Note   The equation $d = 16t^2$ is a specific example of the equation used in physics to find the distance a body falls from rest during a time t. In physics, the equation is $d = \frac{1}{2}gt^2$, where g is the acceleration of gravity. On earth the acceleration of gravity is $g = 32$ feet/second^2. The equation, then, is $d = \frac{1}{2} \cdot 32t^2$ or $d = 16t^2$.

Inverse Variation

If two variables are related so that an *increase* in one produces a proportional *decrease* in the other, then the variables are said to **vary inversely.** If y varies inversely with x, then $y = K\dfrac{1}{x}$ or $y = \dfrac{K}{x}$. We can also say y is **inversely proportional** to x. The constant K is again called the *constant of variation* or *proportionality constant.*

English Phrase	Algebraic Equation
y is inversely proportional to x	$y = \dfrac{K}{x}$
s varies inversely with the square of t	$s = \dfrac{K}{t^2}$
y is inversely proportional to x^4	$y = \dfrac{K}{x^4}$
z varies inversely with the cube root of t	$z = \dfrac{K}{\sqrt[3]{t}}$

▶ **EXAMPLE 3** y varies inversely with the square of x. If y is 4 when x is 5, find y when x is 10.

Solution Since y is inversely proportional to the square of x, we can write

$$y = \frac{K}{x^2}$$

Evaluating K using the information given, we have

$$\text{When} \qquad x = 5$$
$$\text{and} \qquad y = 4$$
$$\text{the equation} \qquad y = \frac{K}{x^2}$$
$$\text{becomes} \qquad 4 = \frac{K}{5^2}$$
$$\text{or} \qquad 4 = \frac{K}{25}$$
$$\text{and} \qquad K = 100$$

Now we write the equation again as

$$y = \frac{100}{x^2}$$

We finish by substituting $x = 10$ into the last equation:

$$y = \frac{100}{10^2} = \frac{100}{100} = 1$$

◀

▶ **EXAMPLE 4**   The volume of a gas is inversely proportional to the pressure of the gas on its container. If a pressure of 48 pounds per square inch corresponds to a volume of 50 cubic feet, what pressure is needed to produce a volume of 100 cubic feet?

Solution   We can represent volume with V and pressure with P:

$$V = \frac{K}{P}$$

Using $P = 48$ and $V = 50$, we have

$$50 = \frac{K}{48}$$

$$K = 50(48)$$

$$K = 2,400$$

The equation that describes the relationship between P and V is

$$V = \frac{2,400}{P}$$

Substituting $V = 100$ into this last equation, we get

$$100 = \frac{2,400}{P}$$

$$100P = 2,400$$

$$P = \frac{2,400}{100}$$

$$P = 24$$

A volume of 100 cubic feet is produced by a pressure of 24 pounds per square inch.   ◀

Note   The relationship between pressure and volume as given in this example is known as Boyle's Law and applies to situations such as those encountered in a piston-cylinder arrangement. It was Robert Boyle who, in 1662, published the results of some of his experiments that showed, among other things, that the volume of a gas decreases as the pressure increases. This is an example of inverse variation.

Joint Variation and Other Variation Combinations

Many times relationships among different quantities are described in terms of more than two variables. If the variable y varies directly with *two* other variables, say x and z, then we say y **varies jointly** with x and z. In addition to joint variation, there are many other combinations of direct and inverse variation involving more than two variables. The following table is a list of some variation statements and their equivalent mathematical forms:

English Phrase	Algebraic Equation
y varies jointly with x and z	$y = Kxz$
z varies jointly with r and the square of s	$z = Krs^2$
V is directly proportional to T and inversely proportional to P	$V = \dfrac{KT}{P}$
F varies jointly with m_1 and m_2 and inversely with the square of r	$F = \dfrac{Km_1 \cdot m_2}{r^2}$

▶ **EXAMPLE 5** y varies jointly with x and the square of z. When x is 5 and z is 3, y is 180. Find y when x is 2 and z is 4.

Solution   The general equation is given by

$$y = Kxz^2$$

Substituting $x = 5$, $z = 3$, and $y = 180$, we have

$$180 = K(5)(3)^2$$
$$180 = 45K$$
$$K = 4$$

The specific equation is

$$y = 4xz^2$$

When $x = 2$ and $z = 4$, the last equation becomes

$$y = 4(2)(4)^2$$
$$y = 128$$ ◀

▶ **EXAMPLE 6**   In electricity, the resistance of a cable is directly proportional to its length and inversely proportional to the square of the diameter. If a 100-foot cable 0.5 inch in diameter has a resistance of 0.2 ohm, what will be the resistance of a cable made from the same material if it is 200 feet long with a diameter of 0.25 inch?

Solution   Let R = resistance, l = length, and d = diameter. The equation is

$$R = \frac{Kl}{d^2}$$

When $R = 0.2$, $l = 100$, and $d = 0.5$, the equation becomes

$$0.2 = \frac{K(100)}{(0.5)^2}$$

or

$$K = 0.0005$$

Using this value of K in our original equation, the result is

$$R = \frac{0.0005l}{d^2}$$

When $l = 200$ and $d = 0.25$, the equation becomes

$$R = \frac{0.0005(200)}{(0.25)^2}$$

$$R = 1.6 \text{ ohms} \qquad \blacktriangleleft$$

Graphing Direct and Inverse Variation Statements

When y varies directly with x, the relationship is always written $y = Kx$. For any value of K, the graph of $y = Kx$ is a straight line with slope K and y-intercept 0 (in slope-intercept form the equation is $y = Kx + 0$). Since the y-intercept is 0, the graph of every direct-variation statement of the form $y = Kx$ will pass through the origin.

▶ **EXAMPLE 7**   Graph the direct-variation statements $y = 2x$ and $y = -2x$ on the same coordinate system.

Solution   The graph of each equation is a straight line that passes through the origin. The graph of $y = 2x$ has a slope of 2, while the graph of $y = -2x$ has a slope of -2. Both graphs are shown in Figure 1.

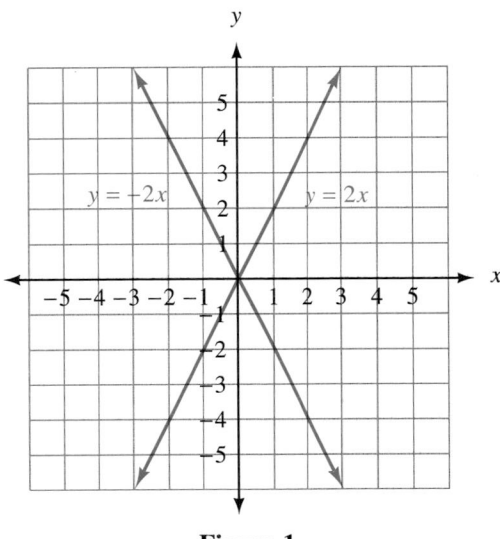

$y = -2x$ $y = 2x$

Figure 1

Next, we will graph two simple inverse variation statements.

▶ **EXAMPLE 8**   Graph the inverse variation statement $y = \dfrac{1}{x}$.

Solution   Since this is the first time we have graphed an equation of this form, we will make a table of values for x and y that satisfy the equation. Before we do, let's make some generalizations about the graph.

First, notice that, since y is equal to 1 divided by x, y will be positive when x is positive. (The quotient of two positive numbers is a positive number.) Likewise, when x is negative, y will be negative. In other words, x and y will always have the same sign. Thus, our graph will appear in quadrants 1 and 3 only, because in those quadrants x and y have the same sign.

Next, notice that the expression $\frac{1}{x}$ will be undefined when x is 0, meaning that there is no value of y corresponding to $x = 0$. Because of this, the graph will not cross the y-axis. Further, the graph will not cross the x-axis either. If we try to find the x-intercept by letting $y = 0$, we have

$$0 = \frac{1}{x}$$

But there is no value of x to divide into 1 to obtain 0. Therefore, since there is no solution to this equation, our graph will not cross the x-axis.

To summarize, we can expect to find the graph in quadrants 1 and 3 only, and the graph will cross neither axis. The graph is shown in Figure 2 on the next page.

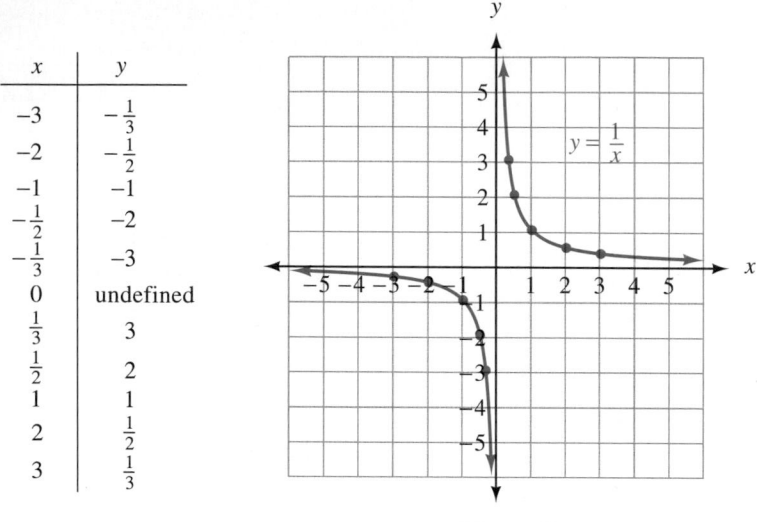

x	y
-3	$-\frac{1}{3}$
-2	$-\frac{1}{2}$
-1	-1
$-\frac{1}{2}$	-2
$-\frac{1}{3}$	-3
0	undefined
$\frac{1}{3}$	3
$\frac{1}{2}$	2
1	1
2	$\frac{1}{2}$
3	$\frac{1}{3}$

Figure 2

▶ **EXAMPLE 9** Graph the inverse variation statement $y = \dfrac{-6}{x}$.

Solution Since y is -6 divided by x, when x is positive, y will be negative (a negative divided by a positive is negative), and when x is negative, y will be positive (a negative divided by a negative). Thus, the graph will appear in quadrants 2 and 4 only. As was the case in Example 8, the graph will not cross either axis.

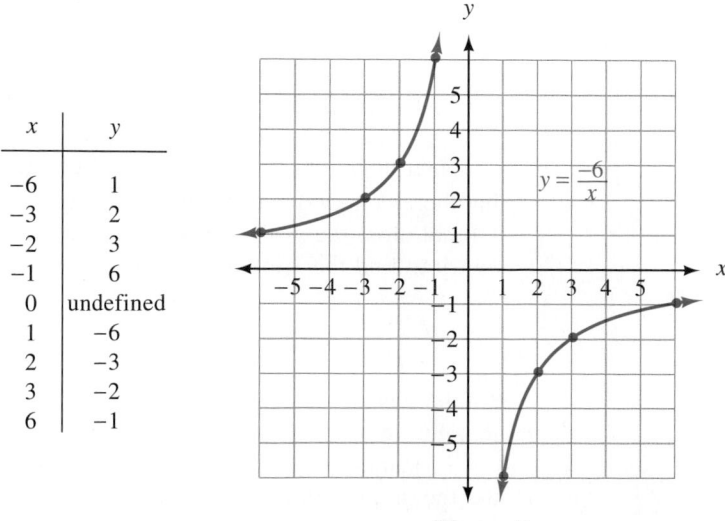

x	y
-6	1
-3	2
-2	3
-1	6
0	undefined
1	-6
2	-3
3	-2
6	-1

Figure 3

In Chapter 4 we found that the first Ferris wheel, designed and built by George Ferris in 1893, had a circumference of 785 feet. In the next example we graph the relationship between the average speed of a person riding this wheel and the amount of time it takes the wheel to complete one revolution.

▶ **EXAMPLE 10**   A Ferris wheel has a circumference of 785 feet. If one complete revolution of the wheel takes from 10 to 30 minutes, then the relationship between the average speed of a rider on the wheel and the amount of time it takes the wheel to complete one revolution is given by the equation

$$r = \frac{785}{t} \quad 10 \leq t \leq 30$$

where r is the average speed and t is the amount of time it takes the wheel to complete one revolution. Graph the equation.

Solution   Since the variables r and t represent speed and time, both must be positive quantities. Therefore, the graph of this equation will lie in the first quadrant only. Below is a table of values of t and r found from the equation, along with the graph of the equation. (Some of the numbers in the table have been rounded to the nearest tenth.)

t	r
10	78.5
15	52.3
20	39.3
25	31.4
30	26.2

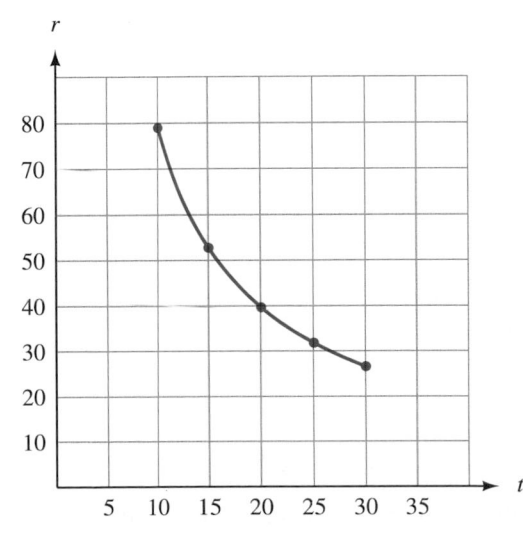

Figure 4

USING TECHNOLOGY: Graphing Calculators

If we use a graphing calculator to graph the equation in Example 10, it is not necessary to construct the table first. In fact, if we graph

$$Y_1 = 785/X \quad \text{in Window: X from 0 to 40, Y from 0 to 90}$$

we can use the Trace and Zoom features together to produce the numbers in the table next to Figure 4. Graph the equation above and Zoom in on the point with x-coordinate 20 until you are convinced that the table values for x and y are correct.

P R O B L E M   S E T   5 . 7

For the following problems, y varies directly with x.

1. If y is 10 when x is 2, find y when x is 6.
2. If y is 20 when x is 5, find y when x is 3.
3. If y is -32 when x is 4, find x when y is -40.
4. If y is -50 when x is 5, find x when y is -70.

For the following problems, r is inversely proportional to s.

5. If r is -3 when s is 4, find r when s is 2.
6. If r is -10 when s is 6, find r when s is -5.
7. If r is 8 when s is 3, find s when r is 48.
8. If r is 12 when s is 5, find s when r is 30.

For the following problems, d varies directly with the square of r.

9. If $d = 10$ when $r = 5$, find d when $r = 10$.
10. If $d = 12$ when $r = 6$, find d when $r = 9$.
11. If $d = 100$ when $r = 2$, find d when $r = 3$.
12. If $d = 50$ when $r = 5$, find d when $r = 7$.

For the following problems, y varies inversely with the square of x.

13. If $y = 45$ when $x = 3$, find y when x is 5.
14. If $y = 12$ when $x = 2$, find y when x is 6.
15. If $y = 18$ when $x = 3$, find x when y is 2.
16. If $y = 45$ when $x = 4$, find x when y is 5.

For the following problems, z varies jointly with x and the square of y.

17. If z is 54 when x and y are 3, find z when $x = 2$ and $y = 4$.
18. If z is 80 when x is 5 and y is 2, find z when $x = 2$ and $y = 5$.
19. If z is 64 when $x = 1$ and $y = 4$, find x when $z = 32$ and $y = 1$.
20. If z is 27 when $x = 6$ and $y = 3$, find x when $z = 50$ and $y = 4$.
21. The length a spring stretches is directly proportional to the force applied. If a force of 5 pounds stretches a spring 3 inches, how much force is necessary to stretch the same spring 10 inches?
22. The weight of a certain material varies directly with the surface area of that material. If 8 square feet weighs half a pound, how much will 10 square feet weigh?
23. The volume of a gas is inversely proportional to the pressure. If a pressure of 36 pounds per square inch corresponds to a volume of 25 cubic feet, what pressure is needed to produce a volume of 75 cubic feet?
24. The frequency of an electromagnetic wave varies inversely with the length. If a wave of length 200 meters has a frequency of 800 kilocycles per second, what frequency will be associated with a wave of length 500 meters?
25. The surface area of a hollow cylinder varies jointly with the height and radius of the cylinder. If a cylinder with radius 3 inches and height

5 inches has a surface area of 94 square inches, what is the surface area of a cylinder with radius 2 inches and height 8 inches?

26. The capacity of a cylinder varies jointly with the height and the square of the radius. If a cylinder with a radius of 3 cm (centimeters) and a height of 6 cm has a capacity of 169.56 cm³ (cubic centimeters), what will be the capacity of a cylinder with radius 4 cm and height 9 cm?

27. The resistance of a wire varies directly with the length and inversely with the square of the diameter. If 100 feet of wire with diameter 0.01 inch has a resistance of 10 ohms, what is the resistance of 60 feet of the same type of wire if its diameter is 0.02 inch?

28. The volume of a gas varies directly with its temperature and inversely with the pressure. If the volume of a certain gas is 30 cubic feet at a temperature of 300 K and a pressure of 20 pounds per square inch, what is the volume of the same gas at 340 K when the pressure is 30 pounds per square inch?

Graph each pair of direct variation statements on the same coordinate system.

29. $y = 4x$ and $y = -4x$
30. $y = \frac{1}{4}x$ and $y = -\frac{1}{4}x$
31. $y = 3x$ and $y = -\frac{1}{3}x$
32. $y = 2x$ and $y = -\frac{1}{2}x$
33. $y = \frac{2}{3}x$ and $y = -\frac{3}{2}x$
34. $y = \frac{3}{5}x$ and $y = -\frac{5}{3}x$

Graph each of the following inverse variation statements.

35. $y = \dfrac{2}{x}$ 36. $y = \dfrac{-2}{x}$

37. $y = \dfrac{-4}{x}$ 38. $y = \dfrac{4}{x}$

39. $y = \dfrac{8}{x}$ 40. $y = \dfrac{-8}{x}$

41. Graph $y = \dfrac{3}{x}$ and $x + y = 4$ on the same coordinate system. At what points do the two graphs intersect?

42. Graph $y = \dfrac{4}{x}$ and $x - y = 3$ on the same coordinate system. At what points do the two graphs intersect?

Applying the Concepts

If the circumference of a Ferris wheel is C, then the relationship between the average speed of a rider and the amount of time for the wheel to complete one revolution is $r = C/t$ where r is the average speed and t is time.

43. A Ferris wheel has a circumference of 204 feet. If a ride once around the wheel takes from 20 to 50 seconds, graph the relationship between the average speed of a rider on the wheel and the amount of time it takes the wheel to complete one revolution.

44. A Ferris wheel has a diameter of 102 feet. If a ride once around the wheel takes from 3 to 5 minutes, graph the relationship between the average speed of a rider on the wheel and the amount of time it takes the wheel to complete one revolution.

Review Problems

The following problems review material we covered in Sections 4.6 and 4.7.

Solve each equation. [4.6]

45. $\dfrac{t}{3} - \dfrac{1}{2} = -1$

46. $\dfrac{x}{x-2} + \dfrac{2}{3} = \dfrac{2}{x-2}$

47. $2 + \dfrac{5}{y} = \dfrac{3}{y^2}$ 48. $1 - \dfrac{1}{y} = \dfrac{12}{y^2}$

Solve each application problem. [4.7]

49. The sum of a number and its reciprocal is $\frac{41}{20}$. Find the number.

50. It takes an inlet pipe 8 hours to fill a tank. The drain can empty the tank in 6 hours. If the tank is full and both the inlet pipe and drain are open, how long will it take to drain the tank?

USING TECHNOLOGY

Graph each pair of equations in the same window. Then find the coordinates of the points where the two graphs intersect.

51. $Y_1 = 2X$, $Y_2 = 2/X$
52. $Y_1 = -2X$, $Y_2 = -2/X$
53. $Y_1 = 0.5X$, $Y_2 = 0.5/X$
54. $Y_1 = -0.5X$, $Y_2 = -0.5/X$

55. Use the results you found in Problems 51–54 to make a statement about the points of intersection of the graphs of

$$y = ax \quad \text{and} \quad y = \frac{a}{x}$$

Prove that your statement is true algebraically.

CHAPTER 5 SUMMARY

Examples

1. The equation $3x + 2y = 6$ is an example of a linear equation in two variables.

Linear Equations in Two Variables [5.1]
A *linear equation in two variables* is any equation that can be put in the form $ax + by = c$. The graph of every linear equation is a straight line.

2. To find the x-intercept for $3x + 2y = 6$, we let $y = 0$ and get

$$3x = 6$$
$$x = 2$$

In this case the x-intercept is 2, and the graph crosses the x-axis at $(2, 0)$.

Intercepts [5.1]
The *x-intercept* of an equation is the x-coordinate of the point where the graph crosses the x-axis. The *y-intercept* is the y-coordinate of the point where the graph crosses the y-axis. We find the y-intercept by substituting $x = 0$ into the equation and solving for y. The x-intercept is found by letting $y = 0$ and solving for x.

3. The slope of the line through $(6, 9)$ and $(1, -1)$ is

$$m = \frac{9 - (-1)}{6 - 1} = \frac{10}{5} = 2$$

The Slope of a Line [5.2]
The *slope* of the line containing points (x_1, y_1) and (x_2, y_2) is given by

$$\text{Slope} = m = \frac{\text{rise}}{\text{run}} = \frac{y_2 - y_1}{x_2 - x_1}$$

Horizontal lines have 0 slope, and vertical lines have no slope. Parallel lines have equal slopes, and perpendicular lines have slopes that are negative reciprocals.

4. The equation of the line with slope 5 and y-intercept 3 is

$$y = 5x + 3$$

The Slope-Intercept Form of a Line [5.3]
The equation of a line with slope m and y-intercept b is given by

$$y = mx + b$$

5. The equation of the line through (3, 2) with slope -4 is

$$y - 2 = -4(x - 3)$$

which can be simplified to

$$y = -4x + 14$$

The Point-Slope Form of a Line [5.3]

The equation of the line through (x_1, y_1) that has slope m can be written as

$$y - y_1 = m(x - x_1)$$

6. The graph of $x - y \leq 3$ is

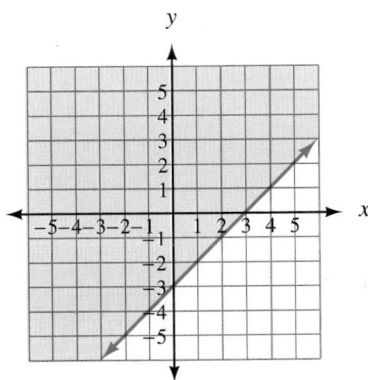

Linear Inequalities in Two Variables [5.4]

An inequality of the form $ax + by < c$ is a *linear inequality in two variables.* The equation for the boundary of the solution set is given by $ax + by = c$. (This equation is found by simply replacing the inequality symbol with an equal sign.)

To graph a linear inequality, first graph the boundary. Next, choose any point not on the boundary and substitute its coordinates into the original inequality. If the resulting statement is true, the graph lies on the same side of the boundary as the test point. A false statement indicates that the solution set lies on the other side of the boundary.

7. The relation $\{(8, 1), (6, 1), (-3, 0)\}$ is also a function since no ordered pairs have the same first coordinates. The domain is $\{8, 6, -3\}$ and the range is $\{1, 0\}$.

Relations and Functions [5.5]

A *relation* is any set of ordered pairs. The set of all first coordinates is called the *domain* of the relation and the set of all second coordinates is the *range* of the relation.

A *function* is a relation in which no two different ordered pairs have the same first coordinates. A function is a rule that pairs each element in one set, called the domain, with exactly one element from a second set, called the range.

If the domain for a relation or a function is not specified, it is assumed to be all real numbers for which the relation (or function) is defined. Since we are concerned only with real number functions, a function is not defined for those values of x that give 0 in the denominator or the square root of a negative number.

8. The graph of $x = y^2$ shown in Figure 3 on page 312 fails the vertical line test. It is not the graph of a function.

Vertical Line Test [5.5]

If a vertical line crosses the graph of a relation in more than one place, the relation cannot be a function. If no vertical line can be found that crosses the graph in more than one place, the relation must be a function.

9. If y varies directly with x, then

$$y = Kx$$

If, also, y is 18 when x is 6, then

$$18 = K \cdot 6$$

or $K = 3$

So the equation can be written more specifically as

$$y = 3x$$

If we want to know what y is when x is 4, we simply substitute:

$$y = 3 \cdot 4$$
$$y = 12$$

Variation [5.7]

If y *varies directly* with x (y is directly proportional to x), then we say

$$y = Kx$$

If y *varies inversely* with x (y is inversely proportional to x), then we say

$$y = \frac{K}{x}$$

If z *varies jointly* with x and y (z is directly proportional to both x and y), then we say

$$z = Kxy$$

In each case, K is called the *constant of variation.*

COMMON MISTAKES

1. When graphing ordered pairs, the most common mistake is to associate the first coordinate with the y-axis and the second with the x-axis. If you make this mistake you would graph (3, 1) by going up 3 and to the right 1, which is just the reverse of what you should do. Remember, the first coordinate is always associated with the horizontal axis, and the second coordinate is always associated with the vertical axis.

2. The two most common mistakes students make when first working with the formula for the slope of a line are

(a) Putting the difference of the x-coordinates over the difference of the y-coordinates.

(b) Subtracting in one order in the numerator and then subtracting in the opposite order in the denominator.

3. When graphing linear inequalities in two variables, remember to graph the boundary with a broken line when the inequality symbol is a $<$ symbol or a $>$ symbol. The only time you use a solid line for the boundary is when the inequality symbol is $\leq$ or $\geq$.

CHAPTER 5 REVIEW

Graph each line. [5.1]

1. $3x + 2y = 6$ **2.** $-3x + 2y = 6$
3. $5x - 2y = 10$ **4.** $y = 2x - 3$
5. $y = -\frac{3}{2}x + 1$ **6.** $x = 3$

Find the slope of the line through the following pairs of points. [5.2]

7. (5, 2), (3, 6)   **8.** (3, −2), (−1, 2)
9. (−4, 2), (3, 2)   **10.** (−6, 4), (−6, 8)

Find x if the line through the two given points has the given slope. [5.2]

11. $(4, x), (1, -3), m = 2$
12. $(-3, x), (-1, -2), m = -2$
13. $(-4, 7), (2, x), m = -\frac{1}{3}$
14. $(-3, 5), (1, x), m = -\frac{1}{2}$

Solve the following problems. [5.2]

15. Find the slope of any line parallel to the line through $(3, 8)$ and $(5, -2)$.
16. Find the slope of any line perpendicular to the line through $(-3, 8)$ and $(5, 2)$.
17. The line through $(5, y^2)$ and $(2, y)$ is parallel to a line with slope 4. What are the possible values of y?
18. The line through $(2, y^2)$ and $(-2, 4y)$ is perpendicular to a line with slope $-\frac{1}{3}$. What are the possible values for y?

Give the equation of the line with the following slope and y-intercept. [5.3]

19. $m = 3, b = 5$ **20.** $m = 5, b = 3$
21. $m = -2, b = 0$ **22.** $m = \frac{1}{3}, b = -\frac{2}{3}$

Give the slope and y-intercept of each equation. [5.3]

23. $3x - y = 6$ **24.** $2x - 3y = 6$
25. $2x - 3y = 9$ **26.** $3x - 2y = 4$

Find the equation of the line that contains the given point and has the given slope. [5.3]

27. $(2, 4), m = 2$ **28.** $(3, 2), m = 3$
29. $(-3, 1), m = -\frac{1}{3}$ **30.** $(\frac{1}{3}, \frac{2}{3}), m = -\frac{1}{2}$

Find the equation of the line that contains the given pair of points. [5.3]

31. $(2, 5), (-3, -5)$ **32.** $(1, 4), (-1, -2)$
33. $(-3, 7), (4, 7)$ **34.** $(-2, 6), (-2, 3)$
35. $(-5, -1), (-3, -4)$
36. $(-8, -2), (1, -3)$
37. Find the equation of the line that is parallel to $2x - y = 4$ and contains the point $(2, -3)$. [5.3]

38. Find the equation of the line that is parallel to the graph of $3x - 2y = 4$ and contains the point $(0, -1)$. [5.3]
39. Find the equation of the line perpendicular to $y = -3x + 1$ that has an x-intercept of 2. [5.3]
40. Find the equation of the line perpendicular to $2x - 4y = 5$ that has an x-intercept of -2. [5.3]

Graph each linear inequality. [5.4]

41. $y \leq 2x - 3$ **42.** $2x - 3y > 6$
43. $x \geq -1$ **44.** $y > 4$

State the domain and range of each relation, and then indicate which relations are also functions. [5.5]

45. $\{(2, 4), (3, 3), (4, 2)\}$
46. $\{(-5, 2), (3, 4), (-5, -2)\}$
47. $\{(6, 3), (-4, 3), (-2, 0)\}$
48. $\{(1, -1), (3, 0), (-2, 0)\}$

Let $f(x) = 3x - 2$ and $g(x) = x^2 - 2x + 4$ and evaluate the following. [5.6]

49. $f(0)$ **50.** $g(-3)$
51. $f(3) - g(1)$ **52.** $g(-1) - f(-1)$

If $f = \{(2, -1), (-3, 0), (4, \frac{1}{2}), (\pi, 2)\}$ and $g = \{(2, 2), (-1, 4), (0, 0)\}$, find the following. [5.6]

53. $f(-3)$ **54.** $g(2)$
55. $f(2) + g(2)$
56. $f(-3) + g(-1) + g(0)$

Let $f(x) = 2x^2 - 4x + 1$ and $g(x) = 3x + 2$ and evaluate each of the following. [5.6]

57. $f(0)$ **58.** $f(-1)$
59. $g(a)$ **60.** $g(a - 2)$
61. $f[g(0)]$ **62.** $g[f(-2)]$
63. $f[g(x)]$ **64.** $g[f(x)]$
65. The length of a rectangle is 3 inches less than 2 times the width. If x is the width of the rectangle, use function notation to write the perimeter $P(x)$ in terms of the width x. [5.6]

66. Graph the function you found in Problem 65. Then draw a vertical line, starting at 2 on the *x*-axis, that corresponds to $P(2)$. [5.6]

The formula for the volume of a right circular cylinder is $V = \pi r^2 h$, where r is the radius of the base and h is the height of the cylinder. Use this information to work Problems 67 and 68. [5.6]

67. If the height is twice the radius, use function notation to write the volume $V(r)$ as a function of just r.

68. If the radius is twice the height, use function notation to write the volume $V(h)$ as a function of just h.

For the following problems, y varies directly with x. [5.7]

69. If y is 6 when x is 2, find y when x is 8.
70. If y is -3 when x is 5, find y when x is -10.

For the following problems, y varies inversely with the square of x. [5.7]

71. If y is 9 when x is 2, find y when x is 3.
72. If y is 4 when x is 5, find y when x is 2.

For the following problems, z varies jointly with x and the cube of y. [5.7]

73. If z is 6 when x is 2 and y is -1, find z when x is 3 and y is 2.
74. If z is -48 when x is 3 and y is 2, find z when x is 2 and y is 3.

75. If z is -20 when x is -5 and y is 1, find y when z is 64 and x is 2.
76. If z is -108 when x is 4 and y is 3, find x when z is 8 and y is -2.

Solve each application problem. [5.7]

77. The tension t in a spring varies directly with the distance d the spring is stretched. If the tension is 42 pounds when the spring is stretched 2 inches, find the tension when the spring is stretched twice as far.

78. The power P in an electric circuit varies directly with the square of the current I. If P is 30 when I is 2, find P when I is 6.

79. The intensity of a light source varies inversely with the square of the distance from the source. Four feet from the source the intensity is 9 foot candles. What is the intensity 3 feet from the source?

80. The weight of a body varies inversely with the square of its distance from the center of the earth. If a man weighs 150 pounds 4,000 miles from the center of the earth, how much will he weigh 5,000 miles from the center of the earth?

Graph each equation. [5.7]

81. $y = 3x$
82. $y = -2x$
83. $y = \dfrac{8}{x}$
84. $y = \dfrac{-4}{x}$

CHAPTER 5 TEST

For each of the following straight lines, identify the *x*-intercept, *y*-intercept, and slope, and sketch the graph. [5.1–5.3]

1. $2x + y = 6$
2. $y = -2x - 3$
3. $y = \frac{3}{2}x + 4$
4. $x = -2$

Find the equation for each line. [5.3]

5. Give the equation of the line through $(-1, 3)$ that has slope $m = 2$.

6. Give the equation of the line through $(-3, 2)$ and $(4, -1)$.
7. Line l contains the point $(5, -3)$ and has a graph parallel to the graph of $2x - 5y = 10$. Find the equation for l.
8. Line l contains the point $(-1, -2)$ and has a graph perpendicular to the graph of $y = 3x - 1$. Find the equation for l.
9. Give the equation of the vertical line through $(4, -7)$.

Graph the following linear inequalities. [5.4]

10. $3x - 4y < 12$ **11.** $y \leq -x + 2$

Specify the domain and range for the following relations and indicate which relations are also functions. [5.5]

12. $\{(-2, 0), (-3, 0), (-2, 1)\}$
13. $y = x^2 - 9$

Let $f(x) = x - 2$, $g(x) = 3x + 4$,   and $h(x) = 3x^2 - 2x - 8$, and find the following. [5.6]

14. $f(3) + g(2)$ **15.** $h(0) + g(0)$
16. $f[g(2)]$ **17.** $f[g(x)]$

A piece of cardboard is in the shape of a square with each side 8 inches long. You are going to make a box out of the piece of cardboard by cutting four equal squares from each corner and then folding up the sides, as shown in Figure 1. Use this information to answer questions 18, 19, and 20. [5.6]

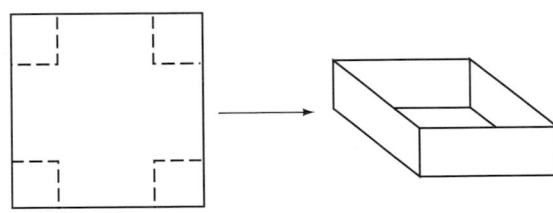

Figure 1

18. If the length of each side of the squares you are cutting from the corners is represented by the variable x, state the restrictions on x.

19. Use function notation to write the volume of the resulting box $V(x)$ as a function of x.
20. Find $V(2)$ and explain what it represents.

A company manufactures and sells typewriter ribbons. If they charge p dollars for each ribbon they sell, they will sell $x = 1{,}200 - 100p$ ribbons per week. The total cost to manufacture x ribbons is $C(x) = 4x + 600$. [5.6]

21. Find $R(x)$.   **22.** Find $R(p)$.
23. Find $P(x)$.   **24.** Find $P(500)$.

Solve the following variation problems. [5.7]

25. Quantity y varies directly with the square of x. If y is 50 when x is 5, find y when x is 3.
26. Quantity z varies jointly with x and the cube of y. If z is 15 when x is 5 and y is 2, find z when x is 2 and y is 3.
27. The maximum load (L) a horizontal beam can safely hold varies jointly with the width (w) and the square of the depth (d) and inversely with the length (l). If a 10-foot beam with width 3 and depth 4 will safely hold up to 800 pounds, how many pounds will a 12-foot beam with width 3 and depth 4 hold?

Graph each of the following. [5.7]

28. $y = \dfrac{4}{x}$ **29.** $y = \dfrac{-4}{x}$

6

RATIONAL EXPONENTS AND ROOTS

INTRODUCTION

Figure 1 shows a square in which each of the four sides is 1 inch long. To find the square of the length of the diagonal c we apply the Pythagorean Theorem:

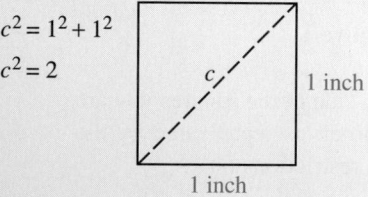

$$c^2 = 1^2 + 1^2$$
$$c^2 = 2$$

1 inch

1 inch

Figure 1

Because we know that c is positive and that its square is 2, we call c the **positive square root** of 2, and we write $c = \sqrt{2}$. The ability to associate numbers, such as $\sqrt{2}$, with the diagonal of a square, or rectangle, allows us to analyze some interesting items from geometry. Two particularly interesting geometric objects that we will study in this chapter are shown in Figures 2 and 3. Both are constructed from right triangles and the numbers associated with the diagonals in each are found from the Pythagorean Theorem.

The Spiral of Roots

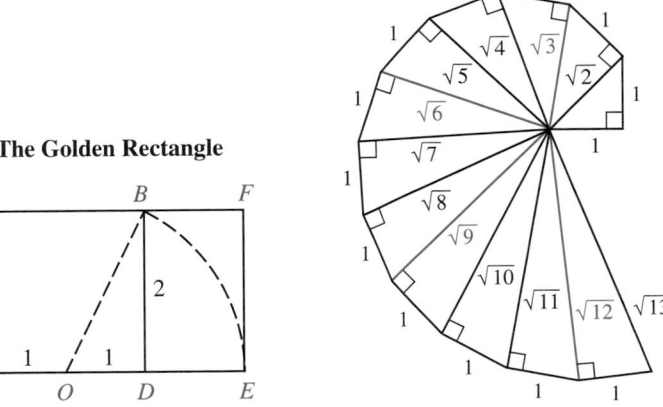

The Golden Rectangle

Figure 2

Figure 3

OVERVIEW

The work we have done previously with exponents and polynomials will be very useful in understanding the concepts presented in this chapter. The square roots, cube roots, and higher roots we will study take us in the reverse direction from exponents. Although roots and expressions involving roots are new to us, you will find that they behave in much the same way as the polynomials we have dealt with in the past. That is, the rules we have developed for working with polynomials can be applied to these new expressions, so that finding the product $(\sqrt{5} + 3)(\sqrt{5} - 2)$ is very similar to finding the product $(x + 3)(x - 2)$.

Also in this chapter is our introduction to a larger set of numbers, the complex numbers. These numbers include the real numbers as well as numbers we have not worked with yet. Complex numbers also behave like polynomials in many situations, so the rules we have for polynomials can be applied to complex numbers as well.

STUDY SKILLS

This is the last chapter in which we will mention study skills. You know by now what works best for you, and what you have to do to achieve your goals for this course. From now on it is simply a matter of sticking with the things that work for you and avoiding the things that do not. It seems simple, but as with anything that takes effort, it is up to you to see that you maintain the skills that get you where you want to be in the course.

If you intend to take more classes in mathematics, and you want to ensure your success in those classes, then you can work towards this goal: *Become the type of student who can learn mathematics on your own.* Most people who have degrees in mathematics were students who could learn mathematics on their own. This doesn't mean that you have to learn it all on your own; it simply means that if you have to, you can learn it on your own. Attaining this goal gives you independence, and puts you in control of your success in any math class you take.

SECTION

6.1   **Rational Exponents**

In Chapter 3, we developed notation (exponents) to give us the square, cube, or any power of a number. For instance, if we wanted the square of 3, we wrote $3^2 = 9$. If we wanted the cube of 3, we wrote $3^3 = 27$. In this section, we will develop notation that will take us in the reverse direction, that is, from the square of a number, say 25, back to the original number, 5.

DEFINITION   If x is a nonnegative real number, then the expression $\sqrt{x}$ is called the **positive square root** of x and is such that

$$(\sqrt{x})^2 = x$$

In words: $\sqrt{x}$ is the positive number we square to get x.

The negative square root of x, $-\sqrt{x}$, is defined in a similar manner.

▶ **EXAMPLE 1**   The positive square root of 64 is 8 because 8 is the positive number with the property $8^2 = 64$. The negative square root of 64 is -8 since -8 is the negative number whose square is 64. We can summarize both of these facts by saying

$$\sqrt{64} = 8 \quad \text{and} \quad -\sqrt{64} = -8 \qquad \blacktriangleleft$$

Note   It is a common mistake to assume that an expression like $\sqrt{25}$ indicates both square roots, 5 and -5. The expression $\sqrt{25}$ indicates only the positive square root of 25, which is 5. If we want the negative square root, we must use a negative sign: $-\sqrt{25} = -5$.

The higher roots, cube roots, fourth roots, and so on, are defined by definitions similar to that of square roots.

DEFINITION If x is a real number, and n is a positive integer, then

Positive square root of x, $\sqrt{x}$, is such that $(\sqrt{x})^2 = x$ $\qquad x \geq 0$

Cube root of x, $\sqrt[3]{x}$, is such that $(\sqrt[3]{x})^3 = x$

Positive fourth root of x, $\sqrt[4]{x}$, is such that $(\sqrt[4]{x})^4 = x$ $\qquad x \geq 0$

Fifth root of x, $\sqrt[5]{x}$, is such that $(\sqrt[5]{x})^5 = x$

$$\vdots \qquad\qquad \vdots \;\; \vdots$$

The **nth root of x,** $\sqrt[n]{x}$, is such that $(\sqrt[n]{x})^n = x$ $\qquad x \geq 0$ if n is even

Note We have restricted the even roots in this definition to nonnegative numbers. Even roots of negative numbers exist, but are not represented by real numbers. That is, $\sqrt{-4}$ is not a real number since there is no real number whose square is -4.

Here is a table of the most common roots used in this book. Any of the roots that are unfamiliar should be memorized.

Square Roots		Cube Roots	Fourth Roots
$\sqrt{0} = 0$	$\sqrt{49} = 7$	$\sqrt[3]{0} = 0$	$\sqrt[4]{0} = 0$
$\sqrt{1} = 1$	$\sqrt{64} = 8$	$\sqrt[3]{1} = 1$	$\sqrt[4]{1} = 1$
$\sqrt{4} = 2$	$\sqrt{81} = 9$	$\sqrt[3]{8} = 2$	$\sqrt[4]{16} = 2$
$\sqrt{9} = 3$	$\sqrt{100} = 10$	$\sqrt[3]{27} = 3$	$\sqrt[4]{81} = 3$
$\sqrt{16} = 4$	$\sqrt{121} = 11$	$\sqrt[3]{64} = 4$	
$\sqrt{25} = 5$	$\sqrt{144} = 12$	$\sqrt[3]{125} = 5$	
$\sqrt{36} = 6$	$\sqrt{169} = 13$		

Notation An expression like $\sqrt[3]{8}$ that involves a root is called a **radical expression**. In the expression $\sqrt[3]{8}$, the 3 is called the **index**, the $\sqrt{}$ is the **radical sign**, and 8 is called the **radicand**. The index of a radical must be a positive integer greater than 1. If no index is written, it is assumed to be 2.

Roots and Negative Numbers

When dealing with negative numbers and radicals, the only restriction concerns negative numbers under even roots. We can have negative signs in front of radicals and negative numbers under odd roots and still obtain real numbers.

Here are some examples to help clarify this. In the last section of this chapter we will see how to deal with even roots of negative numbers.

▶ **EXAMPLES** Simplify each expression, if possible.

2. $\sqrt[3]{-8} = -2$ because $(-2)^3 = -8$

3. $\sqrt{-4}$ is not a real number since there is no real number whose square is -4

4. $-\sqrt{25} = -5$ this is the negative square root of 25

5. $\sqrt[5]{-32} = -2$ because $(-2)^5 = -32$

6. $\sqrt[4]{-81}$ is not a real number since there is no real number we can raise to the fourth power and obtain -81 ◀

Variables under a Radical

From the preceding examples it is clear that we must be careful that we do not try to take an even root of a negative number. For this reason, we will assume that all variables appearing under a radical sign represent nonnegative numbers.

▶ **EXAMPLES** Assume all variables represent nonnegative numbers and simplify each expression as much as possible.

7. $\sqrt{25a^4b^6} = 5a^2b^3$ because $(5a^2b^3)^2 = 25a^4b^6$

8. $\sqrt[3]{x^6y^{12}} = x^2y^4$ because $(x^2y^4)^3 = x^6y^{12}$

9. $\sqrt[4]{81r^8s^{20}} = 3r^2s^5$ because $(3r^2s^5)^4 = 81r^8s^{20}$ ◀

Rational Numbers as Exponents

We will now develop a second kind of notation involving exponents that will allow us to designate square roots, cube roots, and so on in another way.

Consider the equation $x = 8^{1/3}$. Although we have not encountered fractional exponents before, let's assume that all the properties of exponents hold in this case. Cubing both sides of the equation, we have

$$x^3 = (8^{1/3})^3$$
$$x^3 = 8^{(1/3)(3)}$$
$$x^3 = 8^1$$
$$x^3 = 8$$

The last line tells us that x is the number whose cube is 8. It must be true, then, that x is the cube root of 8, $x = \sqrt[3]{8}$. Since we started with $x = 8^{1/3}$, it follows that

$$8^{1/3} = \sqrt[3]{8}$$

It seems reasonable, then, to define fractional exponents as indicating roots. Here is the formal definition.

DEFINITION   If x is a real number and n is a positive integer greater than 1, then

$$x^{1/n} = \sqrt[n]{x} \qquad (x \geq 0 \text{ when } n \text{ is even})$$

In words: The quantity $x^{1/n}$ is the nth root of x.

With this definition we have a way of representing roots with exponents. Here are some examples.

▶ **EXAMPLES**   Write each expression as a root and then simplify, if possible.

10. $8^{1/3} = \sqrt[3]{8} = 2$

11. $36^{1/2} = \sqrt{36} = 6$

12. $-25^{1/2} = -\sqrt{25} = -5$

13. $(-25)^{1/2} = \sqrt{-25}$, which is not a real number

14. $\left(\dfrac{4}{9}\right)^{1/2} = \sqrt{\dfrac{4}{9}} = \dfrac{2}{3}$ ◀

Note   If we were using a scientific calculator to work Example 10, this is how the sequence of key strokes would look:

$$8 \boxed{y^x} \; 3 \; \boxed{1/x} \; \boxed{=}$$

or

$$8 \boxed{y^x} \boxed{(} \; 1 \; \boxed{\div} \; 3 \; \boxed{)} \boxed{=}$$

The properties of exponents developed in Chapter 3 were applied to integer exponents only. We will now extend these properties to include rational exponents also. We do so without proof.

Properties of Exponents

If a and b are real numbers and r and s are rational numbers, and a and b are nonnegative whenever r and s indicate even roots, then

1. $a^r \cdot a^s = a^{r+s}$ 4. $a^{-r} = \dfrac{1}{a^r}$ $(a \neq 0)$

2. $(a^r)^s = a^{rs}$ 5. $\left(\dfrac{a}{b}\right)^r = \dfrac{a^r}{b^r}$ $(b \neq 0)$

3. $(ab)^r = a^r b^r$ 6. $\dfrac{a^r}{a^s} = a^{r-s}$ $(a \neq 0)$

There are times when rational exponents can simplify our work with radicals. Here are Examples 8 and 9 again, but this time we will work them using rational exponents.

▶ **EXAMPLES**   Write each radical with a rational exponent and then simplify.

15. $\sqrt[3]{x^6 y^{12}} = (x^6 y^{12})^{1/3}$
$= (x^6)^{1/3} (y^{12})^{1/3}$
$= x^2 y^4$

16. $\sqrt[4]{81 r^8 s^{20}} = (81 r^8 s^{20})^{1/4}$
$= 81^{1/4} (r^8)^{1/4} (s^{20})^{1/4}$
$= 3 r^2 s^5$ ◀

So far, the numerators of all the rational exponents we have encountered have been 1. The next theorem extends the work we can do with rational exponents to rational exponents with numerators other than 1.

We can extend our properties of exponents with the following theorem.

Theorem 6.1

If a is a nonnegative real number, m is an integer, and n is a positive integer, then

$$a^{m/n} = (a^{1/n})^m = (a^m)^{1/n}$$

PROOF   We can prove Theorem 6.1 using the properties of exponents. Since $\frac{m}{n} = m(\frac{1}{n})$ we have

$$a^{m/n} = a^{m(1/n)} \qquad\qquad a^{m/n} = a^{(1/n)(m)}$$
$$= (a^m)^{1/n} \qquad\qquad\quad = (a^{1/n})^m$$

Here are some examples that illustrate how we use this theorem.

▶ **EXAMPLES**   Simplify as much as possible.

17. $8^{2/3} = (8^{1/3})^2$ Theorem 6.1
$= 2^2$ Definition of fractional exponents
$= 4$ The square of 2 is 4

Note   On a scientific calculator, Example 17 would look like this:

$$8 \boxed{y^x} \boxed{(} 2 \boxed{\div} 3 \boxed{)} \boxed{=}$$

18. $25^{3/2} = (25^{1/2})^3$ Theorem 6.1

$\phantom{25^{3/2}} = 5^3$ Definition of fractional exponents

$\phantom{25^{3/2}} = 125$ The cube of 5 is 125

19. $9^{-3/2} = (9^{1/2})^{-3}$ Theorem 6.1

$\phantom{9^{-3/2}} = 3^{-3}$ Definition of fractional exponents

$\phantom{9^{-3/2}} = \dfrac{1}{3^3}$ Property 4 for exponents

$\phantom{9^{-3/2}} = \dfrac{1}{27}$ The cube of 3 is 27

20. $\left(\dfrac{27}{8}\right)^{-4/3} = \left[\left(\dfrac{27}{8}\right)^{1/3}\right]^{-4}$ Theorem 6.1

$\phantom{\left(\dfrac{27}{8}\right)^{-4/3}} = \left(\dfrac{3}{2}\right)^{-4}$ Definition of fractional exponents

$\phantom{\left(\dfrac{27}{8}\right)^{-4/3}} = \left(\dfrac{2}{3}\right)^{4}$ Property 4 for exponents

$\phantom{\left(\dfrac{27}{8}\right)^{-4/3}} = \dfrac{16}{81}$ $\left(\dfrac{2}{3}\right)^{4} = \dfrac{16}{81}$ ◀

The following examples show the application of the properties of exponents to rational exponents.

▶ **EXAMPLES**   Assume all variables represent positive quantities and simplify as much as possible.

21. $x^{1/3} \cdot x^{5/6} = x^{1/3 + 5/6}$ Property 1

$\phantom{x^{1/3} \cdot x^{5/6}} = x^{2/6 + 5/6}$ LCD is 6

$\phantom{x^{1/3} \cdot x^{5/6}} = x^{7/6}$ Add fractions

22. $(y^{2/3})^{3/4} = y^{(2/3)(3/4)}$ Property 2

$\phantom{(y^{2/3})^{3/4}} = y^{1/2}$ Multiply fractions: $\dfrac{2}{3} \cdot \dfrac{3}{4} = \dfrac{6}{12} = \dfrac{1}{2}$

23. $\dfrac{z^{1/3}}{z^{1/4}} = z^{1/3 - 1/4}$ Property 6

$\phantom{\dfrac{z^{1/3}}{z^{1/4}}} = z^{4/12 - 3/12}$ LCD is 12

$\phantom{\dfrac{z^{1/3}}{z^{1/4}}} = z^{1/12}$ Subtract fractions

24. $\left(\dfrac{a^{-1/3}}{b^{1/2}}\right)^{6} = \dfrac{(a^{-1/3})^6}{(b^{1/2})^6}$ Property 5

$\phantom{\left(\dfrac{a^{-1/3}}{b^{1/2}}\right)^{6}} = \dfrac{a^{-2}}{b^3}$ Property 2

$\phantom{\left(\dfrac{a^{-1/3}}{b^{1/2}}\right)^{6}} = \dfrac{1}{a^2 b^3}$ Property 4

25. $\dfrac{(x^{-3}y^{1/2})^4}{x^{10}y^{3/2}} = \dfrac{(x^{-3})^4(y^{1/2})^4}{x^{10}y^{3/2}}$ Property 3

$= \dfrac{x^{-12}y^2}{x^{10}y^{3/2}}$ Property 2

$= x^{-22}y^{1/2}$ Property 6

$= \dfrac{y^{1/2}}{x^{22}}$ Property 4   ◀

Facts from Geometry: The Pythagorean Theorem (Again) and the Golden Rectangle

Now that we have had some experience working with square roots, we can rewrite the Pythagorean Theorem using a square root. If triangle ABC is a right triangle with $C = 90°$, then the length of the longest side is the *positive square root* of the sum of the squares of the other two sides.

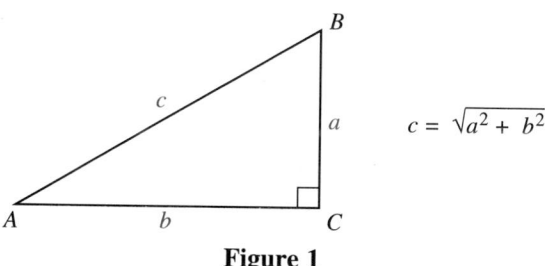

Figure 1

In the introduction to this chapter we mentioned the golden rectangle. Its origins can be traced back over 2,000 years to the Greek civilization that produced Pythagoras, Socrates, Plato, Aristotle, and Euclid. The most important mathematical work to come from that Greek civilization was Euclid's *Elements,* an elegantly written summary of all that was known about geometry at that time in history. Euclid's *Elements,* according to Howard Eves, an authority on the history of mathematics, exercised a greater influence on scientific thinking than any other work. Here is how we construct a golden rectangle from a square of side 2, using the same method that Euclid used in his *Elements.*

Constructing a Golden Rectangle from a Square of Side 2

Step 1: Draw a square with a side of length 2. Connect the midpoint of side CD to corner B. (Note that we have labeled the midpoint of segment CD with the letter O.)

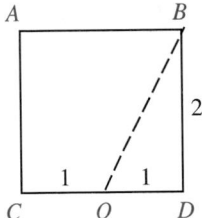

Step 2: Drop the diagonal from step 1 down so it aligns with side *CD*.

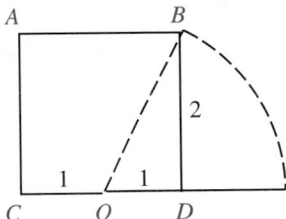

Step 3: Form rectangle *ACEF*. This is a golden rectangle.

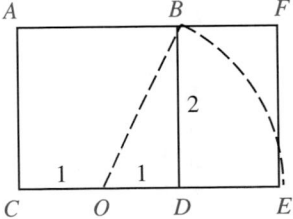

All golden rectangles are constructed from squares. Every golden rectangle, no matter how large or small it is, will have the same shape. To associate a number with the shape of the golden rectangle, we use the ratio of its length to its width. This ratio is called the **golden ratio.** To calculate the golden ratio we must first find the length of the diagonal we used to construct the golden rectangle. Figure 2 on the next page shows the golden rectangle that we constructed from a square of side 2. The length of the diagonal *OB* is found by applying the Pythagorean Theorem to triangle *OBD*.

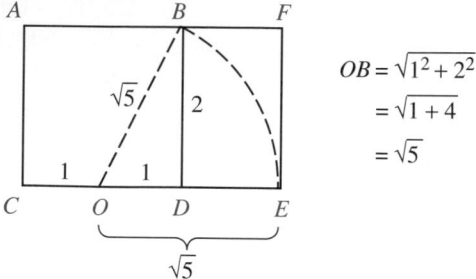

$$OB = \sqrt{1^2 + 2^2}$$
$$= \sqrt{1 + 4}$$
$$= \sqrt{5}$$

Figure 2

The length of segment OE is equal to the length of diagonal OB; both are $\sqrt{5}$. Since the distance from C to O is 1, the length CE of the golden rectangle is $1 + \sqrt{5}$. Now we can find the golden ratio:

$$\text{Golden ratio} = \frac{\text{length}}{\text{width}} = \frac{CE}{EF} = \frac{1 + \sqrt{5}}{2}$$

USING TECHNOLOGY: Graphing Calculators

A Word of Caution   Some graphing calculators give surprising results when evaluating expressions such as $(-8)^{2/3}$. As you know from reading this section, the expression $(-8)^{2/3}$ simplifies to 4, either by taking the cube root first and then squaring the result, or by squaring the base first and then taking the cube root of the result. Here are three different ways to have your calculator evaluate this expression:

1. $(-8)^\wedge(2/3)$ to evaluate $(-8)^{2/3}$
2. $((-8)^\wedge 2)^\wedge(1/3)$ to evaluate $((-8)^2)^{1/3}$
3. $((-8)^\wedge(1/3))^\wedge 2$ to evaluate $((-8)^{1/3})^2$

Note any differences in the results.
   Next, graph each of the following functions, one at a time.

1. $Y_1 = X^{2/3}$
2. $Y_2 = (X^2)^{1/3}$
3. $Y_3 = (X^{1/3})^2$

The correct graph is shown in Figure 3. Note which of your graphs match the correct graph.

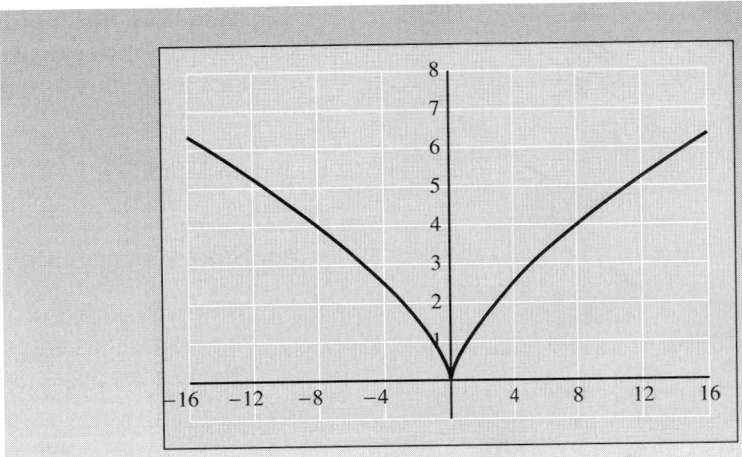

Figure 3

The discussion above is intended to make you aware of any differences that exist in the way your calculator evaluates exponential expressions.

PROBLEM SET 6.1

Find each of the following roots, if possible.

1. $\sqrt{144}$

2. $-\sqrt{144}$

3. $\sqrt{-144}$

4. $\sqrt{-49}$

5. $-\sqrt{49}$

6. $\sqrt{49}$

7. $\sqrt[3]{-27}$

8. $-\sqrt[3]{27}$

9. $\sqrt[4]{16}$

10. $-\sqrt[4]{16}$

11. $\sqrt[4]{-16}$

12. $-\sqrt[4]{-16}$

13. $\sqrt{0.04}$

14. $\sqrt{0.81}$

15. $\sqrt[3]{0.008}$

16. $\sqrt[3]{0.125}$

Simplify each expression. Assume all variables represent nonnegative numbers.

17. $\sqrt{36a^8}$

18. $\sqrt{49a^{10}}$

19. $\sqrt[3]{27a^{12}}$

20. $\sqrt[3]{8a^{15}}$

21. $\sqrt[3]{x^3y^6}$

22. $\sqrt[3]{x^6y^3}$

23. $\sqrt[5]{32x^{10}y^5}$

24. $\sqrt[5]{32x^5y^{10}}$

25. $\sqrt[4]{16a^{12}b^{20}}$

26. $\sqrt[4]{81a^{24}b^8}$

Use the definition of rational exponents to write each of the following with the appropriate root. Then simplify.

27. $36^{1/2}$

28. $49^{1/2}$

29. $-9^{1/2}$

30. $-16^{1/2}$

31. $8^{1/3}$

32. $-8^{1/3}$

33. $(-8)^{1/3}$

34. $-27^{1/3}$

35. $32^{1/5}$

36. $81^{1/4}$

37. $\left(\frac{81}{25}\right)^{1/2}$

38. $\left(\frac{9}{16}\right)^{1/2}$

39. $\left(\frac{64}{125}\right)^{1/3}$

40. $\left(\frac{8}{27}\right)^{1/3}$

Use Theorem 6.1 to simplify each of the following as much as possible.

41. $27^{2/3}$

42. $8^{4/3}$

43. $25^{3/2}$

44. $9^{3/2}$

45. $16^{3/4}$

46. $81^{3/4}$

Simplify each expression. Remember, negative exponents give reciprocals.

47. $27^{-1/3}$

48. $9^{-1/2}$

49. $81^{-3/4}$

50. $4^{-3/2}$

51. $\left(\frac{25}{36}\right)^{-1/2}$

52. $\left(\frac{16}{49}\right)^{-1/2}$

53. $\left(\frac{81}{16}\right)^{-3/4}$

54. $\left(\frac{27}{8}\right)^{-2/3}$

55. $16^{1/2} + 27^{1/3}$

56. $25^{1/2} + 100^{1/2}$

57. $8^{-2/3} + 4^{-1/2}$

58. $49^{-1/2} + 25^{-1/2}$

Use the properties of exponents to simplify each of the following as much as possible. Assume all bases are positive.

59. $x^{3/5} \cdot x^{1/5}$

60. $x^{3/4} \cdot x^{5/4}$

61. $(a^{3/4})^{4/3}$

62. $(a^{2/3})^{3/4}$

63. $\dfrac{x^{1/5}}{x^{3/5}}$

64. $\dfrac{x^{2/7}}{x^{5/7}}$

65. $\dfrac{x^{5/6}}{x^{2/3}}$

66. $\dfrac{x^{7/8}}{x^{8/7}}$

67. $(x^{3/5}y^{5/6}z^{1/3})^{3/5}$

68. $(x^{3/4}y^{1/8}z^{5/6})^{4/5}$

69. $\dfrac{a^{3/4}b^2}{a^{7/8}b^{1/4}}$

70. $\dfrac{a^{1/3}b^4}{a^{3/5}b^{1/3}}$

71. $\dfrac{(y^{2/3})^{3/4}}{(y^{1/3})^{3/5}}$

72. $\dfrac{(y^{5/4})^{2/5}}{(y^{1/4})^{4/3}}$

73. $\left(\dfrac{a^{-1/4}}{b^{1/2}}\right)^8$

74. $\left(\dfrac{a^{-1/5}}{b^{1/3}}\right)^{15}$

75. $\dfrac{(r^{-2}s^{1/3})^6}{r^8s^{3/2}}$

76. $\dfrac{(r^{-5}s^{1/2})^4}{r^{12}s^{5/2}}$

77. $\dfrac{(25a^6b^4)^{1/2}}{(8a^{-9}b^3)^{-1/3}}$

78. $\dfrac{(27a^3b^6)^{1/3}}{(81a^8b^{-4})^{1/4}}$

79. Show that the expression $(a^{1/2} + b^{1/2})^2$ is not equal to $a + b$ by replacing a with 9 and b with 4 in both expressions and then simplifying each.

80. Show that the statement $(a^2 + b^2)^{1/2} = a + b$ is not, in general, true by replacing a with 3 and b with 4 and then simplifying both sides.

81. You may have noticed, if you have been using a calculator to find roots, that you can find the fourth root of a number by pressing the square root button twice. Written in symbols, this fact looks like this:

$$\sqrt{\sqrt{a}} = \sqrt[4]{a} \quad (a \geq 0)$$

Show that this statement is true by rewriting each side with exponents instead of radical notation and then simplifying the left side.

82. Show that the statement below is true by rewriting each side with exponents instead of radical notation and then simplifying the left side.

$$\sqrt[3]{\sqrt{a}} = \sqrt[6]{a} \quad (a \geq 0)$$

Applying the Concepts

83. The maximum speed (v) that an automobile can travel around a curve of radius r without skidding is given by the equation

$$v = \left(\frac{5r}{2}\right)^{1/2}$$

where v is in miles/hour and r is measured in feet. What is the maximum speed a car can travel around a curve with a radius of 250 feet without skidding?

84. The equation $L = \left(1 - \dfrac{v^2}{c^2}\right)^{1/2}$ gives the relativistic length of a 1-foot ruler traveling with velocity v. Find L if $\dfrac{v}{c} = \dfrac{3}{5}$.

85. The golden ratio is the ratio of the length to the width in any golden rectangle. The exact value of this number is $\dfrac{1 + \sqrt{5}}{2}$. Use a calculator to find a decimal approximation to this number and round it to the nearest thousandth.

86. The reciprocal of the golden ratio is $\dfrac{2}{1 + \sqrt{5}}$. Find a decimal approximation to this number that is accurate to the nearest thousandth.

87. Find the next term in the sequence below. Then explain how this sequence is related to the Fibonacci sequence.

$$\frac{3}{2}, \frac{5}{3}, \frac{8}{5}, \cdots$$

88. Write the first 10 terms in the sequence shown in Problem 87. Then find a decimal approximation to each of the 10 terms, rounding each to the nearest thousandth.

Review Problems

The problems that follow review material we covered in Sections 3.4 and 5.1. Reviewing these problems will help you understand the next section.

91. $(x - 3)(x + 5)$ **92.** $(x - 2)(x + 2)$
93. $(x^2 - 5)^2$ **94.** $(x^2 + 5)^2$
95. $(x - 3)(x^2 + 3x + 9)$
96. $(x + 3)(x^2 - 3x + 9)$

Multiply. [3.4]

Find the x- and y-intercepts for each line. [5.1]

89. $x^2(x^4 - x)$ **90.** $5x^2(2x^3 - x)$

97. $5x - 4y = 10$ **98.** $12x - 5y = 15$
99. $y = \frac{2}{3}x + 4$ **100.** $y = \frac{3}{2}x - 6$

USING TECHNOLOGY

Graph the equation $y = x^{3/4}$ and then use the Trace and Zoom feature to approximate each of the following to the nearest tenth.

101. $2^{3/4}$ **102.** $3^{3/4}$
103. $10^{3/4}$ **104.** $16^{3/4}$

Graph $y = x^{3/4}$ and $y = x^{4/3}$ on the same coordinate system, using a Window with X from -4 to 4 and Y from -1 to 4. (If you have not read the word of caution at the end of Section 6.1, do so now.) Use the graphs to answer the following questions.

105. Where do the two graphs intersect?
106. For what values of x is $x^{3/4} \geq x^{4/3}$?

Carbon-14 dating is used extensively in science to find the age of fossils. If at one time a nonliving substance contains an amount A of carbon-14, then t years later the amount of carbon-14 it contains is given by the formula below.

$$A \cdot 2^{-t/5{,}600}$$

Use this formula and a calculator to solve the following problems.

107. A nonliving substance contains 3 micrograms of carbon-14. How much carbon-14 will be left at the end of
(a) 5,000 years?   (b) 10,000 years?
(c) 56,000 years?   (d) 112,000 years?

108. A nonliving substance contains 5 micrograms of carbon-14. How much carbon-14 will be left at the end of
(a) 500 years?   (b) 5,000 years?
(c) 56,000 years?   (d) 112,000 years?

Research Project 7

As we mentioned in the introduction to Chapter 4, the three items shown below are very closely related. You may have done some of the research necessary to show how these items are related if you worked Problems 43 and 44 in Problem Set 4.5 and Problems 85 through 88 in this problem set.

A Continued Fraction

$$1 + \cfrac{1}{1 + \cfrac{1}{1 + \cfrac{1}{1 + \cdots}}}$$

The Fibonacci Sequence

1, 1, 2, 3, 5, . . .

The Golden Rectangle

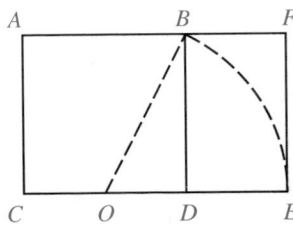

Begin your research by trying to find a decimal approximation to the continued fraction that is accurate to the nearest thousandth. (You will have to go through the Fibonacci sequence to find the decimal approximation.) Once you have this decimal approximation, compare it with what you know about the golden rectangle.

Once you see how the three items are related, write your results in essay form so that anyone who has successfully completed intermediate algebra can use it as a guide to discover the relationship that you have discovered.

More Expressions Involving Rational Exponents

In this section, we will look at multiplication, division, factoring, and simplification of some expressions that resemble polynomials but contain rational exponents. The problems in this section will be of particular interest to you if you are planning to take either an engineering calculus class or a business calculus class. As was the case in the previous section, we will assume all variables represent nonnegative real numbers. That way, we will not have to worry about the possibility of introducing undefined terms—even roots of negative numbers—into any of our examples. Let's begin this section with a look at multiplication of expressions containing rational exponents.

▶ **EXAMPLE 1** Multiply $x^{2/3}(x^{4/3} - x^{1/3})$.

Solution Applying the distributive property and then simplifying the resulting terms, we have

$$x^{2/3}(x^{4/3} - x^{1/3}) = x^{2/3}x^{4/3} - x^{2/3}x^{1/3} \qquad \text{Distributive property}$$
$$= x^{6/3} - x^{3/3} \qquad \text{Add exponents}$$
$$= x^2 - x \qquad \text{Simplify} \qquad ◀$$

▶ **EXAMPLE 2** Multiply $(x^{2/3} - 3)(x^{2/3} + 5)$.

Solution Applying the FOIL method, we multiply as if we were multiplying two binomials:

$$(x^{2/3} - 3)(x^{2/3} + 5) = x^{2/3}x^{2/3} + 5x^{2/3} - 3x^{2/3} - 15$$
$$= x^{4/3} + 2x^{2/3} - 15 \qquad ◀$$

▶ **EXAMPLE 3** Multiply $(3a^{1/3} - 2b^{1/3})(4a^{1/3} - b^{1/3})$.

Solution Again, we use the FOIL method to multiply:

$$(3a^{1/3} - 2b^{1/3})(4a^{1/3} - b^{1/3})$$
$$= 3a^{1/3}4a^{1/3} - 3a^{1/3}b^{1/3} - 2b^{1/3}4a^{1/3} + 2b^{1/3}b^{1/3}$$
$$= 12a^{2/3} - 11a^{1/3}b^{1/3} + 2b^{2/3} \qquad ◀$$

▶ **EXAMPLE 4** Expand $(t^{1/2} - 5)^2$.

Solution We can use the definition of exponents and the FOIL method:

$$\begin{aligned} (t^{1/2} - 5)^2 &= (t^{1/2} - 5)(t^{1/2} - 5) \\ &= t^{1/2}t^{1/2} - 5t^{1/2} - 5t^{1/2} + 25 \\ &= t - 10t^{1/2} + 25 \end{aligned}$$

We can obtain the same result by using the formula for the square of a binomial, $(a - b)^2 = a^2 - 2ab + b^2$.

$$\begin{aligned} (t^{1/2} - 5)^2 &= (t^{1/2})^2 - 2t^{1/2} \cdot 5 + 5^2 \\ &= t - 10t^{1/2} + 25 \end{aligned}$$ ◀

▶ **EXAMPLE 5** Multiply $(x^{3/2} - 2^{3/2})(x^{3/2} + 2^{3/2})$.

Solution This product has the form $(a - b)(a + b)$, which will result in the difference of two squares, $a^2 - b^2$:

$$\begin{aligned} (x^{3/2} - 2^{3/2})(x^{3/2} + 2^{3/2}) &= (x^{3/2})^2 - (2^{3/2})^2 \\ &= x^3 - 2^3 \\ &= x^3 - 8 \end{aligned}$$ ◀

▶ **EXAMPLE 6** Multiply $(a^{1/3} - b^{1/3})(a^{2/3} + a^{1/3}b^{1/3} + b^{2/3})$.

Solution We can find this product by multiplying in columns:

$$\begin{array}{l} a^{2/3} + a^{1/3}b^{1/3} + b^{2/3} \\ \underline{\qquad\qquad a^{1/3} - b^{1/3}} \\ a \quad + a^{2/3}b^{1/3} + a^{1/3}b^{2/3} \\ \underline{\qquad - a^{2/3}b^{1/3} - a^{1/3}b^{2/3} - b} \\ a \qquad\qquad\qquad\qquad\quad - b \end{array}$$

The product is $a - b$. ◀

Our next example involves division with expressions that contain rational exponents. As you will see, this kind of division is very similar to division of a polynomial by a monomial (as shown previously in Section 4.2).

▶ **EXAMPLE 7** Divide $\dfrac{15x^{2/3}y^{1/3} - 20x^{4/3}y^{2/3}}{5x^{1/3}y^{1/3}}$.

Solution We can approach this problem in the same way we approached division by a monomial. We simply divide each term in the numerator by the term in the denominator:

$$\frac{15x^{2/3}y^{1/3} - 20x^{4/3}y^{2/3}}{5x^{1/3}y^{1/3}} = \frac{15x^{2/3}y^{1/3}}{5x^{1/3}y^{1/3}} - \frac{20x^{4/3}y^{2/3}}{5x^{1/3}y^{1/3}}$$
$$= 3x^{1/3} - 4xy^{1/3} \qquad \blacktriangleleft$$

The next three examples involve factoring. In the first example, we are told what to factor from each term of an expression.

▶ **EXAMPLE 8** Factor $3(x - 2)^{1/3}$ from $12(x - 2)^{4/3} - 9(x - 2)^{1/3}$ and then simplify, if possible.

Solution This solution is similar to factoring out the greatest common factor:

$$12(x - 2)^{4/3} - 9(x - 2)^{1/3} = 3(x - 2)^{1/3}[4(x - 2) - 3]$$
$$= 3(x - 2)^{1/3}(4x - 11) \qquad \blacktriangleleft$$

Although an expression containing rational exponents is not a polynomial —remember, a polynomial must have exponents that are whole numbers—we are going to treat the expressions that follow as if they were polynomials.

▶ **EXAMPLE 9** Factor $x^{2/3} - 3x^{1/3} - 10$ as if it were a trinomial.

Solution We can think of $x^{2/3} - 3x^{1/3} - 10$ as if it is a trinomial in which the variable is $x^{1/3}$. To see this, replace $x^{1/3}$ with y to get

$$y^2 - 3y - 10$$

Since this trinomial in y factors as $(y - 5)(y + 2)$, we can factor our original expression similarly:

$$x^{2/3} - 3x^{1/3} - 10 = (x^{1/3} - 5)(x^{1/3} + 2)$$

Remember, with factoring, we can always multiply our factors to check that we have factored correctly. ◀

▶ **EXAMPLE 10** Factor $6x^{2/5} + 11x^{1/5} - 10$ as if it were a trinomial.

Solution We can think of the expression in question as a trinomial in $x^{1/5}$:

$$6x^{2/5} + 11x^{1/5} - 10 = (3x^{1/5} - 2)(2x^{1/5} + 5) \qquad \blacktriangleleft$$

In our next example, we combine two expressions by applying the methods we used to add and subtract fractions or rational expressions in Chapter 4.

▶ **EXAMPLE 11**   Subtract $(x^2 + 4)^{1/2} - \dfrac{x^2}{(x^2 + 4)^{1/2}}$.

Solution   In order to combine these two expressions, we need to find a least common denominator, change to equivalent fractions, and subtract numerators. The least common denominator is $(x^2 + 4)^{1/2}$.

$$(x^2 + 4)^{1/2} - \frac{x^2}{(x^2 + 4)^{1/2}} = \frac{(x^2 + 4)^{1/2}}{1} \cdot \frac{(x^2 + 4)^{1/2}}{(x^2 + 4)^{1/2}} - \frac{x^2}{(x^2 + 4)^{1/2}}$$

$$= \frac{x^2 + 4 - x^2}{(x^2 + 4)^{1/2}}$$

$$= \frac{4}{(x^2 + 4)^{1/2}} \quad ◀$$

▶ **EXAMPLE 12**   If you purchase an investment for P dollars and t years later it is worth A dollars, then the annual rate of return r on that investment is given by the formula

$$r = \left(\frac{A}{P}\right)^{1/t} - 1$$

Find the annual rate of return on a coin collection that was purchased for $500 and sold 4 years later for $800.

Solution   Using $A = 800$, $P = 500$, and $t = 4$ in the formula, we have

$$r = \left(\frac{800}{500}\right)^{1/4} - 1$$

The easiest way to simplify this expression is with a scientific calculator. If we first change $\frac{1}{4}$ to 0.25 and $\frac{800}{500}$ to 1.6, then pressing the keys in the following sequence will give us an approximation to $(\frac{800}{500})^{1/4}$:

$$1.6 \;\boxed{y^x}\; .25 \;\boxed{=}$$

Allowing three decimal places, the result is 1.125. Using this result, we can complete the problem:

$$r = (1.6)^{0.25} - 1$$

$$= 1.125 - 1$$

$$= 0.125 \text{ or } 12.5\%$$

The annual return on the coin collection is approximately 12.5%. To do as well with a savings account, we would have to invest the original $500 in an account that paid 12.5%, compounded annually. ◀

PROBLEM SET 6.2

Multiply. (Assume all variables in this problem set represent nonnegative real numbers.)

1. $x^{2/3}(x^{1/3} + x^{4/3})$ **2.** $x^{2/5}(x^{3/5} - x^{8/5})$
3. $a^{1/2}(a^{3/2} - a^{1/2})$ **4.** $a^{1/4}(a^{3/4} + a^{7/4})$
5. $2x^{1/3}(3x^{8/3} - 4x^{5/3} + 5x^{2/3})$
6. $5x^{1/2}(4x^{5/2} + 3x^{3/2} + 2x^{1/2})$
7. $4x^{1/2}y^{3/5}(3x^{3/2}y^{-3/5} - 9x^{-1/2}y^{7/5})$
8. $3x^{4/5}y^{1/3}(4x^{6/5}y^{-1/3} - 12x^{-4/5}y^{5/3})$
9. $(x^{2/3} - 4)(x^{2/3} + 2)$
10. $(x^{2/3} - 5)(x^{2/3} + 2)$
11. $(a^{1/2} - 3)(a^{1/2} - 7)$
12. $(a^{1/2} - 6)(a^{1/2} - 2)$
13. $(4y^{1/3} - 3)(5y^{1/3} + 2)$
14. $(5y^{1/3} - 2)(4y^{1/3} + 3)$
15. $(5x^{2/3} + 3y^{1/2})(2x^{2/3} + 3y^{1/2})$
16. $(4x^{2/3} - 2y^{1/2})(5x^{2/3} - 3y^{1/2})$
17. $(t^{1/2} + 5)^2$ **18.** $(t^{1/2} - 3)^2$
19. $(x^{3/2} + 4)^2$ **20.** $(x^{3/2} - 6)^2$
21. $(a^{1/2} - b^{1/2})^2$ **22.** $(a^{1/2} + b^{1/2})^2$
23. $(2x^{1/2} - 3y^{1/2})^2$ **24.** $(5x^{1/2} + 4y^{1/2})^2$
25. $(a^{1/2} - 3^{1/2})(a^{1/2} + 3^{1/2})$
26. $(a^{1/2} - 5^{1/2})(a^{1/2} + 5^{1/2})$
27. $(x^{3/2} + y^{3/2})(x^{3/2} - y^{3/2})$
28. $(x^{5/2} + y^{5/2})(x^{5/2} - y^{5/2})$
29. $(t^{1/2} - 2^{3/2})(t^{1/2} + 2^{3/2})$
30. $(t^{1/2} - 5^{3/2})(t^{1/2} + 5^{3/2})$
31. $(2x^{3/2} + 3^{1/2})(2x^{3/2} - 3^{1/2})$
32. $(3x^{1/2} + 2^{3/2})(3x^{1/2} - 2^{3/2})$
33. $(x^{1/3} + y^{1/3})(x^{2/3} - x^{1/3}y^{1/3} + y^{2/3})$
34. $(x^{1/3} - y^{1/3})(x^{2/3} + x^{1/3}y^{1/3} + y^{2/3})$
35. $(a^{1/3} - 2)(a^{2/3} + 2a^{1/3} + 4)$
36. $(a^{1/3} + 3)(a^{2/3} - 3a^{1/3} + 9)$
37. $(2x^{1/3} + 1)(4x^{2/3} - 2x^{1/3} + 1)$
38. $(3x^{1/3} - 1)(9x^{2/3} + 3x^{1/3} + 1)$
39. $(t^{1/4} - 1)(t^{1/4} + 1)(t^{1/2} + 1)$
40. $(t^{1/4} - 2)(t^{1/4} + 2)(t^{1/2} + 4)$

Divide.

41. $\dfrac{18x^{3/4} + 27x^{1/4}}{9x^{1/4}}$ **42.** $\dfrac{25x^{1/4} + 30x^{3/4}}{5x^{1/4}}$

43. $\dfrac{12x^{2/3}y^{1/3} - 16x^{1/3}y^{2/3}}{4x^{1/3}y^{1/3}}$

44. $\dfrac{12x^{4/3}y^{1/3} - 18x^{1/3}y^{4/3}}{6x^{1/3}y^{1/3}}$

45. $\dfrac{21a^{7/5}b^{3/5} - 14a^{2/5}b^{8/5}}{7a^{2/5}b^{3/5}}$

46. $\dfrac{24a^{9/5}b^{3/5} - 16a^{4/5}b^{8/5}}{8a^{4/5}b^{3/5}}$

47. Factor $3(x - 2)^{1/2}$ from $12(x - 2)^{3/2} - 9(x - 2)^{1/2}$.
48. Factor $4(x + 1)^{1/3}$ from $4(x + 1)^{4/3} + 8(x + 1)^{1/3}$.
49. Factor $5(x - 3)^{7/5}$ from $5(x - 3)^{12/5} - 15(x - 3)^{7/5}$.
50. Factor $6(x + 3)^{8/7}$ from $6(x + 3)^{15/7} - 12(x + 3)^{8/7}$.
51. Factor $3(x + 1)^{1/2}$ from $9x(x + 1)^{3/2} + 6(x + 1)^{1/2}$.
52. Factor $4x(x + 1)^{1/2}$ from $4x^2(x + 1)^{1/2} + 8x(x + 1)^{3/2}$.

Factor each of the following as if it were a trinomial.

53. $x^{2/3} - 5x^{1/3} + 6$ **54.** $x^{2/3} - x^{1/3} - 6$
55. $a^{2/5} - 2a^{1/5} - 8$ **56.** $a^{2/5} + 2a^{1/5} - 8$
57. $2y^{2/3} - 5y^{1/3} - 3$ **58.** $3y^{2/3} + 5y^{1/3} - 2$
59. $9t^{2/5} - 25$ **60.** $16t^{2/5} - 49$
61. $4x^{2/7} + 20x^{1/7} + 25$
62. $25x^{2/7} - 20x^{1/7} + 4$

Simplify each of the following to a single fraction.

63. $\dfrac{3}{x^{1/2}} + x^{1/2}$ **64.** $\dfrac{2}{x^{1/2}} - x^{1/2}$

65. $x^{2/3} + \dfrac{5}{x^{1/3}}$ **66.** $x^{3/4} - \dfrac{7}{x^{1/4}}$

67. $\dfrac{3x^2}{(x^3 + 1)^{1/2}} + (x^3 + 1)^{1/2}$

68. $\dfrac{x^3}{(x^2 - 1)^{1/2}} + 2x(x^2 - 1)^{1/2}$

69. $\dfrac{x^2}{(x^2 + 4)^{1/2}} - (x^2 + 4)^{1/2}$

70. $\dfrac{x^5}{(x^2 - 2)^{1/2}} + 4x^3(x^2 - 2)^{1/2}$

Use a calculator to find approximations to each of the following. Round your answers for Problems 75 and 76 to three places past the decimal point.

71. $16^{0.25}$
72. $81^{0.25}$
73. $9^{1.5}$ **74.** $32^{0.4}$
75. $(\frac{1}{2})^{1/5}$ **76.** $(\frac{1}{2})^{1/10}$

Applying the Concepts

77. A coin collection is purchased as an investment for $500 and sold 4 years later for $900. Find the annual rate of return on the investment.
78. An investor buys stock in a company for $800. Five years later, the same stock is worth $1,600. Find the annual rate of return on the stocks.
79. Find the annual rate of return on a home that is purchased for $60,000 and is sold 5 years later for $80,000.
80. Find the annual rate of return on a home that is purchased for $75,000 and is sold 10 years later for $150,000.

Review Problems

The following problems review material we covered in Section 5.2.

Find the slope of the line that contains the following pairs of points.

81. $(-4, -1)$ and $(-2, 5)$
82. $(-2, -3)$ and $(-5, 1)$
83. Find y if the slope of the line through $(5, y)$ and $(4, 2)$ is 3.
84. Find x if the slope of the line through $(4, 9)$ and $(x, -2)$ is $-\frac{7}{3}$.

A line has a slope of $\frac{2}{3}$. Find the slope of any line

85. parallel to it.
86. perpendicular to it.
87. Find the slope of the line with x-intercept 5 and y-intercept -2.
88. Find the slope of the line with x-intercept -3 and y-intercept 6.

SECTION

6.3   ## Simplified Form for Radicals

In this section, we will use radical notation instead of rational exponents. We will begin by stating two properties of radicals. Following this, we will give a definition for simplified form for radical expressions. The examples in this section show how we use the properties of radicals to write radical expressions in simplified form.

For our first two properties of radicals, we will assume a and b are nonnegative real numbers whenever n is an even number.

Property 1 for Radicals
$$\sqrt[n]{ab} = \sqrt[n]{a}\,\sqrt[n]{b}$$
In words: The nth root of a product is the product of the nth roots.

PROOF OF PROPERTY 1

$$\sqrt[n]{ab} = (ab)^{1/n} \qquad \text{Definition of fractional exponents}$$
$$= a^{1/n}b^{1/n} \qquad \text{Exponents distribute over products}$$
$$= \sqrt[n]{a}\,\sqrt[n]{b} \qquad \text{Definition of fractional exponents}$$

Property 2 for Radicals

$$\sqrt[n]{\frac{a}{b}} = \frac{\sqrt[n]{a}}{\sqrt[n]{b}} \qquad (b \neq 0)$$

In words: The nth root of a quotient is the quotient of the nth roots.

The proof of Property 2 is similar to the proof of Property 1.

Note There is no property for radicals that says the nth root of a sum is the sum of the nth roots. That is,

$$\sqrt[n]{a + b} \neq \sqrt[n]{a} + \sqrt[n]{b}$$

These two properties of radicals allow us to change the form of and simplify radical expressions without changing their value.

Simplified Form for Radical Expressions
A radical expression is in **simplified form** if

1. none of the factors of the radicand (the quantity under the radical sign) can be written as powers greater than or equal to the index—that is, no perfect squares can be factors of the quantity under a square root sign, no perfect cubes can be factors of what is under a cube root sign, and so forth;
2. there are no fractions under the radical sign; and
3. there are no radicals in the denominator.

Satisfying the first condition for simplified form actually amounts to taking as much out from under the radical sign as possible. The following examples illustrate the first condition for simplified form.

▶ **EXAMPLE 1** Write $\sqrt{50}$ in simplified form.

Solution The largest perfect square that divides 50 is 25. We write 50 as $25 \cdot 2$ and apply Property 1 for radicals:

$$\sqrt{50} = \sqrt{25 \cdot 2} \qquad 50 = 25 \cdot 2$$

$$= \sqrt{25}\,\sqrt{2} \qquad \text{Property 1}$$
$$= 5\sqrt{2} \qquad \sqrt{25} = 5$$

We have taken as much as possible out from under the radical sign—in this case, factoring 25 from 50 and then writing $\sqrt{25}$ as 5. ◀

▶ **EXAMPLE 2**   Write in simplified form: $\sqrt{48x^4y^3}$, where x, $y \geq 0$.

Solution   The largest perfect square that is a factor of the radicand is $16x^4y^2$. Applying Property 1 again, we have

$$\sqrt{48x^4y^3} = \sqrt{16x^4y^2 \cdot 3y}$$
$$= \sqrt{16x^4y^2}\,\sqrt{3y}$$
$$= 4x^2y\sqrt{3y} \qquad ◀$$

▶ **EXAMPLE 3**   Write $\sqrt[3]{40a^5b^4}$ in simplified form.

Solution   We now want to factor the largest perfect cube from the radicand. We write $40a^5b^4$ as $8a^3b^3 \cdot 5a^2b$ and proceed as we did in Examples 1 and 2.

$$\sqrt[3]{40a^5b^4} = \sqrt[3]{8a^3b^3 \cdot 5a^2b}$$
$$= \sqrt[3]{8a^3b^3}\,\sqrt[3]{5a^2b}$$
$$= 2ab\,\sqrt[3]{5a^2b} \qquad ◀$$

Here are some further examples concerning the first condition for simplified form.

▶ **EXAMPLES**   Write each expression in simplified form.

4. $\sqrt{12x^7y^6} = \sqrt{4x^6y^6 \cdot 3x}$
$\phantom{\sqrt{12x^7y^6}} = \sqrt{4x^6y^6}\,\sqrt{3x}$
$\phantom{\sqrt{12x^7y^6}} = 2x^3y^3\,\sqrt{3x}$

5. $\sqrt[3]{54a^6b^2c^4} = \sqrt[3]{27a^6c^3 \cdot 2b^2c}$
$\phantom{\sqrt[3]{54a^6b^2c^4}} = \sqrt[3]{27a^6c^3}\,\sqrt[3]{2b^2c}$
$\phantom{\sqrt[3]{54a^6b^2c^4}} = 3a^2c\,\sqrt[3]{2b^2c} \qquad ◀$

The second property of radicals is used to simplify a radical that contains a fraction.

▶ **EXAMPLE 6**   Simplify $\sqrt{\dfrac{3}{4}}$.

Solution   Applying Property 2 for radicals, we have

$$\sqrt{\frac{3}{4}} = \frac{\sqrt{3}}{\sqrt{4}} \qquad \text{Property 2}$$

$$= \frac{\sqrt{3}}{2} \qquad \sqrt{4} = 2$$

The last expression is in simplified form because it satisfies all three conditions for simplified form.   ◄

▶ **EXAMPLE 7**   Write $\sqrt{\dfrac{5}{6}}$ in simplified form.

Solution   Proceeding as in Example 6, we have

$$\sqrt{\frac{5}{6}} = \frac{\sqrt{5}}{\sqrt{6}}$$

The resulting expression satisfies the second condition for simplified form since neither radical contains a fraction. It does, however, violate condition 3 since it has a radical in the denominator. Getting rid of the radical in the denominator is called **rationalizing the denominator** and is accomplished, in this case, by multiplying the numerator and denominator by $\sqrt{6}$:

$$\frac{\sqrt{5}}{\sqrt{6}} = \frac{\sqrt{5}}{\sqrt{6}} \cdot \frac{\sqrt{6}}{\sqrt{6}}$$

$$= \frac{\sqrt{30}}{\sqrt{6^2}}$$

$$= \frac{\sqrt{30}}{6} \qquad\qquad\qquad ◄$$

▶ **EXAMPLES**   Rationalize the denominator.

8. $\dfrac{4}{\sqrt{3}} = \dfrac{4}{\sqrt{3}} \cdot \dfrac{\sqrt{3}}{\sqrt{3}}$

   $= \dfrac{4\sqrt{3}}{\sqrt{3^2}}$

   $= \dfrac{4\sqrt{3}}{3}$

9. $\dfrac{2\sqrt{3x}}{\sqrt{5y}} = \dfrac{2\sqrt{3x}}{\sqrt{5y}} \cdot \dfrac{\sqrt{5y}}{\sqrt{5y}}$

$$= \frac{2\sqrt{15xy}}{\sqrt{(5y)^2}}$$

$$= \frac{2\sqrt{15xy}}{5y}$$

◀

When the denominator involves a cube root, we must multiply by a radical that will produce a perfect cube under the cube root sign in the denominator, as our next example illustrates.

▶ **EXAMPLE 10** Rationalize the denominator in $\dfrac{7}{\sqrt[3]{4}}$.

Solution Since $4 = 2^2$, we can multiply both numerator and denominator by $\sqrt[3]{2}$ and obtain $\sqrt[3]{2^3}$ in the denominator:

$$\frac{7}{\sqrt[3]{4}} = \frac{7}{\sqrt[3]{2^2}}$$

$$= \frac{7}{\sqrt[3]{2^2}} \cdot \frac{\sqrt[3]{2}}{\sqrt[3]{2}}$$

$$= \frac{7\sqrt[3]{2}}{\sqrt[3]{2^3}}$$

$$= \frac{7\sqrt[3]{2}}{2}$$

◀

▶ **EXAMPLE 11** Simplify $\sqrt{\dfrac{12x^5y^3}{5z}}$.

Solution We use Property 2 to write the numerator and denominator as two separate radicals:

$$\sqrt{\frac{12x^5y^3}{5z}} = \frac{\sqrt{12x^5y^3}}{\sqrt{5z}}$$

Simplifying the numerator, we have

$$\frac{\sqrt{12x^5y^3}}{\sqrt{5z}} = \frac{\sqrt{4x^4y^2}\,\sqrt{3xy}}{\sqrt{5z}}$$

$$= \frac{2x^2y\sqrt{3xy}}{\sqrt{5z}}$$

To rationalize the denominator we multiply the numerator and denominator by $\sqrt{5z}$:

$$\frac{2x^2y\sqrt{3xy}}{\sqrt{5z}} \cdot \frac{\sqrt{5z}}{\sqrt{5z}} = \frac{2x^2y\sqrt{15xyz}}{\sqrt{(5z)^2}}$$

$$= \frac{2x^2y\sqrt{15xyz}}{5z}$$

◄

The Square Root of a Perfect Square

So far in this chapter, we have assumed that all our variables are nonnegative when they appear under a square root symbol. There are times, however, when this is not the case.

Consider the following two statements:

$$\sqrt{3^2} = \sqrt{9} = 3 \quad \text{and} \quad \sqrt{(-3)^2} = \sqrt{9} = 3$$

Whether we operate on 3 or -3, the result is the same: both expressions simplify to 3. The other operation we have worked with in the past that produces the same result is absolute value. That is,

$$|3| = 3 \quad \text{and} \quad |-3| = 3$$

This leads us to the next property of radicals.

Property 3 for Radicals
If a is a real number, then $\sqrt{a^2} = |a|$.

The result of this discussion and Property 3 is simply this:

If we know a is positive, then $\sqrt{a^2} = a$.
If we know a is negative, then $\sqrt{a^2} = |a|$.
If we don't know if a is positive or negative, then $\sqrt{a^2} = |a|$.

► **EXAMPLES** Simplify each expression. Do *not* assume the variables represent positive numbers.

12. $\sqrt{9x^2} = 3|x|$
13. $\sqrt{x^3} = |x|\sqrt{x}$
14. $\sqrt{x^2 - 6x + 9} = \sqrt{(x-3)^2} = |x-3|$
15. $\sqrt{x^3 - 5x^2} = \sqrt{x^2(x-5)} = |x|\sqrt{x-5}$

◄

As you can see, we must use absolute value symbols when we take a square root of a perfect square, unless we know the base of the perfect square is a positive number. The same idea holds for higher even roots, but not for odd roots. With odd roots, no absolute value symbols are necessary.

▶ **EXAMPLES**   Simplify each expression.

16. $\sqrt[3]{(-2)^3} = \sqrt[3]{-8} = -2$

17. $\sqrt[3]{(-5)^3} = \sqrt[3]{-125} = -5$ ◀

We can extend the discussion above to all roots as follows:

Extending Property 3 for Radicals
If a is a real number, then

$$\sqrt[n]{a^n} = |a| \quad \text{if} \quad n \text{ is even}$$

$$\sqrt[n]{a^n} = a \quad \text{if} \quad n \text{ is odd}$$

The Spiral of Roots

In order to visualize the square roots of the positive integers we can construct the spiral of roots that we mentioned in the introduction to this chapter. To begin, we draw two line segments, each of length 1, at right angles to each other. Then we use the Pythagorean Theorem to find the length of the diagonal. Figure 1 illustrates this procedure.

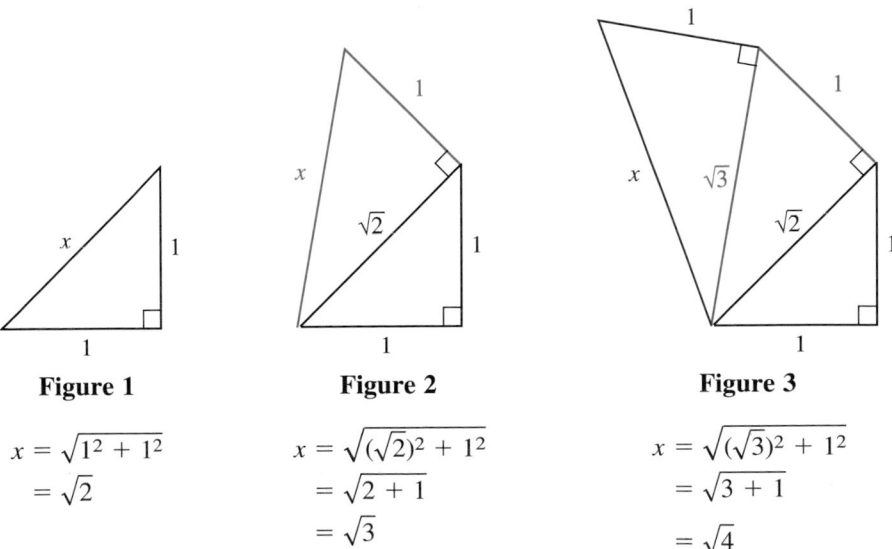

Figure 1   **Figure 2**   **Figure 3**

$$x = \sqrt{1^2 + 1^2} \qquad x = \sqrt{(\sqrt{2})^2 + 1^2} \qquad x = \sqrt{(\sqrt{3})^2 + 1^2}$$

$$= \sqrt{2} \qquad\qquad = \sqrt{2 + 1} \qquad\qquad = \sqrt{3 + 1}$$

$$\qquad\qquad\qquad = \sqrt{3} \qquad\qquad\qquad = \sqrt{4}$$

Next, we construct a second triangle by connecting a line segment of length 1 to the end of the first diagonal so that the angle formed is a right angle. We find the length of the second diagonal using the Pythagorean Theorem. Figure 2 illustrates this procedure. Continuing to draw new triangles by connecting line segments of length 1 to the end of each new diagonal, so that the angle formed is a right angle, the spiral of roots begins to appear (Figure 3).

The Spiral of Roots and Function Notation

Looking over the diagrams and calculations in the discussion above, we see that each diagonal in the spiral of roots is found by using the length of the previous diagonal.

First diagonal: $\sqrt{1^2 + 1^2} = \sqrt{2}$

Second diagonal: $\sqrt{(\sqrt{2})^2 + 1^2} = \sqrt{3}$

Third diagonal: $\sqrt{(\sqrt{3})^2 + 1^2} = \sqrt{4}$

Fourth diagonal: $\sqrt{(\sqrt{4})^2 + 1^2} = \sqrt{5}$

A process like this one, in which the answer to one calculation is used to find the answer to the next calculation, is called a *recursive* process. In this particular case, we can use function notation to model the process. If we let x represent the length of any diagonal, then the length of the next diagonal is given by

$$f(x) = \sqrt{x^2 + 1}$$

To begin the process of finding the diagonals, we let $x = 1$:

$$f(1) = \sqrt{1^2 + 1} = \sqrt{2}$$

To find the next diagonal, we substitute $\sqrt{2}$ for x to obtain

$$f(f(1)) = f(\sqrt{2}) = \sqrt{(\sqrt{2})^2 + 1} = \sqrt{3}$$
$$f(f(f(1))) = f(\sqrt{3}) = \sqrt{(\sqrt{3})^2 + 1} = \sqrt{4}$$

We can describe the process of finding the diagonals of the spiral of roots very concisely this way:

$$f(1), \quad f(f(1)), \quad f(f(f(1))), \; \ldots \qquad \text{where } f(x) = \sqrt{x^2 + 1}$$

The sequence of function values above is a special case of a general category of similar sequences that are closely connected to *fractals* and *chaos,* two topics in mathematics that are currently receiving a good deal of attention.

USING TECHNOLOGY: Graphing Calculators

As our discussion above indicates, the length of each diagonal in the spiral of roots is used to calculate the length of the next diagonal. The ANS key on a graphing calculator can be used very effectively in a situation like this one. To begin, we store the number 1 in the variable ANS. Next, we key in the formula used to produce each diagonal using ANS for the variable. After that, it is simply a matter of pressing ENTER, as many times as we like, to produce the lengths of as many diagonals as we like. Here is a summary of what we do:

Enter This	Display Shows
1 ENT	1.000
$\sqrt{(\text{ANS}^2 + 1)}$ ENT	1.414
ENT	1.732
ENT	2.000
ENT	2.236

If you continue to press the ENT key, you will produce decimal approximations for as many of the diagonals in the spiral of roots as you like.

PROBLEM SET 6.3

Use Property 1 for radicals to write each of the following expressions in simplified form. (Assume all variables are nonnegative through Problem 34.)

1. $\sqrt{8}$
2. $\sqrt{32}$
3. $\sqrt{98}$
4. $\sqrt{75}$
5. $\sqrt{288}$
6. $\sqrt{128}$
7. $\sqrt{80}$
8. $\sqrt{200}$
9. $\sqrt{48}$
10. $\sqrt{27}$
11. $\sqrt{675}$
12. $\sqrt{972}$
13. $\sqrt[3]{54}$
14. $\sqrt[3]{24}$
15. $\sqrt[3]{128}$
16. $\sqrt[3]{162}$
17. $\sqrt[3]{432}$
18. $\sqrt[3]{1,536}$
19. $\sqrt[5]{64}$
20. $\sqrt[4]{48}$
21. $\sqrt{18x^3}$
22. $\sqrt{27x^5}$
23. $\sqrt[4]{32y^7}$
24. $\sqrt[5]{32y^7}$
25. $\sqrt[3]{40x^4y^7}$
26. $\sqrt[3]{128x^6y^2}$
27. $\sqrt{48a^2b^3c^4}$
28. $\sqrt{72a^4b^3c^2}$
29. $\sqrt[3]{48a^2b^3c^4}$
30. $\sqrt[3]{72a^4b^3c^2}$
31. $\sqrt[5]{64x^8y^{12}}$
32. $\sqrt[4]{32x^9y^{10}}$
33. $\sqrt[5]{243x^7y^{10}z^5}$
34. $\sqrt[5]{64x^8y^4z^{11}}$

Substitute the given numbers into the expression $\sqrt{b^2 - 4ac}$ and then simplify.

35. $a = 2, b = -6, c = 3$
36. $a = 6, b = 7, c = -5$
37. $a = 1, b = 2, c = 6$
38. $a = 2, b = 5, c = 3$
39. $a = \frac{1}{2}, b = -\frac{1}{2}, c = -\frac{5}{4}$
40. $a = \frac{7}{4}, b = -\frac{3}{4}, c = -2$

Rationalize the denominator in each of the following expressions. (Assume all variables represent positive numbers through Problem 70.)

41. $\dfrac{2}{\sqrt{3}}$

42. $\dfrac{3}{\sqrt{2}}$

43. $\dfrac{5}{\sqrt{6}}$

44. $\dfrac{7}{\sqrt{5}}$

45. $\sqrt{\dfrac{1}{2}}$

46. $\sqrt{\dfrac{1}{3}}$

47. $\sqrt{\dfrac{1}{5}}$

48. $\sqrt{\dfrac{1}{6}}$

49. $\dfrac{4}{\sqrt[3]{2}}$

50. $\dfrac{5}{\sqrt[3]{3}}$

51. $\dfrac{2}{\sqrt[3]{9}}$

52. $\dfrac{3}{\sqrt[3]{4}}$

53. $\sqrt[4]{\dfrac{3}{2x^2}}$

54. $\sqrt[4]{\dfrac{5}{3x^2}}$

55. $\sqrt[4]{\dfrac{8}{y}}$

56. $\sqrt[4]{\dfrac{27}{y}}$

57. $\sqrt[3]{\dfrac{4x}{3y}}$

58. $\sqrt[3]{\dfrac{7x}{6y}}$

59. $\sqrt[3]{\dfrac{2x}{9y}}$

60. $\sqrt[3]{\dfrac{5x}{4y}}$

61. $\sqrt[4]{\dfrac{1}{8x^3}}$

62. $\sqrt[4]{\dfrac{8}{9x^3}}$

Write each of the following in simplified form.

63. $\sqrt{\dfrac{27x^3}{5y}}$

64. $\sqrt{\dfrac{12x^5}{7y}}$

65. $\sqrt{\dfrac{75x^3y^2}{2z}}$

66. $\sqrt{\dfrac{50x^2y^3}{3z}}$

67. $\sqrt[3]{\dfrac{16a^4b^3}{9c}}$

68. $\sqrt[3]{\dfrac{54a^5b^4}{25c^2}}$

69. $\sqrt[3]{\dfrac{8x^3y^6}{9z}}$

70. $\sqrt[3]{\dfrac{27x^6y^3}{2z^2}}$

Simplify each expression. Do *not* assume the variables represent positive numbers.

71. $\sqrt{25x^2}$

72. $\sqrt{49x^2}$

73. $\sqrt{27x^3y^2}$

74. $\sqrt{40x^3y^2}$

75. $\sqrt{x^2 - 10x + 25}$

76. $\sqrt{x^2 - 16x + 64}$

77. $\sqrt{4x^2 + 12x + 9}$

78. $\sqrt{16x^2 + 40x + 25}$

79. $\sqrt{4a^4 + 16a^3 + 16a^2}$

80. $\sqrt{9a^4 + 18a^3 + 9a^2}$

81. $\sqrt{4x^3 - 8x^2}$ **82.** $\sqrt{18x^3 - 9x^2}$

83. Show that the statement $\sqrt{a + b} = \sqrt{a} + \sqrt{b}$ is not, in general, true by replacing a with 9 and b with 16 and simplifying both sides.

84. Find a pair of values for a and b that will make the statement $\sqrt{a + b} = \sqrt{a} + \sqrt{b}$ true.

Applying the Concepts

85. The distance (d) between opposite corners of a rectangular room with length l and width w is given by

$$d = \sqrt{l^2 + w^2}$$

How far is it between opposite corners of a living room that measures 10 by 15 feet?

86. The radius r of a sphere with volume V can be found by using the formula

$$r = \sqrt[3]{\dfrac{3V}{4\pi}}$$

Find the radius of a sphere with volume 9 cubic feet. Write your answer in simplified form. (Use $\frac{22}{7}$ for π.)

87. Construct your own spiral of roots by using a ruler. Draw the first triangle by using two 1-inch lines. The first diagonal will have a length of $\sqrt{2}$ inches. Each new triangle will be formed by drawing a 1-inch line segment at the end of the previous diagonal so that the angle formed is 90°.

88. Construct a spiral of roots by using line segments of length 2 inches. The length of the first diagonal will be $2\sqrt{2}$ inches. The length of the second diagonal will be $2\sqrt{3}$ inches.

89. If $f(x) = \sqrt{x^2 + 1}$ find the first six terms in the sequence below. Use your results to predict the value of the 10th term and the 100th term.

$$f(1), \quad f(f(1)), \quad f(f(f(1))), \quad \ldots$$

90. If $f(x) = \sqrt{x^2 + 4}$ find the first six terms in the sequence below. Use your results to predict the value of the 10th term and the 100th term. (The numbers in this sequence are the lengths of the diagonals of the spiral you drew in Problem 88.)

$$f(2), \quad f(f(2)), \quad f(f(f(2))), \quad \ldots$$

Review Problems

The problems below review material we covered in Section 5.4.

Graph each inequality.

91. $2x + 3y < 6$ **92.** $2x + y < -5$

93. $y \geq -3x - 4$
94. $y \geq 2x - 1$
95. $x \geq 3$
96. $y > -5$

One Step Further

Factor each radicand into the product of prime factors, then simplify each radical.

97. $\sqrt[3]{8,640}$ **98.** $\sqrt{8,640}$
99. $\sqrt[3]{10,584}$ **100.** $\sqrt{10,584}$

Assume a is a positive number and rationalize each denominator.

101. $\dfrac{1}{\sqrt[10]{a^3}}$ **102.** $\dfrac{1}{\sqrt[12]{a^7}}$

103. $\dfrac{1}{\sqrt[20]{a^{11}}}$ **104.** $\dfrac{1}{\sqrt[15]{a^{13}}}$

USING TECHNOLOGY

105. Show that the two expressions $\sqrt{x^2 + 1}$ and $x + 1$ are not, in general, equal to each other by graphing $y = \sqrt{x^2 + 1}$ and $y = x + 1$ on the same set of axes.

106. Show that the two expressions $\sqrt{x^2 + 9}$ and $x + 3$ are not, in general, equal to each other by graphing $y = \sqrt{x^2 + 9}$ and $y = x + 3$ on the same set of axes.

107. Approximately how far apart are the graphs in Problem 105 when $x = 2$?

108. Approximately how far apart are the graphs in Problem 106 when $x = 2$?

109. For what value of x are the expressions $\sqrt{x^2 + 1}$ and $x + 1$ equal?

110. For what value of x are the expressions $\sqrt{x^2 + 9}$ and $x + 3$ equal?

SECTION
6.4   Addition and Subtraction of Radical Expressions

In Chapter 3 we found that we could add similar terms when combining polynomials. The same idea applies to addition and subtraction of radical expressions.

DEFINITION   Two radicals are said to be **similar radicals** if they have the same index and the same radicand.

The expressions $5\sqrt[3]{7}$ and $-8\sqrt[3]{7}$ are similar since the index is 3 in both cases and the radicands are 7. The expressions $3\sqrt[4]{5}$ and $7\sqrt[3]{5}$ are not similar since they have different indices, while the expressions $2\sqrt[5]{8}$ and $3\sqrt[5]{9}$ are not similar because the radicands are not the same.

We add and subtract radical expressions in the same way we add and subtract polynomials—by combining similar terms under the distributive property.

▶ **EXAMPLE 1** Combine $5\sqrt{3} - 4\sqrt{3} + 6\sqrt{3}$.

Solution All three radicals are similar. We apply the distributive property to get

$$5\sqrt{3} - 4\sqrt{3} + 6\sqrt{3} = (5 - 4 + 6)\sqrt{3}$$
$$= 7\sqrt{3} \qquad \blacktriangleleft$$

▶ **EXAMPLE 2** Combine $3\sqrt{8} + 5\sqrt{18}$.

Solution The two radicals do not seem to be similar. We must write each in simplified form before applying the distributive property.

$$3\sqrt{8} + 5\sqrt{18} = 3\sqrt{4 \cdot 2} + 5\sqrt{9 \cdot 2}$$
$$= 3\sqrt{4}\sqrt{2} + 5\sqrt{9}\sqrt{2}$$
$$= 3 \cdot 2\sqrt{2} + 5 \cdot 3\sqrt{2}$$
$$= 6\sqrt{2} + 15\sqrt{2}$$
$$= (6 + 15)\sqrt{2}$$
$$= 21\sqrt{2} \qquad \blacktriangleleft$$

The result of Example 2 can be generalized to the following rule for sums and differences of radical expressions.

Rule

To add or subtract radical expressions, put each in simplified form and apply the distributive property if possible. We can add only similar radicals. We must write each expression in simplified form for radicals before we can tell if the radicals are similar.

▶ **EXAMPLE 3** Combine $7\sqrt{75xy^3} - 4y\sqrt{12xy}$, where $x, y \geq 0$.

Solution We write each expression in simplified form and combine similar radicals:

$$7\sqrt{75xy^3} - 4y\sqrt{12xy} = 7\sqrt{25y^2}\sqrt{3xy} - 4y\sqrt{4}\sqrt{3xy}$$
$$= 35y\sqrt{3xy} - 8y\sqrt{3xy}$$

$$= (35y - 8y)\sqrt{3xy}$$
$$= 27y\sqrt{3xy} \qquad \blacktriangleleft$$

▶ **EXAMPLE 4**   Combine $10\sqrt[3]{8a^4b^2} + 11a\sqrt[3]{27ab^2}$.

Solution   Writing each radical in simplified form and combining similar terms, we have

$$10\sqrt[3]{8a^4b^2} + 11a\sqrt[3]{27ab^2} = 10\sqrt[3]{8a^3}\sqrt[3]{ab^2} + 11a\sqrt[3]{27}\sqrt[3]{ab^2}$$
$$= 20a\sqrt[3]{ab^2} + 33a\sqrt[3]{ab^2}$$
$$= 53a\sqrt[3]{ab^2} \qquad \blacktriangleleft$$

▶ **EXAMPLE 5**   Combine $\dfrac{\sqrt{3}}{2} + \dfrac{1}{\sqrt{3}}$.

Solution   We begin by writing the second term in simplified form.

$$\frac{\sqrt{3}}{2} + \frac{1}{\sqrt{3}} = \frac{\sqrt{3}}{2} + \frac{1}{\sqrt{3}} \cdot \frac{\sqrt{3}}{\sqrt{3}}$$
$$= \frac{\sqrt{3}}{2} + \frac{\sqrt{3}}{3}$$
$$= \frac{1}{2}\sqrt{3} + \frac{1}{3}\sqrt{3}$$
$$= \left(\frac{1}{2} + \frac{1}{3}\right)\sqrt{3}$$
$$= \frac{5}{6}\sqrt{3} = \frac{5\sqrt{3}}{6} \qquad \blacktriangleleft$$

▶ **EXAMPLE 6**   Construct a golden rectangle from a square of side 4. Then show that the ratio of the length to the width is the golden ratio $\dfrac{1 + \sqrt{5}}{2}$.

Solution   Figure 1 shows the golden rectangle constructed from a square of side 4.

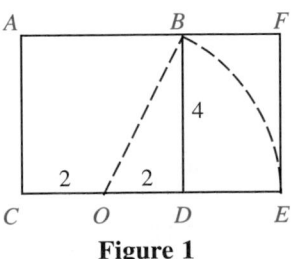

Figure 1

The length of the diagonal OB is found from the Pythagorean Theorem.

$$OB = \sqrt{2^2 + 4^2} = \sqrt{4 + 16} = \sqrt{20} = 2\sqrt{5}$$

The ratio of the length to the width for the rectangle is the golden ratio.

$$\text{Golden ratio} = \frac{CE}{EF} = \frac{2 + 2\sqrt{5}}{4} = \frac{2(1 + \sqrt{5})}{2 \cdot 2} = \frac{1 + \sqrt{5}}{2}$$

As you can see, showing that the ratio of length to width in this rectangle is the golden ratio depends on our ability to write $\sqrt{20}$ as $2\sqrt{5}$ and our ability to reduce to lowest terms by factoring and then dividing out the common factor 2 from the numerator and denominator. ◀

P R O B L E M   S E T   6 . 4

Combine the following expressions. (Assume any variables under an even root are nonnegative.)

1. $3\sqrt{5} + 4\sqrt{5}$

2. $6\sqrt{3} - 5\sqrt{3}$

3. $3x\sqrt{7} - 4x\sqrt{7}$

4. $6y\sqrt{a} + 7y\sqrt{a}$

5. $5\sqrt[3]{10} - 4\sqrt[3]{10}$

6. $6\sqrt[4]{2} + 9\sqrt[4]{2}$

7. $8\sqrt[5]{6} - 2\sqrt[5]{6} + 3\sqrt[5]{6}$

8. $7\sqrt[6]{7} - \sqrt[6]{7} + 4\sqrt[6]{7}$

9. $3x\sqrt{2} - 4x\sqrt{2} + x\sqrt{2}$

10. $5x\sqrt{6} - 3x\sqrt{6} - 2x\sqrt{6}$

11. $\sqrt{20} - \sqrt{80} + \sqrt{45}$ **12.** $\sqrt{8} - \sqrt{32} - \sqrt{18}$

13. $4\sqrt{8} - 2\sqrt{50} - 5\sqrt{72}$

14. $\sqrt{48} - 3\sqrt{27} + 2\sqrt{75}$

15. $5x\sqrt{8} + 3\sqrt{32x^2} - 5\sqrt{50x^2}$

16. $2\sqrt{50x^2} - 8x\sqrt{18} - 3\sqrt{72x^2}$

17. $5\sqrt[3]{16} - 4\sqrt[3]{54}$ **18.** $\sqrt[3]{81} + 3\sqrt[3]{24}$

19. $\sqrt[3]{x^4y^2} + 7x\sqrt[3]{xy^2}$

20. $2\sqrt[3]{x^8y^6} - 3y^2\sqrt[3]{8x^8}$

21. $5a^2\sqrt{27ab^3} - 6b\sqrt{12a^5b}$

22. $9a\sqrt{20a^3b^2} + 7b\sqrt{45a^5}$

23. $b\sqrt[3]{24a^5b} + 3a\sqrt[3]{81a^2b^4}$

24. $7\sqrt[3]{a^4b^3c^2} - 6ab\sqrt[3]{ac^2}$

25. $5x\sqrt[4]{3y^5} + y\sqrt[4]{243x^4y} + \sqrt[4]{48x^4y^5}$

26. $x\sqrt[4]{5xy^8} + y\sqrt[4]{405x^5y^4} + y^2\sqrt[4]{80x^5}$

27. $\dfrac{\sqrt{2}}{2} + \dfrac{1}{\sqrt{2}}$

28. $\dfrac{\sqrt{3}}{3} + \dfrac{1}{\sqrt{3}}$

29. $\dfrac{\sqrt{5}}{3} + \dfrac{1}{\sqrt{5}}$

30. $\dfrac{\sqrt{6}}{2} + \dfrac{1}{\sqrt{6}}$

31. $\sqrt{x} - \dfrac{1}{\sqrt{x}}$

32. $\sqrt{x} + \dfrac{1}{\sqrt{x}}$

33. $\dfrac{\sqrt{18}}{6} + \sqrt{\dfrac{1}{2}} + \dfrac{\sqrt{2}}{2}$

34. $\dfrac{\sqrt{12}}{6} + \sqrt{\dfrac{1}{3}} + \dfrac{\sqrt{3}}{3}$

35. $\sqrt{6} - \sqrt{\dfrac{2}{3}} + \sqrt{\dfrac{1}{6}}$

36. $\sqrt{15} - \sqrt{\dfrac{3}{5}} + \sqrt{\dfrac{5}{3}}$

37. $\sqrt[3]{25} + \dfrac{3}{\sqrt[3]{5}}$ **38.** $\sqrt[4]{8} + \dfrac{1}{\sqrt[4]{2}}$

39. Use a calculator to find a decimal approximation for $\sqrt{12}$ and for $2\sqrt{3}$.

40. Use a calculator to find decimal approximations for $\sqrt{50}$ and $5\sqrt{2}$.

41. Use a calculator to find a decimal approximation for $\sqrt{8} + \sqrt{18}$. Is it equal to the decimal approximation for $\sqrt{26}$ or $\sqrt{50}$?

42. Use a calculator to find a decimal approximation for $\sqrt{3} + \sqrt{12}$. Is it equal to the decimal approximation for $\sqrt{15}$ or $\sqrt{27}$?

Each statement below is false. Correct the right side of each one.

43. $3\sqrt{2x} + 5\sqrt{2x} = 8\sqrt{4x}$

44. $5\sqrt{3} - 7\sqrt{3} = -2\sqrt{9}$

45. $\sqrt{9 + 16} = 3 + 4$

46. $\sqrt{36 + 64} = 6 + 8$

Applying the Concepts

47. Construct a golden rectangle from a square of side 8. Then show that the ratio of the length to the width is the golden ratio $\dfrac{1 + \sqrt{5}}{2}$.

48. Construct a golden rectangle from a square of side 10. Then show that the ratio of the length to the width is the golden ratio $\dfrac{1 + \sqrt{5}}{2}$.

49. Use a ruler to construct a golden rectangle from a square of side 1 inch. Then show that the ratio of the length to the width is the golden ratio.

50. Use a ruler to construct a golden rectangle from a square of side $\frac{2}{3}$ inch. Then show that the ratio of the length to the width is the golden ratio.

51. To show that all golden rectangles have the same ratio of length to width, construct a golden rectangle from a square of side $2x$. Then show that the ratio of the length to the width is the golden ratio.

52. To show that all golden rectangles have the same ratio of length to width, construct a golden rectangle from a square of side x. Then show that the ratio of the length to the width is the golden ratio.

Review Problems

The problems below review material we covered in Section 5.3.

53. Give the slope and y-intercept of the line $2x - 3y = 6$.

54. Give the equation of the line with slope -3 and y-intercept 5.

55. Find the equation of the line with slope $\frac{2}{3}$ that contains the point $(-6, 2)$.

56. Find the equation of the line with slope 5 that contains the point $(3, -2)$.

57. Find the equation of the line through $(1, 3)$ and $(-1, -5)$.

58. Find the equation of the line with x-intercept 3 and y-intercept -2.

59. Find the equation of the line with slope $\frac{3}{4}$ and x-intercept -4.

60. Find the equation of the line through $(-1, 4)$ whose graph is perpendicular to the graph of $y = 2x + 3$.

One Step Further

Assume all variables represent positive numbers, then simplify as much as possible.

61. $5\sqrt{x + 3} + 3\sqrt{4x + 12}$

62. $3x\sqrt{x + 5} + 2\sqrt{x^3 + 5x^2}$

63. $5\sqrt{x^3 + 4x^2} - x\sqrt{25x + 100}$

64. $x\sqrt{9x + 18} - 3\sqrt{x^3 + 2x^2}$

65. $2x\sqrt{x^2 + 10x + 25} + \sqrt{4x^4 + 40x^3 + 100x^2}$

66. $x\sqrt{x^2y^2 + 6xy^2 + 9y^2} + y\sqrt{x^4 + 6x^3 + 9x^2}$

Multiplication and Division of Radical Expressions

In this section, we will look at multiplication and division of expressions that contain radicals. As you will see, multiplication of expressions that contain radicals is very similar to multiplication of polynomials. The division problems in this section are just an extension of the work we did previously when we rationalized denominators.

▶ **EXAMPLE 1**   Multiply $(3\sqrt{5})(2\sqrt{7})$.

Solution   We can rearrange the order and grouping of the numbers in this product by applying the commutative and associative properties. Following this, we apply Property 1 for radicals and multiply:

$$(3\sqrt{5})(2\sqrt{7}) = (3 \cdot 2)(\sqrt{5}\,\sqrt{7}) \qquad \text{Commutative and}$$
$$\text{associative properties}$$
$$= (3 \cdot 2)(\sqrt{5 \cdot 7}) \qquad \text{Property 1 for radicals}$$
$$= 6\sqrt{35} \qquad \text{Multiplication} \qquad ◀$$

In actual practice, it is not necessary to show either of the first two steps.

▶ **EXAMPLE 2**   Multiply $\sqrt{3}(2\sqrt{6} - 5\sqrt{12})$.

Solution   Applying the distributive property, we have

$$\sqrt{3}(2\sqrt{6} - 5\sqrt{12}) = \sqrt{3} \cdot 2\sqrt{6} - \sqrt{3} \cdot 5\sqrt{12}$$
$$= 2\sqrt{18} - 5\sqrt{36}$$

Writing each radical in simplified form gives

$$2\sqrt{18} - 5\sqrt{36} = 2\sqrt{9}\,\sqrt{2} - 5\sqrt{36}$$
$$= 6\sqrt{2} - 30 \qquad ◀$$

▶ **EXAMPLE 3**   Multiply $(\sqrt{3} + \sqrt{5})(4\sqrt{3} - \sqrt{5})$.

Solution   The same principle that applies when multiplying two binomials applies to this product. We must multiply each term in the first expression by each term in the second one. Any convenient method can be used. Let's use the FOIL method.

$$(\sqrt{3} + \sqrt{5})(4\sqrt{3} - \sqrt{5})$$
$$\qquad\quad \overset{\text{F}}{} \quad \overset{\text{O}}{} \quad \overset{\text{I}}{} \quad \overset{\text{L}}{}$$
$$= \sqrt{3} \cdot 4\sqrt{3} - \sqrt{3}\,\sqrt{5} + \sqrt{5} \cdot 4\sqrt{3} - \sqrt{5}\,\sqrt{5}$$
$$= 4 \cdot 3 - \sqrt{15} + 4\sqrt{15} - 5$$
$$= 12 + 3\sqrt{15} - 5$$
$$= 7 + 3\sqrt{15} \qquad ◀$$

▶ **EXAMPLE 4**   Expand and simplify $(\sqrt{x} + 3)^2$.

Solution 1   We can write this problem as a multiplication problem and proceed as we did in Example 3:

$$(\sqrt{x} + 3)^2 = (\sqrt{x} + 3)(\sqrt{x} + 3)$$
$$\qquad\qquad \overset{\text{F}}{} \quad \overset{\text{O}}{} \quad \overset{\text{I}}{} \quad \overset{\text{L}}{}$$
$$= \sqrt{x} \cdot \sqrt{x} + 3\sqrt{x} + 3\sqrt{x} + 3 \cdot 3$$

$$= x + 3\sqrt{x} + 3\sqrt{x} + 9$$
$$= x + 6\sqrt{x} + 9$$

Solution 2   We can obtain the same result by applying the formula for the square of a sum: $(a + b)^2 = a^2 + 2ab + b^2$.

$$(\sqrt{x} + 3)^2 = (\sqrt{x})^2 + 2(\sqrt{x})(3) + 3^2$$
$$= x + 6\sqrt{x} + 9 \qquad \blacktriangleleft$$

▶ **EXAMPLE 5**   Expand $(3\sqrt{x} - 2\sqrt{y})^2$ and simplify the result.

Solution   Let's apply the formula for the square of a difference, $(a - b)^2 = a^2 - 2ab + b^2$.

$$(3\sqrt{x} - 2\sqrt{y})^2 = (3\sqrt{x})^2 - 2(3\sqrt{x})(2\sqrt{y}) + (2\sqrt{y})^2$$
$$= 9x - 12\sqrt{xy} + 4y \qquad \blacktriangleleft$$

▶ **EXAMPLE 6**   Expand and simplify $(\sqrt{x + 2} - 1)^2$.

Solution   Applying the formula $(a - b)^2 = a^2 - 2ab + b^2$, we have

$$(\sqrt{x + 2} - 1)^2 = (\sqrt{x + 2})^2 - 2\sqrt{x + 2}(1) + 1^2$$
$$= x + 2 - 2\sqrt{x + 2} + 1$$
$$= x + 3 - 2\sqrt{x + 2} \qquad \blacktriangleleft$$

▶ **EXAMPLE 7**   Multiply $(\sqrt{6} + \sqrt{2})(\sqrt{6} - \sqrt{2})$.

Solution   We notice the product is of the form $(a + b)(a - b)$, which always gives the difference of two squares, $a^2 - b^2$:

$$(\sqrt{6} + \sqrt{2})(\sqrt{6} - \sqrt{2}) = (\sqrt{6})^2 - (\sqrt{2})^2$$
$$= 6 - 2$$
$$= 4 \qquad \blacktriangleleft$$

In Example 7 the two expressions $(\sqrt{6} + \sqrt{2})$ and $(\sqrt{6} - \sqrt{2})$ are called **conjugates.** In general, the conjugate of $\sqrt{a} + \sqrt{b}$ is $\sqrt{a} - \sqrt{b}$. If a and b are integers, multiplying conjugates of this form always produces a rational number: If a and b are positive integers, then

$$(\sqrt{a} + \sqrt{b})(\sqrt{a} - \sqrt{b})$$
$$= \sqrt{a}\sqrt{a} - \sqrt{a}\sqrt{b} + \sqrt{a}\sqrt{b} - \sqrt{b}\sqrt{b}$$
$$= a - \sqrt{ab} + \sqrt{ab} - b$$
$$= a - b$$

which is an integer (because a and b are integers) and therefore rational.

Division with radical expressions is the same as rationalizing the denominator. In Section 6.3 we were able to divide $\sqrt{3}$ by $\sqrt{2}$ by rationalizing the denominator:

$$\frac{\sqrt{3}}{\sqrt{2}} = \frac{\sqrt{3}}{\sqrt{2}} \cdot \frac{\sqrt{2}}{\sqrt{2}}$$

We can accomplish the same result with expressions such as

$$\frac{6}{\sqrt{5} - \sqrt{3}}$$

by multiplying the numerator and denominator by the conjugate of the denominator.

▶ **EXAMPLE 8** Divide $\dfrac{6}{\sqrt{5} - \sqrt{3}}$. (Rationalize the denominator.)

Solution   Since the product of two conjugates is a rational number, we multiply the numerator and denominator by the conjugate of the denominator.

$$\frac{6}{\sqrt{5} - \sqrt{3}} = \frac{6}{\sqrt{5} - \sqrt{3}} \cdot \frac{(\sqrt{5} + \sqrt{3})}{(\sqrt{5} + \sqrt{3})}$$

$$= \frac{6\sqrt{5} + 6\sqrt{3}}{(\sqrt{5})^2 - (\sqrt{3})^2}$$

$$= \frac{6\sqrt{5} + 6\sqrt{3}}{5 - 3}$$

$$= \frac{6\sqrt{5} + 6\sqrt{3}}{2}$$

The numerator and denominator of this last expression have a factor of 2 in common. We can reduce to lowest terms by factoring 2 from the numerator and then dividing both the numerator and denominator by 2:

$$= \frac{\cancel{2}(3\sqrt{5} + 3\sqrt{3})}{\cancel{2}}$$

$$= 3\sqrt{5} + 3\sqrt{3} \qquad\qquad ◀$$

▶ **EXAMPLE 9** Rationalize the denominator $\dfrac{\sqrt{5} - 2}{\sqrt{5} + 2}$.

Solution   To rationalize the denominator, we multiply the numerator and denominator by the conjugate of the denominator:

$$\frac{\sqrt{5} - 2}{\sqrt{5} + 2} = \frac{\sqrt{5} - 2}{\sqrt{5} + 2} \cdot \frac{(\sqrt{5} - 2)}{(\sqrt{5} - 2)}$$

$$= \frac{5 - 2\sqrt{5} - 2\sqrt{5} + 4}{(\sqrt{5})^2 - 2^2}$$

$$= \frac{9 - 4\sqrt{5}}{5 - 4}$$

$$= \frac{9 - 4\sqrt{5}}{1}$$

$$= 9 - 4\sqrt{5} \qquad \blacktriangleleft$$

▶ **EXAMPLE 10**   A golden rectangle constructed from a square of side 2 is shown in Figure 1. Show that the smaller rectangle *BDEF* is also a golden rectangle by finding the ratio of its length to its width.

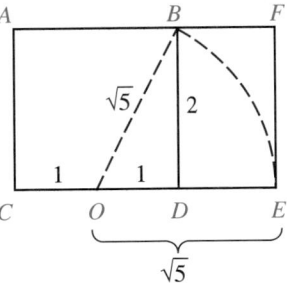

Figure 1

Solution   First we find expressions for the length and width of the smaller rectangle.

$$\text{Length} = EF = 2$$
$$\text{Width} = DE = \sqrt{5} - 1$$

Next, we find the ratio of length to width.

$$\text{Ratio of length to width} = \frac{EF}{DE} = \frac{2}{\sqrt{5} - 1}$$

To show that the small rectangle is a golden rectangle, we must show that the ratio of length to width is the golden ratio. We do so by rationalizing the denominator.

$$\frac{2}{\sqrt{5}-1} = \frac{2}{\sqrt{5}-1} \cdot \frac{\sqrt{5}+1}{\sqrt{5}+1}$$

$$= \frac{2(\sqrt{5}+1)}{5-1}$$

$$= \frac{2(\sqrt{5}+1)}{4}$$

$$= \frac{\sqrt{5}+1}{2} \qquad \text{Divide out common factor 2}$$

Since addition is commutative, this last expression is the golden ratio. Therefore, the small rectangle in Figure 1 is a golden rectangle. ◄

P R O B L E M   S E T   6 . 5

Multiply. (Assume all expressions appearing under a square root symbol represent nonnegative numbers throughout this problem set.)

1. $\sqrt{6}\sqrt{3}$

2. $\sqrt{6}\sqrt{2}$

3. $(2\sqrt{3})(5\sqrt{7})$

4. $(3\sqrt{5})(2\sqrt{7})$

5. $(4\sqrt{6})(2\sqrt{15})(3\sqrt{10})$

6. $(4\sqrt{35})(2\sqrt{21})(5\sqrt{15})$

7. $(3\sqrt[3]{3})(6\sqrt[3]{9})$

8. $(2\sqrt[3]{2})(6\sqrt[3]{4})$

9. $\sqrt{3}(\sqrt{2}-3\sqrt{3})$

10. $\sqrt{2}(5\sqrt{3}+4\sqrt{2})$

11. $6\sqrt[3]{4}(2\sqrt[3]{2}+1)$

12. $7\sqrt[3]{5}(3\sqrt[3]{25}-2)$

13. $(\sqrt{3}+\sqrt{2})(3\sqrt{3}-\sqrt{2})$

14. $(\sqrt{5}-\sqrt{2})(3\sqrt{5}+2\sqrt{2})$

15. $(\sqrt{x}+5)(\sqrt{x}-3)$

16. $(\sqrt{x}+4)(\sqrt{x}+2)$

17. $(3\sqrt{6}+4\sqrt{2})(\sqrt{6}+2\sqrt{2})$

18. $(\sqrt{7}-3\sqrt{3})(2\sqrt{7}-4\sqrt{3})$

19. $(\sqrt{3}+4)^2$

20. $(\sqrt{5}-2)^2$

21. $(\sqrt{x}-3)^2$

22. $(\sqrt{x}+4)^2$

23. $(2\sqrt{a}-3\sqrt{b})^2$

24. $(5\sqrt{a}-2\sqrt{b})^2$

25. $(\sqrt{x-4}+2)^2$

26. $(\sqrt{x-3}+2)^2$

27. $(\sqrt{x-5}-3)^2$

28. $(\sqrt{x-3}-4)^2$

29. $(\sqrt{3}-\sqrt{2})(\sqrt{3}+\sqrt{2})$

30. $(\sqrt{5}-\sqrt{2})(\sqrt{5}+\sqrt{2})$

31. $(\sqrt{a}+7)(\sqrt{a}-7)$

32. $(\sqrt{a}+5)(\sqrt{a}-5)$

33. $(5-\sqrt{x})(5+\sqrt{x})$

34. $(3-\sqrt{x})(3+\sqrt{x})$

35. $(\sqrt{x-4}+2)(\sqrt{x-4}-2)$

36. $(\sqrt{x+3}+5)(\sqrt{x+3}-5)$

37. $(\sqrt{3}+1)^3$

38. $(\sqrt{5}-2)^3$

Rationalize the denominator in each of the following.

39. $\dfrac{\sqrt{2}}{\sqrt{6}-\sqrt{2}}$

40. $\dfrac{\sqrt{5}}{\sqrt{5}+\sqrt{3}}$

41. $\dfrac{\sqrt{5}}{\sqrt{5}+1}$

42. $\dfrac{\sqrt{7}}{\sqrt{7}-1}$

43. $\dfrac{\sqrt{x}}{\sqrt{x}-3}$

44. $\dfrac{\sqrt{x}}{\sqrt{x}+2}$

45. $\dfrac{\sqrt{5}}{2\sqrt{5}-3}$

46. $\dfrac{\sqrt{7}}{3\sqrt{7}-2}$

47. $\dfrac{3}{\sqrt{x}-\sqrt{y}}$

48. $\dfrac{2}{\sqrt{x}+\sqrt{y}}$

49. $\dfrac{\sqrt{6}+\sqrt{2}}{\sqrt{6}-\sqrt{2}}$

50. $\dfrac{\sqrt{5}-\sqrt{3}}{\sqrt{5}+\sqrt{3}}$

51. $\dfrac{\sqrt{7}-2}{\sqrt{7}+2}$

52. $\dfrac{\sqrt{11}+3}{\sqrt{11}-3}$

53. $\dfrac{\sqrt{a}+\sqrt{b}}{\sqrt{a}-\sqrt{b}}$

54. $\dfrac{\sqrt{a}-\sqrt{b}}{\sqrt{a}+\sqrt{b}}$

55. $\dfrac{\sqrt{x}+2}{\sqrt{x}-2}$

56. $\dfrac{\sqrt{x}-3}{\sqrt{x}+3}$

57. $\dfrac{2\sqrt{3} - \sqrt{7}}{3\sqrt{3} + \sqrt{7}}$

58. $\dfrac{5\sqrt{6} + 2\sqrt{2}}{\sqrt{6} - \sqrt{2}}$

59. $\dfrac{3\sqrt{x} + 2}{1 + \sqrt{x}}$

60. $\dfrac{5\sqrt{x} - 1}{2 + \sqrt{x}}$

61. Show that the product
$(\sqrt[3]{2} + \sqrt[3]{3})(\sqrt[3]{4} - \sqrt[3]{6} + \sqrt[3]{9})$ is 5.

62. Show that the product
$(\sqrt[3]{x} + 2)(\sqrt[3]{x^2} - 2\sqrt[3]{x} + 4)$ is $x + 8$.

Each statement below is false. Correct the right side of each one.

63. $5(2\sqrt{3}) = 10\sqrt{15}$

64. $3(2\sqrt{x}) = 6\sqrt{3x}$

65. $(\sqrt{x} + 3)^2 = x + 9$

66. $(\sqrt{x} - 7)^2 = x - 49$

67. $(5\sqrt{3})^2 = 15$

68. $(3\sqrt{5})^2 = 15$

Applying the Concepts

69. If an object is dropped from the top of a 100-foot building, the amount of time t, in seconds, that it takes for the object to be h feet from the ground is given by the formula

$$t = \frac{\sqrt{100 - h}}{4}$$

How long does it take before the object is 50 feet from the ground? How long does it take to reach the ground? (When it is on the ground, h is 0.)

70. Use the formula given in Problem 69 to determine h if t is 1.25 seconds.

71. Rectangle *ACEF* in Figure 2 is a golden rectangle. If side *AC* is 6 inches, show that the smaller rectangle *BDEF* is also a golden rectangle.

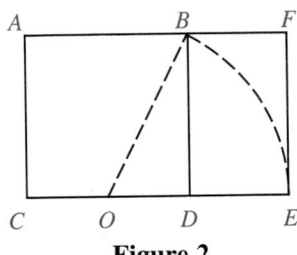

Figure 2

72. Rectangle *ACEF* in Figure 2 is a golden rectangle. If side *AC* is 1 inch, show that the smaller rectangle *BDEF* is also a golden rectangle.

73. If side *AC* in Figure 2 is $2x$, show that rectangle *BDEF* is a golden rectangle.

74. If side *AC* in Figure 2 is x, show that rectangle *BDEF* is a golden rectangle.

Review Problems

The problems that follow review material we covered in Sections 5.1 and 5.7 that will help you with part of the next section.

Graph each line. [5.1]

75. $y + 2x = 4$

76. $y - 2x = 4$

77. $y = -\frac{2}{3}x + 1$

78. $y = \frac{1}{2}x - 3$

Graph each equation. [5.7]

79. $y = \dfrac{12}{x}$

80. $y = \dfrac{-12}{x}$

One Step Further

Rationalize the denominator.

81. $\dfrac{\sqrt{x - 4}}{\sqrt{x - 4} + 2}$

82. $\dfrac{\sqrt{x + 3}}{\sqrt{x + 3} + 5}$

83. $\dfrac{\sqrt{x + 3} + \sqrt{x - 3}}{\sqrt{x + 3} - \sqrt{x - 3}}$

84. $\dfrac{\sqrt{x + 5} + \sqrt{x - 5}}{\sqrt{x + 5} - \sqrt{x - 5}}$

85. $\dfrac{1}{\sqrt[3]{x} + 2}$

86. $\dfrac{1}{\sqrt[3]{x} - 2}$

87. $\dfrac{1}{\sqrt[3]{3} + \sqrt[3]{2}}$

88. $\dfrac{1}{\sqrt[3]{3} - \sqrt[3]{2}}$

Equations with Radicals

This section is concerned with solving equations that involve one or more radicals. The first step in solving an equation that contains a radical is to eliminate the radical from the equation. To do so we need an additional property.

> **Squaring Property of Equality**
> If both sides of an equation are squared, the solutions to the original equation are solutions to the resulting equation.

We will never lose solutions to our equations by squaring both sides. We may, however, introduce **extraneous solutions.** These extraneous solutions satisfy the equation obtained by squaring both sides of the original equation, but do not satisfy the original equation.

We know that if two real numbers a and b are equal, then so are their squares:

$$\text{If} \qquad a = b$$

$$\text{then} \qquad a^2 = b^2$$

On the other hand, extraneous solutions are introduced when we square opposites. That is, even though opposites are not equal, their squares are. For example,

$$5 = -5 \qquad \text{A false statement}$$
$$(5)^2 = (-5)^2 \qquad \text{Square both sides}$$
$$25 = 25 \qquad \text{A true statement}$$

We are free to square both sides of an equation any time it is convenient. We must be aware, however, that doing so may introduce extraneous solutions. We must, therefore, check all our solutions in the original equation if at any time we square both sides of the original equation.

▶ **EXAMPLE 1**   Solve for x: $\sqrt{3x + 4} = 5$.

Solution   We square both sides and proceed as usual:

$$\sqrt{3x + 4} = 5$$
$$(\sqrt{3x + 4})^2 = 5^2$$
$$3x + 4 = 25$$
$$3x = 21$$
$$x = 7$$

Checking $x = 7$ in the original equation, we have

$$\sqrt{3(7) + 4} \overset{?}{=} 5$$
$$\sqrt{21 + 4} \overset{?}{=} 5$$
$$\sqrt{25} \overset{?}{=} 5$$
$$5 = 5$$

The solution $x = 7$ satisfies the original equation.   ◄

▶ **EXAMPLE 2**   Solve $\sqrt{4x - 7} = -3$.

Solution   Squaring both sides, we have

$$\sqrt{4x - 7} = -3$$
$$(\sqrt{4x - 7})^2 = (-3)^2$$
$$4x - 7 = 9$$
$$4x = 16$$
$$x = 4$$

Checking $x = 4$ in the original equation gives

$$\sqrt{4(4) - 7} \overset{?}{=} -3$$
$$\sqrt{16 - 7} \overset{?}{=} -3$$
$$\sqrt{9} \overset{?}{=} -3$$
$$3 = -3 \qquad \text{A false statement}$$

The solution $x = 4$ produces a false statement when checked in the original equation. Since $x = 4$ was the only possible solution, there is no solution to the original equation. The possible solution $x = 4$ is an extraneous solution. It satisfies the equation obtained by squaring both sides of the original equation, but does not satisfy the original equation.   ◄

Note   The fact that there is no solution to the equation in Example 2 was obvious to begin with. Notice that the left side of the equation is the *positive* square root of $4x - 7$, which must be a positive number or 0. The right side of the equation is -3. Since we cannot have a number that is either positive or zero equal to a negative number, there is no solution to the equation.

▶ **EXAMPLE 3**   Solve $\sqrt{5x - 1} + 3 = 7$.

Solution   We must isolate the radical on the left side of the equation. If we attempt to square both sides without doing so, the resulting equation will also contain a radical. Adding -3 to both sides, we have

$$\sqrt{5x - 1} + 3 = 7$$
$$\sqrt{5x - 1} = 4$$

We can now square both sides and proceed as usual:

$$(\sqrt{5x - 1})^2 = 4^2$$

$$5x - 1 = 16$$

$$5x = 17$$

$$x = \frac{17}{5}$$

Checking $x = \frac{17}{5}$, we have

$$\sqrt{5\left(\frac{17}{5}\right) - 1} + 3 \stackrel{?}{=} 7$$

$$\sqrt{17 - 1} + 3 \stackrel{?}{=} 7$$

$$\sqrt{16} + 3 \stackrel{?}{=} 7$$

$$4 + 3 \stackrel{?}{=} 7$$

$$7 = 7 \qquad \blacktriangleleft$$

▶ **EXAMPLE 4** Solve $t + 5 = \sqrt{t + 7}$.

Solution This time, squaring both sides of the equation results in a quadratic equation:

$$(t + 5)^2 = (\sqrt{t + 7})^2 \qquad \text{Square both sides}$$

$$t^2 + 10t + 25 = t + 7$$

$$t^2 + 9t + 18 = 0 \qquad \text{Standard form}$$

$$(t + 3)(t + 6) = 0 \qquad \text{Factor the left side}$$

$$t + 3 = 0 \quad \text{or} \quad t + 6 = 0 \qquad \text{Set factors equal to 0}$$

$$t = -3 \quad \text{or} \qquad t = -6$$

We must check each solution in the original equation:

Check $t = -3$ Check $t = -6$

$$-3 + 5 \stackrel{?}{=} \sqrt{-3 + 7} \qquad -6 + 5 \stackrel{?}{=} \sqrt{-6 + 7}$$

$$2 \stackrel{?}{=} \sqrt{4} \qquad\qquad -1 \stackrel{?}{=} \sqrt{1}$$

$$2 = 2 \qquad\qquad\qquad -1 = 1$$

A true statement   A false statement

Since $t = -6$ does not check, our only solution is $t = -3$.   ◀

▶ **EXAMPLE 5** Solve $\sqrt{x - 3} = \sqrt{x} - 3$.

Solution We begin by squaring both sides. Note carefully what happens when we square the right side of the equation, and compare the square of the right side with the square of the left side. You must convince yourself that these results are correct. (The note that follows this example will help if you are having trouble convincing yourself that what is written below is true.)

$$(\sqrt{x-3})^2 = (\sqrt{x}-3)^2$$
$$x - 3 = x - 6\sqrt{x} + 9$$

Now we still have a radical in our equation, so we will have to square both sides again. Before we do, though, let's isolate the remaining radical.

$$x - 3 = x - 6\sqrt{x} + 9$$

$-3 = -6\sqrt{x} + 9$	Add $-x$ to each side
$-12 = -6\sqrt{x}$	Add -9 to each side
$2 = \sqrt{x}$	Divide each side by -6
$4 = x$	Square each side

Our only possible solution is $x = 4$, which we check in our original equation as follows:

$$\sqrt{4-3} \overset{?}{=} \sqrt{4} - 3$$
$$\sqrt{1} \overset{?}{=} 2 - 3$$

$1 = -1$	A false statement

Substituting 4 for x in the original equation yields a false statement. Since 4 was our only possible solution, there is no solution to our equation. ◀

Note   In reading through Example 5, it is very important that you realize that the square of $(\sqrt{x} - 3)$ is not $x + 9$. Remember, when we square a difference with two terms, we use the formula

$$(a - b)^2 = a^2 - 2ab + b^2$$

Applying this formula to $(\sqrt{x} - 3)^2$, we have

$$(\sqrt{x} - 3)^2 = (\sqrt{x})^2 - 2(\sqrt{x})(3) + 3^2$$
$$= x - 6\sqrt{x} + 9$$

Here is another example of an equation for which we must apply our squaring property twice before all radicals are eliminated.

▶ **EXAMPLE 6**   Solve $\sqrt{x+1} = 1 - \sqrt{2x}$.

Solution   This equation has two separate terms involving radical signs. Squaring both sides gives

$$x + 1 = 1 - 2\sqrt{2x} + 2x$$

$-x = -2\sqrt{2x}$	Add $-2x$ and -1 to both sides
$x^2 = 4(2x)$	Square both sides
$x^2 - 8x = 0$	Standard form

Our equation is a quadratic equation in standard form. To solve for x, we factor the left side and set each factor equal to 0:

$$x(x - 8) = 0 \qquad \text{Factor left side}$$

$$x = 0 \quad \text{or} \quad x - 8 = 0 \qquad \text{Set factors equal to 0}$$

$$x = 8$$

Since we squared both sides of our equation, we have the possibility that one or both of the solutions are extraneous. We must check each one in the original equation:

Check $x = 8$ | Check $x = 0$

$$\sqrt{8 + 1} \stackrel{?}{=} 1 - \sqrt{2 \cdot 8} \qquad \sqrt{0 + 1} \stackrel{?}{=} 1 - \sqrt{2 \cdot 0}$$

$$\sqrt{9} \stackrel{?}{=} 1 - \sqrt{16} \qquad \sqrt{1} \stackrel{?}{=} 1 - \sqrt{0}$$

$$3 \stackrel{?}{=} 1 - 4 \qquad 1 \stackrel{?}{=} 1 - 0$$

$$3 = -3 \qquad 1 = 1$$

A false statement | A true statement

Since $x = 8$ does not check, it is an extraneous solution. Our only solution is $x = 0$. ◀

▶ **EXAMPLE 7** Solve $\sqrt{x + 1} = \sqrt{x + 2} - 1$.

Solution Squaring both sides we have

$$(\sqrt{x + 1})^2 = (\sqrt{x + 2} - 1)^2$$

$$x + 1 = x + 2 - 2\sqrt{x + 2} + 1$$

Once again we are left with a radical in our equation. Before we square each side again, we must isolate the radical on the right side of the equation.

$$x + 1 = x + 3 - 2\sqrt{x + 2} \qquad \text{Simplify the right side}$$

$$1 = 3 - 2\sqrt{x + 2} \qquad \text{Add } -x \text{ to each side}$$

$$-2 = -2\sqrt{x + 2} \qquad \text{Add } -3 \text{ to each side}$$

$$1 = \sqrt{x + 2} \qquad \text{Divide each side by } -2$$

$$1 = x + 2 \qquad \text{Square both sides}$$

$$-1 = x \qquad \text{Add } -2 \text{ to each side}$$

Checking our only possible solution $x = -1$ in our original equation, we have

$$\sqrt{-1 + 1} \stackrel{?}{=} \sqrt{-1 + 2} - 1$$

$$\sqrt{0} \stackrel{?}{=} \sqrt{1} - 1$$

$$0 \overset{?}{=} 1 - 1$$

$$0 = 0 \qquad \text{A true statement}$$

Our solution checks.   ◀

It is also possible to raise both sides of an equation to powers greater than 2. We only need to check for extraneous solutions when we raise both sides of an equation to an even power. Raising both sides of an equation to an odd power will not produce extraneous solutions.

▶ **EXAMPLE 8**   Solve $\sqrt[3]{4x + 5} = 3$.

Solution   Cubing both sides we have

$$(\sqrt[3]{4x + 5})^3 = 3^3$$

$$4x + 5 = 27$$

$$4x = 22$$

$$x = \frac{22}{4}$$

$$x = \frac{11}{2}$$

We do not need to check $x = \frac{11}{2}$ since we raised both sides to an odd power.   ◀

We end this section by looking at the graphs of some equations that contain radicals.

▶ **EXAMPLE 9**   Graph $y = \sqrt{x}$ and $y = \sqrt[3]{x}$.

Solution   The graphs are shown in Figures 1 and 2 on the next page. Notice that the graph of $y = \sqrt{x}$ appears in the first quadrant only, because in the equation $y = \sqrt{x}$, x and y cannot be negative.

The graph of $y = \sqrt[3]{x}$ appears in quadrants 1 and 3 since the cube root of a positive number is also a positive number, and the cube root of a negative number is a negative number. That is, when x is positive, y will be positive and when x is negative, y will be negative.

The graphs of both equations will contain the origin, since $y = 0$ when $x = 0$ in both equations.

x	y
-4	undefined
-1	undefined
0	0
1	1
4	2
9	3
16	4

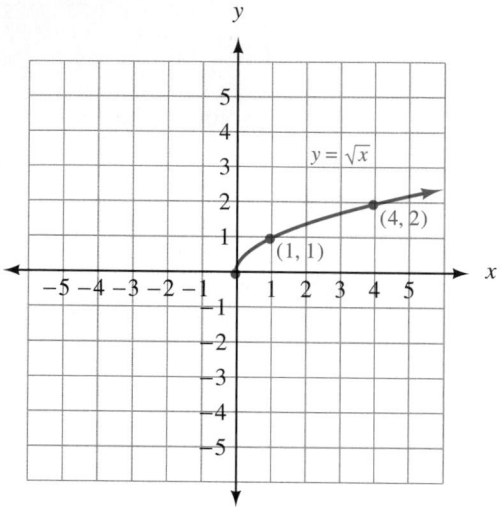

Figure 1

x	y
-27	-3
-8	-2
-1	-1
0	0
1	1
8	2
27	3

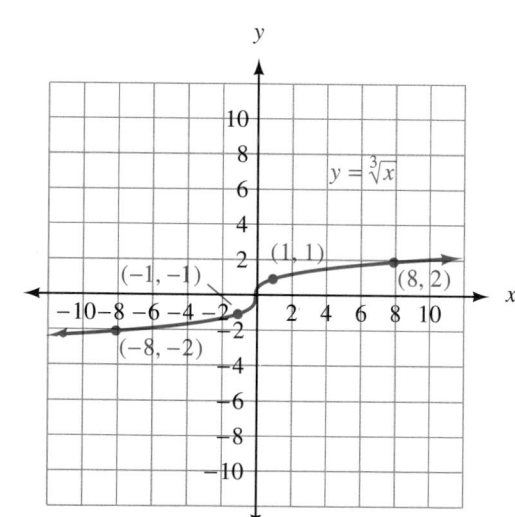

Figure 2

P R O B L E M   S E T   6 . 6

Solve each of the following equations.

1. $\sqrt{2x + 1} = 3$

2. $\sqrt{3x + 1} = 4$

3. $\sqrt{4x + 1} = -5$

4. $\sqrt{6x + 1} = -5$

5. $\sqrt{2y - 1} = 3$

6. $\sqrt{3y - 1} = 2$

7. $\sqrt{5x - 7} = -1$

8. $\sqrt{8x + 3} = -6$

9. $\sqrt{2x - 3} - 2 = 4$

10. $\sqrt{3x + 1} - 4 = 1$

11. $\sqrt{4a + 1} + 3 = 2$

12. $\sqrt{5a - 3} + 6 = 2$

13. $\sqrt[4]{3x + 1} = 2$

14. $\sqrt[4]{4x + 1} = 3$

15. $\sqrt[3]{2x - 5} = 1$ **16.** $\sqrt[3]{5x + 7} = 2$
17. $\sqrt[3]{3a + 5} = -3$ **18.** $\sqrt[3]{2a + 7} = -2$
19. $\sqrt{y - 3} = y - 3$ **20.** $\sqrt{y + 3} = y - 3$
21. $\sqrt{a + 2} = a + 2$ **22.** $\sqrt{a + 10} = a - 2$
23. $\sqrt{2x + 4} = \sqrt{1 - x}$
24. $\sqrt{3x + 4} = -\sqrt{2x + 3}$
25. $\sqrt{4a + 7} = -\sqrt{a + 2}$
26. $\sqrt{7a - 1} = \sqrt{2a + 4}$
27. $\sqrt[4]{5x - 8} = \sqrt[4]{4x - 1}$
28. $\sqrt[4]{6x + 7} = \sqrt[4]{x + 2}$
29. $x + 1 = \sqrt{5x + 1}$ **30.** $x - 1 = \sqrt{6x + 1}$
31. $t + 5 = \sqrt{2t + 9}$ **32.** $t + 7 = \sqrt{2t + 13}$
33. $\sqrt{y - 8} = \sqrt{8 - y}$
34. $\sqrt{2y + 5} = \sqrt{5y + 2}$
35. $\sqrt[3]{3x + 5} = \sqrt[3]{5 - 2x}$
36. $\sqrt[3]{4x + 9} = \sqrt[3]{3 - 2x}$

The following equations will require that you square both sides twice before all the radicals are eliminated. Solve each equation using the methods shown in Examples 5, 6, and 7.

37. $\sqrt{x - 8} = \sqrt{x} - 2$
38. $\sqrt{x + 3} = \sqrt{x} - 3$
39. $\sqrt{x + 1} = \sqrt{x} + 1$
40. $\sqrt{x - 1} = \sqrt{x} - 1$
41. $\sqrt{x + 8} = \sqrt{x - 4} + 2$
42. $\sqrt{x + 5} = \sqrt{x - 3} + 2$
43. $\sqrt{x - 5} - 3 = \sqrt{x - 8}$
44. $\sqrt{x - 3} - 4 = \sqrt{x - 3}$
45. $\sqrt{x + 4} = 2 - \sqrt{2x}$
46. $\sqrt{5x + 1} = 1 + \sqrt{5x}$
47. $\sqrt{2x + 4} = \sqrt{x + 3} + 1$
48. $\sqrt{2x - 1} = \sqrt{x - 4} + 2$

Applying the Concepts

49. Solve the following formula for h:
$$t = \frac{\sqrt{100 - h}}{4}$$

50. Solve the following formula for h:
$$t = \sqrt{\frac{2h - 40t}{g}}$$

51. The length of time (T) in seconds it takes the pendulum of a grandfather clock to swing through one complete cycle is given by the formula
$$T = 2\pi \sqrt{\frac{L}{32}}$$
where L is the length, in feet, of the pendulum, and π is approximately $\frac{22}{7}$. How long must the pendulum be if one complete cycle takes 2 seconds?

52. Solve the formula in Problem 51 for L.

Graph each equation.

53. $y = 2\sqrt{x}$ **54.** $y = -2\sqrt{x}$
55. $y = \sqrt{x} - 2$ **56.** $y = \sqrt{x} + 2$
57. $y = \sqrt{x - 2}$ **58.** $y = \sqrt{x + 2}$
59. $y = 3\sqrt[3]{x}$ **60.** $y = -3\sqrt[3]{x}$
61. $y = \sqrt[3]{x} + 3$ **62.** $y = \sqrt[3]{x} - 3$
63. $y = \sqrt[3]{x + 3}$ **64.** $y = \sqrt[3]{x - 3}$

Review Problems

The problems that follow review material we covered in Section 6.5. Reviewing these problems will help you understand the next section.

Multiply.

65. $\sqrt{2}(\sqrt{3} - \sqrt{2})$
66. $(\sqrt{x} - 4)(\sqrt{x} + 5)$
67. $(\sqrt{x} + 5)^2$
68. $(\sqrt{5} + \sqrt{3})(\sqrt{5} - \sqrt{3})$

Rationalize the denominator.

69. $\dfrac{\sqrt{x}}{\sqrt{x} + 3}$

70. $\dfrac{\sqrt{5} - \sqrt{3}}{\sqrt{5} + \sqrt{3}}$

One Step Further

Solve each equation.

71. $\dfrac{x}{3\sqrt{2x-3}} - \dfrac{1}{\sqrt{2x-3}} = \dfrac{1}{3}$

72. $\dfrac{x}{5\sqrt{2x+10}} + \dfrac{1}{\sqrt{2x+10}} = \dfrac{1}{5}$

73. $x + 1 = \sqrt[3]{4x+4}$

74. $x - 1 = \sqrt[3]{4x-4}$

Solve for y in terms of x.

75. $y + 2 = \sqrt{x^2 + (y-2)^2}$

76. $y + \frac{1}{2} = \sqrt{x^2 + (y - \frac{1}{2})^2}$

USING TECHNOLOGY

77. Use your Y variables list, or write a program, to graph the family of curves $Y = \sqrt{X} + B$ for $B = -3, -2, -1, 0, 1, 2,$ and 3.

78. Use your Y variables list, or write a program, to graph the family of curves $Y = \sqrt{X+B}$ for $B = -3, -2, -1, 0, 1, 2,$ and 3.

79. Summarize the results of Problem 77 by giving a written description of the effect of b on the graph of $y = \sqrt{x} + b$.

80. Summarize the results of Problem 78 by giving a written description of the effect of b on the graph of $y = \sqrt{x+b}$.

81. Use your Y variables list, or write a program, to graph the family of curves $Y = \sqrt[3]{X} + B$ for $B = -3, -2, -1, 0, 1, 2,$ and 3.

82. Use your Y variables list, or write a program, to graph the family of curves $Y = \sqrt[3]{X+B}$ for $B = -3, -2, -1, 0, 1, 2,$ and 3.

83. Summarize the results of Problem 81 by giving a written description of the effect of b on the graph of $y = \sqrt[3]{x} + b$.

84. Summarize the results of Problem 82 by giving a written description of the effect of b on the graph of $y = \sqrt[3]{x+b}$.

85. Use your Y variables list, or write a program to graph the family of curves $Y = A\sqrt{X}$ for $A = -3, -2, -1, 0, 1, 2,$ and 3.

86. Use your Y variables list, or write a program, to graph the family of curves $Y = A\sqrt{X}$ for $A = \frac{1}{4}, \frac{1}{3}, \frac{1}{2}, 1, 2,$ and 3.

87. Summarize the results of Problems 85 and 86 by giving a written description of the effect of a on the graph of $y = a\sqrt{x}$.

SECTION

6.7 Complex Numbers

The equation $x^2 = -9$ has no real number solutions since the square of a real number is always positive. We have been unable to work with square roots of negative numbers like $\sqrt{-25}$ and $\sqrt{-16}$ for the same reason. Complex numbers allow us to expand our work with radicals to include square roots of negative numbers and to solve equations like $x^2 = -9$ and $x^2 = -64$. Our work with complex numbers is based on the following definition.

DEFINITION The **number i** is such that $i = \sqrt{-1}$ (which is the same as saying $i^2 = -1$).

The number i, as we have defined it here, is not a real number. Because of the way we have defined i, we can use it to simplify square roots of negative numbers.

Square Roots of Negative Numbers
If a is a positive number, then $\sqrt{-a}$ can always be written as $i\sqrt{a}$. That is,

$$\sqrt{-a} = i\sqrt{a} \qquad \text{if } a \text{ is a positive number}$$

To justify our rule we simply square the quantity $i\sqrt{a}$ to obtain $-a$. Here is what it looks like when we do so:

$$(i\sqrt{a})^2 = i^2 \cdot (\sqrt{a})^2$$
$$= -1 \cdot a$$
$$= -a$$

Here are some examples that illustrate the use of our new rule.

▶ **EXAMPLES** Write each square root in terms of the number i.

1. $\sqrt{-25} = i\sqrt{25} = i \cdot 5 = 5i$
2. $\sqrt{-49} = i\sqrt{49} = i \cdot 7 = 7i$
3. $\sqrt{-12} = i\sqrt{12} = i \cdot 2\sqrt{3} = 2i\sqrt{3}$
4. $\sqrt{-17} = i\sqrt{17}$ ◀

Note In Examples 3 and 4 we wrote i before the radical simply to avoid confusion. If we were to write the answer to 3 as $2\sqrt{3}i$, some people would think the i was under the radical sign and it is not.

If we assume all the properties of exponents hold when the base is i, we can write any power of i as i, -1, $-i$, or 1. Using the fact that $i^2 = -1$, we have

$$i^1 = i$$
$$i^2 = -1$$
$$i^3 = i^2 \cdot i = -1(i) = -i$$
$$i^4 = i^2 \cdot i^2 = -1(-1) = 1$$

Since $i^4 = 1$, i^5 will simplify to i, and we will begin repeating the sequence $i, -1, -i, 1$ as we simplify higher powers of i: Any power of i simplifies to $i, -1, -i,$ or 1. The easiest way to simplify higher powers of i is to write them in terms of i^2. For instance, to simplify i^{21}, we would write it as

$$(i^2)^{10} \cdot i \qquad \text{because } 2 \cdot 10 + 1 = 21$$

Then, since $i^2 = -1$, we have

$$(-1)^{10} \cdot i = 1 \cdot i = i$$

▶ **EXAMPLES**   Simplify as much as possible.

5. $i^{30} = (i^2)^{15} = (-1)^{15} = -1$
6. $i^{11} = (i^2)^5 \cdot i = (-1)^5 \cdot i = (-1)i = -i$
7. $i^{40} = (i^2)^{20} = (-1)^{20} = 1$ ◀

DEFINITION   A **complex number** is any number that can be put in the form

$$a + bi$$

where a and b are real numbers and $i = \sqrt{-1}$. The form $a + bi$ is called *standard form* for complex numbers. The number a is called the *real part* of the complex number. The number b is called the *imaginary part* of the complex number.

Every real number is also a complex number. The real number 8, for example, can be written as $8 + 0i$; therefore, 8 is also considered a complex number.

Equality for Complex Numbers

Two complex numbers are equal if and only if their real parts are equal and their imaginary parts are equal. That is, for real numbers a, b, c, and d,

$$a + bi = c + di \quad \text{if and only if} \quad a = c \quad \text{and} \quad b = d$$

▶ **EXAMPLE 8**   Find x and y if $3x + 4i = 12 - 8yi$.

Solution   Since the two complex numbers are equal, their real parts are equal and their imaginary parts are equal:

$$3x = 12 \quad \text{and} \quad 4 = -8y$$

$$x = 4 \qquad\qquad y = -\frac{1}{2}$$ ◀

▶ **EXAMPLE 9** Find x and y if $(4x - 3) + 7i = 5 + (2y - 1)i$.

Solution The real parts are $4x - 3$ and 5. The imaginary parts are 7 and $2y - 1$:

$$4x - 3 = 5 \quad \text{and} \quad 7 = 2y - 1$$
$$4x = 8 \qquad\qquad 8 = 2y$$
$$x = 2 \qquad\qquad y = 4 \qquad\qquad \blacktriangleleft$$

Addition and Subtraction of Complex Numbers

To add two complex numbers, add their real parts and add their imaginary parts. That is, if a, b, c, and d are real numbers, then

$$(a + bi) + (c + di) = (a + c) + (b + d)i$$

If we assume that the commutative, associative, and distributive properties hold for the number i, then the definition of addition is simply an extension of these properties.

We define subtraction in a similar manner. If a, b, c, and d are real numbers, then

$$(a + bi) - (c + di) = (a - c) + (b - d)i$$

▶ **EXAMPLES** Add or subtract as indicated.

10. $(3 + 4i) + (7 - 6i) = (3 + 7) + (4 - 6)i = 10 - 2i$
11. $(7 + 3i) - (5 + 6i) = (7 - 5) + (3 - 6)i = 2 - 3i$
12. $(5 - 2i) - (9 - 4i) = (5 - 9) + (-2 + 4)i = -4 + 2i$ ◀

Multiplication of Complex Numbers

Since complex numbers have the same form as binomials, we find the product of two complex numbers the same way we find the product of two binomials.

▶ **EXAMPLE 13** Multiply $(3 - 4i)(2 + 5i)$.

Solution Multiplying each term in the second complex number by each term in the first, we have

$$\overset{\text{F}\qquad\text{O}\qquad\text{I}\qquad\text{L}}{}$$
$$(3 - 4i)(2 + 5i) = 3 \cdot 2 + 3 \cdot 5i - 2 \cdot 4i - 5i(4i)$$
$$= 6 + 15i - 8i - 20i^2$$

Combining similar terms and using the fact that $i^2 = -1$, we can simplify as follows:

$$6 + 15i - 8i - 20i^2 = 6 + 7i - 20(-1)$$
$$= 6 + 7i + 20$$
$$= 26 + 7i$$

The product of the complex numbers $3 - 4i$ and $2 + 5i$ is the complex number $26 + 7i$. ◀

▶ **EXAMPLE 14**   Multiply $2i(4 - 6i)$.

Solution   Applying the distributive property gives us

$$2i(4 - 6i) = 2i \cdot 4 - 2i \cdot 6i$$
$$= 8i - 12i^2$$
$$= 12 + 8i$$ ◀

▶ **EXAMPLE 15**   Expand $(3 + 5i)^2$.

Solution   We treat this like the square of a binomial. Remember: $(a + b)^2 = a^2 + 2ab + b^2$:

$$(3 + 5i)^2 = 3^2 + 2(3)(5i) + (5i)^2$$
$$= 9 + 30i + 25i^2$$
$$= 9 + 30i - 25$$
$$= -16 + 30i$$ ◀

▶ **EXAMPLE 16**   Multiply $(2 - 3i)(2 + 3i)$.

Solution   This product has the form $(a - b)(a + b)$, which we know results in the difference of two squares, $a^2 - b^2$:

$$(2 - 3i)(2 + 3i) = 2^2 - (3i)^2$$
$$= 4 - 9i^2$$
$$= 4 + 9$$
$$= 13$$ ◀

The product of the two complex numbers $2 - 3i$ and $2 + 3i$ is the real number 13. The two complex numbers $2 - 3i$ and $2 + 3i$ are called complex conjugates. The fact that their product is a real number is very useful.

DEFINITION   The complex numbers $a + bi$ and $a - bi$ are called **complex conjugates.** One important property they have is that their product is the real number $a^2 + b^2$. Here's why:

$$(a + bi)(a - bi) = a^2 - (bi)^2$$
$$= a^2 - b^2i^2$$
$$= a^2 - b^2(-1)$$
$$= a^2 + b^2$$

Division with Complex Numbers

The fact that the product of two complex conjugates is a real number is the key to division with complex numbers.

▶ **EXAMPLE 17**   Divide $\dfrac{2 + i}{3 - 2i}$.

Solution   We want a complex number in standard form that is equivalent to the quotient $(2 + i)/(3 - 2i)$. We need to eliminate i from the denominator. Multiplying the numerator and denominator by $3 + 2i$ will give us what we want:

$$\frac{2 + i}{3 - 2i} = \frac{2 + i}{3 - 2i} \cdot \frac{(3 + 2i)}{(3 + 2i)}$$

$$= \frac{6 + 4i + 3i + 2i^2}{9 - 4i^2}$$

$$= \frac{6 + 7i - 2}{9 + 4}$$

$$= \frac{4 + 7i}{13}$$

$$= \frac{4}{13} + \frac{7}{13}i$$

Dividing the complex number $2 + i$ by $3 - 2i$ gives the complex number $\frac{4}{13} + \frac{7}{13}i$.   ◀

▶ **EXAMPLE 18**   Divide $\dfrac{7 - 4i}{i}$.

Solution   The conjugate of the denominator is $-i$. Multiplying numerator and denominator by this number, we have

$$\frac{7 - 4i}{i} = \frac{7 - 4i}{i} \cdot \frac{-i}{-i}$$

$$= \frac{-7i + 4i^2}{-i^2}$$

$$= \frac{-7i + 4(-1)}{-(-1)}$$

$$= -4 - 7i \qquad \blacktriangleleft$$

PROBLEM SET 6.7

Write the following in terms of i and simplify as much as possible.

1. $\sqrt{-36}$ **2.** $\sqrt{-49}$

3. $-\sqrt{-25}$ **4.** $-\sqrt{-81}$

5. $\sqrt{-72}$ **6.** $\sqrt{-48}$

7. $-\sqrt{-12}$ **8.** $-\sqrt{-75}$

Write each of the following as i, -1, $-i$, or 1.

9. i^{28} **10.** i^{31}

11. i^{26} **12.** i^{37}

13. i^{75} **14.** i^{42}

Find x and y so each of the following equations is true.

15. $2x + 3yi = 6 - 3i$

16. $4x - 2yi = 4 + 8i$

17. $2 - 5i = -x + 10yi$

18. $4 + 7i = 6x - 14yi$

19. $2x + 10i = -16 - 2yi$

20. $4x - 5i = -2 + 3yi$

21. $(2x - 4) - 3i = 10 - 6yi$

22. $(4x - 3) - 2i = 8 + yi$

23. $(7x - 1) + 4i = 2 + (5y + 2)i$

24. $(5x + 2) - 7i = 4 + (2y + 1)i$

Combine the following complex numbers.

25. $(2 + 3i) + (3 + 6i)$

26. $(4 + i) + (3 + 2i)$

27. $(3 - 5i) + (2 + 4i)$

28. $(7 + 2i) + (3 - 4i)$

29. $(5 + 2i) - (3 + 6i)$

30. $(6 + 7i) - (4 + i)$

31. $(3 - 5i) - (2 + i)$

32. $(7 - 3i) - (4 + 10i)$

33. $[(3 + 2i) - (6 + i)] + (5 + i)$

34. $[(4 - 5i) - (2 + i)] + (2 + 5i)$

35. $[(7 - i) - (2 + 4i)] - (6 + 2i)$

36. $[(3 - i) - (4 + 7i)] - (3 - 4i)$

37. $(3 + 2i) - [(3 - 4i) - (6 + 2i)]$

38. $(7 - 4i) - [(-2 + i) - (3 + 7i)]$

39. $(4 - 9i) + [(2 - 7i) - (4 + 8i)]$

40. $(10 - 2i) - [(2 + i) - (3 - i)]$

Find the following products.

41. $3i(4 + 5i)$ **42.** $2i(3 + 4i)$

43. $6i(4 - 3i)$ **44.** $11i(2 - i)$

45. $(3 + 2i)(4 + i)$ **46.** $(2 - 4i)(3 + i)$

47. $(4 + 9i)(3 - i)$ **48.** $(5 - 2i)(1 + i)$

49. $(1 + i)^3$ **50.** $(1 - i)^3$

51. $(2 - i)^3$ **52.** $(2 + i)^3$

53. $(2 + 5i)^2$ **54.** $(3 + 2i)^2$

55. $(1 - i)^2$ **56.** $(1 + i)^2$

57. $(3 - 4i)^2$ **58.** $(6 - 5i)^2$

59. $(2 + i)(2 - i)$ **60.** $(3 + i)(3 - i)$

61. $(6 - 2i)(6 + 2i)$ **62.** $(5 + 4i)(5 - 4i)$

63. $(2 + 3i)(2 - 3i)$ **64.** $(2 - 7i)(2 + 7i)$

65. $(10 + 8i)(10 - 8i)$

66. $(11 - 7i)(11 + 7i)$

Find the following quotients. Write all answers in standard form for complex numbers.

67. $\dfrac{2 - 3i}{i}$

68. $\dfrac{3 + 4i}{i}$

69. $\dfrac{5 + 2i}{-i}$

70. $\dfrac{4 - 3i}{-i}$

71. $\dfrac{4}{2 - 3i}$

72. $\dfrac{3}{4 - 5i}$

73. $\dfrac{6}{-3 + 2i}$

74. $\dfrac{-1}{-2 - 5i}$

75. $\dfrac{2 + 3i}{2 - 3i}$

76. $\dfrac{4 - 7i}{4 + 7i}$

77. $\dfrac{5 + 4i}{3 + 6i}$

78. $\dfrac{2 + i}{5 - 6i}$

79. $\dfrac{3 - 7i}{9 - 5i}$

80. $\dfrac{4 + 10i}{3 + 6i}$

Review Problems

The problems that follow review material we covered in Section 5.7.

81. If y varies directly with the square of x, and y is 75 when x is 5, find y when x is 7.

82. Suppose y varies directly with the cube root of x. If y is 14 when x is 8, find x when y is 21.

83. Suppose y varies inversely with the square root of x. If y is 10 when x is 25, find x when y is 5.

84. If y varies inversely with the cube of x, and y is 2 when x is 2, find y when x is 4.

85. Suppose z varies jointly with x and the square of y. If z is 40 when x is 5 and y is 2, find z when x is 2 and y is 5.

86. Suppose z varies jointly with x and the cube root of y. If z is 48 when x is 3 and y is 8, find z when x is 4 and y is $\frac{1}{8}$.

One Step Further

87. Show that $-i$ and $\dfrac{1}{i}$ (the opposite and the reciprocal of i) are the same number.

88. Show that i^{2n+1} is the same as i for all positive even integers n.

89. Show that $x = 1 + i$ is a solution to the equation $x^2 - 2x + 2 = 0$.

90. Show that $x = 1 - i$ is a solution to the equation $x^2 - 2x + 2 = 0$.

91. Show that $x = 2 + i$ is a solution to the equation $x^3 - 11x + 20 = 0$.

92. Show that $x = 2 - i$ is a solution to the equation $x^3 - 11x + 20 = 0$.

CHAPTER 6 SUMMARY

Examples
1. The number 49 has two square roots, 7 and -7. They are written like this:

$$\sqrt{49} = 7 \qquad -\sqrt{49} = -7$$

Square Roots [6.1]
Every positive real number x has two square roots. The *positive square root of x* is written $\sqrt{x}$, while the *negative square root of x* is written $-\sqrt{x}$. Both the positive and the negative square roots of x are numbers we square to get x. That is,

$$\left.\begin{array}{l}(\sqrt{x})^2 = x \\ \text{and} \quad (-\sqrt{x})^2 = x\end{array}\right\} \text{ for } x \geq 0$$

2. $\sqrt[3]{8} = 2$

$\sqrt[3]{-27} = -3$

Higher Roots [6.1]

In the expression $\sqrt[n]{a}$, n is the *index*, a is the *radicand*, and $\sqrt{}$ is the *radical sign*. The expression $\sqrt[n]{a}$ is such that

$$(\sqrt[n]{a})^n = a \qquad a \geq 0 \text{ when } n \text{ is even}$$

3. $25^{1/2} = \sqrt{25} = 5$

$8^{2/3} = (\sqrt[3]{8})^2 = 2^2 = 4$

$9^{3/2} = (\sqrt{9})^3 = 3^3 = 27$

Rational Exponents [6.1, 6.2]

Rational exponents are used to indicate roots. The relationship between rational exponents and roots is as follows:

$$a^{1/n} = \sqrt[n]{a} \quad \text{and} \quad a^{m/n} = (a^{1/n})^m = (a^m)^{1/n}$$

$$a \geq 0 \text{ when } n \text{ is even}$$

4. $\sqrt{4 \cdot 5} = \sqrt{4}\,\sqrt{5} = 2\sqrt{5}$

$\sqrt{\dfrac{7}{9}} = \dfrac{\sqrt{7}}{\sqrt{9}} = \dfrac{\sqrt{7}}{3}$

Properties of Radicals [6.3]

If a and b are nonnegative real numbers whenever n is even, then

1. $\sqrt[n]{ab} = \sqrt[n]{a}\,\sqrt[n]{b}$

2. $\sqrt[n]{\dfrac{a}{b}} = \dfrac{\sqrt[n]{a}}{\sqrt[n]{b}} \qquad (b \neq 0)$

5. $\sqrt{\dfrac{4}{5}} = \dfrac{\sqrt{4}}{\sqrt{5}}$

$= \dfrac{2}{\sqrt{5}} \cdot \dfrac{\sqrt{5}}{\sqrt{5}}$

$= \dfrac{2\sqrt{5}}{5}$

Simplified Form for Radicals [6.3]

A radical expression is said to be in *simplified form*

1. if there is no factor of the radicand that can be written as a power greater than or equal to the index;
2. if there are no fractions under the radical sign; and
3. if there are no radicals in the denominator.

6. $5\sqrt{3} - 7\sqrt{3} = (5 - 7)\sqrt{3}$

$= -2\sqrt{3}$

$\sqrt{20} + \sqrt{45} = 2\sqrt{5} + 3\sqrt{5}$

$= (2 + 3)\sqrt{5}$

$= 5\sqrt{5}$

Addition and Subtraction of Radical Expressions [6.4]

We add and subtract radical expressions by using the distributive property to combine similar radicals. *Similar radicals* are radicals with the same index and the same radicand.

7. $(\sqrt{x} + 2)(\sqrt{x} + 3)$

$= \sqrt{x}\,\sqrt{x} + 3\sqrt{x} + 2\sqrt{x} + 2 \cdot 3$

$= x + 5\sqrt{x} + 6$

Multiplication of Radical Expressions [6.5]

We multiply radical expressions in the same way that we multiply polynomials. We can use the distributive property and the FOIL method.

8.
$$\frac{3}{\sqrt{2}} = \frac{3}{\sqrt{2}} \cdot \frac{\sqrt{2}}{\sqrt{2}} = \frac{3\sqrt{2}}{2}$$

$$\frac{3}{\sqrt{5} - \sqrt{3}} = \frac{3}{\sqrt{5} - \sqrt{3}} \cdot \frac{\sqrt{5} + \sqrt{3}}{\sqrt{5} + \sqrt{3}}$$

$$= \frac{3\sqrt{5} + 3\sqrt{3}}{5 - 3}$$

$$= \frac{3\sqrt{5} + 3\sqrt{3}}{2}$$

9. $\sqrt{2x + 1} = 3$

$(\sqrt{2x + 1})^2 = 3^2$

$2x + 1 = 9$

$x = 4$

10. $3 + 4i$ is a complex number.

Addition

$(3 + 4i) + (2 - 5i) = 5 - i$

Multiplication

$(3 + 4i)(2 - 5i)$

$= 6 - 15i + 8i - 20i^2$

$= 6 - 7i + 20$

$= 26 - 7i$

Division

$$\frac{2}{3 + 4i} = \frac{2}{3 + 4i} \cdot \frac{3 - 4i}{3 - 4i}$$

$$= \frac{6 - 8i}{9 + 16}$$

$$= \frac{6}{25} - \frac{8}{25} i$$

Rationalizing the Denominator [6.3, 6.5]

When a fraction contains a square root in the denominator, we *rationalize the denominator* by multiplying numerator and denominator by

1. the square root itself if there is only one term in the denominator, or
2. the conjugate of the denominator if there are two terms in the denominator.

Rationalizing the denominator can also be called division of radical expressions.

Squaring Property of Equality [6.6]

We may square both sides of an equation any time it is convenient to do so, as long as we check all resulting solutions in the original equation.

Complex Numbers [6.7]

A *complex number* is any number that can be put in the form

$$a + bi$$

where a and b are real numbers and $i = \sqrt{-1}$. The *real part* of the complex number is a, and b is the *imaginary part*.

If a, b, c, and d are real numbers, then we have the following definitions associated with complex numbers:

1. Equality

   $a + bi = c + di$ if and only if $a = c$ and $b = d$

2. Addition and subtraction

   $(a + bi) + (c + di) = (a + c) + (b + d)i$

   $(a + bi) - (c + di) = (a - c) + (b - d)i$

3. Multiplication

   $(a + bi)(c + di) = (ac - bd) + (ad + bc)i$

4. Division is similar to rationalizing the denominator.

COMMON MISTAKES

1. The most common mistake when working with radicals is to assume that the square root of a sum is the sum of the square roots—or

$$\sqrt{x + y} = \sqrt{x} + \sqrt{y} \quad \text{Mistake}$$

The problem with this is it just isn't true. If we try it with 16 and 9, the mistake becomes obvious:

$$\sqrt{16 + 9} \overset{?}{=} \sqrt{16} + \sqrt{9}$$

$$\sqrt{25} \overset{?}{=} 4 + 3$$

$$5 \neq 7$$

2. A common mistake when working with complex numbers is to mistake i for -1. The letter i is not -1; it is the square root of -1. That is, $i = \sqrt{-1}$.

3. When both sides of an equation are squared in the process of solving the equation, a common mistake occurs when the resulting solutions are not checked in the original equation. Remember, every time we square both sides of an equation, there is the possibility we have introduced an extraneous root.

4. When squaring a quantity that has two terms involving radicals, it is a common mistake to omit the middle term in the result. For example,

$$(\sqrt{x + 3} + \sqrt{2x})^2 = (x + 3) + (2x) \quad \text{Mistake}$$

It should look like this:

$$(\sqrt{x + 3} + \sqrt{2x})^2 = (x + 3) + 2\sqrt{2x}\sqrt{x + 3} + (2x)$$

Remember: $(a + b)^2 = a^2 + 2ab + b^2$.

CHAPTER 6 REVIEW

Simplify each expression as much as possible. [6.1]

1. $\sqrt{49}$

2. $\sqrt[3]{8}$

3. $(-27)^{1/3}$

4. $27^{1/3}$

5. $16^{1/4}$

6. $\left(\dfrac{27}{64}\right)^{1/3}$

7. $9^{3/2}$

8. $8^{2/3}$

9. $\sqrt[5]{32x^{15}y^{10}}$

10. $\sqrt[3]{125x^9y^{12}}$

11. $8^{-4/3}$

12. $8^{-2/3} + 25^{-1/2}$

Use the properties of exponents to simplify each expression. Assume all bases represent positive numbers. [6.1]

13. $x^{2/3} \cdot x^{4/3}$

14. $(y^{3/4})^{4/3}$

15. $(a^{2/3}b^{4/3})^3$

16. $x^{2/3} \cdot x^{3/4}$

17. $\dfrac{a^{3/5}}{a^{1/4}}$

18. $(x^{1/3}y^{1/2}z^{5/6})^{6/5}$

19. $\dfrac{a^{2/3}b^3}{a^{1/4}b^{1/3}}$

20. $\dfrac{(y^{3/4})^{8/3}}{(y^{1/3})^{3/2}}$

Multiply. [6.2]

21. $(3x^{1/2} + 5y^{1/2})(4x^{1/2} - 3y^{1/2})$

22. $(4x^{1/2} - 3)(5x^{1/2} + 2)$

23. $(a^{1/3} - 5)^2$

24. $(2t^{1/3} - 1)(4t^{2/3} + 2t^{1/3} + 1)$

Divide. (Assume $x > 0$.) [6.2]

25. $\dfrac{28x^{5/6} + 14x^{7/6}}{7x^{1/3}}$

26. $\dfrac{39a^{5/7}b^{4/5} - 26a^{3/7}b^{6/5}}{13a^{2/7}b^{3/5}}$

27. Factor $2(x - 3)^{1/4}$ from
$8(x - 3)^{5/4} - 2(x - 3)^{1/4}$. [6.2]

28. Factor $6x^{2/5} - 11x^{1/5} - 10$ as if it were a trinomial. [6.2]

Simplify into a single fraction. (Assume $x > 0$.) [6.2]

29. $x^{3/4} + \dfrac{5}{x^{1/4}}$

30. $\dfrac{4x^3}{(x^2 + 1)^{1/2}} + (x^2 + 1)^{1/2}$

Write each expression in simplified form for radicals. (Assume all variables represent nonnegative numbers.) [6.3]

31. $\sqrt{12}$ **32.** $\sqrt{27}$

33. $\sqrt{50}$ **34.** $\sqrt{20}$

35. $\sqrt[3]{16}$ **36.** $\sqrt[3]{32}$

37. $\sqrt{18x^2}$ **38.** $\sqrt{72y^5}$

39. $\sqrt{80a^3b^4c^2}$ **40.** $\sqrt[3]{27x^4y^3}$

41. $\sqrt[4]{32a^4b^5c^6}$ **42.** $\sqrt[4]{162a^6b^5c^4}$

Rationalize the denominator in each expression. [6.3]

43. $\dfrac{3}{\sqrt{2}}$ **44.** $\sqrt{\dfrac{2}{5}}$

45. $\dfrac{6}{\sqrt[3]{2}}$ **46.** $\dfrac{7}{\sqrt[3]{9}}$

Write each expression in simplified form. (Assume all variables represent positive numbers.) [6.3]

47. $\sqrt{\dfrac{48x^3}{7y}}$ **48.** $\sqrt{\dfrac{75x^2y^3}{2z}}$

49. $\sqrt[3]{\dfrac{40x^2y^3}{3z}}$ **50.** $\sqrt[3]{\dfrac{54x^4y^3}{5z^2}}$

Combine the following expressions. (Assume all variables represent positive numbers.) [6.4]

51. $5x\sqrt{6} + 2x\sqrt{6} - 9x\sqrt{6}$

52. $3x\sqrt{7} + 7x\sqrt{7} - 2x\sqrt{7}$

53. $\sqrt{12} + \sqrt{3}$ **54.** $\sqrt{18} + \sqrt{8}$

55. $\dfrac{3}{\sqrt{5}} + \sqrt{5}$ **56.** $\sqrt{15} - \sqrt{\dfrac{3}{5}}$

57. $3\sqrt{8} - 4\sqrt{72} + 5\sqrt{50}$

58. $3\sqrt{48} - 3\sqrt{75} + 2\sqrt{27}$

59. $3b\sqrt{27a^5b} + 2a\sqrt{3a^3b^3}$

60. $4a\sqrt{18a^3b^2} - 2b\sqrt{50a^5}$

61. $2x\sqrt[3]{xy^3z^2} - 6y\sqrt[3]{x^4z^2}$

62. $7x\sqrt[3]{81x^2y^4} - 3y\sqrt[3]{24x^5y}$

Multiply. [6.5]

63. $\sqrt{2}(\sqrt{3} - 2\sqrt{2})$

64. $4\sqrt{5}(2\sqrt{10} - 3\sqrt{5})$

65. $(\sqrt{x} - 2)(\sqrt{x} - 3)$

66. $(\sqrt{6} + 3\sqrt{2})(2\sqrt{6} + \sqrt{2})$

67. $(5\sqrt{6} - 2\sqrt{3})(2\sqrt{6} + \sqrt{3})$

68. $(\sqrt{x} - 2)^2$

69. $(\sqrt{8} - \sqrt{2})(\sqrt{8} + \sqrt{2})$

70. $(3\sqrt{5} + 1)(3\sqrt{5} - 1)$

Rationalize the denominator. (Assume x, $y > 0$.) [6.5]

71. $\dfrac{3}{\sqrt{5} - 2}$ **72.** $\dfrac{3}{\sqrt{6} - \sqrt{3}}$

73. $\dfrac{\sqrt{7} + \sqrt{5}}{\sqrt{7} - \sqrt{5}}$ **74.** $\dfrac{\sqrt{x} + \sqrt{y}}{\sqrt{x} - \sqrt{y}}$

75. $\dfrac{3\sqrt{7}}{3\sqrt{7} - 4}$ **76.** $\dfrac{5\sqrt{6}}{2\sqrt{6} + 7}$

Solve each equation. [6.6]

77. $\sqrt{4a + 1} = 1$ **78.** $\sqrt{7x - 4} = -2$

79. $\sqrt[3]{3x - 8} = 1$ **80.** $\sqrt[3]{8 - 3x} = -1$

81. $\sqrt{3x + 1} - 3 = 1$ **82.** $\sqrt{4x + 8} - 2 = 2$

83. $\sqrt{x + 4} = \sqrt{x} - 2$

84. $\sqrt{x + 11} = \sqrt{x - 5} + 2$

85. $\sqrt{2y - 8} = y - 4$

86. $\sqrt{y + 3} = y + 3$

Graph each equation. [6.6]

87. $y = 3\sqrt{x}$ **88.** $y = \sqrt{x} + 3$

89. $y = \sqrt[3]{x} + 2$ **90.** $y = 4\sqrt[3]{x}$

Write in terms of i and then simplify. [6.7]

91. $\sqrt{-49}$ **92.** $\sqrt{-80}$

Write each of the following as i, -1, $-i$, or 1. [6.7]

93. i^{24} **94.** i^{27}

Find x and y so that each of the following equations is true. [6.7]

95. $3 - 4i = -2x + 8yi$

96. $(3x + 2) - 8i = -4 + 2yi$

Combine the following complex numbers. [6.7]

97. $(3 + 5i) + (6 - 2i)$

98. $(3 + 5i) - (6 - 2i)$

99. $(2 + 5i) - [(3 + 2i) + (6 - i)]$

100. $[(6 + 2i) - (3 - 4i)] - (5 - i)$

Multiply. [6.7]

101. $3i(4 + 2i)$ **102.** $-5i(6 - i)$

103. $(2 + 3i)(4 + i)$ **104.** $(6 - 3i)(2 + 4i)$

105. $(4 + 2i)^2$ **106.** $(1 + i)^2$

107. $(4 + 3i)(4 - 3i)$ **108.** $(3 + i)(3 - i)$

Divide. Write all answers in standard form for complex numbers. [6.7]

109. $\dfrac{3 + i}{i}$ **110.** $\dfrac{2 - i}{-i}$

111. $\dfrac{-3}{2 + i}$ **112.** $\dfrac{3 + 2i}{3 - 2i}$

113. $\dfrac{4 - 3i}{4 + 3i}$ **114.** $\dfrac{7 - i}{3 - 2i}$

CHAPTER 6 TEST

Simplify each of the following. (Assume all variable bases are positive integers and all variable exponents are positive real numbers throughout this test.) [6.1]

1. $27^{-2/3}$ **2.** $\left(\frac{25}{49}\right)^{-1/2}$

3. $a^{3/4} \cdot a^{-1/3}$ **4.** $\dfrac{(x^{2/3}y^{-3})^{1/2}}{(x^{3/4}y^{1/2})^{-1}}$

5. $\sqrt{49x^8y^{10}}$ **6.** $\sqrt[5]{32x^{10}y^{20}}$

7. $\dfrac{(36a^8b^4)^{1/2}}{(27a^9b^6)^{1/3}}$ **8.** $\dfrac{(x^n y^{1/n})^n}{(x^{1/n}y^n)^{n^2}}$

Multiply. [6.2]

9. $2a^{1/2}(3a^{3/2} - 5a^{1/2})$ **10.** $(4a^{3/2} - 5)^2$

Factor. [6.2]

11. $3x^{2/3} + 5x^{1/3} - 2$ **12.** $9x^{2/3} - 49$

Combine. [6.2]

13. $\dfrac{4}{x^{1/2}} + x^{1/2}$

14. $\dfrac{x^2}{(x^2 - 3)^{1/2}} - (x^2 - 3)^{1/2}$

Write in simplified form. [6.3]

15. $\sqrt{125x^3y^5}$ **16.** $\sqrt[3]{40x^7y^8}$

17. $\sqrt{\frac{2}{3}}$ **18.** $\sqrt{\dfrac{12a^4b^3}{5c}}$

Combine. [6.4]

19. $3\sqrt{12} - 4\sqrt{27}$

20. $2\sqrt[3]{24a^3b^3} - 5a\sqrt[3]{3b^3}$

Multiply. [6.5]

21. $(\sqrt{x} + 7)(\sqrt{x} - 4)$ **22.** $(3\sqrt{2} - \sqrt{3})^2$

Rationalize the denominator. [6.5]

23. $\dfrac{5}{\sqrt{3} - 1}$ **24.** $\dfrac{\sqrt{x} - \sqrt{2}}{\sqrt{x} + \sqrt{2}}$

Solve for x. [6.6]

25. $\sqrt{3x + 1} = x - 3$ **26.** $\sqrt[3]{2x + 7} = -1$

27. $\sqrt{x + 3} = \sqrt{x + 4} - 1$

Graph. [6.6]

28. $y = \sqrt{x - 2}$ **29.** $y = \sqrt[3]{x} + 3$

30. Solve for x and y so that the following equation is true [6.7]:

$$(2x + 5) - 4i = 6 - (y - 3)i$$

Perform the indicated operations. [6.7]

31. $(3 + 2i) - [(7 - i) - (4 + 3i)]$

32. $(2 - 3i)(4 + 3i)$

33. $(5 - 4i)^2$

34. $\dfrac{2 - 3i}{2 + 3i}$

35. Show that i^{38} can be written as -1. [6.7]

7

QUADRATIC EQUATIONS

INTRODUCTION

Table 1 below is taken from the trail map given to skiers skiing at Northstar at Tahoe ski resort in Lake Tahoe, California. The table gives the length of each chair lift at Northstar, along with the change in elevation from the beginning of the lift to the end of the lift.

TABLE 1 From the Trail Map for Northstar at Tahoe Ski Resort

	Lift Information	
Lift	Vertical Rise	Length
Big Springs Gondola	480'	4,100'
Bear Paw Double	120'	790'
Echo Triple	710'	4,890'
Aspen Express Quad	900'	5,100'
Forest Double	1,170'	5,750'
Lookout Double	960'	4,330'
Comstock Express Quad	1,250'	5,900'
Rendezvous Triple	650'	2,900'
Schaffer Camp Triple	1,860'	6,150'
Chipmunk Tow Lift	28'	280'
Bear Cub Tow Lift	120'	750'

Right triangles are good mathematical models for chair lifts. In this chapter we will use our knowledge of right triangles, along with the new material developed in the chapter, to solve a variety of problems, some of which will involve chair lifts.

OVERVIEW

We begin this chapter by extending the work we started in Chapter 3 with quadratic equations. In Chapter 3 we solved quadratic equations by factoring. Up to this point, factoring is the only method we have had for solving quadratic equations. The new methods we will develop in this chapter will allow us to solve any quadratic equation, regardless of whether or not it is factorable.

Once we have presented the new methods of solving quadratic equations, we will take a closer look at the relationship between solutions to equations and the equations themselves. The result of doing so will give us a better overall view of equations and their solutions, and allow us to build equations from their solutions.

After our work with equations is finished, we will turn our attention to quadratic inequalities. Our ability to solve a wider variety of equations will increase the types of inequalities we can solve.

In the last two sections of the chapter we broaden our work with graphing equations in two variables. The new graphs we will encounter are called **parabolas,** and they are connected very closely to the quadratic equations presented in this chapter.

To be successful in this chapter, you should have a working knowledge of factoring, binomial squares, square roots, and complex numbers. Also important for your success in this chapter is your ability to graph equations in two variables by constructing tables of ordered pairs and by using intercepts.

SECTION

7.1

Completing the Square

In this section, we will develop the first of our new methods of solving quadratic equations. The new method is called **completing the square.** Completing the square on a quadratic equation allows us to obtain solutions, regardless of whether the equation can be factored. Before we solve equations by completing the square, we need to learn how to solve equations by taking square roots of both sides.

Consider the equation

$$x^2 = 16$$

We could solve it by writing it in standard form, factoring the left side, and proceeding as we did in Chapter 3. However, we can shorten our work considerably if we simply notice that x must be either the positive square root of 16 or the negative square root of 16. That is,

$$\text{If } x^2 = 16$$

$$\text{then} \quad x = \sqrt{16} \quad \text{or} \quad x = -\sqrt{16}$$

$$x = 4 \quad \text{or} \quad x = -4$$

We can generalize this result into a theorem as follows.

Theorem 7.1
If $a^2 = b$ where b is a real number, then $a = \sqrt{b}$ or $a = -\sqrt{b}$.

Notation The expression $a = \sqrt{b}$ or $a = -\sqrt{b}$ can be written in shorthand form as $a = \pm\sqrt{b}$. The symbol $\pm$ is read "plus or minus."

We can apply Theorem 7.1 to some fairly complicated quadratic equations.

▶ **EXAMPLE 1** Solve $(2x - 3)^2 = 25$.

Solution

$$(2x - 3)^2 = 25$$

$$2x - 3 = \pm\sqrt{25} \qquad \text{Theorem 7.1}$$

$$2x - 3 = \pm 5 \qquad \sqrt{25} = 5$$

$$2x = 3 \pm 5 \qquad \text{Add 3 to both sides}$$

$$x = \frac{3 \pm 5}{2} \qquad \text{Divide both sides by 2}$$

The last equation can be written as two separate statements:

$$x = \frac{3 + 5}{2} \quad \text{or} \quad x = \frac{3 - 5}{2}$$

$$= \frac{8}{2} \qquad\qquad = \frac{-2}{2}$$

$$= 4 \qquad \text{or} \qquad = -1$$

The solution set is $\{4, -1\}$. ◀

Notice that we could have solved the equation in Example 1 by expanding the left side, writing the resulting equation in standard form, and then factoring. The problem would look like this:

$$(2x - 3)^2 = 25 \qquad \text{Original equation}$$

$$4x^2 - 12x + 9 = 25 \qquad \text{Expand the left side}$$

$$4x^2 - 12x - 16 = 0 \qquad \text{Add } -25 \text{ to each side}$$

$$4(x^2 - 3x - 4) = 0 \qquad \text{Begin factoring}$$

$$4(x - 4)(x + 1) = 0 \qquad \text{Factor completely}$$

$$x - 4 = 0 \quad \text{or} \quad x + 1 = 0 \qquad \text{Set variable factors equal to 0}$$

$$x = 4 \quad \text{or} \qquad x = -1$$

As you can see, solving the equation by factoring leads to the same two solutions.

▶ **EXAMPLE 2**   Solve for x: $(3x - 1)^2 = -12$.

Solution

$$(3x - 1)^2 = -12$$

$$3x - 1 = \pm\sqrt{-12} \qquad \text{Theorem 7.1}$$

$$3x - 1 = \pm 2i\sqrt{3} \qquad \sqrt{-12} = 2i\sqrt{3}$$

$$3x = 1 \pm 2i\sqrt{3} \qquad \text{Add 1 to both sides}$$

$$x = \frac{1 \pm 2i\sqrt{3}}{3} \qquad \text{Divide both sides by 3}$$

The solution set is $\left\{ \dfrac{1 + 2i\sqrt{3}}{3}, \dfrac{1 - 2i\sqrt{3}}{3} \right\}$

Both solutions are complex. Here is a check of the first solution:

When $x = \dfrac{1 + 2i\sqrt{3}}{3}$

the equation $(3x - 1)^2 = -12$

becomes $\left(3 \cdot \dfrac{1 + 2i\sqrt{3}}{3} - 1 \right)^2 \overset{?}{=} -12$

or $(1 + 2i\sqrt{3} - 1)^2 \overset{?}{=} -12$

$$(2i\sqrt{3})^2 \overset{?}{=} -12$$

$$4 \cdot i^2 \cdot 3 \overset{?}{=} -12$$

$$12(-1) \overset{?}{=} -12$$

$$-12 = -12 \qquad \blacktriangleleft$$

Note   We cannot solve the equation in Example 2 by factoring. If we expand the left side and write the resulting equation in standard form, we are left with a quadratic equation that does not factor:

$$(3x - 1)^2 = -12 \qquad \text{Equation from Example 2}$$

$$9x^2 - 6x + 1 = -12 \qquad \text{Expand the left side}$$

$$9x^2 - 6x + 13 = 0 \qquad \text{Standard form, but not factorable}$$

▶ **EXAMPLE 3**   Solve $x^2 + 6x + 9 = 12$.

Solution   We can solve this equation as we have the equations in Examples 1 and 2 if we first write the left side as $(x + 3)^2$.

Note   We can use a calculator to get decimal approximations to these solutions. If $\sqrt{33} \approx 5.74$, then

$$\frac{-5 + 5.74}{2} = 0.37$$

$$\frac{-5 - 5.74}{2} = -5.37$$

◀

▶ **EXAMPLE 5**   Solve for x: $3x^2 - 8x + 7 = 0$.

Solution

$$3x^2 - 8x + 7 = 0$$

$$3x^2 - 8x = -7 \qquad \text{Add } -7 \text{ to both sides}$$

We cannot complete the square on the left side because the leading coefficient is not 1. We take an extra step and divide both sides by 3:

$$\frac{3x^2}{3} - \frac{8x}{3} = -\frac{7}{3}$$

$$x^2 - \frac{8}{3}x = -\frac{7}{3}$$

Half of $\frac{8}{3}$ is $\frac{4}{3}$, the square of which is $\frac{16}{9}$:

$$x^2 - \frac{8}{3}x + \mathbf{\frac{16}{9}} = -\frac{7}{3} + \mathbf{\frac{16}{9}} \qquad \text{Add } \mathbf{\frac{16}{9}} \text{ to both sides}$$

$$\left(x - \frac{4}{3}\right)^2 = -\frac{5}{9} \qquad \text{Simplify right side}$$

$$x - \frac{4}{3} = \pm\sqrt{-\frac{5}{9}} \qquad \text{Theorem 7.1}$$

$$x - \frac{4}{3} = \pm\frac{i\sqrt{5}}{3} \qquad \sqrt{-\frac{5}{9}} = \frac{\sqrt{-5}}{3} = \frac{i\sqrt{5}}{3}$$

$$x = \frac{4}{3} \pm \frac{i\sqrt{5}}{3} \qquad \text{Add } \frac{4}{3} \text{ to both sides}$$

$$x = \frac{4 \pm i\sqrt{5}}{3}$$

The solution set is $\left\{\dfrac{4 + i\sqrt{5}}{3}, \dfrac{4 - i\sqrt{5}}{3}\right\}$.

◀

> **Strategy for Solving a Quadratic Equation by Completing the Square**
>
> To summarize the method used in the preceding two examples, we list the following steps:
>
> *Step 1:* Write the equation in the form $ax^2 + bx = c$.
>
> *Step 2:* If the leading coefficient is not 1, divide both sides by the coefficient so that the resulting equation has a leading coefficient of 1. That is, if $a \neq 1$, then divide both sides by a.
>
> *Step 3:* Add the square of half the coefficient of the linear term to both sides of the equation.
>
> *Step 4:* Write the left side of the equation as the square of a binomial and simplify the right side if possible.
>
> *Step 5:* Apply Theorem 7.1 and solve as usual.

Facts from Geometry: More Special Triangles

The triangles shown in Figures 1 and 2 occur frequently in mathematics.

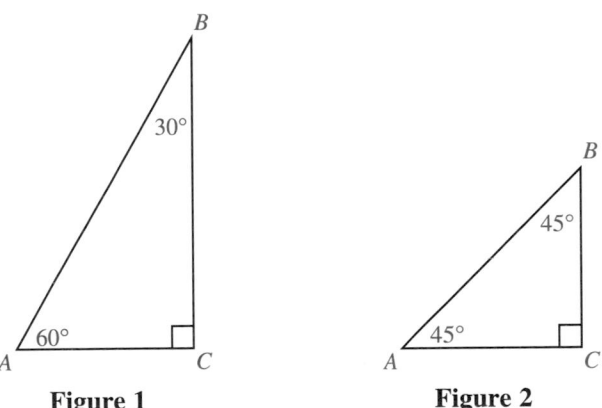

Figure 1   Figure 2

Note that both of the triangles are right triangles. We refer to the triangle in Figure 1 as a $30°-60°-90°$ triangle, and the triangle in Figure 2 as a $45°-45°-90°$ triangle.

▶ **EXAMPLE 6**   If the shortest side in a $30°-60°-90°$ triangle is 1 inch, find the lengths of the other two sides.

Solution   In Figure 3 on the next page, triangle ABC is a $30°-60°-90°$ triangle in which the shortest side AC is 1 inch long. Triangle DBC is also a $30°-60°-90°$ triangle in which the shortest side DC is 1 inch long.

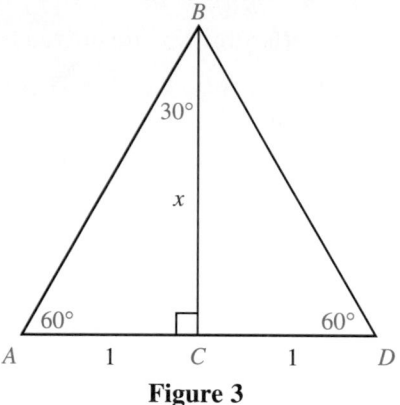

Figure 3

Notice that the large triangle *ABD* is an equilateral triangle because each of its interior angles is 60°. Each side of triangle *ABD* is 2 inches long. Therefore, side *AB* in triangle *ABC* is 2 inches. To find the length of side *BC*, we use the Pythagorean Theorem.

$$BC^2 + AC^2 = AB^2$$

$$x^2 + 1^2 = 2^2$$

$$x^2 + 1 = 4$$

$$x^2 = 3$$

$$x = \sqrt{3} \text{ inches}$$

Note that we write only the positive square root because *x* is the length of a side in a triangle and is therefore a positive number. ◀

▶ **EXAMPLE 7**   Table 1 in the introduction to this chapter gives the vertical rise of the Forest Double chair lift as 1,170 feet and the length of the chair lift as 5,750 feet. To the nearest foot, find the horizontal distance covered by a person riding this lift.

Solution   The figure below is a model of the Forest Double chair lift. A rider gets on the lift at point *A* and exits at point *B*. The length of the lift is *AB*.

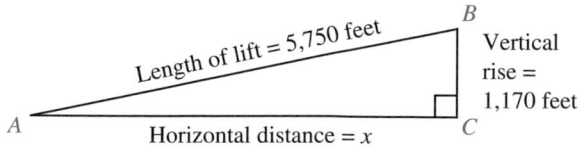

To find the horizontal distance covered by a person riding the chair lift we use the Pythagorean Theorem.

$$5,750^2 = x^2 + 1,170^2 \qquad \text{Pythagorean Theorem}$$
$$33,062,500 = x^2 + 1,368,900 \qquad \text{Simplify squares}$$
$$x^2 = 33,062,500 - 1,368,900 \qquad \text{Solve for } x^2$$
$$x^2 = 31,693,600 \qquad \text{Simplify the right side}$$
$$x = \sqrt{31,693,600} \qquad \text{Theorem 7.1}$$
$$= 5,630 \text{ feet} \quad \text{to the nearest foot}$$

A rider getting on the lift at point A and riding to point B will cover a horizontal distance of approximately 5,630 feet. ◀

PROBLEM SET 7.1

Solve the following equations.

1. $x^2 = 25$
2. $x^2 = 16$
3. $a^2 = -9$
4. $a^2 = -49$
5. $y^2 = \frac{3}{4}$
6. $y^2 = \frac{5}{9}$
7. $x^2 + 12 = 0$
8. $x^2 + 8 = 0$
9. $4a^2 - 45 = 0$
10. $9a^2 - 20 = 0$
11. $(2y - 1)^2 = 25$
12. $(3y + 7)^2 = 1$
13. $(2a + 3)^2 = -9$
14. $(3a - 5)^2 = -49$
15. $(5x + 2)^2 = -8$
16. $(6x - 7)^2 = -75$
17. $x^2 + 8x + 16 = -27$
18. $x^2 - 12x + 36 = -8$
19. $4a^2 - 12a + 9 = -4$
20. $9a^2 - 12a + 4 = -9$

Simplify the left side of each equation and then solve for x.

21. $(x + 5)^2 + (x - 5)^2 = 52$
22. $(2x + 1)^2 + (2x - 1)^2 = 10$
23. $(2x + 3)^2 + (2x - 3)^2 = 26$
24. $(3x + 2)^2 + (3x - 2)^2 = 26$
25. $(3x + 4)(3x - 4) - (x + 2)(x - 2) = -4$
26. $(5x + 2)(5x - 2) - (x + 3)(x - 3) = 29$

Copy each of the following and fill in the blanks so that the left side of each is a perfect square trinomial. That is, complete the square.

27. $x^2 + 12x + \underline{\ \ } = (x + \underline{\ \ })^2$
28. $x^2 + 6x + \underline{\ \ } = (x + \underline{\ \ })^2$
29. $x^2 - 4x + \underline{\ \ } = (x - \underline{\ \ })^2$
30. $x^2 - 2x + \underline{\ \ } = (x - \underline{\ \ })^2$
31. $a^2 - 10a + \underline{\ \ } = (a - \underline{\ \ })^2$
32. $a^2 - 8a + \underline{\ \ } = (a - \underline{\ \ })^2$
33. $x^2 + 5x + \underline{\ \ } = (x + \underline{\ \ })^2$
34. $x^2 + 3x + \underline{\ \ } = (x + \underline{\ \ })^2$
35. $y^2 - 7y + \underline{\ \ } = (y - \underline{\ \ })^2$
36. $y^2 - y + \underline{\ \ } = (y - \underline{\ \ })^2$

Solve each of the following quadratic equations by completing the square.

37. $x^2 + 4x = 12$
38. $x^2 - 2x = 8$
39. $x^2 + 12x = -27$
40. $x^2 - 6x = 16$
41. $a^2 - 2a + 5 = 0$
42. $a^2 + 10a + 22 = 0$
43. $y^2 - 8y + 1 = 0$
44. $y^2 + 6y - 1 = 0$
45. $x^2 - 5x - 3 = 0$
46. $x^2 - 5x - 2 = 0$
47. $2x^2 - 4x - 8 = 0$
48. $3x^2 - 9x - 12 = 0$
49. $3t^2 - 8t + 1 = 0$
50. $5t^2 + 12t - 1 = 0$
51. $4x^2 - 3x + 5 = 0$
52. $7x^2 - 5x + 2 = 0$

Applying the Concepts

53. If the shortest side in a 30°–60°–90° triangle is $\frac{1}{2}$ inch long, find the lengths of the other two sides.

54. If the shortest side in a 30°–60°–90° triangle is 3 feet long, find the lengths of the other two sides.

55. If the shortest side of a 30°–60°–90° triangle is x in length, find the lengths of the other two sides in terms of x.

56. If the longest side of a 30°–60°–90° triangle is x in length, find the lengths of the other two sides in terms of x.

57. If the shorter sides of a 45°–45°–90° triangle are each 1 inch long, find the length of the hypotenuse.

58. If the shorter sides of a 45°–45°–90° triangle are each 3 feet long, find the length of the hypotenuse.

59. If the hypotenuse of a 45°–45°–90° triangle is 1 inch long, find the length of the shorter sides.

60. If the hypotenuse of a 45°–45°–90° triangle is 2 feet long, find the length of the shorter sides.

61. If the shorter sides of a 45°–45°–90° triangle are each x in length, find the length of the hypotenuse, in terms of x.

62. If the hypotenuse of a 45°–45°–90° triangle is x in length, find the length of the shorter sides, in terms of x.

63. Use Table 1 from the introduction to this chapter to find the horizontal distance covered by a person riding the Bear Paw Double chair lift. Round your answer to the nearest foot.

64. Use Table 1 from the introduction to this chapter to find the horizontal distance covered by a person riding the Big Springs Gondola lift. Round your answer to the nearest foot.

65. Using a right triangle to model the Forest Double chair lift, find the slope of the lift to the nearest hundredth.

66. Using a right triangle to model the Echo Triple chair lift, find the slope of the lift to the nearest hundredth.

Review Problems

The problems that follow review material we covered in Section 6.3. Reviewing these problems will help you with the next section.

Write each of the following in simplified form for radicals.

67. $\sqrt{45}$ **68.** $\sqrt{24}$

69. $\sqrt{27y^5}$ **70.** $\sqrt{8y^3}$

71. $\sqrt[3]{54x^6y^5}$ **72.** $\sqrt[3]{16x^9y^7}$

73. Simplify $\sqrt{b^2-4ac}$ when $a=6$, $b=7$, and $c=-5$.

74. Simplify $\sqrt{b^2-4ac}$ when $a=2$, $b=-6$, and $c=3$.

Rationalize the denominator.

75. $\dfrac{3}{\sqrt{2}}$ **76.** $\dfrac{5}{\sqrt{3}}$

77. $\dfrac{2}{\sqrt[3]{4}}$ **78.** $\dfrac{3}{\sqrt[3]{2}}$

One Step Further

Solve for x.

79. $(x+a)^2 + (x-a)^2 = 10a^2$
80. $(ax+1)^2 + (ax-1)^2 = 10$

Assume a is a positive number and solve for x by completing the square on x.

81. $x^2 + 2ax = -a^2$
82. $x^2 + 2ax = -4a^2$
83. $x^2 + 2ax = 0$
84. $x^2 + ax = 0$

Assume p and q are positive numbers and solve for x by completing the square on x.

85. $x^2 + px + q = 0$
86. $x^2 - px + q = 0$

The Quadratic Formula

In this section, we will use the method of completing the square from the preceding section to derive the quadratic formula. The **quadratic formula** is a very useful tool in mathematics. It allows us to solve all types of quadratic equations.

> **The Quadratic Theorem**
>
> For any quadratic equation in the form $ax^2 + bx + c = 0$, where $a \neq 0$, the two solutions are
>
> $$x = \frac{-b + \sqrt{b^2 - 4ac}}{2a} \quad \text{and} \quad x = \frac{-b - \sqrt{b^2 - 4ac}}{2a}$$

PROOF We will prove the quadratic theorem by completing the square on $ax^2 + bx + c = 0$:

$$ax^2 + bx + c = 0$$

$$ax^2 + bx = -c \qquad \text{Add } -c \text{ to both sides}$$

$$x^2 + \frac{b}{a}x = -\frac{c}{a} \qquad \text{Divide both sides by } a$$

To complete the square on the left side, we add the square of $\frac{1}{2}$ of $\frac{b}{a}$ to both sides. $\left(\frac{1}{2} \text{ of } \frac{b}{a} \text{ is } \frac{b}{2a}. \right)$

$$x^2 + \frac{b}{a}x + \left(\frac{b}{2a} \right)^2 = -\frac{c}{a} + \left(\frac{b}{2a} \right)^2$$

We now simplify the right side as a separate step. We square the second term and combine the two terms by writing each with the least common denominator $4a^2$:

$$-\frac{c}{a} + \left(\frac{b}{2a} \right)^2 = -\frac{c}{a} + \frac{b^2}{4a^2} = \frac{4a}{4a}\left(\frac{-c}{a} \right) + \frac{b^2}{4a^2} = \frac{-4ac + b^2}{4a^2}$$

It is convenient to write this last expression as

$$\frac{b^2 - 4ac}{4a^2}$$

Continuing with the proof, we have

$$x^2 + \frac{b}{a}x + \left(\frac{b}{2a}\right)^2 = \frac{b^2 - 4ac}{4a^2}$$

$$\left(x + \frac{b}{2a}\right)^2 = \frac{b^2 - 4ac}{4a^2} \qquad \text{Write left side as a binomial square}$$

$$x + \frac{b}{2a} = \pm\frac{\sqrt{b^2 - 4ac}}{2a} \qquad \text{Theorem 7.1}$$

$$x = -\frac{b}{2a} \pm \frac{\sqrt{b^2 - 4ac}}{2a} \qquad \text{Add } -\frac{b}{2a} \text{ to both sides}$$

$$x = \frac{-b \pm \sqrt{b^2 - 4ac}}{2a}$$

Our proof is now complete. What we have is this: If our equation is in the form $ax^2 + bx + c = 0$ (standard form), where $a \neq 0$, the two solutions are always given by the formula

$$x = \frac{-b \pm \sqrt{b^2 - 4ac}}{2a}$$

This formula is known as the **quadratic formula.** If we substitute the coefficients a, b, and c of any quadratic equation in standard form into the formula, we need only perform some basic arithmetic to arrive at the solution set.

▶ **EXAMPLE 1**   Use the quadratic formula to solve $6x^2 + 7x - 5 = 0$.

Solution   Using $a = 6$, $b = 7$, and $c = -5$ in the formula

$$x = \frac{-b \pm \sqrt{b^2 - 4ac}}{2a}$$

we have

$$x = \frac{-7 \pm \sqrt{49 - 4(6)(-5)}}{2(6)}$$

or

$$x = \frac{-7 \pm \sqrt{49 + 120}}{12}$$

$$= \frac{-7 \pm \sqrt{169}}{12}$$

$$= \frac{-7 \pm 13}{12}$$

We separate the last equation into the two statements

$$x = \frac{-7 + 13}{12} \quad \text{or} \quad x = \frac{-7 - 13}{12}$$

$$x = \frac{1}{2} \qquad \text{or} \quad x = -\frac{5}{3}$$

The solution set is $\{\frac{1}{2}, -\frac{5}{3}\}$.   ◀

Whenever the solutions to a quadratic equation are rational numbers, as they are in Example 1, it means that the original equation was solvable by factoring. To illustrate, let's solve the equation from Example 1 again, but this time by factoring:

$$6x^2 + 7x - 5 = 0 \qquad \text{Equation in standard form}$$

$$(3x + 5)(2x - 1) = 0 \qquad \text{Factor the left side}$$

$$3x + 5 = 0 \quad \text{or} \quad 2x - 1 = 0 \qquad \text{Set factors equal to 0}$$

$$x = -\frac{5}{3} \quad \text{or} \qquad x = \frac{1}{2}$$

When an equation can be solved by factoring, then factoring is usually the faster method of solution. It is best to try to factor first, and then if you have trouble factoring, go to the quadratic formula. It always works.

▶ **EXAMPLE 2** Solve $\dfrac{x^2}{3} - x = -\dfrac{1}{2}$.

Solution Multiplying through by 6 and writing the result in standard form, we have

$$2x^2 - 6x + 3 = 0$$

the left side of which is not factorable. Therefore, we use the quadratic formula with $a = 2$, $b = -6$, and $c = 3$. The two solutions are given by

$$x = \frac{-(-6) \pm \sqrt{36 - 4(2)(3)}}{2(2)}$$

$$= \frac{6 \pm \sqrt{12}}{4}$$

$$= \frac{6 \pm 2\sqrt{3}}{4} \qquad \sqrt{12} = \sqrt{4 \cdot 3} = \sqrt{4}\sqrt{3} = 2\sqrt{3}$$

We can reduce this last expression to lowest terms by factoring 2 from the numerator and denominator and then dividing the numerator and denominator by 2:

$$x = \frac{\cancel{2}(3 \pm \sqrt{3})}{\cancel{2} \cdot 2} = \frac{3 \pm \sqrt{3}}{2} \qquad\qquad ◀$$

▶ **EXAMPLE 3** Solve $\dfrac{1}{x+2} - \dfrac{1}{x} = \dfrac{1}{3}$.

Solution To solve this equation, we must first put it in standard form. To do so, we must clear the equation of fractions by multiplying each side by the LCD for all the denominators, which is $3x(x+2)$. Multiplying both sides by the LCD, we have

$$3x(x+2)\left(\frac{1}{x+2} - \frac{1}{x}\right) = \frac{1}{3} \cdot 3x(x+2) \qquad \text{Multiply each by the LCD}$$

$$3x(x+2) \cdot \frac{1}{x+2} - 3x(x+2) \cdot \frac{1}{x} = \frac{1}{3} \cdot 3x(x+2)$$

$$3x - 3(x+2) = x(x+2)$$

$$3x - 3x - 6 = x^2 + 2x \qquad \text{Multiplication}$$

$$-6 = x^2 + 2x \qquad \text{Simplify left side}$$

$$0 = x^2 + 2x + 6 \qquad \text{Add 6 to each side}$$

Since the right side of our last equation is not factorable, we use the quadratic formula. From our last equation, we have $a = 1$, $b = 2$, and $c = 6$. Using these numbers for a, b, and c in the quadratic formula gives us

$$x = \frac{-2 \pm \sqrt{4 - 4(1)(6)}}{2(1)}$$

$$= \frac{-2 \pm \sqrt{4 - 24}}{2} \qquad \text{Simplify inside the radical}$$

$$= \frac{-2 \pm \sqrt{-20}}{2} \qquad 4 - 24 = -20$$

$$= \frac{-2 \pm 2i\sqrt{5}}{2} \qquad \sqrt{-20} = i\sqrt{20} = i\sqrt{4}\sqrt{5} = 2i\sqrt{5}$$

$$= \frac{2(-1 \pm i\sqrt{5})}{2} \qquad \text{Factor 2 from the numerator}$$

$$= -1 \pm i\sqrt{5} \qquad \text{Divide numerator and denominator by 2}$$

Since neither of the two solutions, $-1 + i\sqrt{5}$ nor $-1 - i\sqrt{5}$, will make any of the denominators in our original equation 0, they are both solutions. ◀

Although the equation in our next example is not a quadratic equation, we solve it by using both factoring and the quadratic formula.

▶ **EXAMPLE 4** Solve $27t^3 - 8 = 0$.

Solution It would be a mistake to add 8 to each side of this equation and then take the cube root of each side because we would lose two of our solutions.

Instead, we factor the left side and then set the factors equal to 0:

$$27t^3 - 8 = 0 \qquad \text{Equation in standard form}$$

$$(3t - 2)(9t^2 + 6t + 4) = 0 \qquad \begin{array}{l}\text{Factor as the difference} \\ \text{of two cubes}\end{array}$$

$$3t - 2 = 0 \quad \text{or} \quad 9t^2 + 6t + 4 = 0 \qquad \text{Set each factor equal to 0}$$

The first equation leads to a solution of $t = \frac{2}{3}$. The second equation does not factor, so we use the quadratic formula with $a = 9$, $b = 6$, and $c = 4$:

$$t = \frac{-6 \pm \sqrt{36 - 4(9)(4)}}{2(9)}$$

$$= \frac{-6 \pm \sqrt{36 - 144}}{18}$$

$$= \frac{-6 \pm \sqrt{-108}}{18}$$

$$= \frac{-6 \pm 6i\sqrt{3}}{18} \qquad \sqrt{-108} = i\sqrt{36 \cdot 3} = 6i\sqrt{3}$$

$$= \frac{\cancel{6}(-1 \pm i\sqrt{3})}{\cancel{6} \cdot 3} \qquad \begin{array}{l}\text{Factor 6 from the numerator} \\ \text{and denominator}\end{array}$$

$$= \frac{-1 \pm i\sqrt{3}}{3} \qquad \text{Divide out common factor 6}$$

The three solutions to our original equation are

$$\frac{2}{3}, \frac{-1 + i\sqrt{3}}{3}, \text{ and } \frac{-1 - i\sqrt{3}}{3} \qquad \blacktriangleleft$$

▶ **EXAMPLE 5** If an object is thrown downward with an initial velocity of 20 feet/second, the distance s it travels in an amount of time t is given by the equation $s = 20t + 16t^2$. How long does it take the object to fall 40 feet?

Solution We let $s = 40$ and solve for t:

When $\qquad\qquad\qquad\qquad\qquad\qquad s = 40$

the equation $\qquad\qquad\qquad\qquad\quad s = 20t + 16t^2$

becomes $\qquad\qquad\qquad\qquad 40 = 20t + 16t^2$

or $\qquad\qquad\qquad 16t^2 + 20t - 40 = 0$

$\qquad\qquad\qquad\qquad 4t^2 + 5t - 10 = 0 \qquad \text{Divide by 4}$

Using the quadratic formula, we have

$$t = \frac{-5 \pm \sqrt{25 - 4(4)(-10)}}{2(4)}$$

$$t = \frac{-5 \pm \sqrt{185}}{8}$$

$$t = \frac{-5 + \sqrt{185}}{8} \quad \text{or} \quad t = \frac{-5 - \sqrt{185}}{8}$$

The second solution is impossible since it is a negative number and t must be positive.

It takes

$$t = \frac{-5 + \sqrt{185}}{8}$$

or approximately

$$\frac{-5 + 13.60}{8} \approx 1.08 \text{ seconds}$$

for the object to fall 40 feet. ◀

Recall from Chapter 3 that the relationship between profit, revenue, and cost is given by the formula

$$P(x) = R(x) - C(x)$$

where $P(x)$ is the profit, $R(x)$ is the total revenue, and $C(x)$ is the total cost of producing and selling x items.

▶ **EXAMPLE 6** A company produces and sells copies of an accounting program for home computers. The total weekly cost (in dollars) to produce x copies of the program is $C(x) = 8x + 500$, while the weekly revenue for selling all x programs is $R(x) = 35x - 0.1x^2$. How many programs must it sell each week for its weekly profit to be $1,200?

Solution Substituting the given expressions for $R(x)$ and $C(x)$ in the equation $P(x) = R(x) - C(x)$, we have a polynomial in x that represents the weekly profit $P(x)$:

$$
\begin{aligned}
P(x) &= R(x) - C(x) \\
&= 35x - 0.1x^2 - (8x + 500) \\
&= 35x - 0.1x^2 - 8x - 500 \\
&= -500 + 27x - 0.1x^2
\end{aligned}
$$

Setting this expression equal to 1,200, we have a quadratic equation to solve that will give us the number of programs x that need to be sold each week to bring in a profit of $1,200:

$$1,200 = -500 + 27x - 0.1x^2$$

We can write this equation in standard form by adding the opposite of each term on the right side of the equation to both sides of the equation. Doing so produces the following equation:

$$0.1x^2 - 27x + 1,700 = 0$$

Applying the quadratic formula to this equation with $a = 0.1$, $b = -27$, and $c = 1,700$, we have

$$x = \frac{27 \pm \sqrt{(-27)^2 - 4(0.1)(1,700)}}{2(0.1)}$$

$$= \frac{27 \pm \sqrt{729 - 680}}{0.2}$$

$$= \frac{27 \pm \sqrt{49}}{0.2}$$

$$= \frac{27 \pm 7}{0.2}$$

Writing this last expression as two separate expressions, we have our two solutions:

$$x = \frac{27 + 7}{0.2} \quad \text{or} \quad x = \frac{27 - 7}{0.2}$$

$$= \frac{34}{0.2} \qquad\qquad = \frac{20}{0.2}$$

$$= 170 \qquad\qquad = 100$$

The weekly profit will be $1,200 if they produce and sell 100 items or if they produce and sell 170 items. ◀

What is interesting about the equation we solved in Example 6 is that it has rational solutions, meaning it could have been solved by factoring. But looking back to the equation, factoring does not seem like a reasonable method of solution because the coefficients are either large or small. So, there are times when using the quadratic formula is a faster method of solution, even though the equation you are solving is factorable.

USING TECHNOLOGY: Graphing Calculators

More about Example 5 We can solve the problem in Example 5 by graphing the function $Y_1 = 20X + 16X^2$ in a Window with X from 0 to 4 (because X is taking the place of t and we know t is a positive quantity) and Y from 0 to 50 (because we are looking for X when Y_1 is 40). Graphing Y_1 gives a graph similar to the graph in Figure 1 below. Using the Zoom and Trace features at $Y_1 = 40$ gives us X = 1.08 to the nearest hundredth, matching the results we obtained by solving the original equation algebraically.

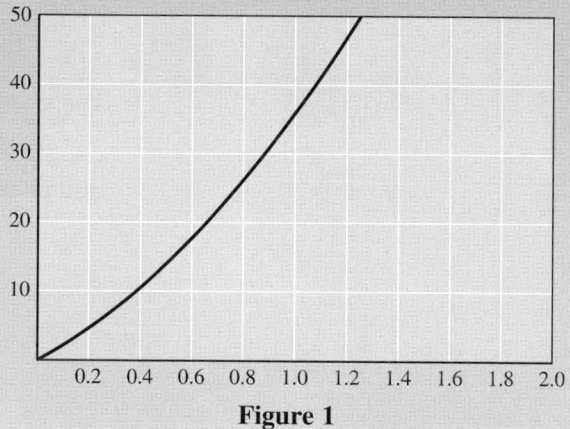

Figure 1

More about Example 6 To visualize the functions in Example 6, we set up our calculators this way:

$$Y_1 = 35X - .1X^2 \qquad \text{Revenue function}$$

$$Y_2 = 8X + 500 \qquad \text{Cost function}$$

$$Y_3 = Y_1 - Y_2 \qquad \text{Profit function}$$

Window: X from 0 to 350, Y from 0 to 3500

Graphing these functions produces graphs similar to the ones shown in Figure 2. The lower graph is the graph of the profit function. Using Zoom and Trace on the lower graph at $Y_3 = 1,200$ produces two corresponding values of X, 170 and 100, which match the results in Example 6.

We will continue this discussion of the relationship between graphs of functions and solutions to equations in the Using Technology material in the next section.

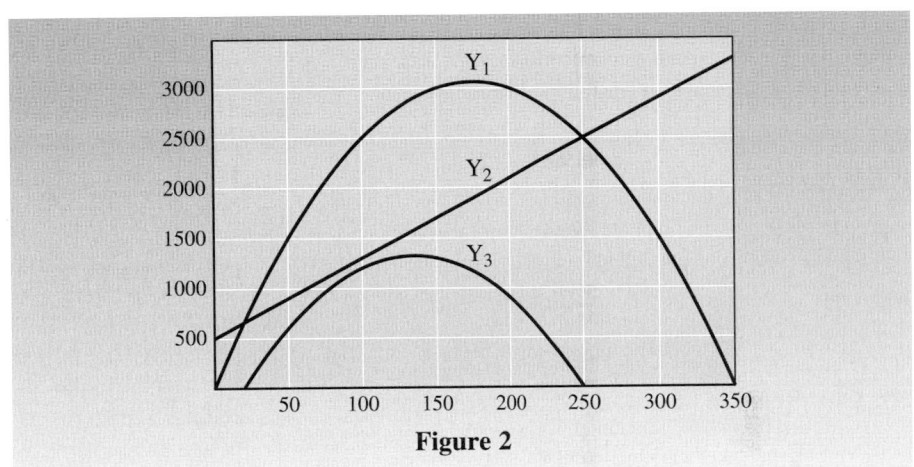

Figure 2

PROBLEM SET 7.2

Solve each equation. Use factoring or the quadratic formula, whichever is appropriate. (Try factoring first. If you have any difficulty factoring, then go right to the quadratic formula.)

1. $x^2 + 5x + 6 = 0$

2. $x^2 + 5x - 6 = 0$

3. $a^2 - 4a + 1 = 0$

4. $a^2 + 4a + 1 = 0$

5. $\frac{1}{6}x^2 - \frac{1}{2}x + \frac{1}{3} = 0$

6. $\frac{1}{4}x^2 + \frac{1}{4}x - \frac{1}{2} = 0$

7. $\frac{x^2}{2} + 1 = \frac{2x}{3}$

8. $\frac{x^2}{2} + \frac{2}{3} = -\frac{2x}{3}$

9. $y^2 - 5y = 0$

10. $2y^2 + 10y = 0$

11. $30x^2 + 40x = 0$

12. $50x^2 - 20x = 0$

13. $\frac{2t^2}{3} - t = -\frac{1}{6}$

14. $\frac{t^2}{3} - \frac{t}{2} = -\frac{3}{2}$

15. $0.01x^2 + 0.06x - 0.08 = 0$

16. $0.02x^2 - 0.03x + 0.05 = 0$

17. $2x + 3 = -2x^2$

18. $2x - 3 = 3x^2$

19. $100x^2 - 200x + 100 = 0$

20. $100x^2 - 600x + 900 = 0$

21. $\frac{1}{2}r^2 = \frac{1}{6}r - \frac{2}{3}$

22. $\frac{1}{4}r^2 = \frac{2}{5}r + \frac{1}{10}$

23. $(x - 3)(x - 5) = 1$

24. $(x - 3)(x + 1) = -6$

25. $(x + 3)^2 + (x - 8)(x - 1) = 16$

26. $(x - 4)^2 + (x + 2)(x + 1) = 9$

27. $\frac{x^2}{3} - \frac{5x}{6} = \frac{1}{2}$

28. $\frac{x^2}{6} + \frac{5}{6} = -\frac{x}{3}$

Multiply both sides of each equation by its LCD. Then solve the resulting equation.

29. $\frac{1}{x + 1} - \frac{1}{x} = \frac{1}{2}$

30. $\frac{1}{x + 1} + \frac{1}{x} = \frac{1}{3}$

31. $\frac{1}{y - 1} + \frac{1}{y + 1} = 1$

32. $\frac{2}{y + 2} + \frac{3}{y - 2} = 1$

33. $\frac{1}{x + 2} + \frac{1}{x + 3} = 1$

34. $\frac{1}{x + 3} + \frac{1}{x + 4} = 1$

35. $\frac{6}{r^2 - 1} - \frac{1}{2} = \frac{1}{r + 1}$

36. $2 + \frac{5}{r - 1} = \frac{12}{(r - 1)^2}$

Solve each equation. In each case you will have three solutions.

37. $x^3 - 8 = 0$

38. $x^3 - 27 = 0$

39. $8a^3 + 27 = 0$

40. $27a^3 + 8 = 0$

41. $125t^3 - 1 = 0$

42. $64t^3 + 1 = 0$

Each of the following equations has three solutions. Look first for the greatest common factor, then use the quadratic formula to find all solutions.

43. $2x^3 + 2x^2 + 3x = 0$

44. $6x^3 - 4x^2 + 6x = 0$

45. $3y^4 = 6y^3 - 6y^2$

46. $4y^4 = 16y^3 - 20y^2$

47. $6t^5 + 4t^4 = -2t^3$ **48.** $8t^5 + 2t^4 = -10t^3$

49. One solution to a quadratic equation is $\dfrac{-3 + 2i}{5}$. What is the other solution?

50. One solution to a quadratic equation is $\dfrac{-2 + 3i\sqrt{2}}{5}$. What is the other solution?

Applying the Concepts

Problems 51–58 may be solved using a graphing calculator.

51. An object is thrown downward with an initial velocity of 5 feet/second. The relationship between the distance s it travels and time t is given by $s(t) = 5t + 16t^2$. How long does it take the object to fall 74 feet?

52. The distance an object falls from rest is given by the equation $s(t) = 16t^2$, where $s =$ distance and $t =$ time. How long does it take an object dropped from a 100-foot cliff to hit the ground?

53. An object is thrown upward with an initial velocity of 20 feet/second. The equation that gives the height h of the object at any time t is $h(t) = 20t - 16t^2$. At what times will the object be 4 feet off the ground?

54. An object is propelled upward with an initial velocity of 32 feet/second from a height of 16 feet above the ground. The equation giving the object's height h at any time t is $h(t) = 16 + 32t - 16t^2$. Does the object ever reach a height of 32 feet?

55. The total cost (in dollars) for a company to manufacture and sell x items per week is $C(x) = 60x + 300$, while the revenue brought in by selling all x items is $R(x) =$ $100x - 0.5x^2$. How many items must be sold to obtain a weekly profit of $300?

56. The total cost (in dollars) for a company to produce and sell x items per week is $C(x) = 200x + 1,600$, while the revenue brought in by selling all x items is $R(x) = 300x - 0.5x^2$. How many items must be sold in order for the weekly profit to be $2,150?

57. Suppose it costs a company selling patterns $C(x) = 800 + 6.5x$ dollars to produce and sell x patterns a month. If the revenue obtained by selling x patterns is $R(x) = 10x - 0.002x^2$, how many patterns must it sell each month if it wants a monthly profit of $700?

58. Suppose a company manufactures and sells x picture frames each month with a total cost of $C(x) = 1,200 + 3.5x$ dollars. If the revenue obtained by selling x frames is $R(x) = 9x - 0.002x^2$, find the number of frames it must sell each month if its monthly profit is to be $2,300.

Review Problems

The problems that follow review material we covered in Sections 4.2 and 6.1. Reviewing the problems from Section 4.2 will help you with the next section.

Divide, using long division. [4.2]

59. $\dfrac{8y^2 - 26y - 9}{2y - 7}$ **60.** $\dfrac{6y^2 + 7y - 18}{3y - 4}$

61. $\dfrac{x^3 + 9x^2 + 26x + 24}{x + 2}$

62. $\dfrac{x^3 + 6x^2 + 11x + 6}{x + 3}$

Simplify each expression. (Assume $x, y > 0$.) [6.1]

63. $25^{1/2}$ **64.** $8^{1/3}$

65. $\left(\dfrac{9}{25}\right)^{3/2}$ **66.** $\left(\dfrac{16}{81}\right)^{3/4}$

67. $8^{-2/3}$ **68.** $4^{-3/2}$

69. $\dfrac{(49x^8y^{-4})^{1/2}}{(27x^{-3}y^9)^{-1/3}}$ **70.** $\dfrac{(x^{-2}y^{1/3})^6}{x^{-10}y^{3/2}}$

One Step Further

So far, all the equations we have solved have had coefficients that were rational numbers. Here are some equations that have irrational coefficients and some that have complex coefficients. Solve each equation. (Remember, $i^2 = -1$.)

71. $x^2 + \sqrt{3}x - 6 = 0$

72. $x^2 - \sqrt{5}x - 5 = 0$

73. $\sqrt{2}x^2 + 2x - \sqrt{2} = 0$

74. $\sqrt{7}x^2 + 2\sqrt{2}x - \sqrt{7} = 0$

75. $x^2 + ix + 2 = 0$

76. $x^2 + 3ix - 2 = 0$

77. $ix^2 + 3x + 4i = 0$

78. $4ix^2 + 5x + 9i = 0$

SECTION

7.3

Additional Items Involving Solutions to Equations

In this section, we will do two things. First, we will define the discriminant and use it to find the kind of solutions a quadratic equation has without solving the equation. Second, we will use the zero-factor property to build equations from their solutions.

The Discriminant

The quadratic formula

$$x = \frac{-b \pm \sqrt{b^2 - 4ac}}{2a}$$

gives the solutions to any quadratic equation in standard form. There are times, when working with quadratic equations, when it is important only to know what kind of solutions the equation has.

> **DEFINITION** The expression under the radical in the quadratic formula is called the **discriminant:**
>
> $$\text{Discriminant} = D = b^2 - 4ac$$

The discriminant indicates the number and type of solutions to a quadratic equation, when the original equation has integer coefficients. For example, if we were to use the quadratic formula to solve the equation $2x^2 + 2x + 3 = 0$, we would find the discriminant to be

$$b^2 - 4ac = 2^2 - 4(2)(3) = -20$$

Since the discriminant appears under a square root symbol, we have the square root of a negative number in the quadratic formula. Our solutions would therefore be complex numbers. Similarly, if the discriminant were 0, the quadratic formula would yield

$$x = \frac{-b \pm \sqrt{0}}{2a} = \frac{-b \pm 0}{2a} = \frac{-b}{2a}$$

and the equation would have one rational solution: the number $\dfrac{-b}{2a}$.

The following table gives the relationship between the discriminant and the type of solutions to the equation.

For the equation $ax^2 + bx + c = 0$ where a, b, and c are integers and $a \neq 0$:

If the discriminant $b^2 - 4ac$ is	Then the equation will have
Negative	Two complex solutions containing i
Zero	One rational solution
A positive number that is also a perfect square	Two rational solutions
A positive number that is not a perfect square	Two irrational solutions

In the second and third cases, when the discriminant is 0 or a positive perfect square, the solutions are rational numbers. The quadratic equations in these two cases are the ones that can be factored.

▶ **EXAMPLES**   For each equation, give the number and kind of solutions.

1. $x^2 - 3x - 40 = 0$

Solution   Using $a = 1$, $b = -3$, and $c = -40$ in $b^2 - 4ac$, we have $(-3)^2 - 4(1)(-40) = 9 + 160 = 169$.

The discriminant is a perfect square. Therefore, the equation has two rational solutions.

2. $2x^2 - 3x + 4 = 0$

Solution   Using $a = 2$, $b = -3$, and $c = 4$, we have

$$b^2 - 4ac = (-3)^2 - 4(2)(4) = 9 - 32 = -23$$

The discriminant is negative, implying the equation has two complex solutions that contain i.

3. $4x^2 - 12x + 9 = 0$

Solution   Using $a = 4$, $b = -12$, and $c = 9$, the discriminant is

$$b^2 - 4ac = (-12)^2 - 4(4)(9) = 144 - 144 = 0$$

Since the discriminant is 0, the equation will have one rational solution.

4. $x^2 + 6x = 8$

Solution   We must first put the equation in standard form by adding -8 to each side. If we do so, the resulting equation is

$$x^2 + 6x - 8 = 0$$

Now we identify a, b, and c as 1, 6, and -8, respectively:

$$b^2 - 4ac = 6^2 - 4(1)(-8) = 36 + 32 = 68$$

The discriminant is a positive number, but not a perfect square. Therefore, the equation will have two irrational solutions.   ◀

▶ **EXAMPLE 5**   Find an appropriate k so that the equation $4x^2 - kx = -9$ has exactly one rational solution.

Solution   We begin by writing the equation in standard form:

$$4x^2 - kx + 9 = 0$$

Using $a = 4$, $b = -k$, and $c = 9$, we have

$$b^2 - 4ac = (-k)^2 - 4(4)(9)$$
$$= k^2 - 144$$

An equation has exactly one rational solution when the discriminant is 0. We set the discriminant equal to 0 and solve:

$$k^2 - 144 = 0$$
$$k^2 = 144$$
$$k = \pm 12$$

Choosing k to be 12 or -12 will result in an equation with one rational solution.   ◀

Building Equations from Their Solutions

Suppose we know that the solutions to an equation are $x = 3$ and $x = -2$. We can find equations with these solutions by using the zero-factor property. First, let's write our solutions as equations with 0 on the right side:

If	$x = 3$	First solution
then	$x - 3 = 0$	Add -3 to each side
and if	$x = -2$	Second solution
then	$x + 2 = 0$	Add 2 to each side

Now, since both $x - 3$ and $x + 2$ are 0, their product must be 0 also. Therefore, we can write

$$(x - 3)(x + 2) = 0 \qquad \text{Zero-factor property}$$

$$x^2 - x - 6 = 0 \qquad \text{Multiply out the left side}$$

There are many other equations that have 3 and -2 as solutions. For example, any constant multiple of $x^2 - x - 6 = 0$, such as $5x^2 - 5x - 30 = 0$, also has 3 and -2 as solutions. Similarly, any equation built from positive integer powers of the factors $x - 3$ and $x + 2$ will also have 3 and -2 as solutions. One such equation is

$$(x - 3)^2(x + 2) = 0$$

$$(x^2 - 6x + 9)(x + 2) = 0$$

$$x^3 - 4x^2 - 3x + 18 = 0$$

In mathematics, we distinguish between the solutions to this last equation and those to the equation $x^2 - x - 6 = 0$ by saying $x = 3$ is a solution of *multiplicity* 2 in the equation $x^3 - 4x^2 - 3x + 18 = 0$, and a solution of *multiplicity* 1 in the equation $x^2 - x - 6 = 0$.

▶ **EXAMPLE 6** Find an equation that has solutions $t = 5$, $t = -5$, and $t = 3$.

Solution First, we use the given solutions to write equations that have 0 on their right sides:

$$\text{If} \qquad t = 5 \qquad t = -5 \qquad t = 3$$

$$\text{then} \quad t - 5 = 0 \qquad t + 5 = 0 \qquad t - 3 = 0$$

Since $t - 5$, $t + 5$, and $t - 3$ are all 0, their product is also 0 by the zero-factor property. An equation with solutions of 5, -5, and 3 is

$$(t - 5)(t + 5)(t - 3) = 0 \qquad \text{Zero-factor property}$$

$$(t^2 - 25)(t - 3) = 0 \qquad \text{Multiply first two binomials}$$

$$t^3 - 3t^2 - 25t + 75 = 0 \qquad \text{Complete the multiplication}$$

The last line gives us an equation with solutions of 5, -5, and 3. Remember, there are many other equations with these same solutions. ◀

▶ **EXAMPLE 7** Find an equation with solutions $x = -\frac{2}{3}$ and $x = \frac{4}{5}$.

Solution The solution $x = -\frac{2}{3}$ can be rewritten as $3x + 2 = 0$ as follows:

$$x = -\frac{2}{3} \qquad \text{The first solution}$$

$$3x = -2 \qquad \text{Multiply each side by 3}$$

$$3x + 2 = 0 \qquad \text{Add 2 to each side}$$

Similarly, the solution $x = \frac{4}{5}$ can be rewritten as $5x - 4 = 0$:

$$x = \frac{4}{5} \qquad \text{The second solution}$$

$$5x = 4 \qquad \text{Multiply each side by 5}$$

$$5x - 4 = 0 \qquad \text{Add } -4 \text{ to each side}$$

Since both $3x + 2$ and $5x - 4$ are 0, their product is 0 also, giving us the equation we are looking for:

$$(3x + 2)(5x - 4) = 0 \qquad \text{Zero-factor property}$$

$$15x^2 - 2x - 8 = 0 \qquad \text{Multiplication} \qquad \blacktriangleleft$$

From Example 6 and the discussion that preceded it, we know that if $x = a$ is a solution to an equation, then $x - a$ must be a factor of the same equation. We can use this fact to solve equations that we would otherwise be unable to solve.

▶ **EXAMPLE 8**   Find all solutions to the equation

$$x^3 + 9x^2 + 26x + 24 = 0$$

if $x = -2$ is one of its solutions.

Solution   Since $x = -2$ is a solution to the equation, $x + 2$ must be a factor of the left side of the equation—meaning $x + 2$ divides the left side evenly. In Example 9 of Section 4.2, we used long division to divide $x^3 + 9x^2 + 26x + 24$ by $x + 2$ and found that quotient to be $x^2 + 7x + 12$. (You may want to look back to Section 4.2 to see what we are talking about.) Without showing that division problem again, here is how we use it to find all solutions to the equation in question:

$$x^3 + 9x^2 + 26x + 24 = 0$$

$$(x + 2)(x^2 + 7x + 12) = 0 \qquad \text{From long division by } x + 2$$

$$(x + 2)(x + 3)(x + 4) = 0 \qquad \text{Factoring the trinomial}$$

$$x + 2 = 0 \quad \text{or} \quad x + 3 = 0 \quad \text{or} \quad x + 4 = 0 \qquad \begin{array}{l}\text{Setting each}\\\text{factor to 0}\end{array}$$

$$x = -2 \quad \text{or} \quad x = -3 \quad \text{or} \quad x = -4 \qquad \begin{array}{l}\text{The three}\\\text{solutions} \quad \blacktriangleleft\end{array}$$

USING TECHNOLOGY: Graphing Calculators

....

Solving Equations

Now that we have explored the relationship between equations and their solutions, we can look at how a graphing calculator can be used in the solution process. To begin, let's solve the equation $x^2 = x + 2$ using techniques from algebra: writing it in standard form, factoring, and then setting each factor equal to zero.

$$x^2 - x - 2 = 0 \qquad \text{Standard form}$$

$$(x - 2)(x + 1) = 0 \qquad \text{Factor}$$

$$x - 2 = 0 \quad \text{or} \quad x + 1 = 0 \qquad \text{Set each factor equal to zero}$$

$$x = 2 \quad \text{or} \qquad x = -1 \qquad \text{Solve}$$

Our original equation, $x^2 = x + 2$, has two solutions, $x = 2$ and $x = -1$. To solve our equation using a graphing calculator, we need to associate it with an equation, or equations, in two variables. One way to do this is to associate the left side with the equation $y = x^2$ and the right side of the equation with $y = x + 2$. To do so, we set up the functions list in our calculator this way:

$$Y_1 = X^2$$

$$Y_2 = X + 2$$

Window: X from -5 to 5, Y from -5 to 5

Graphing these functions, in this window, will produce a graph similar to the one shown in Figure 1.

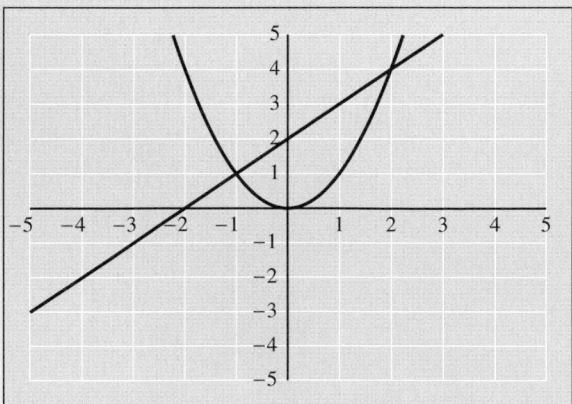

Figure 1

If we use the Trace feature to find the coordinates of the points of intersection, we find the two curves intersect at $(-1, 1)$ and $(2, 4)$. We note that the x-coordinates of these two points match the solutions to the equation $x^2 = x + 2$, which we found using algebraic techniques. This makes sense because if two graphs intersect at a point (x, y), then the coordinates of that point satisfy both equations. If a point (x, y) satisfies both $y = x^2$ and $y = x + 2$, then, for that particular point, $x^2 = x + 2$. From this we conclude that the x-coordinates of the points of intersection are solutions to our original equation. Here is a summary of what we have discovered:

CONCLUSION 1   If the graphs of two functions $y = f(x)$ and $y = g(x)$ intersect in the coordinate plane, then the x-coordinates of the points of intersection are solutions to the equation $f(x) = g(x)$.

A second method of solving our original equation $x^2 = x + 2$ graphically, requires the use of one function instead of two. To begin, we write the equation in standard form as $x^2 - x - 2 = 0$. Next, we graph the function $y = x^2 - x - 2$. The x-intercepts of the graph are the points with y-coordinates of 0. Therefore they satisfy the equation $0 = x^2 - x - 2$, which is our original equation. The graph in Figure 2 shows $Y_1 = X^2 - X - 2$ in a Window with X from -5 to 5 and Y from -5 to 5.

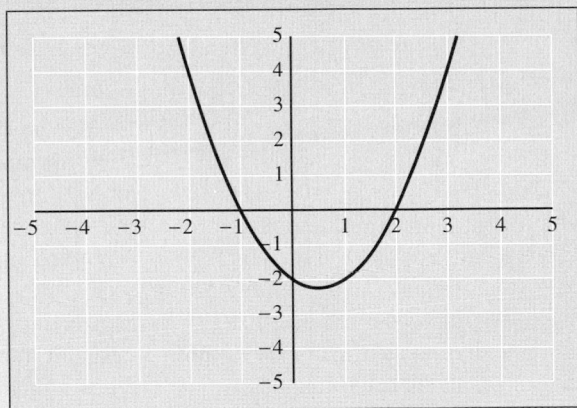

Figure 2

Using the Trace feature, we find that the x-intercepts of the graph are $x = -1$ and $x = 2$, which match the solutions to our original equation

$x^2 = x + 2$. We can summarize the relationship between solutions to an equation and the intercepts of its associated graph this way:

CONCLUSION 2   If $y = f(x)$ is a function, then any x-intercept on the graph of $y = f(x)$ is a solution to the equation $f(x) = 0$.

Solving Equations with Complex Solutions

There are limitations to using a graph to solve an equation. To illustrate, suppose we solve the equation $4x^3 - 6x^2 + 2x - 3 = 0$ by factoring:

$$4x^3 - 6x^2 + 2x - 3 = 0 \qquad \text{Original equation}$$

$$2x^2(2x - 3) + (2x - 3) = 0$$

$$(2x^2 + 1)(2x - 3) = 0 \qquad \text{Factor by grouping}$$

$$2x^2 + 1 = 0 \quad \text{or} \quad 2x - 3 = 0 \qquad \text{Set factors equal to zero}$$

$$2x^2 = -1 \quad \text{or} \quad 2x = 3$$

$$x^2 = -\frac{1}{2} \quad \text{or} \quad x = \frac{3}{2} \qquad \text{Solve the resulting equations}$$

$$x = \pm\frac{1}{\sqrt{2}}\, i = \pm\frac{\sqrt{2}}{2}\, i$$

We have three solutions, $-\dfrac{\sqrt{2}}{2}\, i$, $\dfrac{\sqrt{2}}{2}\, i$, and $\frac{3}{2}$. Figure 3 shows the graph of $y = 4x^3 - 6x^2 + 2x - 3$. As you can see, the graph crosses the x-axis exactly once at $x = \frac{3}{2}$, which we expect. The rectangular coordinate system consists of ordered pairs of *real numbers,* so our complex solutions cannot appear on the graph.

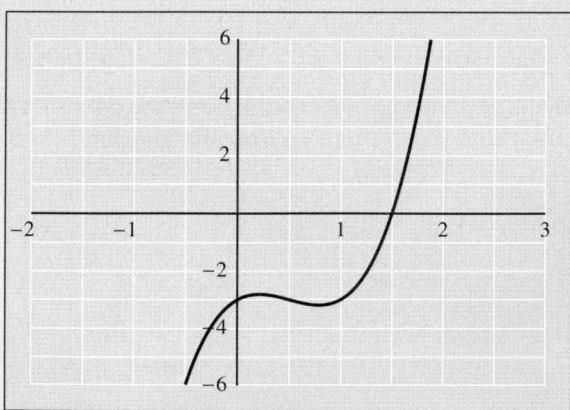

Figure 3

Every cubic equation will have at least one real number solution. If we are interested in exact values for all solutions to our equation, including complex solutions as well as irrational solutions, we can take our one real solution and, with the aid of long division and the quadratic formula, find the other solutions, just as we did in Example 8 on page 431. But how
do we find the one real solution in the first place? We use our graphing calculators.

Suppose we want to solve the equation $x^3 = 15x + 4$ completely. We can use our graphing calculator to graph the function $y = x^3 - 15x - 4$, and note that it crosses the x-axis at $x = 4$. The graph is shown in Figure 4.

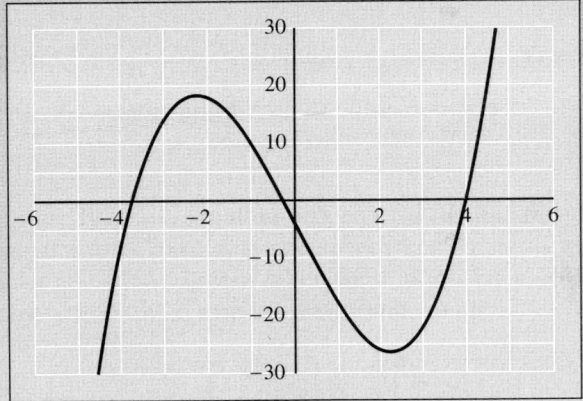

Figure 4

If we Zoom and Trace on the other two solutions, we find they are approximated by -3.732 and -0.268 to the nearest thousandth. To find exact values for these last two solutions, we reason that if $x = 4$ is a solution, then $x - 4$ must be a factor of the equation $x^3 - 15x - 4 = 0$. That is, $x - 4$ divides $x^3 - 15x - 4$ without a remainder. Here is the division problem:

$$
\begin{array}{r}
x^2 + 4x + 1 \\
x - 4 \overline{\smash{)}x^3 + 0x^2 - 15x - 4} \\
\underline{x^3 - 4x^2} \\
4x^2 - 15x \\
\underline{4x^2 - 16x} \\
x - 4 \\
\underline{x - 4} \\
0
\end{array}
$$

We find that it factors into $(x - 4)(x^2 + 4x + 1)$. This allows us to find all solutions to our original equation.

$$x^3 - 15x - 4 = 0$$

$$(x - 4)(x^2 + 4x + 1) = 0$$

$$x - 4 = 0 \quad \text{or} \quad x^2 + 4x + 1 = 0$$

$$x = 4 \quad \text{or} \qquad x = \frac{-4 \pm \sqrt{16 - 4(1)(1)}}{2(1)}$$

$$= \frac{-4 \pm \sqrt{12}}{2}$$

$$= \frac{-4 \pm 2\sqrt{3}}{2}$$

$$= -2 \pm \sqrt{3}$$

The three exact solutions to our equation are 4, $-2 + \sqrt{3}$, and $-2 - \sqrt{3}$.

PROBLEM SET 7.3

Use the discriminant to find the number and kind of solutions for each of the following equations.

1. $x^2 - 6x + 5 = 0$ **2.** $x^2 - x - 12 = 0$
3. $4x^2 - 4x = -1$ **4.** $9x^2 + 12x = -4$
5. $x^2 + x - 1 = 0$ **6.** $x^2 - 2x + 3 = 0$
7. $2y^2 = 3y + 1$ **8.** $3y^2 = 4y - 2$
9. $x^2 - 9 = 0$ **10.** $4x^2 - 81 = 0$
11. $5a^2 - 4a = 5$ **12.** $3a = 4a^2 - 5$

Determine k so that each of the following has exactly one real solution.

13. $x^2 - kx + 25 = 0$ **14.** $x^2 + kx + 25 = 0$
15. $x^2 = kx - 36$ **16.** $x^2 = kx - 49$
17. $4x^2 - 12x + k = 0$
18. $9x^2 + 30x + k = 0$
19. $kx^2 - 40x = 25$ **20.** $kx^2 - 2x = -1$
21. $3x^2 - kx + 2 = 0$ **22.** $5x^2 + kx + 1 = 0$

For each of the following problems, find an equation that has the given solutions.

23. $x = 5, x = 2$ **24.** $x = -5, x = -2$
25. $t = -3, t = 6$ **26.** $t = -4, t = 2$

27. $y = 2, y = -2, y = 4$
28. $y = 1, y = -1, y = 3$
29. $x = \frac{1}{2}, x = 3$ **30.** $x = \frac{1}{3}, x = 5$
31. $t = -\frac{3}{4}, t = 3$ **32.** $t = -\frac{4}{5}, t = 2$
33. $x = 3, x = -3, x = \frac{5}{6}$
34. $x = 5, x = -5, x = \frac{2}{3}$
35. $a = -\frac{1}{2}, a = \frac{3}{5}$
36. $a = -\frac{1}{3}, a = \frac{4}{7}$
37. $x = -\frac{2}{3}, x = \frac{2}{3}, x = 1$
38. $x = -\frac{4}{5}, x = \frac{4}{5}, x = -1$
39. $x = 2, x = -2, x = 3, x = -3$
40. $x = 1, x = -1, x = 5, x = -5$
41. Find an equation that has a solution of $x = 3$ of multiplicity 1 and a solution $x = -5$ of multiplicity 2.
42. Find an equation that has a solution of $x = 5$ of multiplicity 1 and a solution $x = -3$ of multiplicity 2.
43. Find an equation that has solutions $x = 3$ and $x = -3$, both of multiplicity 2.
44. Find an equation that has solutions $x = 4$ and $x = -4$, both of multiplicity 2.

45. Find all solutions to $x^3 + 6x^2 + 11x + 6 = 0$ if $x = -3$ is one of its solutions.

46. Find all solutions to $x^3 + 10x^2 + 29x + 20 = 0$ if $x = -4$ is one of its solutions.

47. One solution to $y^3 + 5y^2 - 2y - 24 = 0$ is $y = -3$. Find all solutions.

48. One solution to $y^3 + 3y^2 - 10y - 24 = 0$ is $y = -2$. Find all solutions.

49. If $x = 3$ is one solution to $x^3 - 5x^2 + 8x = 6$, find the other solutions.

50. If $x = 2$ is one solution to $x^3 - 6x^2 + 13x = 10$, find the other solutions.

51. Find all solutions to $t^3 = 13t^2 - 65t + 125$ if $t = 5$ is one of the solutions.

52. Find all solutions to $t^3 = 8t^2 - 25t + 26$ if $t = 2$ is one of the solutions.

Review Problems

The problems that follow review material we covered in Section 6.2. Reviewing these problems will help you with the next section.

Multiply.

53. $a^4(a^{3/2} - a^{1/2})$
54. $(a^{1/2} - 5)(a^{1/2} + 3)$
55. $(x^{3/2} - 3)^2$
56. $(x^{1/2} - 8)(x^{1/2} + 8)$

Divide.

57. $\dfrac{30x^{3/4} - 25x^{5/4}}{5x^{1/4}}$

58. $\dfrac{45x^{5/3} y^{7/3} - 36x^{8/3}y^{4/3}}{9x^{2/3} y^{1/3}}$

59. Factor $5(x - 3)^{1/2}$ from $10(x - 3)^{3/2} - 15(x - 3)^{1/2}$.

60. Factor $2(x + 1)^{1/3}$ from $8(x + 1)^{4/3} - 2(x + 1)^{1/3}$.

Factor each of the following as if they were trinomials.

61. $2x^{2/3} - 11x^{1/3} + 12$
62. $9x^{2/3} + 12x^{1/3} + 4$

One Step Further

63. Find all solutions to $x^4 + x^3 - x^2 + x - 2 = 0$ if $x = -2$ is one solution.

64. Find all solutions to $x^4 - x^3 + 2x^2 - 4x - 8 = 0$ if $x = 2$ is one solution.

65. Find all solutions to $x^3 + 3ax^2 + 3a^2x + a^3 = 0$ if $x = -a$ is one solution.

66. Find all solutions to $x^3 + 6ax^2 + 12a^2x + 8a^3 = 0$ if $x = -2a$ is one solution.

USING TECHNOLOGY

Find all solutions to the following equations. Solve using algebra and by graphing. If rounding is necessary, round to the nearest hundredth.

67. $x^2 = 4x + 5$ **68.** $4x^2 = 8x + 5$
69. $x^2 - 1 = 2x$ **70.** $4x^2 - 1 = 4x$

Find all solutions to each equation. If rounding is necessary, round to the nearest hundredth.

71. $2x^3 - x^2 - 2x + 1 = 0$

72. $3x^3 - 2x^2 - 3x + 2 = 0$
73. $2x^3 + 2 = x^2 + 4x$
74. $3x^3 - 9x = 2x^2 - 6$

Each equation below has only one real solution. Find it by graphing. Then use long division and factoring or the quadratic formula to find the remaining solutions.

75. $3x^3 - 8x^2 + 10x - 4 = 0$
76. $10x^3 + 6x^2 + x - 2 = 0$

Equations Quadratic in Form

We are now in a position to put our knowledge of quadratic equations to work to solve a variety of equations.

▶ **EXAMPLE 1**   Solve $(x + 3)^2 - 2(x + 3) - 8 = 0$.

Solution   We can see that this equation is quadratic in form by replacing $x + 3$ with another variable, say y. Replacing $x + 3$ with y we have

$$y^2 - 2y - 8 = 0$$

We can solve this equation by factoring the left side and then setting each factor equal to 0.

$$y^2 - 2y - 8 = 0$$
$$(y - 4)(y + 2) = 0 \qquad \text{Factor}$$
$$y - 4 = 0 \quad \text{or} \quad y + 2 = 0 \qquad \text{Set factors to 0}$$
$$y = 4 \quad \text{or} \qquad y = -2$$

Since our original equation was written in terms of the variable x, we would like our solutions in terms of x also. Replacing y with $x + 3$ and then solving for x we have

$$x + 3 = 4 \quad \text{or} \quad x + 3 = -2$$
$$x = 1 \quad \text{or} \qquad x = -5$$

The solutions to our original equation are 1 and -5.

The method we have just shown lends itself well to other types of equations that are quadratic in form, as we will see. In this example, however, there is another method that works just as well. Let's solve our original equation again, but this time, let's begin by expanding $(x + 3)^2$ and $2(x + 3)$.

$$(x + 3)^2 - 2(x + 3) - 8 = 0$$
$$x^2 + 6x + 9 - 2x - 6 - 8 = 0 \qquad \text{Multiply}$$
$$x^2 + 4x - 5 = 0 \qquad \text{Combine similar terms}$$
$$(x - 1)(x + 5) = 0 \qquad \text{Factor}$$
$$x - 1 = 0 \quad \text{or} \quad x + 5 = 0 \qquad \text{Set factors to 0}$$
$$x = 1 \quad \text{or} \qquad x = -5$$

As you can see, either method produces the same result.   ◀

▶ **EXAMPLE 2**   Solve $4x^4 + 7x^2 = 2$.

Solution   This equation is quadratic in x^2. We can make it easier to look at by using the substitution $y = x^2$. (The choice of the letter y is arbitrary. We could just as easily use the substitution $m = x^2$.) Making the substitution $y = x^2$ and then solving the resulting equation we have

$$4y^2 + 7y = 2$$

$$4y^2 + 7y - 2 = 0 \qquad \text{Standard form}$$

$$(4y - 1)(y + 2) = 0 \qquad \text{Factor}$$

$$4y - 1 = 0 \quad \text{or} \quad y + 2 = 0 \qquad \text{Set factors to 0}$$

$$y = \frac{1}{4} \quad \text{or} \qquad y = -2$$

Now we replace y with x^2 in order to solve for x:

$$x^2 = \frac{1}{4} \qquad \text{or} \quad x^2 = -2$$

$$x = \pm\sqrt{\frac{1}{4}} \quad \text{or} \quad x = \pm\sqrt{-2} \qquad \text{Theorem 7.1}$$

$$x = \pm\frac{1}{2} \qquad \text{or} \quad x = \pm i\sqrt{2}$$

The solution set is $\{\frac{1}{2}, -\frac{1}{2}, i\sqrt{2}, -i\sqrt{2}\}$.   ◀

▶ **EXAMPLE 3**   Solve for x: $x + \sqrt{x} - 6 = 0$.

Solution   To see that this equation is quadratic in form, we have to notice that $(\sqrt{x})^2 = x$. That is, the equation can be rewritten as

$$(\sqrt{x})^2 + \sqrt{x} - 6 = 0$$

Replacing $\sqrt{x}$ with y and solving as usual we have

$$y^2 + y - 6 = 0$$

$$(y + 3)(y - 2) = 0$$

$$y + 3 = 0 \quad \text{or} \quad y - 2 = 0$$

$$y = -3 \quad \text{or} \qquad y = 2$$

Again, to find x we replace y with $\sqrt{x}$ and solve:

$$\sqrt{x} = -3 \quad \text{or} \quad \sqrt{x} = 2$$

$$x = 9 \qquad\qquad x = 4 \qquad \text{Square both sides of each equation}$$

Since we squared both sides of each equation, we have the possibility of obtaining extraneous solutions. We have to check both solutions in our original equation.

When $x = 9$ When $x = 4$

the equation $x + \sqrt{x} - 6 = 0$ the equation $x + \sqrt{x} - 6 = 0$

becomes $9 + \sqrt{9} - 6 \overset{?}{=} 0$ becomes $4 + \sqrt{4} - 6 \overset{?}{=} 0$

$9 + 3 - 6 \overset{?}{=} 0$ $4 + 2 - 6 \overset{?}{=} 0$

$6 \neq 0$ $0 = 0$

This means 9 is extraneous.   This means 4 is a solution.

The only solution to the equation $x + \sqrt{x} - 6 = 0$ is $x = 4$. ◀

We should note here that the two possible solutions, 9 and 4, to the equation in Example 3 can be obtained by another method. Instead of substituting for $\sqrt{x}$, we can isolate it on one side of the equation and then square both sides to clear the equation of radicals.

$x + \sqrt{x} - 6 = 0$

$\sqrt{x} = -x + 6$ Isolate $\sqrt{x}$

$x = x^2 - 12x + 36$ Square both sides

$0 = x^2 - 13x + 36$ Add $-x$ to both sides

$0 = (x - 4)(x - 9)$ Factor

$x - 4 = 0$ or $x - 9 = 0$

$x = 4$ $x = 9$

We obtain the same two possible solutions. Since we squared both sides of the equation to find them, we would have to check each one in the original equation. As was the case in Example 3, only $x = 4$ is a solution; $x = 9$ is extraneous.

▶ **EXAMPLE 4** If an object is tossed into the air with an upward velocity of 12 feet/second from the top of a building h feet high, the time it takes for the object to hit the ground below is given by the formula

$$16t^2 - 12t - h = 0$$

Solve this formula for t.

Solution The formula is in standard form and is quadratic in t. The coefficients a, b, and c that we need to apply to the quadratic formula are $a = 16$, $b = -12$, and $c = -h$. Substituting these quantities into the quadratic formula, we have

$$t = \frac{12 \pm \sqrt{144 - 4(16)(-h)}}{2(16)}$$

$$t = \frac{12 \pm \sqrt{144 + 64h}}{32}$$

We can factor the perfect square 16 from the two terms under the radical and simplify our radical somewhat:

$$t = \frac{12 \pm \sqrt{16(9 + 4h)}}{32}$$

$$t = \frac{12 \pm 4\sqrt{9 + 4h}}{32}$$

Now we can reduce to lowest terms by factoring a 4 from the numerator and denominator:

$$t = \frac{\cancel{4}(3 \pm \sqrt{9 + 4h})}{\cancel{4} \cdot 8}$$

$$t = \frac{3 \pm \sqrt{9 + 4h}}{8}$$

If we were given a value of h, we would find that one of the solutions to this last formula would be a negative number. Since time is always measured in positive units, we wouldn't use that solution. ◀

More about the Golden Ratio

In Section 6.1 we derived the golden ratio $\dfrac{1 + \sqrt{5}}{2}$ by finding the ratio of length to width for a golden rectangle. The golden ratio was actually discovered before the golden rectangle, by the Greeks who lived before Euclid. The early Greeks found the golden ratio by dividing a line segment into two parts so that the ratio of the shorter part to the longer part was the same as the ratio of the longer part to the whole segment. When they divided a line segment in this manner they said it was divided in "extreme and mean ratio." Figure 1 illustrates a line segment divided this way.

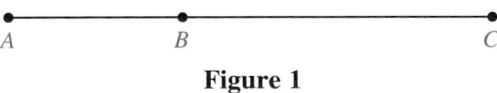

Figure 1

If point B divides segment AC in "extreme and mean ratio," then

$$\frac{\text{Length of shorter segment}}{\text{Length of longer segment}} = \frac{\text{length of longer segment}}{\text{length of whole segment}}$$

$$\frac{AB}{BC} = \frac{BC}{AC}$$

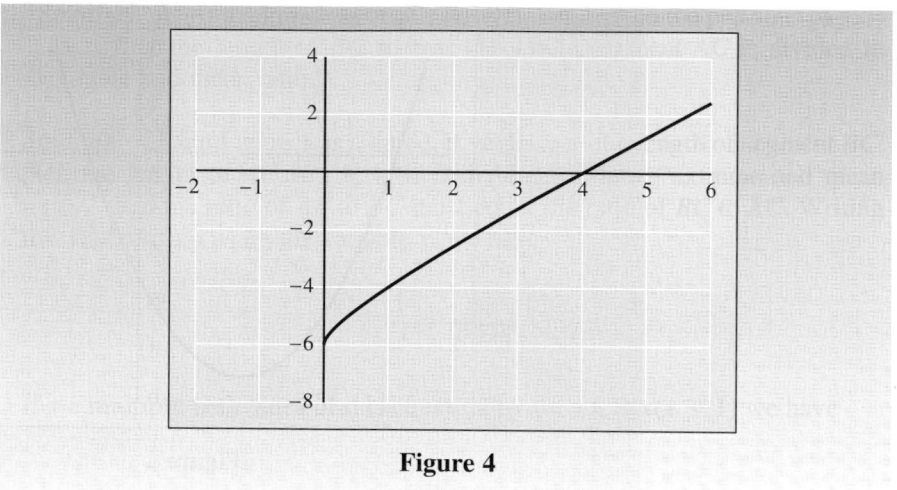

Figure 4

PROBLEM SET 7.4

Solve each equation.

1. $(x - 3)^2 + 3(x - 3) + 2 = 0$
2. $(x + 4)^2 - (x + 4) - 6 = 0$
3. $2(x + 4)^2 + 5(x + 4) - 12 = 0$
4. $3(x - 5)^2 + 14(x - 5) - 5 = 0$
5. $x^4 - 6x^2 - 27 = 0$ **6.** $x^4 + 2x^2 - 8 = 0$
7. $x^4 + 9x^2 = -20$ **8.** $x^4 - 11x^2 = -30$
9. $(2a - 3)^2 - 9(2a - 3) = -20$
10. $(3a - 2)^2 + 2(3a - 2) = 3$
11. $2(4a + 2)^2 = 3(4a + 2) + 20$
12. $6(2a + 4)^2 = (2a + 4) + 2$
13. $6t^4 = -t^2 + 5$
14. $3t^4 = -2t^2 + 8$
15. $9x^4 - 49 = 0$
16. $25x^4 - 9 = 0$

Solve each of the following equations. Remember, if you square both sides of an equation in the process of solving it, you have to check all solutions in the original equation.

17. $x - 7\sqrt{x} + 10 = 0$ **18.** $x - 6\sqrt{x} + 8 = 0$
19. $t - 2\sqrt{t} - 15 = 0$ **20.** $t - 3\sqrt{t} - 10 = 0$
21. $6x + 11\sqrt{x} = 35$ **22.** $2x + \sqrt{x} = 15$
23. $(a - 2) - 11\sqrt{a - 2} + 30 = 0$
24. $(a - 3) - 9\sqrt{a - 3} + 20 = 0$

25. $(2x + 1) - 8\sqrt{2x + 1} + 15 = 0$
26. $(2x - 3) - 7\sqrt{2x - 3} + 12 = 0$

Applying the Concepts

27. An object is tossed into the air with an upward velocity of 8 feet/second from the top of a building h feet high. The time it takes for the object to hit the ground below is given by the formula $16t^2 - 8t - h = 0$. Solve this formula for t.

28. An object is tossed into the air with an upward velocity of 6 feet/second from the top of a building h feet high. The time it takes for the object to hit the ground below is given by the formula $16t^2 - 6t - h = 0$. Solve this formula for t.

29. An object is tossed into the air with an upward velocity of v feet/second from the top of a building 20 feet high. The time it takes for the object to hit the ground below is given by the formula $16t^2 - vt - 20 = 0$. Solve this formula for t.

30. An object is tossed into the air with an upward velocity of v feet/second from the top of a building 40 feet high. The time it takes for the object to hit the ground below is given by the formula $16t^2 - vt - 40 = 0$. Solve this formula for t.

Use Figure 1 from this section as a guide to working the following problems.

31. If AB in Figure 1 is 4 inches, and B divides AC in "extreme and mean ratio," find BC and then show that BC is four times the golden ratio.

32. If AB in Figure 1 is $\frac{1}{2}$ inch, and B divides AC in "extreme and mean ratio," find BC and then show that BC is half the golden ratio.

33. If AB in Figure 1 is 2 inches, and B divides AC in "extreme and mean ratio," find BC and then show that the ratio of BC to AB is the golden ratio.

34. If AB in Figure 1 is $\frac{1}{2}$ inch, and B divides AC in "extreme and mean ratio," find BC and then show that the ratio of BC to AB is the golden ratio.

35. Solve the formula $16t^2 - vt - h = 0$ for t.

36. Solve the formula $16t^2 + vt + h = 0$ for t.

37. Solve the formula $kx^2 + 8x + 4 = 0$ for x.

38. Solve the formula $k^2x^2 + kx + 4 = 0$ for x.

39. Solve $x^2 + 2xy + y^2 = 0$ for x by using the quadratic formula with $a = 1$, $b = 2y$, and $c = y^2$.

40. Solve $x^2 - 2xy + y^2 = 0$ for x.

Review Problems

The problems that follow review material we covered in Sections 6.4 and 6.5.

Combine, if possible. [6.4]

41. $5\sqrt{7} - 2\sqrt{7}$ **42.** $6\sqrt{2} - 9\sqrt{2}$

43. $\sqrt{18} - \sqrt{8} + \sqrt{32}$ **44.** $\sqrt{50} + \sqrt{72} - \sqrt{8}$

45. $9x\sqrt{20x^3y^2} + 7y\sqrt{45x^5}$

46. $5x^2\sqrt{27xy^3} - 6y\sqrt{12x^5y}$

Multiply. [6.5]

47. $(\sqrt{5} - 2)(\sqrt{5} + 8)$

48. $(2\sqrt{3} - 7)(2\sqrt{3} + 7)$

49. $(\sqrt{x} + 2)^2$ **50.** $(3 - \sqrt{x})(3 + \sqrt{x})$

Rationalize the denominator. [6.5]

51. $\dfrac{\sqrt{7}}{\sqrt{7} - 2}$

52. $\dfrac{\sqrt{5} - \sqrt{2}}{\sqrt{5} + \sqrt{2}}$

One Step Further

Find the x- and y-intercepts.

53. $y = x^3 - 4x$

54. $y = x^4 - 10x^2 + 9$

55. $y = 3x^3 + x^2 - 27x - 9$

56. $y = 2x^3 + x^2 - 8x - 4$

57. The graph of $y = 2x^3 - 7x^2 - 5x + 4$ crosses the x-axis at $x = 4$. Where else does it cross the x-axis?

58. The graph of $y = 6x^3 + x^2 - 12x + 5$ crosses the x-axis at $x = 1$. Where else does it cross the x-axis?

Research Project 8

If you successfully completed Research Project 7 in the previous chapter, then you used number sequences and inductive reasoning to conclude that the continued fraction

$$1 + \cfrac{1}{1 + \cfrac{1}{1 + \cfrac{1}{1 + \cdots}}}$$

is equal to the golden ratio.

The same conclusion can be found with the quadratic formula by using the following conditional statement.

$$\text{If } x = 1 + \cfrac{1}{1 + \cfrac{1}{1 + \cfrac{1}{1 + \cdots}}}, \text{ then } x = 1 + \frac{1}{x}$$

Work with this conditional statement until you see that it is true. Then solve the equation $x = 1 + \frac{1}{x}$. Write an essay in which you explain in your own words why the conditional statement is true. Then show the details of the solution to the equation $x = 1 + \frac{1}{x}$. The message that you want to get across in your essay is that the continued fraction shown above is actually the same as the golden ratio.

Quadratic Inequalities

Quadratic inequalities in one variable are inequalities of the form

$$ax^2 + bx + c < 0 \qquad ax^2 + bx + c > 0$$
$$ax^2 + bx + c \leq 0 \qquad ax^2 + bx + c \geq 0$$

where a, b, and c are constants, with $a \neq 0$. The technique we will use to solve inequalities of this type involves graphing. Suppose, for example, we wish to find the solution set for the inequality $x^2 - x - 6 > 0$. We begin by factoring the left side to obtain

$$(x - 3)(x + 2) > 0$$

We have two real numbers $x - 3$ and $x + 2$ whose product $(x - 3)(x + 2)$ is greater than zero. That is, their product is positive. The only way the product can be positive is either if both factors, $(x - 3)$ and $(x + 2)$, are positive or if they are both negative. To help visualize where $x - 3$ is positive and where it is negative, we draw a real number line and label it accordingly:

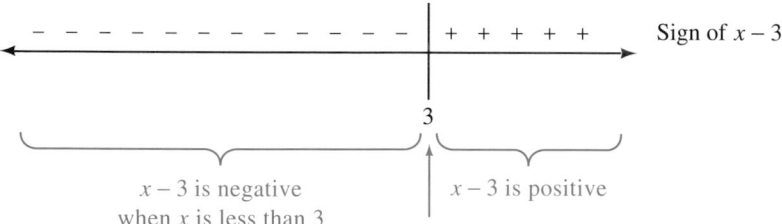

Here is a similar diagram showing where the factor $x + 2$ is positive and where it is negative:

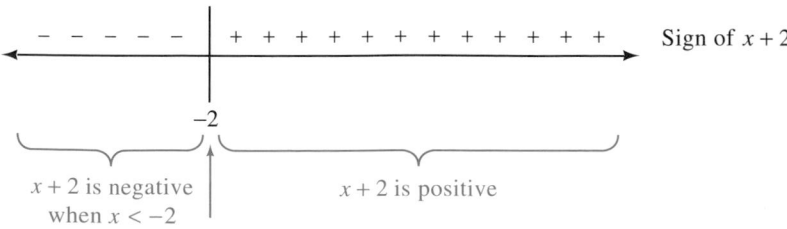

Drawing the two number lines together, we have

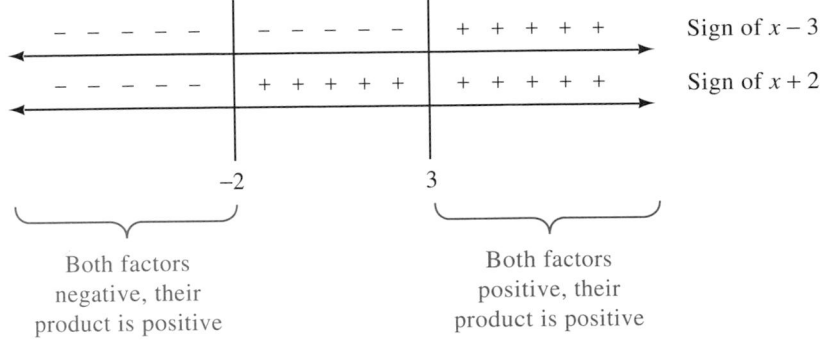

We can see from the diagram above that the graph of the solution to $x^2 - x - 6 > 0$ is

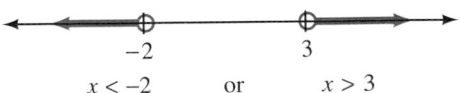

$x < -2 \qquad$ or $\qquad x > 3$

▶ **EXAMPLE 1** Solve for x: $x^2 - 2x - 8 \leq 0$.

Solution We begin by factoring:

$$x^2 - 2x - 8 \leq 0$$

$$(x - 4)(x + 2) \leq 0$$

The product $(x - 4)(x + 2)$ is negative or zero. The factors must have opposite signs. We draw a diagram showing where each factor is positive and where each factor is negative:

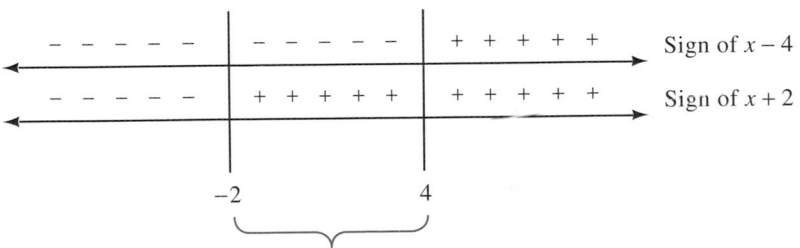

From the diagram we have the graph of the solution set:

$-2 \leq x \leq 4$

▶ **EXAMPLE 2** Solve for x: $6x^2 - x \geq 2$.

Solution

$$6x^2 - x \geq 2$$

$$6x^2 - x - 2 \geq 0 \leftarrow \text{Standard form}$$

$$(3x - 2)(2x + 1) \geq 0$$

The product is positive, so the factors must agree in sign. Here is the diagram showing where that occurs:

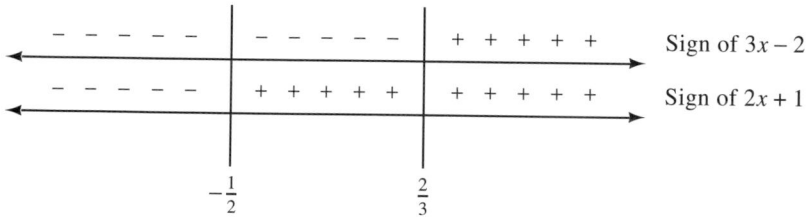

Since the factors agree in sign below $-\frac{1}{2}$ and above $\frac{2}{3}$, the graph of the solution set is

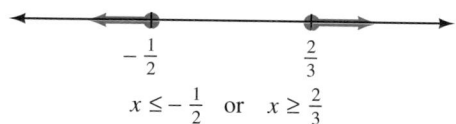

$$x \leq -\frac{1}{2} \quad \text{or} \quad x \geq \frac{2}{3}$$

◀

▶ **EXAMPLE 3**   Solve $x^2 - 6x + 9 \geq 0$.

Solution

$$x^2 - 6x + 9 \geq 0$$

$$(x - 3)^2 \geq 0$$

This is a special case in which both factors are the same. Since $(x - 3)^2$ is always positive or zero, the solution set is all real numbers. That is, any real number that is used in place of x in the original inequality will produce a true statement.

◀

▶ **EXAMPLE 4**   Solve $\dfrac{x - 4}{x + 1} \leq 0$.

Solution   The inequality indicates that the quotient of $(x - 4)$ and $(x + 1)$ is negative or 0 (less than or equal to 0). We can use the same reasoning we used to solve the first three examples, because quotients are positive or negative under the same conditions that products are positive or negative. Here is the diagram that shows where each factor is positive and where each factor is negative.

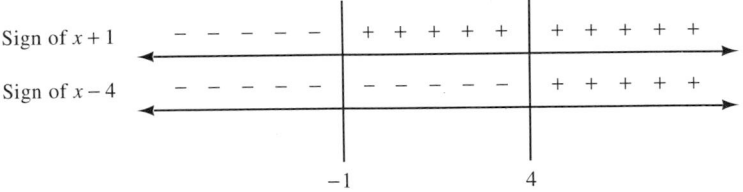

The region between -1 and 4 is where the solutions lie, since the original inequality indicates the quotient $\dfrac{x-4}{x+1}$ is negative, and between -1 and 4 the factors have opposite signs, making the quotient negative. The solution set and its graph are

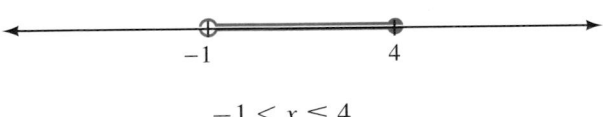

$$-1 < x \le 4$$

Notice the left endpoint is open—that is, not included in the solution set— because $x = -1$ would make the original inequality undefined. It is important to check all endpoints of solution sets to inequalities that involve rational expressions. ◀

▶ **EXAMPLE 5** Solve $\dfrac{3}{x-2} - \dfrac{2}{x-3} > 0$.

Solution We begin by adding the two rational expressions on the left side. The common denominator is $(x-2)(x-3)$:

$$\frac{3}{x-2} \cdot \frac{(x-3)}{(x-3)} - \frac{2}{x-3} \cdot \frac{(x-2)}{(x-2)} > 0$$

$$\frac{3x - 9 - 2x + 4}{(x-2)(x-3)} > 0$$

$$\frac{x-5}{(x-2)(x-3)} > 0$$

This time the quotient involves three factors. Here is the diagram that shows the signs of the three factors.

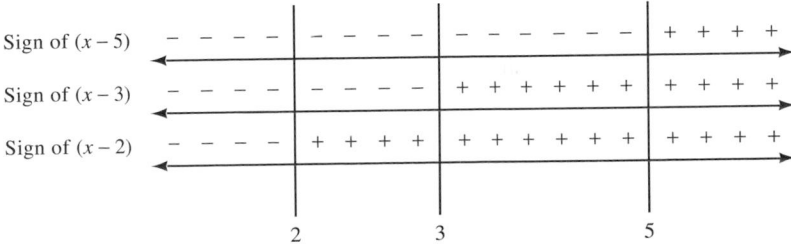

The original inequality indicates that the quotient is positive. In order for this to happen, of the three factors either all must be positive, or exactly two must be negative. Looking back to our diagram, we see the regions that satisfy these conditions are between 2 and 3 or above 5. Here is our solution set.

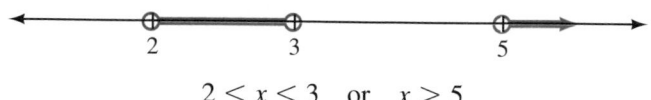

$$2 < x < 3 \quad \text{or} \quad x > 5 \qquad \blacktriangleleft$$

▶ **EXAMPLE 6**   Find the values of x for which y is positive and the values of x for which y is negative if $y = x^2 - 9$.

Solution   Since y is equal to $x^2 - 9$, y will be positive whenever $x^2 - 9$ is positive, and y will be negative whenever $x^2 - 9$ is negative. Since $x^2 - 9 = (x + 3)(x - 3)$, a sign chart for the product $(x + 3)(x - 3)$ will give us the intervals over which that quantity is positive and the intervals over which it is negative:

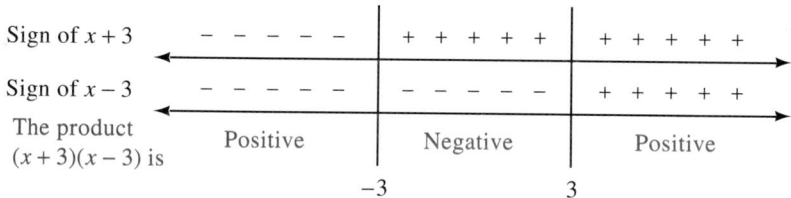

Since $y = x^2 - 9 = (x + 3)(x - 3)$, y is positive for $x < -3$ and for $x > 3$; y is negative for x between -3 and 3, which is the interval $-3 < x < 3$.   ◀

P R O B L E M   S E T   7 . 5

Solve each of the following inequalities and graph the solution set.

1. $x^2 + x - 6 > 0$ **2.** $x^2 + x - 6 < 0$
3. $x^2 - x - 12 \leq 0$ **4.** $x^2 - x - 12 \geq 0$
5. $x^2 + 5x \geq -6$ **6.** $x^2 - 5x > 6$
7. $6x^2 < 5x - 1$ **8.** $4x^2 \geq -5x + 6$
9. $x^2 - 9 < 0$ **10.** $x^2 - 16 \geq 0$
11. $4x^2 - 9 \geq 0$ **12.** $9x^2 - 4 < 0$
13. $2x^2 - x - 3 < 0$ **14.** $3x^2 + x - 10 \geq 0$
15. $x^2 - 4x + 4 \geq 0$ **16.** $x^2 - 4x + 4 < 0$
17. $x^2 - 10x + 25 < 0$
18. $x^2 - 10x + 25 > 0$
19. $(x - 2)(x - 3)(x - 4) > 0$
20. $(x - 2)(x - 3)(x - 4) < 0$
21. $(x + 1)(x + 2)(x + 3) \leq 0$

22. $(x + 1)(x + 2)(x + 3) \geq 0$

23. $\dfrac{x - 1}{x + 4} \leq 0$ **24.** $\dfrac{x + 4}{x - 1} \leq 0$

25. $\dfrac{3x}{x + 6} - \dfrac{8}{x + 6} < 0$

26. $\dfrac{5x}{x + 1} - \dfrac{3}{x + 1} < 0$

27. $\dfrac{4}{x - 6} + 1 > 0$

28. $\dfrac{2}{x - 3} + 1 \geq 0$

29. $\dfrac{x - 2}{(x + 3)(x - 4)} < 0$

30. $\dfrac{x-1}{(x+2)(x-5)} < 0$

31. $\dfrac{2}{x-4} - \dfrac{1}{x-3} > 0$

32. $\dfrac{4}{x+3} - \dfrac{3}{x+2} > 0$

33. $\dfrac{x+7}{2x+12} + \dfrac{6}{x^2-36} \le 0$

34. $\dfrac{x+1}{2x-2} - \dfrac{2}{x^2-1} \le 0$

Find the intervals over which y is positive and the intervals over which y is negative if

35. $y = x^2 - 25$ **36.** $y = x^2 - 36$
37. $y = x^2 - 6x + 5$ **38.** $y = x^2 - 2x - 3$
39. $y = x^2 + 4$ **40.** $y = x^2 + 49$

Applying the Concepts

41. The length of a rectangle is 3 inches more than twice the width. If the area is to be at least 44 square inches, what are the possibilities for the width?
42. The length of a rectangle is 5 inches less than three times the width. If the area is to be less than 12 square inches, what are the possibilities for the width?
43. A manufacturer of portable radios knows that the weekly revenue produced by selling x radios is given by the equation $R = 1{,}300p - 100p^2$ where p is the price of each radio. What price should she charge for each radio if she wants her weekly revenue to be at least $4,000?

44. A manufacturer of small calculators knows that the weekly revenue produced by selling x calculators is given by the equation $R = 1{,}700p - 100p^2$ where p is the price of each calculator. What price should be charged for each calculator if the revenue is to be at least $7,000 each week?

Review Problems

The problems that follow review material we covered in Sections 5.1 and 6.6. Reviewing these problems will help you understand parts of the next section.

Graph each line by finding the x- and y-intercepts. [5.1]

45. $5x - 3y = 15$ **46.** $-2x + 4y = 6$
47. $\frac{1}{2}x + \frac{1}{3}y = 1$ **48.** $\frac{2}{3}x - \frac{4}{3}y = 4$

Solve each equation. [6.6]

49. $\sqrt{3t-1} = 2$ **50.** $\sqrt{4t+5} + 7 = 3$
51. $\sqrt{x+3} = x - 3$ **52.** $\sqrt{x+3} = \sqrt{x} - 3$

Graph each equation. [6.6]

53. $y = \sqrt[3]{x-1}$ **54.** $y = \sqrt[3]{x} - 1$

One Step Further

Graph the solution set for each inequality.

55. $x^2 - 2x - 1 < 0$
56. $x^2 - 6x + 7 < 0$
57. $x^2 - 8x + 13 > 0$
58. $x^2 - 10x + 18 > 0$

SECTION

7.6   ## Graphing Parabolas

Recall from Section 5.5 that the solution set to equations such as

$$y = x^2 - 3$$

will consist of ordered pairs. One method of graphing the solution set is to find a number of ordered pairs that satisfy the equation and to graph them. We can obtain some ordered pairs that are solutions to $y = x^2 - 3$ by use of a table as follows:

x	$y = x^2 - 3$	y	Solutions
-3	$y = (-3)^2 - 3 = 9 - 3 = 6$	6	$(-3, 6)$
-2	$y = (-2)^2 - 3 = 4 - 3 = 1$	1	$(-2, 1)$
-1	$y = (-1)^2 - 3 = 1 - 3 = -2$	-2	$(-1, -2)$
0	$y = 0^2 - 3 = 0 - 3 = -3$	-3	$(0, -3)$
1	$y = 1^2 - 3 = 1 - 3 = -2$	-2	$(1, -2)$
2	$y = 2^2 - 3 = 4 - 3 = 1$	1	$(2, 1)$
3	$y = 3^2 - 3 = 9 - 3 = 6$	6	$(3, 6)$

Graphing these solutions and then connecting them with a smooth curve, we have the graph of $y = x^2 - 3$. (See Figure 1.)

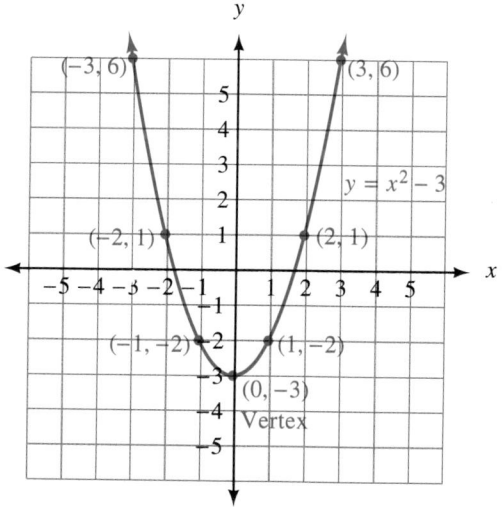

Figure 1

This graph is an example of a **parabola.** All equations of the form $y = ax^2 + bx + c, a \neq 0$, have parabolas for graphs.

Although it is always possible to graph parabolas by making a table of values of x and y that satisfy the equation, there are other methods that are faster and, in some cases, more accurate.

The important points associated with the graph of a parabola are the highest (or lowest) point on the graph and the x-intercepts. The y-intercepts can also be useful.

Intercepts for Parabolas

The graph of the equation $y = ax^2 + bx + c$ will cross the y-axis at $y = c$, since substituting $x = 0$ into $y = ax^2 + bx + c$ yields $y = c$.

Since the graph will cross the x-axis when $y = 0$, the x-intercepts are those values of x that are solutions to the quadratic equation $0 = ax^2 + bx + c$.

The Vertex of a Parabola

The highest or lowest point on a parabola is called the **vertex**. The vertex for the graph of $y = ax^2 + bx + c$ will always occur when

$$x = \frac{-b}{2a}$$

To see this, we must transform the right side of $y = ax^2 + bx + c$ into an expression that contains x in just one of its terms. This is accomplished by completing the square on the first two terms. Here is what it looks like:

$$y = ax^2 + bx + c$$

$$y = a\left(x^2 + \frac{b}{a}x \right) + c$$

$$y = a\left[x^2 + \frac{b}{a}x + \left(\frac{b}{2a} \right)^2 \right] + c - a\left(\frac{b}{2a} \right)^2$$

$$y = a\left(x + \frac{b}{2a} \right)^2 + \frac{4ac - b^2}{4a}$$

It may not look like it, but this last line indicates that the vertex of the graph of $y = ax^2 + bx + c$ has an x-coordinate of $\dfrac{-b}{2a}$. Since a, b, and c are constants, the only quantity that is varying in the last expression is the x in $\left(x + \dfrac{b}{2a} \right)^2$.

Since the quantity $\left(x + \dfrac{b}{2a} \right)^2$ is the square of $x + \dfrac{b}{2a}$, the smallest it will ever be is 0, and that will happen when $x = \dfrac{-b}{2a}$.

We can use the vertex point along with the x- and y-intercepts to sketch the graph of any equation of the form $y = ax^2 + bx + c$. Here is a summary of the preceding information.

Graphing Parabolas

The graph of $y = ax^2 + bx + c$, $a \neq 0$, will have

1. a y-intercept at $y = c$
2. x-intercepts (if they exist) at

$$x = \frac{-b \pm \sqrt{b^2 - 4ac}}{2a}$$

3. a vertex when $x = \dfrac{-b}{2a}$

▶ **EXAMPLE 1**   Sketch the graph of $y = x^2 - 6x + 5$.

Solution   To find the x-intercepts we let $y = 0$ and solve for x:

$$0 = x^2 - 6x + 5$$
$$0 = (x - 5)(x - 1)$$
$$x = 5 \quad \text{or} \quad x = 1$$

To find the coordinates of the vertex we first find

$$x = \frac{-b}{2a} = \frac{-(-6)}{2(1)} = 3$$

The x-coordinate of the vertex is 3. To find the y-coordinate we substitute 3 for x in our original equation:

$$y = 3^2 - 6(3) + 5 = 9 - 18 + 5 = -4$$

The graph crosses the x-axis at 1 and 5 and has its vertex at $(3, -4)$. Plotting these points and connecting them with a smooth curve, we have the graph shown in Figure 2. The graph is a parabola that opens up, so we say the graph is **concave up.** The vertex is the lowest point on the graph. (Note that the graph crosses the y-axis at 5, which is the value of y we obtain when we let $x = 0$.)

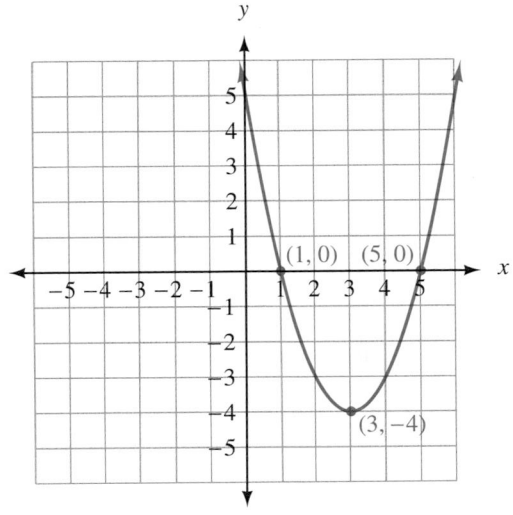

Figure 2

Finding the Vertex by Completing the Square

Another way to locate the vertex of the parabola in Example 1 is by completing the square on the first two terms on the right side of the equation $y = x^2 - 6x + 5$. In this case, we would do so by adding 9 to and subtracting 9 from the right side of the equation. This amounts to adding 0 to the equation, so we know we haven't changed its solutions. This is what it looks like:

$$y = (x^2 - 6x \qquad) + 5$$
$$y = (x^2 - 6x + \mathbf{9}) + 5 - \mathbf{9}$$
$$y = (x - 3)^2 - 4$$

You may have to look at this last equation awhile to see this, but when $x = 3$, then $y = (x - 3)^2 - 4 = 0^2 - 4 = -4$ is the smallest y will ever be. And that is why the vertex is at $(3, -4)$. As a matter of fact, this is the same kind of reasoning we used when we derived the formula $x = \dfrac{-b}{2a}$ for the x-coordinate of the vertex.

▶ **EXAMPLE 2** Graph $y = -x^2 - 2x + 3$.

Solution To find the x-intercepts, we let $y = 0$:

$$0 = -x^2 - 2x + 3$$
$$0 = x^2 + 2x - 3 \qquad \text{Multiply each side by} -1$$

$$0 = (x + 3)(x - 1)$$

$$x = -3 \quad \text{or} \quad x = 1$$

The x-coordinate of the vertex is given by

$$x = \frac{-b}{2a} = \frac{-(-2)}{2(-1)} = \frac{2}{-2} = -1$$

To find the y-coordinate of the vertex, we substitute -1 for x in our original equation to get

$$y = -(-1)^2 - 2(-1) + 3 = -1 + 2 + 3 = 4$$

Our parabola has x-intercepts at -3 and 1, and a vertex at $(-1, 4)$. Figure 3 shows the graph. We say the graph is **concave down** since it opens downward.

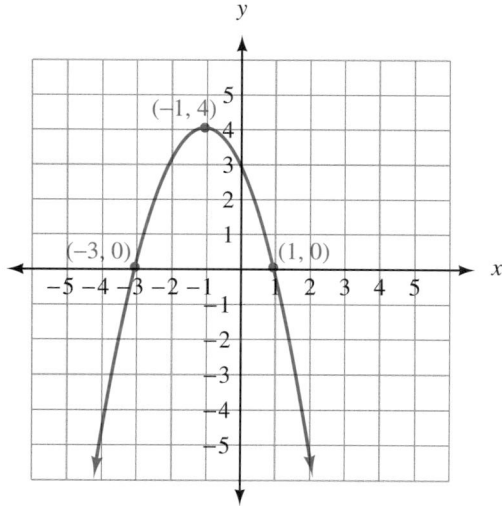

Figure 3

Again, we could have obtained the coordinates of the vertex by completing the square on the first two terms on the right side of our equation. To do so we must first factor -1 from the first two terms. (Remember, the leading coefficient must be 1 in order to complete the square.) When we complete the square, we add 1 inside the parentheses, which actually decreases the right side of the equation by -1 since everything in the parentheses is multiplied by -1. To make up for it, we add 1 outside the parentheses.

$$y = -1(x^2 + 2x \qquad) + 3$$

$$y = -1(x^2 + 2x + \mathbf{1}) + 3 + \mathbf{1}$$

$$y = -1(x + 1)^2 + 4$$

The last line tells us that the *largest* value of y will be 4, and that will occur when $x = -1$. ◀

▶ **EXAMPLE 3**   Graph $y = 3x^2 - 6x + 1$.

Solution   To find the x-intercepts, we let $y = 0$ and solve for x:

$$0 = 3x^2 - 6x + 1$$

Since the right side of this equation does not factor, we can look at the discriminant to see what kind of solutions are possible. The discriminant for this equation is

$$b^2 - 4ac = 36 - 4(3)(1) = 24$$

Since the discriminant is a positive number but not a perfect square, the equation will have irrational solutions. This means that the x-intercepts are irrational numbers and will have to be approximated with decimals using the quadratic formula. Rather than use the quadratic formula, we will find some other points on the graph, but first let's find the vertex.

Here are both methods of finding the vertex:

Using the formula that gives us the x-coordinate of the vertex, we have:	To complete the square on the right side of the equation, we factor 3 from the first two terms, add 1 inside the parentheses, and add -3 outside the parentheses (this amounts to adding 0 to the right side):

$$x = \frac{-b}{2a} = \frac{-(-6)}{2(3)} = 1$$

Substituting 1 for x in the equation gives us the y-coordinate of the vertex:

$$y = 3 \cdot 1^2 - 6 \cdot 1 + 1 = -2$$

$$y = 3(x^2 - 2x \quad\ ) + 1$$
$$y = 3(x^2 - 2x + \mathbf{1}) + 1 - \mathbf{3}$$
$$y = 3(x - 1)^2 - 2$$

In either case, the vertex is $(1, -2)$.

If we can find two points, one on each side of the vertex, we can sketch the graph. Let's let $x = 0$ and $x = 2$, since each of these numbers is the same distance from $x = 1$, and $x = 0$ will give us the y-intercept.

When $x = 0$	When $x = 2$
$y = 3(0)^2 - 6(0) + 1$	$y = 3(2)^2 - 6(2) + 1$
$= 0 - 0 + 1$	$= 12 - 12 + 1$
$= 1$	$= 1$

The two points just found are $(0, 1)$ and $(2, 1)$. Plotting these two points along with the vertex $(1, -2)$, we have the graph shown in Figure 4 on the next page.

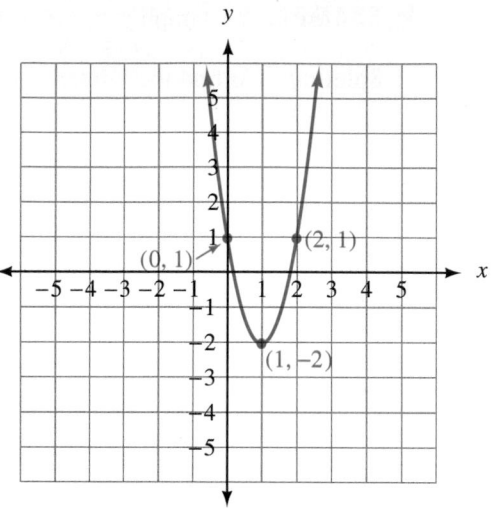

Figure 4

◀

▶ **EXAMPLE 4** Graph $y = -2x^2 + 6x - 5$.

Solution Letting $y = 0$, we have

$$0 = -2x^2 + 6x - 5$$

Again, the right side of this equation does not factor. The discriminant is $b^2 - 4ac = 36 - 4(-2)(-5) = -4$, which indicates that the solutions are complex numbers. This means that our original equation does not have x-intercepts. The graph does not cross the x-axis.

Let's find the vertex.

Using our formula for the x-coordinate of the vertex, we have

$$x = \frac{-b}{2a} = \frac{-6}{2(-2)} = \frac{6}{4} = \frac{3}{2}$$

To find the y-coordinate, we let $x = \frac{3}{2}$:

$$y = -2\left(\frac{3}{2}\right)^2 + 6\left(\frac{3}{2}\right) - 5$$

$$= \frac{-18}{4} + \frac{18}{2} - 5$$

$$= \frac{-18 + 36 - 20}{4}$$

$$= -\frac{1}{2}$$

Finding the vertex by completing the square is a more complicated matter. In order to make the coefficient of x^2 a 1, we must factor -2 from the first two terms. To complete the square inside the parentheses, we add $\frac{9}{4}$. Since each term inside the parentheses is multiplied by -2, we add $\frac{9}{2}$ outside the parentheses so that the net result is the same as adding 0 to the right side:

$$y = -2(x^2 - 3x \quad) - 5$$

$$y = -2\left(x^2 - 3x + \frac{9}{4}\right) - 5 + \frac{9}{2}$$

$$y = -2\left(x - \frac{3}{2}\right)^2 - \frac{1}{2}$$

The vertex is $(\frac{3}{2}, -\frac{1}{2})$. Since this is the only point we have so far, we must find two others. Let's let $x = 3$ and $x = 0$, since each point is the same distance from $x = \frac{3}{2}$ and on either side:

When $x = 3$ When $x = 0$

$$y = -2(3)^2 + 6(3) - 5 \qquad\qquad y = -2(0)^2 + 6(0) - 5$$
$$= -18 + 18 - 5 \qquad\qquad\qquad = 0 + 0 - 5$$
$$= -5 \qquad\qquad\qquad\qquad\quad = -5$$

The two additional points on the graph are $(3, -5)$ and $(0, -5)$. Figure 5 shows the graph.

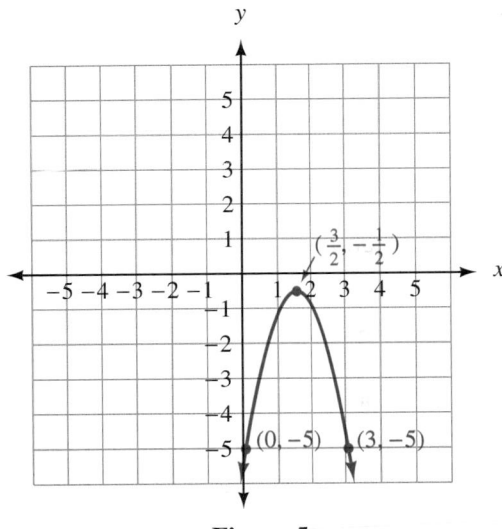

Figure 5

The graph is concave down. The vertex is the highest point on the graph. ◀

By looking at the equations and graphs in Examples 1 through 4, we can conclude that the graph of $y = ax^2 + bx + c$ will be concave up when a is positive, and concave down when a is negative. Taking this even further, if $a > 0$, then the vertex is the lowest point on the graph, and if $a < 0$, the vertex is the highest point on the graph. We can use this information to solve some problems in which we are interested in finding the largest or smallest value of a variable.

▶ **EXAMPLE 5** A company selling copies of an accounting program for home computers finds that it will make a weekly profit of P dollars from selling x copies of the program, according to the equation

$$P(x) = -0.1x^2 + 27x - 500$$

How many copies of the program should it sell to make the largest possible profit, and what is the largest possible profit?

Solution   Since the coefficient of x^2 is negative, we know the graph of this parabola will be concave down, meaning that the vertex is the highest point of the curve. We find the vertex by first finding its x-coordinate:

$$x = \frac{-b}{2a} = \frac{-27}{2(-0.1)} = \frac{27}{0.2} = 135$$

This represents the number of programs the company needs to sell each week in order to make a maximum profit. To find the maximum profit, we substitute 135 for x in the original equation. (A calculator is helpful for these kinds of calculations.)

$$
\begin{aligned}
P(135) &= -0.1(135)^2 + 27(135) - 500 \\
&= -0.1(18{,}225) + 3{,}645 - 500 \\
&= -1{,}822.5 + 3{,}645 - 500 \\
&= 1{,}322.5
\end{aligned}
$$

The maximum weekly profit is $1,322.50 and is obtained by selling 135 programs a week.   ◀

▶ **EXAMPLE 6**   An art supply store finds that they can sell x sketch pads each week at p dollars each, according to the equation $x = 900 - 300p$. Graph the revenue equation $R = xp$. Then use the graph to find the price p that will bring in the maximum revenue. Finally, find the maximum revenue.

Solution   As it stands, the revenue equation contains three variables. Since we are asked to find the value of p that gives us the maximum value of R, we rewrite the equation using just the variables R and p. Since $x = 900 - 300p$, we have

$$R = xp = (900 - 300p)p$$

The graph of this equation is shown in Figure 6. The graph appears in the first quadrant only, since R and p are both positive quantities. From the graph we see that the maximum value of R occurs when $p = \$1.50$. We can calculate the maximum value of R from the equation:

When $p = 1.5$

the equation $R = (900 - 300p)p$

becomes $\begin{aligned} R &= (900 - 300 \cdot 1.5)1.5 \\ &= (900 - 450)1.5 \\ &= 450 \cdot 1.5 \\ &= 675 \end{aligned}$

The maximum revenue is $675. It is obtained by setting the price of each sketch pad at $p = \$1.50$.

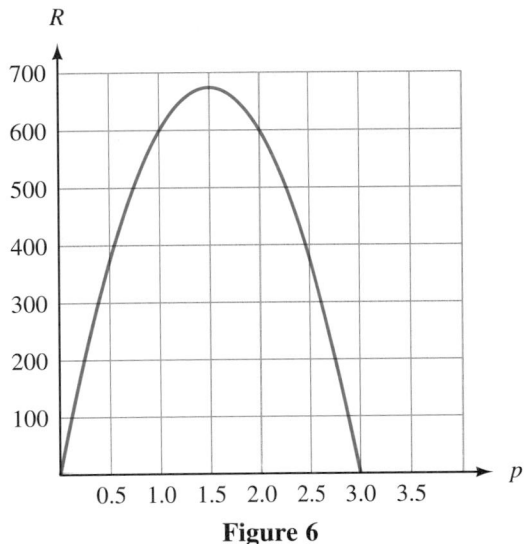

Figure 6

USING TECHNOLOGY: Graphing Calculators

If you have been using a graphing calculator for some of the material in this course, you are well aware that your calculator can draw all the graphs in this section very easily. However, it is important that you are able to recognize and sketch the graph of any parabola by hand. It is a skill that all successful intermediate algebra students should possess, even if they are proficient in the use of a graphing calculator. My suggestion is that you work the problems in this section and problem set without your calculator. Then use your calculator to check your results.

PROBLEM SET 7.6

For each of the following equations, give the x-intercepts and the coordinates of the vertex and sketch the graph.

1. $y = x^2 + 2x - 3$
2. $y = x^2 - 2x - 3$
3. $y = -x^2 - 4x + 5$
4. $y = x^2 + 4x - 5$
5. $y = x^2 - 1$
6. $y = x^2 - 4$
7. $y = -x^2 + 9$
8. $y = -x^2 + 1$
9. $y = 2x^2 - 4x - 6$
10. $y = 2x^2 + 4x - 6$
11. $y = x^2 - 2x - 4$
12. $y = x^2 - 2x - 2$

Find the vertex and any two convenient points to sketch the graphs of the following.

13. $y = x^2 - 4x - 4$
14. $y = x^2 - 2x + 3$
15. $y = -x^2 + 2x - 5$
16. $y = -x^2 + 4x - 2$
17. $y = x^2 + 1$
18. $y = x^2 + 4$
19. $y = -x^2 - 3$
20. $y = -x^2 - 2$
21. $y = 3x^2 + 4x + 1$
22. $y = 2x^2 + 4x + 3$

For each of the following equations, find the coordinates of the vertex and indicate whether the vertex is the highest point on the graph or the lowest point on the graph. (Do not graph.)

23. $y = x^2 - 6x + 5$

24. $y = -x^2 + 6x - 5$

25. $y = -x^2 + 2x + 8$

26. $y = x^2 - 2x - 8$

27. $y = 12 + 4x - x^2$

28. $y = -12 - 4x + x^2$

29. $y = -x^2 - 8x$

30. $y = x^2 + 8x$

Applying the Concepts

31. A company earns a weekly profit of P dollars by selling x items, according to the equation $P(x) = -0.5x^2 + 40x - 300$. Find the number of items the company must sell each week in order to obtain the largest possible profit. Then, find the largest possible profit.

32. A company earns a weekly profit of P dollars by selling x items, according to the equation $P(x) = -0.5x^2 + 100x - 1,600$. Find the number of items the company must sell each week in order to obtain the largest possible profit. Then, find the largest possible profit.

33. A company finds that it can make a profit of P dollars each month by selling x patterns, according to the formula $P(x) = -0.002x^2 + 3.5x - 800$. How many patterns must it sell each month in order to have a maximum profit? What is the maximum profit?

34. A company selling picture frames finds that it can make a profit of P dollars each month by selling x frames, according to the formula $P(x) = -0.002x^2 + 5.5x - 1,200$. How many frames must it sell each month in order to have a maximum profit? What is the maximum profit?

35. An arrow is shot straight up into the air with an initial velocity of 128 feet/second. If h represents the height of the arrow at any time t, then the equation that gives h in terms of t is $h(t) = 128t - 16t^2$. Find the maximum height attained by the arrow.

36. A ball is projected into the air with an upward velocity of 32 feet/second. The equation that gives the height h of the ball at any time t is $h(t) = 32t - 16t^2$. Find the maximum height attained by the ball.

37. A company that manufactures typewriter ribbons knows that the number of ribbons they can sell each week x is related to the price of each ribbon p by the equation $x = 1,200 - 100p$. Graph the revenue equation $R = xp$. Then use the graph to find the price p that will bring in the maximum revenue. Finally, find the maximum revenue.

38. A company that manufactures diskettes for home computers finds that they can sell x diskettes each day at p dollars per diskette according to the equation $x = 800 - 100p$. Graph the revenue equation $R = xp$. Then use the graph to find the price p that will bring in the maximum revenue. Finally, find the maximum revenue.

39. The relationship between the number of calculators a company sells each day x and the price of each calculator p is given by the equation $x = 1,700 - 100p$. Graph the revenue equation $R = xp$ and use the graph to find the price p that will bring in the maximum revenue. Then find the maximum revenue.

40. The relationship between the number of pencil sharpeners a company sells each week and the price of each sharpener is given by the equation $x = 1,800 - 100p$. Graph the revenue equation $R = xp$ and use the graph to find the price p that will bring in the maximum revenue. Then find the maximum revenue.

Review Problems

The problems that follow review material we covered in Section 6.7.

Perform the indicated operations.

41. $(3 - 5i) - (2 - 4i)$

42. $2i(5 - 6i)$

43. $(3 + 2i)(7 - 3i)$

44. $(4 + 5i)^2$

45. $\dfrac{i}{3 + i}$

46. $\dfrac{2 + 3i}{2 - 3i}$

Parabolas That Open Left and Right

In this section, we extend the work we did in the previous section by considering parabolas that open left and right instead of up and down.

In Example 1 of the previous section we graphed the equation $y = x^2 - 6x + 5$. That graph is shown in Figure 1.

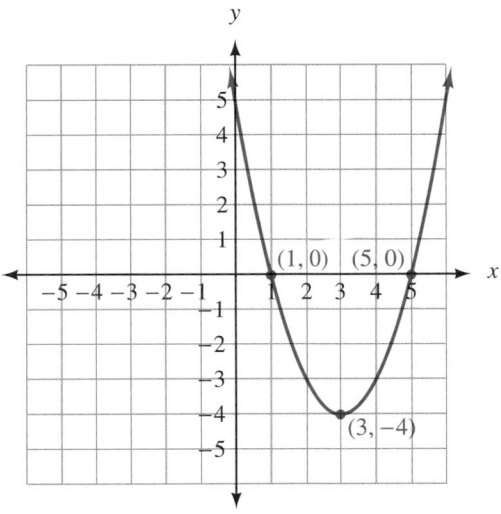

Figure 1

Note that the x-intercepts are 1 and 5, and the vertex is at $(3, -4)$.

If we interchange x and y in the equation $y = x^2 - 6x + 5$, we obtain the equation

$$x = y^2 - 6y + 5$$

We can expect the graph of this equation to be similar to the graph shown in Figure 1, except that the role of x and y will be interchanged. Example 1 gives more details.

▶ **EXAMPLE 1**   Graph $x = y^2 - 6y + 5$.

Solution   Since the x-intercepts for $y = x^2 - 6x + 5$ are $x = 1$ and $x = 5$, we expect the y-intercepts for $x = y^2 - 6y + 5$ to be $y = 1$ and $y = 5$. To see that this is true, we let $x = 0$ in $x = y^2 - 6y + 5$, and solve for y:

$$0 = y^2 - 6y + 5$$

$$0 = (y - 1)(y - 5)$$

$$y - 1 = 0 \quad \text{or} \quad y - 5 = 0$$

$$y = 1 \quad \text{or} \quad y = 5$$

The graph of $x = y^2 - 6y + 5$ will cross the y-axis at 1 and 5.

Since the graph in Figure 1 has its vertex at $x = 3, y = -4$, we expect the graph of $x = y^2 - 6y + 5$ to have its vertex at $y = 3, x = -4$. Completing the square on y shows that this is true.

$$x = (y^2 - 6y + \mathbf{9}) + 5 - \mathbf{9}$$

$$x = (y - 3)^2 - 4$$

This last line tells us the smallest value of x is -4, and it occurs when y is 3. Therefore, the vertex must be at the point $(-4, 3)$. Here is that graph:

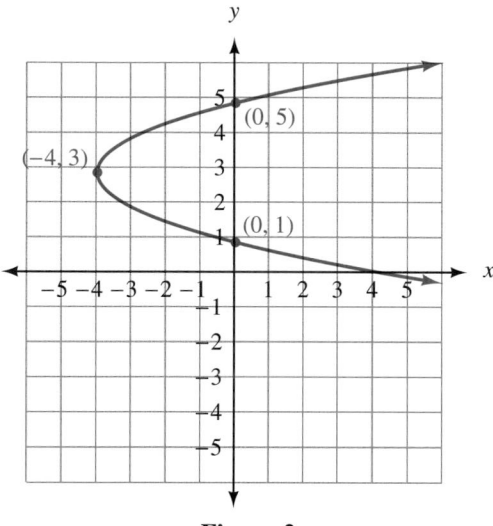

Figure 2

A summary of the discussion above is given below.

Parabolas That Open Left and Right

The graph of $x = ay^2 + by + c$ will have

1. an x-intercept at $x = c$
2. y-intercepts (if they exist) at

$$y = \frac{-b \pm \sqrt{b^2 - 4ac}}{2a}$$

3. a vertex when $y = \dfrac{-b}{2a}$

The equations we will graph in Examples 2 and 3 that follow are the equations graphed in Examples 2 and 3 of the previous section with the variables x and y interchanged.

▶ **EXAMPLE 2**   Graph $x = -y^2 - 2y + 3$.

Solution   The intercepts and vertex are found as shown below. The results are used to sketch the graph in Figure 3.

x-intercept: When $y = 0$, $x = 3$
y-intercepts: When $x = 0$, we have $-y^2 - 2y + 3 = 0$, which yields solutions $y = -3$ and $y = 1$.

Vertex: $y = \dfrac{-b}{2a} = \dfrac{-(-2)}{2(-1)} = -1$. Substituting -1 for y in the equation gives us $x = 4$. The vertex is at $(4, -1)$.

The graph of our equation is shown below on the right. Note the similarities to the graph of $y = -x^2 - 2x + 3$ shown on the left.

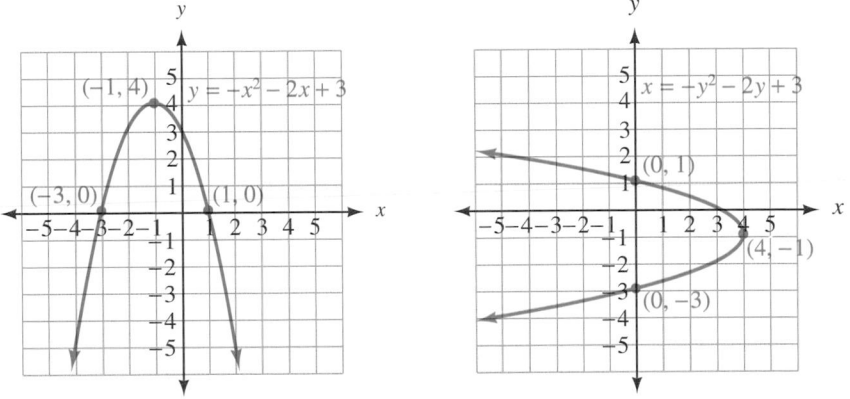

Figure 3

▶ **EXAMPLE 3**   Graph $x = 3y^2 - 6y + 1$.

Solution   For this last example, we simply show the graph of our equation alongside the graph of the equation $y = 3x^2 - 6x + 1$. Both are shown in Figure 4 on the next page.

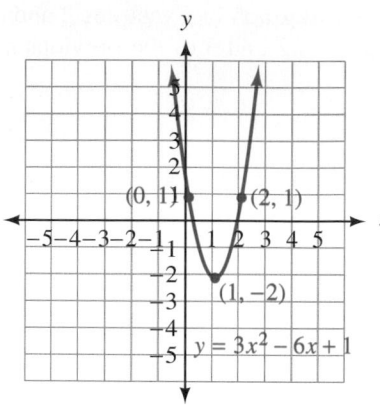
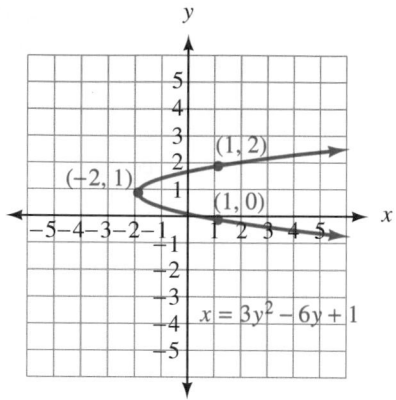

Figure 4

PROBLEM SET 7.7

Graph each equation.

1. $x = y^2 + 2y - 3$ **2.** $x = y^2 - 2y - 3$
3. $x = -y^2 - 4y + 5$ **4.** $x = y^2 + 4y - 5$
5. $x = y^2 - 1$ **6.** $x = y^2 - 4$
7. $x = -y^2 + 9$ **8.** $x = -y^2 + 1$
9. $x = 2y^2 - 4y - 6$ **10.** $x = 2y^2 + 4y - 6$
11. $x = y^2 - 2y - 4$ **12.** $x = y^2 - 2y + 3$

Review Problems

The problems that follow review material we covered in Section 6.7.

Write in terms of i and simplify.

13. $\sqrt{-20}$ **14.** $-\sqrt{-20}$

Write each of the following as i, -1, $-i$, or 1.

15. i^{20} **16.** i^{22}

Find the following quotients. Write all answers in standard form for complex numbers.

17. $\dfrac{3 - 4i}{i}$ **18.** $\dfrac{-3 + 4i}{i}$

CHAPTER 7 SUMMARY

Examples

1. If $(x - 3)^2 = 25$
then $x - 3 = \pm 5$
$x = 3 \pm 5$
$x = 8$ or $x = -2$

Theorem 7.1 [7.1]
If $a^2 = b$, where b is a real number, then
$$a = \sqrt{b} \quad \text{or} \quad a = -\sqrt{b} \quad \text{or} \quad a = \pm\sqrt{b}$$

2. Solve $x^2 - 6x - 6 = 0$.

$$x^2 - 6x = 6$$

$$x^2 - 6x + \mathbf{9} = 6 + \mathbf{9}$$

$$(x - 3)^2 = 15$$

$$x - 3 = \pm\sqrt{15}$$

$$x = 3 \pm \sqrt{15}$$

To Solve a Quadratic Equation by Completing the Square [7.1]

Step 1: Write the equation in the form $ax^2 + bx = c$.

Step 2: If $a \neq 1$, divide through by the constant a so the coefficient of x^2 is 1.

Step 3: Complete the square on the left side by adding the square of $\frac{1}{2}$ the coefficient of x to both sides.

Step 4: Write the left side of the equation as the square of a binomial. Simplify the right side if possible.

Step 5: Apply Theorem 7.1 and solve as usual.

3. If $2x^2 + 3x - 4 = 0$, then

$$x = \frac{-3 \pm \sqrt{9 - 4(2)(-4)}}{2(2)}$$

$$= \frac{-3 \pm \sqrt{41}}{4}$$

The Quadratic Theorem [7.2]

For any quadratic equation in the form $ax^2 + bx + c = 0$, $a \neq 0$, the two solutions are

$$x = \frac{-b \pm \sqrt{b^2 - 4ac}}{2a}$$

This last expression is known as the *quadratic formula.*

4. The discriminant for

$$x^2 + 6x + 9 = 0$$

is $D = 36 - 4(1)(9) = 0$, which means the equation has one rational solution.

The Discriminant [7.3]

The expression $b^2 - 4ac$ which appears under the radical sign in the quadratic formula is known as the *discriminant.*

We can classify the solutions to $ax^2 + bx + c = 0$:

The solutions are	When the discriminant is
Two complex numbers containing i	Negative
One rational number	Zero
Two rational numbers	A positive perfect square
Two irrational numbers	A positive number, but not a perfect square

5. The equation $x^4 - x^2 - 12 = 0$ is quadratic in x^2. Letting $y = x^2$ we have

$$y^2 - y - 12 = 0$$

$$(y - 4)(y + 3) = 0$$

$$y = 4 \quad \text{or} \quad y = -3$$

Resubstituting x^2 for y we have

$$x^2 = 4 \quad \text{or} \quad x^2 = -3$$

$$x = \pm 2 \quad \text{or} \quad x = \pm i\sqrt{3}$$

Equations Quadratic in Form [7.4]

There are a variety of equations whose form is quadratic. We solve most of them by making a substitution so the equation becomes quadratic, and then solving that equation by factoring or the quadratic formula. For example,

The equation	*is quadratic in*
$(2x - 3)^2 + 5(2x - 3) - 6 = 0$	$2x - 3$
$4x^4 - 7x^2 - 2 = 0$	x^2
$2x - 7\sqrt{x} + 3 = 0$	$\sqrt{x}$

6. Solve $x^2 - 2x - 8 > 0$. We factor and draw the sign diagram.

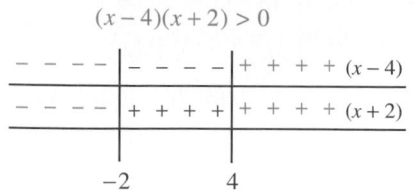

$$(x-4)(x+2) > 0$$

The solution is $x < -2$ or $x > 4$.

7. The graph of $y = x^2 - 4$ will be a parabola. It will cross the x-axis at 2 and -2, and the vertex will be $(0, -4)$.

Quadratic Inequalities [7.5]
We solve quadratic inequalities by manipulating the inequality to get 0 on the right side and then factoring the left side. We then make a diagram that indicates where the factors are positive and where they are negative. From this sign diagram and the original inequality we graph the appropriate solution set.

Graphing Parabolas [7.6]
The graph of any equation of the form

$$y = ax^2 + bx + c \qquad a \neq 0$$

is a *parabola*. The graph is *concave up* if $a > 0$, and *concave down* if $a < 0$. The highest or lowest point on the graph is called the *vertex* and will always have an x-coordinate of $x = \dfrac{-b}{2a}$.

CHAPTER 7 REVIEW

Solve each equation. [7.1]

1. $(2t - 5)^2 = 25$
2. $(3t - 2)^2 = 4$
3. $(3y - 4)^2 = -49$
4. $(8y + 1)^2 = -36$
5. $(2x + 6)^2 = 12$
6. $(3x - 4)^2 = 18$

Solve by completing the square. [7.1]

7. $2x^2 + 6x - 20 = 0$
8. $3x^2 + 15x = -18$
9. $a^2 + 9 = 6a$
10. $a^2 + 4 = 4a$
11. $2y^2 + 6y = -3$
12. $3y^2 + 3 = 9y$

Solve each equation. [7.2]

13. $\frac{1}{6}x^2 + \frac{1}{2}x - \frac{5}{3} = 0$
14. $\frac{2}{15}x^2 + \frac{1}{3}x + \frac{1}{5} = 0$
15. $8x^2 - 18x = 0$
16. $8x^2 - 18 = 0$
17. $4t^2 - 8t + 19 = 0$
18. $4t^2 - 4t + 9 = 0$
19. $100x^2 - 200x = 100$
20. $10x^2 - 60x = -70$
21. $0.06a^2 + 0.05a = 0.04$
22. $0.06a^2 - 0.01a = 0.02$
23. $9 - 6x = -x^2$
24. $2x - 1 = x^2$

25. $(2x + 1)(x - 5) - (x + 3)(x - 2) = -17$
26. $(2x - 3)^2 - (3x + 2)(x - 2) = 2x + 6$
27. $2y^3 + 2y = 10y^2$
28. $3y^3 - y = 5y^2$
29. $5x^2 = -2x + 3$
30. $2x^2 = -3x + 7$
31. $x^3 - 27 = 0$
32. $8a^3 + 27 = 0$
33. $3 - \dfrac{2}{x} + \dfrac{1}{x^2} = 0$
34. $\dfrac{2}{x^2} + 1 = \dfrac{4}{x}$
35. $\dfrac{1}{x - 3} + \dfrac{1}{x + 2} = 1$
36. $\dfrac{1}{x + 3} + \dfrac{1}{x - 2} = 1$

37. The total cost (in dollars) for a company to produce x items per week is $C(x) = 7x + 400$. The revenue for selling all x items is $R(x) = 34x - 0.1x^2$. How many items must it produce and sell each week for its weekly profit to be $1,300? [7.2]

38. The total cost (in dollars) for a company to produce x items per week is $C(x) = 70x + 300$. The revenue for selling all x items is

$R(x) = 110x - 0.5x^2$. How many items must it produce and sell each week for its weekly profit to be \$300? [7.2]

Use the discriminant to find the number and kind of solutions for each equation. [7.3]

39. $2x^2 - 8x = -8$ **40.** $4x^2 - 8x = -4$
41. $2x^2 + x - 3 = 0$ **42.** $5x^2 + 11x = 12$
43. $x^2 - x = 1$ **44.** $x^2 - 5x = -5$
45. $3x^2 + 5x = -4$ **46.** $4x^2 - 3x = -6$

Determine k so that each equation has exactly one real solution. [7.3]

47. $25x^2 - kx + 4 = 0$
48. $4x^2 + kx + 25 = 0$
49. $kx^2 + 12x + 9 = 0$
50. $kx^2 - 16x + 16 = 0$
51. $9x^2 + 30x + k = 0$
52. $4x^2 + 28x + k = 0$

For each of the following problems, find an equation that has the given solutions. [7.3]

53. $x = 3, x = 5$ **54.** $y = \frac{1}{2}, y = -4$
55. $t = 3, t = -3, t = 5$
56. $t = 2, t = -2, t = 6$
57. Find all solutions to $x^3 - 4x^2 + x + 6 = 0$ if $x = 2$ is one solution. [7.3]
58. Find all solutions to $x^3 - 3x^2 - 6x + 8 = 0$ if $x = 1$ is one solution. [7.3]

Find all solutions. [7.4]

59. $(x - 2)^2 - 4(x - 2) - 60 = 0$
60. $(x + 3)^2 - 3(x + 3) - 70 = 0$
61. $6(2y + 1)^2 - (2y + 1) - 2 = 0$
62. $3(4y - 1)^2 + (4y - 1) - 10 = 0$
63. $x^4 - x^2 = 12$ **64.** $x^4 - 7x^2 = 18$
65. $x - \sqrt{x} - 2 = 0$ **66.** $x - 3\sqrt{x} + 2 = 0$
67. $2x - 11\sqrt{x} = -12$ **68.** $3x + \sqrt{x} = 2$
69. $\sqrt{x + 5} = \sqrt{x} + 1$ **70.** $\sqrt{x - 2} = 2 - \sqrt{x}$
71. $\sqrt{y + 21} + \sqrt{y} = 7$
72. $\sqrt{y - 3} - \sqrt{y} = -1$
73. $\sqrt{y + 9} - \sqrt{y - 6} = 3$

74. $\sqrt{y + 7} - \sqrt{y + 2} = 1$
75. An object is tossed into the air with an upward velocity of 10 feet/second from the top of a building h feet high. The time it takes for the object to hit the ground below is given by the formula $16t^2 - 10t - h = 0$. Solve this formula for t. [7.4]
76. An object is tossed into the air with an upward velocity of v feet/second from the top of a 10-foot wall. The time it takes for the object to hit the ground below is given by the formula $16t^2 - vt - 10 = 0$. Solve this formula for t. [7.4]

Solve each inequality and graph the solution set. [7.5]

77. $x^2 - x - 2 < 0$
78. $3x^2 - 14x + 8 \le 0$
79. $2x^2 + 5x - 12 \ge 0$
80. $x^2 - 4x + 4 < 0$
81. $(x + 2)(x - 3)(x + 4) > 0$
82. $x^3 + 2x^2 - 9x - 18 < 0$

83. $\dfrac{x}{x - 3} + \dfrac{2}{x - 3} > 0$

84. $\dfrac{x}{x + 4} - \dfrac{2}{x + 4} > 0$

85. $\dfrac{x - 3}{(x - 2)(x - 4)} > 0$

86. $\dfrac{x + 2}{(x + 1)(x - 1)} > 0$

Find the x-intercepts, if they exist, and the vertex for each parabola. Then use them to sketch the graph. [7.6]

87. $y = x^2 - 6x + 8$
88. $y = x^2 - x - 2$
89. $y = x^2 - 4$
90. $y = x^2 + 6x + 9$

CHAPTER 7 TEST

Solve each equation. [7.1, 7.2]

1. $(2x + 4)^2 = 25$ **2.** $(2x - 6)^2 = -8$
3. $y^2 - 10y + 25 = -4$
4. $(y + 1)(y - 3) = -6$

5. $8t^3 - 125 = 0$ **6.** $\dfrac{1}{a + 2} - \dfrac{1}{3} = \dfrac{1}{a}$

7. Solve the formula $64(1 + r)^2 = A$ for r. [7.1]
8. Solve $x^2 - 4x = -2$ by completing the square. [7.1]
9. An object projected upward with an initial velocity of 32 feet/second will rise and fall according to the equation $s(t) = 32t - 16t^2$, where s is its distance above the ground at time t. At what times will the object be 12 feet above the ground? [7.2]
10. The total weekly cost for a company to make x ceramic coffee cups is given by the formula $C(x) = 2x + 100$. If the weekly revenue from selling all x cups is $R(x) = 25x - 0.2x^2$, how many cups must it sell a week to make a profit of $200 a week? [7.2]
11. Find k so that $kx^2 = 12x - 4$ has one rational solution. [7.3]
12. Use the discriminant to identify the number and kind of solutions to $2x^2 - 5x = 7$. [7.3]

Find equations that have the given solutions. [7.3]

13. $x = 5, x = -\frac{2}{3}$
14. $x = 2, x = -2, x = 7$

15. Find all the solutions to the equation $4x^3 - 16x^2 + 17x - 2 = 0$ if $x = 2$ is one solution. [7.3]

Solve each equation. [7.4]

16. $4x^4 - 7x^2 - 2 = 0$
17. $(2t + 1)^2 - 5(2t + 1) + 6 = 0$
18. $2t - 7\sqrt{t} + 3 = 0$
19. An object is tossed into the air with an upward velocity of 14 feet/second from the top of a building h feet high. The time it takes for the object to hit the ground below is given by the formula $16t^2 - 14t - h = 0$. Solve this formula for t. [7.4]

Graph each of the following inequalities. [7.5]

20. $x^2 - x - 6 \le 0$ **21.** $2x^2 + 5x > 3$

22. $\dfrac{x + 2}{x - 1} \ge 0$

Sketch the graph of each of the following. Give the coordinates of the vertex in each case. [7.6]

23. $y = x^2 - 2x - 3$ **24.** $y = -x^2 + 2x + 8$
25. Find the maximum weekly profit for a company with weekly costs of $C(x) = 5x + 100$ and weekly revenue of $R(x) = 25x - 0.1x^2$. [7.6]

**C H A P T E R**

SYSTEMS OF LINEAR EQUATIONS

INTRODUCTION

Table 1 gives some corresponding temperatures on the Fahrenheit and Celsius temperature scales. Each pair of numbers is obtained by measuring the temperature of water using two different thermometers. The initial reading of 25° C and 77° F is the water at room temperature. The last reading, 100° C and 212° F, is boiling water. The numbers in between these pairs are obtained as the water is heated from room temperature to boiling.

TABLE 1 Corresponding Temperatures

In Degrees Fahrenheit	In Degrees Celsius
77	25
95	35
167	75
212	100

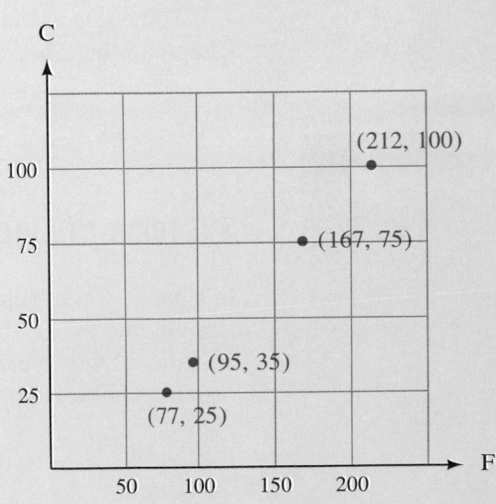

If we graph these pairs of numbers as ordered pairs (F, C), we find that they lie in a straight line—that is, the relationship between the variables C and F is a linear relationship. Although we could use the methods developed in Chapter 5

to find the formula that gives the exact relationship between C and F, the relationship can also be found as a system of two equations in two variables, which is one of the main topics of this chapter.

Systems of linear equations are used extensively in many different disciplines. They can be used to solve multiple-loop circuit problems in electronics, kinship patterns in anthropology, genetics problems in biology, and profit-and-cost problems in economics. There are many other applications as well.

OVERVIEW

In Chapter 5 we worked with linear equations in two variables. In this chapter we will extend our work with linear equations to include systems of linear equations in two variables. A system of linear equations in two variables is two linear equations considered at the same time. The solution to this type of system of equations is all the points that lie on both lines. That is, to solve a system of equations we look for all the points the individual equations have in common.

After we have introduced systems of equations in two variables, we will turn our attention to systems of equations in three variables. One of the ways in which we solve systems of equations in three variables is by reducing them to systems of equations in two variables.

Also in this chapter is our introduction to determinants. There are many places in mathematics where determinants are found. For our introduction to them we will see how they can be used to solve systems of linear equations.

To be successful in this chapter you should be familiar with the concepts in Chapter 5 that are concerned with linear equations in two variables.

SECTION 8.1 Systems of Linear Equations in Two Variables

In Chapter 5 we found the graph of an equation of the form $ax + by = c$ to be a straight line. Since the graph is a straight line, the equation is said to be a linear equation. Two linear equations considered together form a **linear system** of equations. For example,

$$3x - 2y = 6$$
$$2x + 4y = 20$$

is a linear system. The solution set to the system is the set of all ordered pairs that satisfy both equations. If we graph each equation on the same set of axes, we can see the solution set (see Figure 1).

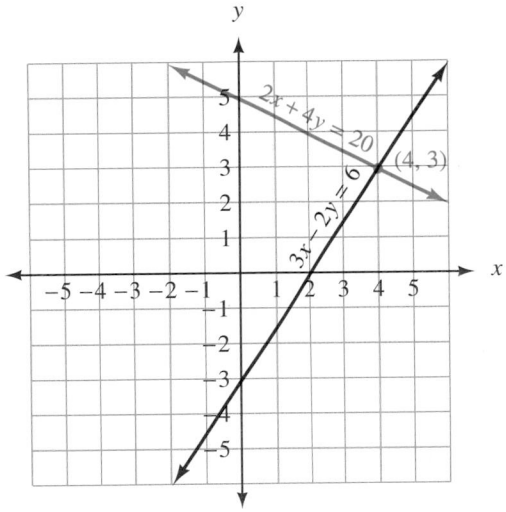

Figure 1

The point $(4, 3)$ lies on both lines and therefore must satisfy both equations. It is obvious from the graph that it is the only point that does so. The solution set for the system is $\{(4, 3)\}$.

More generally, if $a_1x + b_1y = c_1$ and $a_2x + b_2y = c_2$ are linear equations, then the solution set for the system

$$a_1x + b_1y = c_1$$

$$a_2x + b_2y = c_2$$

can be illustrated through one of the following graphs (Discussion on next page.):

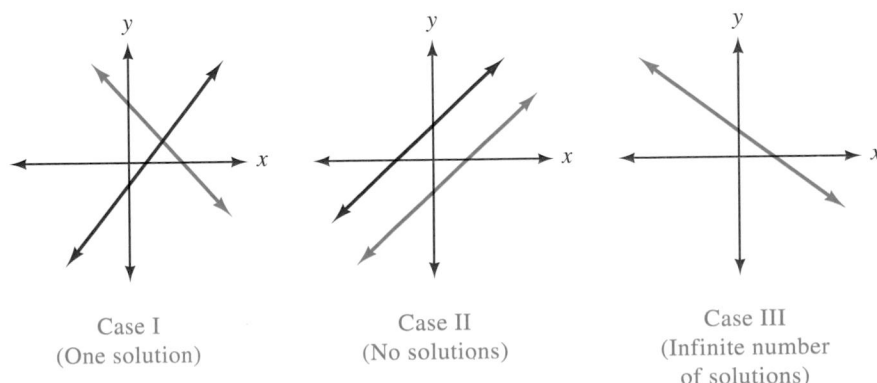

Case I
(One solution)

Case II
(No solutions)

Case III
(Infinite number
of solutions)

Figure 2

Case I The two lines intersect at one and only one point. The coordinates of the point give the solution to the system. This is what usually happens.

Case II The lines are parallel and therefore have no points in common. The solution set to the system is the empty set, Ø. In this case, we say the equations are **inconsistent.**

Case III The lines coincide. That is, their graphs represent the same line. The solution set consists of all ordered pairs that satisfy either equation. In this case, the equations are said to be **dependent.**

In the beginning of this section we found the solution set for the system

$$3x - 2y = 6$$

$$2x + 4y = 20$$

by graphing each equation and then reading the solution set from the graph. Solving a system of linear equations by graphing is the least accurate method. If the coordinates of the point of intersection are not integers, it can be very difficult to read the solution set from the graph. There is another method of solving a linear system that does not depend on the graph. It is called the **addition method.**

▶ **EXAMPLE 1** Solve the system

$$4x + 3y = 10$$

$$2x + y = 4$$

Solution If we multiply the bottom equation by -3, the coefficients of y in the resulting equation and the top equation will be opposites:

$$4x + 3y = 10 \xrightarrow{\text{No change}} 4x + 3y = 10$$

$$2x + y = 4 \xrightarrow[\text{Multiply by } -3]{} -6x - 3y = -12$$

Adding the left and right sides of the resulting equations, we have

$$
\begin{array}{r}
4x + 3y = 10 \\
-6x - 3y = -12 \\
\hline
-2x = -2
\end{array}
$$

The result is a linear equation in one variable. We have eliminated the variable y from the equations by addition. (It is for this reason we call this method of solving a linear system the *addition method.*) Solving $-2x = -2$ for x, we have

$$x = 1$$

This is the x-coordinate of the solution to our system. To find the y-coordinate, we substitute $x = 1$ into any of the equations containing both the variables x and y. Let's try the second equation in our original system:

$$2(1) + y = 4$$

$$2 + y = 4$$

$$y = 2$$

This is the y-coordinate of the solution to our system. The ordered pair $(1, 2)$ is the solution to the system.   ◀

▶ **EXAMPLE 2**   Solve the system

$$3x - 5y = -2$$

$$2x - 3y = 1$$

Solution   We can eliminate either variable. Let's decide to eliminate the variable x. We can do so by multiplying the top equation by 2 and the bottom equation by -3, and then adding the left and right sides of the resulting equations:

$$
\begin{array}{rl}
3x - 5y = -2 \xrightarrow{\text{Multiply by 2}} & 6x - 10y = -4 \\
2x - 3y = 1 \xrightarrow[\text{Multiply by } -3]{} & \underline{-6x + 9y = -3} \\
& -y = -7 \\
& y = 7
\end{array}
$$

The y-coordinate of the solution to the system is 7. Substituting this value of y into any of the equations with both x- and y-variables gives $x = 11$. The solution to the system is $(11, 7)$. It is the only ordered pair that satisfies both equations.   ◀

▶ **EXAMPLE 3**   Solve the system

$$2x - 3y = 4$$

$$4x + 5y = 3$$

Solution   We can eliminate x by multiplying the top equation by -2 and adding it to the bottom equation:

$$
\begin{array}{rl}
2x - 3y = 4 \xrightarrow{\text{Multiply by } -2} & -4x + 6y = -8 \\
4x + 5y = 3 \xrightarrow[\text{No change}]{} & \underline{4x + 5y = 3} \\
& 11y = -5 \\
& y = -\dfrac{5}{11}
\end{array}
$$

The y-coordinate of our solution is $-\frac{5}{11}$. If we were to substitute this value of y back into either of our original equations, we would find the arithmetic necessary to solve for x cumbersome. For this reason, it is probably best to go back to the original system and solve it a second time—for x instead of y. Here is how we do that:

$$2x - 3y = 4 \xrightarrow{\text{Multiply by 5}} 10x - 15y = 20$$
$$4x + 5y = 3 \xrightarrow[\text{Multiply by 3}]{} \underline{12x + 15y = 9}$$
$$22x = 29$$
$$x = \frac{29}{22}$$

The solution to our system is $(\frac{29}{22}, -\frac{5}{11})$. ◀

▶ **EXAMPLE 4** Solve the system

$$5x - 2y = 1$$
$$-10x + 4y = 3$$

Solution We can eliminate y by multiplying the first equation by 2 and adding the result to the second equation:

$$5x - 2y = 1 \xrightarrow{\text{Multiply by 2}} 10x - 4y = 2$$
$$-10x + 4y = 3 \xrightarrow[\text{No change}]{} \underline{-10x + 4y = 3}$$
$$0 = 5$$

The result is the false statement $0 = 5$, which indicates there is no solution to the system. If we were to graph the two lines, we would find that they are parallel. In a case like this, we say the system is **inconsistent.** Whenever both variables have been eliminated and the resulting statement is false, the solution set for the system will be the empty set, $\varnothing$. ◀

▶ **EXAMPLE 5** Solve the system

$$4x - 3y = 2$$
$$8x - 6y = 4$$

Solution Multiplying the top equation by -2 and adding, we can eliminate the variable x:

$$4x - 3y = 2 \xrightarrow{\text{Multiply by } -2} -8x + 6y = -4$$
$$8x - 6y = 4 \xrightarrow[\text{No change}]{} \underline{8x - 6y = 4}$$
$$0 = 0$$

Both variables have been eliminated and the resulting statement $0 = 0$ is true. In this case the lines coincide and the system is said to be **dependent.** The solution set consists of all ordered pairs that satisfy either equation. We can write the solution set as $\{(x, y) \,|\, 4x - 3y = 2\}$ or $\{(x, y) \,|\, 8x - 6y = 4\}$. ◀

The previous two examples illustrate the two special cases in which the graphs of the equations in the system either coincide or are parallel. In both cases the left-hand sides of the equations were multiples of one another. In the case of the dependent equations, the right-hand sides were also multiples. We can generalize these observations as follows:

The equations in the system

$$a_1 x + b_1 y = c_1$$

$$a_2 x + b_2 y = c_2$$

will be inconsistent (their graphs are parallel lines) if

$$\frac{a_1}{a_2} = \frac{b_1}{b_2} \neq \frac{c_1}{c_2}$$

and will be dependent (their graphs will coincide) if

$$\frac{a_1}{a_2} = \frac{b_1}{b_2} = \frac{c_1}{c_2}$$

▶ **EXAMPLE 6**   Solve the system

$$\frac{1}{2} x - \frac{1}{3} y = 2$$

$$\frac{1}{4} x + \frac{2}{3} y = 6$$

Solution   Although we could solve this system without clearing the equations of fractions, there is probably less chance for error if we have only integer coefficients to work with. So let's begin by multiplying both sides of the top equation by 6, and both sides of the bottom equation by 12, to clear each equation of fractions:

$$\frac{1}{2} x - \frac{1}{3} y = 2 \xrightarrow{\text{Times 6}} 3x - 2y = 12$$

$$\frac{1}{4} x + \frac{2}{3} y = 6 \xrightarrow[\text{Times 12}]{} 3x + 8y = 72$$

Now we can eliminate x by multiplying the top equation by -1 and leaving the bottom equation unchanged:

$$\begin{aligned} 3x - 2y = 12 &\xrightarrow{\text{Times } -1} -3x + 2y = -12 \\ 3x + 8y = 72 &\xrightarrow[\text{No change}]{} \underline{3x + 8y = 72} \\ & 10y = 60 \\ & y = 6 \end{aligned}$$

We can substitute $y = 6$ into any equation that contains both x and y. Let's use $3x - 2y = 12$.

$$3x - 2(6) = 12$$
$$3x - 12 = 12$$
$$3x = 24$$
$$x = 8$$

The solution to the system is (8, 6). ◄

We end this section by considering another method of solving a linear system. The method is called the **substitution method** and is shown in the following examples.

▶ **EXAMPLE 7** Solve the system

$$2x - 3y = -6$$
$$y = 3x - 5$$

Solution The second equation tells us y is $3x - 5$. Substituting the expression $3x - 5$ for y in the first equation, we have

$$2x - 3(3x - 5) = -6$$

The result of the substitution is the elimination of the variable y. Solving the resulting linear equation in x as usual, we have

$$2x - 9x + 15 = -6$$
$$-7x + 15 = -6$$
$$-7x = -21$$
$$x = 3$$

Putting $x = 3$ into the second equation in the original system, we have

$$y = 3(3) - 5$$
$$= 9 - 5$$
$$= 4$$

The solution to the system is (3, 4). ◄

▶ **EXAMPLE 8** Solve by substitution

$$2x + 3y = 5$$
$$x - 2y = 6$$

Solution   In order to use the substitution method we must solve one of the two equations for x or y. We can solve for x in the second equation by adding $2y$ to both sides:

$$x - 2y = 6$$

$$x = 2y + 6 \qquad \text{Add } 2y \text{ to both sides}$$

Substituting the expression $2y + 6$ for x in the first equation of our system, we have

$$2(2y + 6) + 3y = 5$$

$$4y + 12 + 3y = 5$$

$$7y + 12 = 5$$

$$7y = -7$$

$$y = -1$$

Using $y = -1$ in either equation in the original system, we find $x = 4$. The solution is $(4, -1)$. ◀

Both the substitution method and the addition method can be used to solve any system of linear equations in two variables. However, systems like the one in Example 7 are easier to solve using the substitution method, since one of the variables is already written in terms of the other. A system like the one in Example 2 is easier to solve using the addition method, since solving for one of the variables would lead to an expression involving fractions. The system in Example 8 could be solved easily by either method, since solving the second equation for x is a one-step process.

USING TECHNOLOGY: Graphing Calculators

Solving Systems That Intersect in Exactly One Point
A graphing calculator can be used to solve a system of equations in two variables if the equations intersect in exactly one point. To solve the system shown in Example 3, we first solve each equation for y. Here is the result:

$$2x - 3y = 4 \quad \text{becomes} \quad y = \frac{4 - 2x}{-3}$$

$$4x + 5y = 3 \quad \text{becomes} \quad y = \frac{3 - 4x}{5}$$

Note that we do not have to write the equations in simplest form since it does not matter to the calculator whether there is a negative number in the denominator or not. Graphing these two functions on the calculator gives a diagram similar to the one in Figure 3 on the next page.

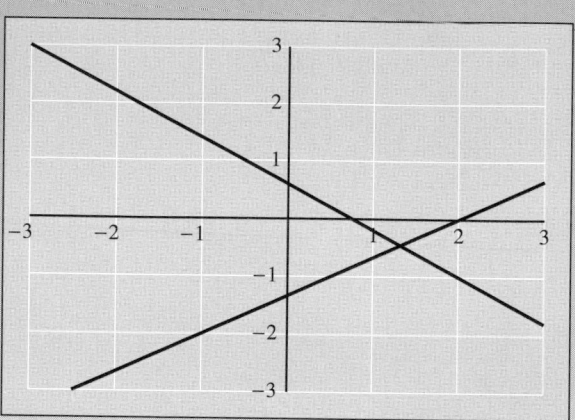

Figure 3

Using Trace and Zoom until we are accurate to the nearest hundredth, we find the two lines intersect at $x = 1.32$ and $y = -0.45$, which are the decimal equivalents (accurate to the nearest hundredth) of the fractions found in Example 3.

Special Cases

We cannot assume that two lines that look parallel on the calculator display are in fact parallel. If you graph the functions $y = x - 5$ and $y = 0.99x + 2$ in a Window where x and y range from -10 to 10, the lines look parallel. However, we know this is not the case, since their slopes are different. As we Zoom out repeatedly, the lines begin to look as if they coincide. We know this is not the case, because the two lines have different y-intercepts. To summarize: If we graph two lines on a calculator and the graphs look as if they are parallel or coincide, we should use algebraic methods, not the calculator, to determine the solution to the system.

P R O B L E M   S E T   8 . 1

Solve each system by graphing both equations on the same set of axes and then reading the solution from the graph.

1. $3x - 2y = 6$
$\quad\; x - \; y = 1$

2. $5x - 2y = 10$
$\quad\; x - \; y = -1$

3. $\quad\; y = \frac{3}{5}x - 3$
$\quad 2x - y = -4$

4. $\quad\; y = \frac{1}{2}x - 2$
$\quad 2x - y = -1$

5. $y = \frac{1}{2}x$
$\quad y = -\frac{3}{4}x + 5$

6. $y = \frac{2}{3}x$
$\quad y = -\frac{1}{3}x + 6$

7. $3x + 3y = -2$
$\quad\quad\; y = -x + 4$

8. $2x - 2y = 6$
$\quad\quad\; y = x - 3$

9. $2x - y = 5$
$\quad\quad y = 2x - 5$

10. $x + 2y = 5$
$\quad\quad y = -\frac{1}{2}x + 3$

Solve each of the following systems by the addition method.

11. $x + y = 5$
$3x - y = 3$

12. $x - y = 4$
$-x + 2y = -3$

13. $3x + y = 4$
$4x + y = 5$

14. $6x - 2y = -10$
$6x + 3y = -15$

15. $3x - 2y = 6$
$6x - 4y = 12$

16. $4x + 5y = -3$
$-8x - 10y = 3$

17. $x + 2y = 0$
$2x - 6y = 5$

18. $x + 3y = 3$
$2x - 9y = 1$

19. $2x - 5y = 16$
$4x - 3y = 11$

20. $5x - 3y = -11$
$7x + 6y = -12$

21. $6x + 3y = -1$
$9x + 5y = 1$

22. $5x + 4y = -1$
$7x + 6y = -2$

23. $4x + 3y = 14$
$9x - 2y = 14$

24. $7x - 6y = 13$
$6x - 5y = 11$

25. $2x - 5y = 3$
$-4x + 10y = 3$

26. $3x - 2y = 1$
$-6x + 4y = -2$

27. $\frac{1}{4}x - \frac{1}{6}y = -2$
$-\frac{1}{6}x + \frac{1}{5}y = 4$

28. $-\frac{1}{3}x + \frac{1}{4}y = 0$
$\frac{1}{5}x - \frac{1}{10}y = 1$

29. $\frac{1}{2}x + \frac{1}{3}y = 13$
$\frac{2}{5}x + \frac{1}{4}y = 10$

30. $\frac{1}{2}x + \frac{1}{3}y = \frac{2}{3}$
$\frac{2}{3}x + \frac{2}{5}y = \frac{14}{15}$

31. $\frac{2}{3}x + \frac{2}{5}y = 4$
$\frac{1}{3}x - \frac{1}{2}y = -\frac{1}{3}$

32. $\frac{1}{2}x - \frac{1}{3}y = \frac{5}{6}$
$-\frac{2}{5}x + \frac{1}{2}y = -\frac{9}{10}$

Solve each of the following systems by the substitution method.

33. $7x - y = 24$
$x = 2y + 9$

34. $3x - y = -8$
$y = 6x + 3$

35. $6x - y = 10$
$y = -\frac{3}{4}x - 1$

36. $2x - y = 6$
$y = -\frac{4}{3}x + 1$

37. $x - y = 4$
$2x - 3y = 6$

38. $x + y = 3$
$2x + 3y = -4$

39. $y = 3x - 2$
$y = 4x - 4$

40. $y = 5x - 2$
$y = -2x + 5$

41. $2x - y = 5$
$4x - 2y = 10$

42. $-10x + 8y = -6$
$y = \frac{5}{4}x$

43. $\frac{1}{3}x - \frac{1}{2}y = 0$
$x = \frac{3}{2}y$

44. $\frac{2}{5}x - \frac{2}{3}y = 0$
$y = \frac{3}{5}x$

You may want to read Example 3 again before solving the systems that follow.

45. $4x - 7y = 3$
$5x + 2y = -3$

46. $3x - 4y = 7$
$6x - 3y = 5$

47. $9x - 8y = 4$
$2x + 3y = 6$

48. $4x - 7y = 10$
$-3x + 2y = -9$

49. $3x - 5y = 2$
$7x + 2y = 1$

50. $4x - 3y = -1$
$5x + 8y = 2$

51. Multiply both sides of the second equation in the following system by 100 and then solve as usual.

$$x + y = 10{,}000$$
$$0.06x + 0.05y = 560$$

52. Multiply both sides of the second equation in the following system by 10 and then solve as usual.

$$x + y = 12$$
$$0.20x + 0.50y = 0.30(12)$$

53. What value of c will make the following system a dependent system (one in which the lines coincide)?

$$6x - 9y = 3$$
$$4x - 6y = c$$

54. What value of c will make the following system a dependent system?

$$5x - 7y = c$$
$$-15x + 21y = 9$$

Applying the Concepts

Problems 55–56 may be solved using a graphing calculator.

55. One telephone company charges 41 cents for the first minute and 32 cents for each additional minute for a certain long-distance phone call. If the number of additional minutes after the first minute is x and the cost, in cents, for the call is y, then the equation that

gives the total cost, in cents, for the call is
$y = 32x + 41$.

(a) How much does it cost to make a 10-minute long-distance call under these conditions?

(b) If a second phone company charges 45 cents for the first minute and 30 cents for each additional minute, write the equation that gives the total cost y of a call in terms of the number of additional minutes x.

(c) After how many additional minutes will the two companies charge an equal amount? (What is the x-coordinate of the point of intersection of the two lines?)

56. In a certain city, a taxi ride costs 75¢ for the first $\frac{1}{7}$ of a mile and 10¢ for every additional $\frac{1}{7}$ of a mile after the first seventh. If x is the number of additional sevenths of a mile, then the total cost y of a taxi ride is

$$y = 10x + 75$$

(a) How much does it cost to ride a taxi for 10 miles in this city?

(b) Suppose a taxi ride in another city costs 50¢ for the first $\frac{1}{7}$ of a mile, and 15¢ for each additional $\frac{1}{7}$ of a mile. Write an equation that gives the total cost y, in cents, to ride x sevenths of a mile past the first seventh, in this city.

(c) Solve the two equations given above simultaneously (as a system of equations)

and explain in words what your solution represents.

Review Problems

The problems that follow review material we covered in Section 7.1.

Solve each equation.

57. $(2x - 1)^2 = 25$ **58.** $(3x + 5)^2 = -12$

59. What number would you add to $x^2 - 10x$ to make it a perfect square trinomial?

60. What number would you add to $x^2 - 5x$ to make it a perfect square trinomial?

Solve by completing the square.

61. $x^2 - 10x + 8 = 0$ **62.** $x^2 - 5x + 4 = 0$

63. $3x^2 - 6x + 6 = 0$

64. $4x^2 - 16x - 8 = 0$

One Step Further

65. Find a and b so that the line $ax + by = 7$ passes through the points $(1, -2)$ and $(3, 1)$.

66. Find a and b so that the line $ax + by = 2$ passes through the points $(2, 2)$ and $(6, 7)$.

67. Find a and b so the parabola $y = ax^2 + bx$ will pass through the points $(-1, 3)$ and $(3, 3)$. Then find the vertex of the parabola.

68. Find a and b so the parabola $y = ax^2 + bx$ will pass through the points $(1, 3)$ and $(-3, 3)$. Then find the vertex of the parabola.

SECTION

8.2   # Systems of Linear Equations in Three Variables

A solution to an equation in three variables such as

$$2x + y - 3z = 6$$

is an ordered triple of numbers (x, y, z). For example, the ordered triples $(0, 0, -2)$, $(2, 2, 0)$, and $(0, 9, 1)$ are solutions to the equation $2x + y - 3z = 6$, since they produce a true statement when their coordinates are replaced for x, y, and z in the equation.

> **DEFINITION** The **solution set** for a system of three linear equations in three variables is the set of ordered triples that satisfy all three equations.

▶ **EXAMPLE 1** Solve the system

$$x + y + z = 6 \qquad (1)$$
$$2x - y + z = 3 \qquad (2)$$
$$x + 2y - 3z = -4 \qquad (3)$$

Solution We want to find the ordered triple (x, y, z) that satisfies all three equations. We have numbered the equations so it will be easier to keep track of where they are and what we are doing.

There are many ways to proceed. The main idea is to take two different pairs of equations and eliminate the same variable from each pair. We begin by adding equations (1) and (2) to eliminate the y-variable. The resulting equation is numbered (4):

$$
\begin{array}{ll}
x + y + z = 6 & (1) \\
\underline{2x - y + z = 3} & (2) \\
3x + 2z = 9 & (4)
\end{array}
$$

Adding twice equation (2) to equation (3) will also eliminate the variable y. The resulting equation is numbered (5):

$$
\begin{array}{ll}
4x - 2y + 2z = 6 & \text{Twice (2)} \\
\underline{x + 2y - 3z = -4} & (3) \\
5x - z = 2 & (5)
\end{array}
$$

Equations (4) and (5) form a linear system in two variables. By multiplying equation (5) by 2 and adding the result to equation (4), we will succeed in eliminating the variable z from the new pair of equations:

$$
\begin{array}{ll}
3x + 2z = 9 & (4) \\
\underline{10x - 2z = 4} & \text{Twice (5)} \\
13x = 13 & \\
x = 1 &
\end{array}
$$

Substituting $x = 1$ into equation (4), we have

$$
\begin{aligned}
3(1) + 2z &= 9 \\
2z &= 6 \\
z &= 3
\end{aligned}
$$

Using $x = 1$ and $z = 3$ in equation (1) gives us

$$1 + y + 3 = 6$$
$$y + 4 = 6$$
$$y = 2$$

The solution is the ordered triple (1, 2, 3).   ◀

▶ **EXAMPLE 2**   Solve the system

$$2x + y - z = 3 \qquad (1)$$
$$3x + 4y + z = 6 \qquad (2)$$
$$2x - 3y + z = 1 \qquad (3)$$

Solution   It is easiest to eliminate z from the equations. The equation produced by adding (1) and (2) is

$$5x + 5y = 9 \qquad (4)$$

The equation that results from adding (1) and (3) is

$$4x - 2y = 4 \qquad (5)$$

Equations (4) and (5) form a linear system in two variables. We can eliminate the variable y from this system as follows:

$$
\begin{aligned}
5x + 5y = 9 \xrightarrow{\text{Multiply by 2}} & 10x + 10y = 18 \\
4x - 2y = 4 \xrightarrow[\text{Multiply by 5}]{} & \underline{20x - 10y = 20} \\
& 30x \qquad\quad = 38
\end{aligned}
$$

$$x = \frac{38}{30}$$

$$x = \frac{19}{15}$$

Substituting $x = \frac{19}{15}$ into equation (5) or equation (4) and solving for y gives

$$y = \frac{8}{15}$$

Using $x = \frac{19}{15}$ and $y = \frac{8}{15}$ in equation (1), (2), or (3) and solving for z results in

$$z = \frac{1}{15}$$

The ordered triple that satisfies all three equations is $(\frac{19}{15}, \frac{8}{15}, \frac{1}{15})$.   ◀

▶ **EXAMPLE 3**   Solve the system

$$2x + 3y - z = 5 \qquad (1)$$
$$4x + 6y - 2z = 10 \qquad (2)$$
$$x - 4y + 3z = 5 \qquad (3)$$

Solution   Multiplying equation (1) by -2 and adding the result to equation (2) looks like this:

$$
\begin{array}{ll}
-4x - 6y + 2z = -10 & -2 \text{ times (1)} \\
\underline{4x + 6y - 2z = 10} & (2) \\
\qquad\qquad 0 = 0 &
\end{array}
$$

All three variables have been eliminated, and we are left with a true statement. As was the case in Section 8.1, this finding implies that the two equations are dependent.

With a system of three equations in three variables, however, a dependent system can have no solution or an infinite number of solutions. After we have concluded the examples in this section, we will discuss the geometry behind these systems. Doing so will give you some additional insight into dependent systems.   ◀

▶ **EXAMPLE 4**   Solve the system

$$x - 5y + 4z = 8 \qquad (1)$$
$$3x + y - 2z = 7 \qquad (2)$$
$$-9x - 3y + 6z = 5 \qquad (3)$$

Solution   Multiplying equation (2) by 3 and adding the result to equation (3) produces

$$
\begin{array}{ll}
9x + 3y - 6z = 21 & 3 \text{ times (2)} \\
\underline{-9x - 3y + 6z = \ \ 5} & (3) \\
\qquad\qquad 0 = 26 &
\end{array}
$$

In this case all three variables have been eliminated, and we are left with a false statement. The two equations are inconsistent; there are no ordered triples that satisfy both equations. The solution set for the system is the empty set, $\varnothing$. If equations (2) and (3) have no ordered triples in common, then certainly (1), (2), and (3) do not either.   ◀

▶ **EXAMPLE 5** Solve the system

$$x + 3y = 5 \qquad (1)$$
$$6y + z = 12 \qquad (2)$$
$$x - 2z = -10 \qquad (3)$$

Solution It may be helpful to rewrite the system as

$$x + 3y \qquad = 5 \qquad (1)$$
$$6y + z = 12 \qquad (2)$$
$$x \qquad - 2z = -10 \qquad (3)$$

Equation (2) does not contain the variable x. If we multiply equation (3) by -1 and add the result to equation (1), we will be left with another equation that does not contain the variable x:

$$\begin{array}{ll} x + 3y \qquad = 5 & (1) \\ -x \qquad + 2z = 10 & -1 \text{ times } (3) \\ \hline 3y + 2z = 15 & (4) \end{array}$$

Equations (2) and (4) form a linear system in two variables. Multiplying equation (2) by -2 and adding the result to equation (4) eliminates the variable z:

$$\begin{array}{l} 6y + z = 12 \xrightarrow{\text{Multiply by } -2} -12y - 2z = -24 \\ 3y + 2z = 15 \xrightarrow[\text{No change}]{} \underline{ 3y + 2z = 15} \\ -9y = -9 \\ y = 1 \end{array}$$

Using $y = 1$ in equation (4) and solving for z, we have

$$z = 6$$

Substituting $y = 1$ into equation (1) gives

$$x = 2$$

The ordered triple that satisfies all three equations is (2, 1, 6). ◀

The Geometry Behind Equations in Three Variables

We can graph an ordered triple on a coordinate system with three axes. The graph will be a point in space. The coordinate system is drawn in perspective; you have to imagine that the x-axis comes out of the paper and is perpendicular to both the y-axis and the z-axis. To graph the point (3, 4, 5) we move 3 units in the x-direction, 4 units in the y-direction, and then 5 units in the z-direction, as shown in Figure 1.

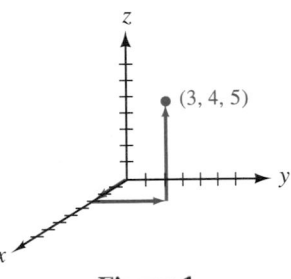

(3, 4, 5)

Figure 1

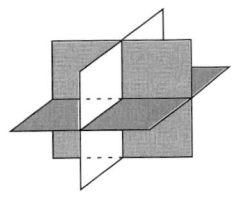

Case 1 The three planes have exactly one point in common, as in Examples 1, 2, and 5.

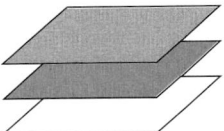

Case 2 The three planes have no points in common. The system they represent is an inconsistent system.

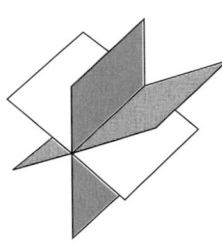

Case 3 The three planes intersect in a line. Any point on the line is a solution to the system of equations represented by the planes, so there is an infinite number of solutions to the system. This is an example of a dependent system.

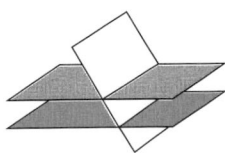

Case 4 Two of the planes are parallel, while the third plane intersects each of the parallel planes. There is no solution to the system; it is an inconsistent system.

Although in actual practice it is sometimes difficult to graph equations in three variables, if we were to graph a linear equation in three variables we would find that the graph was a plane in space. A system of three equations in three variables is represented by three planes in space.

There are a number of possible ways in which these three planes can intersect, some of which are shown in the margin on this page. And there are still other possibilities that are not among those shown in the margin.

In Example 3, we found that equations 1 and 2 were dependent equations. They represent the same plane. That is, they have all their points in common. But the system of equations that they came from has either no solution or an infinite number of solutions. It all depends on the third plane. If the third plane coincides with the first two, then the solution to the system is a plane. If the third plane is parallel to the first two, then there is no solution to the system. And finally, if the third plane intersects the first two, but does not coincide with them, then the solution to the system is that line of intersection.

In Example 4 we found that trying to eliminate a variable from the second and third equations resulted in a false statement. This means that the two planes represented by these equations are parallel. It makes no difference where the third plane is; there is no solution to the system in Example 4. (If we were to graph the three planes from Example 4, we would obtain a diagram similar to Case 2 or Case 4 in the margin.)

If, in the process of solving a system of linear equations in three variables, we eliminate all the variables from a pair of equations and are left with a false statement, we will say the system is inconsistent. If we eliminate all the variables and are left with a true statement, then we will say the system is a dependent one.

PROBLEM SET 8.2

Solve the following systems.

1. $x + y + z = 4$
$x - y + 2z = 1$
$x - y - 3z = -4$

2. $x - y - 2z = -1$
$x + y + z = 6$
$x + y - z = 4$

3. $x + y + z = 6$
$x - y + 2z = 7$
$2x - y - 4z = -9$

4. $x + y + z = 0$
$x + y - z = 6$
$x - y + 2z = -7$

5. $x + 2y + z = 3$
$2x - y + 2z = 6$
$3x + y - z = 5$

6. $2x + y - 3z = -14$
$x - 3y + 4z = 22$
$3x + 2y + z = 0$

7. $2x + 3y - 2z = 4$
$x + 3y - 3z = 4$
$3x - 6y + z = -3$

8. $4x + y - 2z = 0$
$2x - 3y + 3z = 9$
$-6x - 2y + z = 0$

9. $-x + 4y - 3z = 2$
$2x - 8y + 6z = 1$
$3x - y + z = 0$

10. $4x + 6y - 8z = 1$
$-6x - 9y + 12z = 0$
$x - 2y - 2z = 3$

11. $\frac{1}{2}x - y + z = 0$
$2x + \frac{1}{3}y + z = 2$
$x + y + z = -4$

12. $\frac{1}{3}x + \frac{1}{2}y + z = -1$
$x - y + \frac{1}{5}z = 1$
$x + y + z = 5$

13. $2x - y - 3z = 1$
$x + 2y + 4z = 3$
$4x - 2y - 6z = 2$

14. $3x + 2y + z = 3$
$x - 3y + z = 4$
$-6x - 4y - 2z = 1$

15. $2x - y + 3z = 4$
$x + 2y - z = -3$
$4x + 3y + 2z = -5$

16. $6x - 2y + z = 5$
$3x + y + 3z = 7$
$x + 4y - z = 4$

17. $x + y = 9$
$y + z = 7$
$x - z = 2$

18. $x - y = -3$
$x + z = 2$
$y - z = 7$

19. $2x + y = 2$
$y + z = 3$
$4x - z = 0$

20. $2x + y = 6$
$3y - 2z = -8$
$x + z = 5$

21. $2x - 3y = 0$
$6y - 4z = 1$
$x + 2z = 1$

22. $3x + 2y = 3$
$y + 2z = 2$
$6x - 4z = 1$

23. $\frac{1}{2}x + \frac{2}{3}y = \frac{5}{2}$
$\frac{1}{5}x - \frac{1}{2}z = -\frac{3}{10}$
$\frac{1}{3}y - \frac{1}{4}z = \frac{3}{4}$

24. $\frac{1}{2}x - \frac{1}{3}y = \frac{1}{6}$
$\frac{1}{3}y - \frac{1}{3}z = 1$
$\frac{1}{5}x - \frac{1}{2}z = -\frac{4}{5}$

25. $\frac{1}{2}x - \frac{1}{4}y + \frac{1}{2}z = -2$
$\frac{1}{4}x - \frac{1}{12}y - \frac{1}{3}z = \frac{1}{4}$
$\frac{1}{6}x + \frac{1}{3}y - \frac{1}{2}z = \frac{3}{2}$

26. $\frac{1}{2}x + \frac{1}{2}y + z = \frac{1}{2}$
$\frac{1}{2}x - \frac{1}{4}y - \frac{1}{4}z = 0$
$\frac{1}{4}x + \frac{1}{12}y + \frac{1}{6}z = \frac{1}{6}$

Applying the Concepts

27. In the following diagram of an electrical circuit, x, y, and z represent the amount of current (in amperes) flowing across the 5-ohm, 20-ohm, and 10-ohm resistors, respectively. (In circuit diagrams resistors are represented by $-W$ and potential differences by $\dashv\vdash$.)

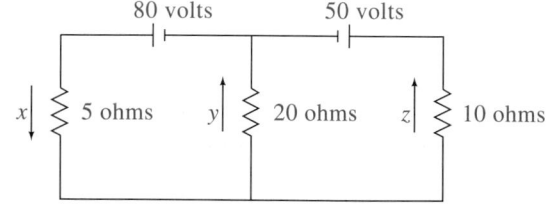

The system of equations used to find the three currents x, y, and z is

$$x - y - z = 0$$
$$5x + 20y = 80$$
$$20y - 10z = 50$$

Solve the system for all variables.

28. If a car rental company charges $10 a day and 8¢ a mile to rent one of its cars, then the cost z, in dollars, to rent a car for x days and drive y miles can be found from the equation

$$z = 10x + 0.08y$$

(a) How much does it cost to rent a car for 2 days and drive it 200 miles under these conditions?

(b) A second company charges $12 a day and 6¢ a mile for the same car. Write an equation that gives the cost z, in dollars, to rent a car from this company for x days and drive it y miles.

(c) A car is rented from each of the companies mentioned above for 2 days. To find the mileage at which the cost of renting the cars from each of the two companies will be equal, solve the following system for y:

$$z = 10x + 0.08y$$
$$z = 12x + 0.06y$$
$$x = 2$$

Review Problems

The problems below review material we covered in Sections 2.2 and 7.2. Reviewing the problems from Section 2.2 will help you with some of the next section.

Solve each equation for y. [2.2]

29. $2x - 3y = 6$ **30.** $3x + 2y = 6$

31. $2x - 3y = 5$ **32.** $3x - 2y = 5$

Solve. [7.2]

33. $2x^2 + 4x - 3 = 0$ **34.** $3x^2 + 4x - 2 = 0$
35. $(2y - 3)(2y - 1) = -4$
36. $(y - 1)(3y - 3) = 10$
37. $t^3 - 125 = 0$ **38.** $8t^3 + 1 = 0$
39. $4x^5 - 16x^4 = 20x^3$ **40.** $3x^4 + 6x^2 = 6x^3$

41. $\dfrac{1}{x - 3} + \dfrac{1}{x + 2} = 1$

42. $\dfrac{1}{x + 3} + \dfrac{1}{x - 2} = 1$

One Step Further

43. Find a, b, and c so that the parabola $y = ax^2 + bx + c$ passes through the points $(1, 0)$, $(3, -4)$, and $(5, 0)$.

44. Find a, b, and c so that the parabola $y = ax^2 + bx + c$ passes through the points $(0, 3)$, $(1, 0)$, and $(-1, 4)$.

Solve each system for the solution (x, y, z, w).

45.
$$\begin{aligned} x + \ y + z + \ w &= \ \ 10 \\ x + 2y - z + \ w &= \ \ \ 6 \\ x - \ y - z + 2w &= \ \ \ 4 \\ x - 2y + z - 3w &= -12 \end{aligned}$$

46.
$$\begin{aligned} x + \ y + \ z + \ w &= \ 16 \\ x - \ y + 2z - \ w &= \ \ 1 \\ x + 3y - \ z - \ w &= -2 \\ x - 3y - 2z + 2w &= -4 \end{aligned}$$

SECTION

8.3   Introduction to Determinants

In this section we will expand and evaluate **determinants**. The purpose of this section is simply to be able to find the value of a given determinant. As we will see in the next section, determinants are very useful in solving systems of linear equations. Before we apply determinants to systems of linear equations, however, we must practice calculating the value of some determinants.

DEFINITION The value of the **2 × 2** (2 by 2) **determinant**

$$\begin{vmatrix} a & c \\ b & d \end{vmatrix}$$

is given by

$$\begin{vmatrix} a & c \\ b & d \end{vmatrix} = ad - bc$$

From the preceding definition we see that a determinant is simply a square array of numbers with two vertical lines enclosing it. The value of a 2 × 2 determinant is found by cross-multiplying on the diagonals and then subtracting, a diagram of which looks like

$$\begin{vmatrix} a & c \\ b & d \end{vmatrix} = ad - bc$$

▶ **EXAMPLE 1** Find the value of the following 2 × 2 determinants:

(a) $\begin{vmatrix} 1 & 2 \\ 3 & 4 \end{vmatrix} = 1(4) - 3(2) = 4 - 6 = -2$

(b) $\begin{vmatrix} 3 & 5 \\ -2 & 7 \end{vmatrix} = 3(7) - (-2)5 = 21 + 10 = 31$ ◀

▶ **EXAMPLE 2** Solve for x if

$$\begin{vmatrix} x^2 & 2 \\ x & 1 \end{vmatrix} = 8$$

Solution We expand the determinant on the left side to get

$$x^2(1) - x(2) = 8$$
$$x^2 - 2x = 8$$
$$x^2 - 2x - 8 = 0$$
$$(x - 4)(x + 2) = 0$$
$$x - 4 = 0 \quad \text{or} \quad x + 2 = 0$$
$$x = 4 \quad \text{or} \quad x = -2$$ ◀

We now turn our attention to 3 × 3 determinants. A 3 × 3 determinant is also a square array of numbers, the value of which is given by the following definition.

DEFINITION   The value of the **3 × 3 determinant**

$$\begin{vmatrix} a_1 & b_1 & c_1 \\ a_2 & b_2 & c_2 \\ a_3 & b_3 & c_3 \end{vmatrix}$$

is given by

$$\begin{vmatrix} a_1 & b_1 & c_1 \\ a_2 & b_2 & c_2 \\ a_3 & b_3 & c_3 \end{vmatrix} = a_1 b_2 c_3 + a_3 b_1 c_2 + a_2 b_3 c_1 - a_3 b_2 c_1 - a_1 b_3 c_2 - a_2 b_1 c_3$$

At first glance, the expansion of a 3 × 3 determinant looks a little complicated. There are actually two different methods used to find the six products given above that simplify matters somewhat.

Method 1   We begin by writing the determinant with the first two columns repeated on the right:

$$\begin{vmatrix} a_1 & b_1 & c_1 \\ a_2 & b_2 & c_2 \\ a_3 & b_3 & c_3 \end{vmatrix} \begin{matrix} a_1 & b_1 \\ a_2 & b_2 \\ a_3 & b_3 \end{matrix}$$

The positive products in the definition come from multiplying down the three full diagonals:

The negative products come from multiplying up the three full diagonals:

▶ **EXAMPLE 3**   Find the value of

$$\begin{vmatrix} 1 & 3 & -2 \\ 2 & 0 & 1 \\ 4 & -1 & 1 \end{vmatrix}$$

Solution Repeating the first two columns and then finding the products down the diagonals and the products up the diagonals as given in Method 1, we have

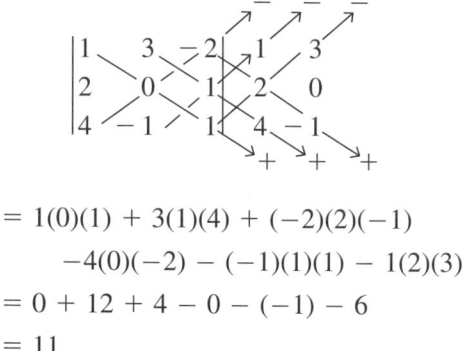

$$= 1(0)(1) + 3(1)(4) + (-2)(2)(-1)$$
$$-4(0)(-2) - (-1)(1)(1) - 1(2)(3)$$
$$= 0 + 12 + 4 - 0 - (-1) - 6$$
$$= 11 \quad \blacktriangleleft$$

Method 2 The second method of evaluating a 3×3 determinant is called **expansion by minors.**

DEFINITION The **minor** for an element in a 3×3 determinant is the determinant consisting of the elements remaining when the row and column to which the element belongs are deleted. For example, in the determinant

$$\begin{vmatrix} a_1 & b_1 & c_1 \\ a_2 & b_2 & c_2 \\ a_3 & b_3 & c_3 \end{vmatrix}$$

Minor for element $a_1 = \begin{vmatrix} b_2 & c_2 \\ b_3 & c_3 \end{vmatrix}$

Minor for element $b_2 = \begin{vmatrix} a_1 & c_1 \\ a_3 & c_3 \end{vmatrix}$

Minor for element $c_3 = \begin{vmatrix} a_1 & b_1 \\ a_2 & b_2 \end{vmatrix}$

Before we can evaluate a 3×3 determinant by Method 2, we must first define what is known as the *sign array* for a 3×3 determinant.

DEFINITION   The **sign array** for a 3×3 determinant is a 3×3 array of signs in the following pattern:

$$\begin{vmatrix} + & - & + \\ - & + & - \\ + & - & + \end{vmatrix}$$

The sign array begins with a $+$ sign in the upper left-hand corner. The signs then alternate between $+$ and $-$ across every row and down every column.

Strategy for Evaluating a 3×3 Determinant by Expansion of Minors

We can evaluate a 3×3 determinant by expanding across any row or down any column as follows:

Step 1: Choose a row or column to expand about.
Step 2: Write the product of each element in the row or column chosen in step 1 with its minor.
Step 3: Connect the three products in step 2 with the signs in the corresponding row or column in the sign array.

To illustrate the procedure, we will use the same determinant we used in Example 3.

▶ **EXAMPLE 4**   Expand across the first row:

$$\begin{vmatrix} 1 & 3 & -2 \\ 2 & 0 & 1 \\ 4 & -1 & 1 \end{vmatrix}$$

Solution   The products of the three elements in row 1 with their minors are

$$1 \begin{vmatrix} 0 & 1 \\ -1 & 1 \end{vmatrix} \quad 3 \begin{vmatrix} 2 & 1 \\ 4 & 1 \end{vmatrix} \quad (-2) \begin{vmatrix} 2 & 0 \\ 4 & -1 \end{vmatrix}$$

Connecting these three products with the signs from the first row of the sign array, we have

$$+1 \begin{vmatrix} 0 & 1 \\ -1 & 1 \end{vmatrix} - 3 \begin{vmatrix} 2 & 1 \\ 4 & 1 \end{vmatrix} + (-2) \begin{vmatrix} 2 & 0 \\ 4 & -1 \end{vmatrix}$$

We complete the problem by evaluating each of the three 2×2 determinants and then simplifying the resulting expression:

$$+1[0 - (-1)] - 3(2 - 4) + (-2)(-2 - 0)$$
$$= 1(1) - 3(-2) + (-2)(-2)$$
$$= 1 + 6 + 4$$
$$= 11$$

The results of Examples 3 and 4 match. It makes no difference which method we use — the value of a 3 × 3 determinant is unique. ◀

Note The method shown in Example 4 is actually more valuable than our first method, because it will work with any size determinant from 3 × 3 to 4 × 4 to any higher order determinant. Method 1 works only on 3 × 3 determinants. It cannot be used on a 4 × 4 determinant.

▶ **EXAMPLE 5** Expand down column 2:

$$\begin{vmatrix} 2 & 3 & -2 \\ 1 & 4 & 1 \\ 1 & 5 & -1 \end{vmatrix}$$

Solution We connect the products of elements in column 2 and their minors with the signs from the second column in the sign array:

$$\begin{vmatrix} 2 & 3 & -2 \\ 1 & 4 & 1 \\ 1 & 5 & -1 \end{vmatrix} = -3\begin{vmatrix} 1 & 1 \\ 1 & -1 \end{vmatrix} + 4\begin{vmatrix} 2 & -2 \\ 1 & -1 \end{vmatrix} - 5\begin{vmatrix} 2 & -2 \\ 1 & 1 \end{vmatrix}$$
$$= -3(-1 - 1) + 4[-2 - (-2)] - 5[2 - (-2)]$$
$$= -3(-2) + 4(0) - 5(4)$$
$$= 6 + 0 - 20$$
$$= -14$$ ◀

PROBLEM SET 8.3

Find the value of the following 2 × 2 determinants.

1. $\begin{vmatrix} 1 & 0 \\ 2 & 3 \end{vmatrix}$

2. $\begin{vmatrix} 5 & 4 \\ 3 & 2 \end{vmatrix}$

3. $\begin{vmatrix} 2 & 1 \\ 3 & 4 \end{vmatrix}$

4. $\begin{vmatrix} 4 & 1 \\ 5 & 2 \end{vmatrix}$

5. $\begin{vmatrix} 0 & 1 \\ 1 & 0 \end{vmatrix}$

6. $\begin{vmatrix} 1 & 0 \\ 0 & 1 \end{vmatrix}$

7. $\begin{vmatrix} -3 & 2 \\ 6 & -4 \end{vmatrix}$

8. $\begin{vmatrix} 8 & -3 \\ -2 & 5 \end{vmatrix}$

9. $\begin{vmatrix} -3 & -1 \\ 4 & -2 \end{vmatrix}$

10. $\begin{vmatrix} 5 & 3 \\ 7 & -6 \end{vmatrix}$

Solve each of the following for x.

11. $\begin{vmatrix} 2x & 1 \\ x & 3 \end{vmatrix} = 10$

12. $\begin{vmatrix} 3x & -2 \\ 2x & 3 \end{vmatrix} = 26$

13. $\begin{vmatrix} 1 & 2x \\ 2 & -3x \end{vmatrix} = 21$

14. $\begin{vmatrix} -5 & 4x \\ 1 & -x \end{vmatrix} = 27$

15. $\begin{vmatrix} 2x & -4 \\ 2 & x \end{vmatrix} = -8x$

16. $\begin{vmatrix} 3x & 2 \\ 2 & x \end{vmatrix} = -11x$

17. $\begin{vmatrix} x^2 & 3 \\ x & 1 \end{vmatrix} = 10$ **18.** $\begin{vmatrix} x^2 & -2 \\ x & 1 \end{vmatrix} = 35$

Find the value of each of the following 3 × 3 determinants by using Method 1 of this section.

19. $\begin{vmatrix} 1 & 2 & 0 \\ 0 & 2 & 1 \\ 1 & 1 & 1 \end{vmatrix}$ **20.** $\begin{vmatrix} -1 & 0 & 2 \\ 3 & 0 & 1 \\ 0 & 1 & 3 \end{vmatrix}$

21. $\begin{vmatrix} 1 & 2 & 3 \\ 3 & 2 & 1 \\ 1 & 1 & 1 \end{vmatrix}$ **22.** $\begin{vmatrix} -1 & 2 & 0 \\ 3 & -2 & 1 \\ 0 & 5 & 4 \end{vmatrix}$

Find the value of each determinant by using Method 2 and expanding across the first row.

23. $\begin{vmatrix} 0 & 1 & 2 \\ 1 & 0 & 1 \\ -1 & 2 & 0 \end{vmatrix}$ **24.** $\begin{vmatrix} 3 & -2 & 1 \\ 0 & -1 & 0 \\ 2 & 0 & 1 \end{vmatrix}$

25. $\begin{vmatrix} 3 & 0 & 2 \\ 0 & -1 & -1 \\ 4 & 0 & 0 \end{vmatrix}$ **26.** $\begin{vmatrix} 1 & 1 & 1 \\ 1 & -1 & 1 \\ 1 & 1 & -1 \end{vmatrix}$

Find the value of each of the following determinants.

27. $\begin{vmatrix} 2 & -1 & 0 \\ 1 & 0 & -2 \\ 0 & 1 & 2 \end{vmatrix}$ **28.** $\begin{vmatrix} 5 & 0 & -4 \\ 0 & 1 & 3 \\ -1 & 2 & -1 \end{vmatrix}$

29. $\begin{vmatrix} 1 & 3 & 7 \\ -2 & 6 & 4 \\ 3 & 7 & -1 \end{vmatrix}$ **30.** $\begin{vmatrix} 2 & 1 & 5 \\ 6 & -3 & 4 \\ 8 & 9 & -2 \end{vmatrix}$

Applying the Concepts

31. Show that the following determinant equation is another way to write the slope-intercept form of the equation of a line.

$$\begin{vmatrix} y & x \\ m & 1 \end{vmatrix} = b$$

32. Show that the following determinant equation is another way to write the equation $F = \frac{9}{5}C + 32$.

$$\begin{vmatrix} C & F & 1 \\ 5 & 41 & 1 \\ -10 & 14 & 1 \end{vmatrix} = 0$$

Review Problems

The problems that follow review material we covered in Section 7.3.

Use the discriminant to find the number and kind of solutions to the following equations.

33. $2x^2 - 5x + 4 = 0$ **34.** $4x^2 - 12x = -9$

For each of the following problems, find an equation with the given solutions.

35. $x = -3, x = 5$
36. $x = 2, x = -2, x = 1$
37. $y = \frac{2}{3}, y = 3$ **38.** $y = -\frac{3}{5}, y = 2$
39. Find all solutions to $x^3 - 8x^2 + 21x - 18 = 0$ if $x = 3$ is one solution.
40. Find all solutions to $3x^3 - 2x^2 - 10x + 4 = 0$ if $x = 2$ is one solution.

One Step Further

A 4 × 4 determinant can be evaluated only by using Method 2, expansion by minors; Method 1 will not work. Below is a 4 × 4 determinant and its associated sign array.

$$\begin{vmatrix} 2 & 0 & 1 & -3 \\ -1 & 2 & 0 & 1 \\ -3 & 0 & 1 & 0 \\ 1 & 1 & 0 & 0 \end{vmatrix} \quad \begin{vmatrix} + & - & + & - \\ - & + & - & + \\ + & - & + & - \\ - & + & - & + \end{vmatrix}$$

4 × 4 determinant   4 × 4 sign array

41. Use expansion by minors to evaluate the 4 × 4 determinant above by expanding it across row 1.
42. Evaluate the determinant above by expanding it down column 4.
43. Use expansion by minors down column 3 to evaluate the determinant above.
44. Evaluate the determinant above by expanding it across row 4.

8.4   Cramer's Rule

We begin this section with a look at how determinants can be used to solve a system of linear equations in two variables. The method we use is called **Cramer's rule.** We state it here as a theorem without proof.

Theorem 8.1 (Cramer's Rule)

The solution to the system

$$a_1 x + b_1 y = c_1$$
$$a_2 x + b_2 y = c_2$$

is given by

$$x = \frac{D_x}{D}, \quad y = \frac{D_y}{D}$$

where

$$D = \begin{vmatrix} a_1 & b_1 \\ a_2 & b_2 \end{vmatrix} \qquad D_x = \begin{vmatrix} c_1 & b_1 \\ c_2 & b_2 \end{vmatrix} \qquad D_y = \begin{vmatrix} a_1 & c_1 \\ a_2 & c_2 \end{vmatrix} \qquad (D \neq 0)$$

The determinant D is made up of the coefficients of x and y in the original system. The determinants D_x and D_y are found by replacing the coefficients of x or y by the constant terms in the original system. Notice also that Cramer's rule does not apply if $D = 0$. In this case the equations are either inconsistent or dependent.

▶ **EXAMPLE 1**   Use Cramer's rule to solve

$$2x - 3y = 4$$
$$4x + 5y = 3$$

Solution   We begin by calculating the determinants D, D_x, and D_y:

$$D = \begin{vmatrix} 2 & -3 \\ 4 & 5 \end{vmatrix} = 2(5) - 4(-3) = 22$$

$$D_x = \begin{vmatrix} 4 & -3 \\ 3 & 5 \end{vmatrix} = 4(5) - 3(-3) = 29$$

$$D_y = \begin{vmatrix} 2 & 4 \\ 4 & 3 \end{vmatrix} = 2(3) - 4(4) = -10$$

$$x = \frac{D_x}{D} = \frac{29}{22} \quad \text{and} \quad y = \frac{D_y}{D} = \frac{-10}{22} = -\frac{5}{11}$$

The solution set for the system is $\{(\frac{29}{22}, -\frac{5}{11})\}$. ◀

Cramer's rule can be applied to systems of linear equations in three variables also.

Theorem 8.2 (also Cramer's Rule)

The solution set to the system

$$a_1 x + b_1 y + c_1 z = d_1$$
$$a_2 x + b_2 y + c_2 z = d_2$$
$$a_3 x + b_3 y + c_3 z = d_3$$

is given by

$$x = \frac{D_x}{D}, \quad y = \frac{D_y}{D}, \quad \text{and} \quad z = \frac{D_z}{D}$$

where

$$D = \begin{vmatrix} a_1 & b_1 & c_1 \\ a_2 & b_2 & c_2 \\ a_3 & b_3 & c_3 \end{vmatrix} \qquad D_x = \begin{vmatrix} d_1 & b_1 & c_1 \\ d_2 & b_2 & c_2 \\ d_3 & b_3 & c_3 \end{vmatrix} \qquad (D \neq 0)$$

$$D_y = \begin{vmatrix} a_1 & d_1 & c_1 \\ a_2 & d_2 & c_2 \\ a_3 & d_3 & c_3 \end{vmatrix} \qquad D_z = \begin{vmatrix} a_1 & b_1 & d_1 \\ a_2 & b_2 & d_2 \\ a_3 & b_3 & d_3 \end{vmatrix}$$

Again the determinant D consists of the coefficients of x, y, and z in the original system. The determinants D_x, D_y, and D_z are found by replacing the coefficients of x, y, and z, respectively, with the constant terms from the original system. If $D = 0$, there is no unique solution to the system.

▶ **EXAMPLE 2** Use Cramer's rule to solve

$$x + y + z = 6$$
$$2x - y + z = 3$$
$$x + 2y - 3z = -4$$

Solution This is the same system used in Example 1 in Section 8.2. We begin by setting up and evaluating $D, D_x, D_y,$ and D_z. (Recall that there are a number of ways to evaluate a 3×3 determinant. Since we have four of these determinants, we can use both Methods 1 and 2 from the previous section.) We evaluate D using Method 1 from Section 8.3.

$$D = \begin{vmatrix} 1 & 1 & 1 \\ 2 & -1 & 1 \\ 1 & 2 & -3 \end{vmatrix}$$

$$= 3 + 1 + 4 - (-1) - (2) - (-6) = 13$$

We evaluate D_x using Method 2 from Section 8.3 and expanding across row 1:

$$D_x = \begin{vmatrix} 6 & 1 & 1 \\ 3 & -1 & 1 \\ -4 & 2 & -3 \end{vmatrix} = 6 \begin{vmatrix} -1 & 1 \\ 2 & -3 \end{vmatrix} - 1 \begin{vmatrix} 3 & 1 \\ -4 & -3 \end{vmatrix} + 1 \begin{vmatrix} 3 & -1 \\ -4 & 2 \end{vmatrix}$$

$$= 6(1) - 1(-5) + 1(2)$$

$$= 13$$

Find D_y by expanding across row 2:

$$D_y = \begin{vmatrix} 1 & 6 & 1 \\ 2 & 3 & 1 \\ 1 & -4 & -3 \end{vmatrix} = -2 \begin{vmatrix} 6 & 1 \\ -4 & -3 \end{vmatrix} + 3 \begin{vmatrix} 1 & 1 \\ 1 & -3 \end{vmatrix} - 1 \begin{vmatrix} 1 & 6 \\ 1 & -4 \end{vmatrix}$$

$$= -2(-14) + 3(-4) - 1(-10)$$

$$= 26$$

Find D_z by expanding down column 1:

$$D_z = \begin{vmatrix} 1 & 1 & 6 \\ 2 & -1 & 3 \\ 1 & 2 & -4 \end{vmatrix} = 1 \begin{vmatrix} -1 & 3 \\ 2 & -4 \end{vmatrix} - 2 \begin{vmatrix} 1 & 6 \\ 2 & -4 \end{vmatrix} + 1 \begin{vmatrix} 1 & 6 \\ -1 & 3 \end{vmatrix}$$

$$= 1(-2) - 2(-16) + 1(9)$$

$$= 39$$

$$x = \frac{D_x}{D} = \frac{13}{13} = 1 \qquad y = \frac{D_y}{D} = \frac{26}{13} = 2 \qquad z = \frac{D_z}{D} = \frac{39}{13} = 3$$

The solution set is $\{(1, 2, 3)\}$. ◀

▶ **EXAMPLE 3**   Use Cramer's rule to solve

$$x + y = -1$$
$$2x - z = 3$$
$$y + 2z = -1$$

Solution   It is helpful to rewrite the system using zeros for the coefficients of those variables not shown:

$$x + y + 0z = -1$$
$$2x + 0y - z = 3$$
$$0x + y + 2z = -1$$

The four determinants used in Cramer's rule are

$$D = \begin{vmatrix} 1 & 1 & 0 \\ 2 & 0 & -1 \\ 0 & 1 & 2 \end{vmatrix} = -3$$

$$D_x = \begin{vmatrix} -1 & 1 & 0 \\ 3 & 0 & -1 \\ -1 & 1 & 2 \end{vmatrix} = -6$$

$$D_y = \begin{vmatrix} 1 & -1 & 0 \\ 2 & 3 & -1 \\ 0 & -1 & 2 \end{vmatrix} = 9$$

$$D_z = \begin{vmatrix} 1 & 1 & -1 \\ 2 & 0 & 3 \\ 0 & 1 & -1 \end{vmatrix} = -3$$

$$x = \frac{D_x}{D} = \frac{-6}{-3} = 2, \qquad y = \frac{D_y}{D} = \frac{9}{-3} = -3, \qquad z = \frac{D_z}{D} = \frac{-3}{-3} = 1$$

The solution set is $\{(2, -3, 1)\}$.   ◀

Finally, we should mention the possible situations that can occur when the determinant D is 0, when we are using Cramer's rule.

If $D = 0$ and at least one of the other determinants, D_x or D_y (or D_z), is not 0, then the system is inconsistent. In this case, there is no solution to the system.

On the other hand, if $D = 0$ and both D_x and D_y (and D_z in a system of three equations in three variables) are 0, then the system is a dependent one.

PROBLEM SET 8.4

Solve each of the following systems using Cramer's rule.

1. $2x - 3y = 3$
$4x - 2y = 10$

2. $3x + y = -2$
$-3x + 2y = -4$

3. $5x - 2y = 4$
$-10x + 4y = 1$

4. $-4x + 3y = -11$
$5x + 4y = 6$

5. $4x - 7y = 3$
$5x + 2y = -3$

6. $3x - 4y = 7$
$6x - 2y = 5$

7. $9x - 8y = 4$
$2x + 3y = 6$

8. $4x - 7y = 10$
$-3x + 2y = -9$

9. $x + y + z = 4$
$x - y - z = 2$
$2x + 2y - z = 2$

10. $-x + y + 3z = 6$
$x + y + 2z = 7$
$2x + 3y + z = 4$

11. $x + y - z = 2$
$-x + y + z = 3$
$x + y + z = 4$

12. $-x - y + z = 1$
$x - y + z = 3$
$x + y - z = 4$

13. $3x - y + 2z = 4$
$6x - 2y + 4z = 8$
$x - 5y + 2z = 1$

14. $2x - 3y + z = 1$
$3x - y - z = 4$
$4x - 6y + 2z = 3$

15. $2x - y + 3z = 4$
$x - 5y - 2z = 1$
$-4x - 2y + z = 3$

16. $4x - y + 5z = 1$
$2x + 3y + 4z = 5$
$x + y + 3z = 2$

17. $-x - 7y = 1$
$x + 3z = 11$
$2y + z = 0$

18. $x + y = 2$
$-x + 3z = 0$
$2y + z = 3$

19. $x - y = 2$
$3x + z = 11$
$y - 2z = -3$

20. $4x + 5y = -1$
$2y + 3z = -5$
$x + 2z = -1$

Applying the Concepts

21. If a company has fixed costs of $100 per week and each item it produces costs $10 to manufacture, then the total cost y per week to produce x items is

$$y = 10x + 100$$

If the company sells each item it manufactures for $12, then the total amount of money y the company brings in for selling x items is

$$y = 12x$$

Use Cramer's rule to solve the system

$$y = 10x + 100$$

$$y = 12x$$

for x to find the number of items the company must sell per week in order to break even.

22. Suppose a company has fixed costs of $200 per week, and each item it produces costs $20 to manufacture.
(a) Write an equation that gives the total cost per week, y, to manufacture x items.
(b) If each item sells for $25, write an equation that gives the total amount of money y the company brings in for selling x items.
(c) Use Cramer's rule to find the number of items the company must sell each week to break even.

Review Problems

The problems below review material we covered in Section 7.4.

Solve each equation.

23. $x^4 - 2x^2 - 8 = 0$

24. $x^4 - 8x^2 - 9 = 0$

25. $(3x - 1)^2 - 7(3x - 1) = -10$

26. $(2x - 3)^2 + 6(2x - 3) = 7$

27. $2x - 5\sqrt{x} + 3 = 0$

28. $3x - 8\sqrt{x} + 4 = 0$

29. $(3x + 1) - 6\sqrt{3x + 1} + 8 = 0$

30. $(2x - 1) - 2\sqrt{2x - 1} - 15 = 0$

31. Solve $kx^2 + 4x - k = 0$ for x.

32. Solve $4x^2 - 4x + k = 0$ for x.

One Step Further

Solve for x and y using Cramer's rule. Your answers will contain the constants a and b.

33. $ax + by = -1$
$bx + ay = 1$

34. $ax + y = b$
$bx + y = a$

35. $a^2x + by = 1$ **36.** $ax + by = a$
     $b^2x + ay = 1$ $bx + ay = a$

37. Name the system of equations for which Cramer's rule yields the following determinants.

$$D = \begin{vmatrix} 1 & 2 \\ 3 & 4 \end{vmatrix} \quad D_x = \begin{vmatrix} 1 & 2 \\ 0 & 4 \end{vmatrix}$$

38. Name the system of equations for which Cramer's rule yields the following determinants.

$$D = \begin{vmatrix} 1 & 3 & 2 \\ -1 & 0 & 4 \\ 2 & 5 & -1 \end{vmatrix} \quad D_y = \begin{vmatrix} 1 & 1 & 2 \\ -1 & 3 & 4 \\ 2 & 5 & -1 \end{vmatrix}$$

SECTION

8.5

Matrix Solutions to Linear Systems

In mathematics, a **matrix** is a rectangular array of elements considered as a whole. We can use matrices to represent systems of linear equations. To do so, we write the coefficients of the variables and the constant terms in the same position in the matrix as they are in the system of equations. To show where the coefficients end and the constant terms begin, we use vertical lines instead of equal signs. For example, the system

$$2x + 5y = -4$$
$$x - 3y = 9$$

can be represented by the matrix

$$\left[\begin{array}{cc|c} 2 & 5 & -4 \\ 1 & -3 & 9 \end{array} \right]$$

which is called an **augmented matrix** because it includes both the coefficients of the variables and the constant terms.

In order to solve a system of linear equations by using the augmented matrix for that system, we need the following row operations as the tools of that solution process. The row operations tell us what we can do to an augmented matrix that may change the numbers in the matrix, but will always produce a matrix that represents a system of equations with the same solution as that of our original system.

Row Operations

1. We can interchange any two rows of a matrix.
2. We can multiply any row by a nonzero constant.
3. We can add to any row a constant multiple of another row.

The three row operations are simply a list of the properties we use to solve systems of linear equations, translated to fit an augmented matrix. For instance, the second operation in our list is actually just another way to state the multiplication property of equality.

We solve a system of linear equations by transforming the augmented matrix into a matrix that has 1's down the diagonal of the coefficient matrix, and 0's below it. For instance, we will have solved the system

$$2x + 5y = -4$$
$$x - 3y = 9$$

When the matrix

$$\left[\begin{array}{rr|r} 2 & 5 & -4 \\ 1 & -3 & 9 \end{array}\right]$$

has been transformed, using the row operations listed above, into a matrix of the form

$$\left[\begin{array}{rr|r} 1 & - & - \\ 0 & 1 & - \end{array}\right]$$

To accomplish this, we begin with the first column and try to produce a 1 in the first position and a 0 below it. Interchanging rows 1 and 2 gives us a 1 in the top position of the first column:

$$\downarrow \quad \text{Interchange rows 1 and 2}$$

$$\left[\begin{array}{rr|r} 1 & -3 & 9 \\ 2 & 5 & -4 \end{array}\right]$$

Multiplying row 1 by -2 and adding the result to row 2 gives us a zero where we want it.

$$\downarrow \quad \begin{array}{l}\text{Multiply row 1 by } -2 \text{ and} \\ \text{add the result to row 2}\end{array}$$

$$\left[\begin{array}{rr|r} 1 & -3 & 9 \\ 0 & 11 & -22 \end{array}\right]$$

$$\downarrow \quad \text{Multiply row 2 by } \frac{1}{11}$$

$$\left[\begin{array}{rr|r} 1 & -3 & 9 \\ 0 & 1 & -2 \end{array}\right]$$

Taking this last matrix and writing the system of equations it represents, we have

$$x - 3y = 9$$
$$y = -2$$

Substituting -2 for y in the top equation gives us

$$x = 3$$

The solution to our system is $(3, -2)$.

▶ **EXAMPLE 1**   Solve the system below using an augmented matrix:

$$x + y - z = 2$$
$$2x + 3y - z = 7$$
$$3x - 2y + z = 9$$

Solution   We begin by writing the system in terms of an augmented matrix:

$$\begin{vmatrix} 1 & 1 & -1 & 2 \\ 2 & 3 & -1 & 7 \\ 3 & -2 & 1 & 9 \end{vmatrix}$$

Next, we want to produce 0's in the second two positions of column 1:

$$\downarrow \qquad \text{Multiply row 1 by } -2 \text{ and}$$
$$\text{add the result to row 2}$$

$$\begin{bmatrix} 1 & 1 & -1 & 2 \\ 0 & 1 & 1 & 3 \\ 3 & -2 & 1 & 9 \end{bmatrix}$$

$$\downarrow \qquad \text{Multiply row 1 by } -3 \text{ and}$$
$$\text{add the result to row 3}$$

$$\begin{bmatrix} 1 & 1 & -1 & 2 \\ 0 & 1 & 1 & 3 \\ 0 & -5 & 4 & 3 \end{bmatrix}$$

Note that we could have done these two steps in one single step. As you become more familiar with this method of solving systems of equations, you will do just that.

$$\downarrow \qquad \text{Multiply row 2 by 5 and}$$
$$\text{add the result to row 3}$$

$$\begin{bmatrix} 1 & 1 & -1 & 2 \\ 0 & 1 & 1 & 3 \\ 0 & 0 & 9 & 18 \end{bmatrix}$$

$$\downarrow \qquad \text{Multiply row 3 by } \dfrac{1}{9}$$

$$\begin{bmatrix} 1 & 1 & -1 & 2 \\ 0 & 1 & 1 & 3 \\ 0 & 0 & 1 & 2 \end{bmatrix}$$

Converting back to a system of equations, we have

$$x + y - z = 2$$
$$y + z = 3$$
$$z = 2$$

This system is equivalent to our first one, but much easier to solve. Substituting $z = 2$ into the second equation, we have

$$y = 1$$

Substituting $z = 2$ and $y = 1$ into the first equation, we have

$$x = 3$$

The solution to our original system is (3, 1, 2). It satisfies each of our original equations. You can check this, if you like.   ◀

PROBLEM SET 8.5

Solve the following systems of equations by using matrices.

1. $x + y = 5$
  $3x - y = 3$

2. $x + y = -2$
  $2x - y = -10$

3. $3x - 5y = 7$
  $-x + y = -1$

4. $2x - y = 4$
  $x + 3y = 9$

5. $2x - 8y = 6$
  $3x - 8y = 13$

6. $3x - 6y = 3$
  $-2x + 3y = -4$

7. $x + y + z = 4$
  $x - y + 2z = 1$
  $x - y - z = -2$

8. $x - y - 2z = -1$
  $x + y + z = 6$
  $x + y - z = 4$

9. $x + 2y + z = 3$
  $2x - y + 2z = 6$
  $3x + y - z = 5$

10. $x - 3y + 4z = -4$
  $2x + y - 3z = 14$
  $3x + 2y + z = 10$

11. $x + 2y = 3$
  $y + z = 3$
  $4x - z = 2$

12. $x + y = 2$
  $3y - 2z = -8$
  $x + z = 5$

13. $x + 3y = 7$
  $3x - 4z = -8$
  $5y - 2z = -5$

14. $x + 4y = 13$
  $2x - 5z = -3$
  $4y - 3z = 9$

Solve each system using matrices. Remember, multiplying a row by a nonzero constant will not change the solution to the system.

15. $\frac{1}{3}x + \frac{1}{5}y = 2$
  $\frac{1}{3}x - \frac{1}{2}y = -\frac{1}{3}$

16. $\frac{1}{2}x + \frac{1}{3}y = 13$
  $\frac{1}{5}x + \frac{1}{8}y = 5$

The systems that follow are inconsistent systems. In both cases, the lines are parallel. Try solving each system using matrices and see what happens.

17. $2x - 3y = 4$
  $4x - 6y = 4$

18. $10x - 15y = 5$
  $-4x + 6y = -4$

The systems that follow are dependent systems, In each case, the lines coincide. Try solving each system using matrices and see what happens.

19. $-6x + 4y = 8$
  $-3x + 2y = 4$

20. $x + 2y = 5$
  $-x - 2y = -5$

Review Problems

The problems below review material we covered in Sections 2.3 and 2.4.

21. The sum of two numbers is 11. If one of them is 1 less than twice the other, find the two numbers.

22. Stephanie has 18 coins totaling $3.45. If she has only dimes and quarters, how many of each type does she have?

These problems are taken from the book *A First Course in Algebra,* written by Wallace C. Boyden and published by Silver, Burdett and Company in 1894.

23. A man bought 12 pairs of boots and 6 suits of clothes for $168. If a suit of clothes costs $2 less than four times as much as a pair of boots, what was the price of each?

24. A farmer pays just as much for 4 horses as he does for 6 cows. If a cow costs 15 dollars less than a horse, what is the cost of each?

25. Two men whose wages differ by 8 dollars receive together a total of $44 per month. How much does each receive?

26. Mr. Ames builds 3 houses. The first costs $2,000 more than the second, and the third twice as much as the first. If they all together cost $18,000, what was the cost of each house?

SECTION

8.6 **Applications**

Many times word problems involve more than one unknown quantity. If a problem is stated in terms of two unknowns and we represent each unknown quantity with a different variable, then we must write the relationships between the variables with two equations. The two equations written in terms of the two variables form a system of linear equations that we solve using the methods developed in this chapter. If we find a problem that relates three unknown quantities, then we need three different equations in order to form a linear system we can solve.

▶ **EXAMPLE 1** One number is 2 more than 3 times another. Their sum is 26. Find the two numbers.

Solution If we let x and y represent the two numbers, then the translation of the first sentence in the problem into an equation would be

$$y = 3x + 2$$

The second sentence gives us a second equation:

$$x + y = 26$$

The linear system that describes the situation is

$$x + y = 26$$
$$y = 3x + 2$$

Substituting the expression for y from the second equation into the first and solving for x yields

$$x + (3x + 2) = 26$$
$$4x + 2 = 26$$
$$4x = 24$$
$$x = 6$$

Using $x = 6$ in $y = 3x + 2$ gives the second number:

$$y = 3(6) + 2$$
$$y = 20$$

The two numbers are 6 and 20. Their sum is 26, and the second is 2 more than 3 times the first.   ◀

▶ **EXAMPLE 2**   Suppose 850 tickets were sold for a game for a total of $1,100. If adult tickets cost $1.50 and children's tickets cost $1.00, how many of each kind of ticket were sold?

Solution   If we let x = the number of adult tickets and y = the number of children's tickets, then

$$x + y = 850$$

since a total of 850 tickets were sold. Since each adult ticket costs $1.50 and each children's ticket costs $1.00 and the total amount of money paid for tickets was $1,100, a second equation is

$$1.50x + 1.00y = 1,100$$

The same information can also be obtained by summarizing the problem with a table. One such table follows. Notice that the two equations we obtained previously are given by the two rows of the table.

	Adult tickets	Children's tickets	Total
Number	x	y	850
Value	$1.50x$	$1.00y$	1,100

Whether we use a table to summarize the information in the problem, or just talk our way through the problem, the system of equations that describes the situation is

$$x + y = 850$$
$$1.50x + 1.00y = 1,100$$

If we multiply the second equation by 10 to clear it of decimals, we have the system

$$x + y = 850$$
$$15x + 10y = 11,000$$

Multiplying the first equation by -10 and adding the result to the second equation eliminates the variable y from the system:

$$-10x - 10y = -8,500$$
$$\underline{15x + 10y = 11,000}$$
$$5x \qquad = 2,500$$
$$x = 500$$

The number of adult tickets sold was 500. To find the number of children's tickets, we substitute $x = 500$ into $x + y = 850$ to get

$$500 + y = 850$$
$$y = 350$$

The number of children's tickets sold was 350. ◀

▶ **EXAMPLE 3** Suppose a person invests a total of $10,000 in two accounts. One account earns 8% annually and the other earns 9% annually. If the total interest earned from both accounts in a year is $860, how much is invested in each account?

Solution The form of the solution to this problem is very similar to that of Example 2. We let x equal the amount invested at 9% and y be the amount invested at 8%. Since the total investment is $10,000, one relationship between x and y can be written as

$$x + y = 10,000$$

The total interest earned from both accounts is $860. The amount of interest earned on x dollars at 9% is $0.09x$, while the amount of interest earned on y dollars at 8% is $0.08y$. This relationship is represented by the equation

$$0.09x + 0.08y = 860$$

The two equations we have just written can also be found by first summarizing the information from the problem in a table. Again, the two rows of the table yield the two equations just written. See the table on the next page.

	Dollars at 9%	Dollars at 8%	Total
Number	x	y	10,000
Interest	$0.09x$	$0.08y$	860

The system of equations that describes this situation is given by

$$x + y = 10{,}000$$
$$0.09x + 0.08y = 860$$

Multiplying the second equation by 100 will clear it of decimals. The system that results after doing so is

$$x + y = 10{,}000$$
$$9x + 8y = 86{,}000$$

We can eliminate y from this system by multiplying the first equation by -8 and adding the result to the second equation.

$$-8x - 8y = -80{,}000$$
$$\underline{9x + 8y = 86{,}000}$$
$$x = 6{,}000$$

The amount of money invested at 9% is $6,000. Since the total investment was $10,000, the amount invested at 8% must be $4,000. ◀

▶ **EXAMPLE 4** How much 20% alcohol solution and 50% alcohol solution must be mixed to get 12 gallons of 30% alcohol solution?

Solution To solve this problem we must first understand that a 20% alcohol solution is 20% alcohol and 80% water.

Let $x =$ the number of gallons of 20% alcohol solution needed, and $y =$ the number of gallons of 50% alcohol solution needed. Since we must end up with a total of 12 gallons of solution, one equation for the system is

$$x + y = 12$$

The amount of alcohol in the x gallons of 20% solution is $0.20x$, while the amount of alcohol in the y gallons of 50% solution is $0.50y$. Since the total amount of alcohol in the 20% and 50% solutions must add up to the amount of alcohol in the 12 gallons of 30% solution, the second equation in our system can be written as

$$0.20x + 0.50y = 0.30(12)$$

Again, let's make a table that summarizes the information we have to this point in the problem.

	20% solution	50% solution	Final solution
Total number of gallons	x	y	12
Gallons of alcohol	$0.20x$	$0.50y$	$0.30(12)$

Our system of equations is

$$x + \quad y = 12$$

$$0.20x + 0.50y = 0.30(12) = 3.6$$

Multiplying the second equation by 10 gives us an equivalent system:

$$x + \ y = 12$$

$$2x + 5y = 36$$

Multiplying the top equation by -2 to eliminate the x-variable, we have

$$
\begin{array}{r}
-2x - 2y = -24 \\
2x + 5y = 36 \\
\hline
3y = 12 \\
\end{array}
$$

$$y = \quad 4$$

Substituting $y = 4$ into $x + y = 12$, we solve for x:

$$x + 4 = 12$$

$$x = 8$$

It takes 8 gallons of 20% alcohol solution and 4 gallons of 50% alcohol solution to produce 12 gallons of 30% alcohol solution.   ◀

▶ **EXAMPLE 5**   It takes 2 hours for a boat to travel 28 miles downstream. The same boat can travel 18 miles upstream in 3 hours. What is the speed of the boat in still water and the speed of the current of the river?

Solution   Let $x =$ the speed of the boat in still water and $y =$ the speed of the current. Using a table as we did in Section 4.7, we have

	d	r	t
Upstream	18	$x - y$	3
Downstream	28	$x + y$	2

Since $d = r \cdot t$, the system we need to solve the problem is

$$18 = (x - y) \cdot 3$$
$$28 = (x + y) \cdot 2$$

which is equivalent to

$$6 = x - y$$
$$14 = x + y$$

Adding the two equations, we have

$$20 = 2x$$
$$x = 10$$

Substituting $x = 10$ into $14 = x + y$, we see that

$$y = 4$$

The speed of the boat in still water is 10 mph and the speed of the current is 4 mph. ◀

▶ **EXAMPLE 6** A coin collection consists of 14 coins with a total value of $1.35. If the coins are nickels, dimes, and quarters, and the number of nickels is three less than twice the number of dimes, how many of each coin is there in the collection?

Solution Since we have three types of coins we will have to use three variables. Let's let $x =$ the number of nickels, $y =$ the number of dimes, and $z =$ the number of quarters. Since the total number of coins is 14, we have our first equation

$$x + y + z = 14$$

Since the number of nickels is three less than twice the number of dimes, we have a second equation.

$$x = 2y - 3 \quad \text{which is equivalent to} \quad x - 2y = -3$$

Our last equation is obtained by considering the value of each coin and the total value of the collection. Let's write the equation in terms of cents, so we won't have to clear it of decimals later.

$$5x + 10y + 25z = 135$$

Here is our system, with the equations numbered for reference:

$$x + y + z = 14 \qquad (1)$$
$$x - 2y = -3 \qquad (2)$$
$$5x + 10y + 25z = 135 \qquad (3)$$

Let's begin by eliminating x from the first and second equations, and the first and third equations. Adding -1 times the second equation to the first equation gives us an equation in only y and z. We call this equation (4).

$$3y + z = 17 \quad (4)$$

Adding -5 times equation (1) to equation (3) gives us

$$5y + 20z = 65 \quad (5)$$

We can eliminate z from equations (4) and (5) by adding -20 times (4) to (5). Here is the result:

$$-55y = -275$$

$$y = 5$$

Substituting $y = 5$ into equation (4) gives us $z = 2$. Substituting $y = 5$ and $z = 2$ into equation (1) gives us $x = 7$. The collection consists of 7 nickels, 5 dimes, and 2 quarters. ◀

If you go on to take a chemistry class, you may see the next example (or one much like it) again.

▶ **EXAMPLE 7** In a chemistry lab, students record the temperature of water at room temperature and find that it is 77° on the Fahrenheit temperature scale and 25° on the Celsius temperature scale. The water is then heated until it boils. The temperature of the boiling water is 212° F and 100° C. Assume that the relationship between the two temperature scales is a linear one, then use the data above to find the formula that gives the Celsius temperature C in terms of the Fahrenheit temperature F.

Solution The data is summarized in Table 1.

TABLE 1 Corresponding Temperatures

In Degrees Fahrenheit	In Degrees Celsius
77	25
212	100

If we assume the relationship is linear, then the formula that relates the two temperature scales can be written in slope-intercept form as

$$C = mF + b$$

Substituting C = 25 and F = 77 into this formula gives us

$$25 = 77m + b$$

Substituting C = 100 and F = 212 into the formula yields

$$100 = 212m + b$$

Together, the two equations form a system of equations, which we can solve using the addition method.

$$
\begin{array}{lll}
25 = 77m + b & \xrightarrow{\text{Times } -1} & -25 = -77m - b \\
100 = 212m + b & \xrightarrow{\text{No change}} & 100 = 212m + b \\
\hline
& & 75 = 135m
\end{array}
$$

$$m = \frac{75}{135} = \frac{5}{9}$$

To find the value of b we substitute $m = \frac{5}{9}$ into $25 = 77m + b$ and solve for b.

$$25 = 77\left(\frac{5}{9}\right) + b$$

$$25 = \frac{385}{9} + b$$

$$b = 25 - \frac{385}{9} = \frac{225}{9} - \frac{385}{9} = -\frac{160}{9}$$

The equation that gives C in terms of F is

$$C = \frac{5}{9}F - \frac{160}{9}$$

◄

PROBLEM SET 8.6

Number Problems

1. One number is 3 more than twice another. The sum of the numbers is 18. Find the two numbers.

2. The sum of two numbers is 32. One of the numbers is 4 less than 5 times the other. Find the two numbers.

3. The difference of two numbers is 6. Twice the smaller is 4 more than the larger. Find the two numbers.

4. The larger of two numbers is 5 more than twice the smaller. If the smaller is subtracted from the larger, the result is 12. Find the two numbers.

5. The sum of three numbers is 8. Twice the smallest is 2 less than the largest, while the sum of the largest and smallest is 5. Use a linear system in three variables to find the three numbers.

6. The sum of three numbers is 14. The largest is 4 times the smallest, while the sum of the small-

est and twice the largest is 18. Use a linear system in three variables to find the three numbers.

Ticket and Interest Problems

7. A total of 925 tickets were sold for a game for a total of $1,150. If adult tickets sold for $2.00 and children's tickets sold for $1.00, how many of each kind of ticket were sold?

8. If tickets for a show cost $2.00 for adults and $1.50 for children, how many of each kind of ticket were sold if a total of 300 tickets were sold for $525?

9. Mr. Jones has $20,000 to invest. He invests part at 6% and the rest at 7%. If he earns $1,280 in interest after one year, how much did he invest at each rate?

10. A man invests $17,000 in two accounts. One account earns 5% interest per year and the other 6.5%. If his total interest after one year is $970, how much does he invest at each rate?

11. Susan invests twice as much money at 7.5% as she does at 6%. If her total interest after a year is $840, how much does she have invested at each rate?

12. A woman earns $1,350 in interest from two accounts in a year. If she has three times as much invested at 7% as she does at 6%, how much does she have in each account?

13. A man invests $2,200 in three accounts that pay 6%, 8%, and 9% in annual interest. He has three times as much invested at 9% as he does at 6%. If his total interest for the year is $178, how much is invested at each rate?

14. A student has money in three accounts that pay 5%, 7%, and 8% in annual interest. She has three times as much invested at 8% as she does at 5%. If the total amount she has invested is $1,600 and her interest for the year comes to $115, how much money does she have in each account?

Mixture Problems

15. How many gallons of 20% alcohol solution and 50% alcohol solution must be mixed to get 9 gallons of 30% alcohol solution?

16. How many ounces of 30% hydrochloric acid solution and 80% hydrochloric acid solution must be mixed to get 10 ounces of 50% hydrochloric acid solution?

17. A mixture of 16% disinfectant solution is to be made from 20% and 14% disinfectant solutions. How much of each solution should be used if 15 gallons of the 16% solution are needed?

18. How much 25% antifreeze and 50% antifreeze should be combined to give 40 gallons of 30% antifreeze?

Rate Problems

19. It takes a boat 2 hours to travel 24 miles downstream and 3 hours to travel 18 miles upstream. What is the speed of the boat in still water? What is the speed of the current of the river?

20. A boat on a river travels 20 miles downstream in only 2 hours. It takes the same boat 6 hours to travel 12 miles upstream. What are the speed of the boat and the speed of the current?

21. An airplane flying with the wind can cover a certain distance in 2 hours. The return trip against the wind takes $2\frac{1}{2}$ hours. How fast is the plane and what is the speed of the air, if the distance is 600 miles?

22. An airplane covers a distance of 1,500 miles in 3 hours when it flies with the wind and $3\frac{1}{3}$ hours when it flies against the wind. What is the speed of the plane in still air?

Coin Problems

23. Bob has 20 coins totaling $1.40. If he has only dimes and nickels, how many of each coin does he have?

24. If Amy has 15 coins totaling $2.70, and the coins are quarters and dimes, how many of each coin does she have?

25. A collection of nickels, dimes, and quarters consists of 9 coins with a total value of $1.20. If the number of dimes is equal to the number of nickels, find the number of each type of coin.

26. A coin collection consists of 12 coins with a total value of $1.20. If the collection consists only of nickels, dimes, and quarters, and the

number of dimes is two more than twice the number of nickels, how many of each type of coin are in the collection?

Finding the Equation Problems

27. A manufacturing company finds that they can sell 300 items if the price per item is $2.00, and 400 items if the price is $1.50 per item. If the relationship between the number of items sold x and the price per item p is a linear one, find a formula that gives x in terms of p. Then use the formula to find the number of items they will sell if the price per item is $3.00.

28. A company manufactures and sells bracelets. They have found from past experience that they can sell 300 bracelets each week if the price per bracelet is $2.00, but only 150 bracelets are sold if the price is $2.50 per bracelet. If the relationship between the number of bracelets sold x and the price per bracelet p is a linear one, find a formula that gives x in terms of p. Then use the formula to find the number of bracelets they will sell at $3.00 each.

29. Five Cities Garbage charges a flat monthly fee for their services plus a certain amount for each bag of trash they pick up. A customer notices that the January bill for picking up 5 bags of trash was $25.60, while the February bill for 7 bags of trash was $27.10. Assume the relationship between the total monthly charges C and the number of bags of trash picked up x is a linear relationship. Use the data to find the formula that gives C in terms of x. Then use the formula to predict the cost for picking up 12 bags of trash in one month.

30. A bottled water company charges a flat fee each month for the use of their water dispenser plus a certain amount for each gallon of water delivered. Suppose that the company delivers 10 gallons of water in March and the March bill is $18. Then, in April, 15 gallons of water are delivered for a total charge of $23.50. Assume the relationship between the total monthly

charges C and the number of gallons of water delivered x is a linear relationship. Use the data to find the formula that gives C in terms of x. Then use the formula to predict the cost if 20 gallons of water are delivered in one month.

31. A ball is tossed into the air so that the height after 1, 3, and 5 seconds is as given in the table below. If the relationship between the height of the ball h and the time t is quadratic, then the relationship can be written as

$$h = at^2 + bt + c$$

Use the information in the table to write a system of three equations in three variables a, b, and c. Solve the system to find the exact relationship between h and t. Then find the maximum height attained by the ball.

t (in seconds)	h (in feet)
1	128
3	128
5	0

32. A ball is tossed into the air and its height above the ground after 1, 3, and 4 seconds is recorded as shown in the table on the next page. The relationship between the height of the ball h and the time t is quadratic and can be written as

$$h = at^2 + bt + c$$

Use the information in the table to write a system of three equations in three variables a, b, and c. Solve the system to find the exact relationship between the variables h and t. Then find the maximum height attained by the ball.

t (in seconds)	h (in feet)
1	96
3	64
4	0

Review Problems

The problems below review material we covered in Section 7.6.

Find the vertex for each of the following parabolas and then indicate if it is the highest or lowest point on the graph.

33. $y = 2x^2 + 8x - 15$
34. $y = 3x^2 - 9x - 10$
35. $y = 12x - 4x^2$
36. $y = 18x - 6x^2$
37. An object is projected into the air with an initial upward velocity of 64 feet/second. Its height h at any time t is given by the formula $h(t) = 64t - 16t^2$. Find the time at which the object reaches its maximum height. Then, find the maximum height.
38. An object is projected into the air with an initial upward velocity of 64 feet/second from the top of a building 40 feet high. If the height h of the object t seconds after it is projected into the air is $h(t) = 40 + 64t - 16t^2$, find the time at which the object reaches its maximum height. Then, find the maximum height it attains.

Research Project 9

Zeno of Elea was born at about the same time that Pythagoras died. He is responsible for three paradoxes that have come to be known as Zeno's paradoxes. One of the three has to do with a race between Achilles and a tortoise. Achilles is much faster than the tortoise, but the tortoise has a head start. According to Zeno's method of reasoning, Achilles can never pass the tortoise because each time he reaches the place where the tortoise was, the tortoise is gone. Research Zeno's paradox concerning Achilles and the tortoise. Put your findings into essay form that begins with a definition for the word "paradox." Then use Zeno's method of reasoning to describe a race between Achilles and the tortoise if Achilles runs at 10 miles per hour, the tortoise at 1 mile per hour, and the tortoise has a 1-mile head start. Next, use the methods shown in this section to find the distance at which Achilles reaches the tortoise and the time at which Achilles reaches the tortoise. Conclude your essay by summarizing what you have done and showing how the two results you have obtained form a paradox.

CHAPTER 8 SUMMARY

Examples

1. The solution to the system

$$x + 2y = 4$$
$$x - y = 1$$

is the ordered pair (2, 1). It is the only ordered pair that satisfies both equations.

Systems of Linear Equations [8.1, 8.2]

A *system of linear equations* consists of two or more linear equations considered simultaneously. The solution set to a linear system in two variables is the set of ordered pairs that satisfy both equations. The solution set to a linear system in three variables consists of the ordered triples that satisfy all three equations in the system.

9

RELATIONS, FUNCTIONS, AND CONIC SECTIONS

INTRODUCTION AND OVERVIEW

In the first half of this chapter we will continue our study of relations and functions. Relations and functions have many applications in the real world, as we have seen in the examples in previous chapters. When we say, "the price of gasoline is increasing because there is more demand for it this year," we are expressing a relationship between the price of gasoline and the demand for it. We are implying the price of gasoline is a function of the demand for it. Mathematics becomes a part of this problem when we express, with an equation, the exact relationship between the two quantities.

In the second half of this chapter we take an introductory look at conic sections. Conic sections include circles, which you are familiar with on an intuitive level, as well as parabolas, which you have already been introduced to in Chapter 7. Ellipses and hyperbolas are the other two conic sections, and they each can be observed in the world around us. For instance, the path the earth travels in its orbit around the sun is an ellipse, while a comet that comes in contact with the gravitational field of the sun may travel on a hyperbolic path.

SECTION

9.1   More about Relations and Functions

We begin this section with a review of the basic concepts associated with relations and functions. First, let us review the definitions for relations and functions, along with the definitions for domain and range.

DEFINITION A **relation** is a set of ordered pairs. The set of all first coordinates forms the **domain** of the relation, while the **range** is the set of all second coordinates. A **function** is a relation in which no ordered pairs have the same first coordinates. The formal definition is stated this way: A function is a rule or correspondence that pairs each element in one set, called the domain, with exactly one element from a second set, called the range.

Relations and functions are both sets of ordered pairs. What makes them different from each other is that the ordered pairs of a function will never repeat their first coordinates. Every function is a relation, but not all relations are functions.

If we have the graph of a relation, we can check it visually to see if it is also a function by using the vertical line test. If any vertical line can be found that crosses a graph in more than one place, then that graph cannot be the graph of a function. Each point on a vertical line has the same first coordinate as every other point on the line. If a vertical line crosses a graph in two places, then we have two points with the same first coordinates. Therefore, the graph is not the graph of a function. Example 1 illustrates.

▶ **EXAMPLE 1** Use the vertical line test to see which of the following is a function: (a) $y = x^2 - 2$; (b) $x = y^2 - 2$.

Solution The graph of each relation is given in Figure 1. The equation $y = x^2 - 2$ is a function, since there are no vertical lines that cross its graph in more than one place. The equation $x = y^2 - 2$ does not represent a function, since we can find a vertical line that crosses its graph in more than one place.

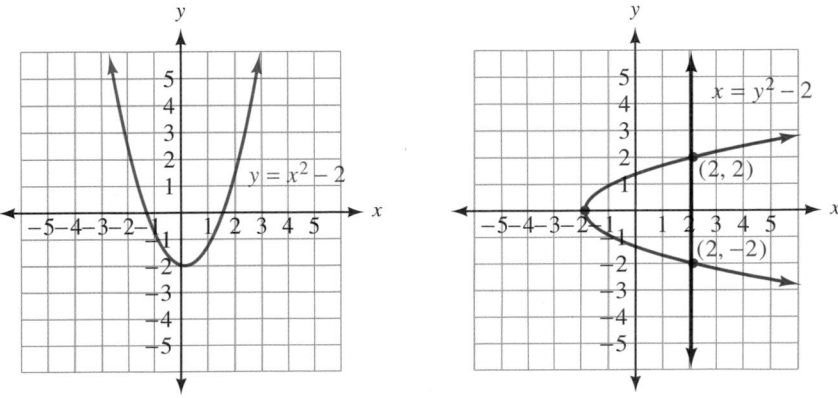

Figure 1

The Domain and Range of a Function

When a function (or relation) is given in terms of an equation, the domain is the set of all possible replacements for the variable x. If the domain of a function (or relation) is not specified, it is assumed to be all real numbers that do not give undefined terms in the equation. That is, we cannot use values of x in the domain that will produce 0 in a denominator or an even root of a negative number.

▶ **EXAMPLE 2**   Specify the domain for $y = \dfrac{6}{x - 2}$.

Solution   The domain can be any real number that does not produce an undefined term. If $x = 2$, the denominator on the right side will be 0. Hence, the domain is all real numbers except 2.   ◀

To visualize what it means for the domain of this relation to include all real numbers except 2, we can look at its graph.

▶ **EXAMPLE 3**   Graph $y = \dfrac{6}{x - 2}$.

Solution   This graph will have a shape similar to the graphs we considered in Section 5.7, when we graphed $y = \dfrac{1}{x}$ and $y = -\dfrac{6}{x}$. Unlike those graphs, however, this graph will cross the y-axis. To find the y-intercept, we let x equal 0.

$$\text{When } x = 0, \quad y = \frac{6}{0 - 2} = \frac{6}{-2} = -3 \quad y\text{-intercept}$$

On the other hand, the graph will not cross the x-axis. If it did, we would have a solution to the equation

$$0 = \frac{6}{x - 2}$$

which has no solution because there is no number to divide 6 by to obtain 0.

The graph of our equation is shown in Figure 2 along with a table giving values of x and y that satisfy the equation. Since y is undefined when x is 2, the graph will not cross the vertical line $x = 2$. (If it did, there would be a value of y for $x = 2$.) The line $x = 2$ is called a **vertical asymptote** of the graph. The graph will get very close to the vertical asymptote, but will never touch or cross it.

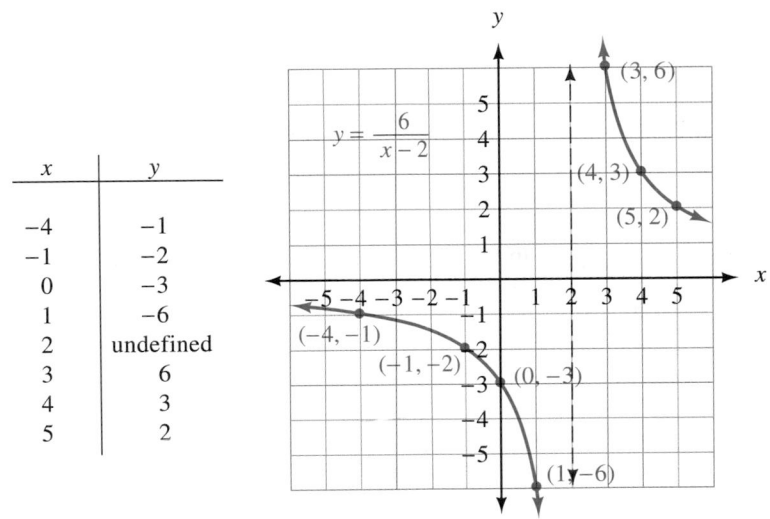

x	y
−4	−1
−1	−2
0	−3
1	−6
2	undefined
3	6
4	3
5	2

Figure 2

If you compare the graphs in Section 5.7 with Figure 2, you can see that the graph of $y = \dfrac{6}{x - 2}$ is the graph of $y = \dfrac{6}{x}$ with all points shifted 2 units to the right. Note also that the graph in Figure 2 is the graph of a function because no vertical line crosses the graph in more than one place. ◀

▶ **EXAMPLE 4**   Give the domain for $y = \sqrt{x + 4}$.

Solution   Since the domain must consist of real numbers, the quantity under the radical will have to be greater than or equal to 0:

$$x + 4 \geq 0$$
$$x \geq -4$$

The domain in this case is $\{x \mid x \geq -4\}$. To visualize this fact we have the graph of $y = \sqrt{x + 4}$ in Figure 3 on page 526. By the vertical line test we can conclude that this is the graph of a function.

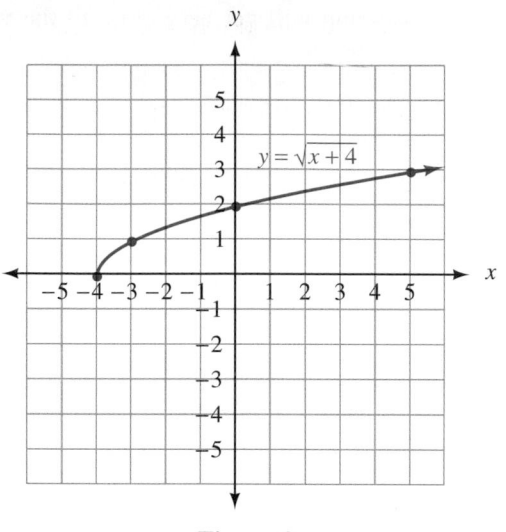

Figure 3 ◄

If we look closely at the graph in Figure 3, not only do we see that the domain is $\{x \mid x \geq -4\}$, but we also see that the range must be $\{y \mid y \geq 0\}$ because the graph starts on the x-axis and rises continuously from there. In fact, it is sometimes easiest to identify the domain *and* range of a function (or relation) by looking at its graph. The next example continues this discussion.

► **EXAMPLE 5** Give the domain and range for $y = x^2 - 3$.

Solution The graph is the parabola shown in Figure 4.

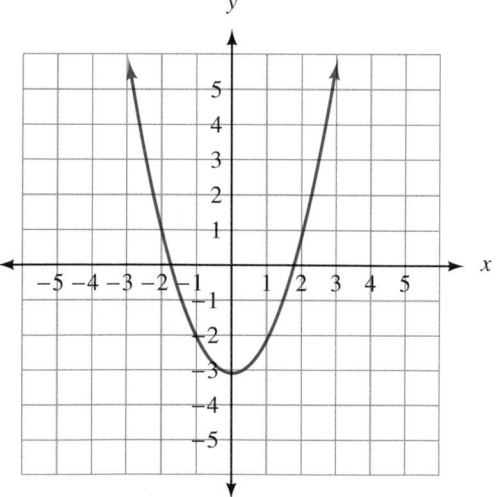

Figure 4

The domain is all real numbers, and the range is the set $\{y \mid y \geq -3\}$. ◄

Function Notation

Recall that, when we are working with functions, we can use the notation $f(x)$ in place of the variable y. That is, $y = f(x)$, which we read as "y is equal to f of x." If $f(x) = 2x - 4$, we can find $f(0)$ by substituting 0 for x in the expression $2x - 4$. Here is that problem, plus some others:

$$f(0) = 2(0) - 4 = -4$$

$$f(2) = 2(2) - 4 = 4 - 4 = 0$$

$$f\left(\frac{1}{2}\right) = 2\left(\frac{1}{2}\right) - 4 = 1 - 4 = -3$$

$$f(a) = 2(a) - 4 = 2a - 4$$

We say, "the value of the function f at 0 is -4, the value of f at 2 is 0, and the value of f at $\frac{1}{2}$ is -3."

▶ **EXAMPLE 6** If $f(x) = 2x - 4$, find $\dfrac{f(x) - f(a)}{x - a}$.

Solution

$$\frac{f(x) - f(a)}{x - a} = \frac{(2x - 4) - (2a - 4)}{x - a}$$

$$= \frac{2x - 2a}{x - a}$$

$$= \frac{2(x - a)}{x - a}$$

$$= 2 \qquad \blacktriangleleft$$

▶ **EXAMPLE 7** If $f(x) = 2x - 3$, find $\dfrac{f(x + h) - f(x)}{h}$.

Solution The expression $f(x + h)$ is given by

$$f(x + h) = 2(x + h) - 3$$
$$= 2x + 2h - 3$$

Using this result gives us

$$\frac{f(x + h) - f(x)}{h} = \frac{(2x + 2h - 3) - (2x - 3)}{h}$$

$$= \frac{2h}{h}$$

$$= 2 \qquad \blacktriangleleft$$

Algebra with Functions

If we are given two functions, f and g, with a common domain, we can define four other functions as follows.

DEFINITION

$(f + g)(x) = f(x) + g(x)$ The function $f + g$ is the sum of the functions f and g.

$(f - g)(x) = f(x) - g(x)$ The function $f - g$ is the difference of the functions f and g.

$(fg)(x) = f(x)g(x)$ The function fg is the product of the functions f and g.

$\dfrac{f}{g}(x) = \dfrac{f(x)}{g(x)}$ The function $\dfrac{f}{g}$ is the quotient of the functions f and g, where $g(x) \neq 0$.

▶ **EXAMPLE 8**   Let $f(x) = 4x - 3$, $g(x) = 4x^2 - 7x + 3$, and $h(x) = x - 1$. Find $f + g, fh, fg,$ and $\dfrac{g}{f}$.

Solution   The function $f + g$, the sum of functions f and g, is defined by

$$(f + g)(x) = f(x) + g(x)$$
$$= (4x - 3) + (4x^2 - 7x + 3)$$
$$= 4x^2 - 3x$$

The function fh, the product of functions f and h, is defined by

$$(fh)(x) = f(x)h(x)$$
$$= (4x - 3)(x - 1)$$
$$= 4x^2 - 7x + 3$$
$$= g(x)$$

The product of the functions f and g, fg, is given by

$$(fg)(x) = f(x)g(x)$$
$$= (4x - 3)(4x^2 - 7x + 3)$$
$$= 16x^3 - 28x^2 + 12x - 12x^2 + 21x - 9$$
$$= 16x^3 - 40x^2 + 33x - 9$$

The quotient of the functions g and f, $\dfrac{g}{f}$, is defined as

$$\frac{g}{f}(x) = \frac{g(x)}{f(x)}$$

$$= \frac{4x^2 - 7x + 3}{4x - 3}$$

Factoring the numerator, we can reduce to lowest terms:

$$\frac{g}{f}(x) = \frac{(4x - 3)(x - 1)}{4x - 3}$$

$$= x - 1$$

$$= h(x)$$

◄

P R O B L E M   S E T   9 . 1
. . . .

Give the domain for each of the following functions.

1. $y = \sqrt{2x - 1}$ **2.** $y = \sqrt{3x - 2}$

3. $y = \sqrt{1 - 4x}$ **4.** $y = \sqrt{2 + 3x}$

5. $y = \dfrac{x + 2}{x - 5}$ **6.** $y = \dfrac{x - 3}{x + 4}$

7. $y = \dfrac{3}{2x^2 + 5x - 3}$

8. $y = \dfrac{-1}{3x^2 - 5x + 2}$

Problems 9–22 may be solved using a graphing calculator. Graph each of the following relations. Use the graph to find the domain and range, and indicate which relations are also functions.

9. $y = x^2 - 4$ **10.** $y = x^2 + 4$

11. $y = x^2 - 4x$ **12.** $y = x^2 + 4x$

13. $x = y^2 - 4$ **14.** $x = y^2 - 4y$

15. $y = \sqrt{x - 2}$ **16.** $y = \sqrt{x} - 2$

17. $y = \sqrt{x} + 1$ **18.** $y = \sqrt{x + 1}$

19. $x = y^2 + 2$ **20.** $x = y^2 - 1$

21. $x = y^2 - 2y + 1$ **22.** $x = y^2 + 4y + 4$

For each of the following functions, evaluate the quantity $\dfrac{f(x) - f(a)}{x - a}$.

23. $f(x) = 3x$ **24.** $f(x) = -2x$
25. $f(x) = 4x - 5$ **26.** $f(x) = 3x + 1$
27. $f(x) = x^2$ **28.** $f(x) = 2x^2$
29. $y = x^2 + 1$ **30.** $y = x^2 - 1$

For each of the following functions, evaluate the quantity $\dfrac{f(x + h) - f(x)}{h}$.

31. $f(x) = 2x + 3$ **32.** $f(x) = 3x - 2$
33. $y = -4x - 1$ **34.** $y = -x + 4$
35. $y = 3x^2 - 2$ **36.** $y = 4x^2 + 3$

If the functions f, g, and h are defined by $f(x) = 3x - 5$, $g(x) = x - 2$, and $h(x) = 3x^2 - 11x + 10$, write a formula for each of the following functions.

37. $g + f$ **38.** $f + h$
39. $g + h$ **40.** $f - g$
41. $g - f$ **42.** $h - g$
43. fg **44.** gf
45. fh **46.** gh
47. $\dfrac{h}{f}$ **48.** $\dfrac{h}{g}$

49. $\dfrac{f}{h}$ **50.** $\dfrac{g}{h}$

51. $f + g + h$ **52.** $h - g + f$
53. $h + fg$ **54.** $h - fg$

Review Problems

The following problems review material we covered in Section 8.1.

Solve each system by the addition method.

55. $4x + 3y = 10$ **56.** $3x - 5y = -2$
       $2x + y = 4$ $2x - 3y = 1$
57. $4x + 5y = 5$ **58.** $4x + 2y = -2$
       $\frac{6}{5}x + y = 2$ $\frac{1}{2}x + y = 0$

Solve each system by the substitution method.

59. $x + y = 3$ **60.** $x + y = 6$
           $y = x + 3$ $y = x - 4$
61. $2x - 3y = -6$ **62.** $7x - y = 24$
           $y = 3x - 5$ $x = 2y + 9$

One Step Further

To get a clearer picture of the relationship between some of the graphs you have drawn and their vertical asymptotes, use a calculator to answer the following problems.

63. If $y = \dfrac{6}{x - 2}$, find the values of y that correspond to $x = 1.5, 1.7, 1.9, 2.1, 2.3$, and 2.5.

64. If $y = \dfrac{1}{x - 3}$, find the values of y that correspond to $x = 2.5, 2.7, 2.9, 3.1, 3.3$, and 3.5.

65. If $y = \dfrac{2}{x - 4}$, find the values of y that correspond to $x = 3.9, 3.99, 4.01$, and 4.1.

66. If $y = \dfrac{4}{x + 2}$, find the values of y that correspond to $x = -2.1, -2.01, -2.001, -1.999, -1.99$, and -1.9.

SECTION

9.2   Classification of Functions

Much of the work we have done in previous chapters has involved functions. All linear equations in two variables, except those with vertical lines for graphs, are functions. The parabolas we worked with in Section 7.6 are graphs of functions.

Constant Functions

Any function that can be written in the form

$$f(x) = c$$

where c is a real number, is called a **constant function.** The graph of every constant function is a horizontal line.

▶ **EXAMPLE 1**   The function $f(x) = 3$ is an example of a constant function. Since all ordered pairs belonging to f have a y-coordinate of 3, the graph is the horizontal line given by $y = 3$. Remember, y and $f(x)$ are equivalent—that is, $y = f(x)$.   ◀

Linear Functions

Any function that can be written in the form

$$f(x) = ax + b$$

where a and b are real numbers, $a \neq 0$, is called a **linear function.** The graph of every linear function is a straight line. In the past we have written linear functions in the form

$$y = mx + b$$

▶ **EXAMPLE 2** The function $f(x) = 2x - 3$ is an example of a linear function. The graph of this function is a straight line with slope 2 and y-intercept -3. ◀

Quadratic Functions

A **quadratic function** is any function that can be written in the form

$$f(x) = ax^2 + bx + c$$

where a, b, and c are real numbers and $a \neq 0$. The graph of every quadratic function is a parabola. We considered parabolic graphs of this type in Section 7.6. At that time the quadratic functions were written as $y = ax^2 + bx + c$ using y instead of $f(x)$.

▶ **EXAMPLE 3** The function $f(x) = 2x^2 - 4x - 6$ is an example of a quadratic function. Its graph is a parabola. ◀

Much of the work we have done previously with quadratic functions can be written in terms of function notation. For example, since y and $f(x)$ are equivalent, the x-intercepts for the graph of the function $f(x) = ax^2 + bx + c$ are values of x for which $f(x) = 0$.

▶ **EXAMPLE 4** Find the values of x for which $f(x) = 0$ if $f(x) = x^2 - 4x + 1$.

Solution If $f(x) = x^2 - 4x + 1$ and $f(x) = 0$, then

$$x^2 - 4x + 1 = 0$$

This equation does not factor, so in order to solve it we must use the quadratic formula:

$$x = \frac{4 \pm \sqrt{16 - 4(1)(1)}}{2(1)}$$

$$= \frac{4 \pm \sqrt{12}}{2}$$

$$= \frac{4 \pm 2\sqrt{3}}{2}$$

$$= 2 \pm \sqrt{3}$$

If we were to graph the equation $y = x^2 - 4x + 1$, we would find that it crossed the x-axis at $2 + \sqrt{3}$ and also at $2 - \sqrt{3}$, which are approximately 3.7 and 0.3.   ◀

Cubic Functions

A **cubic function** is any function that can be written in the form

$$f(x) = ax^3 + bx^2 + cx + d$$

where a, b, c, and d are real numbers and $a \neq 0$.

Cubic functions, linear functions, and quadratic functions all belong to a larger class of functions known as **polynomial functions.**

▶ **EXAMPLE 5**   Graph the cubic function $y = \frac{1}{2}x^3$.

Solution   The graph is shown in Figure 1. The table next to the graph shows some ordered pairs that satisfy the equation.

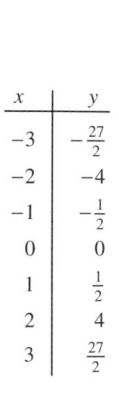

x	y
-3	$-\frac{27}{2}$
-2	-4
-1	$-\frac{1}{2}$
0	0
1	$\frac{1}{2}$
2	4
3	$\frac{27}{2}$

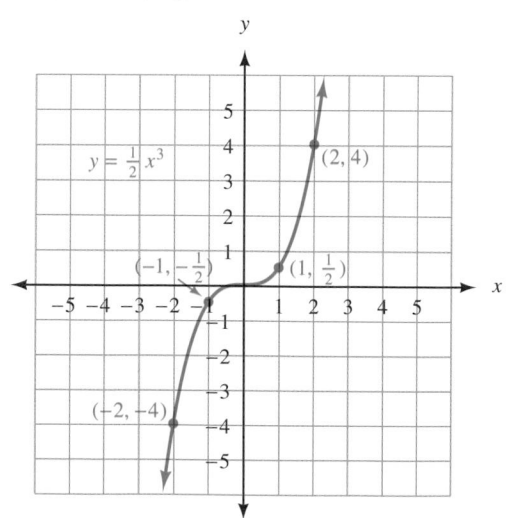

Figure 1   ◀

If we take all the functions that come from the ratio of two polynomials, we get the next category of functions.

Rational Functions

A **rational function** is any function that can be written in the form

$$f(x) = \frac{P(x)}{Q(x)}$$

where $P(x)$ and $Q(x)$ are polynomials and $Q(x) \neq 0$.

▶ **EXAMPLE 6** Graph the rational function $y = \dfrac{x - 4}{x - 2}$.

Solution In addition to making a table to find some points on the graph, we can analyze the graph as follows:

1. The graph will have a y-intercept of 2, because when $x = 0$, $y = -4/-2 = 2$.
2. To find the x-intercept, we let $y = 0$ to get

$$0 = \frac{x - 4}{x - 2}$$

   The only way this expression can be 0 is if the numerator is 0, which happens when $x = 4$. (If you want to solve the equation above, multiply both sides by $x - 2$. You will get the same solution, $x = 4$.)
3. The graph will have a vertical asymptote at $x = 2$, since $x = 2$ will make the denominator of the function 0, meaning y is undefined when x is 2.
4. The graph will have a horizontal asymptote at $y = 1$ because for very large values of x, $\dfrac{x - 4}{x - 2}$ is very close to 1. The larger x is, the closer $\dfrac{x - 4}{x - 2}$ is to 1. The same is true for very small values of x, such as -1000 and $-10,000$.

Putting this information together with the ordered pairs in the table next to the figure, we have the graph shown in Figure 2 on page 534.

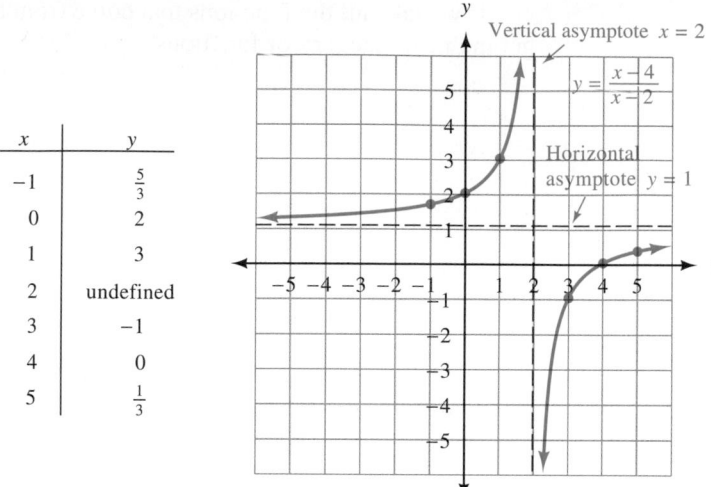

x	y
-1	$\frac{5}{3}$
0	2
1	3
2	undefined
3	-1
4	0
5	$\frac{1}{3}$

Figure 2

◀

Exponential Functions

An **exponential function** is any function that can be written in the form

$$f(x) = b^x$$

where b is a positive real number other than 1.

Each of the following is an exponential function:

$$f(x) = 2^x, \quad y = 3^x, \quad f(x) = \left(\frac{1}{4}\right)^x$$

The first step in becoming familiar with exponential functions is to find some values for specific exponential functions.

▶ **EXAMPLE 7** If the exponential functions f and g are defined by

$$f(x) = 2^x \quad \text{and} \quad g(x) = 3^x$$

then

$$f(0) = 2^0 = 1 \qquad\qquad g(0) = 3^0 = 1$$

$$f(1) = 2^1 = 2 \qquad\qquad g(1) = 3^1 = 3$$

$$f(2) = 2^2 = 4 \qquad\qquad g(2) = 3^2 = 9$$

$$f(3) = 2^3 = 8 \qquad\qquad g(3) = 3^3 = 27$$

$$f(-2) = 2^{-2} = \frac{1}{2^2} = \frac{1}{4} \qquad g(-2) = 3^{-2} = \frac{1}{3^2} = \frac{1}{9}$$

$$f(-3) = 2^{-3} = \frac{1}{2^3} = \frac{1}{8} \qquad g(-3) = 3^{-3} = \frac{1}{3^3} = \frac{1}{27}$$ ◀

We will now turn our attention to the graphs of exponential functions. Since the notation y is easier to use when graphing, and $y = f(x)$, for convenience we will write the exponential functions as

$$y = b^x$$

▶ **EXAMPLE 8** Sketch the graph of the exponential function $y = 2^x$.

Solution Using the results of Example 7, we have the table shown below. Graphing the ordered pairs given in the table and connecting them with a smooth curve, we have the graph of $y = 2^x$ shown in Figure 3.

x	y
-3	$\frac{1}{8}$
-2	$\frac{1}{4}$
-1	$\frac{1}{2}$
0	1
1	2
2	4
3	8

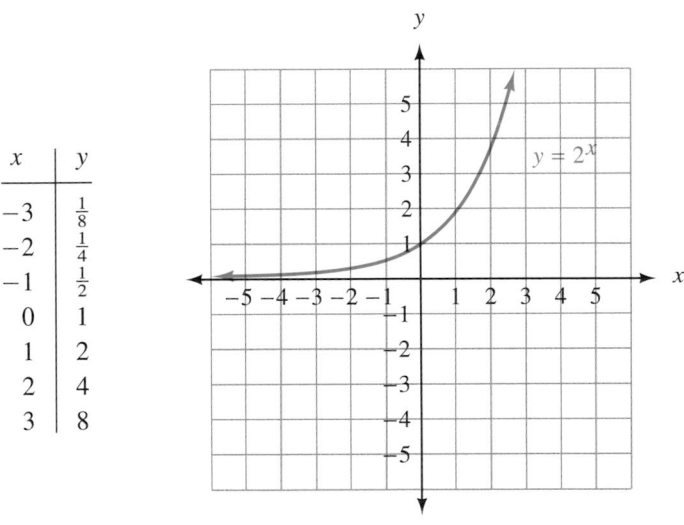

Figure 3

Notice that the graph does not cross the x-axis. It *approaches* the x-axis—in fact, we can get it as close to the x-axis as we want without it actually intersecting the x-axis. In order for the graph of $y = 2^x$ to intersect the x-axis, we would have to find a value of x that would make $2^x = 0$. Because no such value of x exists, the graph of $y = 2^x$ cannot intersect the x-axis. ◀

▶ **EXAMPLE 9** Sketch the graph of $y = (\frac{1}{3})^x$.

Solution   We can make a table that will give some ordered pairs that satisfy the equation:

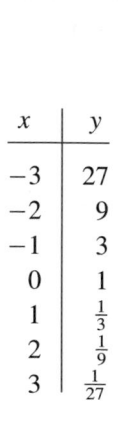

x	y
-3	27
-2	9
-1	3
0	1
1	$\frac{1}{3}$
2	$\frac{1}{9}$
3	$\frac{1}{27}$

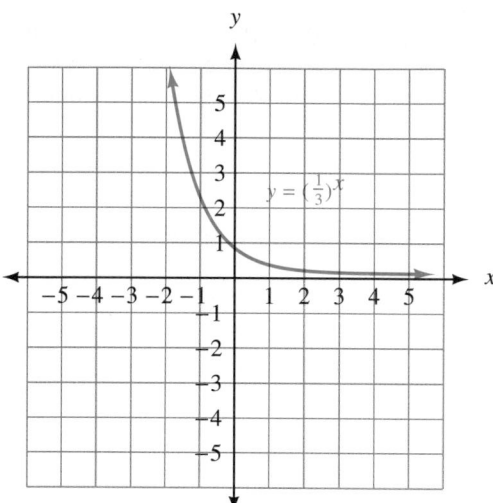

Figure 4

Using the ordered pairs from the table, we have the graph shown in Figure 4.

◀

The graphs of all exponential functions have two things in common: (1) each crosses the y-axis at $(0, 1)$, since $b^0 = 1$; and (2) none can cross the x-axis, since $b^x = 0$ is impossible because of the restrictions on b.

PROBLEM SET 9.2

Sketch the graph of each of the following functions. Identify each as a constant function, linear function, quadratic function, cubic function, or rational function.

1. $f(x) = x^2 - 3$ **2.** $g(x) = 2x^2$

3. $g(x) = 4x - 1$ **4.** $f(x) = 3x + 2$

5. $f(x) = 5$ **6.** $f(x) = -3$

7. $f(x) = x^2 + 4x - 5$

8. $f(x) = -x^2 - 4x + 5$

9. $y = x^3$ **10.** $y = 2x^3$

11. $y = x^3 - 2$ **12.** $y = x^3 + 3$

13. $y = (x - 2)^3$ **14.** $y = (x + 3)^3$

15. $y = \dfrac{x - 3}{x - 1}$ **16.** $y = \dfrac{x + 4}{x - 2}$

17. $y = \dfrac{x + 3}{x - 1}$ **18.** $y = \dfrac{x - 2}{x - 1}$

19. $y = \dfrac{x - 3}{x + 1}$ **20.** $y = \dfrac{x - 2}{x + 1}$

For each function below, find all values of x for which $f(x) = 0$.

21. $f(x) = 3x - 9$ **22.** $f(x) = -2x + 12$

23. $f(x) = 6x^2 - x - 15$

24. $f(x) = 12x^2 - 5x - 2$
25. $f(x) = x^2 + 4x + 1$
26. $f(x) = x^2 - 4x - 1$
27. $f(x) = 2x^3 + x^2 - 18x - 9$
28. $f(x) = 3x^3 + x^2 - 12x - 4$
29. The number of people in a store t hours after it opens at 9:00 A.M. is given by the function $N(t) = 60t - 10t^2$. At what time is the number of people in the store at a maximum? How many people are in the store at that time?
30. The number of students in an elementary school who will get the flu t days after the first student shows symptoms is given by the formula $N(t) = 40t - 5t^2$. If a fourth grader comes down with the flu on Monday, on what day will the largest number of students have the flu? How many students will have the flu on that day?

Let $f(x) = 3^x$ and $g(x) = (\frac{1}{2})^x$ and evaluate each of the following.

31. $g(0)$ **32.** $f(0)$
33. $g(-1)$ **34.** $g(-4)$
35. $f(-3)$ **36.** $f(-1)$
37. $f(2) + g(-2)$ **38.** $f(2) - g(-2)$

Graph each of the following functions.

39. $y = 4^x$ **40.** $y = 2^{-x}$
41. $y = 3^{-x}$ **42.** $y = (\frac{1}{3})^{-x}$
43. $y = 2^{x+1}$ **44.** $y = 2^{x-3}$
45. $y = 2^{2x}$ **46.** $y = 3^{2x}$
47. Suppose it takes 1 day for a certain strain of bacteria to reproduce by dividing in half. If there are 100 bacteria present to begin with, then the total number present after x days will be $f(x) = 100 \cdot 2^x$. Find the total number

present after 1 day, 2 days, 3 days, and 4 days. How many days must elapse before there are over 100,000 bacteria present?
48. Suppose it takes 12 hours for a certain strain of bacteria to reproduce by dividing in half. If there are 50 bacteria present to begin with, then the total number present after x days will be $f(x) = 50 \cdot 4^x$. Find the total number present after 1 day, 2 days, and 3 days.

Review Problems

The problems below review material we covered in Section 8.2.

Solve each system. [8.2].

49. $x + y + z = 6$
     $2x - y + z = 3$
      $x + 2y - 3z = -4$
50. $x + y + z = 6$
      $x - y + 2z = 7$
     $2x - y - z = 0$
51. $3x + 4y = 15$ **52.** $x + 3y = 5$
    $2x - 5z = -3$ $6y + z = 12$
     $4y - 3z = 9$ $x - 2z = -10$

One Step Further

Graph each cubic function. First find the x- and y-intercepts, then make a table that gives four or five ordered pairs that satisfy the equation.

53. $y = x^3 - x$ **54.** $y = x^3 - 9x$
55. $y = x^3 - 3x$ **56.** $y = 4x^3 - 9x$
57. $y = x^3 + 3x^2 - x - 3$
58. $y = x^3 - 2x^2 - x + 2$
59. $y = x^3 + x^2 - 9x - 9$
60. $y = x^3 + 4x^2 - x - 4$

SECTION

9.3   **The Inverse of a Function**

Suppose the function f is given by

$$f = \{(1, 4), (2, 5), (3, 6), (4, 7)\}$$

The inverse of f is obtained by reversing the order of the coordinates in each ordered pair in f. The inverse of f is the relation given by

$$g = \{(4, 1), (5, 2), (6, 3), (7, 4)\}$$

It is obvious that the domain of f is now the range of g, and the range of f is now the domain of g. Every function (or relation) has an inverse that is obtained from the original function by interchanging the components of each ordered pair.

Suppose a function f is defined with an equation instead of a list of ordered pairs. We can obtain the equation of the inverse of f by interchanging the role of x and y in the equation for f.

▶ **EXAMPLE 1**   If the function f is defined by $f(x) = 2x - 3$, find the equation that represents the inverse of f.

Solution   Since the inverse of f is obtained by interchanging the components of all the ordered pairs belonging to f, and each ordered pair in f satisfies the equation $y = 2x - 3$, we simply exchange x and y in the equation $y = 2x - 3$ to get the formula for the inverse of f:

$$x = 2y - 3$$

We now solve this equation for y in terms of x:

$$x + 3 = 2y$$

$$y = \frac{x + 3}{2}$$

The last line gives the equation that defines the inverse of f. Let's compare the graphs of f and its inverse as given above. (See Figure 1.)

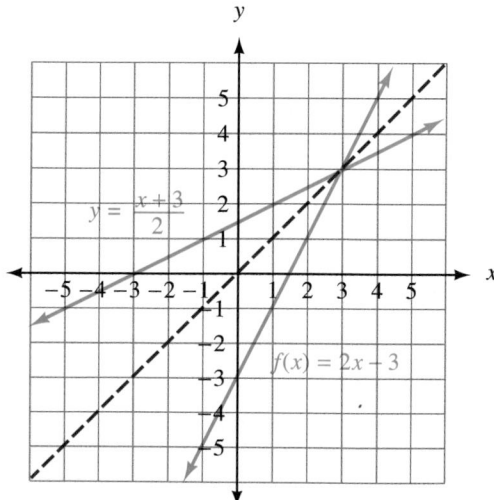

Figure 1

The graphs of f and its inverse have symmetry about the line $y = x$. This is a reasonable result since the one function was obtained from the other by interchanging x and y in the equation. The ordered pairs (a, b) and (b, a) always have symmetry about the line $y = x$. ◀

▶ **EXAMPLE 2** Graph the function $y = x^2 - 2$ and its inverse. Give the equation for the inverse.

Solution We can obtain the graph of the inverse of $y = x^2 - 2$ by graphing $y = x^2 - 2$ by the usual methods, and then reflecting the graph about the line $y = x$.

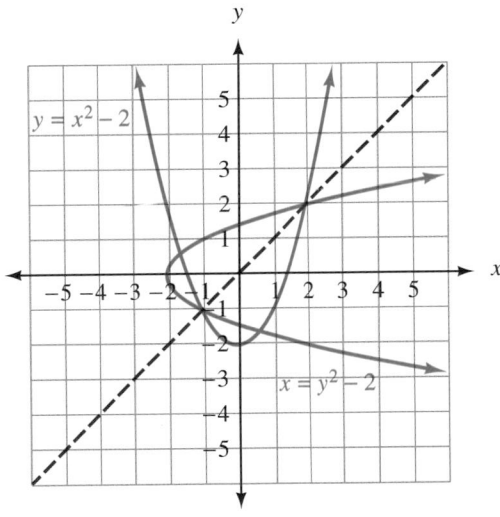

Figure 2

The equation that corresponds to the inverse of $y = x^2 - 2$ is obtained by interchanging x and y to get $x = y^2 - 2$.

$$x = y^2 - 2$$
$$x + 2 = y^2$$
$$y = \pm\sqrt{x + 2}$$ ◀

Comparing the graphs from Examples 1 and 2, we observe that the inverse of a function is not always a function. In Example 1, both f and its inverse have graphs that are nonvertical straight lines and therefore both represent functions. In Example 2, the inverse of function f is not a function, since a vertical line crosses it in more than one place.

We can distinguish between those functions with inverses that are also functions and those functions with inverses that are not functions with the following definition.

One-to-One Functions

> **DEFINITION** A function is a **one-to-one function** if every element in the range comes from exactly one element in the domain.

This definition indicates that a one-to-one function will yield a set of ordered pairs in which no two different ordered pairs have the same second coordinates. For example, the function

$$f = \{(2, 3), (-1, 3), (5, 8)\}$$

is not one-to-one because the element 3 in the range comes from both 2 and -1 in the domain. On the other hand, the function

$$g = \{(5, 7), (3, -1), (4, 2)\}$$

is a one-to-one function because every element in the range comes from only one element in the domain.

If we have the graph of a function, we can determine if the function is one-to-one with the following test.

Horizontal Line Test

If a horizontal line crosses the graph of a function in more than one place, then the function is not a one-to-one function because the points at which the horizontal line crosses the graph will be points with the same y-coordinates, but different x-coordinates. Therefore, the function will have an element in the range (the y-coordinate) that comes from more than one element in the domain (the x-coordinates).

Of the functions we have covered previously, all the linear functions and exponential functions are one-to-one functions because no horizontal lines can be found that will cross their graphs in more than one place.

Functions Whose Inverses Are Also Functions

Because one-to-one functions do not repeat second coordinates, when we reverse the order of the ordered pairs in a one-to-one function, we obtain a relation in which no two ordered pairs have the same first coordinate—by definition, this relation must be a function. In other words, every one-to-one function has an inverse that is itself a function. Because of this, we can use function notation to represent that inverse.

Inverse Function Notation

If $y = f(x)$ is a one-to-one function, then the inverse of f is also a function and can be denoted by $y = f^{-1}(x)$. To illustrate, in Example 1 we found the inverse of $f(x) = 2x - 3$ was the function $y = \dfrac{x + 3}{2}$. We can write this inverse function with inverse function notation as

$$f^{-1}(x) = \frac{x + 3}{2}$$

On the other hand, the inverse of the function in Example 2 is not itself a function, so we do not use the notation $f^{-1}(x)$ to represent it.

Note   The notation f^{-1} does not represent the reciprocal of f. That is, the -1 in this notation is not an exponent. The notation f^{-1} is defined as representing the inverse function for a one-to-one function.

▶ **EXAMPLE 3**   Find the inverse of $g(x) = \dfrac{x - 4}{x - 2}$.

Solution   To find the inverse for g, we begin by replacing $g(x)$ with y to obtain

$$y = \frac{x - 4}{x - 2} \qquad \text{(the original function)}$$

To find an equation for the inverse, we exchange x and y.

$$x = \frac{y - 4}{y - 2} \qquad \text{(the inverse of the original function)}$$

To solve for y we first multiply each side by $y - 2$ to obtain

$$x(y - 2) = y - 4$$

$$xy - 2x = y - 4 \qquad \text{Distributive property}$$

$$xy - y = 2x - 4 \qquad \text{Collect all terms containing}$$
$$\qquad\qquad\qquad\qquad\qquad y \text{ on the left side}$$

$$y(x - 1) = 2x - 4 \qquad \text{Factor } y \text{ from each term on the left side}$$

$$y = \frac{2x - 4}{x - 1} \qquad \text{Divide each side by } x - 1$$

Since our original function is one-to-one (see Figure 2 in Section 9.2), its inverse is also a function. Therefore we can use inverse function notation to write

$$g^{-1}(x) = \frac{2x - 4}{x - 1}$$

◀

▶ **EXAMPLE 4**   Graph the function $y = 2^x$ and its inverse $x = 2^y$.

Solution   We graphed $y = 2^x$ in the preceding section. We simply reflect its graph about the line $y = x$ to obtain the graph of its inverse $x = 2^y$. (See Figure 3.)

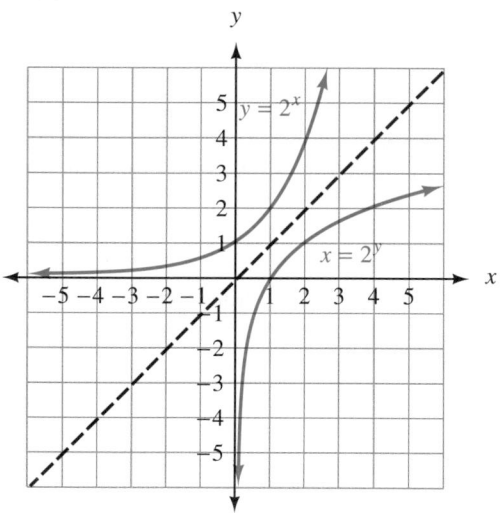

Figure 3

As you can see from the graph, $x = 2^y$ is a function. However, we do not have the mathematical tools to solve this equation for y. Therefore, we are unable to use the inverse function notation to represent this function. In Chapter 10 we will give a definition that solves this problem. For now, we simply leave the equation as $x = 2^y$.   ◀

Here is a summary of some of the things we know about functions, relations, and their inverses:

1. Every function is a relation, but not every relation is a function.
2. Every function has an inverse, but only one-to-one functions have inverses that are also functions.
3. The domain of a function is the range of its inverse and the range of a function is the domain of its inverse.
4. If $y = f(x)$ is a one-to-one function, then we can use the notation $y = f^{-1}(x)$ to represent its inverse function.
5. The graph of a function and its inverse have symmetry about the line $y = x$.
6. If (a, b) belongs to the function f, then the point (b, a) belongs to its inverse.

P R O B L E M   S E T   9 . 3

For each of the following one-to-one functions, find the equation of the inverse. Write the inverse using the notation $f^{-1}(x)$.

1. $f(x) = 3x - 1$
2. $f(x) = 2x - 5$
3. $f(x) = x^3$
4. $f(x) = x^3 - 2$

5. $f(x) = \dfrac{x - 3}{x - 1}$
6. $f(x) = \dfrac{x - 2}{x - 3}$

7. $f(x) = \dfrac{x - 3}{4}$
8. $f(x) = \dfrac{x + 7}{2}$

9. $f(x) = \frac{1}{2} x - 3$
10. $f(x) = \frac{1}{3} x + 1$

11. $f(x) = \dfrac{2x + 1}{3x + 1}$

12. $f(x) = \dfrac{3x + 2}{5x + 1}$

For each of the following relations, sketch the graph of the relation and its inverse, and write an equation for the inverse.

13. $y = 2x - 1$
14. $y = 3x + 1$
15. $y = x^2 - 3$
16. $y = x^2 + 1$
17. $y = x^2 - 2x - 3$
18. $y = x^2 + 2x - 3$
19. $y = 3^x$
20. $y = (\frac{1}{2})^x$
21. $y = 4$
22. $y = -2$
23. $y = \frac{1}{2} x^3$
24. $y = x^3 - 2$
25. $y = \frac{1}{2} x + 2$
26. $y = \frac{1}{3} x - 1$
27. $y = \sqrt{x + 2}$
28. $y = \sqrt{x} + 2$

29. If $f(x) = 3x - 2$, then $f^{-1}(x) = \dfrac{x + 2}{3}$. Use these two functions to find

    (a) $f(2)$ (b) $f^{-1}(2)$
    (c) $f[f^{-1}(2)]$ (d) $f^{-1}[f(2)]$

30. If $f(x) = \frac{1}{2} x + 5$, then $f^{-1}(x) = 2x - 10$. Use these two functions to find

    (a) $f(-4)$ (b) $f^{-1}(-4)$
    (c) $f[f^{-1}(-4)]$ (d) $f^{-1}[f(-4)]$

31. Let $f(x) = \dfrac{1}{x}$, and find $f^{-1}(x)$.

32. Let $f(x) = \dfrac{a}{x}$, and find $f^{-1}(x)$. (a is a real number constant.)

Review Problems

The following problems review material we covered in Section 8.6. They are taken from the book *Algebra for the Practical Man,* written by J. E. Thompson and published by D. Van Nostrand Company in 1931.

33. A man spent $112.80 for 108 geese and ducks, each goose costing 14 dimes and each duck 6 dimes. How many of each did he buy?

34. If 15 pounds of tea and 10 pounds of coffee together cost $15.50, while 25 pounds of tea and 13 pounds of coffee at the same prices cost $24.55, find the price per pound of each.

35. A number of oranges at the rate of three for ten cents and apples at fifteen cents a dozen cost, together, $6.80. Five times as many oranges and one fourth as many apples at the same rates would have cost $25.45. How many of each were bought?

36. An estate is divided among three persons A, B, and C. A's share is three times that of B and B's share is twice that of C. If A receives $9,000 more than C, how much does each receive?

One Step Further

For each of the following functions, find $f^{-1}(x)$. Then show that $f(f^{-1}(x)) = x$.

37. $f(x) = 3x + 5$
38. $f(x) = 6 - 8x$
39. $f(x) = x^3 + 1$
40. $f(x) = x^3 - 8$

41. $f(x) = \dfrac{x - 4}{x - 2}$

42. $f(x) = \dfrac{x - 3}{x - 1}$

The Circle

Before we find the general equation of a circle, we must first derive what is known as the **distance formula.**

Suppose (x_1, y_1) and (x_2, y_2) are any two points in the first quadrant. (Actually, we could choose the two points to be anywhere on the coordinate plane. It is just more convenient to have them in the first quadrant.) We can name the points P_1 and P_2, respectively, and draw the diagram shown in Figure 1.

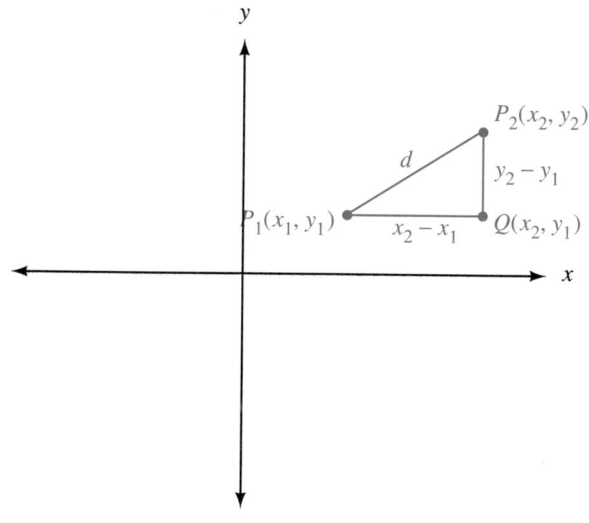

Figure 1

Notice the coordinates of point Q. The x-coordinate is x_2 since Q is directly below point P_2. The y-coordinate of Q is y_1 since Q is directly across from point P_1. It is evident from the diagram that the length of P_2Q is $y_2 - y_1$ and the length of P_1Q is $x_2 - x_1$. Using the Pythagorean Theorem, we have

$$(P_1P_2)^2 = (P_1Q)^2 + (P_2Q)^2$$

or

$$d^2 = (x_2 - x_1)^2 + (y_2 - y_1)^2$$

Taking the square root of both sides, we have

$$d = \sqrt{(x_2 - x_1)^2 + (y_2 - y_1)^2}$$

We know this is the positive square root, since d is the distance from P_1 to P_2 and must therefore be positive. This formula is called the **distance formula.**

▶ **EXAMPLE 1** Find the distance between (3, 5) and (2, −1).

Solution If we let (3, 5) be (x_1, y_1) and (2, −1) be (x_2, y_2) and apply the distance formula, we have

$$\begin{aligned} d &= \sqrt{(2-3)^2 + (-1-5)^2} \\ &= \sqrt{(-1)^2 + (-6)^2} \\ &= \sqrt{1 + 36} \\ &= \sqrt{37} \end{aligned}$$

◀

▶ **EXAMPLE 2** Find x if the distance from $(x, 5)$ to (3, 4) is $\sqrt{2}$.

Solution Using the distance formula, we have

$$\begin{aligned} \sqrt{2} &= \sqrt{(x-3)^2 + (5-4)^2} \\ 2 &= (x-3)^2 + 1^2 \\ 2 &= x^2 - 6x + 9 + 1 \\ 0 &= x^2 - 6x + 8 \\ 0 &= (x-4)(x-2) \\ x &= 4 \quad \text{or} \quad x = 2 \end{aligned}$$

The two solutions are 4 and 2, which indicates there are two points, (4, 5) and (2, 5), which are $\sqrt{2}$ units from (3, 4).   ◀

We can use the distance formula to derive the equation of a circle.

Theorem 9.1

The equation of the circle with center at (a, b) and radius r is given by

$$(x - a)^2 + (y - b)^2 = r^2$$

PROOF By definition, all points on the circle are a distance r from the center (a, b). If we let (x, y) represent any point on the circle, then (x, y) is r units from (a, b). Applying the distance formula, we have

$$r = \sqrt{(x - a)^2 + (y - b)^2}$$

Squaring both sides of this equation gives the equation of the circle:

$$(x - a)^2 + (y - b)^2 = r^2$$

We can use Theorem 9.1 to find the equation of a circle given its center and radius, or to find its center and radius given the equation.

▶ **EXAMPLE 3**   Find the equation of the circle with center at $(-3, 2)$ having a radius of 5.

Solution   We have $(a, b) = (-3, 2)$ and $r = 5$. Applying Theorem 9.1 yields

$$[x - (-3)]^2 + (y - 2)^2 = 5^2$$

$$(x + 3)^2 + (y - 2)^2 = 25$$ ◀

▶ **EXAMPLE 4**   Give the equation of the circle with radius 3 whose center is at the origin.

Solution   The coordinates of the center are $(0, 0)$, and the radius is 3. The equation must be

$$(x - 0)^2 + (y - 0)^2 = 3^2$$

$$x^2 + y^2 = 9$$ ◀

We can see from Example 4 that the equation of any circle with its center at the origin and radius r will be $x^2 + y^2 = r^2$.

▶ **EXAMPLE 5**   Find the center and radius, and sketch the graph, of the circle whose equation is

$$(x - 1)^2 + (y + 3)^2 = 4$$

Solution   Writing the equation in the form

$$(x - a)^2 + (y - b)^2 = r^2$$

we have

$$(x - 1)^2 + [y - (-3)]^2 = 2^2$$

The center is at $(1, -3)$ and the radius is 2. (See Figure 2.)

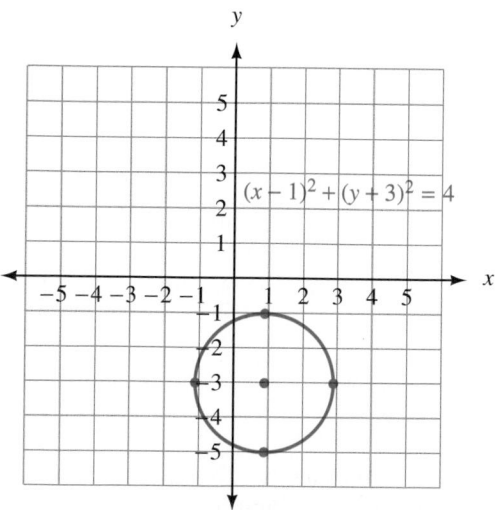

Figure 2   ◀

▶ **EXAMPLE 6**   Sketch the graph of $x^2 + y^2 = 9$.

Solution   Since the equation can be written in the form

$$(x - 0)^2 + (y - 0)^2 = 3^2$$

it must have its center at $(0, 0)$ and a radius of 3. (See Figure 3.)

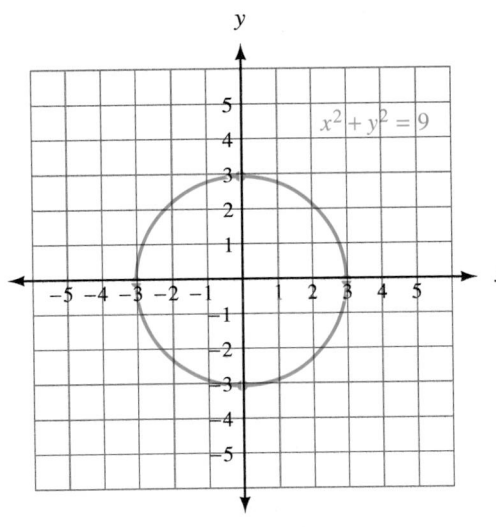

Figure 3   ◀

▶ **EXAMPLE 7**   Sketch the graph of $x^2 + y^2 + 6x - 4y - 12 = 0$.

Solution   To sketch the graph we must find the center and radius. The center and radius can be identified if the equation has the form

$$(x - a)^2 + (y - b)^2 = r^2$$

The original equation can be written in this form by completing the squares on x and y:

$$x^2 + y^2 + 6x - 4y - 12 = 0$$
$$x^2 + 6x \quad + y^2 - 4y \quad = 12$$
$$x^2 + 6x + \mathbf{9} + y^2 - 4y + \mathbf{4} = 12 + \mathbf{9} + \mathbf{4}$$
$$(x + 3)^2 + (y - 2)^2 = 25$$
$$(x + 3)^2 + (y - 2)^2 = 5^2$$

From the last line it is apparent that the center is at $(-3, 2)$ and the radius is 5. (See Figure 4 on the next page.)

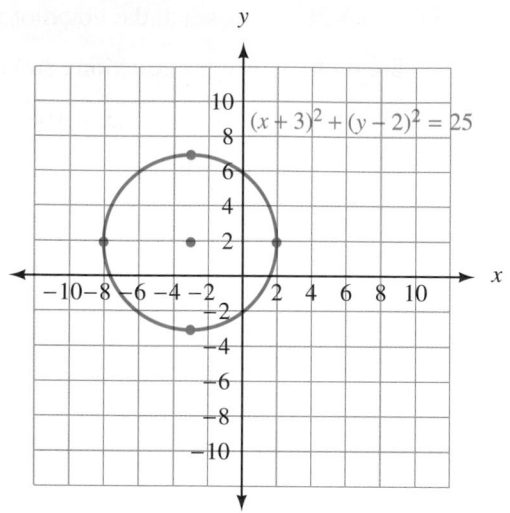

Figure 4

P R O B L E M   S E T   9 . 4

Find the distance between the following points.

1. (3, 7) and (6, 3) **2.** (4, 7) and (8, 1)
3. (0, 9) and (5, 0) **4.** (−3, 0) and (0, 4)
5. (3, −5) and (−2, 1)
6. (−8, 9) and (−3, −2)
7. (−1, −2) and (−10, 5)
8. (−3, −8) and (−1, 6)
9. Find x so the distance between $(x, 2)$ and $(1, 5)$ is $\sqrt{13}$.
10. Find x so the distance between $(-2, 3)$ and $(x, 1)$ is 3.
11. Find y so the distance between $(7, y)$ and $(8, 3)$ is 1.
12. Find y so the distance between $(3, -5)$ and $(3, y)$ is 9.

Write the equation of the circle with the given center and radius.

13. Center (2, 3); $r = 4$
14. Center (3, −1); $r = 5$
15. Center (3, −2); $r = 3$
16. Center (−2, 4); $r = 1$
17. Center (−5, −1); $r = \sqrt{5}$
18. Center (−7, −6); $r = \sqrt{3}$

19. Center (0, −5); $r = 1$
20. Center (0, −1); $r = 7$
21. Center (0, 0); $r = 2$
22. Center (0, 0); $r = 5$

Give the center and radius, and sketch the graph of each of the following circles.

23. $x^2 + y^2 = 4$ **24.** $x^2 + y^2 = 16$
25. $(x - 1)^2 + (y - 3)^2 = 25$
26. $(x - 4)^2 + (y - 1)^2 = 36$
27. $(x + 2)^2 + (y - 4)^2 = 8$
28. $(x - 3)^2 + (y + 1)^2 = 12$
29. $(x + 1)^2 + (y + 1)^2 = 1$
30. $(x + 3)^2 + (y + 2)^2 = 9$
31. $x^2 + y^2 - 6y = 7$
32. $x^2 + y^2 - 4y = 5$
33. $x^2 + y^2 + 2x = 1$
34. $x^2 + y^2 + 10x = 0$
35. $x^2 + y^2 - 4x - 6y = -4$
36. $x^2 + y^2 - 4x + 2y = 4$
37. $x^2 + y^2 + 2x + y = \frac{11}{4}$
38. $x^2 + y^2 - 6x - y = -\frac{1}{4}$
39. Find the equation of the circle with center at the origin that contains the point (3, 4).

40. Find the equation of the circle with center at the origin that contains the point $(-5, 12)$.
41. Find the equation of the circle with center at the origin and x-intercepts 3 and -3.
42. Find the equation of the circle with y-intercepts 4 and -4, and center at the origin.
43. A circle with center at $(-1, 3)$ passes through the point $(4, 3)$. Find the equation.
44. A circle with center at $(2, 5)$ passes through the point $(-1, 4)$. Find the equation.
45. If we solved the equation $x^2 + y^2 = 9$ for y, we would obtain the equation $y = \pm\sqrt{9 - x^2}$. This last equation is equivalent to the two equations $y = \sqrt{9 - x^2}$, in which y is always positive, and $y = -\sqrt{9 - x^2}$, in which y is always negative. Look at the graph of $x^2 + y^2 = 9$ in Example 6 of this section and indicate what part of the graph each of the two equations corresponds to.
46. Solve the equation $x^2 + y^2 = 9$ for x, and then indicate what part of the graph in Example 6 each of the two resulting equations corresponds to.
47. The formula for the circumference of a circle is $C = 2\pi r$. If the units of the coordinate system used in Problem 25 are in meters, what is the circumference of that circle?
48. The formula for the area of a circle is $A = \pi r^2$. What is the area of the circle mentioned in Problem 47?

Review Problems

The following problems review material we covered in Section 8.4.

Solve each system by using Cramer's rule.

49. $4x - 7y = 3$
$5x + 2y = -3$
50. $9x - 8y = 4$
$2x + 3y = 6$
51. $3x + 4y = 15$
$2x - 5z = -3$
$4y - 3z = 9$
52. $x + 3y = 5$
$6y + z = 12$
$x - 2z = -10$

One Step Further

A circle is **tangent** to a line if it touches, but does not cross, the line.

53. Find the equation of the circle with center at $(2, 3)$ if the circle is tangent to the y-axis.
54. Find the equation of the circle with center at $(3, 2)$ if the circle is tangent to the x-axis.
55. Find the equation of the circle with center at $(2, 3)$ if the circle is tangent to the vertical line $x = 4$.
56. Find the equation of the circle with center at $(3, 2)$ if the circle is tangent to the horizontal line $y = 6$.

Find the distance from the origin to the center of each circle given below.

57. $x^2 + y^2 - 6x + 8y = 144$
58. $x^2 + y^2 - 8x + 6y = 144$
59. $x^2 + y^2 - 6x - 8y = 144$
60. $x^2 + y^2 + 8x + 6y = 144$

SECTION

9.5 **Ellipses and Hyperbolas**

This section is concerned with the graphs of ellipses and hyperbolas. To simplify matters somewhat we will begin by considering those graphs that are centered about the origin.

Suppose we want to graph the equation

$$\frac{x^2}{25} + \frac{y^2}{9} = 1$$

We can find the y-intercepts by letting $x = 0$, and the x-intercepts by letting $y = 0$:

When	$x = 0$	When	$y = 0$
	$\dfrac{0^2}{25} + \dfrac{y^2}{9} = 1$		$\dfrac{x^2}{25} + \dfrac{0^2}{9} = 1$
	$y^2 = 9$		$x^2 = 25$
	$y = \pm 3$		$x = \pm 5$

The graph crosses the y-axis at $(0, 3)$ and $(0, -3)$ and the x-axis at $(5, 0)$ and $(-5, 0)$. Graphing these points and then connecting them with a smooth curve gives the graph shown in Figure 1.

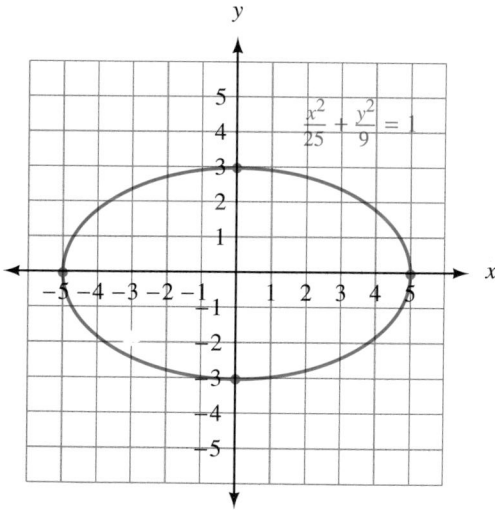

Figure 1

We can find other ordered pairs on the graph by substituting in values for x (or y) and then solving for y (or x). For example, if we let $x = 3$ then

$$\frac{3^2}{25} + \frac{y^2}{9} = 1$$

$$\frac{9}{25} + \frac{y^2}{9} = 1$$

$$0.36 + \frac{y^2}{9} = 1$$

$$\frac{y^2}{9} = 0.64$$

$$y^2 = 5.76$$

$$y = \pm 2.4$$

This would give us the two ordered pairs $(3, -2.4)$ and $(3, 2.4)$.

A graph of the type shown in Figure 1 is called an **ellipse.** If we were to find some other ordered pairs that satisfy our original equation, we would find that their graphs lie on the ellipse. Also, the coordinates of any point on the ellipse will satisfy the equation. We can generalize these results as follows.

Ellipses Centered at the Origin

The graph of any equation of the form

$$\frac{x^2}{a^2} + \frac{y^2}{b^2} = 1 \qquad \text{Standard form}$$

will be an **ellipse** centered at the origin. The ellipse will cross the x-axis at $(a, 0)$ and $(-a, 0)$. It will cross the y-axis at $(0, b)$ and $(0, -b)$. When a and b are equal, the ellipse will be a circle. Each of the points $(a, 0)$, $(-a, 0)$, $(0, b)$, and $(0, -b)$ is a vertex of the graph.

The most convenient way to graph an ellipse is to locate the intercepts.

▶ **EXAMPLE 1** Sketch the graph of $4x^2 + 9y^2 = 36$.

Solution To write the equation in the form

$$\frac{x^2}{a^2} + \frac{y^2}{b^2} = 1$$

we must divide both sides by 36:

$$\frac{4x^2}{36} + \frac{9y^2}{36} = \frac{36}{36}$$

$$\frac{x^2}{9} + \frac{y^2}{4} = 1$$

The graph crosses the x-axis at $(3, 0)$, $(-3, 0)$ and the y-axis at $(0, 2)$, $(0, -2)$. (See Figure 2 on the next page.)

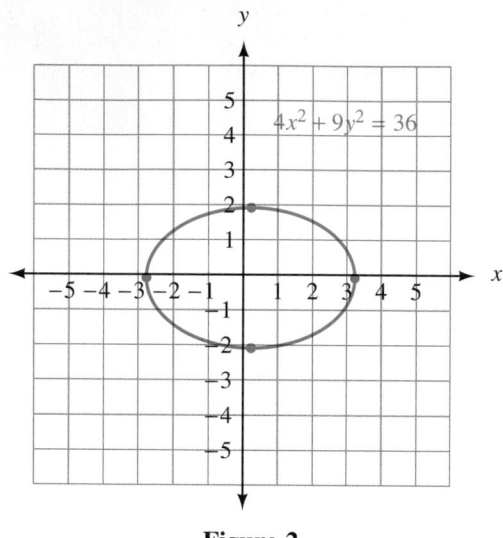

Figure 2

Consider the equation

$$\frac{x^2}{9} - \frac{y^2}{4} = 1$$

If we were to find a number of ordered pairs that are solutions to the equation and connect their graphs with a smooth curve, we would have Figure 3.

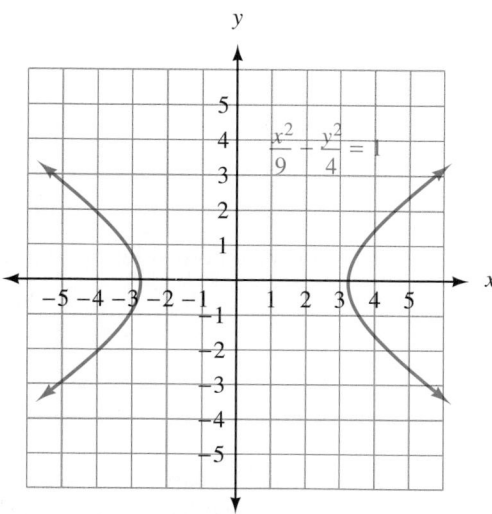

Figure 3

This graph is an example of a **hyperbola.** Notice that the graph has x-intercepts at $(3, 0)$ and $(-3, 0)$. The graph has no y-intercepts and hence does not cross the y-axis, since substituting $x = 0$ into the equation yields

$$\frac{0^2}{9} - \frac{y^2}{4} = 1$$

$$-y^2 = 4$$

$$y^2 = -4$$

for which there is no real solution. We can, however, use the number below y^2 to help sketch the graph. If we draw a rectangle that has its sides parallel to the x- and y-axes and that passes through the x-intercepts and the points on the y-axis corresponding to the square roots of the number below y^2, $+2$, and -2, it looks like the rectangle in Figure 4. The lines that connect opposite corners of the rectangle are called **asymptotes.** The graph of the hyperbola

$$\frac{x^2}{9} - \frac{y^2}{4} = 1$$

will approach these lines. Figure 4 is the graph.

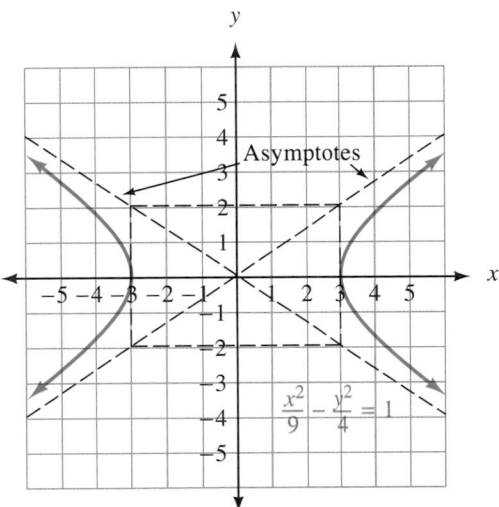

Figure 4

▶ **EXAMPLE 2** Graph the equation $\dfrac{y^2}{9} - \dfrac{x^2}{16} = 1$.

Solution   In this case the y-intercepts are 3 and -3, and the x-intercepts do not exist. We can use the square roots of the number below x^2, however, to find the asymptotes associated with the graph. The sides of the rectangle used to draw the asymptotes must pass through 3 and -3 on the y-axis, and 4 and -4 on the x-axis. (See Figure 5.)

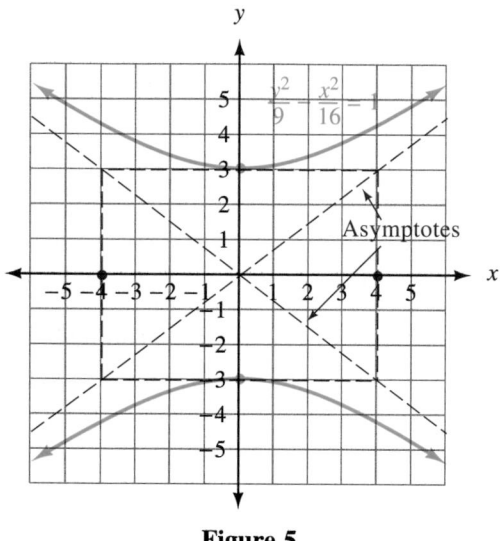

Figure 5

Here is a summary of what we have for hyperbolas.

Hyperbolas Centered at the Origin

The graph of the equation

$$\frac{x^2}{a^2} - \frac{y^2}{b^2} = 1$$

will be a hyperbola centered at the origin. The graph will have x-intercepts (vertices) at $-a$ and a.

The graph of the equation

$$\frac{y^2}{a^2} - \frac{x^2}{b^2} = 1$$

will be a hyperbola centered at the origin. The graph will have y-intercepts (vertices) at $-a$ and a.

As an aid in sketching either of the preceding equations, the asymptotes can be found by drawing lines through opposite corners of the rectangle whose sides pass through $-a$, a, $-b$, and b on the axes.

Ellipses and Hyperbolas Not Centered at the Origin

The equation below is the equation of an ellipse with its center at the point $(4, 1)$.

$$\frac{(x - 4)^2}{9} + \frac{(y - 1)^2}{4} = 1$$

To see why the center is at (4, 1) we substitute x' (read "x prime") for $x - 4$ and y' for $y - 1$ in the equation. That is,

if $\qquad\qquad\qquad x' = x - 4$

and $\qquad\qquad\quad\; y' = y - 1$

the equation $\qquad \dfrac{(x - 4)^2}{9} + \dfrac{(y - 1)^2}{4} = 1$

becomes $\qquad\quad \dfrac{(x')^2}{9} + \dfrac{(y')^2}{4} \qquad = 1$

which is the equation of an ellipse in a coordinate system with an x'-axis and a y'-axis. We call this new coordinate system the $x'y'$-coordinate system. The center of our ellipse is at the origin in the $x'y'$-coordinate system. The question is this: What are the coordinates of the center of this ellipse in the original xy-coordinate system? To answer this question we go back to our original substitutions:

$$x' = x - 4$$
$$y' = y - 1$$

In the $x'y'$-coordinate system, the center of our ellipse is at $x' = 0$, $y' = 0$ (the origin of the $x'y'$ system). Substituting these numbers for x' and y', we have

$$0 = x - 4$$
$$0 = y - 1$$

Solving these equations for x and y will give us the coordinates of the center of our ellipse in the xy-coordinate system. As you can see, the solutions are $x = 4$ and $y = 1$. Therefore, in the xy-coordinate system, the center of our ellipse is at the point (4, 1). Figure 6 illustrates this discussion.

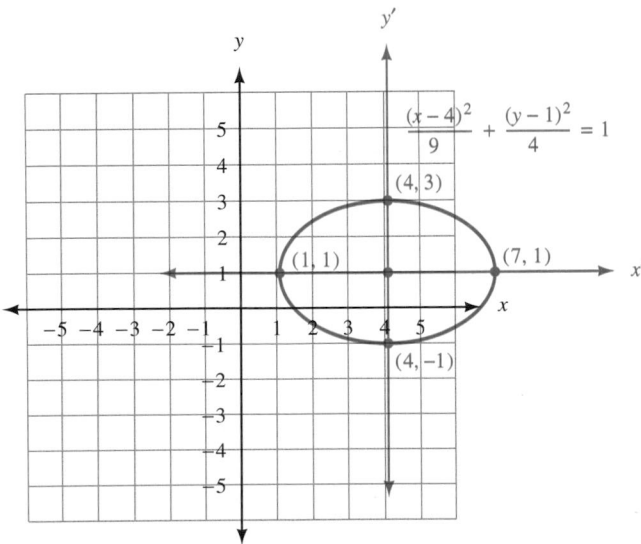

Figure 6

The coordinates of all points labeled in Figure 6 are given with respect to the xy-coordinate system. The x' and y' axes are shown simply for reference in our discussion. Note that the horizontal distance from the center to the vertices is 3—the square root of the number below $x - 4$. Likewise, the vertical distance from the center to the other vertices is 2—the square root of the number below $y - 1$. We summarize the information above with the following:

An Ellipse with Center at (h, k)

The graph of the equation

$$\frac{(x - h)^2}{a^2} + \frac{(y - k)^2}{b^2} = 1$$

will be an ellipse with center at (h, k). The vertices of the ellipse will be at the points $(h + a, k)$, $(h - a, k)$, $(h, k + b)$, and $(h, k - b)$.

▶ **EXAMPLE 3**   Graph the ellipse $x^2 + 9y^2 + 4x - 54y + 76 = 0$.

Solution   In order to identify the coordinates of the center, we must complete the square on x and also on y. To begin, we rearrange the terms so that the terms containing x are together, the terms containing y are together, and the constant term is on the other side of the equal sign. Doing so gives us the following equation:

$$x^2 + 4x \quad + 9y^2 - 54y \quad = -76$$

Before we can complete the square on y we must factor 9 from each term containing y.

$$x^2 + 4x \quad + 9(y^2 - 6y \quad) = -76$$

To complete the square on x, we add 4 to each side of the equation. To complete the square on y, we add 9 inside the parentheses. This increases the left side of the equation by 81 since each term within the parentheses is multiplied by 9. Therefore, we must add 81 to the right side of the equation also.

$$x^2 + 4x + \mathbf{4} + 9(y^2 - 6y + \mathbf{9}) = -76 + \mathbf{4} + \mathbf{81}$$

$$(x + 2)^2 + 9(y - 3)^2 = 9$$

To identify the distances to the vertices, we divide each term on both sides by 9.

$$\frac{(x + 2)^2}{9} + \frac{9(y - 3)^2}{9} = \frac{9}{9}$$

$$\frac{(x + 2)^2}{9} + \frac{(y - 3)^2}{1} = 1$$

The graph is an ellipse with center at $(-2, 3)$, as shown in Figure 7.

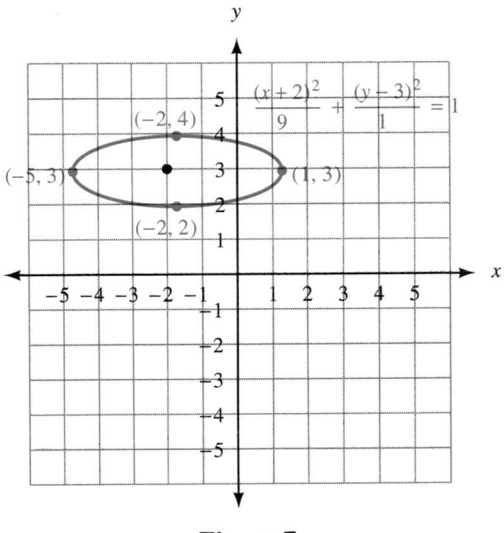

Figure 7   ◄

The ideas associated with graphing hyperbolas whose centers are not at the origin parallel the ideas just presented about graphing ellipses whose centers have been moved off the origin. Without showing the justification for doing so, we state the following guidelines for graphing hyperbolas:

Hyperbolas with Centers at (h, k)
The graphs of the equations

$$\frac{(x - h)^2}{a^2} - \frac{(y - k)^2}{b^2} = 1 \quad \text{and} \quad \frac{(y - k)^2}{b^2} - \frac{(x - h)^2}{a^2} = 1$$

will be hyperbolas with their centers at (h, k). The vertices of the graph of the first equation will be at the points $(h + a, k)$ and $(h - a, k)$, while the vertices for the graph of the second equation will be at $(h, k + b)$ and $(h, k - b)$. In either case, the asymptotes can be found by connecting opposite corners of the rectangle that contains the four vertices mentioned previously.

▶ **EXAMPLE 4** Graph the hyperbola $4x^2 - y^2 + 4y - 20 = 0$.

Solution In order to identify the coordinates of the center of the hyperbola, we need to complete the square on y. (Since there is no linear term in x, we do not need to complete the square on x. The x-coordinate of the center will be $x = 0$.)

$$4x^2 - y^2 + 4y - 20 = 0$$

$$4x^2 - y^2 + 4y \qquad = 20 \qquad \text{Add 20 to each side}$$

$$4x^2 - 1(y^2 - 4y \quad) = 20 \qquad \text{Factor } -1 \text{ from each term containing } y$$

To complete the square on y, we add 4 to the terms inside the parentheses. Doing so adds -4 to the left side of the equation since everything inside the parentheses is multiplied by -1. To keep from changing the equation we must add -4 to the right side also.

$$4x^2 - 1(y^2 - 4y + \mathbf{4}) = 20 - \mathbf{4}$$

$$4x^2 - 1(y - 2)^2 = 16$$

$$\frac{4x^2}{16} - \frac{(y - 2)^2}{16} = \frac{16}{16}$$

$$\frac{x^2}{4} - \frac{(y - 2)^2}{16} = 1$$

This is the equation of a hyperbola with center at $(0, 2)$. The graph opens to the right and left as shown in Figure 8.

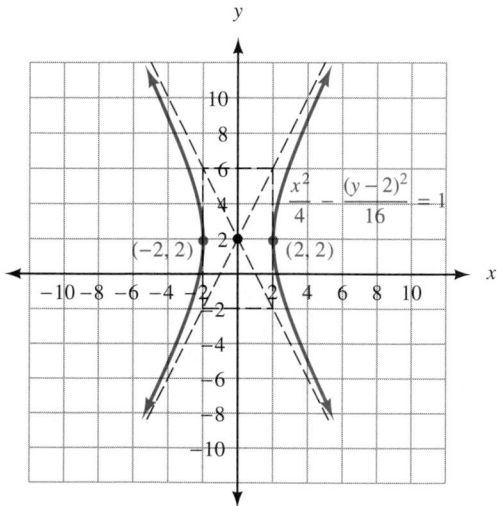

Figure 8

PROBLEM SET 9.5

Graph each of the following. Be sure to label both the x- and y-intercepts.

1. $\dfrac{x^2}{9} + \dfrac{y^2}{16} = 1$ **2.** $\dfrac{x^2}{25} + \dfrac{y^2}{4} = 1$

3. $\dfrac{x^2}{16} + \dfrac{y^2}{9} = 1$ **4.** $\dfrac{x^2}{4} + \dfrac{y^2}{25} = 1$

5. $\dfrac{x^2}{3} + \dfrac{y^2}{4} = 1$ **6.** $\dfrac{x^2}{4} + \dfrac{y^2}{3} = 1$

7. $4x^2 + 25y^2 = 100$ **8.** $4x^2 + 9y^2 = 36$
9. $x^2 + 8y^2 = 16$ **10.** $12x^2 + y^2 = 36$

Graph each of the following. Show the intercepts and the asymptotes in each case.

11. $\dfrac{x^2}{9} - \dfrac{y^2}{16} = 1$ **12.** $\dfrac{x^2}{25} - \dfrac{y^2}{4} = 1$

13. $\dfrac{x^2}{16} - \dfrac{y^2}{9} = 1$ **14.** $\dfrac{x^2}{4} - \dfrac{y^2}{25} = 1$

15. $\dfrac{y^2}{9} - \dfrac{x^2}{16} = 1$ **16.** $\dfrac{y^2}{25} - \dfrac{x^2}{4} = 1$

17. $\dfrac{y^2}{36} - \dfrac{x^2}{4} = 1$ **18.** $\dfrac{y^2}{4} - \dfrac{x^2}{36} = 1$

19. $x^2 - 4y^2 = 4$ **20.** $y^2 - 4x^2 = 4$
21. $16y^2 - 9x^2 = 144$
22. $4y^2 - 25x^2 = 100$

Find the x- and y-intercepts, if they exist, for each of the following. Do not graph.

23. $0.4x^2 + 0.9y^2 = 3.6$
24. $1.6x^2 + 0.9y^2 = 14.4$

25. $\dfrac{x^2}{0.04} - \dfrac{y^2}{0.09} = 1$

26. $\dfrac{y^2}{0.16} - \dfrac{x^2}{0.25} = 1$

27. $\dfrac{25x^2}{9} + \dfrac{25y^2}{4} = 1$

28. $\dfrac{16x^2}{9} + \dfrac{16y^2}{25} = 1$

Graph each of the following ellipses. In each case, label the coordinates of the center and the vertices.

29. $\dfrac{(x-4)^2}{4} + \dfrac{(y-2)^2}{9} = 1$

30. $\dfrac{(x-2)^2}{4} + \dfrac{(y-4)^2}{9} = 1$

31. $4x^2 + y^2 - 4y - 12 = 0$
32. $4x^2 + y^2 - 24x - 4y + 36 = 0$
33. $x^2 + 9y^2 + 4x - 54y + 76 = 0$
34. $4x^2 + y^2 - 16x + 2y + 13 = 0$

Graph each of the following hyperbolas. In each case, label the coordinates of the center and the vertices, and show the asymptotes.

35. $\dfrac{(x-2)^2}{16} - \dfrac{y^2}{4} = 1$ **36.** $\dfrac{(y-2)^2}{16} - \dfrac{x^2}{4} = 1$

37. $9y^2 - x^2 - 4x + 54y + 68 = 0$
38. $4x^2 - y^2 - 24x + 4y + 28 = 0$
39. $4y^2 - 9x^2 - 16y + 72x - 164 = 0$
40. $4x^2 - y^2 - 16x - 2y + 11 = 0$

41. Find y when x is 4 in the equation $\dfrac{x^2}{25} + \dfrac{y^2}{9} = 1$.

42. Find x when y is 3 in the equation $\dfrac{x^2}{4} + \dfrac{y^2}{25} = 1$.

43. Find y when x is 1.8 in $16x^2 + 9y^2 = 144$.
44. Find y when x is 1.6 in $49x^2 + 4y^2 = 196$.
45. Give the equations of the two asymptotes in the graph you found in Problem 15.
46. Give the equations of the two asymptotes in the graph you found in Problem 16.
47. For the ellipses you have graphed in this section, the longer line segment connecting opposite intercepts is called the **major axis** of the ellipse. Give the length of the major axis of the ellipse you graphed in Problem 3.
48. For the ellipses you have graphed in this section, the shorter line segment connecting opposite intercepts is called the **minor axis** of the ellipse. Give the length of the minor axis of the ellipse you graphed in Problem 3.

▶ **EXAMPLE 3**   Graph $4y^2 - 9x^2 < 36$.

Solution   The boundary is the hyperbola $4y^2 - 9x^2 = 36$ and is not included in the solution set. Testing $(0, 0)$ in the original inequality yields a true statement, which means that the region containing the origin is the solution set. (See Figure 4.)

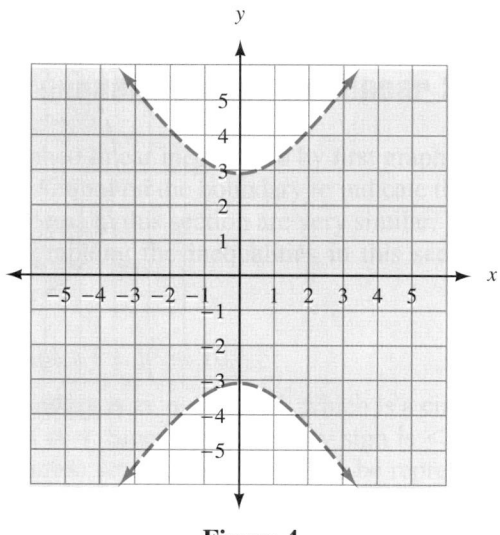

Figure 4

Next we solve systems of equations that contain at least one second-degree equation. The most convenient method of solving a system that contains one or two second-degree equations is by substitution, although the addition method can be used at times.

▶ **EXAMPLE 4**   Solve the system

$$x^2 + y^2 = 4$$

$$x - 2y = 4$$

Solution   In this case the substitution method is the most convenient. Solving the second equation for x in terms of y, we have

$$x - 2y = 4$$

$$x = 2y + 4$$

We now substitute $2y + 4$ for x in the first equation in our original system and proceed to solve for y:

$$(2y + 4)^2 + y^2 = 4$$

$$4y^2 + 16y + 16 + y^2 = 4$$

$$5y^2 + 16y + 12 = 0$$

$$(5y + 6)(y + 2) = 0$$

$$5y + 6 = 0 \quad \text{or} \quad y + 2 = 0$$

$$y = -\frac{6}{5} \quad \text{or} \quad y = -2$$

These are the y-coordinates of the two solutions to the system. Substituting $y = -\frac{6}{5}$ into $x - 2y = 4$ and solving for x gives us $x = \frac{8}{5}$. Using $y = -2$ in the same equation yields $x = 0$. The two solutions to our system are $(\frac{8}{5}, -\frac{6}{5})$ and $(0, -2)$. Although graphing the system is not necessary, it does help us visualize the situation. (See Figure 5.)

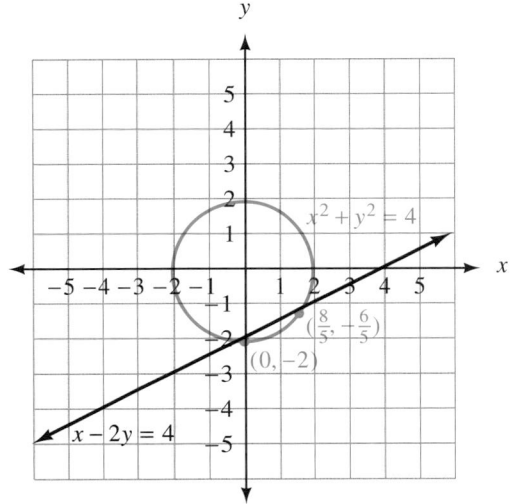

Figure 5 ◀

▶ **EXAMPLE 5**   Solve the system

$$16x^2 - 4y^2 = 64$$

$$x^2 + y^2 = 9$$

Solution   Since each equation is of the second degree in both x and y, it is easier to solve this system by eliminating one of the variables by addition. To eliminate y we multiply the bottom equation by 4 and add the result to the top equation:

$$\begin{array}{r} 16x^2 - 4y^2 = 64 \\ 4x^2 + 4y^2 = 36 \\ \hline 20x^2 = 100 \end{array}$$

$$x^2 = 5$$

$$x = \pm\sqrt{5}$$

Solution   The boundary for the top inequality is a circle with center at the origin and a radius of 3. The solution set lies inside the boundary. The boundary for the second inequality is an ellipse. In this case the solution set lies outside the boundary. For both graphs, the boundary is also part of the solution set.

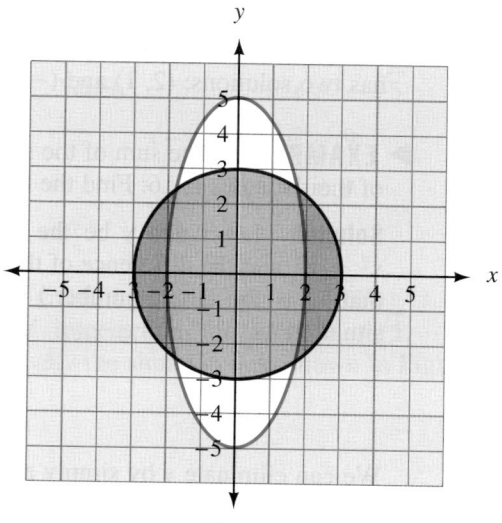

Figure 7

The solution set is the intersection of the two individual solution sets.   ◄

▶ **EXAMPLE 9**   Graph the solution set for the system below.

$$x - 2y \le 4$$

$$x + y \le 4$$

$$x \ge -1$$

Solution   We have three linear inequalities, representing three sections of the coordinate plane. The graph of the solution set for this system will be the intersection of these three sections. The graph of $x - 2y \le 4$ is the section above and including the boundary $x - 2y = 4$. The graph of $x + y \le 4$ is the section below and including the boundary line $x + y = 4$. The graph of $x \ge -1$ is all the points to the right of, and including, the vertical line $x = -1$. The intersection of these three graphs is shown in Figure 8.

Figure 8

PROBLEM SET 9.6

Graph each of the following inequalities.

1. $x^2 + y^2 \leq 49$ **2.** $x^2 + y^2 < 49$
3. $(x - 2)^2 + (y + 3)^2 < 16$
4. $(x + 3)^2 + (y - 2)^2 \geq 25$
5. $y < x^2 - 6x + 7$ **6.** $y \geq x^2 + 2x - 8$
7. $\dfrac{x^2}{25} - \dfrac{y^2}{9} \geq 1$ **8.** $\dfrac{x^2}{25} - \dfrac{y^2}{9} \leq 1$
9. $4x^2 + 25y^2 \leq 100$ **10.** $25x^2 - 4y^2 > 100$

Graph the solution sets to the following systems.

11. $x^2 + y^2 < 9$
$\quad\quad y \geq x^2 - 1$
12. $x^2 + y^2 \leq 16$
$\quad\quad y < x^2 + 2$
13. $\dfrac{x^2}{9} + \dfrac{y^2}{25} \leq 1$
$\quad \dfrac{x^2}{4} - \dfrac{y^2}{9} > 1$
14. $\dfrac{x^2}{4} + \dfrac{y^2}{16} \geq 1$
$\quad \dfrac{x^2}{9} - \dfrac{y^2}{25} < 1$
15. $4x^2 + 9y^2 \leq 36$
$\quad\quad y > x^2 + 2$
16. $9x^2 + 4y^2 \geq 36$
$\quad\quad y < x^2 + 1$

17. $x + y \leq 3$
$\quad x - 3y \leq 3$
$\quad\quad x \geq -2$
18. $x - y \leq 4$
$\quad x + 2y \leq 4$
$\quad\quad x \geq -1$
19. $x + y \leq 2$
$\quad -x + y \leq 2$
$\quad\quad y \geq -2$
20. $x - y \leq 3$
$\quad -x - y \leq 3$
$\quad\quad y \leq -1$
21. $x + y \leq 4$
$\quad x \geq 0$
$\quad y \geq 0$
22. $x - y \leq 2$
$\quad x \geq 0$
$\quad y \leq 0$

Solve each of the following systems of equations.

23. $x^2 + y^2 = 9$
$\quad 2x + y = 3$
24. $x^2 + y^2 = 9$
$\quad x + 2y = 3$
25. $x^2 + y^2 = 16$
$\quad x + 2y = 8$
26. $x^2 + y^2 = 16$
$\quad x - 2y = 8$
27. $x^2 + y^2 = 25$
$\quad x^2 - y^2 = 25$
28. $x^2 + y^2 = 4$
$\quad 2x^2 - y^2 = 5$
29. $x^2 + y^2 = 9$
$\quad y = x^2 - 3$
30. $x^2 + y^2 = 4$
$\quad y = x^2 - 2$
31. $x^2 + y^2 = 16$
$\quad y = x^2 - 4$
32. $x^2 + y^2 = 1$
$\quad y = x^2 - 1$
33. $3x + 2y = 10$
$\quad y = x^2 - 5$

7.

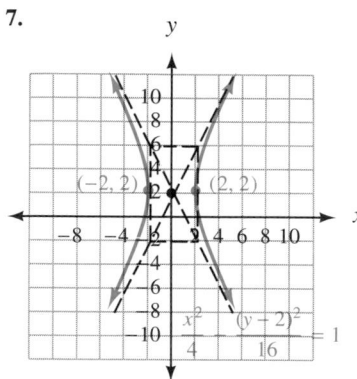

Hyperbolas with Centers at (h, k) [9.5]

The graphs of the equations

$$\frac{(x - h)^2}{a^2} - \frac{(y - k)^2}{b^2} = 1 \quad \text{and} \quad \frac{(y - k)^2}{b^2} - \frac{(x - h)^2}{a^2} = 1$$

will be hyperbolas with their centers at (h, k). The vertices of the graph of the first equation will be at the points $(h + a, k)$ and $(h - a, k)$, while the vertices for the graph of the second equation will be at $(h, k + b)$ and $(h, k - b)$. In either case, the asymptotes can be found by connecting opposite corners of the rectangle that contains the four vertices mentioned previously.

8. The graph of the inequality

$$x^2 + y^2 < 9$$

is all points inside the circle with center at the origin and radius 3. The circle itself is not part of the solution and is, therefore, shown with a broken curve.

Second-Degree Inequalities in Two Variables [9.6]

We graph second-degree inequalities in two variables in much the same way that we graphed linear inequalities. That is, we begin by graphing the boundary, using a solid curve if the boundary is included in the solution (this happens when the inequality symbol is $\geq$ or $\leq$), or a broken curve if the boundary is not included in the solution (when the inequality symbol is $>$ or $<$). After we have graphed the boundary, we choose a test point that is not on the boundary and try it in the original inequality. A true statement indicates we are in the region of the solution. A false statement indicates we are not in the region of the solution.

9. We can solve the system

$$x^2 + y^2 = 4$$

$$x = 2y + 4$$

by substituting $2y + 4$ from the second equation for x in the first equation:

$$(2y + 4)^2 + y^2 = 4$$

$$4y^2 + 16y + 16 + y^2 = 4$$

$$5y^2 + 16y + 12 = 0$$

$$(5y + 6)(y + 2) = 0$$

$$y = -\frac{6}{5} \quad \text{or} \quad y = -2$$

Substituting these values of y into the second equation in our system gives $x = \frac{8}{5}$ and $x = 0$. The solutions are $\left(\frac{8}{5}, -\frac{6}{5}\right)$ and $(0, -2)$.

Systems of Nonlinear Equations [9.6]

A system of nonlinear equations is two equations, at least one of which is not linear, considered at the same time. The solution set for the system consists of all ordered pairs that satisfy both equations. In most cases, we use the substitution method to solve these systems; however, the addition method can be used if like variables are raised to the same power in both equations. It is sometimes helpful to graph each equation in the system on the same set of axes in order to anticipate the number and approximate positions of the solutions.

COMMON MISTAKE

1. A common mistake occurs when the expression $f^{-1}(x)$ is interpreted as meaning the reciprocal of $f(x)$. The notation $f^{-1}(x)$ is used to denote the *inverse* of $f(x)$:

$$f^{-1}(x) \neq \frac{1}{f(x)} \qquad f^{-1}(x) = \text{inverse of } f$$

CHAPTER 9 REVIEW

Give the domain of each function. [9.1]

1. $y = \sqrt{x + 2}$

2. $y = \sqrt{3 - 2x}$

3. $y = \dfrac{x + 3}{x - 6}$

4. $y = \dfrac{4}{x^2 - 2x - 8}$

For each function below, evaluate the quantity $\dfrac{f(x + h) - f(x)}{h}$. [9.1]

5. $f(x) = 2x + 1$

6. $f(x) = x^2 - 1$

For each function below, evaluate the quantity $\dfrac{f(x) - f(a)}{x - a}$. [9.1]

7. $f(x) = 2x - 1$

8. $f(x) = x^2$

Let $f(x) = x + 3$, let $g(x) = 2x - 4$, and let $h(x) = 2x^2 + 2x - 12$. Write a formula for each function below. [9.1]

9. $f + g$

10. fg

11. $\dfrac{h}{f}$

12. $h + gf$

13. gg

14. $2f + 3g$

Find the following if $f(x) = x + 3$, $g(x) = 2x - 4$, and $h(x) = 2x^2 + 2x - 12$. [9.1]

15. $f(2) + g(2)$
16. $f(0) - g(0)$
17. $f(-1) \cdot g(-1)$
18. $h(3)/f(3)$

Find all values of x for which $f(x) = 0$. [9.2]

19. $f(x) = 2x - 5$
20. $f(x) = x^2 - 5x - 14$
21. $f(x) = x^3 + 2x^2 - 9x - 18$
22. $f(x) = x^3 - 3x^2 - 4x + 12$

Identify each of the following as a constant function, linear function, quadratic function, exponential function, cubic function, or rational function. [9.2]

23. $f(x) = 3x - 1$
24. $f(x) = 4x^3$
25. $f(x) = (\frac{1}{3})^x$
26. $y = -3$
27. $y = x^2 - x - 2$
28. $f(x) = \dfrac{x - 3}{x + 2}$

Graph each function. [9.2]

29. $y = 2^x$
30. $y = 2^{-x}$
31. $y = \frac{1}{4} x^3$
32. $y = \dfrac{x - 3}{x + 1}$

Let $f(x) = 2^x$ and $g(x) = (\frac{1}{3})^x$ and evaluate each of the following. [9.2]

33. $f(4)$
34. $f(-1)$
35. $g(2)$
36. $f(2) - g(-2)$
37. $f(-1) + g(1)$
38. $g(-1) + f(2)$

For each function below, find the equation of the inverse. Write the inverse using the notation $f^{-1}(x)$ if the inverse is itself a function. [9.3]

39. $f(x) = 2x + 3$
40. $f(x) = x^2 - 1$
41. $f(x) = \frac{1}{2}x + 2$
42. $f(x) = 4 - 2x^2$

For each relation that follows, sketch the graph of the relation and its inverse, and write an equation for the inverse. [9.3]

43. $y = 2x + 1$

44. $y = x^2 - 4$

Find the distance between the following points. [9.4]

45. $(2, 6), (-1, 5)$

46. $(3, -4), (1, -1)$

47. $(0, 3), (-4, 0)$

48. $(-3, 7), (-3, -2)$

49. Find x so that the distance between $(x, -1)$ and $(2, -4)$ is 5. [9.4]

50. Find y so that the distance between $(3, -4)$ and $(-3, y)$ is 10. [9.4]

Write the equation of the circle with the given center and radius. [9.4]

51. Center $(3, 1)$, $r = 2$

52. Center $(3, -1)$, $r = 4$

53. Center $(-5, 0)$, $r = 3$

54. Center $(-3, 4)$, $r = 3\sqrt{2}$

Find the equation of each circle. [9.4]

55. Center at the origin, x-intercepts ± 5

56. Center at the origin, y-intercepts ± 3

57. Center at $(-2, 3)$ and passing through the point $(2, 0)$

58. Center at $(-6, 8)$ and passing through the origin

Give the center and radius of each circle and then sketch the graph. [9.4]

59. $x^2 + y^2 = 4$

60. $(x - 3)^2 + (y + 1)^2 = 16$

61. $x^2 + y^2 - 6x + 4y = -4$

62. $x^2 + y^2 + 4x - 2y = 4$

Graph each of the following. Label the x- and y-intercepts. [9.5]

63. $\dfrac{x^2}{4} + \dfrac{y^2}{9} = 1$

64. $4x^2 + y^2 = 16$

Graph the following. Show the asymptotes. [9.5]

65. $\dfrac{x^2}{4} - \dfrac{y^2}{9} = 1$

66. $4x^2 - y^2 = 16$

Graph each equation. [9.5]

67. $\dfrac{(x + 2)^2}{9} + \dfrac{(y - 3)^2}{1} = 1$

68. $\dfrac{(x - 2)^2}{16} - \dfrac{y^2}{4} = 1$

69. $9y^2 - x^2 - 4x + 54y + 68 = 0$

70. $9x^2 + 4y^2 - 72x - 16y + 124 = 0$

Graph each of the following inequalities. [9.6]

71. $x^2 + y^2 < 9$

72. $(x + 2)^2 + (y - 1)^2 \leq 4$

73. $y \geq x^2 - 1$

74. $9x^2 + 4y^2 \leq 36$

Graph the solution set for each system. [9.6]

75. $x^2 + y^2 < 16$
$y > x^2 - 4$

76. $x + y \leq 2$
$-x + y \leq 2$
$y \geq -2$

Solve each system of equations. [9.6]

77. $x^2 + y^2 = 16$
$2x + y = 4$

78. $x^2 + y^2 = 4$
$y = x^2 - 2$

79. $9x^2 - 4y^2 = 36$
$9x^2 + 4y^2 = 36$

80. $2x^2 - 4y^2 = 8$
$x^2 + 2y^2 = 10$

CHAPTER 9 TEST

Indicate any restrictions on the domain of the following. [9.1]

1. $y = \sqrt{x - 4}$

2. $y = \dfrac{x - 1}{x^2 - 2x - 8}$

Let $f(x) = x - 2$, $g(x) = 3x + 4$, and $h(x) = 3x^2 - 2x - 8$, and find the following. [9.1]

3. $(fg)(x)$

4. $(f - g)(x)$

5. $(f + g)(x)$ **6.** $\left(\dfrac{h}{g}\right)(x)$

Graph each function. [9.2]

7. $y = \frac{1}{4}x^3$ **8.** $y = \dfrac{x + 4}{x - 1}$

9. $y = 2^x$ **10.** $y = 3^{-x}$

Find an equation for the inverse of each of the following functions. Sketch the graph of the function and its inverse on the same set of axes. [9.3]

11. $f(x) = 2x - 3$
12. $f(x) = x^2 - 4$
13. Find x so that $(x, 2)$ is $2\sqrt{5}$ units from $(-1, 4)$. [9.4]
14. Give the equation of the circle with center at $(-2, 4)$ and radius 3. [9.4]

15. Give the equation of the circle with center at the origin that contains the point $(-3, -4)$. [9.4]
16. Find the center and radius of the circle $x^2 + y^2 - 10x + 6y = 5$. [9.4]

Graph each of the following. [9.5, 9.6]

17. $4x^2 - y^2 = 16$ **18.** $\dfrac{x^2}{25} + \dfrac{y^2}{4} = 1$

19. $(x - 2)^2 + (y + 1)^2 \leq 9$
20. $9x^2 + 4y^2 - 72x - 16y + 124 = 0$

Solve the following systems. [9.6]

21. $x^2 + y^2 = 25$ **22.** $x^2 + y^2 = 16$
      $2x + y = 5$ $y = x^2 - 4$

10

LOGARITHMS

INTRODUCTION AND OVERVIEW

This chapter is mainly concerned with applications of a new notation for exponents. **Logarithms** are exponents. The properties of logarithms are actually the properties of exponents. There are many applications of logarithms to both science and higher mathematics. For example, the pH of a liquid is defined in terms of logarithms. (That's the same pH that is given on the label of many hair conditioners.) The Richter scale for measuring earthquake intensity is a logarithmic scale, as is the decibel scale used for measuring the intensity of sound.

We will begin this chapter with the definition of logarithms and the three main properties of logarithms. The rest of the chapter involves applications of the definition and properties. An understanding of the section on inverse functions from Chapter 9 will be very useful to you in getting started in this chapter.

Logarithms Are Exponents

As you know from your work in the previous chapter, equations of the form

$$y = b^x \qquad (b > 0, b \neq 1)$$

are called exponential functions. Since the equation of the inverse of a function can be obtained by exchanging x and y in the equation of the original function, the inverse of an exponential function must have the form

$$x = b^y \qquad (b > 0, b \neq 1)$$

Now, this last equation is actually the equation of a logarithmic function, as the following definition indicates:

DEFINITION The expression $y = \log_b x$ is read "y is the logarithm to the base b of x" and is equivalent to the expression

$$x = b^y \qquad (b > 0, b \neq 1)$$

In words, we say "y is the number we raise b to in order to get x."

Notation When an expression is in the form $x = b^y$, it is said to be in *exponential form*. On the other hand, if an expression is in the form $y = \log_b x$, it is said to be in *logarithmic form*.

Here are some equivalent statements written in both forms.

Exponential Form		Logarithmic Form
$8 = 2^3$	$\Leftrightarrow$	$\log_2 8 = 3$
$25 = 5^2$	$\Leftrightarrow$	$\log_5 25 = 2$
$0.1 = 10^{-1}$	$\Leftrightarrow$	$\log_{10} 0.1 = -1$
$\dfrac{1}{8} = 2^{-3}$	$\Leftrightarrow$	$\log_2 \dfrac{1}{8} = -3$
$r = z^s$	$\Leftrightarrow$	$\log_z r = s$

▶ **EXAMPLE 1** Solve for x: $\log_3 x = -2$.

Solution In exponential form the equation looks like this:

$$x = 3^{-2}$$

or

$$x = \frac{1}{9}$$

The solution is $\frac{1}{9}$. ◀

▶ **EXAMPLE 2** Solve $\log_x 4 = 3$.

Solution Again, we use the definition of logarithms to write the expression in exponential form:

$$4 = x^3$$

Taking the cube root of both sides, we have

$$\sqrt[3]{4} = \sqrt[3]{x^3}$$
$$x = \sqrt[3]{4}$$

The solution set is $\{\sqrt[3]{4}\}$.   ◀

▶ **EXAMPLE 3**   Solve $\log_8 4 = x$.

Solution   We write the expression again in exponential form:

$$4 = 8^x$$

Since both 4 and 8 can be written as powers of 2, we write them in terms of powers of 2:

$$2^2 = (2^3)^x$$
$$2^2 = 2^{3x}$$

The only way the left and right sides of this last line can be equal is if the exponents are equal — that is, if

$$2 = 3x$$

or

$$x = \frac{2}{3}$$

The solution is $\frac{2}{3}$. We check as follows:

$$\log_8 4 = \frac{2}{3} \Leftrightarrow 4 = 8^{2/3}$$
$$4 = (\sqrt[3]{8})^2$$
$$4 = 2^2$$
$$4 = 4$$

The solution checks when used in the original equation.   ◀

Graphing Logarithmic Functions

Graphing logarithmic functions can be done using the graphs of exponential functions and the fact that the graphs of inverse functions have symmetry about the line $y = x$. Here's an example to illustrate.

▶ **EXAMPLE 4**   Graph the equation $y = \log_2 x$.

Solution   The equation $y = \log_2 x$ is, by definition, equivalent to the exponential equation

$$x = 2^y$$

which is the equation of the inverse of the function

$$y = 2^x$$

The graph of $y = 2^x$ was given in Figure 3 of Section 9.2. We simply reflect the graph of $y = 2^x$ about the line $y = x$ to get the graph of $x = 2^y$, which is also the graph of $y = \log_2 x$. (See Figure 1.)

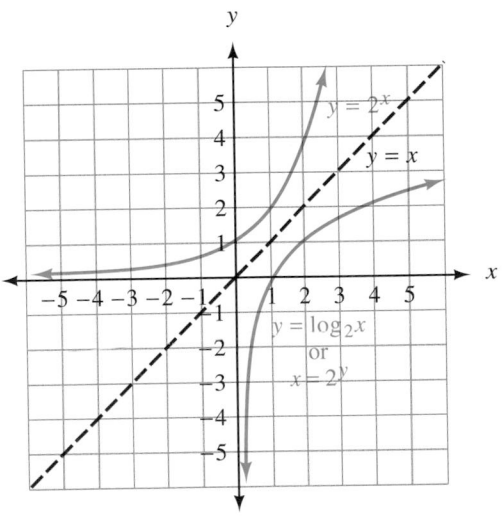

Figure 1

It is apparent from the graph that $y = \log_2 x$ is a function, since no vertical line will cross its graph in more than one place. The same is true for all logarithmic equations of the form $y = \log_b x$ where b is a positive number other than 1. Note also that the graph of $y = \log_b x$ will always appear to the right of the y-axis, meaning that x will always be positive in the expression $y = \log_b x$. ◀

Two Special Identities

If b is a positive real number other than 1, then each of the following is a consequence of the definition of a logarithm:

$$(1) \quad b^{\log_b x} = x \quad \text{and} \quad (2) \quad \log_b b^x = x$$

The justifications for these identities are similar. Let's consider only the first one. Consider the expression

$$y = \log_b x$$

By definition, it is equivalent to

$$x = b^y$$

Substituting $\log_b x$ for y in the last line gives us

$$x = b^{\log_b x}$$

The next examples in this section show how these two special properties can be used to simplify expressions involving logarithms.

▶ **EXAMPLE 5** Simplify $\log_2 8$.

Solution Substitute 2^3 for 8:

$$\log_2 8 = \log_2 2^3$$
$$= 3 \qquad \blacktriangleleft$$

▶ **EXAMPLE 6** Simplify $\log_{10} 10{,}000$.

Solution 10,000 can be written as 10^4:

$$\log_{10} 10{,}000 = \log_{10} 10^4$$
$$= 4 \qquad \blacktriangleleft$$

▶ **EXAMPLE 7** Simplify $\log_b b$ ($b > 0, b \neq 1$).

Solution Since $b^1 = b$, we have

$$\log_b b = \log_b b^1$$
$$= 1 \qquad \blacktriangleleft$$

▶ **EXAMPLE 8** Simplify $\log_b 1$ ($b > 0, b \neq 1$).

Solution Since $1 = b^0$, we have

$$\log_b 1 = \log_b b^0$$
$$= 0 \qquad \blacktriangleleft$$

▶ **EXAMPLE 9** Simplify $\log_4(\log_5 5)$.

Solution Since $\log_5 5 = 1$,

$$\log_4(\log_5 5) = \log_4 1$$
$$= 0 \qquad \blacktriangleleft$$

One application of logarithms is in measuring the magnitude of an earthquake. If an earthquake has a shockwave T times greater than the smallest shockwave that can be measured on a seismograph, then the magnitude M of the earthquake, as measured on the Richter scale, is given by the formula

$$M = \log_{10} T$$

(When we talk about the size of a shockwave, we are talking about its amplitude. The amplitude of a wave is half the difference between its highest point and its lowest point.)

To illustrate the discussion, an earthquake that produces a shockwave that is 10,000 times greater than the smallest shockwave measurable on a seismograph will have a magnitude M on the Richter scale of

$$M = \log_{10} 10,000 = 4$$

▶ **EXAMPLE 10**   If an earthquake has a magnitude of $M = 5$ on the Richter scale, what can you say about the size of its shockwave?

Solution   To answer this question, we put $M = 5$ into the formula $M = \log_{10} T$ to obtain

$$5 = \log_{10} T$$

Writing this expression in exponential form, we have

$$T = 10^5 = 100,000$$

We can say that an earthquake that measures 5 on the Richter scale has a shockwave 100,000 times greater than the smallest shockwave measurable on a seismograph.   ◀

From Example 10 and the discussion that preceded it, we find that an earthquake of magnitude 5 has a shockwave that is 10 times greater than that of an earthquake of magnitude 4, because 100,000 is 10 times 10,000.

P R O B L E M   S E T   1 0 . 1

Write each of the following expressions in logarithmic form.

1. $2^4 = 16$ **2.** $3^2 = 9$

3. $125 = 5^3$ **4.** $16 = 4^2$

5. $0.01 = 10^{-2}$ **6.** $0.001 = 10^{-3}$

7. $2^{-5} = \frac{1}{32}$ **8.** $4^{-2} = \frac{1}{16}$

9. $\left(\frac{1}{2}\right)^{-3} = 8$ **10.** $\left(\frac{1}{3}\right)^{-2} = 9$

11. $27 = 3^3$ **12.** $81 = 3^4$

Write each of the following expressions in exponential form.

13. $\log_{10} 100 = 2$

14. $\log_2 8 = 3$

15. $\log_2 64 = 6$

16. $\log_2 32 = 5$

17. $\log_8 1 = 0$

18. $\log_9 9 = 1$

19. $\log_{10} 0.001 = -3$

20. $\log_{10} 0.0001 = -4$

21. $\log_6 36 = 2$

22. $\log_7 49 = 2$

23. $\log_5 \frac{1}{25} = -2$

24. $\log_3 \frac{1}{81} = -4$

Solve each of the following equations for x.

25. $\log_3 x = 2$ **26.** $\log_4 x = 3$

27. $\log_5 x = -3$ **28.** $\log_2 x = -4$

29. $\log_2 16 = x$ **30.** $\log_3 27 = x$

31. $\log_8 2 = x$ **32.** $\log_{25} 5 = x$

33. $\log_x 4 = 2$ **34.** $\log_x 16 = 4$

35. $\log_x 5 = 3$ **36.** $\log_x 8 = 2$

Sketch the graph of each of the following logarithmic equations.

37. $y = \log_3 x$

38. $y = \log_{1/2} x$

39. $y = \log_{1/3} x$

40. $y = \log_4 x$

41. $y = \log_5 x$

42. $y = \log_{1/5} x$

43. $y = \log_{10} x$

44. $y = \log_{1/4} x$

Simplify each of the following.

45. $\log_2 16$

46. $\log_3 9$

47. $\log_{25} 125$

48. $\log_9 27$

49. $\log_{10} 1,000$

50. $\log_{10} 10,000$

51. $\log_3 3$

52. $\log_4 4$

53. $\log_5 1$

54. $\log_{10} 1$

55. $\log_3(\log_6 6)$

56. $\log_5(\log_3 3)$

57. $\log_4[\log_2(\log_2 16)]$

58. $\log_4[\log_3(\log_2 8)]$

Applying the Concepts

In chemistry, the pH of a solution is defined in terms of logarithms as pH $= -\log_{10}[H^+]$ where $[H^+]$ is the concentration of the hydrogen ion in solution. An acid solution has a pH below 7, and a basic solution has a pH higher than 7.

59. In distilled water, the concentration of the hydrogen ion is $[H^+] = 10^{-7}$. What is the pH?

60. Find the pH of a bottle of vinegar, if the concentration of the hydrogen ion is $[H^+] = 10^{-3}$.

61. A hair conditioner has a pH of 6. Find the concentration of the hydrogen ion, $[H^+]$, in the conditioner.

62. If a glass of orange juice has a pH of 4, what is the concentration of the hydrogen ion, $[H^+]$, in the orange juice?

63. Find the magnitude M of an earthquake with a shockwave that measures $T = 100$ on a seismograph.

64. Find the magnitude M of an earthquake with a shockwave that measures $T = 100,000$ on a seismograph.

65. If an earthquake has a magnitude of 8 on the Richter scale, how many times greater is its shockwave than the smallest shockwave measurable on a seismograph?

66. If an earthquake has a magnitude of 6 on the Richter scale, how many times greater is its shockwave than the smallest shockwave measurable on a seismograph?

Review Problems

The problems below review material from Section 9.1.

State the domain for each of the following functions.

67. $y = \sqrt{3x + 1}$

68. $y = \dfrac{-4}{x^2 + 2x - 35}$

If $f(x) = 2x^2 - 18$ and $g(x) = 2x - 6$, find

69. $f(0)$

70. $g[f(0)]$

71. $\dfrac{g(x + h) - g(x)}{h}$

72. $\dfrac{g}{f}(x)$

Graph each function.

73. $y = \dfrac{6}{x + 1}$

74. $y = \dfrac{6}{x - 1}$

SECTION

10.2 **Properties of Logarithms**

For the following three properties, x, y, and b are all positive real numbers, $b \neq 1$, and r is any real number.

> **Property 1** $\log_b(xy) = \log_b x + \log_b y$
>
> *In words:* The logarithm of a product is the sum of the logarithms.
>
> **Property 2** $\log_b\left(\dfrac{x}{y}\right) = \log_b x - \log_b y$
>
> *In words:* The logarithm of a quotient is the difference of the logarithms.
>
> **Property 3** $\log_b x^r = r \log_b x$
>
> *In words:* The logarithm of a number raised to a power is the product of the power and the logarithm of the number.

PROOF OF PROPERTY 1   To prove Property 1, we simply apply the first identity for logarithms given at the end of the preceding section:

$$b^{\log_b xy} = xy = (b^{\log_b x})(b^{\log_b y}) = b^{\log_b x + \log_b y}$$

Since the first and last expressions are equal and the bases are the same, the exponents $\log_b xy$ and $\log_b x + \log_b y$ must be equal. Therefore,

$$\log_b xy = \log_b x + \log_b y$$

The proofs of Properties 2 and 3 proceed in much the same manner, so we will omit them here. The examples that follow show how the three properties can be used.

▶ **EXAMPLE 1**   Expand, using the properties of logarithms: $\log_5 \dfrac{3xy}{z}$.

Solution   Applying Property 2, we can write the quotient of $3xy$ and z in terms of a difference:

$$\log_5 \frac{3xy}{z} = \log_5 3xy - \log_5 z$$

Applying Property 1 to the product $3xy$, we write it in terms of addition:

$$\log_5 \frac{3xy}{z} = \log_5 3 + \log_5 x + \log_5 y - \log_5 z \qquad \blacktriangleleft$$

▶ **EXAMPLE 2**   Expand, using the properties of logarithms:

$$\log_2 \frac{x^4}{\sqrt{y} \cdot z^3}$$

Solution We write $\sqrt{y}$ as $y^{1/2}$ and apply the properties:

$$\log_2 \frac{x^4}{\sqrt{y} \cdot z^3} = \log_2 \frac{x^4}{y^{1/2}z^3} \qquad\qquad \sqrt{y} = y^{1/2}$$

$$= \log_2 x^4 - \log_2(y^{1/2} \cdot z^3) \qquad \text{Property 2}$$

$$= \log_2 x^4 - (\log_2 y^{1/2} + \log_2 z^3) \qquad \text{Property 1}$$

$$= \log_2 x^4 - \log_2 y^{1/2} - \log_2 z^3 \qquad \text{Remove parentheses}$$

$$= 4 \log_2 x - \frac{1}{2} \log_2 y - 3 \log_2 z \qquad \text{Property 3}$$ ◀

We can also use the three properties to write an expression in expanded form as just one logarithm.

▶ **EXAMPLE 3** Write as a single logarithm:

$$2 \log_{10} a + 3 \log_{10} b - \frac{1}{3} \log_{10} c$$

Solution We begin by applying Property 3:

$$2 \log_{10} a + 3 \log_{10} b - \frac{1}{3} \log_{10} c$$

$$= \log_{10} a^2 + \log_{10} b^3 - \log_{10} c^{1/3} \qquad \text{Property 3}$$

$$= \log_{10}(a^2 \cdot b^3) - \log_{10} c^{1/3} \qquad \text{Property 1}$$

$$= \log_{10} \frac{a^2b^3}{c^{1/3}} \qquad\qquad \text{Property 2}$$

$$= \log_{10} \frac{a^2b^3}{\sqrt[3]{c}} \qquad\qquad c^{1/3} = \sqrt[3]{c}$$ ◀

The properties of logarithms along with the definition of logarithms are useful in solving equations that involve logarithms.

▶ **EXAMPLE 4** Solve for x: $\log_2(x + 2) + \log_2 x = 3$.

Solution Applying Property 1 to the left side of the equation allows us to write it as a single logarithm:

$$\log_2(x + 2) + \log_2 x = 3$$

$$\log_2[(x + 2)(x)] = 3$$

The last line can be written in exponential form using the definition of logarithms:

$$(x + 2)(x) = 2^3$$

Solve as usual:

$$x^2 + 2x = 8$$
$$x^2 + 2x - 8 = 0$$
$$(x + 4)(x - 2) = 0$$
$$x + 4 = 0 \quad \text{or} \quad x - 2 = 0$$
$$x = -4 \quad \text{or} \quad x = 2$$

In the previous section, we noted the fact that x in the expression $y = \log_b x$ cannot be a negative number. Since substitution of $x = -4$ into the original equation gives

$$\log_2(-2) + \log_2(-4) = 3$$

which contains logarithms of negative numbers, we cannot use -4 as a solution. The solution set is $\{2\}$. ◄

P R O B L E M   S E T   1 0 . 2

Use the three properties of logarithms given in this section to expand each expression as much as possible.

1. $\log_3 4x$

2. $\log_2 5x$

3. $\log_6 \dfrac{5}{x}$

4. $\log_3 \dfrac{x}{5}$

5. $\log_2 y^5$

6. $\log_7 y^3$

7. $\log_9 \sqrt[3]{z}$

8. $\log_8 \sqrt{z}$

9. $\log_6 x^2 y^4$

10. $\log_{10} x^2 y^4$

11. $\log_5 \sqrt{x} \cdot y^4$

12. $\log_8 \sqrt[3]{xy^6}$

13. $\log_b \dfrac{xy}{z}$

14. $\log_b \dfrac{3x}{y}$

15. $\log_{10} \dfrac{4}{xy}$

16. $\log_{10} \dfrac{5}{4y}$

17. $\log_{10} \dfrac{x^2 y}{\sqrt{z}}$

18. $\log_{10} \dfrac{\sqrt{x} \cdot y}{z^3}$

19. $\log_{10} \dfrac{x^3 \sqrt{y}}{z^4}$

20. $\log_{10} \dfrac{x^4 \sqrt[3]{y}}{\sqrt{z}}$

21. $\log_b \sqrt[3]{\dfrac{x^2 y}{z^4}}$

22. $\log_b \sqrt[4]{\dfrac{x^4 y^3}{z^5}}$

Write each expression as a single logarithm.

23. $\log_b x + \log_b z$

24. $\log_b x - \log_b z$

25. $2 \log_3 x - 3 \log_3 y$

26. $4 \log_2 x + 5 \log_2 y$

27. $\frac{1}{2} \log_{10} x + \frac{1}{3} \log_{10} y$

28. $\frac{1}{3} \log_{10} x - \frac{1}{4} \log_{10} y$

29. $3 \log_2 x + \frac{1}{2} \log_2 y - \log_2 z$

30. $2 \log_3 x + 3 \log_3 y - \log_3 z$

31. $\frac{1}{2} \log_2 x - 3 \log_2 y - 4 \log_2 z$

32. $3 \log_{10} x - \log_{10} y - \log_{10} z$

33. $\frac{3}{2} \log_{10} x - \frac{3}{4} \log_{10} y - \frac{4}{5} \log_{10} z$

34. $3 \log_{10} x - \frac{4}{3} \log_{10} y - 5 \log_{10} z$

Solve each of the following equations.

35. $\log_2 x + \log_2 3 = 1$
36. $\log_3 x + \log_3 3 = 1$
37. $\log_3 x - \log_3 2 = 2$
38. $\log_3 x + \log_3 2 = 2$
39. $\log_3 x + \log_3(x - 2) = 1$
40. $\log_6 x + \log_6(x - 1) = 1$
41. $\log_3(x + 3) - \log_3(x - 1) = 1$
42. $\log_4(x - 2) - \log_4(x + 1) = 1$
43. $\log_2 x + \log_2(x - 2) = 3$
44. $\log_4 x + \log_4(x + 6) = 2$
45. $\log_8 x + \log_8(x - 3) = \frac{2}{3}$
46. $\log_{27} x + \log_{27}(x + 8) = \frac{2}{3}$
47. $\log_5 \sqrt{x} + \log_5 \sqrt{6x + 5} = 1$
48. $\log_2 \sqrt{x} + \log_2 \sqrt{6x + 5} = 1$

Applying the Concepts

49. The formula $M = 0.21(\log_{10} a - \log_{10} b)$ is used in the food processing industry to find the number of minutes M of heat processing a certain food should undergo at 250°F to reduce the probability of survival of *C. botulinum* spores. The letter a represents the number of spores per can before heating, and b represents the number of spores per can after heating. Find M if $a = 1$ and $b = 10^{-12}$. Then, find M using the same values for a and b in the formula

$$M = 0.21 \log_{10} \frac{a}{b}.$$

50. The formula $N = \log_{10} \dfrac{P_1}{P_2}$ is used in radio electronics to find the ratio of the acoustic powers of two electric circuits in terms of their electric powers. Find N if P_1 is 100 and P_2 is 1. Then, use the same two values of P_1 and P_2 to find N in the formula $N = \log_{10} P_1 - \log_{10} P_2$.

51. Use the properties of logarithms to show that $\log_{10}(8.43 \times 10^2)$ can be written as $2 + \log_{10} 8.43$.

52. Use the properties of logarithms to show that $\log_{10}(2.76 \times 10^3)$ can be written as $3 + \log_{10} 2.76$.

53. Use the properties of logarithms to show that the formula $\log_{10} A = \log_{10} 100(1.06)^t$ can be written as $\log_{10} A = 2 + t \log_{10} 1.06$.

54. Use the properties of logarithms to show that the formula $\log_{10} A = \log_{10} 3(2)^{t/5,600}$ can be written as $\log_{10} A = \log_{10} 3 + \dfrac{t}{5,600} \log_{10} 2$.

Review Problems

The problems that follow review material we covered in Sections 9.2 and 9.3.

Graph each of the following functions. [9.2]

55. $f(x) = 2x - 3$
56. $f(x) = 3$
57. $f(x) = 2x^2 - 4x - 6$
58. $f(x) = 3^x$

Find all values of x for which $f(x) = 0$. [9.2]

59. $f(x) = 4x - 3$
60. $f(x) = -3x + 1$
61. $f(x) = x^2 + 2x - 1$
62. $f(x) = x^2 - 8x - 1$

Find the equation of the inverse of each of the following functions. Write the inverse using the notation $f^{-1}(x)$, if the inverse is itself a function. [9.3]

63. $f(x) = 2x + 3$
64. $f(x) = 3x - 2$
65. $f(x) = x^2 - 4$
66. $f(x) = x^2 + 2$
67. $f(x) = \dfrac{x - 3}{5}$
68. $f(x) = \dfrac{x + 2}{3}$

Common Logarithms and Natural Logarithms

There are two kinds of logarithms that occur more frequently than other logarithms. Logarithms with a base of 10 are very common because our number system is a base-10 number system. For this reason, we call base-10 logarithms **common logarithms.**

Common Logarithms

DEFINITION A **common logarithm** is a logarithm with a base of 10. Since common logarithms are used so frequently, it is customary, in order to save time, to omit notating the base. That is,

$$\log_{10} x = \log x$$

When the base is not shown, it is assumed to be 10.

Common logarithms of powers of 10 are very simple to evaluate. We need only recognize that $\log 10 = \log_{10} 10 = 1$ and apply the third property of logarithms: $\log_b x^r = r \log_b x$.

$$\log 1{,}000 = \log 10^3 \quad = 3 \log 10 \quad = 3(1) \quad = 3$$
$$\log 100 \quad = \log 10^2 \quad = 2 \log 10 \quad = 2(1) \quad = 2$$
$$\log 10 \quad = \log 10^1 \quad = 1 \log 10 \quad = 1(1) \quad = 1$$
$$\log 1 \quad = \log 10^0 \quad = 0 \log 10 \quad = 0(1) \quad = 0$$
$$\log 0.1 \quad = \log 10^{-1} = -1 \log 10 = -1(1) = -1$$
$$\log 0.01 \quad = \log 10^{-2} = -2 \log 10 = -2(1) = -2$$
$$\log 0.001 = \log 10^{-3} = -3 \log 10 = -3(1) = -3$$

To find common logarithms of numbers that are not powers of 10, we use a calculator with a $\boxed{\log}$ key or a table of logarithms. We will assume the use of a calculator for the rest of this chapter.

Check the following logarithms to be sure you know how to use your calculator. (These answers have been rounded to the nearest ten thousandth.)

$$\log 7.02 = 0.8463$$
$$\log 1.39 = 0.1430$$
$$\log 6.00 = 0.7782$$
$$\log 9.99 = 0.9996$$

▶ **EXAMPLE 1**   Use a calculator to find log 2,760.

Solution

$$\log 2{,}760 = 3.4409$$

To work this problem on a scientific calculator, we simply enter the number 2,760 and press the key labeled $\boxed{\log}$. To work the problem on a graphing calculator we press the $\boxed{\log}$ button first, then enter the number 2,760, then we press the $\boxed{\text{ENT}}$ key.

$$2{,}760\ \boxed{\log}$$

The 3 in the answer is called the **characteristic,** while the decimal part of the logarithm is called the **mantissa.**   ◀

▶ **EXAMPLE 2**   Find log 0.0391.

Solution $\log 0.0391 = -1.4078$ ◀

▶ **EXAMPLE 3**   Find log 0.00523.

Solution $\log 0.00523 = -2.2815$ ◀

▶ **EXAMPLE 4**   Find x if log $x = 3.8774$.

Solution   We are looking for the number whose logarithm is 3.8774. On a scientific calculator, we enter 3.8774 and press the key labeled $\boxed{10^x}$. On a graphing calculator we press $\boxed{10^x}$ first, then we enter the number 3.8774. The result is 7,540 to four significant digits.

$$\text{If} \qquad \log x = 3.8774$$
$$\text{then} \qquad x = 10^{3.8774}$$
$$x = 7{,}540$$

The number 7,540 is called the **antilogarithm** or just **antilog** of 3.8774. That is, 7,540 is the number whose logarithm is 3.8774.   ◀

▶ **EXAMPLE 5**   Find x if log $x = -2.4179$.

Solution   Using the $\boxed{10^x}$ key on a calculator, with $x = 2.4179$, the result is 0.00382.

$$\text{If} \qquad \log x = -2.4179$$
$$\text{then} \qquad x = 10^{-2.4179}$$
$$= 0.00382$$

The antilog of -2.4179 is 0.00382. That is, the logarithm of 0.00382 is -2.4179.   ◀

In Section 10.1, we found that the magnitude M of an earthquake that produces a shockwave T times larger than the smallest shockwave that can be measured on a seismograph is given by the formula

$$M = \log_{10} T$$

We can rewrite this formula using our shorthand notation for common logarithms as

$$M = \log T$$

▶ **EXAMPLE 6**   The San Francisco earthquake of 1906 measured 8.3 on the Richter scale. The San Fernando earthquake of 1971 measured 6.6 on the Richter scale. Find T for each earthquake and then give some indication of how much stronger the 1906 earthquake was than the 1971 earthquake.

Solution   For the 1906 earthquake:

$$\text{If } \log T = 8.3, \text{ then } T = 2.00 \times 10^8$$

For the 1971 earthquake:

$$\text{If } \log T = 6.6, \text{ then } T = 3.98 \times 10^6$$

Dividing the two values of T and rounding our answer to the nearest whole number, we have

$$\frac{2.00 \times 10^8}{3.98 \times 10^6} = 50$$

The shockwave for the 1906 earthquake was approximately 50 times as large as the shockwave for the 1971 earthquake.   ◀

Natural Logarithms

The next kind of logarithms we want to discuss are **natural logarithms.** In order to give a definition for natural logarithms, we need to talk about a special number that is denoted by the letter e. The number e is a number like π. It is irrational and occurs in many formulas that describe the world around us. Like π, it can be approximated with a decimal number. Whereas π is approximately 3.1416, e is approximately 2.7183. (If you have a calculator with a key labeled $\boxed{e^x}$, you can use it to find e' to see a more accurate approximation to e.) We cannot give a more precise definition of the number e without using some of the topics taught in calculus. For the work we are going to do with the number e, we need know only that it is an irrational number that is approximately 2.7183. (If this bothers you, try to think of the last time you saw a precise definition for the number π. Even though you may not know the definition of π, you are still able to work problems that use the number π, simply by knowing that it is an irrational number that is approximately 3.1416.) Here is our definition for natural logarithms.

DEFINITION   A **natural logarithm** is a logarithm with a base of e. The natural logarithm of x is denoted by $\ln x$. That is,

$$\ln x = \log_e x$$

We can assume that all our properties of exponents and logarithms hold for expressions with a base of e, since e is a real number. Here are some examples intended to make you more familiar with the number e and natural logarithms.

▶ **EXAMPLE 7**   Simplify each of the following expressions.

(a) $e^0 = 1$
(b) $e^1 = e$
(c) $\ln e = 1$ In exponential form, $e^1 = e$
(d) $\ln 1 = 0$ In exponential form, $e^0 = 1$
(e) $\ln e^3 = 3$
(f) $\ln e^{-4} = -4$
(g) $\ln e^t = t$ ◀

▶ **EXAMPLE 8**   Use the properties of logarithms to expand the expression $\ln Ae^{5t}$.

Solution   Since the properties of logarithms hold for natural logarithms, we have

$$\ln Ae^{5t} = \ln A + \ln e^{5t}$$
$$= \ln A + 5t \ln e$$
$$= \ln A + 5t \qquad \text{Because } \ln e = 1 \qquad ◀$$

▶ **EXAMPLE 9**   If $\ln 2 = 0.6931$ and $\ln 3 = 1.0986$, find

(a) $\ln 6$ (b) $\ln 0.5$ (c) $\ln 8$

Solution
(a)   Since $6 = 2 \cdot 3$, we have

$$\ln 6 = \ln 2 \cdot 3$$
$$= \ln 2 + \ln 3$$
$$= 0.6931 + 1.0986$$
$$= 1.7917$$

(b)   Writing 0.5 as $\frac{1}{2}$ and applying property 2 for logarithms gives us

$$\ln 0.5 = \ln \frac{1}{2}$$

$$= \ln 1 - \ln 2$$
$$= 0 - 0.6931$$
$$= -0.6931$$

(c) Writing 8 as 2^3 and applying property 3 for logarithms, we have

$$\ln 8 = \ln 2^3$$
$$= 3 \ln 2$$
$$= 3(0.6931)$$
$$= 2.0793$$ ◄

PROBLEM SET 10.3

Find the following logarithms.

1. log 378
2. log 426
3. log 37.8
4. log 42,600
5. log 3,780
6. log 0.4260
7. log 0.0378
8. log 0.0426
9. log 37,800
10. log 4,900
11. log 600
12. log 900
13. log 2,010
14. log 10,200
15. log 0.00971
16. log 0.0312
17. log 0.0314
18. log 0.00052
19. log 0.399
20. log 0.111

Find x in the following equations.

21. $\log x = 2.8802$
22. $\log x = 4.8802$
23. $\log x = -2.1198$
24. $\log x = -3.1198$
25. $\log x = 3.1553$
26. $\log x = 5.5911$
27. $\log x = -5.3497$
28. $\log x = -1.5670$
29. $\log x = -7.0372$
30. $\log x = -4.2000$
31. $\log x = 10$
32. $\log x = -1$
33. $\log x = -10$
34. $\log x = 1$
35. $\log x = 20$
36. $\log x = -20$
37. $\log x = -2$
38. $\log x = 4$
39. $\log x = \log_2 8$
40. $\log x = \log_3 9$
41. $\log x = \log 5$
42. $\log x = \log 10$

Applying the Concepts

In Problem Set 10.1, we indicated that the pH of a solution is defined in terms of logarithms as

$$pH = -\log[H^+]$$

where $[H^+]$ is the concentration of the hydrogen ion in that solution.

43. Find the pH of orange juice if the concentration of the hydrogen ion in the juice is $[H^+] = 6.50 \times 10^{-4}$.
44. Find the pH of milk if the concentration of the hydrogen ion in milk is $[H^+] = 1.88 \times 10^{-6}$.
45. Find the concentration of hydrogen ions in a glass of wine if the pH is 4.75.
46. Find the concentration of hydrogen ions in a bottle of vinegar if the pH is 5.75.

Find the relative size T of the shockwave of earthquakes with the following magnitudes, as measured on the Richter scale.

47. 5.5
48. 6.6
49. 8.3
50. 8.7
51. How much larger is the shockwave of an earthquake that measures 6.5 on the Richter scale than one that measures 5.5 on the same scale?
52. How much larger is the shockwave of an earthquake that measures 8.5 on the Richter scale than one that measures 5.5 on the same scale?

The annual rate of depreciation, r, on a car that is purchased for P dollars and is worth W dollars t years later can be found from the formula

$$\log(1 - r) = \frac{1}{t} \log \frac{W}{P}$$

53. Find the annual rate of depreciation on a car that is purchased for $9,000 and sold 5 years later for $4,500.

54. Find the annual rate of depreciation on a car that is purchased for $9,000 and sold 4 years later for $3,000.

55. Find the annual rate of depreciation on a car that is purchased for $7,550 and sold 5 years later for $5,750.

56. Find the annual rate of depreciation on a car that is purchased for $7,550 and sold 3 years later for $5,750.

Simplify each of the following expressions.

57. $\ln e$ **58.** $\ln 1$
59. $\ln e^5$ **60.** $\ln e^{-3}$
61. $\ln e^x$ **62.** $\ln e^y$

Use the properties of logarithms to expand each of the following expressions.

63. $\ln 10e^{3t}$
64. $\ln 10e^{4t}$
65. $\ln Ae^{-2t}$
66. $\ln Ae^{-3t}$

If $\ln 2 = 0.6931$, $\ln 3 = 1.0986$, and $\ln 5 = 1.6094$, find each of the following.

67. $\ln 15$ **68.** $\ln 10$
69. $\ln \frac{1}{3}$ **70.** $\ln \frac{1}{5}$
71. $\ln 9$ **72.** $\ln 25$
73. $\ln 16$ **74.** $\ln 81$

Review Problems

The problems that follow review material we covered in Sections 9.4 and 9.5.

Solve the following. [9.4]

75. Find x so the distance between $(x, 3)$ and $(1, 6)$ is $\sqrt{13}$.

76. Find y so the distance between $(2, y)$ and $(1, 2)$ is $\sqrt{2}$.

Give the center and radius of each of the following circles. [9.4]

77. $x^2 + y^2 + 6x - 4y = 3$
78. $x^2 + y^2 - 8x + 2y = 8$

Graph each equation. [9.5]

79. $25x^2 + 4y^2 = 100$ **80.** $4x^2 - 25y^2 = 100$

SECTION

10.4   Exponential Equations and Change of Base

Logarithms are very important in solving equations in which the variable appears as an exponent. The equation

$$5^x = 12$$

is an example of one such equation. Equations of this form are called **exponential equations.** Since the quantities 5^x and 12 are equal, so are their common logarithms. We begin our solution by taking the logarithm of both sides:

$$\log 5^x = \log 12$$

We now apply Property 3 for logarithms, $\log x^r = r \log x$, to turn x from an exponent into a coefficient:

$$x \log 5 = \log 12$$

Dividing both sides by log 5 gives us

$$x = \frac{\log 12}{\log 5}$$

If we want a decimal approximation to the solution, we can find log 12 and log 5 on a calculator and divide:

$$x = \frac{1.0792}{0.6990}$$

$$= 1.5439$$

The complete problem looks like this:

$$5^x = 12$$

$$\log 5^x = \log 12$$

$$x \log 5 = \log 12$$

$$x = \frac{\log 12}{\log 5}$$

$$= \frac{1.0792}{0.6990}$$

$$= 1.5439$$

Here is another example of solving an exponential equation using logarithms.

▶ **EXAMPLE 1**   Solve for x: $25^{2x+1} = 15$.

Solution   Taking the logarithm of both sides and then writing the exponent $(2x + 1)$ as a coefficient, we proceed as follows:

$$25^{2x+1} = 15$$

$$\log 25^{2x+1} = \log 15 \qquad\qquad \text{Take the log of both sides}$$

$$(2x + 1) \log 25 = \log 15 \qquad\qquad \text{Property 3}$$

$$2x + 1 = \frac{\log 15}{\log 25} \qquad\qquad \text{Divide by log 25}$$

$$2x = \frac{\log 15}{\log 25} - 1 \qquad\qquad \text{Add } -1 \text{ to both sides}$$

$$x = \frac{1}{2}\left(\frac{\log 15}{\log 25} - 1\right) \qquad\qquad \text{Multiply both sides by } \frac{1}{2}$$

PROOF   We begin by writing the identity

$$a^{\log_a x} = x$$

Taking the logarithm base b of both sides and writing the exponent $\log_a x$ as a coefficient, we have

$$\log_b a^{\log_a x} = \log_b x$$

$$\log_a x \log_b a = \log_b x$$

Dividing both sides by $\log_b a$, we have the desired result:

$$\frac{\log_a x \log_b a}{\log_b a} = \frac{\log_b x}{\log_b a}$$

$$\log_a x = \frac{\log_b x}{\log_b a}$$

We can use this property to find logarithms we could not otherwise compute on our calculators—that is, logarithms with bases other than 10 or e. The next example illustrates the use of this property.

▶ **EXAMPLE 4**   Find $\log_8 24$.

Solution   Since we do not have base-8 logarithms on our calculators, we can change this expression to an equivalent expression that contains only base-10 logarithms:

$$\log_8 24 = \frac{\log 24}{\log 8} \qquad \text{Property 4}$$

Don't be confused. We did not just drop the base, we changed to base 10. We could have written the last line like this:

$$\log_8 24 = \frac{\log_{10} 24}{\log_{10} 8}$$

From our calculators, we write

$$\log_8 24 = \frac{1.3802}{0.9031}$$

$$= 1.5283$$

Here is the complete calculator solution to Example 4:

Scientific calculator   **Graphing calculator**

24 $\boxed{\log}$ $\boxed{\div}$ 8 $\boxed{\log}$ $\boxed{=}$ $\boxed{\log}$ 24 $\boxed{\div}$ $\boxed{\log}$ 8 $\boxed{\text{ENT}}$ ◀

▶ **EXAMPLE 5**   Suppose that the population in a small city is 32,000 in the beginning of 1994 and that the city council assumes that the population size t years later can be estimated by the equation

$$P = 32{,}000e^{0.05t}$$

Approximately when will the city have a population of 50,000?

Solution We substitute 50,000 for P in the equation and solve for t:

$$50{,}000 = 32{,}000e^{0.05t}$$

$$1.56 = e^{0.05t} \qquad \frac{50{,}000}{32{,}000} \text{ is approximately } 1.56$$

To solve this equation for t, we can take the natural logarithm of each side:

$$\ln 1.56 = \ln e^{0.05t}$$

$$\ln 1.56 = 0.05t \ln e \qquad \text{Property 3 for logarithms}$$

$$\ln 1.56 = 0.05t \qquad \text{Because } \ln e = 1$$

$$t = \frac{\ln 1.56}{0.05} \qquad \text{Divide each side by } 0.05$$

$$= \frac{0.4447}{0.05}$$

$$= 8.89 \text{ years}$$

We can estimate that the population will reach 50,000 toward the end of 2002. ◀

PROBLEM SET 10.4

Solve each exponential equation. Use a calculator to write the answer in decimal form.

1. $3^x = 5$

2. $4^x = 3$

3. $5^x = 3$

4. $3^x = 4$

5. $5^{-x} = 12$

6. $7^{-x} = 8$

7. $12^{-x} = 5$

8. $8^{-x} = 7$

9. $8^{x+1} = 4$

10. $9^{x+1} = 3$

11. $4^{x-1} = 4$

12. $3^{x-1} = 9$

13. $3^{2x+1} = 2$

14. $2^{2x+1} = 3$

15. $3^{1-2x} = 2$

16. $2^{1-2x} = 3$

17. $15^{3x-4} = 10$

18. $10^{3x-4} = 15$

19. $6^{5-2x} = 4$

20. $9^{7-3x} = 5$

Applying the Concepts

21. If $5,000 is placed in an account with an annual interest rate of 12% compounded once a year, how much money will be in the account 10 years later?

22. If $5,000 is placed in an account with an annual interest rate of 12% compounded four times a year, how much money will be in the account 10 years later?

23. If $200 is placed in an account with an annual interest rate of 8% compounded twice a year, how much money will be in the account 10 years later?

24. If $200 is placed in an account with an annual interest rate of 8% compounded once a year, how much money will be in the account 10 years later?

25. How long will it take for $500 to double if it is invested at 6% annual interest compounded twice a year?

26. How long will it take for $500 to double if it is invested at 6% annual interest compounded twelve times a year?

Evaluate each of the following. [10.1, 10.3, 10.4]

5. $\log_8 4$

6. $\log_7 21$

7. $\log 23{,}400$

8. $\log 0.0123$

9. $\ln 46.2$

10. $\ln 0.0462$

Use the properties of logarithms to expand each expression. [10.2]

11. $\log_2 \dfrac{8x^2}{y}$

12. $\log \dfrac{\sqrt{x}}{y^4 \sqrt[5]{z}}$

Write each expression as a single logarithm. [10.2]

13. $2 \log_3 x - \frac{1}{2} \log_3 y$

14. $\frac{1}{3} \log x - \log y - 2 \log z$

Use a calculator to find x. [10.3]

15. $\log x = 4.8476$

16. $\log x = -2.6478$

Solve for x. [10.2, 10.4]

17. $4^x = 8$

18. $27^x = 9$

19. $5 = 3^x$

20. $4^{2x-1} = 8$

21. $\log_5 x - \log_5 3 = 1$

22. $\log_2 x + \log_2(x - 7) = 3$

23. Find the pH of a solution in which $[H^+] = 6.6 \times 10^{-7}$. [10.3]

24. If $400 is deposited in an account that earns 10% annual interest compounded twice a year, how much money will be in the account after 5 years? [10.4]

25. How long will it take $600 to become $1,800 if the $600 is deposited in an account that earns 8% annual interest compounded four times a year? [10.4]

11

SEQUENCES
AND SERIES

INTRODUCTION AND OVERVIEW

In this chapter, we will introduce two closely related topics—**sequences** and **series.** These topics are frequently used to describe the world around us. For example, the principal left on a loan after each monthly payment is made forms a sequence of numbers. The same is true for the resale value of a car at yearly intervals after it is purchased. In the first two sections of this chapter, we will work with sequences and series in general. After that, we will look at some specific categories of sequences and series.

Sequences

Many of the sequences in this chapter will be familiar to you on an intuitive level because you have worked with them for some time now. Here are some of those sequences:

The sequence of odd numbers

$$1, 3, 5, 7, \ldots$$

The sequence of even numbers

$$2, 4, 6, 8, \ldots$$

The sequence of squares

$$1^2, 2^2, 3^2, 4^2, \ldots = 1, 4, 9, 16, \ldots$$

The sequence of factorials

$$1!, 2!, 3!, 4!, \ldots = 1, 2, 6, 24, \ldots$$

The numbers in each of these sequences can be found from the formulas that define functions. For example, the sequence of even numbers can be found from the function

$$f(x) = 2x$$

by finding $f(1), f(2), f(3), f(4)$, and so forth. This gives us justification for the formal definition of a sequence.

DEFINITION   A **sequence** is a function whose domain is the set of positive integers $\{1, 2, 3, 4, \ldots\}$.

As you can see, sequences are simply functions with a specific domain. If we want to form a sequence from the function $f(x) = 3x + 5$ we simply find $f(1)$, $f(2), f(3)$, and so on. Doing so gives us the sequence

$$8, 11, 14, 17, \ldots$$

because $f(1) = 3(1) + 5 = 8, \; f(2) = 3(2) + 5 = 11, \; f(3) = 3(3) + 5 = 14,$ and $f(4) = 3(4) + 5 = 17$.

Notation   Since the domain for a sequence is always the set $\{1, 2, 3, \ldots\}$, we can simplify the notation we use to represent the terms of a sequence. Using the letter a instead of f, and subscripts instead of numbers enclosed by parentheses, we can represent the sequence from the discussion above as follows:

$$a_n = 3n + 5$$

Instead of $f(1)$ we write a_1 for the *first term* of the sequence
Instead of $f(2)$ we write a_2 for the *second term* of the sequence
Instead of $f(3)$ we write a_3 for the *third term* of the sequence
Instead of $f(4)$ we write a_4 for the *fourth term* of the sequence
Instead of $f(n)$ we write a_n for the *nth term* of the sequence

The *n*th term is also called the **general term** of the sequence. The general term is used to define the other terms of the sequence. That is, if we are given the formula for the general term, a_n, we can find any other term in the sequence. The following examples illustrate.

▶ **EXAMPLE 1**   Find the first four terms of the sequence whose general term is given by $a_n = 2n - 1$.

Solution   The subscript notation a_n works the same way function notation works. To find the first, second, third, and fourth terms of this sequence, we simply substitute 1, 2, 3, and 4 for n in the formula $2n - 1$:

If   the general term is $a_n = 2n - 1$

then   the first term is $a_1 = 2(1) - 1 = 1$

the second term is $a_2 = 2(2) - 1 = 3$

the third term is $a_3 = 2(3) - 1 = 5$

the fourth term is $a_4 = 2(4) - 1 = 7$

The first four terms of this sequence are the odd numbers 1, 3, 5, and 7. The whole sequence can be written as

$$1, 3, 5, \ldots, 2n - 1, \ldots$$

Since each term in this sequence is larger than the preceding term, we say the sequence is an **increasing sequence.** ◀

▶ **EXAMPLE 2** Write the first four terms of the sequence defined by

$$a_n = \frac{1}{n + 1}$$

Solution Replacing n with 1, 2, 3, and 4, we have, respectively, the first four terms:

$$\text{First term} = a_1 = \frac{1}{1 + 1} = \frac{1}{2}$$

$$\text{Second term} = a_2 = \frac{1}{2 + 1} = \frac{1}{3}$$

$$\text{Third term} = a_3 = \frac{1}{3 + 1} = \frac{1}{4}$$

$$\text{Fourth term} = a_4 = \frac{1}{4 + 1} = \frac{1}{5}$$

The sequence defined by

$$a_n = \frac{1}{n + 1}$$

can be written as

$$\frac{1}{2}, \frac{1}{3}, \frac{1}{4}, \ldots, \frac{1}{n + 1}, \ldots$$

Since each term in the sequence is smaller than the term preceding it, the sequence is said to be a **decreasing sequence.** ◀

▶ **EXAMPLE 3** Find the fifth and sixth terms of the sequence whose general term is given by $a_n = \dfrac{(-1)^n}{n^2}$.

Solution For the fifth term, we replace n with 5. For the sixth term, we replace n with 6:

$$\text{Fifth term} = a_5 = \frac{(-1)^5}{5^2} = \frac{-1}{25}$$

$$\text{Sixth term} = a_6 = \frac{(-1)^6}{6^2} = \frac{1}{36} \qquad \blacktriangleleft$$

The sequence in Example 3 can be written as

$$-1, \frac{1}{4}, -\frac{1}{9}, \frac{1}{16}, \ldots, \frac{(-1)^n}{n^2}, \ldots$$

Since the terms alternate in sign—if one term is positive, then the next term is negative—we call this an **alternating sequence.** The first three examples all illustrate how we work with a sequence in which we are given a formula for the general term. The next example gives us another way to write the general term.

Recursion Formulas

Let's go back to one of the first sequences we looked at in this section:

$$8, 11, 14, 17, \ldots$$

Each term in the sequence can be found by simply substituting positive integers for n in the formula $a_n = 3n + 5$. However, another way to look at this sequence is to notice that each term can be found by adding 3 to the preceding term; so, we could give all the terms of this sequence by simply saying

Start with 8, then add 3 to each term to get the next term.

The same idea, expressed in symbols, looks like this:

$$a_1 = 8 \quad \text{and} \quad a_n = a_{n-1} + 3 \qquad \text{for } n > 1$$

The formula above is called a **recursion formula** because each term is written *recursively* in terms of the term or terms that precede it.

▶ **EXAMPLE 4** Write the first four terms of the sequence given recursively by

$$a_1 = 4 \quad \text{and} \quad a_n = 5a_{n-1} \qquad \text{for } n > 1$$

Solution The formula tells us to start the sequence with the number 4, then multiply each term by 5 to get the next term. Therefore,

$$a_1 = 4$$
$$a_2 = 5a_1 = 5(4) = 20$$

$$a_3 = 5a_2 = 5(20) = 100$$

$$a_4 = 5a_3 = 5(100) = 500$$

The sequence is 4, 20, 100, 500, ... ◀

Finding the General Term

In the first four examples, we found some terms of a sequence after being given the general term. In the next two examples, we will do the reverse. That is, given some terms of a sequence, we will find the formula for the general term.

▶ **EXAMPLE 5**   Find a formula for the nth term of the sequence 2, 8, 18, 32, ...

Solution   Solving a problem like this involves some guessing. Looking over the first four terms, we see each is twice a perfect square:

$$2 = 2(1)$$

$$8 = 2(4)$$

$$18 = 2(9)$$

$$32 = 2(16)$$

If we write each square with an exponent of 2, the formula for the nth term becomes obvious:

$$a_1 = 2 \ = 2(1)^2$$

$$a_2 = 8 \ = 2(2)^2$$

$$a_3 = 18 = 2(3)^2$$

$$a_4 = 32 = 2(4)^2$$

$$\cdot$$

$$\cdot$$

$$\cdot$$

$$a_n = \qquad 2(n)^2 = 2n^2$$

The general term of the sequence 2, 8, 18, 32, . . . is $a_n = 2n^2$. ◀

▶ **EXAMPLE 6**   Find the general term for the sequence $2, \frac{3}{8}, \frac{4}{27}, \frac{5}{64}, \ldots$.

Solution   The first term can be written as $\frac{2}{1}$. The denominators are all perfect cubes. The numerators are all 1 more than the base of the cubes in the denominators:

$$a_1 = \frac{2}{1} = \frac{1+1}{1^3}$$

$$a_2 = \frac{3}{8} = \frac{2+1}{2^3}$$

$$a_3 = \frac{4}{27} = \frac{3+1}{3^3}$$

$$a_4 = \frac{5}{64} = \frac{4+1}{4^3}$$

Observing this pattern, we recognize the general term to be

$$a_n = \frac{n+1}{n^3}$$

◄

Note   Finding the *n*th term of a sequence from the first few terms is not always automatic. That is, it sometimes takes a while to recognize the pattern. Don't be afraid to guess at the formula for the general term. Many times an incorrect guess leads to the correct formula.

PROBLEM SET 11.1

Write the first five terms of the sequences with the following general terms.

1. $a_n = 3n + 1$

2. $a_n = 2n + 3$

3. $a_n = 4n - 1$

4. $a_n = n + 4$

5. $a_n = n$

6. $a_n = -n$

7. $a_n = n^2 + 3$

8. $a_n = n^3 + 1$

9. $a_n = \dfrac{n}{n+3}$

10. $a_n = \dfrac{n}{n+2}$

11. $a_n = \dfrac{n+1}{n+2}$

12. $a_n = \dfrac{n+3}{n+4}$

13. $a_n = \dfrac{1}{n^2}$

14. $a_n = \dfrac{1}{n^3}$

15. $a_n = 2^n$

16. $a_n = 3^n$

17. $a_n = 3^{-n}$

18. $a_n = 2^{-n}$

19. $a_n = 1 + \dfrac{1}{n}$

20. $a_n = 1 - \dfrac{1}{n}$

21. $a_n = n - \dfrac{1}{n}$

22. $a_n = n + \dfrac{1}{n}$

23. $a_n = (-2)^n$

24. $a_n = (-3)^n$

Write the first five terms of the sequences defined by the following recursion formulas.

25. $a_1 = 3$ $a_n = -3a_{n-1}$ $n > 1$

26. $a_1 = -3$ $a_n = 3a_{n-1}$ $n > 1$

27. $a_1 = 3$ $a_n = a_{n-1} - 3$ $n > 1$

28. $a_1 = -3$ $a_n = a_{n-1} - 3$ $n > 1$

29. $a_1 = 1$ $a_n = 2a_{n-1} + 3$ $n > 1$

30. $a_1 = 1$ $a_n = 3a_{n-1} + 2$ $n > 1$

31. $a_1 = 1$ $a_n = a_{n-1} + n$ $n > 1$

32. $a_1 = 2$ $a_n = a_{n-1} - n$ $n > 1$

Determine the general term for each of the following sequences.

33. 2, 3, 4, 5, . . .

34. 3, 6, 9, 12, . . .

35. 4, 8, 12, 16, 20, . . .

36. 3, 4, 5, 6, . . .

37. 7, 10, 13, 16, . . .

38. 4, 9, 14, 19, . . .

39. 1, 4, 9, 16, . . .

40. 1, 8, 27, 64, . . .

41. 3, 12, 27, 48, . . .

42. 2, 16, 54, 128, . . .

43. 4, 8, 16, 32, . . .

44. 3, 9, 27, 81, . . .

45. −2, 4, −8, 16, . . .

46. −3, 9, −27, 81, . . .

47. $\frac{1}{4}, \frac{1}{8}, \frac{1}{16}, \frac{1}{32}, \ldots$ **48.** $\frac{1}{3}, \frac{1}{9}, \frac{1}{27}, \frac{1}{81}, \ldots$

49. $\frac{1}{4}, \frac{2}{9}, \frac{3}{16}, \frac{4}{25}, \ldots$ **50.** $\frac{1}{4}, \frac{2}{10}, \frac{3}{28}, \frac{4}{82}, \ldots$

Review Problems

The problems that follow review material we covered in Section 10.1.

Write each expression in logarithmic form.

51. $100 = 10^2$ **52.** $4^{3/2} = 8$

Write each expression in exponential form.

53. $\log_3 81 = 4$ **54.** $-2 = \log_{10} 0.01$

Find x in each of the following.

55. $\log_9 x = \frac{3}{2}$ **56.** $\log_x \frac{1}{4} = -2$

Simplify each expression.

57. $\log_2 32$ **58.** $\log_{10} 10,000$

59. $\log_3 [\log_2 8]$ **60.** $\log_5 [\log_6 6]$

One Step Further

61. As n increases, the terms in the sequence $a_n = \left(1 + \dfrac{1}{n}\right)^n$ get closer and closer to the number e (that's the same e we used in defining natural logarithms). However, it takes some fairly large values of n before we can see this happening. Use a calculator to find $a_{100}, a_{1,000}, a_{10,000},$ and $a_{100,000}$ and compare them to the decimal approximation we gave for the number e.

62. The sequence $a_n = \left(1 + \dfrac{1}{n}\right)^{-n}$ gets close to the number $\dfrac{1}{e}$ as n becomes large. Use a calculator to find approximations for a_{100} and $a_{1,000}$, and then compare them to $\dfrac{1}{2.7183}$.

63. Write the first 10 terms of the sequence defined by the recursion formula
$$a_1 = 1, a_2 = 1, a_n = a_{n-1} + a_{n-2} \quad n > 2$$

64. Write the first 10 terms of the sequence defined by the recursion formula
$$a_1 = 2, a_2 = 2, a_n = a_{n-1} + a_{n-2} \quad n > 2$$

65. Simplify each complex fraction in the sequence below and then compare this sequence with the sequence you wrote in Problem 63.

$$1 + \cfrac{1}{1+1}, 1 + \cfrac{1}{1 + \cfrac{1}{1+1}}, 1 + \cfrac{1}{1 + \cfrac{1}{1 + \cfrac{1}{1+1}}} \cdots$$

66. Write the first five terms of the sequence given by the recursion formula below. Compare this sequence with the sequence in Problem 65.

$$a_1 = \frac{3}{2}, a_n = 1 + \frac{1}{a_{n-1}} \quad n > 1$$

11.2   Series

There is an interesting relationship between the sequence of odd numbers and the sequence of squares that is found by adding the terms in the sequence of odd numbers.

$$
\begin{aligned}
1 &= 1 \\
1 + 3 &= 4 \\
1 + 3 + 5 &= 9 \\
1 + 3 + 5 + 7 &= 16
\end{aligned}
$$

When we add the terms of a sequence the result is called a **series.**

DEFINITION The sum of a number of terms in a sequence is called a **series.**

A sequence can be finite or infinite depending on whether or not the sequence ends at the nth term. For example,

$$1, 3, 5, 7, 9$$

is a finite sequence, while

$$1, 3, 5, \ldots$$

is an infinite sequence. Associated with each of the above sequences is a series found by adding the terms of the sequence:

$$1 + 3 + 5 + 7 + 9 \qquad \text{Finite series}$$

$$1 + 3 + 5 + \ \ldots \qquad \text{Infinite series}$$

In this section we will consider only finite series. We can introduce a new kind of notation here that is a compact way of indicating a finite series. The notation is called **summation notation,** or **sigma notation** since it is written using the Greek letter sigma. The expression

$$\sum_{i=1}^{4} (8i - 10)$$

is an example of an expression that uses summation notation. The summation notation in this expression is used to indicate the sum of all the expressions $8i - 10$ from $i = 1$ up to and including $i = 4$. That is,

$$\sum_{i=1}^{4} (8i - 10) = (8 \cdot 1 - 10) + (8 \cdot 2 - 10) + (8 \cdot 3 - 10) + (8 \cdot 4 - 10)$$

$$= -2 + 6 + 14 + 22$$

$$= 40$$

The letter i as used here is called the **index of summation,** or just **index** for short.

Here are some examples illustrating the use of summation notation.

▶ **EXAMPLE 1**   Expand and simplify $\sum\limits_{i=1}^{5} (i^2 - 1)$.

Solution   We replace i in the expression $i^2 - 1$ with all consecutive integers from 1 up to 5, including 1 and 5:

$$\sum_{i=1}^{5} (i^2 - 1) = (1^2 - 1) + (2^2 - 1) + (3^2 - 1) + (4^2 - 1) + (5^2 - 1)$$

$$= 0 + 3 + 8 + 15 + 24$$
$$= 50 \qquad \blacktriangleleft$$

▶ **EXAMPLE 2**   Expand and simplify $\sum\limits_{i=3}^{6} (-2)^i$.

Solution   We replace i in the expression $(-2)^i$ with the consecutive integers beginning at 3 and ending at 6:

$$\sum_{i=3}^{6} (-2)^i = (-2)^3 + (-2)^4 + (-2)^5 + (-2)^6$$

$$= -8 + 16 + (-32) + 64$$
$$= 40 \qquad \blacktriangleleft$$

▶ **EXAMPLE 3**   Expand $\sum\limits_{i=2}^{5} (x^i - 3)$.

Solution   We must be careful not to confuse the letter x with i. The index i is the quantity we replace by the consecutive integers from 2 to 5, not x:

$$\sum_{i=2}^{5} (x^i - 3) = (x^2 - 3) + (x^3 - 3) + (x^4 - 3) + (x^5 - 3) \qquad \blacktriangleleft$$

In the first three examples, we were given an expression with summation notation and asked to expand it. The next examples in this section illustrate how we can write an expression in expanded form as an expression involving summation notation.

▶ **EXAMPLE 4**   Write with summation notation $1 + 3 + 5 + 7 + 9$.

Solution   A formula that gives us the terms of this sum is

$$a_i = 2i - 1$$

where i ranges from 1 up to and including 5. Notice we are using the subscript i in exactly the same way we used the subscript n in the last section—to indicate the general term. Writing the sum

$$1 + 3 + 5 + 7 + 9$$

with summation notation looks like this:

$$\sum_{i=1}^{5} (2i - 1)$$

◀

▶ **EXAMPLE 5** Write with summation notation $3 + 12 + 27 + 48$.

Solution We need a formula, in terms of i, that will give each term in the sum. Writing the sum as

$$3 \cdot 1^2 + 3 \cdot 2^2 + 3 \cdot 3^2 + 3 \cdot 4^2$$

we see the formula

$$a_i = 3 \cdot i^2$$

where i ranges from 1 up to and including 4. Using this formula and summation notation, we can represent the sum

$$3 + 12 + 27 + 48$$

as

$$\sum_{i=1}^{4} 3i^2$$

◀

▶ **EXAMPLE 6** Write with summation notation

$$\frac{x + 3}{x^3} + \frac{x + 4}{x^4} + \frac{x + 5}{x^5} + \frac{x + 6}{x^6}$$

Solution A formula that gives each of these terms is

$$a_i = \frac{x + i}{x^i}$$

where i assumes all integer values between 3 and 6, including 3 and 6. The sum can be written as

$$\sum_{i=3}^{6} \frac{x + i}{x^i}$$

◀

P R O B L E M   S E T   1 1 . 2

Expand and simplify each of the following.

1. $\displaystyle\sum_{i=1}^{4} (2i + 4)$

2. $\displaystyle\sum_{i=1}^{5} (3i - 1)$

3. $\displaystyle\sum_{i=1}^{3} (2i - 1)$

4. $\displaystyle\sum_{i=1}^{4} (2i - 1)$

5. $\displaystyle\sum_{i=2}^{3} (i^2 - 1)$

6. $\displaystyle\sum_{i=3}^{6} (i^2 + 1)$

7. $\displaystyle\sum_{i=1}^{4} \frac{i}{1 + i}$

8. $\displaystyle\sum_{i=1}^{4} \frac{i^2}{1 + i}$

9. $\displaystyle\sum_{i=1}^{3} \frac{i^2}{2i - 1}$

10. $\displaystyle\sum_{i=3}^{5} (i^3 + 4)$

11. $\displaystyle\sum_{i=1}^{4} (-3)^i$

12. $\displaystyle\sum_{i=1}^{4} \left(-\frac{1}{3}\right)^i$

13. $\displaystyle\sum_{i=3}^{6} (-2)^i$

14. $\displaystyle\sum_{i=4}^{6} \left(-\frac{1}{2}\right)^i$

Expand the following.

15. $\displaystyle\sum_{i=1}^{5} (x + i)$

16. $\displaystyle\sum_{i=3}^{6} (x - i)$

17. $\displaystyle\sum_{i=2}^{7} (x + 1)^i$

18. $\displaystyle\sum_{i=1}^{4} (x + 3)^i$

19. $\displaystyle\sum_{i=1}^{5} \frac{x + i}{x - 1}$

20. $\displaystyle\sum_{i=1}^{6} \frac{x - 3i}{x + 3i}$

21. $\displaystyle\sum_{i=3}^{8} (x + i)^i$

22. $\displaystyle\sum_{i=4}^{7} (x - 2i)^i$

23. $\displaystyle\sum_{i=1}^{5} (x + i)^{i+1}$

24. $\displaystyle\sum_{i=2}^{6} (x + i)^{i-1}$

Write each of the following sums with summation notation.

25. $2 + 4 + 8 + 16$

26. $3 + 5 + 7 + 9 + 11$

27. $4 + 8 + 16 + 32 + 64$

28. $1 + 3 + 5$

29. $5 + 10 + 17 + 26 + 37$

30. $3 + 8 + 15 + 24$

31. $\frac{3}{4} + \frac{4}{5} + \frac{5}{6} + \frac{6}{7} + \frac{7}{8}$ **32.** $\frac{1}{2} + \frac{2}{3} + \frac{3}{4} + \frac{4}{5}$

33. $\frac{1}{3} + \frac{2}{5} + \frac{3}{7} + \frac{4}{9}$ **34.** $\frac{3}{1} + \frac{5}{3} + \frac{7}{5} + \frac{9}{7}$

35. $(x - 3) + (x - 4) + (x - 5) + (x - 6)$

36. $x^2 + x^3 + x^4 + x^5 + x^6$

37. $\dfrac{x}{x + 3} + \dfrac{x}{x + 4} + \dfrac{x}{x + 5}$

38. $\dfrac{x - 3}{x^3} + \dfrac{x - 4}{x^4} + \dfrac{x - 5}{x^5} + \dfrac{x - 6}{x^6}$

39. $x^2(x + 2) + x^3(x + 3) + x^4(x + 4)$

40. $x(x + 2)^2 + x(x + 3)^3 + x(x + 4)^4$

Applying the Concepts

41. A skydiver jumps from a plane and falls 16 feet the first second, 48 feet the second second, and 80 feet the third second. If he continues to fall in the same manner, how far will he fall the seventh second? What is the distance he falls in 7 seconds?

42. After 1 day, a colony of 50 bacteria reproduces to become 200 bacteria. After 2 days, they reproduce to become 800 bacteria. If they continue to reproduce at this rate, how many bacteria will be present after 4 days?

Review Problems

The problems below review material we covered in Section 10.2.

Use the properties of logarithms to expand each of the following expressions.

43. $\log_2 x^3 y$

44. $\log_7 \dfrac{x^2}{y^4}$

45. $\log_{10} \dfrac{\sqrt[3]{x}}{y^2}$

46. $\log_{10} \sqrt[3]{\dfrac{x}{y^2}}$

Write each expression as a single logarithm.

47. $\log_{10} x - \log_{10} y^2$ **48.** $\log_{10} x^2 + \log_{10} y^2$

49. $2 \log_3 x - 3 \log_3 y - 4 \log_3 z$

50. $\frac{1}{2} \log_6 x + \frac{1}{3} \log_6 y + \frac{1}{4} \log_6 z$

Solve each equation.

51. $\log_4 x - \log_4 5 = 2$ **52.** $\log_3 6 + \log_3 x = 4$

53. $\log_2 x + \log_2(x - 7) = 3$

54. $\log_5 (x + 1) + \log_5 (x - 3) = 1$

SECTION

11.3   Arithmetic Sequences

In this and the following section, we will review and extend two major types of sequences, which we have worked with previously—**arithmetic sequences** and **geometric sequences.**

> **DEFINITION**   An **arithmetic sequence** is a sequence of numbers in which each term is obtained from the preceding term by adding the same amount each time. An arithmetic sequence is also called an **arithmetic progression.**

The sequence

$$2, 6, 10, 14 \ldots$$

is an example of an arithmetic sequence, since each term is obtained from the preceding term by adding 4 each time. The amount we add each time—in this case, 4—is called the **common difference,** since it can be obtained by subtracting any two consecutive terms. (The term with the larger subscript must be written first.) The common difference is denoted by d.

▶ **EXAMPLE 1**   Give the common difference d for the arithmetic sequence 4, 10, 16, 22, . . .

Solution   Since each term can be obtained from the preceding term by adding 6, the common difference is 6. That is, $d = 6$.   ◀

▶ **EXAMPLE 2**   Give the common difference for 100, 93, 86, 79, ...

Solution   The common difference in this case is $d = -7$, since adding -7 to any term always produces the next consecutive term.   ◀

▶ **EXAMPLE 3**   Give the common difference for $\frac{1}{2}$, 1, $\frac{3}{2}$, 2, ...

Solution   The common difference is $d = \frac{1}{2}$.   ◀

The General Term

The general term, a_n, of an arithmetic progression can always be written in terms of the first term a_1 and the common difference d. Consider the sequence from Example 1:

$$4, 10, 16, 22, \ldots$$

We can write each term in terms of the first term 4 and the common difference 6:

$$4, \quad 4 + (1 \cdot 6), \quad 4 + (2 \cdot 6), \quad 4 + (3 \cdot 6), \quad \ldots$$
$$a_1, \quad\quad a_2, \quad\quad\quad a_3, \quad\quad\quad a_4, \quad\quad\quad \ldots$$

Observing the relationship between the subscript on the terms in the second line and the coefficients of the 6's in the first line, we write the general term for the sequence as

$$a_n = 4 + (n - 1)6$$

We generalize this result to include the general term of any arithmetic sequence.

Arithmetic Sequences

The *general term* of an arithmetic progression with first term a_1 and common difference d is given by

$$a_n = a_1 + (n - 1)d$$

▶ **EXAMPLE 4**   Find the general term for the sequence

$$7, 10, 13, 16, \ldots$$

Solution   The first term is $a_1 = 7$, and the common difference is $d = 3$. Substituting these numbers into the formula given above, we have

$$a_n = 7 + (n - 1)3$$

which we can simplify, if we choose, to

$$a_n = 7 + 3n - 3$$
$$= 3n + 4 \qquad\qquad ◀$$

▶ **EXAMPLE 5**   Find the general term of the arithmetic progression whose third term, a_3, is 7 and whose eighth term, a_8, is 17.

Solution   According to the formula for the general term, the third term can be written as $a_3 = a_1 + 2d$, and the eighth term can be written as

$a_8 = a_1 + 7d$. Since these terms are also equal to 7 and 17, respectively, we can write

$$a_3 = a_1 + 2d = 7$$
$$a_8 = a_1 + 7d = 17$$

To find a_1 and d, we simply solve the system:

$$a_1 + 2d = 7$$
$$a_1 + 7d = 17$$

We add the opposite of the top equation to the bottom equation. The result is

$$5d = 10$$
$$d = 2$$

To find a_1, we simply substitute 2 for d in either of the original equations and get

$$a_1 = 3$$

The general term for this progression is

$$a_n = 3 + (n-1)2$$

which we can simplify to

$$a_n = 2n + 1 \qquad \blacktriangleleft$$

The sum of the first n terms of an arithmetic sequence is denoted by S_n. The following theorem gives the formula for finding S_n, which is sometimes called the **nth partial sum.**

Theorem 11.1

The sum of the first n terms of an arithmetic sequence whose first term is a_1 and whose nth term is a_n is given by

$$S_n = \frac{n}{2}(a_1 + a_n)$$

PROOF We can write S_n in expanded form as

$$S_n = a_1 + [a_1 + d] + [a_1 + 2d] + \cdots + [a_1 + (n-1)d]$$

We can arrive at this same series by starting with the last term, a_n, and subtracting d each time. Writing S_n this way, we have

$$S_n = a_n + [a_n - d] + [a_n - 2d] + \cdots + [a_n - (n-1)d]$$

If we add the preceding two expressions term by term, we have

$$2S_n = (a_1 + a_n) + (a_1 + a_n) + (a_1 + a_n) + \cdots + (a_1 + a_n)$$

$$2S_n = n(a_1 + a_n)$$

$$S_n = \frac{n}{2}(a_1 + a_n)$$

▶ **EXAMPLE 6**   Find the sum of the first 10 terms of the arithmetic progression 2, 10, 18, 26, ...

Solution   The first term is 2 and the common difference is 8. The tenth term is

$$a_{10} = 2 + 9(8)$$
$$= 2 + 72$$
$$= 74$$

Substituting $n = 10$, $a_1 = 2$, and $a_{10} = 74$ into the formula

$$S_n = \frac{n}{2}(a_1 + a_n)$$

we have

$$S_{10} = \frac{10}{2}(2 + 74)$$
$$= 5(76)$$
$$= 380$$

The sum of the first 10 terms is 380.   ◀

P R O B L E M   S E T   1 1 . 3

Determine which of the following sequences are arithmetic progressions. For those that are arithmetic progressions, identify the common difference d.

1. 1, 2, 3, 4, . . .
2. 4, 6, 8, 10, . . .
3. 1, 2, 4, 7, . . .
4. 1, 2, 4, 8, . . .
5. 50, 45, 40, . . .
6. $1, \frac{1}{2}, \frac{1}{4}, \frac{1}{8}, \ldots$
7. 1, 4, 9, 16 . . .
8. 5, 7, 9, 11, . . .
9. $\frac{1}{3}, 1, \frac{5}{3}, \frac{7}{3}, \ldots$
10. 5, 11, 17, . . .

Each of the following problems refers to arithmetic sequences.

11. If $a_1 = 3$ and $d = 4$, find a_n and a_{24}.
12. If $a_1 = 5$ and $d = 10$, find a_n and a_{100}.
13. If $a_1 = 6$ and $d = -2$, find a_{10} and S_{10}.
14. If $a_1 = 7$ and $d = -1$, find a_{24} and S_{24}.
15. If $a_6 = 17$ and $a_{12} = 29$, find the first term a_1, the common difference d, and then find a_{30}.

16. If $a_5 = 23$ and $a_{10} = 48$, find the first term a_1, the common difference d, and then find a_{40}.
17. If the third term is 16 and the eighth term is 26, find the first term, the common difference, and then find a_{20} and S_{20}.
18. If the third term is 16 and the eighth term is 51, find the first term, the common difference, and then find a_{50} and S_{50}.
19. Find the sum of the first 100 terms of the sequence 5, 9, 13, 17, ...
20. Find the sum of the first 50 terms of the sequence 8, 11, 14, 17, ...
21. Find a_{35} for the sequence 12, 7, 2, -3, ...
22. Find a_{45} for the sequence 25, 20, 15, 10, ...
23. Find the tenth term and the sum of the first ten terms of the sequence $\frac{1}{2}$, 1, $\frac{3}{2}$, 2, ...
24. Find the fifteenth term and the sum of the first fifteen terms of the sequence $-\frac{1}{3}$, 0, $\frac{1}{3}$, $\frac{2}{3}$, ...

Applying the Concepts

25. Suppose a woman earns $18,000 the first year she works and then gets a raise of $850 every year after that. Write a sequence that gives her salary for each of the first 5 years she works. What is the general term of this sequence? At this rate, how much will she be making the tenth year she works?
26. Suppose a school teacher makes $16,500 the first year he works and then gets a $900 raise every year after that. Write a sequence that gives his salary for the first 5 years he works. What is the general term of this sequence? How much will he be making the twentieth year he works?

Review Problems

The problems that follow review material we covered in Section 10.3.

Find the following common logarithms.

27. log 576
28. log 57,600
29. log 0.0576
30. log 0.000576

Find x.

31. $\log x = 2.6484$
32. $\log x = 7.9832$
33. $\log x = -7.3516$
34. $\log x = -2.0168$

SECTION

11.4 Geometric Sequences

This section is concerned with the second major classification of sequences, called **geometric sequences.** The problems in this section are very similar to the problems in the preceding section.

> **DEFINITION** A sequence of numbers in which each term is obtained from the previous term by multiplying by the same amount each time is called a **geometric sequence.** Geometric sequences are also called **geometric progressions.**

The sequence

$$3, 6, 12, 24, \ . \ . \ .$$

is an example of a geometric progression. Each term is obtained from the previous term by multiplying by 2. The amount by which we multiply each time—in this case, 2—is called the **common ratio.** The common ratio is denoted by r and can be found by taking the ratio of any two consecutive terms. (The term with the larger subscript must be in the numerator.)

▶ **EXAMPLE 1** Find the common ratio for the geometric progression

$$\frac{1}{2}, \frac{1}{4}, \frac{1}{8}, \frac{1}{16}, \ldots$$

Solution Since each term can be obtained from the term before it by multiplying by $\frac{1}{2}$, the common ratio is $\frac{1}{2}$. That is $r = \frac{1}{2}$. ◀

▶ **EXAMPLE 2** Find the common ratio for $\sqrt{3}, 3, 3\sqrt{3}, 9, \ldots$.

Solution If we take the ratio of the third term to the second term, we have

$$\frac{3\sqrt{3}}{3} = \sqrt{3}$$

The common ratio is $r = \sqrt{3}$. ◀

Geometric Sequences

The *general term, a_n,* of a geometric sequence with first term a_1 and common ratio r is given by

$$a_n = a_1 r^{n-1}$$

To see how we arrive at this formula, consider the following geometric progression whose common ratio is 3:

$$2, 6, 18, 54, \ldots$$

We can write each term of the sequence in terms of the first term 2 and the common ratio 3:

$$2 \cdot 3^0, \quad 2 \cdot 3^1, \quad 2 \cdot 3^2, \quad 2 \cdot 3^3, \quad \ldots$$
$$a_1, \quad\quad a_2, \quad\quad a_3, \quad\quad a_4, \quad \ldots$$

Observing the relationship between the two lines written above, we find we can write the general term of this progression as

$$a_n = 2 \cdot 3^{n-1}$$

Since the first term can be designated by a_1 and the common ratio by r, the formula

$$a_n = 2 \cdot 3^{n-1}$$

coincides with the formula

$$a_n = a_1 r^{n-1}$$

▶ **EXAMPLE 3** Find the general term for the geometric progression

$$5, 10, 20, \ldots$$

Solution The first term is $a_1 = 5$, and the common ratio is $r = 2$. Using these values in the formula

$$a_n = a_1 r^{n-1}$$

we have

$$a_n = 5 \cdot 2^{n-1} \qquad \blacktriangleleft$$

▶ **EXAMPLE 4** Find the tenth term of the sequence $3, \frac{3}{2}, \frac{3}{4}, \frac{3}{8}, \ldots$.

Solution The sequence is a geometric progression with first term $a_1 = 3$ and common ratio $r = \frac{1}{2}$. The tenth term is

$$a_{10} = 3 \left(\frac{1}{2} \right)^9 = \frac{3}{512} \qquad \blacktriangleleft$$

▶ **EXAMPLE 5** Find the general term for the geometric progression whose fourth term is 16 and whose seventh term is 128.

Solution The fourth term can be written as $a_4 = a_1 r^3$, and the seventh term can be written as $a_7 = a_1 r^6$.

$$a_4 = a_1 r^3 = 16$$

$$a_7 = a_1 r^6 = 128$$

We can solve for r by using the ratio a_7 / a_4.

$$\frac{a_7}{a_4} = \frac{a_1 r^6}{a_1 r^3} = \frac{128}{16}$$

$$r^3 = 8$$

$$r = 2$$

The common ratio is 2. To find the first term we substitute $r = 2$ into either of the original two equations. The result is

$$a_1 = 2$$

The general term for this progression is

$$a_n = 2 \cdot 2^{n-1}$$

which we can simplify by adding exponents, since the bases are equal:

$$a_n = 2^n \qquad \blacktriangleleft$$

As was the case in the preceding section, the sum of the first n terms of a geometric progression is denoted by S_n, which is called the ***n*th partial sum** of the progression.

Theorem 11.2
The sum of the first n terms of a geometric progression with first term a_1 and common ratio r is given by the formula

$$S_n = \frac{a_1(r^n - 1)}{r - 1}$$

PROOF   We can write the sum of the first n terms in expanded form:

$$S_n = a_1 + a_1r + a_1r^2 + \cdots + a_1r^{n-1} \tag{1}$$

Then multiplying both sides by r, we have

$$rS_n = a_1r + a_1r^2 + a_1r^3 + \cdots + a_1r^n \tag{2}$$

If we subtract the left side of equation (1) from the left side of equation (2) and do the same for the right sides, we end up with

$$rS_n - S_n = a_1r^n - a_1$$

We factor S_n from both terms on the left side and a_1 from both terms on the right side of this equation:

$$S_n(r - 1) = a_1(r^n - 1)$$

Dividing both sides by $r - 1$ gives the desired result:

$$S_n = \frac{a_1(r^n - 1)}{r - 1}$$

▶ **EXAMPLE 6**   Find the sum of the first 10 terms of the geometric progression $5, 15, 45, 135, \ldots$.

Solution   The first term is $a_1 = 5$, and the common ratio is $r = 3$. Substituting these values into the formula for S_{10}, we have the sum of the first 10 terms of the sequence:

$$S_{10} = \frac{5(3^{10} - 1)}{3 - 1}$$

$$= \frac{5(3^{10} - 1)}{2}$$

The answer can be left in this form. A calculator will give the result as 147,620.   ◀

Infinite Geometric Series

Suppose the common ratio for a geometric sequence is a number whose absolute value is less than 1, for instance $\frac{1}{2}$. The sum of the first n terms is given by the formula

$$S_n = \frac{a_1\left[\left(\dfrac{1}{2}\right)^n - 1\right]}{\dfrac{1}{2} - 1}$$

As n becomes larger and larger, the term $(\frac{1}{2})^n$ will become closer and closer to 0. That is, for $n = 10$, 20, and 30, we have the following approximations:

$$\left(\frac{1}{2}\right)^{10} \approx 0.001$$

$$\left(\frac{1}{2}\right)^{20} \approx 0.000001$$

$$\left(\frac{1}{2}\right)^{30} \approx 0.000000001$$

so that for large values of n, there is very little difference between the expression

$$\frac{a_1(r^n - 1)}{r - 1}$$

and the expression

$$\frac{a_1(0 - 1)}{r - 1} = \frac{-a_1}{r - 1} = \frac{a_1}{1 - r} \quad \text{if} \quad |r| < 1$$

In fact, the sum of the terms of a geometric sequence in which $|r| < 1$ actually becomes the expression

$$\frac{a_1}{1 - r}$$

as n approaches infinity. To summarize, we have the following:

The Sum of an Infinite Geometric Series

If a geometric sequence has first term a_1 and common ratio r such that $|r| < 1$, then the following is called an **infinite geometric series**:

$$S = \sum_{i=0}^{\infty} a_1 r^i = a_1 + a_1 r + a_1 r^2 + a_1 r^3 + \cdots$$

Its sum is given by the formula

$$S = \frac{a_1}{1 - r}$$

▶ **EXAMPLE 7**   Find the sum of the infinite geometric series

$$\frac{1}{5} + \frac{1}{10} + \frac{1}{20} + \frac{1}{40} + \cdots$$

Solution   The first term is $a_1 = \frac{1}{5}$ and the common ratio is $r = \frac{1}{2}$, which has an absolute value less than 1. Therefore, the sum of this series is

$$S = \frac{a_1}{1 - r} = \frac{\dfrac{1}{5}}{1 - \dfrac{1}{2}} = \frac{\dfrac{1}{5}}{\dfrac{1}{2}} = \frac{2}{5}$$

◀

▶ **EXAMPLE 8**   Show that 0.999 . . . is equal to 1.

Solution   We begin by writing 0.999 . . . as an infinite geometric series:

$$0.999 \ldots = 0.9 + 0.09 + 0.009 + 0.0009 + \cdots$$

$$= \frac{9}{10} + \frac{9}{100} + \frac{9}{1{,}000} + \frac{9}{10{,}000} + \cdots$$

$$= \frac{9}{10} + \frac{9}{10}\left(\frac{1}{10}\right) + \frac{9}{10}\left(\frac{1}{10}\right)^2 + \frac{9}{10}\left(\frac{1}{10}\right)^3 + \cdots$$

As the last line indicates, we have an infinite geometric series with $a_1 = \frac{9}{10}$ and $r = \frac{1}{10}$. The sum of this series is given by

$$S = \frac{a_1}{1 - r} = \frac{\dfrac{9}{10}}{1 - \dfrac{1}{10}} = \frac{\dfrac{9}{10}}{\dfrac{9}{10}} = 1$$

◀

P R O B L E M   S E T   1 1 . 4

Identify those sequences that are geometric progressions. For those that are geometric, give the common ratio r.

1. 1, 5, 25, 125, . . .   **2.** 6, 12, 24, 48, . . .   **5.** 4, 9, 16, 25, . . .   **6.** $-1, \frac{1}{3}, -\frac{1}{9}, \frac{1}{27}, \ldots$

3. $\frac{1}{2}, \frac{1}{6}, \frac{1}{18}, \frac{1}{54}, \ldots$ **4.** 5, 10, 15, 20, . . .   **7.** $-2, 4, -8, 16, \ldots$ **8.** 1, 8, 27, 64, . . .

9. 4, 6, 8, 10, . . .
10. 1, −3, 9, −27, . . .

Each of the following problems gives some information about a specific geometric progression.

11. If $a_1 = 4$ and $r = 3$, find a_n.
12. If $a_1 = 5$ and $r = 2$, find a_n.
13. If $a_1 = -2$ and $r = -\frac{1}{2}$, find a_6.
14. If $a_1 = 25$ and $r = -\frac{1}{5}$, find a_6.
15. If $a_1 = 3$ and $r = -1$, find a_{20}.
16. If $a_1 = -3$ and $r = -1$, find a_{20}.
17. If $a_1 = 10$ and $r = 2$, find S_{10}.
18. If $a_1 = 8$ and $r = 3$, find S_5.
19. If $a_1 = 1$ and $r = -1$, find S_{20}.
20. If $a_1 = 1$ and $r = -1$, find S_{21}.
21. Find a_8 for $\frac{1}{5}, \frac{1}{10}, \frac{1}{20}, \ldots$
22. Find a_8 for $\frac{1}{2}, \frac{1}{10}, \frac{1}{50}, \ldots$
23. Find S_5 for $-\frac{1}{2}, -\frac{1}{4}, -\frac{1}{8}, \ldots$
24. Find S_6 for $-\frac{1}{2}, 1, -2, \ldots$
25. Find a_{10} and S_{10} for $\sqrt{2}, 2, 2\sqrt{2}, \ldots$
26. Find a_8 and S_8 for $\sqrt{3}, 3, 3\sqrt{3}, \ldots$
27. Find a_6 and S_6 for 100, 10, 1, ...
28. Find a_6 and S_6 for 100, −10, 1, ...
29. If $a_4 = 40$ and $a_6 = 160$, find r.
30. If $a_5 = \frac{1}{8}$ and $a_8 = \frac{1}{64}$, find r.

Find the sum of each geometric series.

31. $\frac{1}{2} + \frac{1}{4} + \frac{1}{8} + \cdots$ **32.** $\frac{1}{3} + \frac{1}{9} + \frac{1}{27} + \cdots$
33. $4 + 2 + 1 + \cdots$ **34.** $8 + 4 + 2 + \cdots$
35. $\frac{2}{5} + \frac{4}{25} + \frac{8}{125} + \cdots$
36. $\frac{3}{4} + \frac{9}{16} + \frac{27}{64} + \cdots$
37. $\frac{3}{4} + \frac{1}{4} + \frac{1}{12} + \cdots$
38. $\frac{5}{3} + \frac{1}{3} + \frac{1}{15} + \cdots$

39. Show that 0.444 . . . is the same as $\frac{4}{9}$.
40. Show that 0.333 . . . is the same as $\frac{1}{3}$.
41. Show that 0.272727 . . . is the same as $\frac{3}{11}$.
42. Show that 0.545454 . . . is the same as $\frac{6}{11}$.

Review Problems

The problems below review material we covered in Section 3.4. Reviewing these problems will help you understand the next section.

Expand and multiply.

43. $(x + 5)^2$ **44.** $(x + y)^2$
45. $(x + y)^3$ **46.** $(x - 2)^3$
47. $(x + y)^4$ **48.** $(x - 1)^4$

One Step Further

Use a calculator to find the given term or partial sum.

49. Find a_{20} if $a_1 = 100$ and $r = 2$.
50. Find a_{20} if $a_1 = 81$ and $r = \frac{1}{3}$.
51. Find S_{18} if $a_1 = 100$ and $r = \frac{1}{2}$.
52. Find S_{22} if $a_1 = 64$ and $r = -\frac{1}{2}$.

53. Find the sum for $a + \dfrac{a}{2} + \dfrac{a}{4} + \cdots$ if $a > 1$.

54. Find the sum for $\dfrac{1}{a} + \dfrac{1}{a^2} + \dfrac{1}{a^3} + \cdots$ if $a > 1$.

55. Find the sum for $\dfrac{a}{b} + \dfrac{a^2}{b^2} + \dfrac{a^3}{b^3} + \cdots$ if $\left| \dfrac{a}{b} \right| < 1$.

The Binomial Expansion

The purpose of this section is to write and apply the formula for the expansion of expressions of the form $(x + y)^n$, where n is any positive integer. In order to write the formula, we must generalize the information in the following chart:

$$
\begin{aligned}
(x + y)^1 &= & & & & & x & + & y \\
(x + y)^2 &= & & & x^2 & + & 2xy & + & y^2 \\
(x + y)^3 &= & & x^3 & + & 3x^2y & + & 3xy^2 & + & y^3 \\
(x + y)^4 &= & x^4 & + & 4x^3y & + & 6x^2y^2 & + & 4xy^3 & + & y^4 \\
(x + y)^5 &= x^5 & + & 5x^4y & + & 10x^3y^2 & + & 10x^2y^3 & + & 5xy^4 & + & y^5
\end{aligned}
$$

Note The polynomials to the right have been found by expanding the binomials on the left—we just haven't shown the work.

There are a number of similarities to notice among the polynomials on the right. Here is a list:

1. In each polynomial, the sequence of exponents on the variable x decreases to zero from the exponent on the binomial at the left. (The exponent 0 is not shown, since $x^0 = 1$.)
2. In each polynomial, the exponents on the variable y increase from 0 to the exponent on the binomial at the left. (Since $y^0 = 1$, it is not shown in the first term.)
3. The sum of the exponents on the variables in any single term is equal to the exponent on the binomial at the left.

The pattern in the coefficients of the polynomials on the right can best be seen by writing the right side again without the variables. It looks like this:

$$
\begin{array}{ccccccccc}
 & & & & 1 & & 1 \\
 & & & 1 & & 2 & & 1 \\
 & & 1 & & 3 & & 3 & & 1 \\
 & 1 & & 4 & & 6 & & 4 & & 1 \\
1 & & 5 & & 10 & & 10 & & 5 & & 1
\end{array}
$$

This triangle-shaped array of coefficients is called **Pascal's triangle.** Each entry in the triangular array is obtained by adding the two numbers above it. Each row begins and ends with the number 1. If we were to continue Pascal's triangle, the next two rows would be

$$
\begin{array}{ccccccccccccc}
 & 1 & & 6 & & 15 & & 20 & & 15 & & 6 & & 1 \\
1 & & 7 & & 21 & & 35 & & 35 & & 21 & & 7 & & 1
\end{array}
$$

The coefficients for the terms in the expansion of $(x + y)^n$ are given in the nth row of Pascal's triangle.

There is an alternate method of finding these coefficients that does not involve Pascal's triangle. The alternate method involves **factorial notation.**

DEFINITION The expression *n*! is read "*n* factorial" and is the product of all the consecutive integers from *n* down to 1. For example,

$$1! = 1$$
$$2! = 2 \cdot 1 = 2$$
$$3! = 3 \cdot 2 \cdot 1 = 6$$
$$4! = 4 \cdot 3 \cdot 2 \cdot 1 = 24$$
$$5! = 5 \cdot 4 \cdot 3 \cdot 2 \cdot 1 = 120$$

The expression 0! is defined to be 1. We use factorial notation to define **binomial coefficients** as follows.

DEFINITION The expression $\binom{n}{r}$ is called a **binomial coefficient** and is defined by

$$\binom{n}{r} = \frac{n!}{r!(n-r)!}$$

▶ **EXAMPLE 1** Calculate the following binomial coefficients:

$$\binom{7}{5}, \binom{6}{2}, \binom{3}{0}$$

Solution We simply apply the definition for binomial coefficients:

$$\binom{7}{5} = \frac{7!}{5!(7-5)!}$$

$$= \frac{7!}{5! \cdot 2!}$$

$$= \frac{7 \cdot 6 \cdot 5 \cdot 4 \cdot 3 \cdot 2 \cdot 1}{(5 \cdot 4 \cdot 3 \cdot 2 \cdot 1)(2 \cdot 1)}$$

$$= \frac{42}{2}$$

$$= 21$$

$$\binom{6}{2} = \frac{6!}{2!(6-2)!}$$

$$= \frac{6!}{2! \cdot 4!}$$

$$= \frac{6 \cdot 5 \cdot \cancel{4} \cdot \cancel{3} \cdot \cancel{2} \cdot \cancel{1}}{(2 \cdot 1)(\cancel{4} \cdot \cancel{3} \cdot \cancel{2} \cdot \cancel{1})}$$

$$= \frac{30}{2}$$

$$= 15$$

$$\binom{3}{0} = \frac{3!}{0!(3-0)!}$$

$$= \frac{3!}{0! \cdot 3!}$$

$$= \frac{\cancel{3} \cdot \cancel{2} \cdot \cancel{1}}{(1)(\cancel{3} \cdot \cancel{2} \cdot \cancel{1})}$$

$$= 1$$

◀

If we were to calculate all the binomial coefficients in the following array, we would find they match exactly with the numbers in Pascal's triangle. That is why they are called binomial coefficients—because they are the coefficients of the expansion of $(x + y)^n$:

$$
\begin{array}{ccccccccccc}
& & & & \binom{1}{0} & & \binom{1}{1} & & & & \\
& & & \binom{2}{0} & & \binom{2}{1} & & \binom{2}{2} & & & \\
& & \binom{3}{0} & & \binom{3}{1} & & \binom{3}{2} & & \binom{3}{3} & & \\
& \binom{4}{0} & & \binom{4}{1} & & \binom{4}{2} & & \binom{4}{3} & & \binom{4}{4} & \\
\binom{5}{0} & & \binom{5}{1} & & \binom{5}{2} & & \binom{5}{3} & & \binom{5}{4} & & \binom{5}{5}
\end{array}
$$

Using the new notation to represent the entries in Pascal's triangle, we can summarize everything we have noticed about the expansion of binomial powers of the form $(x + y)^n$.

3. For the sequence 3, 7, 11, 15, . . . , $a_1 = 3$ and $d = 4$. The general term is

$$a_n = 3 + (n - 1)4$$
$$= 4n - 1$$

Using this formula to find the tenth term, we have

$$a_{10} = 4(10) - 1 = 39$$

The sum of the first 10 terms is

$$S_{10} = \frac{10}{2}(3 + 39) = 210$$

Arithmetic Sequences [11.3]

An *arithmetic sequence* is a sequence in which each term comes from the preceding term by adding a constant amount each time. If the first term of an arithmetic sequence is a_1 and the amount we add each time (called the *common difference*) is d, then the nth term of the progression is given by

$$a_n = a_1 + (n - 1)d$$

The sum of the first n terms of an arithmetic sequence is

$$S_n = \frac{n}{2}(a_1 + a_n)$$

S_n is called the *nth partial sum.*

4. For the geometric progression 3, 6, 12, 24, . . . , $a_1 = 3$ and $r = 2$. The general term is

$$a_n = 3 \cdot 2^{n-1}$$

The sum of the first 10 terms is

$$S_{10} = \frac{3(2^{10} - 1)}{2 - 1} = 3{,}069$$

Geometric Sequences [11.4]

A *geometric sequence* is a sequence of numbers in which each term comes from the previous term by multiplying by a constant amount each time. The constant by which we multiply each term to get the next term is called the *common ratio*. If the first term of a geometric sequence is a_1 and the common ratio is r, then the formula that gives the general term, a_n, is

$$a_n = a_1 r^{n-1}$$

The sum of the first n terms of a geometric sequence is given by the formula

$$S_n = \frac{a_1(r^n - 1)}{r - 1}$$

5. The sum of the series

$$\frac{1}{3} + \frac{1}{6} + \frac{1}{12} + \cdots$$

is

$$S = \frac{\frac{1}{3}}{1 - \frac{1}{2}} = \frac{\frac{1}{3}}{\frac{1}{2}} = \frac{2}{3}$$

The Sum of an Infinite Geometric Series [11.4]

If a geometric sequence has first term a_1 and common ratio r such that $|r| < 1$, then the following is called an *infinite geometric series:*

$$S = \sum_{i=0}^{\infty} a_1 r^i = a_1 + a_1 r + a_1 r^2 + a_1 r^3 + \cdots$$

Its sum is given by the formula

$$S = \frac{a_1}{1 - r}$$

Factorials [11.5]

The notation $n!$ is called *n factorial* and is defined to be the product of each consecutive integer from n down to 1. That is,

$$0! = 1 \qquad \text{(By definition)}$$
$$1! = 1$$
$$2! = 2 \cdot 1$$
$$3! = 3 \cdot 2 \cdot 1$$
$$4! = 4 \cdot 3 \cdot 2 \cdot 1$$

and so on

6. $\dbinom{7}{3} = \dfrac{7!}{3!(7-3)!}$

$= \dfrac{7!}{3! \cdot 4!}$

$= \dfrac{7 \cdot 6 \cdot 5 \cdot \cancel{4 \cdot 3 \cdot 2 \cdot 1}}{(3 \cdot 2 \cdot 1)(\cancel{4 \cdot 3 \cdot 2 \cdot 1})}$

$= 35$

Binomial Coefficients [11.5]

The notation $\dbinom{n}{r}$ is called a *binomial coefficient* and is defined by

$$\binom{n}{r} = \frac{n!}{r!(n-r)!}$$

Binomial coefficients can be found by using the formula above or by *Pascal's triangle,* which is

$$
\begin{array}{ccccccccc}
& & & & 1 & & 1 & & \\
& & & 1 & & 2 & & 1 & \\
& & 1 & & 3 & & 3 & & 1 \\
& 1 & & 4 & & 6 & & 4 & & 1 \\
1 & & 5 & & 10 & & 10 & & 5 & & 1
\end{array}
$$

and so on

7. $(x + 2)^4$

$= x^4 + 4x^3 \cdot 2 + 6x^2 \cdot 2^2$
$\quad + 4x \cdot 2^3 + 2^4$

$= x^4 + 8x^3 + 24x^2 + 32x + 16$

Binomial Expansion [11.5]

If n is a positive integer, then the formula for expanding $(x + y)^n$ is given by

$$(x+y)^n = \binom{n}{0}x^n y^0 + \binom{n}{1}x^{n-1}y^1 + \binom{n}{2}x^{n-2}y^2$$
$$+ \cdots + \binom{n}{n}x^0 y^n$$

CHAPTER 11 REVIEW

Write the first four terms of the sequence with the following general terms. [11.1]

1. $a_n = 2n + 5$

2. $a_n = 3n - 2$

3. $a_n = n^2 - 1$

4. $a_n = \dfrac{n+3}{n+2}$

5. $a_1 = 4, a_n = 4a_{n-1}, n > 1$

6. $a_1 = \frac{1}{4}$, $a_n = \frac{1}{4}a_{n-1}$, $n > 1$

Determine the general term for each of the following sequences. [11.1]

7. 2, 5, 8, 11, . . .
8. −3, −1, 1, 3, 5, . . .
9. 1, 16, 81, 256, . . .
10. 2, 5, 10, 17, . . .
11. $\frac{1}{2}$, $\frac{1}{4}$, $\frac{1}{8}$, $\frac{1}{16}$, . . .
12. 2, $\frac{3}{4}$, $\frac{4}{9}$, $\frac{5}{16}$, $\frac{6}{25}$, . . .

Expand and simplify each of the following. [11.2]

13. $\displaystyle\sum_{i=1}^{4} (2i + 3)$ **14.** $\displaystyle\sum_{i=1}^{3} (2i^2 - 1)$

15. $\displaystyle\sum_{i=2}^{3} \frac{i^2}{i + 2}$ **16.** $\displaystyle\sum_{i=1}^{4} (-2)^{i-1}$

17. $\displaystyle\sum_{i=3}^{5} (4i + i^2)$ **18.** $\displaystyle\sum_{i=4}^{6} \frac{i + 2}{i}$

Write each of the following sums with summation notation. [11.2]

19. 3 + 6 + 9 + 12
20. 3 + 7 + 11 + 15
21. 5 + 7 + 9 + 11 + 13
22. 4 + 9 + 16
23. $\frac{1}{3} + \frac{1}{4} + \frac{1}{5} + \frac{1}{6}$
24. $\frac{1}{3} + \frac{2}{9} + \frac{3}{27} + \frac{4}{81} + \frac{5}{243}$
25. $(x - 2) + (x - 4) + (x - 6)$

26. $\dfrac{x}{x + 1} + \dfrac{x}{x + 2} + \dfrac{x}{x + 3} + \dfrac{x}{x + 4}$

Determine which of the following sequences are arithmetic progressions, geometric progressions, or neither. [11.3, 11.4]

27. 1, −3, 9, −27, . . .
28. 7, 9, 11, 13, . . .
29. 5, 11, 17, 23, . . . **30.** $\frac{1}{2}$, $\frac{1}{3}$, $\frac{1}{4}$, $\frac{1}{5}$, . . .
31. 4, 8, 16, 32, . . . **32.** $\frac{1}{2}$, $\frac{1}{4}$, $\frac{1}{8}$, $\frac{1}{16}$, . . .
33. 12, 9, 6, 3, . . . **34.** 2, 5, 9, 14, . . .

Each of the following problems refers to arithmetic progressions. [11.3]

35. If $a_1 = 2$ and $d = 3$, find a_n and a_{20}.
36. If $a_1 = 5$ and $d = -3$, find a_n and a_{16}.
37. If $a_1 = -2$ and $d = 4$, find a_{10} and S_{10}.
38. If $a_1 = 3$ and $d = 5$, find a_{16} and S_{16}.
39. If $a_5 = 21$ and $a_8 = 33$, find the first term a_1, the common difference d, and then find a_{10}.
40. If $a_3 = 14$ and $a_7 = 26$, find the first term a_1, the common difference d, and then find a_9 and S_9.
41. If $a_4 = -10$ and $a_8 = -18$, find the first term a_1, the common difference d, and then find a_{20} and S_{20}.
42. Find the sum of the first 100 terms of the sequence 3, 7, 11, 15, 19, ...
43. Find a_{40} for the sequence 100, 95, 90, 85, 80, ...

Each of the following problems refers to infinite geometric progressions. [11.4]

44. If $a_1 = 3$ and $r = 2$, find a_n and a_{20}.
45. If $a_1 = 5$ and $r = -2$, find a_n and a_{16}.
46. If $a_1 = 4$ and $r = \frac{1}{2}$, find a_n and a_{10}.
47. If $a_1 = -2$ and $r = \frac{1}{3}$, find the sum.
48. If $a_1 = 4$ and $r = \frac{1}{2}$, find the sum.
49. If $a_3 = 12$ and $a_4 = 24$, find the first term a_1, the common ratio r, and then find a_6.
50. Find the tenth term of the sequence 3, $3\sqrt{3}$, 9, $9\sqrt{3}$, ...

Evaluate each of the following. [11.5]

51. $\dbinom{8}{2}$ **52.** $\dbinom{7}{4}$

53. $\dbinom{6}{3}$ **54.** $\dbinom{9}{2}$

55. $\dbinom{10}{8}$ **56.** $\dbinom{100}{3}$

Use the binomial formula to expand each of the following. [11.5]

57. $(x - 2)^4$ **58.** $(2x + 3)^4$

59. $(3x + 2y)^3$ **60.** $(x^2 - 2)^5$

61. $\left(\dfrac{x}{2} + 3\right)^4$ **62.** $\left(\dfrac{x}{3} - \dfrac{y}{2}\right)^3$

Use the binomial formula to write the first two terms in the expansion of the following. [11.5]

67. $(x - 2y)^{16}$ **68.** $(x + 2y)^{32}$
69. $(x - 1)^{50}$ **70.** $(x + y)^{150}$
71. Find the sixth term in $(x - 3)^{10}$.
72. Find the fourth term in $(2x + 1)^9$.

Use the binomial formula to write the first three terms in the expansion of the following. [11.5]

63. $(x + 3y)^{10}$ **64.** $(x - 3y)^9$
65. $(x + y)^{11}$ **66.** $(x - 2y)^{12}$

CHAPTER 11 TEST

Write the first five terms of the sequences with the following general terms. [11.1]

1. $a_n = 3n - 5$
2. $a_1 = 3, a_n = a_{n-1} + 4, n > 1$
3. $a_n = n^2 + 1$ **4.** $a_n = 2n^3$

5. $a_n = \dfrac{n + 1}{n^2}$

6. $a_1 = 4, a_n = -2a_{n-1}, n > 1$

Give the general term for each sequence. [11.1]

7. 6, 10, 14, 18, . . .
8. 1, 2, 4, 8, . . .
9. $\frac{1}{2}, \frac{1}{4}, \frac{1}{8}, \frac{1}{16}, \ldots$
10. $-3, 9, -27, 81, \ldots$
11. Expand and simplify each of the following. [11.2]

(a) $\displaystyle\sum_{i=1}^{5} (5i + 3)$ (b) $\displaystyle\sum_{i=3}^{5} (2^i - 1)$

(c) $\displaystyle\sum_{i=2}^{6} (i^2 + 2i)$

12. Find the first term of an arithmetic progression if $a_5 = 11$ and $a_9 = 19$. [11.3]
13. Find the second term of a geometric progression for which $a_3 = 18$ and $a_5 = 162$. [11.4]

Find the sum of the first 10 terms of the following arithmetic progressions. [11.3]

14. 5, 11, 17, . . .   **15.** 25, 20, 15, . . .
16. Write a formula for the sum of the first 50 terms of the geometric progression 3, 6, 12, ... [11.4]
17. Find the sum of $\frac{1}{2} + \frac{1}{6} + \frac{1}{18} + \frac{1}{54} + \cdots$. [11.4]

Use the binomial formula to expand each of the following. [11.5]

18. $(x - 3)^4$ **19.** $(2x - 1)^5$
20. Find the first 3 terms in the expansion of $(x - 1)^{20}$. [11.5]
21. Find the sixth term in $(2x - 3y)^8$. [11.5]

APPENDIX

SYNTHETIC DIVISION

Synthetic division is a short form of long division with polynomials. We will consider synthetic division only for those cases in which the divisor is of the form $x + k$, where k is a constant.

Let's begin by looking over an example of long division with polynomials as done in Section 4.2:

$$
\begin{array}{r}
3x^2 - 2x + 4 \\
x + 3 \overline{\smash{\big)}\, 3x^3 + 7x^2 - 2x - 4} \\
\underline{3x^3 + 9x^2} \\
-2x^2 - 2x \\
\underline{-2x^2 - 6x} \\
4x - 4 \\
\underline{4x + 12} \\
-16
\end{array}
$$

We can rewrite the problem without showing the variable, since the variable is written in descending powers and similar terms are in alignment. It looks like this:

$$
\begin{array}{r}
3 \quad -2 \quad +4 \\
1 + 3 \overline{\smash{\big)}\, 3 \quad\ 7 \quad -2 \quad -4} \\
\underline{(3) + 9} \\
-2 \ (-2) \\
\underline{(-2) \ -6} \\
4 \ (-4) \\
\underline{(4) \ \ 12} \\
-16
\end{array}
$$

We have used parentheses to enclose the numbers that are repetitions of the numbers above them. We can compress the problem by eliminating all repetitions, except the first one:

$$
\begin{array}{r}
3 \quad -2 \quad\ \ 4 \\
1 + 3 \overline{\smash{\big)}\, 3 \quad\ 7 \quad -2 \quad -4} \\
\underline{9 \quad -6 \quad\ 12} \\
3 \quad -2 \quad\ \ 4 \quad -16
\end{array}
$$

10. $\dfrac{3x^3 - 2x + 1}{x - 5}$

11. $\dfrac{x^4 + 2x^2 + 1}{x + 4}$

12. $\dfrac{x^4 - 3x^2 + 1}{x - 4}$

13. $\dfrac{x^5 - 2x^4 + x^3 - 3x^2 - x + 1}{x - 2}$

14. $\dfrac{2x^5 - 3x^4 + x^3 - x^2 + 2x + 1}{x + 2}$

15. $\dfrac{x^2 + x + 1}{x - 1}$

16. $\dfrac{x^2 + x + 1}{x + 1}$

17. $\dfrac{x^4 - 1}{x + 1}$

18. $\dfrac{x^4 + 1}{x - 1}$

19. $\dfrac{x^3 - 1}{x - 1}$

20. $\dfrac{x^3 - 1}{x + 1}$

ANSWERS TO ODD-NUMBERED PROBLEMS, CHAPTER REVIEWS, AND CHAPTER TESTS

CHAPTER 1

Problem Set 1.1

1. $x + 5$ **3.** $6 - x$ **5.** $2t < y$ **7.** $\dfrac{3x}{2y} > 6$ **9.** $x + y < x - y$ **11.** $3(x - 5) > y$

13. $s - t \neq s + t$ **15.** $2(t + 3) \not> t - 6$ **17.** 36   **19.** 100   **21.** 8   **23.** 16   **25.** 10,000
27. 121   **29.** 19   **31.** 27   **33.** 42   **35.** 50   **37.** 16   **39.** 12   **41.** 18   **43.** 56
45. 1,064   **47.** 64   **49.** 34   **51.** 64   **53.** 33   **55.** 33   **57.** 23   **59.** 41   **61.** 65
63. 39   **65.** 7   **67.** 5   **69.** 10   **71.** 1   **73.** 5,431   **75.** 50   **77.** 6   **79.** 32   **81.** 24
83. 41   **85.** 95   **87.** 138   **89.** 78   **91.** 152   **93.** 48   **95.** $2 \cdot 8 + 5 = 21$
97. $36 \div 2 - 3 = 15$ **99.** $9 - 4 + 2 = 7$ **101.** $\{0, 1, 2, 3, 4, 5, 6\}$ **103.** $\{2, 4\}$ **105.** $\{1, 3, 5\}$
107. $\{0, 1, 2, 3, 4, 5, 6\}$ **109.** $\{0, 2\}$ **111.** $\{0, 6\}$ **113.** $\{0, 1, 2, 3, 4, 5, 6, 7\}$ **115.** $\{1, 2, 4, 5\}$
117. 9   **119.** 13

Problem Set 1.2

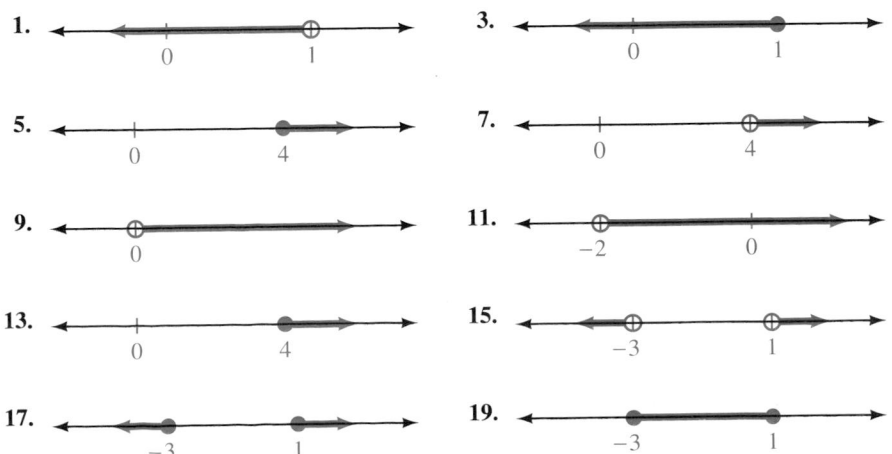

109.
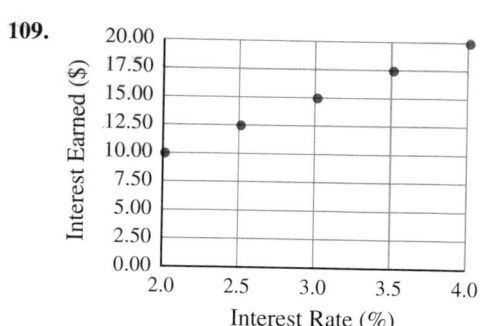

110.
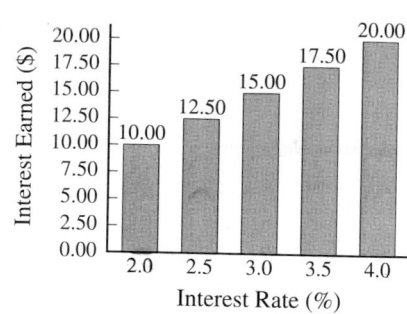

111. Yes, this is a linear relationship   **112.** $22.50, extrapolation

Chapter 1 Test

1. $2(3x + 4y)$ **2.** $2a - 3b < 2a + 3b$ **3.** 57   **4.** 10   **5.** 16   **6.** 0   **7.** $\{2, 4\}$ **8.** $\varnothing$
9. $3, -\frac{1}{3}$ **10.** $-\frac{4}{3}, \frac{3}{4}$ **11.** 3   **12.** -2 **13.** $-5, 0, 1, 4$ **14.** $-5, -4.1, -3.75, -\frac{5}{6}, 0, 1, 1.8, 4$
15. $-\sqrt{2}, \sqrt{3}$ **16.** $3^2 \cdot 5 \cdot 13$

17.

18.

19. Commutative property of addition   **20.** Multiplicative identity property
21. Commutative and associative properties of multiplication
22. Associative and commutative properties of addition   **23.** -19 **24.** 14   **25.** -149 **26.** 213
27. 2   **28.** 0   **29.** $\frac{59}{72}$ **30.** $-4x$ **31.** $-5x - 8$ **32.** $4y - 10$ **33.** $3x - 17$ **34.** $11a - 10$
35. $-\frac{7}{3}$ **36.** $-\frac{5}{2}$
37. Converse: If Emily does not study, then she goes out at night
   Inverse: If Emily does not go out at night, then she studies
   Contrapositive: If Emily studies, then she does not go out at night
38. If $|x| \neq 3$, then $x \neq -3$ **39.** -320, geometric   **40.** -2, arithmetic   **41.** 17   **42.** $\frac{1}{125}$, geometric

43.
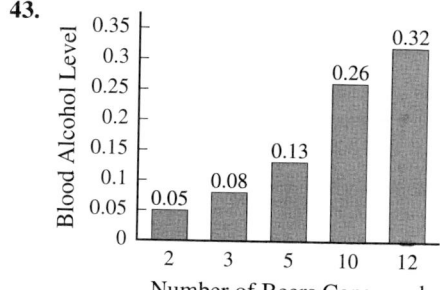

44.
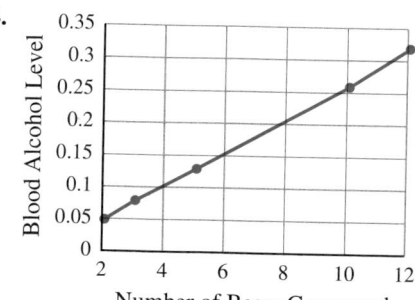

45. 0.10, interpolation   **46.** 0.35, extrapolation

CHAPTER 2

Problem Set 2.1

1. 8   **3.** 5   **5.** 2   **7.** -7 **9.** $-\frac{9}{2}$ **11.** -4 **13.** $-\frac{4}{3}$ **15.** -4 **17.** $\frac{7}{2}$ **19.** $-\frac{11}{5}$
21. 12   **23.** -10 **25.** 7   **27.** -4 **29.** -2 **31.** $\frac{3}{4}$ **33.** 3   **35.** $\frac{4}{5}$ **37.** 2   **39.** 1
41. 2   **43.** 0   **45.** -3 **47.** 4   **49.** 0   **51.** 17   **53.** 2   **55.** -3 **57.** 6   **59.** $-\frac{4}{3}$
61. 3   **63.** $-\frac{3}{2}$ **65.** $\frac{5}{3}$ **67.** 1   **69.** 6,000   **71.** 5,000   **73.** $\frac{3}{2}$ or 1.5   **75.** 1
77. Any method of solution results in a false statement
79. Every attempt at solving the equation results in a true statement   **81.** No solution
83. All real numbers are solutions   **85.** No solution   **87.** Commutative   **89.** Associative
91. Commutative and associative   **93.** Multiplicative identity   **95.** Commutative   **97.** Additive identity
99. 3   **101.** 2   **103.** $-\frac{3}{2}$

Problem Set 2.2

1. -3 **3.** 0   **5.** $\frac{3}{2}$ **7.** 4   **9.** \$5.00   **11.** \$10.00   **13.** 2 centimeters   **15.** 2 inches   **17.** 5 feet

19. $l = \dfrac{A}{w}$ **21.** $t = \dfrac{I}{pr}$ **23.** $T = \dfrac{PV}{nR}$ **25.** $b = y - mx$ **27.** $c = 2s - a - b$ **29.** $r = \dfrac{A - P}{Pt}$

31. $F = \frac{9}{5}C + 32$ **33.** $v = \dfrac{h - 16t^2}{t}$ **35.** $d = \dfrac{A - a}{n - 1}$ **37.** $y = -\frac{2}{3}x + 2$ **39.** $y = \frac{3}{5}x + 3$

41. $y = 3x - 2$ **43.** $y = \frac{1}{3}x + 2$ **45.** $x = \frac{5}{4}y + 2$ **47.** $x = sz + \mu$ **49.** $x = \dfrac{5}{a - b}$

51. $P = \dfrac{A}{1 + rt}$ **53.** $x = \dfrac{d - b}{a - c}$ **55.** $y = -\frac{1}{4}x + 2$ **57.** $y = \frac{3}{5}x - 3$ **59.** 20.52   **61.** 500

63. 25%   **65.** 925   **67.** 4, 7, 10, 13, 16   **69.** 4, 7, 12, 19, 28   **71.** $\frac{1}{4}, \frac{2}{5}, \frac{1}{2}, \frac{4}{7}, \frac{5}{8}$
73. $\frac{1}{1}, \frac{1}{4}, \frac{1}{9}, \frac{1}{16}, \frac{1}{25} \rightarrow 1, \frac{1}{4}, \frac{1}{9}, \frac{1}{16}, \frac{1}{25}$ **75.** 2, 4, 8, 16, 32   **77.** $2, \frac{3}{2}, \frac{4}{3}, \frac{5}{4}, \frac{6}{5}$ **79.** $-\frac{1}{49}$ **81.** $-\frac{1}{17}$
83. $-1, \frac{1}{8}, -\frac{1}{27}, \frac{1}{64}$ **85.** $2(x + 3)$ **87.** $2(x + 3) = 16$ **89.** $5(x - 3)$

91. $3x + 2 = x - 4$ **93.** $x = -\dfrac{a}{b}y + a$ **95.** $a = \dfrac{bc}{b - c}$ **97.** $R = \dfrac{abc}{bc + ac + ab}$ **99.** \$673.68

101. \$674.92   **103.** \$760.05

Problem Set 2.3

Along with the answers to the odd-numbered problems in this problem set, we are including most of the equations used to solve each problem. Be sure that you try the problems on your own before looking here to see what the correct equations are.
1. $3(x + 4) = 3$; -3 **3.** $2(2x + 1) = 3(x - 5)$; -17 **5.** $5x + 2 = 3x + 8$; 3
7. $x + (x + 2) = 18$; 8 and 10   **9.** $x + (x + 1) = 3x - 1$; 2 and 3   **11.** $2x + (x + 1) = 7$; 2
13. $2x - (x + 2) = 5$; 7 and 9   **15.** The width is x, the length is $2x$; $2x + 4x = 60$; 10 feet by 20 feet
17. The length of a side is x; $4x = 28$; 7 feet
19. The shortest side is x, the medium side is $x + 3$, the longest side is $2x$; $x + (x + 3) + 2x = 23$; 5 inches
21. The width is x, the length is $2x - 3$; $2(2x - 3) + 2x = 18$; 4 meters   **23.** \$92.00   **25.** \$9,339.00

43.

Width	L = 6 − W	Length	Area
0	L = 6 − 0	6	0
0.5	L = 6 − 0.5	5.5	2.75
1	L = 6 − 1	5	5
1.5	L = 6 − 1.5	4.5	6.75
2	L = 6 − 2	4	8
2.5	L = 6 − 2.5	3.5	8.75
3	L = 6 − 3	3	9
3.5	L = 6 − 3.5	2.5	8.75
4	L = 6 − 4	2	8
4.5	L = 6 − 4.5	1.5	6.75
5	L = 6 − 5	1	5
5.5	L = 6 − 5.5	0.5	2.75
6	L = 6 − 6	0	0

9 square inches

45.

Time in Months	Amount in Account
0	$100.00
1	$100.50
2	$101.00
3	$101.51
4	$102.02
5	$102.53
6	$103.04
7	$103.55
8	$104.07
9	$104.59
10	$105.11
11	$105.64
12	$106.17
13	$106.70
14	$107.23
15	$107.77
16	$108.31
17	$108.85
18	$109.39
19	$109.94
20	$110.49
21	$111.04
22	$111.60
23	$112.16
24	$112.72

Problem Set 2.5

1. $x \leq \frac{3}{2}$

3. $x > 4$

5. $x \geq -5$

7. $x < 4$

9. $x \geq -6$

11. $x \geq 4$
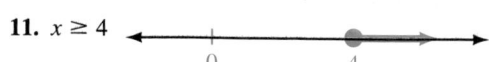

13. $x < -3$

15. $m \geq -1$

17. $x \geq -3$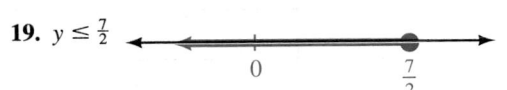

19. $y \leq \frac{7}{2}$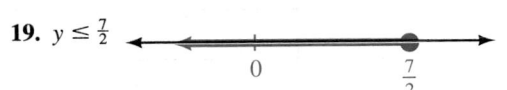

21. $x < 6$

23. $y \geq -52$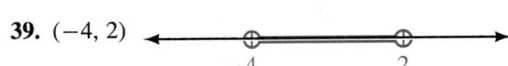

25. $(-\infty, -2]$ **27.** $[1, \infty)$ **29.** $(-\infty, 3)$ **31.** $(-\infty, -1]$ **33.** $[-17, \infty)$ **35.** $(-\infty, -5)$

37. $[3, 7]$

39. $(-4, 2)$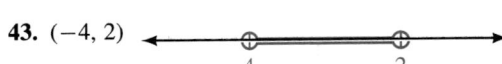

41. $[4, 6]$

43. $(-4, 2)$

45. $(-3, 3)$

47. $(-\infty, -7] \cup [-3, \infty)$

49. $(-\infty, -1] \cup [\frac{3}{5}, \infty)$

51. $(-\infty, -10) \cup (6, \infty)$

53. $p \leq 2$; set the price at \$2.00 or less per pad   **55.** $p > 1.25$; charge more than \$1.25 per pad
57. $y < -\frac{3}{2}x + 3$ **59.** $y \leq \frac{4}{5}x - 4$ **61.** 35° to 45° Celsius; $35° \leq C \leq 45°$
63. $-25°$ to $-10°$ Celsius; $-25° \leq C \leq -10°$ **65.** $-5 < -x < -2$ **67.** No   **69.** -2 **71.** 11

73. The distance between x and 0 on the number line   **75.** $x < \dfrac{c-b}{a}$ **77.** $\dfrac{-c-b}{a} < x < \dfrac{c-b}{a}$

79. $-1.96s + \mu < x < 1.96s + \mu$ **81.** $x > \dfrac{c-b}{a}$

Problem Set 2.6

1. $-4, 4$ **3.** $-2, 2$ **5.** $\varnothing$ **7.** $-1, 1$ **9.** $\varnothing$ **11.** $-6, 6$ **13.** $-3, 7$ **15.** $\frac{17}{3}, \frac{7}{3}$ **17.** 2, 4
19. $-\frac{5}{2}, \frac{5}{6}$ **21.** $-1, 5$ **23.** $\varnothing$ **25.** $20, -4$ **27.** $-4, 8$ **29.** $\frac{2}{3}, -\frac{10}{3}$ **31.** $\varnothing$ **33.** $\frac{3}{2}, -1$
35. $5, 25$ **37.** $-30, 26$ **39.** $-12, 28$ **41.** $-\frac{16}{3}, -6$ **43.** $-3, 4$ **45.** $-3, -2$ **47.** $-5, \frac{3}{5}$
49. $1, \frac{1}{9}$ **51.** $-\frac{1}{2}$ **53.** 0   **55.** $-\frac{1}{2}$ **57.** $-\frac{1}{6}, -\frac{7}{4}$ **59.** All real numbers   **61.** All real numbers

63. $|a - b| = |4 - (-7)| = |11| = 11$
$|b - a| = |-7 - 4| = |-11| = 11$ and $\begin{array}{l}|a - b| = |-5 - (-8)| = |3| = 3\\ |b - a| = |-8 - (-5)| = |-3| = 3\end{array}$

65. $-2, -1, 0, 1, 2, 3, 4$

67. $x \geq 2$

69.

n	a_n
1	2
2	1
3	0
4	1
5	2

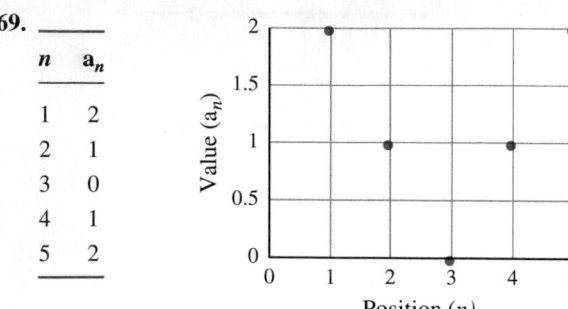

71.

n	a_n
1	4
2	2
3	0
4	2
5	4

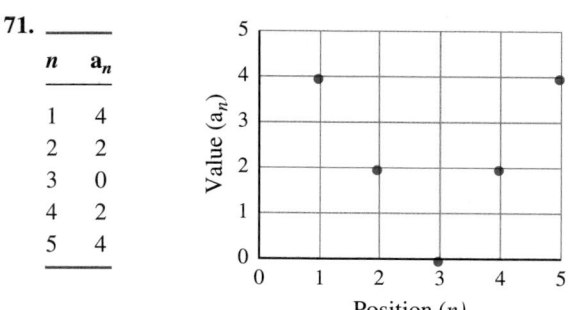

73. **75.**

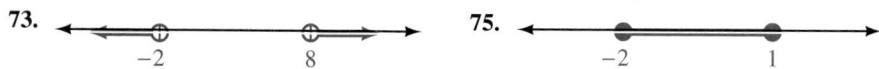

77. $t \leq -\frac{3}{2}$ **79.** $x < -5$ **81.** $\frac{2}{3} < a < \frac{4}{5}$ **83.** $x = a - b$ or $x = a + b$

85. $x = \dfrac{-b - c}{a}$ or $x = \dfrac{-b + c}{a}$ **87.** $x = -\dfrac{a}{b}y - a$ or $x = -\dfrac{a}{b}y + a$

Problem Set 2.7

1. $-3 < x < 3$ **3.** $x \leq -2$ or $x \geq 2$

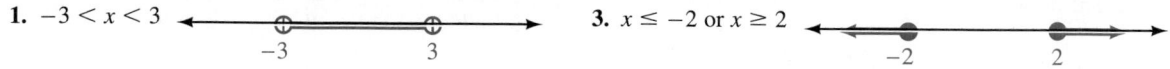

5. $-3 < x < 3$ **7.** $t < -7$ or $t > 7$

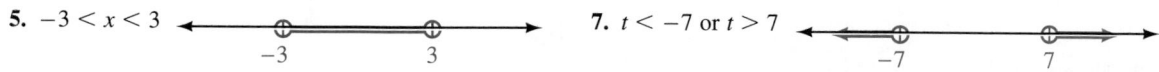

9. $\varnothing$ **11.** All real numbers   **13.** $-4 < x < 10$

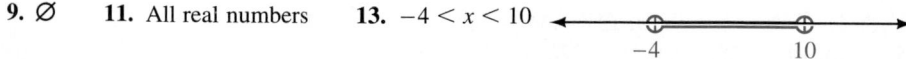

15. $a \leq -9$ or $a \geq -1$ **17.** $\varnothing$

19. $-1 < x < 5$

21. $y \le -5$ or $y \ge -1$

23. $k \le -5$ or $k \ge 2$

25. $-1 < x < 7$

27. $a \le -2$ or $a \ge 1$

29. $-6 < x < \frac{8}{3}$

31. $x < 2$ or $x > 8$

33. $x \le -3$ or $x \ge 12$

35. $x < 2$ or $x > 6$ **37.** $0.99 < x < 1.01$

39. $x \le -\frac{3}{5}$ or $x \ge -\frac{2}{5}$ **41.** $-\frac{1}{6} \le x \le \frac{3}{2}$ **43.** $-0.05 < x < 0.25$ **45.** $|x| \le 4$ **47.** $|x - 5| \le 1$
49. -6 **51.** 66 **53.** -13 **55.** 11 **57.** -4 **59.** $a - b < x < a + b$
61. $x < \dfrac{b - c}{a}$ or $x > \dfrac{b + c}{a}$ **63.** $-a - \dfrac{a}{b}y < x < a - \dfrac{a}{b}y$ **65.** $-ac - \dfrac{a}{b}y < x < ac - \dfrac{a}{b}y$

Problem Set 2.8

1. -1 **3.** -11 **5.** 2 **7.** 4 **9.** 35 **11.** -13 **13.** 1 **15.** -9 **17.** 8 **19.** 19
21. 16 **23.** 0 **25.** $3a^2 - 4a + 1$ **27.** $f(x) = 8.5x$ for $0 \le x \le 40$ **29.** 230
31. $P(x) = 2x + 2(2x + 3) = 6x + 6$, where $x > 0$
33. $A(2) = 3.14(4) = 12.56; A(5) = 3.14(25) = 78.5; A(10) = 3.14(100) = 314$
35. (a) $2.49   (b) $1.53 for a 6-minute call   (c) 5 minutes   **37.** $-3 < x < 3$ **39.** All real numbers
41. $t \le -\frac{4}{3}$ or $t \ge 2$

Chapter 2 Review

1. 10 **2.** -1 **3.** 2 **4.** 3 **5.** 2 **6.** 8 **7.** -3 **8.** 10 **9.** -3 **10.** $-\frac{1}{2}$ **11.** 0
12. 0 **13.** $\frac{2}{3}$ **14.** -1 **15.** $\frac{10}{13}$ **16.** $\frac{14}{13}$ **17.** $\frac{5}{2}$ **18.** 1 **19.** 1 **20.** $\frac{3}{2}$ **21.** $-\frac{5}{11}$
22. $\frac{21}{20}$ **23.** $-\frac{2}{9}$ **24.** $\frac{13}{4}$ **25.** 400 **26.** 600 **27.** $h = 17$ **28.** $b = 40$ **29.** $t = 20$
30. $t = 10$ **31.** $n = 5$ **32.** $n = 4$ **33.** $p = \dfrac{I}{rt}$ **34.** $t = \dfrac{I}{pr}$ **35.** $x = \dfrac{y - b}{m}$ **36.** $m = \dfrac{y - b}{x}$

37. $y = \frac{4}{3}x - 4$ **38.** $x = \frac{3}{4}y + 3$ **39.** $v = \dfrac{d - 16t^2}{t}$ **40.** $v = \dfrac{d + 16t^2}{t}$ **41.** $F = \frac{9}{5}C + 32$

42. $C = \frac{5}{9}(F - 32)$ **43.** 10   **44.** -11 **45.** $-2, -1$ **46.** 8, 10   **47.** 4 feet by 12 feet
48. 5 inches by 11 inches   **49.** 3 meters, 4 meters, 5 meters   **50.** 6 yards, 8 yards, 10 yards
51. $17°, 85°, 78°$ **52.** $80°, 80°, 20°$ **53.** 4, 12   **54.** 5, 20   **55.** 15 dimes, 8 nickels
56. 20 dimes, 3 quarters   **57.** \$400 at 8%, \$500 at 7%   **58.** \$600 at 7%, \$800 at 6%   **59.** $(-\infty, \frac{1}{2})$
60. $(-\infty, \frac{1}{3})$ **61.** $(-\infty, 8]$ **62.** $(-\infty, 13]$ **63.** $(-\infty, 12]$ **64.** $(-\infty, 25]$ **65.** $(-1, \infty)$
66. $(-1, \infty)$ **67.** $[2, 6]$ **68.** $[-3, 2]$ **69.** $[0, 1]$ **70.** $(1, 3)$ **71.** $(-\infty, -\frac{3}{2}] \cup [3, \infty)$
72. $(-\infty, \frac{4}{3}] \cup [2, \infty)$ **73.** $(-\infty, 1) \cup [4, \infty)$ **74.** $(-\infty, -15) \cup [9, \infty)$ **75.** $-2, 2$ **76.** $-3, 3$
77. $-4, 4$ **78.** $-5, 5$ **79.** 2, 4   **80.** $-1, 5$ **81.** $-1, 4$ **82.** $-\frac{5}{3}, 3$ **83.** $-\frac{3}{2}, 3$
84. $-\frac{5}{3}, \frac{7}{3}$ **85.** 5, 9   **86.** 3, 13   **87.** $-1, 1$ **88.** $-\frac{3}{5}, 5$ **89.** 0   **90.** 0

91.

92.

93.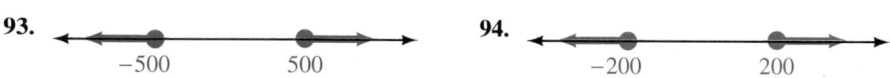

94.

95. $x = 0$

96.

97.

98.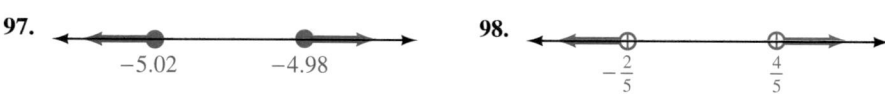

99.

100.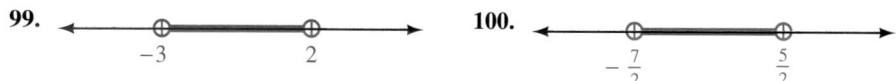

101. $\varnothing$ **102.** $\varnothing$ **103.** $\varnothing$ **104.** All real numbers   **105.** $\varnothing$ **106.** $\varnothing$
107. All real numbers except 0   **108.** All real numbers   **109.** $\varnothing$ **110.** $\varnothing$ **111.** All real numbers
112. All real numbers   **113.** -5 **114.** 1   **115.** 19   **116.** 4   **117.** -8 **118.** 3   **119.** -4
120. $\frac{3}{2}$

Chapter 2 Test

1. 12   **2.** $-\frac{4}{3}$ **3.** 28   **4.** -3 **5.** $-\frac{7}{4}$ **6.** 2   **7.** $w = \dfrac{P - 2l}{2}$ **8.** $B = \dfrac{2A}{h} - b$ **9.** 8, 10

10. 6 inches, 12 inches   **11.** $47°, 47°, 86°$ **12.** 5 and 14   **13.** 6 nickels, 8 dimes
14. \$2,000 at 10%, \$6,000 at 8%

15. $t \geq -6$ $[-6, \infty)$

16. $x < 4$ $(-\infty, 4)$

17. $x < 6$ $(-\infty, 6)$

18. $y \geq -52$

$[-52, \infty)$

19. 6, 2   **20.** 3, -15 **21.** $\varnothing$ **22.** $-5, 1$

23. $x < -1$ or $x > \frac{4}{3}$

24. $-\frac{2}{3} \leq x \leq 4$

25. All real numbers   **26.** $\varnothing$

27. 0   **28.** 0   **29.** 4   **30.** 2

CHAPTER 3

Problem Set 3.1

1. 16   **3.** -16 **5.** -0.027 **7.** 32   **9.** $\frac{1}{8}$ **11.** $\frac{25}{36}$ **13.** x^9 **15.** 64   **17.** $-\frac{8}{27}x^6$ **19.** $-6a^6$
21. $-36x^{11}$ **23.** $\frac{1}{4}n^{34}$ **25.** $\frac{1}{9}$ **27.** $-\frac{1}{32}$ **29.** $\frac{1}{9}$ **31.** $\frac{16}{9}$ **33.** 17   **35.** -4 **37.** x^3

39. $\dfrac{a^6}{b^{15}}$ **41.** $\dfrac{3}{4x^5}$ **43.** $\dfrac{8}{125y^{18}}$ **45.** $\frac{1}{5}$ **47.** $4a$ **49.** $\dfrac{14y^2}{x^4}$ **51.** $15x^9y^3z^2$ **53.** $\dfrac{24a^{12}c^6}{b^3}$

55. $\dfrac{8x^{22}}{81y^{23}}$ **57.** $\dfrac{r^{26}}{9s^{16}}$ **59.** 3.78×10^5 **61.** 4.9×10^3 **63.** 3.7×10^{-4} **65.** 4.95×10^{-3}

67. 5.62×10^{-1} **69.** 5,340   **71.** 7,800,000   **73.** 0.00344   **75.** 0.49   **77.** 8   **79.** 25
81. $(2^2)^3 = 4^3 = 64, 2^{2^3} = 2^8 = 256$ **83.** $g = 32$ **85.** $(2 + 4)^{-1} = 6^{-1} = \frac{1}{6}, 2^{-1} + 4^{-1} = \frac{1}{2} + \frac{1}{4} = \frac{3}{4}$

87. 2.37×10^6 **89.** 6.3×10^8 **91.** 22   **93.** 14   **95.** 14   **97.** -1 **99.** 5   **101.** $\dfrac{1}{x^3}$

103. y^3 **105.** 36

Problem Set 3.2

1. $\dfrac{x^6}{y^4}$ **3.** $\dfrac{4b^2}{a^4}$ **5.** $\frac{1}{9}$ **7.** 8   **9.** $\dfrac{1}{x^{10}}$ **11.** a^{10} **13.** $\dfrac{1}{t^6}$ **15.** x^{12} **17.** a^5 **19.** $\dfrac{1}{a^5}$ **21.** x^{18}

23. x^8 **25.** $\dfrac{1}{a^8}$ **27.** $\dfrac{1}{m}$ **29.** $\dfrac{1}{x^{22}}$ **31.** $3x^2y^4$ **33.** $\dfrac{a^3b^7}{4}$ **35.** $\dfrac{3y^2}{x^4z^3}$ **37.** $\dfrac{4s^3t^2}{r^2}$ **39.** $32x^9y^{11}$

41. $\dfrac{y^{38}}{x^{16}}$ **43.** $\dfrac{32a^{11}}{b^2c}$ **45.** $\dfrac{16y^{16}}{x^8}$ **47.** x^4y^6 **49.** $\dfrac{4x^{18}}{y^{10}}$ **51.** $\dfrac{b^3}{a^4c^3}$ **53.** $\dfrac{x^{10}}{y^{15}}$ **55.** $8x^6y^7$

57. $\dfrac{189x^9}{y^{17}}$ **59.** 9×10^{10} **61.** 2×10^{-4} **63.** 2.5×10^{-6} **65.** 1.8×10^{-7} **67.** 4.98×10^1

69. 2×10^8 **71.** 2×10^4 **73.** 6   **75.** 9   **77.** -7 **79.** 1.003×10^{19} miles   **81.** $2x + 12$
83. 3   **85.** 0   **87.** $-4x + 38$ **89.** 0   **91.** $\frac{119}{84} = \frac{17}{12}$ **93.** 1.1×10^{20} **95.** 1.75×10^{42} **97.** x^5
99. a **101.** 1   **103.** $4x^{2r+2}y^{-s-5}$

Problem Set 3.3

1. Trinomial, 2, 5   **3.** Binomial, 1, 3   **5.** Trinomial, 2, 8   **7.** Polynomial, 3, 4   **9.** Monomial, 0, $-\frac{3}{4}$
11. Trinomial, 3, 6   **13.** $7x + 1$ **15.** $2x^2 + 7x - 15$ **17.** $12a^2 - 7ab - 10b^2$ **19.** $x^2 - 13x + 3$
21. $\frac{1}{4}x^2 - \frac{7}{12}x - \frac{1}{4}$ **23.** $-y^3 - y^2 - 4y + 7$ **25.** $2x^3 + x^2 - 3x - 17$ **27.** $\frac{1}{14}x^2 + \frac{1}{7}xy + \frac{5}{7}y^2$
29. $-3a^3 + 6a^2b - 5ab^2$ **31.** $-3x$ **33.** $3x^2 - 12xy$ **35.** $17x^5 - 12$ **37.** $14a^2 - 2ab + 8b^2$
39. $-x + 2$ **41.** $10x - 5$ **43.** $9x - 35$ **45.** $9y - 4x$ **47.** $9a + 2$ **49.** -2 **51.** 208
53. -15 **55.** $(3 + 4)^2 = 49$; $3^2 + 16 = 25$; $3^2 + 8 \cdot 3 + 16 = 49$ **57.** 240 feet for both
59. $P(x) = -300 + 40x - 0.5x^2$; $P(60) = \$300$ **61.** $P(x) = -800 + 3.5x - 0.002x^2$; $P(1,000) = \$700$
63. $20x^5$ **65.** $2a^3b^3$ **67.** $6x^3$ **69.** 2   **71.** 18   **73.** -2

Problem Set 3.4

1. $12x^3 - 10x^2 + 8x$ **3.** $-3a^5 + 18a^4 - 21a^2$ **5.** $2a^5b - 2a^3b^2 + 2a^2b^4$
7. $-28x^4y^3 + 12x^3y^4 - 24x^2y^5$ **9.** $3r^6s^2 - 6r^5s^3 + 9r^4s^4 + 3r^3s^5$
11. $x^2 - 2x - 15$ **13.** $6x^4 - 19x^2 + 15$ **15.** $x^3 + 9x^2 + 23x + 15$ **17.** $6a^4 + a^3 - 12a^2 + 5a$
19. $a^3 - b^3$ **21.** $8x^3 + y^3$ **23.** $2a^3 - a^2b - ab^2 - 3b^3$ **25.** $18x^3 - 15x^2y - 16y^3$
27. $6x^4 - 44x^3 + 70x^2$ **29.** $6x^3 - 11x^2 - 14x + 24$ **31.** $x^2 + x - 6$ **33.** $x^4 + x^2 - 6$
35. $6a^2 + 13a + 6$ **37.** $20 - 2t - 6t^2$ **39.** $x^6 - 2x^3 - 15$ **41.** $20a^2 + 9a + 1$
43. $20x^2 - 9xy - 18y^2$ **45.** $18t^2 - \frac{2}{9}$ **47.** $b^2 - a^2b - 12a^4$ **49.** $25x^2 + 20xy + 4y^2$
51. $25 - 30t^3 + 9t^6$ **53.** $4a^2 - 9b^2$ **55.** $9r^4 - 49s^2$ **57.** $\frac{1}{9}x^2 - \frac{4}{25}$ **59.** $x^3 - 6x^2 + 12x - 8$
61. $8x^3 - 12x^2 + 6x - 1$ **63.** $x^3 - \frac{3}{2}x^2 + \frac{3}{4}x - \frac{1}{8}$ **65.** $3x^3 - 18x^2 + 33x - 18$
67. $a^2b^2 + b^2 + 8a^2 + 8$ **69.** $3xy^2 + 4x - 6y^2 - 8$ **71.** $3x^2 + 12x + 14$ **73.** $24x$
75. $-6x^2 - 2$ **77.** $(x + y)^2 + (x + y) - 20 = x^2 + 2xy + y^2 + x + y - 20$
79. $(x + y)^2 + 2z(x + y) + z^2 = x^2 + 2xy + y^2 + 2xz + 2yz + z^2$ **81.** $\frac{3}{8}$
83. $2^4 - 3^4 = -65$; $(2 - 3)^4 = 1$; $(2^2 + 3^2)(2 + 3)(2 - 3) = -65$
85. $R(p) = 230p - 20p^2$; $R(6.5) = \$650$ **87.** $R(p) = 350p - 10p^2$; $R(28.5) = \$1,852.50$
89. $S(x) = 2x(x + 1) + 2x(x + 2) + 2(x + 1)(x + 2) = 6x^2 + 12x + 4$
91. $A = 100 + 400r + 600r^2 + 400r^3 + 100r^4$ **93.** $4, -\frac{2}{3}$ **95.** $1, -2$ **97.** $\varnothing$ **99.** $\frac{7}{2}$
101. $x^{2n} - 5x^n + 6$ **103.** $x^{4n} - 9$ **105.** $10x^{2n} + 13x^n - 3$ **107.** $x^{2n} + 10x^n + 25$
109. $x^{3n} + 3x^{2n} + 3x^n + 1$

Problem Set 3.5

1. $5x^2(2x - 3)$ **3.** $9y^3(y^3 + 2)$ **5.** $3ab(3a - 2b)$ **7.** $7xy^2(3y^2 + x)$ **9.** $3(a^2 - 7a + 10)$
11. $4x(x^2 - 4x - 5)$ **13.** $10x^2y^2(x^2 + 2xy - 3y^2)$ **15.** $xy(-x + y - xy)$ **17.** $2xy^2z(2x^2 - 4xz + 3z^2)$
19. $5abc(4abc - 6b + 5ac)$ **21.** $(a - 2b)(5x - 3y)$ **23.** $3(x + y)^2(x^2 - 2y^2)$ **25.** $(x + 5)(2x^2 + 7x + 6)$
27. $(x + 1)(3y + 2a)$ **29.** $(xy + 1)(x + 3)$ **31.** $(x - 2)(3y^2 + 4)$ **33.** $(x - a)(x - b)$
35. $(b + 5)(a - 1)$ **37.** $(b^2 + 1)(a^4 - 5)$ **39.** $(x + 3)(x^2 - 4)$ **41.** $(x + 2)(x^2 - 25)$
43. $(2x + 3)(x^2 - 4)$ **45.** $(x + 3)(4x^2 - 9)$ **47.** 6
49. $P(1 + r) + P(1 + r)r = (1 + r)(P + Pr) = (1 + r)P(1 + r) = P(1 + r)^2$ **51.** $p = 11.5 - 0.05x$; $\$5.25$
53. $p = 35 - 0.1x$; $\$28.50$ **55.** $x^2 + 5x + 6$ **57.** $6y^2 + y - 35$ **59.** $20 - 19a + 3a^2$
61. $x > -6$ **63.** $y \leq 7$ **65.** $t > -2$ **67.** $-1 < t < 2$ **69.** $x^n(x^2 + x + 1)$ **71.** $x^{n+3}(x^2 + x + 1)$

Problem Set 3.6

1. $(x + 3)(x + 4)$ **3.** $(x + 3)(x - 4)$ **5.** $(y + 3)(y - 2)$ **7.** $(2 - x)(8 + x)$ **9.** $(6 + x)(2 + x)$
11. $3(a - 2)(a - 5)$ **13.** $4x(x + 1)(x - 5)$ **15.** $(x + 2y)(x + y)$ **17.** $(a + 6b)(a - 3b)$
19. $(x - 8a)(x + 6a)$ **21.** $(x - 6b)^2$ **23.** $3(x + y)(x - 3y)$ **25.** $2a^3(a^2 + 2ab + 2b^2)$
27. $10x^2y^2(x + 3y)(x - y)$ **29.** $(2x - 3)(x + 5)$ **31.** $(2x - 5)(x + 3)$ **33.** $(2x - 3)(x - 5)$
35. $(2x - 5)(x - 3)$ **37.** Prime   **39.** $(2 + 3a)(1 + 2a)$ **41.** $15(4y + 3)(y - 1)$

43. $x^2(3x - 2)(2x + 1)$ **45.** $10r(2r - 3)^2$ **47.** $(4x + y)(x - 3y)$ **49.** $(5x + 6a)(2x - 3a)$
51. $(6a - 7b)(3a + 4b)$ **53.** $200(1 + 2t)(3 - 2t)$ **55.** $y^2(3y - 2)(3y + 5)$ **57.** $2a^2(4 - 3a)(3 + 2a)$
59. $2x^2y^2(4x + 3y)(x - y)$ **61.** $100(3x^2 + 1)(x^2 + 3)$ **63.** $(4a^2 + 5)(5a^2 + 3)$
65. $3(3 + 4r^2)(1 - r^2)$ **67.** $(x + 5)(2x + 3)(x + 2)$ **69.** $(2x + 3)(x + 2)(x + 5)$ **71.** $9x^2 - 25y^2$
73. $a + 250$ **75.** $y = 2(2x - 1)(x + 5)$; $y = 0$ when $x = \frac{1}{2}$ or $x = -5$, $y = 42$ when $x = 2$
77. $h(t) = 16(6 - t)(1 + t)$; $h(6) = 0$ feet; $h(3) = 192$ feet   **79.** $4x^2 - 9$ **81.** $4x^2 - 12x + 9$

83. $8x^3 - 27$ **85.** $x \le -9$ or $x \ge -1$

87. All real numbers   **89.** $-6 < y < \frac{8}{3}$

91. $(2x^3 + 5y^2)(4x^3 + 3y^2)$ **93.** $(3x - 5)(x + 100)$ **95.** $(\frac{1}{4}x + 1)(\frac{1}{2}x + 2)$ **97.** $(2x + 0.5)(x + 0.5)$
99. $(4x^n + 5)(3x^n + 4)$ **101.** $(20x^n + 1)(x^n - 6)$

Problem Set 3.7

1. $(x - 3)^2$ **3.** $(a - 6)^2$ **5.** $(5 - t)^2$ **7.** $(\frac{1}{3}x + 3)^2$ **9.** $(2y^2 - 3)^2$ **11.** $(4a + 5b)^2$
13. $(\frac{1}{5} + \frac{1}{4}t^2)^2$ **15.** $4(2x - 3)^2$ **17.** $3a(5a + 1)^2$ **19.** $(x + 2 + 3)^2 = (x + 5)^2$
21. $(x + 3)(x - 3)$ **23.** $(7x + 8y)(7x - 8y)$ **25.** $(2a + \frac{1}{2})(2a - \frac{1}{2})$ **27.** $(x + \frac{3}{5})(x - \frac{3}{5})$
29. $(3x + 4y)(3x - 4y)$ **31.** $10(5 + t)(5 - t)$ **33.** $(x^2 + 9)(x + 3)(x - 3)$ **35.** $(3x^3 + 1)(3x^3 - 1)$
37. $(4a^2 + 9)(2a + 3)(2a - 3)$ **39.** $\left(\dfrac{1}{9} + \dfrac{y^2}{4}\right)\left(\dfrac{1}{3} + \dfrac{y}{2}\right)\left(\dfrac{1}{3} - \dfrac{y}{2}\right)$
41. $(x - y)(x + y)(x^2 + xy + y^2)(x^2 - xy + y^2)$ **43.** $2a(a - 2)(a + 2)(a^2 + 2a + 4)(a^2 - 2a + 4)$
45. $(x + 1)(x - 5)$ **47.** $y(y + 8)$ **49.** $(x - 5 + y)(x - 5 - y)$ **51.** $(a + 4 + b)(a + 4 - b)$
53. $(x + y + a)(x + y - a)$ **55.** $(x + 3)(x + 2)(x - 2)$ **57.** $(x + 2)(x + 5)(x - 5)$
59. $(2x + 3)(x + 2)(x - 2)$ **61.** $(x + 3)(2x + 3)(2x - 3)$ **63.** $(x - y)(x^2 + xy + y^2)$
65. $(a + 2)(a^2 - 2a + 4)$ **67.** $(3 + x)(9 - 3x + x^2)$ **69.** $(y - 1)(y^2 + y + 1)$
71. $10(r - 5)(r^2 + 5r + 25)$ **73.** $(4 + 3a)(16 - 12a + 9a^2)$ **75.** $(2x - 3y)(4x^2 + 6xy + 9y^2)$
77. $(t + \frac{1}{3})(t^2 - \frac{1}{3}t + \frac{1}{9})$ **79.** $(3x - \frac{1}{3})(9x^2 + x + \frac{1}{9})$ **81.** $(4a + 5b)(16a^2 - 20ab + 25b^2)$
83. 30 and -30 **85.** \$112.36   **87.** 0 **89.** $y = -\frac{3}{2}x + 6$ **91.** $t = \dfrac{A - P}{Pr}$
93. $(x^n - y^n)(x^n + y^n)$ **95.** $(x^n + y^m)^2$ **97.** $(x^n - 2)(x^{2n} + 2x^n + 4)$ **99.** $(x^n - y^n)(x^{2n} + x^n y^n + y^{2n})$

Problem Set 3.8

1. $(x + 9)(x - 9)$ **3.** $(x + 5)(x - 3)$ **5.** $(x + 2)(x + 3)^2$ **7.** $(x^2 + 2)(y^2 + 1)$ **9.** $2ab(a^2 + 3a + 1)$
11. Does not factor: prime   **13.** $3(2a + 5)(2a - 5)$ **15.** $(3x - 2y)^2$ **17.** $(5 - t)^2$ **19.** $4x(x^2 + 4y^2)$
21. $2y(y + 5)^2$ **23.** $a^4(a + 2b)(a^2 - 2ab + 4b^2)$ **25.** $(t + 3 + x)(t + 3 - x)$ **27.** $(x + 5)(x + 3)(x - 3)$
29. $5(a + b)^2$ **31.** Does not factor: prime   **33.** $3(x + 2y)(x + 3y)$ **35.** $(3a + \frac{1}{3})^2$ **37.** $(x - 3)(x - 7)^2$
39. $(x + 8)(x - 8)$ **41.** $(4 + 3x)(2 - 5x)$ **43.** $a^5(7a + 3)(7a - 3)$ **45.** $(r + \frac{1}{5})(r - \frac{1}{5})$
47. Does not factor: prime   **49.** $100(x - 3)(x + 2)$ **51.** $a(5a + 1)(5a + 3)$ **53.** $(3x^2 + 1)(x^2 - 5)$
55. $3a^2b(2a - 1)(4a^2 + 2a + 1)$ **57.** $(4 - r)(16 + 4r + r^2)$ **59.** $5x^2(2x + 3)(2x - 3)$
61. $100(2t + 3)(2t - 3)$ **63.** $2x^3(4x - 5)(2x - 3)$ **65.** $(y + 1)(y^2 - y + 1)(y - 1)(y^2 + y + 1)$
67. $2(5 + a)(5 - a)$ **69.** $3x^2y^2(2x + 3y)^2$ **71.** $(x - 2 + y)(x - 2 - y)$ **73.** $8, 10$
75. (a) $20°, 70°$ (b) $38°, 142°$ **77.** 8 quarters, 13 nickels

Problem Set 3.9

1. $6, -1$ **3.** $0, 2, 3$ **5.** $\frac{1}{3}, -4$ **7.** $\frac{2}{3}, \frac{3}{2}$ **9.** $5, -5$ **11.** $0, -3, 7$ **13.** $-4, \frac{5}{2}$ **15.** $0, \frac{4}{3}$
17. $-\frac{1}{5}, \frac{1}{3}$ **19.** $0, -\frac{4}{3}, \frac{4}{3}$ **21.** $-10, 0$ **23.** $-5, 1$ **25.** $1, 2$ **27.** $-2, 3$ **29.** $-2, \frac{1}{4}$
31. $-3, -2, 2$ **33.** $-2, -5, 5$ **35.** $-\frac{3}{2}, -2, 2$ **37.** $-3, -\frac{3}{2}, \frac{3}{2}$ **39.** $-5, -3$ or $3, 5$
41. $-5, -4$ or $4, 5$ **43.** $5, 13$ **45.** $6, 8, 10$ **47.** 2 feet, 8 feet **49.** 18 inches, 4 inches
51. 0 and 2 seconds **53.** 1 and 2 seconds **55.** 0 and $\frac{3}{2}$ seconds **57.** 2 and 3 seconds **59.** \$4 or \$8
61. \$7 or \$10 **63.** $-2, 2$ **65.** $1, \frac{7}{3}$ **67.** $-1, 11$ **69.** $-0.01 < x < 0.01$ **71.** All real numbers
73. $t \le -\frac{4}{3}$ or $t \ge 2$

Chapter 3 Review

1. x^{10} **2.** x^9 **3.** $25x^6$ **4.** $36x^8$ **5.** $-32x^{18}y^8$ **6.** $-72x^{17}y^9$ **7.** $\frac{1}{8}$ **8.** $\frac{1}{9}$ **9.** $\frac{9}{4}$ **10.** $\frac{27}{8}$

11. $\frac{1}{2}$ **12.** $\frac{2}{9}$ **13.** 3.45×10^7 **14.** 3.57×10^{-3} **15.** 44,500 **16.** 0.000445 **17.** $\frac{1}{a^9}$ **18.** $3x^4$

19. $\frac{x^{12}}{4}$ **20.** $\frac{9}{x^2}$ **21.** x^2 **22.** x^3 **23.** 8×10^{-2} **24.** 1.8×10^3 **25.** 4×10^{-10} **26.** 6×10^{-3}

27. $2x^2 - 5x + 7$ **28.** $\frac{2}{15}x^2 - \frac{2}{3}x + \frac{8}{15}$ **29.** $2x^3 - 2x^2 - 2x - 4$ **30.** $x^4 - x^2 + x - 3$
31. $x^2 - 2x - 3$ **32.** $2x^2 + x - 7$ **33.** $30x + 12$ **34.** $-35x + 120$ **35.** 15 **36.** -21
37. $12x^3 - 6x^2 + 3x$ **38.** $10x^3 - 2x^2 + 6x$ **39.** $2a^4b^3 + 4a^3b^4 + 2a^2b^5$
40. $3a^5b^2 - 9a^4b^3 + 6a^3b^4$ **41.** $18 - 9y + y^2$ **42.** $14 + 5y - y^2$ **43.** $6x^4 + 5x^2 - 4$
44. $6x^4 - 5x^2 - 6$ **45.** $2t^3 - 4t^2 - 6t$ **46.** $3t^3 + 3t^2 - 6t$ **47.** $x^3 + 27$ **48.** $x^3 - 8$
49. $8x^3 - 27$ **50.** $27x^3 - 8$ **51.** $a^4 - 4a^2 + 4$ **52.** $a^6 - 10a^3 + 25$ **53.** $9x^2 + 30x + 25$
54. $4x^2 + 24x + 36$ **55.** $16x^2 - 24xy + 9y^2$ **56.** $49x^2 - 28xy + 4y^2$ **57.** $x^2 - \frac{1}{9}$ **58.** $x^2 - \frac{1}{16}$
59. $4a^2 - b^2$ **60.** $9a^2 - b^2$ **61.** $x^3 - 3x^2 + 3x - 1$ **62.** $x^3 + 6x^2 + 12x + 8$ **63.** $x^{2m} - 4$
64. $x^{2m} + 10x^m + 25$ **65.** $3xy(2x^3 - 3y^3 + 6x^2y^2)$ **66.** $10x^2y^2(4x + 3xy - 2)$
67. $4(x + y)^2(x^2 - 2y^2)$ **68.** $5x^2(2x - 3y)(3x - 5)$ **69.** $(4x^2 + 5)(2 - y)$
70. $(1 - y)(1 + y)(x^3 + 8b^2)$ **71.** $(x - 2)(x - 3)$ **72.** $(x - 6)(x - 1)$ **73.** $2x(x + 5)(x - 3)$
74. $5x(x + 5)(x + 4)$ **75.** $(5a - 4b)(4a - 5b)$ **76.** $(5a + b)(4a + 5b)$ **77.** $x^2(3x + 2)(2x - 5)$
78. $x(3x + 2)(2x + 5)$ **79.** $3y(4x + 5)(2x - 3)$ **80.** $y^2(4x + 3)(2x + 5)$ **81.** $(x^2 + 4)(x + 2)(x - 2)$
82. $(x^2 + 9)(x + 3)(x - 3)$ **83.** $3(a^2 + 3)^2$ **84.** $4(a^2 - 2)^2$ **85.** $(a - 2)(a^2 + 2a + 4)$
86. $(a + 3)(a^2 - 3a + 9)$ **87.** $5x(x + 3y)^2$ **88.** $3y(2x - 3y)^2$ **89.** $3ab(a - 3b)(a + 3b)$
90. $2ab(b - 2a)(b + 2a)$ **91.** $(x - 5 + y)(x - 5 - y)$ **92.** $(x + 3 + y)(x + 3 - y)$
93. $(6 - 5a)(6 + 5a)$ **94.** $(7 - 5a)(7 + 5a)$ **95.** $(x + 3)(x - 3)(x + 4)$ **96.** $(x + 2)(x - 2)(x + 5)$
97. $-3, -2$ **98.** $-5, -1$ **99.** $-\frac{1}{2}, \frac{4}{5}$ **100.** $\frac{1}{2}, \frac{5}{4}$ **101.** $-\frac{5}{3}, \frac{5}{3}$ **102.** $0, -\frac{3}{4}, \frac{3}{4}$ **103.** $0, -2$
104. $0, 3$ **105.** $-3, 6$ **106.** $-5, 4$ **107.** $-4, 3, -3$ **108.** $-5, -2, 2$ **109.** $-10, -8$ or $8, 10$
110. $-11, -9$ or $9, 11$ **111.** $-5, -4$ or $4, 5$ **112.** $-7, -5$ or $5, 7$ **113.** $3, 4, 5$ **114.** $6, 8, 10$

Chapter 3 Test

1. x^8 **2.** $\frac{1}{32}$ **3.** $\frac{16}{9}$ **4.** $32x^{12}y^{11}$ **5.** a^2 **6.** x^6 **7.** $\frac{2a^{12}}{b^{15}}$ **8.** 6.53×10^6 **9.** 8.7×10^{-4}

10. 8.7×10^7 **11.** 3×10^8 **12.** $\frac{3}{4}x^3 - \frac{5}{4}x^2 - 2x - 1$ **13.** $4x + 75$ **14.** $6y^2 + y - 35$
15. $2x^3 + 3x^2 - 26x + 15$ **16.** $64 - 48t^3 + 9t^6$ **17.** $1 - 36y^2$ **18.** $4x^3 - 2x^2 - 30x$
19. $10t^4 - \frac{1}{10}$ **20.** $(x + 4)(x - 3)$ **21.** $2(3x^2 - 1)(2x^2 + 5)$ **22.** $(4a^2 + 9y^2)(2a + 3y)(2a - 3y)$
23. $(7a - b^2)(x^2 - 2y)$ **24.** $(t + \frac{1}{2})(t^2 - \frac{1}{2}t + \frac{1}{4})$ **25.** $4a^3b(a - 8b)(a + 2b)$
26. $(x - 5 + b)(x - 5 - b)$ **27.** $(9 + x^2)(3 + x)(3 - x)$ **28.** $-\frac{1}{3}, 2$ **29.** $0, 5$ **30.** $-5, 2$
31. $-2, -4, 4$ **32.** $-5, -2$ or $2, 5$ **33.** $6, 8, 10$ inches **34.** $R = 25x - 0.2x^2$
35. $P = -100 + 23x - 0.2x^2$ **36.** \$500 **37.** \$300 **38.** \$200

CHAPTER 4

Problem Set 4.1

1. $-\frac{1}{3}$ **3.** $\frac{b^3}{2}$ **5.** $-\frac{3y^3}{2x}$ **7.** $\frac{18c^2}{7a^2}$ **9.** $\frac{x-4}{6}$ **11.** $\frac{4x-3y}{x(x+y)}$ **13.** $(a^2+9)(a+3)$ **15.** $\frac{a-6}{a+6}$

17. $\frac{2y+3}{y+1}$ **19.** $\frac{5x-2}{x+2}$ **21.** $\frac{2-x}{1-x}$ or $\frac{x-2}{x-1}$ **23.** $\frac{x-3}{x+2}$ **25.** $\frac{a^2-ab+b^2}{a-b}$ **27.** $\frac{2(x-1)}{x}$

29. $\frac{2x+3y}{2x+y}$ **31.** $\frac{x+3}{y-4}$ **33.** $\frac{x+b}{x-2b}$ **35.** $x+2$ **37.** $\frac{1}{x-3}$ **39.** $\frac{2x^2-5}{3x-2}$ **41.** -1

43. $-(y+6)$ **45.** $\frac{-(3a+1)}{3a-1}$ or $-\frac{3a+1}{3a-1}$ **47.** 0 **49.** $4x$

51. and **53.** You can divide out only common factors. You cannot divide out common terms.   **55.** 13

57.

Distance above Surface (in feet)	Intensity in Lumens/Square Foot
1	179.0
2	44.7
3	19.9
4	11.2
5	7.2
6	5.0

59.

Distance above Surface (in feet)	Intensity in Lumens/Square Foot
1	62.0
1.5	27.6
2	15.5
2.5	9.9
3	6.9
3.5	5.1

61. 0.1 mile/minute   **63.** 10.8 miles/gallon   **65.** 6.8 feet/second
67. 3,768 inches/minute and 2,826 inches/minute   **69.** $\frac{10}{3}$ ohms   **71.** 2, 2   **73.** undefined; 4

75. 1, 1   **77.** 3, 3   **79.** $3x^2-7x+4$ **81.** 9   **83.** $8x^2$ **85.** $\frac{2(3x-2)}{2x+3}$ **87.** $\frac{2x^n+3}{3x^n+7}$

89. x^n+3 **91.** $x+7$

Problem Set 4.2

1. $2x^2-4x+3$ **3.** $-2x^2-3x+4$ **5.** $2y^2+\frac{5}{2}-\frac{3}{2y^2}$ **7.** $-\frac{5}{2}x+4+\frac{3}{x}$ **9.** $4ab^3+6a^2b$

11. $-xy+2y^2+3xy^2$ **13.** $x+2$ **15.** $a-3$ **17.** $5x+6y$ **19.** x^2+xy+y^2

21. $(y^2+4)(y+2)$ **23.** $(x+2)(x+5)$ **25.** $(2x+3)(2x-3)$ **27.** $x-7+\frac{7}{x+2}$

29. $2x+5+\frac{2}{3x-4}$ **31.** $2x^2-5x+1+\frac{4}{x+1}$ **33.** $y^2-3y-13$ **35.** $x-3$

37. $3y^2+6y+8+\frac{37}{2y-4}$ **39.** $a^3+2a^2+4a+6+\frac{17}{a-2}$ **41.** y^3+2y^2+4y+8

43. x^2-2x+1 **45.** $(x+3)(x+2)(x+1)$ **47.** $(x+3)(x+4)(x-2)$ **49.** Yes   **51.** 7

A-22 ANSWERS

53. $\frac{21}{10}$ **55.** $\frac{11}{8}$ **57.** $\frac{1}{18}$ **59.** 32 **61.** 17 **63.** $\frac{x^{16}}{y^{22}}$ **65.** $4x^3 - x^2 + 3$

67. $0.5x^2 - 0.4x + 0.3$ **69.** $\frac{3}{2}x - \frac{5}{2} + \frac{1}{2x+4}$ **71.** $\frac{2}{3}x + \frac{1}{3} + \frac{2}{3x-1}$

Problem Set 4.3

1. $\frac{1}{6}$ **3.** $\frac{9}{4}$ **5.** $\frac{1}{2}$ **7.** $\frac{15y}{x^2}$ **9.** $\frac{b}{a}$ **11.** $\frac{2y^5}{z^3}$ **13.** $\frac{x+3}{x+2}$ **15.** $y+1$ **17.** $\frac{3(x+4)}{x-2}$

19. 1 **21.** $\frac{(a-2)(a+2)}{a-5}$ **23.** $\frac{9t^2-6t+4}{4t^2-2t+1}$ **25.** $\frac{x+3}{x+4}$ **27.** $\frac{5a-b}{9a^2+15ab+25b^2}$ **29.** 2

31. $\frac{x(x-1)}{x^2+1}$ **33.** $\frac{(a+4b)(a-3b)}{(a-4b)(a+5b)}$ **35.** $\frac{2y-1}{2y-3}$ **37.** $\frac{(y-2)(y+1)}{(y+2)(y-1)}$ **39.** $\frac{x-1}{x+1}$ **41.** $\frac{x-2}{x+3}$

43. $3x$ **45.** $2(x+5)$ **47.** $x-2$ **49.** $-(y-4)$ **51.** $(a-5)(a+1)$ **53.** $10x^5 + 8x^3 - 6x^2$
55. $12a^2 + 11a - 5$ **57.** $x^4 - 81$ **59.** $16y^2 - 40y + 25$ **61.** $12xy - 6x + 28y - 14$
63. $9 - 6t^2 + t^4$ **65.** $3x^3 + 18x^2 + 33x + 18$ **67.** $\frac{47}{105}$ **69.** $\frac{1}{35}$ **71.** 3 **73.** $\frac{101}{2}$

Problem Set 4.4

1. $\frac{5}{4}$ **3.** $\frac{1}{3}$ **5.** $\frac{41}{24}$ **7.** $\frac{19}{144}$ **9.** $\frac{31}{24}$ **11.** 1 **13.** -1 **15.** $\frac{1}{x+y}$ **17.** 1 **19.** $\frac{a^2+2a-3}{a^3}$

21. 1 **23.** $\frac{4-3t}{2t^2}$ **25.** $\frac{1}{2}$ **27.** $\frac{1}{5}$ **29.** $\frac{x+3}{2(x+1)}$ **31.** $\frac{a-b}{a^2+ab+b^2}$ **33.** $\frac{2y-3}{4y^2+6y+9}$

35. $\frac{2(2x-3)}{(x-3)(x-2)}$ **37.** $\frac{1}{2t-7}$ **39.** $\frac{4}{(a-3)(a+1)}$ **41.** $\frac{-4x^2}{(2x+1)(2x-1)(4x^2+2x+1)}$

43. $\frac{2}{(2x+3)(4x+3)}$ **45.** $\frac{a}{(a+5)(a+4)}$ **47.** $\frac{x+1}{(x+3)(x-2)}$ **49.** $\frac{x-1}{(x+2)(x+1)}$

51. $\frac{1}{(x+1)(x+2)}$ **53.** $\frac{1}{(x+2)(x+3)}$ **55.** $\frac{4x+5}{2x+1}$ **57.** $\frac{22-5t}{4-t}$ **59.** $\frac{2x^2+3x-4}{2x+3}$

61. $\frac{2x-3}{2x}$ **63.** $\frac{1}{2}$ **65.** $\frac{51}{10} = 5.1$ **67.** $(3+4)^{-1} = 7^{-1} = \frac{1}{7}; 3^{-1} + 4^{-1} = \frac{1}{3} + \frac{1}{4} = \frac{7}{12}$ **69.** $\frac{4}{3}$

71. $\frac{x+1}{x}$ **73.** $x + \frac{4}{x} = \frac{x^2+4}{x}$ **75.** $\frac{1}{x} + \frac{1}{x+1} = \frac{2x+1}{x(x+1)}$ **77.** 5.4×10^4 **79.** 3.4×10^{-4}

81. 6,440 **83.** 0.00644 **85.** 1.2×10^4 **87.** 2×10^3 **89.** $\frac{x-1}{x+3}$ **91.** $\frac{x-1}{x+50}$ **93.** $\frac{x+1}{x-100}$

Problem Set 4.5

1. $\frac{9}{8}$ **3.** $\frac{2}{15}$ **5.** $\frac{119}{20}$ **7.** $\frac{1}{x+1}$ **9.** $\frac{a+1}{a-1}$ **11.** $\frac{y-x}{y+x}$ **13.** $\frac{1}{(x-2)(x+5)}$ **15.** $\frac{1}{a^2-a+1}$

17. $\frac{x+3}{x+2}$ **19.** $\frac{a+3}{a-2}$ **21.** $\frac{x-3}{x}$ **23.** $\frac{x+4}{x+2}$ **25.** $\frac{x-3}{x+3}$ **27.** $\frac{a-1}{a+1}$ **29.** $-\frac{x}{3}$

31. $\dfrac{y^2 + 1}{2y}$ **33.** $\dfrac{-x^2 + x - 1}{x - 1}$ **35.** $\frac{5}{3}$ **37.** $\dfrac{2x - 1}{2x + 3}$

39. $(a^{-1} + b^{-1})^{-1} = \left(\dfrac{1}{a} + \dfrac{1}{b}\right)^{-1} = \left(\dfrac{a + b}{ab}\right)^{-1} = \dfrac{ab}{a + b}$ **41.** $\dfrac{1 - x^{-1}}{1 + x^{-1}} = \dfrac{1 - \frac{1}{x}}{1 + \frac{1}{x}} = \dfrac{\frac{x - 1}{x}}{\frac{x + 1}{x}} = \dfrac{x - 1}{x + 1}$

43. First three terms: $\frac{3}{2}, \frac{5}{3}, \frac{8}{5}$ **45.** $\frac{4}{3}, \frac{10}{7}, \frac{24}{17}$ **53.** -15 **55.** 5
    Next three terms: $\frac{13}{8}, \frac{21}{13}, \frac{34}{21}$

57. 1   **59.** $-3, 4$ **61.** $-2, \frac{5}{3}$ **63.** 2, 3   **65.** $\dfrac{x + 3}{(x + 5)(x - 3)}$ **67.** $\dfrac{x - a}{x + a}$

Problem Set 4.6

1. $-\frac{35}{3}$ **3.** $-\frac{18}{5}$ **5.** $\frac{36}{11}$ **7.** 2   **9.** 5   **11.** 2   **13.** $-3, 4$ **15.** $1, -\frac{4}{3}$
17. Possible solution -1, which does not check; $\varnothing$ **19.** 5   **21.** $-\frac{1}{2}, \frac{5}{3}$ **23.** $\frac{2}{3}$ **25.** 18
27. Possible solution 4, which does not check; $\varnothing$ **29.** Possible solutions 3 and -4; only -4 checks; -4
31. -6 **33.** -5 **35.** $\frac{53}{17}$ **37.** Possible solutions 1 and 2; only 2 checks; 2

39. Possible solution 3, which does not check; $\varnothing$ **41.** $\frac{22}{3}$ **43.** 2   **45.** 1, 5   **47.** $x = \dfrac{ab}{a - b}$

49. $R = \dfrac{R_1 R_2}{R_1 + R_2}$ **51.** $y = \dfrac{x - 3}{x - 1}$ **53.** $y = \dfrac{1 - x}{3x - 2}$ **55.** 5   **57.** 8 meters, 13 meters
59. $-6, -5$ or 5, 6   **61.** 3, 4, 5   **63.** $-\frac{2}{3}, -\frac{5}{2}, \frac{5}{2}$ **65.** $-2, 2, 3$

Problem Set 4.7

As you can see, in addition to the answers to the problems we have included some of the equations used to solve the problems. Remember, you should attempt the problems on your own before looking here to check your answers or equations.

1. $\dfrac{1}{x} + \dfrac{1}{3x} = \dfrac{20}{3}; \dfrac{1}{5}$ and $\dfrac{3}{5}$ **3.** $x + \dfrac{1}{x} = \dfrac{10}{3}$; 3 or $\dfrac{1}{3}$ **5.** $\dfrac{1}{x} + \dfrac{1}{x + 1} = \dfrac{7}{12}$; 3, 4   **7.** $\dfrac{7 + x}{9 + x} = \dfrac{5}{6}$; 3

9. Let x = speed of current; $\dfrac{1.5}{5 - x} = \dfrac{3}{5 + x}; \dfrac{5}{3}$ miles/hour   **11.** 6 miles/hour

13. Train A—75 miles/hour, Train B—60 miles/hour   **15.** 540 miles/hour   **17.** 54 miles/hour

19. Let x = time it takes them together; $\dfrac{1}{3} + \dfrac{1}{6} = \dfrac{1}{x}$; 2 days   **21.** 9 hours

23. Let x = time to fill the tank with both open; $\dfrac{1}{8} - \dfrac{1}{16} = \dfrac{1}{x}$; 16 hours   **25.** 15 hours

27. 5.25 minutes   **29.** 51.1 acres   **31.** 5.9 miles/hour   **33.** 20.7 miles/hour   **35.** 4.6 miles/hour

37. 3.6 miles/hour   **39.** $\dfrac{2}{3a}$ **41.** $(x - 3)(x + 2)$ **43.** 1   **45.** $\dfrac{3 - x}{3 + x}$

47. Possible solution 3, which does not check; $\varnothing$

Chapter 4 Review

1. $\dfrac{25x^2}{7y^3}$ **2.** $\dfrac{2z}{x}$ **3.** $\dfrac{a(a-b)}{4}$ **4.** $\dfrac{3}{4a+3b}$ **5.** $\dfrac{x-5}{x+5}$ **6.** $\dfrac{x-5}{x-3}$ **7.** $\dfrac{a+1}{a-1}$ **8.** $x+3$

9. $-\dfrac{x+3}{x-3}$ **10.** -1 **11.** $3x+2+\dfrac{4}{x}$ **12.** $2x^2-x+3$ **13.** $-9b+5a-7a^2b^2$

14. $-2a^2+1-3b^2$ **15.** $x^{3n}-x^{2n}$ **16.** $2x^{2n}-3x^{3n}$ **17.** $x+2$ **18.** $x-3$ **19.** $5x+6y$

20. $x-6y$ **21.** $(y^2+4)(y+2)$ **22.** $(y^2+9)(y+3)$ **23.** $4x+1-\dfrac{2}{2x-7}$

24. $3x+7+\dfrac{10}{3x-4}$ **25.** $y^2-3y-13$ **26.** y^2-5y-1 **27.** $a^3+2a^2+4a+6+\dfrac{17}{a-2}$

28. $a^3-a^2+2a-4+\dfrac{7}{a+2}$ **29.** $\frac{9}{5}$ **30.** 3 **31.** $\dfrac{3x}{4y^2}$ **32.** $\dfrac{6}{y^2}$ **33.** $\dfrac{x-1}{x^2+1}$

34. $x+2$ **35.** 1 **36.** $\dfrac{a-3}{2a-1}$ **37.** $\dfrac{x+2}{x-2}$ **38.** $\dfrac{x+1}{x-1}$ **39.** $(2x-3)(x+3)$

40. $(3x+5)(x+5)$ **41.** $x+3$ **42.** $x-2$ **43.** $\frac{31}{30}$ **44.** $\frac{17}{8}$ **45.** -1 **46.** $-\dfrac{1}{x+y}$

47. $\dfrac{x^2+x+1}{x^3}$ **48.** $\dfrac{x^2-x-1}{x^3}$ **49.** $\dfrac{1}{(y+4)(y+3)}$ **50.** $\dfrac{1}{(y+3)(y+2)}$ **51.** $\dfrac{x-1}{2(x+1)(x+2)}$

52. $\dfrac{x-1}{2(x+1)(x+2)}$ **53.** $\dfrac{15x-2}{5x-2}$ **54.** $\dfrac{49x-23}{7x-3}$ **55.** $\dfrac{1}{(x+2)(x+3)}$ **56.** $\dfrac{1}{(x+3)(x+5)}$

57. $\dfrac{12}{(3x+2)(3x-2)}$ **58.** $\dfrac{24}{(4x+3)(4x-3)}$ **59.** 5 **60.** $\frac{1}{7}$ **61.** $\dfrac{1}{a^2-a+1}$ **62.** $\dfrac{1}{a^2+a+1}$

63. $\dfrac{x^2+x+1}{x^2+1}$ **64.** $\dfrac{x^2+x+1}{x+1}$ **65.** $\dfrac{x+3}{x+2}$ **66.** $\dfrac{2x-1}{2x+1}$ **67.** 6 **68.** $\frac{3}{2}$ **69.** 1 **70.** $-\frac{1}{2}$

71. -6 **72.** 5 **73.** Possible solution -3, which does not check; $\varnothing$
74. Possible solution -1, which does not check; $\varnothing$ **75.** $\frac{22}{3}$ **76.** $-\frac{12}{5}$
77. Possible solutions 4 and -5; only 4 checks   **78.** Possible solutions 2 and 3; only 2 checks
79. $-6, -7$ **80.** Possible solutions -3 and -1; only -1 checks   **81.** $-2, 3$ **82.** $\frac{3}{2}, 4$ **83.** 3, 6
84. 2, 8   **85.** $\frac{60}{11}$ minutes   **86.** 12 minutes   **87.** 4 or $\frac{1}{2}$ **88.** 3 or 1   **89.** 3 mph
90. car 30 mph, truck 20 mph   **91.** 7.5 miles/hour   **92.** 742 miles/hour

Chapter 4 Test

1. $x+y$ **2.** $\dfrac{x-1}{x+1}$ **3.** $6x^2+3xy-4y^2$ **4.** $x^2-4x-2+\dfrac{8}{2x-1}$ **5.** $2(a+4)$ **6.** $4(a+3)$

7. $x+3$ **8.** $\frac{38}{105}$ **9.** $\frac{7}{8}$ **10.** $\dfrac{1}{a-3}$ **11.** $\dfrac{3(x-1)}{x(x-3)}$ **12.** $\dfrac{x}{(x+4)(x+5)}$ **13.** $\dfrac{x+4}{(x+1)(x+2)}$

14. $\dfrac{3a+8}{3a+10}$ **15.** $\dfrac{x-3}{x-2}$ **16.** $-\frac{3}{5}$ **17.** Possible solution 3, which does not check; $\varnothing$

18. $\frac{3}{13}$ **19.** $-2, 3$ **20.** -7 **21.** 6 miles/hour   **22.** 15 hours   **23.** 2.7 miles
24. 1,012 miles/hour

CHAPTER 5

Problem Set 5.1

1–15. (odd)   **17.** $(-\frac{5}{2}, \frac{9}{2})$ **19.** $(-3, \frac{5}{2})$

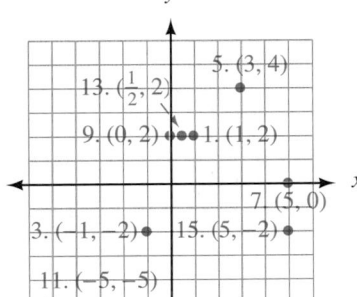

**Odd-Numbered
Answers**

21. $(-2, 0)$ **23.** $(-3, -2)$ **25.** $(-3, -3)$ **27.** $(3, -4)$

29.

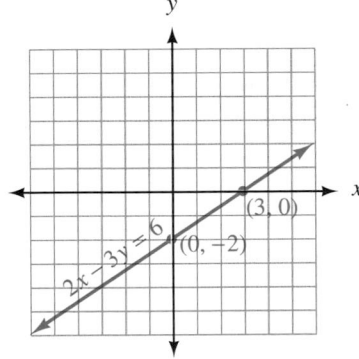

31.

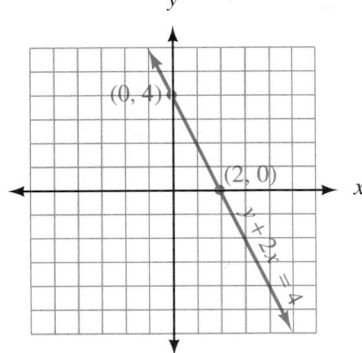

33.

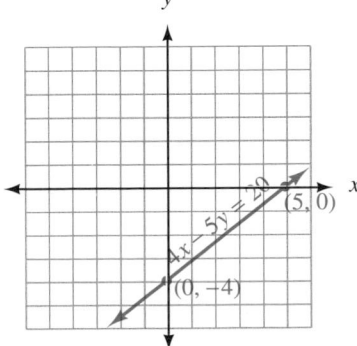

35.

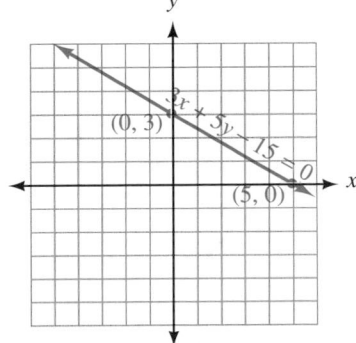

59.

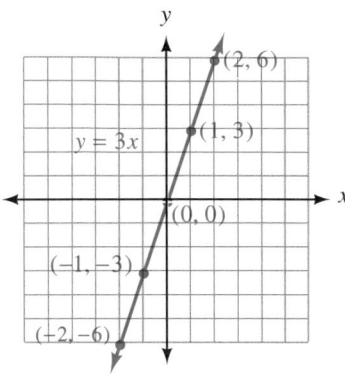

61.

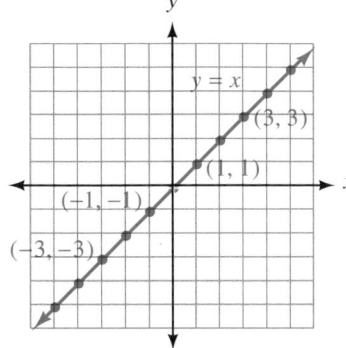

63.

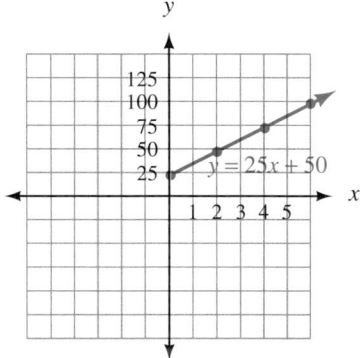

65.

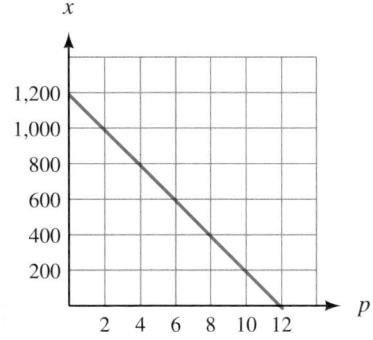

Note that the graph appears in QI only, since x and y represent positive numbers.

67.

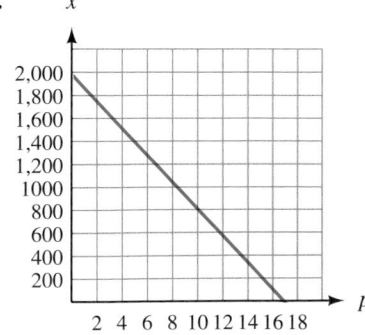

69. $\dfrac{1}{x^2 + 9}$ **71.** $3x - 4x^3y$ **73.** $5x - 4$ **75.** $x^2 + 5x + 25$

77. x-intercept $= \dfrac{c}{a}$, y-intercept $= \dfrac{c}{b}$

79. x-intercept $= a$, y-intercept $= b$ **81.**

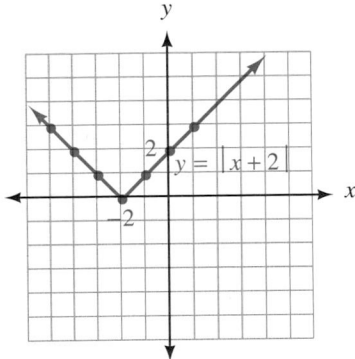

83. (a)

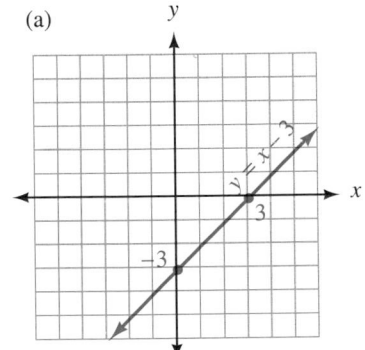

(b)

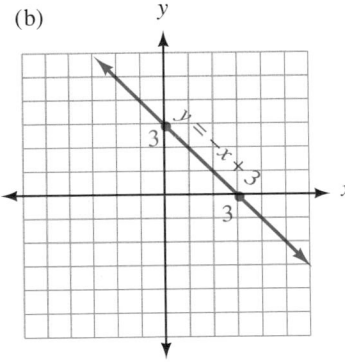

(c)

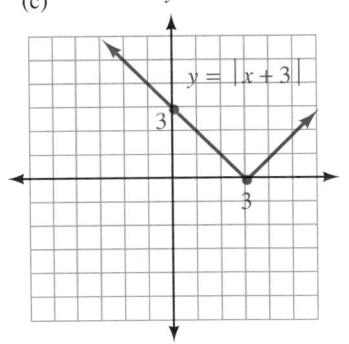

85.

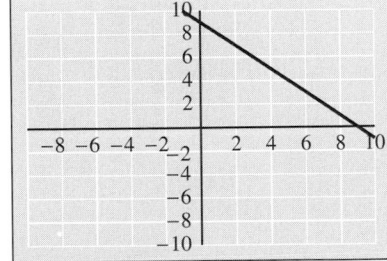

87.

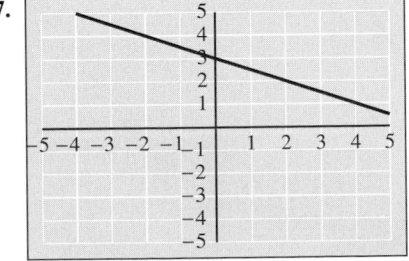

89.
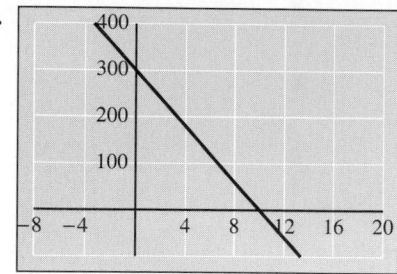

91.
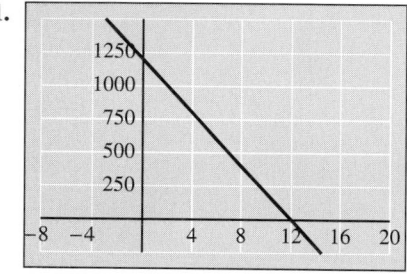

Problem Set 5.2

1. $\frac{3}{2}$ **3.** No slope **5.** $\frac{2}{3}$ **7.**
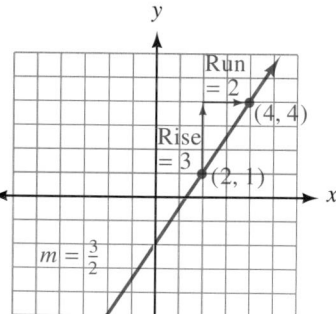

9.
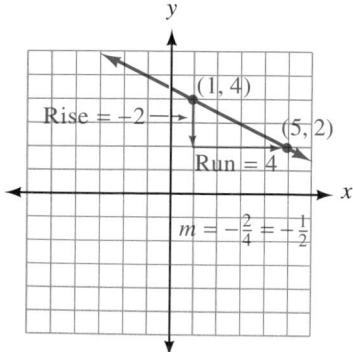

11.
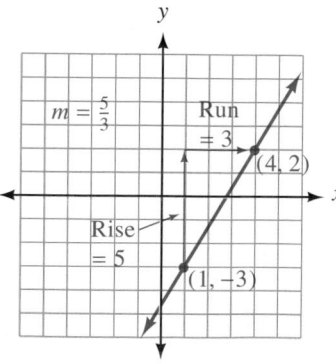

13.

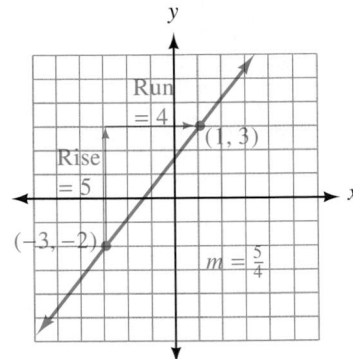

15.

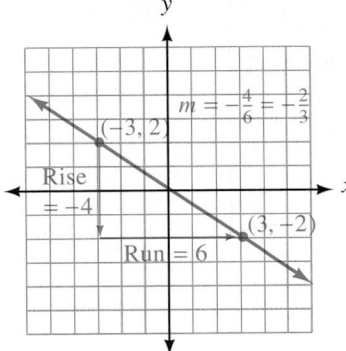

17.

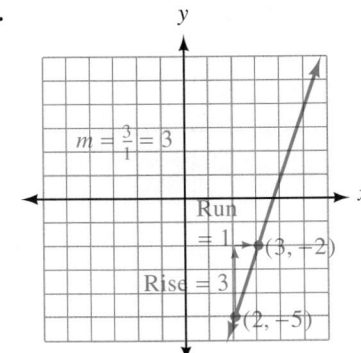

19. 5   **21.** -1 **23.** $-2, 3$ **25.** 2, 1

27.

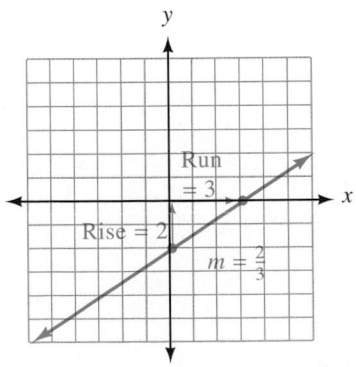

29.

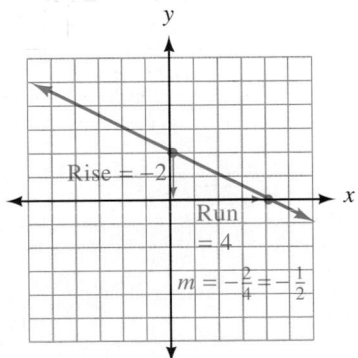

31. $\frac{1}{5}$ **33.** 0   **35.** 8   **37.** $-2, 3$ **39.** 24 feet   **41.** $y = \frac{2}{3}x - 2$ **43.** $y = \frac{2}{3}x - \frac{5}{3}$ **45.** $\dfrac{y^3}{x^2}$

47. 1   **49.** $\dfrac{4x^2 - 6x + 9}{9x^2 - 3x + 1}$ **51.** $-\frac{39}{4}$ **53.** $-\frac{5}{3}$ **55.** $\dfrac{1}{a - b}$

57.

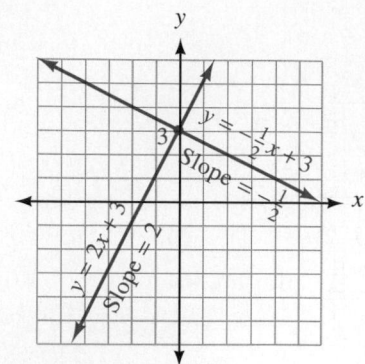

59.

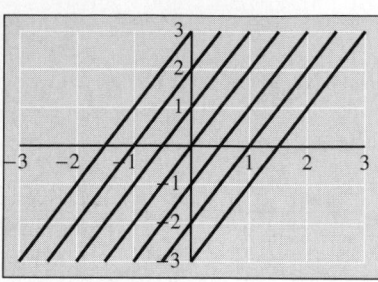

61.

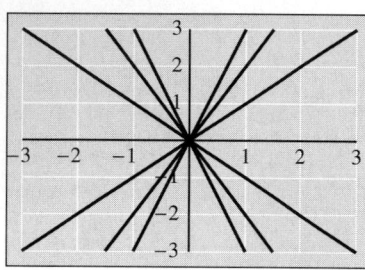

63.

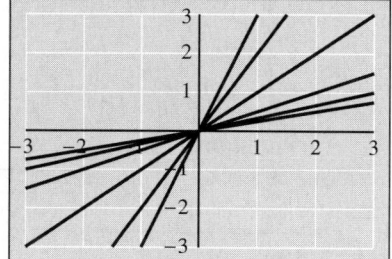

65. For the graph of any line in the form $y = ax + b$, a is the slope of the line and b is the y-intercept.

Problem Set 5.3

1. $y = 2x + 3$ **3.** $y = x - 5$ **5.** $y = \frac{1}{2}x + \frac{3}{2}$ **7.** $y = 4$
9. Slope $= 3$,   **11.** Slope $= \frac{2}{3}$,
   y-intercept $= -2$, y-intercept $= -4$,
   perpendicular slope $= -\frac{1}{3}$ perpendicular slope $= -\frac{3}{2}$

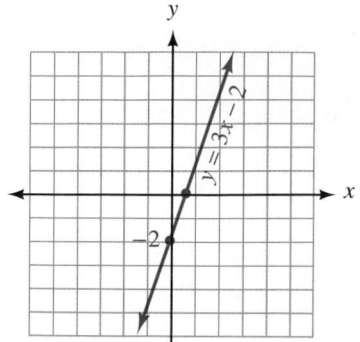

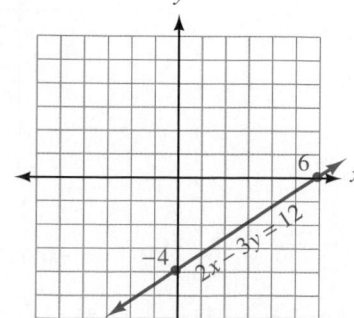

13. Slope $= -\frac{4}{5}$,
   y-intercept $= 4$,
   perpendicular slope $= \frac{5}{4}$

15.

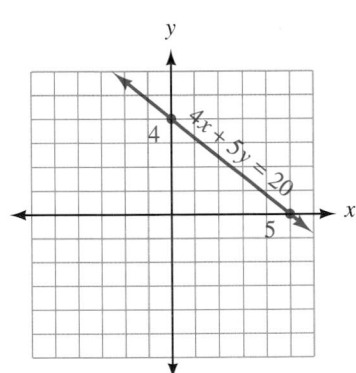

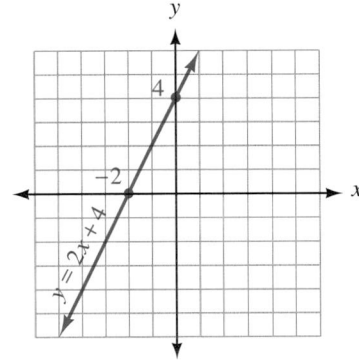

17.

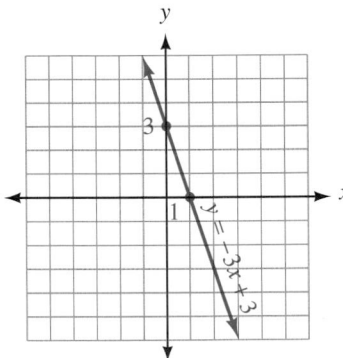

19.

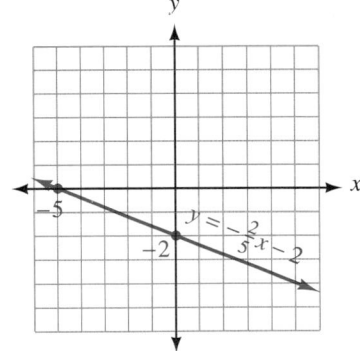

21. $y = 2x - 1$ **23.** $y = -\frac{1}{2}x - 1$ **25.** $y = \frac{3}{2}x - 6$ **27.** $y = -3x + 1$ **29.** $y = x + \frac{11}{10}$

31. $x - y = 2$ **33.** $2x - y = 3$ **35.** $4x - 3y = -6$ **37.** $6x - 5y = 3$ **39.** $3x + 8y = -1$

41. Slope $= 0$, **43.** $y = 3x + 7$ **45.** $y = -\frac{5}{2}x - 13$
   y-intercept $= -2$

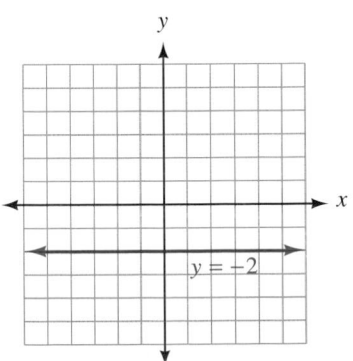

47. $y = \frac{1}{4}x + \frac{1}{4}$ **49.** $y = -\frac{2}{3}x + 2$ **51.** $y = \frac{1}{2}x - \frac{1}{4}$ **53.** 1   **55.** $\dfrac{13 - 3t}{3 - t}$ **57.** $\dfrac{6}{4x + 3}$

59. $\dfrac{x - y}{x^2 + xy + y^2}$ **61.**

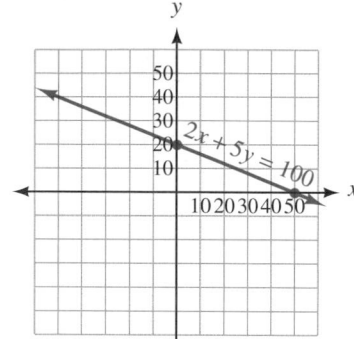

63.

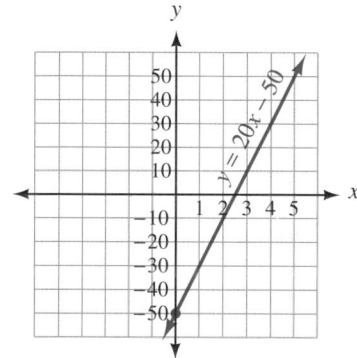

65. $y = -\frac{3}{2}x + 3$: slope $= -\frac{3}{2}$, y-intercept $= 3$, x-intercept $= 2$
67. $y = \frac{3}{2}x + 3$: slope $= \frac{3}{2}$, y-intercept $= 3$, x-intercept $= -2$
69. $y = -\dfrac{b}{a}x + b$: slope $= -\dfrac{b}{a}$, y-intercept $= b$, x-intercept $= a$

Problem Set 5.4

1.

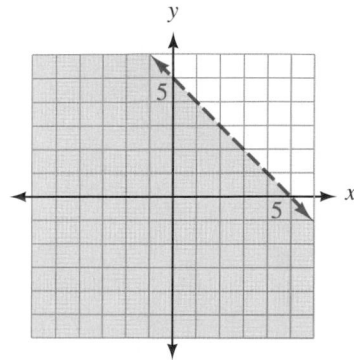

3.

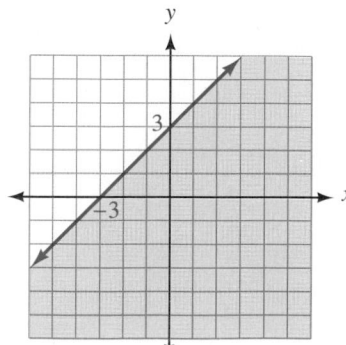

5.

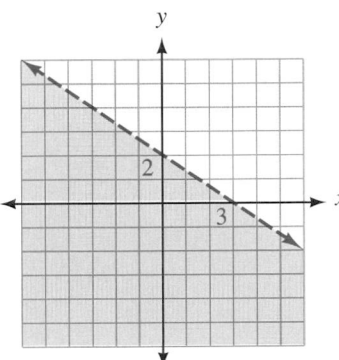

7.

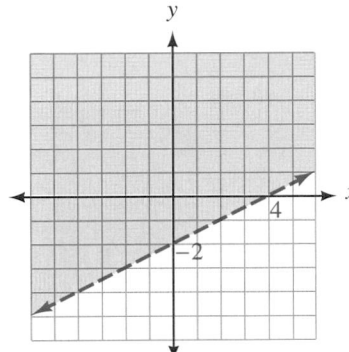

9.

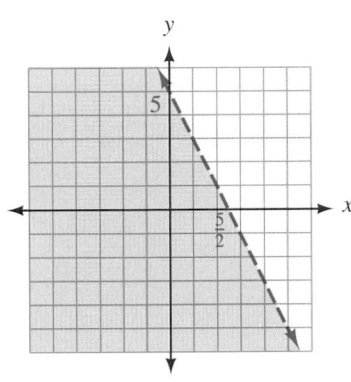

11.

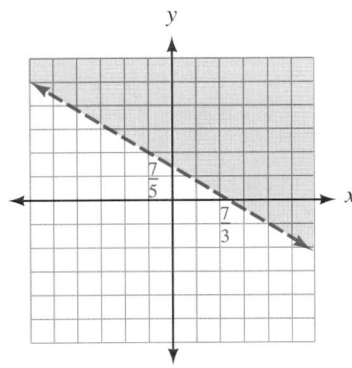

13.

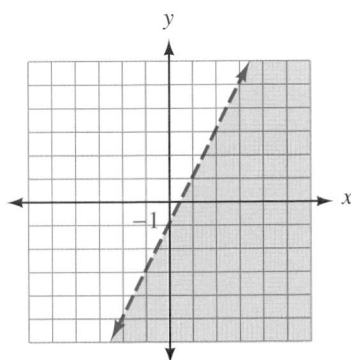

15.

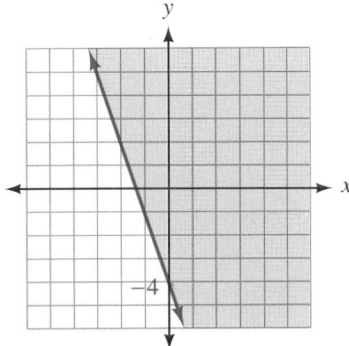

17.

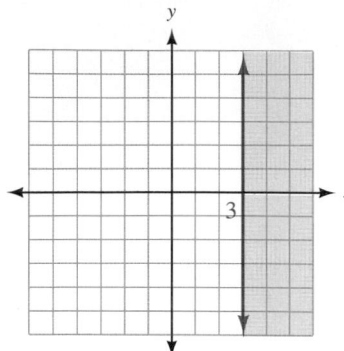

19.

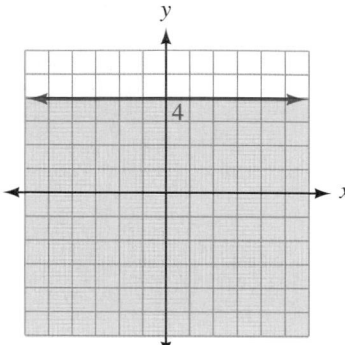

21.

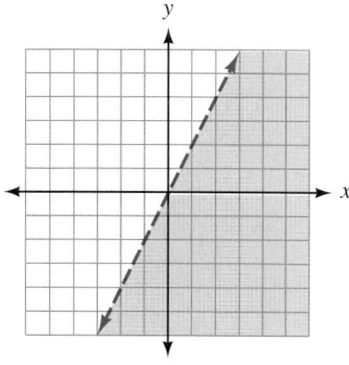

23.

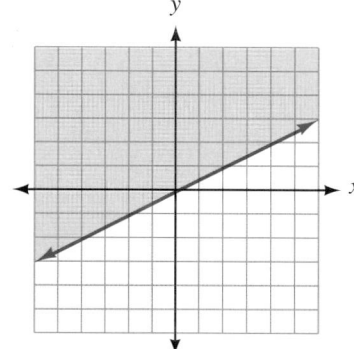

25.

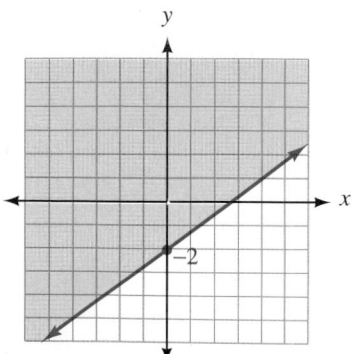

27.

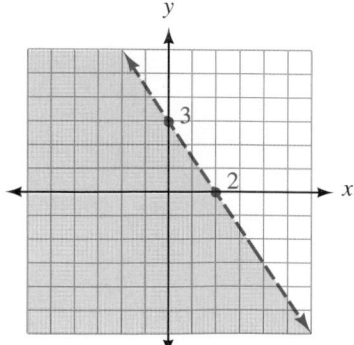

29.

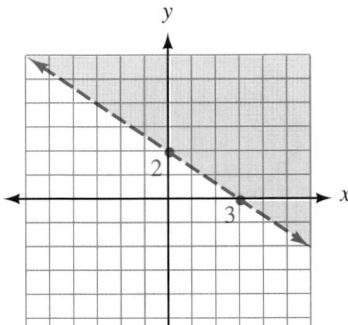

31.

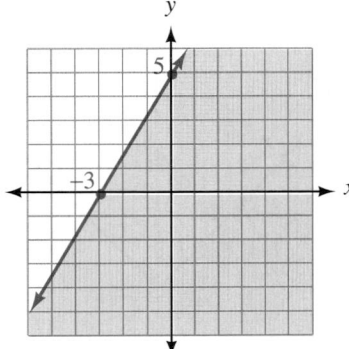

33. $-\frac{1}{8}$ **35.** $\dfrac{y-2}{y+2}$ **37.** $\dfrac{2x+1}{2x-1}$

39.

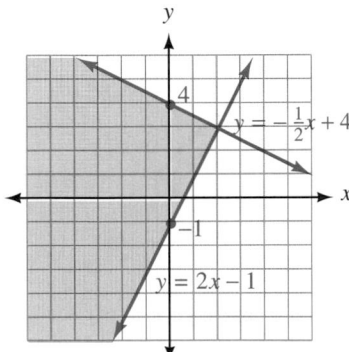

41.

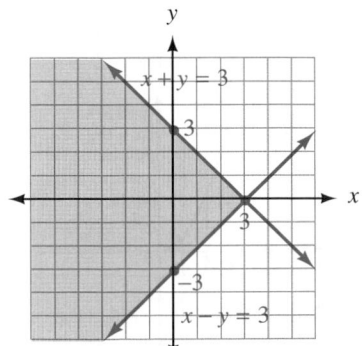

43.

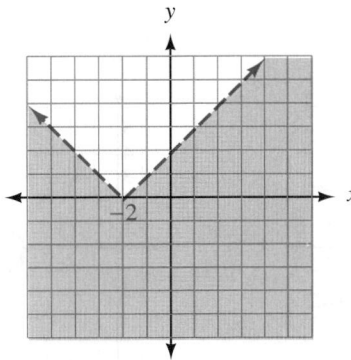

45. Let x = the width, then $x + 2$ is the length. $P = 2x + 2(x + 2) = 4x + 4$.
The variable x must be positive.
47. $A = x(x + 2) = x^2 + 2x$, where $x > 0$.
49. (a) Yes
  (b) Domain = $\{t \mid 0 \le t \le 6\}$, range = $\{h \mid 0 \le h \le 60\}$
  (c) $t = 3$
  (d) $h = 60$
  (e) $t = 6$
51. 10   **53.** -14 **55.** 1   **57.** -3 **59.** 130   **61.** y-intercept: -9, x-intercepts: -3, 3
63. y-intercept: -6, x-intercepts: -2, 3   **65.** y-intercept: 25, x-intercept: -5

67. $x = 3.236$
  $y = 16.944$

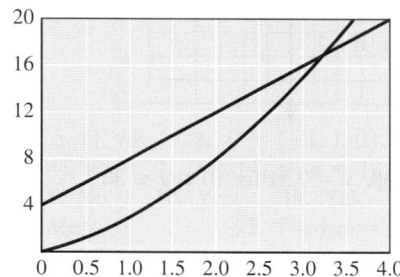

69. 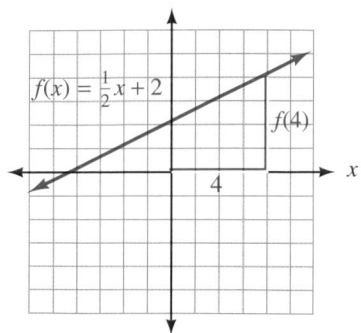   **71.**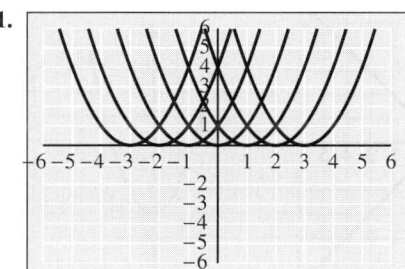

Problem Set 5.6

1. 7   **3.** -18 **5.** -8 **7.** 0   **9.** 24   **11.** 17   **13.** 4   **15.** 0   **17.** 2   **19.** 2   **21.** -8
23. -1 **25.** $2a^2 - 8$ **27.** $2(a + 3)^2 - 8 = 2a^2 + 12a + 10$ **29.** 0   **31.** -2 **33.** -3

35.   **37.** $x = 4$

39.

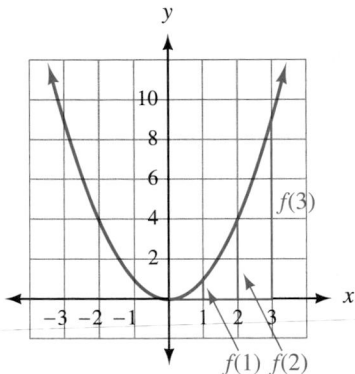

41. $x = 0$ and $x = 1$ **43.** $f(x) = 7.25x$, where $0 \le x \le 40$

45.

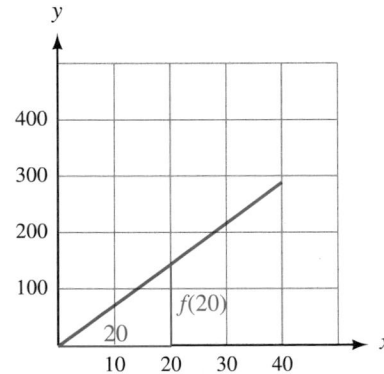

47. $200 + 0.055(500) = \$227.50$

49. $f(x) = 200 + 0.055x$, where $x \ge 0$ **51.** $R(p) = 800p - 100p^2$, $R(x) = 8x - 0.01x^2$

53. $P(x) = -0.01x^2 + 6x - 200$, $P(40) = 24$ **55.** $V(5) = 524$ inches^3, $S(5) = 314$ inches^2

61. $x = 3$ or $x = -3$ **63.** No solutions, $\varnothing$ **65.** $t = 2$ or $t = -\frac{4}{3}$ **67.** $f(-x) = (-x)^2 - 4 = x^2 - 4 = f(x)$

69. $f(-x) = (-x)^{-2} = \dfrac{1}{(-x)^2} = \dfrac{1}{x^2} = x^{-2} = f(x)$ **71.** $f(-x) = 3(-x) = -3x = -f(x)$

73. $f(-x) = (-x)^3 - (-x) = -x^3 + x = -(x^3 - x) = -f(x)$

75.

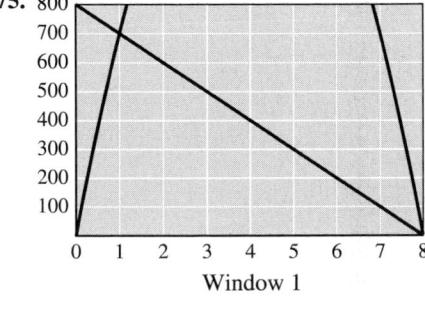

Window 1

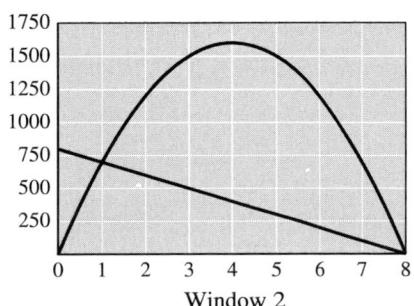

Window 2

Diskette revenue $R(P)$ is at a maximum of 1,600 when the price of diskettes is 4.

Problem Set 5.7

1. 30   **3.** 5   **5.** −6   **7.** $\frac{1}{2}$ **9.** 40   **11.** 225   **13.** $\frac{81}{5}$ **15.** ±9   **17.** 64   **19.** 8
21. $\frac{50}{3}$ pounds   **23.** 12 lb/inch^2 **25.** $\frac{1,504}{15}$ inches^2 **27.** 1.5 ohms

29.

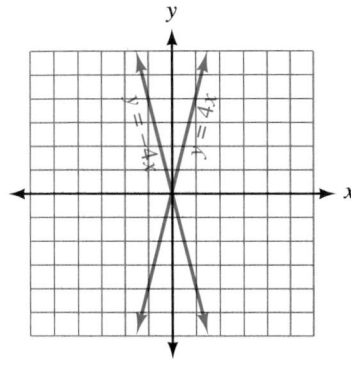

31.

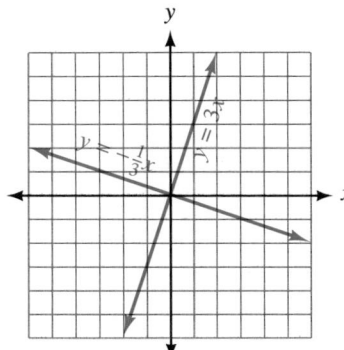

33.

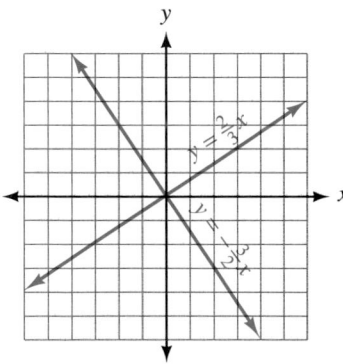

35.

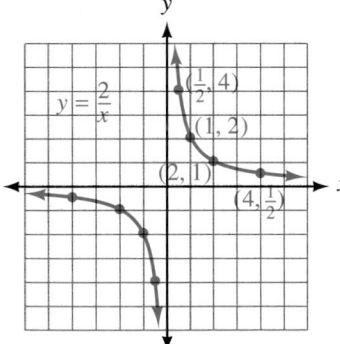

37.

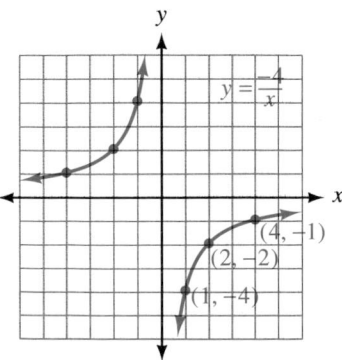

39.

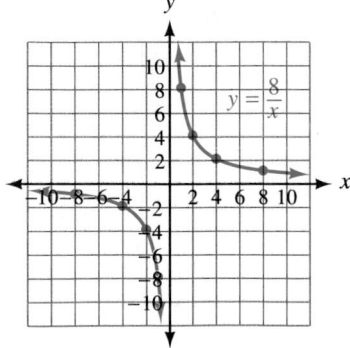

41.

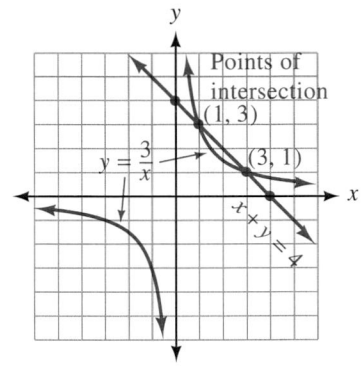

43.

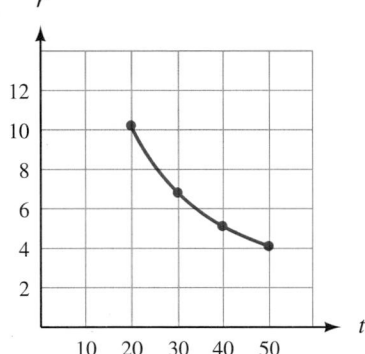

45. $-\frac{3}{2}$ **47.** $-3, \frac{1}{2}$ **49.** $\frac{5}{4}$ or $\frac{4}{5}$

51. $(-1, -2), (1, 2)$ **53.** $(-1, -0.5), (1, 0.5)$

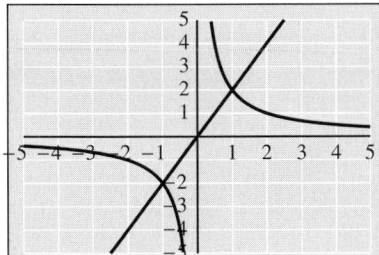

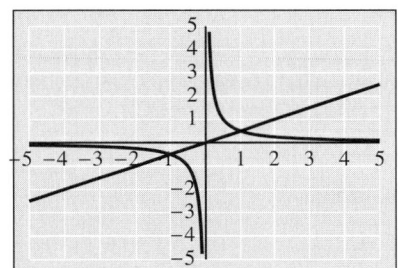

55. The graphs of $y = ax$ and $y = \dfrac{a}{x}$ will intersect when $x = 1$ or $x = -1$, at the points $(1, a)$ and $(-1, -a)$.

$$ax = \frac{a}{x}$$

$$ax^2 = a$$

$$x^2 = \frac{a}{a} \rightarrow x^2 = 1 \rightarrow x = \pm 1$$

Chapter 5 Review

1.

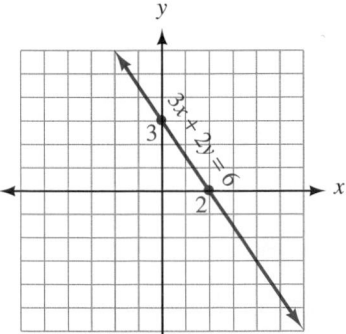

2.

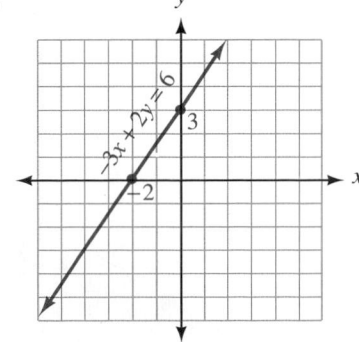

3.

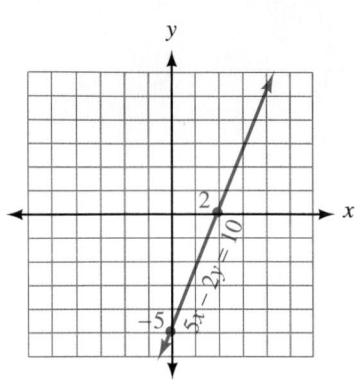

4.

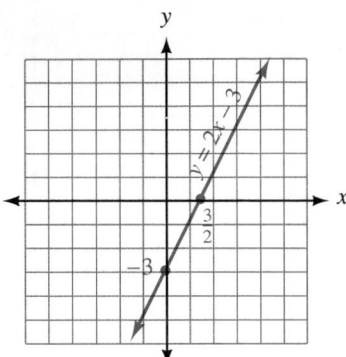

5.

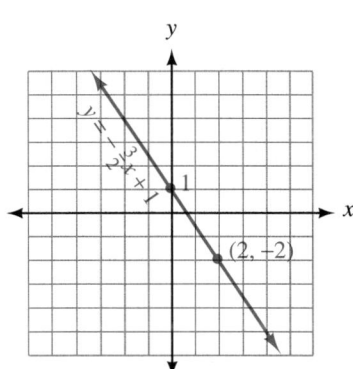

6.

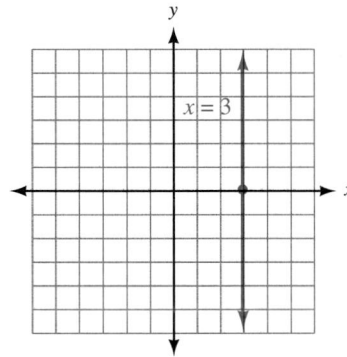

7. -2 **8.** -1 **9.** 0 **10.** No slope   **11.** 3 **12.** 2 **13.** 5 **14.** 3 **15.** -5 **16.** $\frac{4}{3}$
17. $-3, 4$ **18.** $-2, 6$ **19.** $y = 3x + 5$ **20.** $y = 5x + 3$ **21.** $y = -2x$ **22.** $y = \frac{1}{3}x - \frac{2}{3}$
23. $m = 3, b = -6$ **24.** $m = \frac{2}{3}, b = -2$ **25.** $m = \frac{2}{3}, b = -3$ **26.** $m = \frac{3}{2}, b = -2$ **27.** $y = 2x$
28. $y = 3x - 7$ **29.** $y = -\frac{1}{3}x$ **30.** $y = -\frac{1}{2}x + \frac{5}{6}$ **31.** $y = 2x + 1$ **32.** $y = 3x + 1$ **33.** $y = 7$
34. $x = -2$ **35.** $y = -\frac{3}{2}x - \frac{17}{2}$ **36.** $y = -\frac{1}{9}x - \frac{26}{9}$ **37.** $y = 2x - 7$ **38.** $y = \frac{3}{2}x - 1$
39. $y = \frac{1}{3}x - \frac{2}{3}$ **40.** $y = -2x - 4$

41.

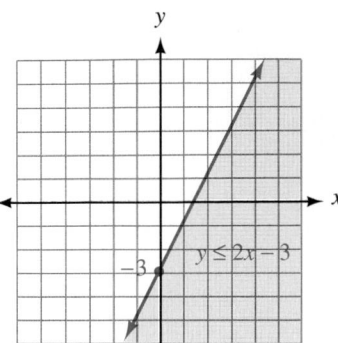

42.

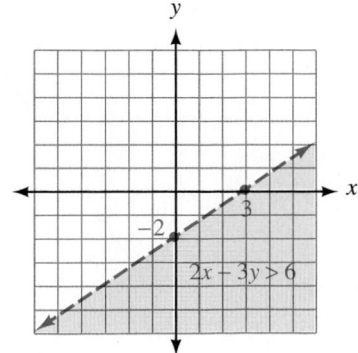

43.

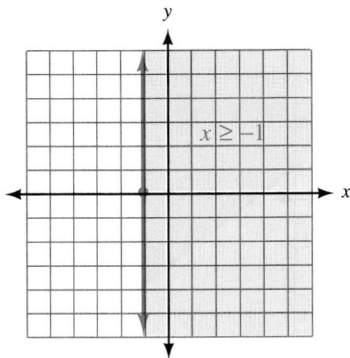

44.

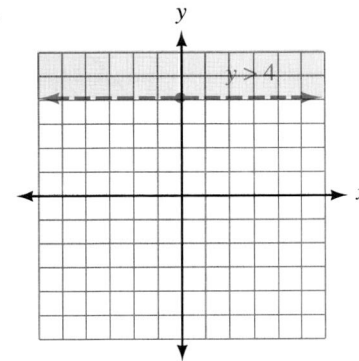

45. $D = \{2, 3, 4\}$, $R = \{4, 3, 2\}$, a function **46.** $D = \{-5, 3\}$, $R = \{2, 4, -2\}$, not a function
47. $D = \{6, -4, -2\}$, $R = \{3, 0\}$, a function **48.** $D = \{1, 3, -2\}$, $R = \{-1, 0\}$, a function
49. -2 **50.** 19 **51.** 4 **52.** 12 **53.** 0 **54.** 2 **55.** 1 **56.** 4 **57.** 1 **58.** 7
59. $3a + 2$ **60.** $3a - 4$ **61.** 1 **62.** 53 **63.** $18x^2 + 12x + 1$ **64.** $6x^2 - 12x + 5$
65. $P(x) = 6x - 6$, where $x > 0$

66.

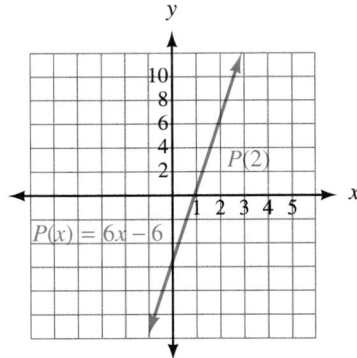

67. $V(r) = 2\pi r^3$ **68.** $V(h) = 4\pi h^3$ **69.** 24 **70.** 6 **71.** 4 **72.** 25 **73.** -72 **74.** -108
75. 2 **76.** 1 **77.** 84 lbs **78.** 270 **79.** 16 foot candles **80.** 96 lbs

81.

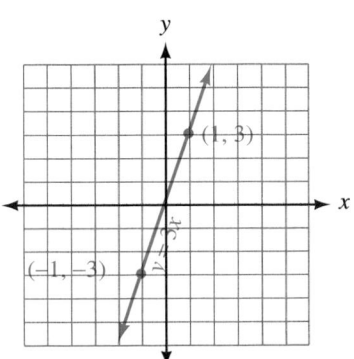

82.

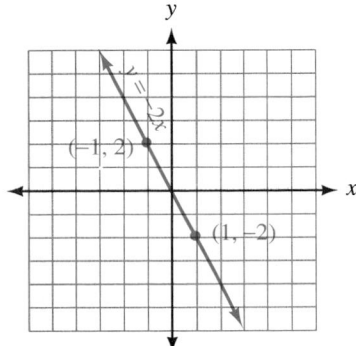

87. $\frac{13}{8}$, numerator and denominator are consecutive members of the Fibonacci sequence   **89.** $x^6 - x^3$
91. $x^2 + 2x - 15$ **93.** $x^4 - 10x^2 + 25$ **95.** $x^3 - 27$ **97.** $x = 2, y = -\frac{5}{2}$ **99.** $x = -6, y = 4$

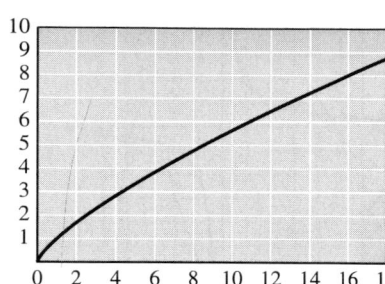

101. When $x = 2, y = 1.7$
103. When $x = 10, y = 5.6$

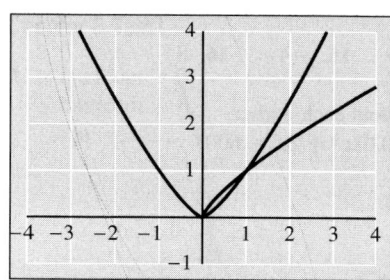

105. Graphs intersect at $x = 1, y = 1$ and $x = 0, y = 0$.

107. (a) 1.62 micrograms
(b) 0.87 micrograms
(c) 0.00293 micrograms
(d) 0.0000029 micrograms

Problem Set 6.2

1. $x + x^2$ **3.** $a^2 - a$ **5.** $6x^3 - 8x^2 + 10x$ **7.** $12x^2 - 36y^2$ **9.** $x^{4/3} - 2x^{2/3} - 8$
11. $a - 10a^{1/2} + 21$ **13.** $20y^{2/3} - 7y^{1/3} - 6$ **15.** $10x^{4/3} + 21x^{2/3}y^{1/2} + 9y$ **17.** $t + 10t^{1/2} + 25$
19. $x^3 + 8x^{3/2} + 16$ **21.** $a - 2a^{1/2}b^{1/2} + b$ **23.** $4x - 12x^{1/2}y^{1/2} + 9y$ **25.** $a - 3$ **27.** $x^3 - y^3$
29. $t - 8$ **31.** $4x^3 - 3$ **33.** $x + y$ **35.** $a - 8$ **37.** $8x + 1$ **39.** $t - 1$ **41.** $2x^{1/2} + 3$
43. $3x^{1/3} - 4y^{1/3}$ **45.** $3a - 2b$ **47.** $3(x - 2)^{1/2}(4x - 11)$ **49.** $5(x - 3)^{7/5}(x - 6)$
51. $3(x + 1)^{1/2}(3x^2 + 3x + 2)$ **53.** $(x^{1/3} - 2)(x^{1/3} - 3)$ **55.** $(a^{1/5} - 4)(a^{1/5} + 2)$

57. $(2y^{1/3} + 1)(y^{1/3} - 3)$ **59.** $(3t^{1/5} + 5)(3t^{1/5} - 5)$ **61.** $(2x^{1/7} + 5)^2$ **63.** $\dfrac{3 + x}{x^{1/2}}$ **65.** $\dfrac{x + 5}{x^{1/3}}$

67. $\dfrac{x^3 + 3x^2 + 1}{(x^3 + 1)^{1/2}}$ **69.** $\dfrac{-4}{(x^2 + 4)^{1/2}}$ **71.** 2   **73.** 27   **75.** 0.871   **77.** 15.8%   **79.** 5.9%
81. 3   **83.** 5   **85.** $\frac{2}{3}$ **87.** $\frac{2}{5}$

Problem Set 6.3

1. $2\sqrt{2}$ **3.** $7\sqrt{2}$ **5.** $12\sqrt{2}$ **7.** $4\sqrt{5}$ **9.** $4\sqrt{3}$ **11.** $15\sqrt{3}$ **13.** $3\sqrt[3]{2}$ **15.** $4\sqrt[3]{2}$ **17.** $6\sqrt[3]{2}$
19. $2\sqrt[5]{2}$ **21.** $3x\sqrt{2x}$ **23.** $2y\sqrt[4]{2y^3}$ **25.** $2xy^2\sqrt[3]{5xy}$ **27.** $4abc^2\sqrt{3b}$ **29.** $2bc\sqrt[3]{6a^2c}$

31. $2xy^2 \sqrt[5]{2x^3y^2}$ **33.** $3xy^2z \sqrt[5]{x^2}$ **35.** $2\sqrt{3}$ **37.** $\sqrt{-20}$, which is not a real number   **39.** $\dfrac{\sqrt{11}}{2}$

41. $\dfrac{2\sqrt{3}}{3}$ **43.** $\dfrac{5\sqrt{6}}{6}$ **45.** $\dfrac{\sqrt{2}}{2}$ **47.** $\dfrac{\sqrt{5}}{5}$ **49.** $2\sqrt[3]{4}$ **51.** $\dfrac{2\sqrt[3]{3}}{3}$ **53.** $\dfrac{\sqrt[4]{24x^2}}{2x}$ **55.** $\dfrac{\sqrt[4]{8y^3}}{y}$

57. $\dfrac{\sqrt[3]{36xy^2}}{3y}$ **59.** $\dfrac{\sqrt[3]{6xy^2}}{3y}$ **61.** $\dfrac{\sqrt[4]{2x}}{2x}$ **63.** $\dfrac{3x\sqrt{15xy}}{5y}$ **65.** $\dfrac{5xy\sqrt{6xz}}{2z}$ **67.** $\dfrac{2ab\sqrt[3]{6ac^2}}{3c}$

69. $\dfrac{2xy^2\sqrt[3]{3z^2}}{3z}$ **71.** $5|x|$ **73.** $3|xy|\sqrt{3x}$ **75.** $|x-5|$ **77.** $|2x+3|$ **79.** $2|a(a+2)|$

81. $2|x|\sqrt{x-2}$ **83.** $\sqrt{9+16} = \sqrt{25} = 5$; $\sqrt{9} + \sqrt{16} = 3 + 4 = 7$ **85.** $5\sqrt{13}$ feet

89. $\sqrt{2}, \sqrt{3}, \sqrt{4}, \sqrt{5}, \sqrt{6}, \sqrt{7}, \ldots$
    10th term $= \sqrt{11}$
    100th term $= \sqrt{101}$

91.

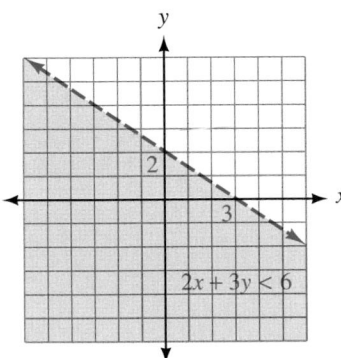

93.

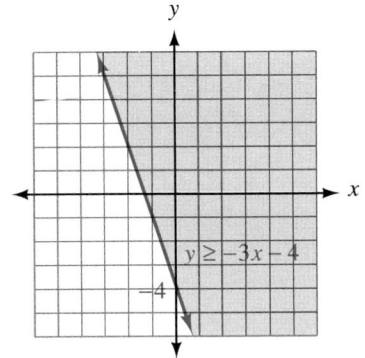

95.

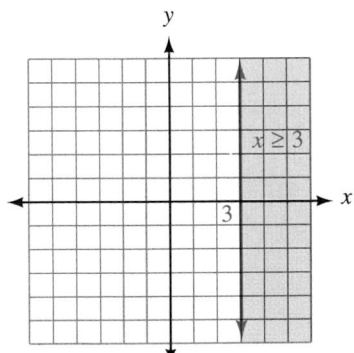

97. $12\sqrt[3]{5}$ **99.** $6\sqrt[3]{49}$ **101.** $\dfrac{\sqrt[10]{a^7}}{a}$

103. $\dfrac{\sqrt[20]{a^9}}{a}$ **105.**

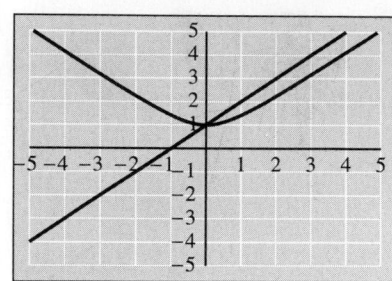

107. About $\frac{3}{4}$ of a unit apart when $x = 2$ **109.** $x = 0$

Problem Set 6.4

1. $7\sqrt{5}$ **3.** $-x\sqrt{7}$ **5.** $\sqrt[3]{10}$ **7.** $9\sqrt[5]{6}$ **9.** 0 **11.** $\sqrt{5}$ **13.** $-32\sqrt{2}$ **15.** $-3x\sqrt{2}$

17. $-2\sqrt[3]{2}$ **19.** $8x\sqrt[3]{xy^2}$ **21.** $3a^2b\sqrt{3ab}$ **23.** $11ab\sqrt[3]{3a^2b}$ **25.** $10xy\sqrt[4]{3y}$ **27.** $\sqrt{2}$ **29.** $\dfrac{8\sqrt{5}}{15}$

31. $\dfrac{(x-1)\sqrt{x}}{x}$ **33.** $\dfrac{3\sqrt{2}}{2}$ **35.** $\dfrac{5\sqrt{6}}{6}$ **37.** $\dfrac{8\sqrt[3]{25}}{5}$ **39.** $\sqrt{12} \approx 3.464; \ 2\sqrt{3} \approx 2(1.732) = 3.464$

41. $\sqrt{8} + \sqrt{18} \approx 2.828 + 4.243 = 7.071; \ \sqrt{50} \approx 7.071; \ \sqrt{26} \approx 5.099$ **43.** $8\sqrt{2x}$ **45.** 5

53. $m = \frac{2}{3}, \ b = -2$ **55.** $y = \frac{2}{3}x + 6$ **57.** $y = 4x - 1$ **59.** $y = \frac{3}{4}x + 3$ **61.** $11\sqrt{x+3}$ **63.** 0

65. $4x(x + 5)$

Problem Set 6.5

1. $3\sqrt{2}$ **3.** $10\sqrt{21}$ **5.** 720 **7.** 54 **9.** $\sqrt{6} - 9$ **11.** $24 + 6\sqrt[3]{4}$ **13.** $7 + 2\sqrt{6}$

15. $x + 2\sqrt{x} - 15$ **17.** $34 + 20\sqrt{3}$ **19.** $19 + 8\sqrt{3}$ **21.** $x - 6\sqrt{x} + 9$ **23.** $4a - 12\sqrt{ab} + 9b$

25. $x + 4\sqrt{x-4}$ **27.** $x - 6\sqrt{x-5} + 4$ **29.** 1 **31.** $a - 49$ **33.** $25 - x$ **35.** $x - 8$

37. $10 + 6\sqrt{3}$ **39.** $\dfrac{\sqrt{3}+1}{2}$ **41.** $\dfrac{5 - \sqrt{5}}{4}$ **43.** $\dfrac{x + 3\sqrt{x}}{x - 9}$ **45.** $\dfrac{10 + 3\sqrt{5}}{11}$ **47.** $\dfrac{3\sqrt{x} + 3\sqrt{y}}{x - y}$

49. $2 + \sqrt{3}$ **51.** $\dfrac{11 - 4\sqrt{7}}{3}$ **53.** $\dfrac{a + 2\sqrt{ab} + b}{a - b}$ **55.** $\dfrac{x + 4\sqrt{x} + 4}{x - 4}$ **57.** $\dfrac{5 - \sqrt{21}}{4}$

59. $\dfrac{\sqrt{x} - 3x + 2}{1 - x}$ **63.** $10\sqrt{3}$ **65.** $x + 6\sqrt{x} + 9$ **67.** 75 **69.** $\dfrac{5\sqrt{2}}{4}$ seconds and $\dfrac{5}{2}$ seconds

75.

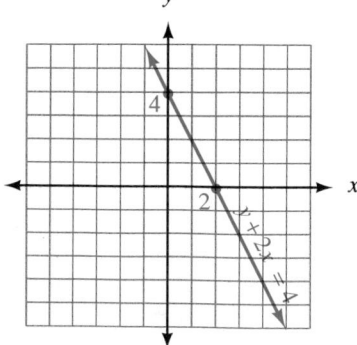

77.

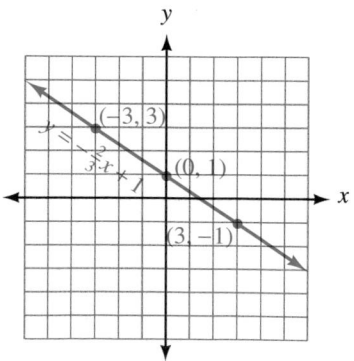

79.

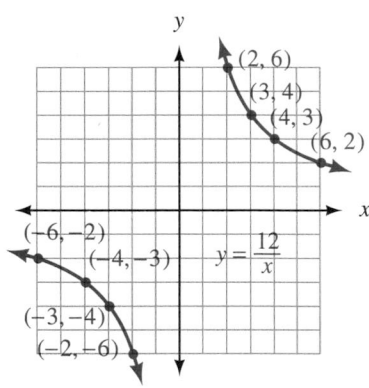

81. $\dfrac{x - 4 - 2\sqrt{x - 4}}{x - 8}$ **83.** $\dfrac{x + \sqrt{x^2 - 9}}{3}$

85. $\dfrac{\sqrt[3]{x^2} - 2\sqrt[3]{x} + 4}{x + 8}$ **87.** $\dfrac{\sqrt[3]{9} - \sqrt[3]{6} + \sqrt[3]{4}}{5}$

Problem Set 6.6

1. 4   **3.** ∅   **5.** 5   **7.** ∅   **9.** $\frac{39}{2}$ **11.** ∅   **13.** 5   **15.** 3   **17.** $-\frac{32}{3}$ **19.** 3, 4
21. −1, −2   **23.** −1   **25.** ∅   **27.** 7   **29.** 0, 3   **31.** −4   **33.** 8   **35.** 0   **37.** 9   **39.** 0
41. 8   **43.** Possible solution 9, which does not check; ∅   **45.** Possible solutions 0 and 32; only 0 checks; 0
47. Possible solutions −2 and 6; only 6 checks; 6   **49.** $h = 100 - 16t^2$ **51.** $\frac{392}{121} \approx 3.24$ feet

53.

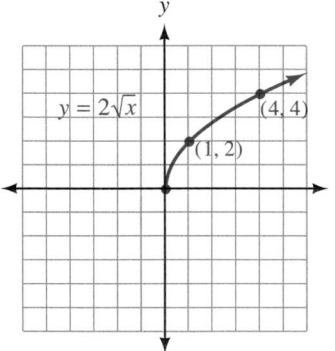

55.

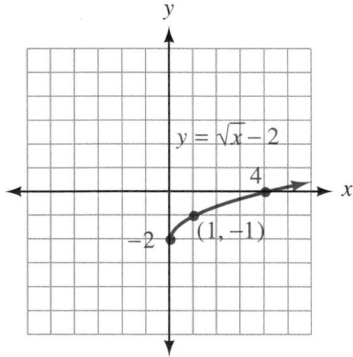

57.

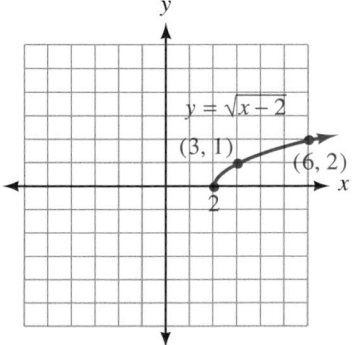

59.

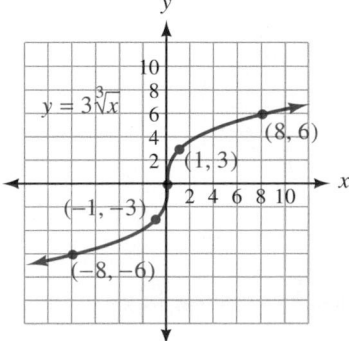

61.

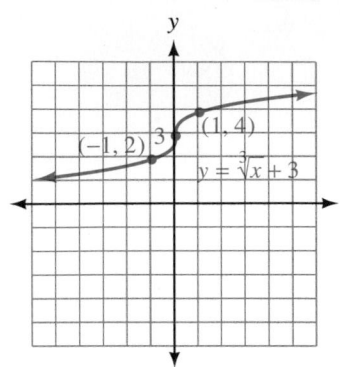

63.

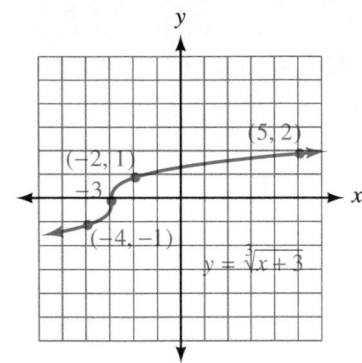

65. $\sqrt{6} - 2$ **67.** $x + 10\sqrt{x} + 25$ **69.** $\dfrac{x - 3\sqrt{x}}{x - 9}$ **71.** Possible solutions 2 and 6; only 6 checks; 6

73. $1, -1, -3$ **75.** $y = \frac{1}{8}x^2$

77.

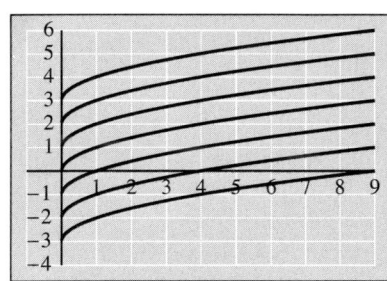

79. The value of b shifts the curve b units along the y-axis.

81.

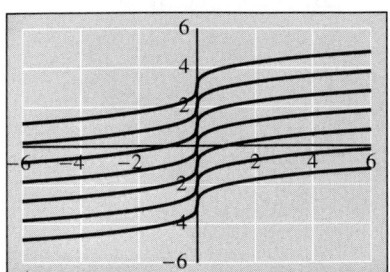

83. The value of b shifts the curve b units along the y-axis.

85.

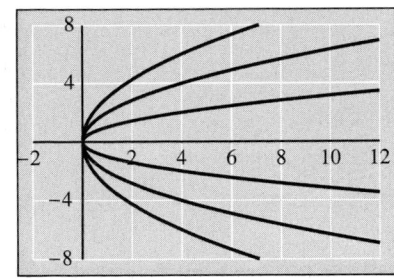

87. The smaller the absolute value of a, the more slowly the graph rises. If a is negative, the graph lies below the x-axis.

Problem Set 6.7

1. $6i$ **3.** $-5i$ **5.** $6i\sqrt{2}$ **7.** $-2i\sqrt{3}$ **9.** 1 **11.** -1 **13.** $-i$ **15.** $x = 3, y = -1$
17. $x = -2, y = -\frac{1}{2}$ **19.** $x = -8, y = -5$ **21.** $x = 7, y = \frac{1}{2}$ **23.** $x = \frac{3}{7}, y = \frac{2}{5}$ **25.** $5 + 9i$
27. $5 - i$ **29.** $2 - 4i$ **31.** $1 - 6i$ **33.** $2 + 2i$ **35.** $-1 - 7i$ **37.** $6 + 8i$ **39.** $2 - 24i$
41. $-15 + 12i$ **43.** $18 + 24i$ **45.** $10 + 11i$ **47.** $21 + 23i$ **49.** $-2 + 2i$ **51.** $2 - 11i$
53. $-21 + 20i$ **55.** $-2i$ **57.** $-7 - 24i$ **59.** 5 **61.** 40 **63.** 13 **65.** 164 **67.** $-3 - 2i$
69. $-2 + 5i$ **71.** $\frac{8}{13} + \frac{12}{13}i$ **73.** $-\frac{18}{13} - \frac{12}{13}i$ **75.** $-\frac{5}{13} + \frac{12}{13}i$ **77.** $\frac{13}{15} - \frac{2}{5}i$ **79.** $\frac{31}{53} - \frac{24}{53}i$
81. 147 **83.** 100 **85.** 100

Chapter 6 Review

1. 7 **2.** 2 **3.** -3 **4.** 3 **5.** 2 **6.** $\frac{3}{4}$ **7.** 27 **8.** 4 **9.** $2x^3y^2$ **10.** $5x^3y^4$
11. $\frac{1}{16}$ **12.** $\frac{9}{20}$ **13.** x^2 **14.** y **15.** a^2b^4 **16.** $x^{17/12}$ **17.** $a^{7/20}$ **18.** $x^{2/5}y^{3/5}z$
19. $a^{5/12}b^{8/3}$ **20.** $y^{3/2}$ **21.** $12x + 11x^{1/2}y^{1/2} - 15y$ **22.** $20x - 7x^{1/2} - 6$ **23.** $a^{2/3} - 10a^{1/3} + 25$
24. $8t - 1$ **25.** $4x^{1/2} + 2x^{5/6}$ **26.** $3a^{3/7}b^{1/5} - 2a^{1/7}b^{3/5}$ **27.** $2(x - 3)^{1/4}(4x - 13)$

28. $(3x^{1/5} + 2)(2x^{1/5} - 5)$ **29.** $\dfrac{x + 5}{x^{1/4}}$ **30.** $\dfrac{4x^3 + x^2 + 1}{(x^2 + 1)^{1/2}}$ **31.** $2\sqrt{3}$ **32.** $3\sqrt{3}$ **33.** $5\sqrt{2}$

34. $2\sqrt{5}$ **35.** $2\sqrt[3]{2}$ **36.** $2\sqrt[3]{4}$ **37.** $3x\sqrt{2}$ **38.** $6y^2\sqrt{2y}$ **39.** $4ab^2c\sqrt{5a}$ **40.** $3xy\sqrt[3]{x}$

41. $2abc\sqrt[4]{2bc^2}$ **42.** $3abc\sqrt[4]{2a^2b}$ **43.** $\dfrac{3\sqrt{2}}{2}$ **44.** $\dfrac{\sqrt{10}}{5}$ **45.** $3\sqrt[3]{4}$ **46.** $\dfrac{7\sqrt[3]{3}}{3}$

47. $\dfrac{4x\sqrt{21xy}}{7y}$ **48.** $\dfrac{5xy\sqrt{6yz}}{2z}$ **49.** $\dfrac{2y\sqrt[3]{45x^2z^2}}{3z}$ **50.** $\dfrac{3xy\sqrt[3]{50xz}}{5z}$ **51.** $-2x\sqrt{6}$ **52.** $8x\sqrt{7}$

53. $3\sqrt{3}$ **54.** $5\sqrt{2}$ **55.** $\dfrac{8\sqrt{5}}{5}$ **56.** $\dfrac{4\sqrt{15}}{5}$ **57.** $7\sqrt{2}$ **58.** $3\sqrt{3}$ **59.** $11a^2b\sqrt{3ab}$

60. $2a^2b\sqrt{2a}$ **61.** $-4xy\sqrt[3]{xz^2}$ **62.** $15xy\sqrt[3]{3x^2y}$ **63.** $\sqrt{6} - 4$ **64.** $40\sqrt{2} - 60$
65. $x - 5\sqrt{x} + 6$ **66.** $18 + 14\sqrt{3}$ **67.** $54 + 3\sqrt{2}$ **68.** $x - 4\sqrt{x} + 4$ **69.** 6

70. 44 **71.** $3\sqrt{5} + 6$ **72.** $\sqrt{6} + \sqrt{3}$ **73.** $6 + \sqrt{35}$ **74.** $\dfrac{x + 2\sqrt{xy} + y}{x - y}$ **75.** $\dfrac{63 + 12\sqrt{7}}{47}$

76. $\dfrac{7\sqrt{6} - 12}{5}$ **77.** 0 **78.** $\varnothing$ **79.** 3 **80.** 3 **81.** 5 **82.** 2 **83.** $\varnothing$ **84.** 14 **85.** $4, 6$

86. $-3, -2$

87.

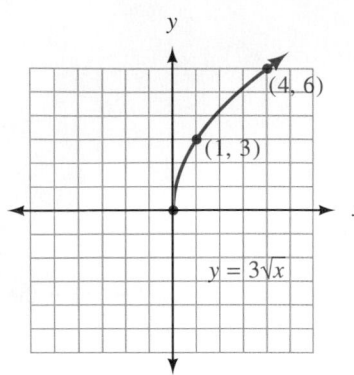

88.

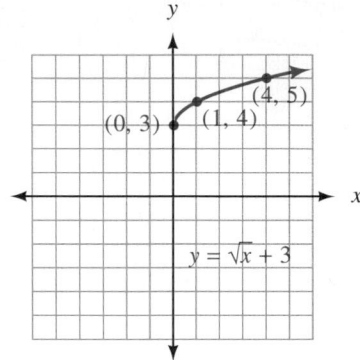

89.

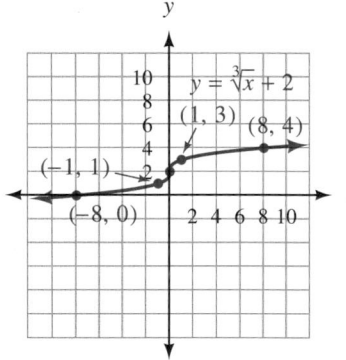

90.

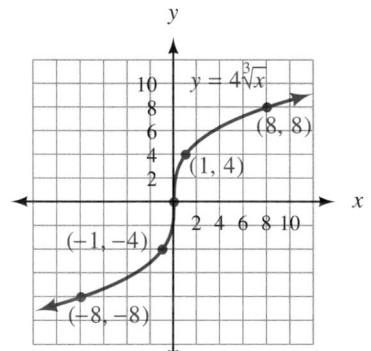

91. $7i$ **92.** $4i\sqrt{5}$ **93.** 1 **94.** $-i$ **95.** $x = -\frac{3}{2}, y = -\frac{1}{2}$ **96.** $x = -2, y = -4$ **97.** $9 + 3i$
98. $-3 + 7i$ **99.** $-7 + 4i$ **100.** $-2 + 7i$ **101.** $-6 + 12i$ **102.** $-5 - 30i$ **103.** $5 + 14i$
104. $24 + 18i$ **105.** $12 + 16i$ **106.** $2i$ **107.** 25 **108.** 10 **109.** $1 - 3i$ **110.** $1 + 2i$
111. $-\frac{6}{5} + \frac{3}{5}i$ **112.** $\frac{5}{13} + \frac{12}{13}i$ **113.** $\frac{7}{25} - \frac{24}{25}i$ **114.** $\frac{23}{13} + \frac{11}{13}i$

Chapter 6 Test

1. $\frac{1}{9}$ **2.** $\frac{7}{5}$ **3.** $a^{5/12}$ **4.** $\dfrac{x^{13/12}}{y}$ **5.** $7x^4y^5$ **6.** $2x^2y^4$ **7.** $2a$ **8.** $x^{n^2-n}y^{1-n^3}$ **9.** $6a^2 - 10a$

10. $16a^3 - 40a^{3/2} + 25$ **11.** $(3x^{1/3} - 1)(x^{1/3} + 2)$ **12.** $(3x^{1/3} - 7)(3x^{1/3} + 7)$ **13.** $\dfrac{x + 4}{x^{1/2}}$

14. $\dfrac{3}{(x^2 - 3)^{1/2}}$ **15.** $5xy^2\sqrt{5xy}$ **16.** $2x^2y^2\sqrt[3]{5xy^2}$ **17.** $\dfrac{\sqrt{6}}{3}$ **18.** $\dfrac{2a^2b\sqrt{15bc}}{5c}$ **19.** $-6\sqrt{3}$

20. $-ab\sqrt[3]{3}$ **21.** $x + 3\sqrt{x} - 28$ **22.** $21 - 6\sqrt{6}$ **23.** $\dfrac{5 + 5\sqrt{3}}{2}$ **24.** $\dfrac{x - 2\sqrt{2x} + 2}{x - 2}$

25. Possible solutions 1 and 8; only 8 checks; 8 **26.** -4 **27.** -3

28.

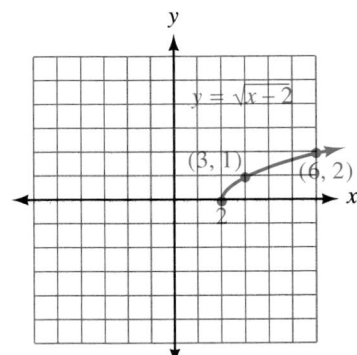

29.

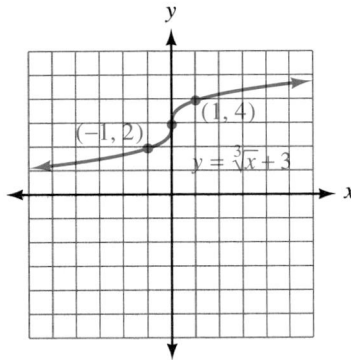

30. $x = \frac{1}{2}$, $y = 7$ **31.** $6i$ **32.** $17 - 6i$ **33.** $9 - 40i$ **34.** $-\frac{5}{13} - \frac{12}{13}i$

35. $i^{38} = (i^2)^{19} = (-1)^{19} = -1$

CHAPTER 7

Problem Set 7.1

1. ± 5 **3.** $\pm 3i$ **5.** $\pm \frac{\sqrt{3}}{2}$ **7.** $\pm 2i\sqrt{3}$ **9.** $\pm \frac{3\sqrt{5}}{2}$ **11.** $-2, 3$ **13.** $\frac{-3 \pm 3i}{2}$

15. $\frac{-2 \pm 2i\sqrt{2}}{5}$ **17.** $-4 \pm 3i\sqrt{3}$ **19.** $\frac{3 \pm 2i}{2}$ **21.** ± 1 **23.** ± 1 **25.** ± 1

27. $x^2 + 12x + 36 = (x + 6)^2$ **29.** $x^2 - 4x + 4 = (x - 2)^2$ **31.** $a^2 - 10a + 25 = (a - 5)^2$

33. $x^2 + 5x + \frac{25}{4} = (x + \frac{5}{2})^2$ **35.** $y^2 - 7y + \frac{49}{4} = (y - \frac{7}{2})^2$ **37.** $-6, 2$ **39.** $-3, -9$ **41.** $1 \pm 2i$

43. $4 \pm \sqrt{15}$ **45.** $\frac{5 \pm \sqrt{37}}{2}$ **47.** $1 \pm \sqrt{5}$ **49.** $\frac{4 \pm \sqrt{13}}{3}$ **51.** $\frac{3 \pm i\sqrt{71}}{8}$ **53.** $\frac{\sqrt{3}}{2}$ inch, 1 inch

55. $x\sqrt{3}, 2x$ **57.** $\sqrt{2}$ inches   **59.** $\frac{\sqrt{2}}{2}$ inch   **61.** $x\sqrt{2}$ **63.** 781 feet

65. $\frac{1,170}{5,630} = 0.21$ to the nearest hundredth   **67.** $3\sqrt{5}$ **69.** $3y^2\sqrt{3y}$ **71.** $3x^2y\sqrt[3]{2y^2}$ **73.** 13

75. $\frac{3\sqrt{2}}{2}$ **77.** $\sqrt[3]{2}$ **79.** $x = \pm 2a$ **81.** $x = -a$ **83.** $x = 0, -2a$ **85.** $x = \frac{-p \pm \sqrt{p^2 - 4q}}{2}$

Problem Set 7.2

1. $-2, -3$ **3.** $2 \pm \sqrt{3}$ **5.** $1, 2$ **7.** $\frac{2 \pm i\sqrt{14}}{3}$ **9.** $0, 5$ **11.** $0, -\frac{4}{3}$ **13.** $\frac{3 \pm \sqrt{5}}{4}$

15. $-3 \pm \sqrt{17}$ **17.** $\frac{-1 \pm i\sqrt{5}}{2}$ **19.** 1   **21.** $\frac{1 \pm i\sqrt{47}}{6}$ **23.** $4 \pm \sqrt{2}$ **25.** $\frac{1}{2}, 1$ **27.** $-\frac{1}{2}, 3$

29. $\frac{-1 \pm i\sqrt{7}}{2}$ **31.** $1 \pm \sqrt{2}$ **33.** $\frac{-3 \pm \sqrt{5}}{2}$ **35.** $-5, 3$ **37.** $2, -1 \pm i\sqrt{3}$ **39.** $-\frac{3}{2}, \frac{3 \pm 3i\sqrt{3}}{4}$

41. $\frac{1}{5}, \dfrac{-1 \pm i\sqrt{3}}{10}$ **43.** $0, \dfrac{-1 \pm i\sqrt{5}}{2}$ **45.** $0, 1 \pm i$ **47.** $0, \dfrac{-1 \pm i\sqrt{2}}{3}$ **49.** $\dfrac{-3 - 2i}{5}$

51. 2 seconds **53.** $\frac{1}{4}$ second and 1 second **55.** $40 \pm 20 = 20$ or 60 items

57. $\dfrac{3.5 \pm 0.5}{0.004} = 750$ or 1,000 patterns **59.** $4y + 1 + \dfrac{-2}{2y - 7}$ **61.** $x^2 + 7x + 12$ **63.** 5 **65.** $\frac{27}{125}$

67. $\frac{1}{4}$ **69.** $21x^3 y$ **71.** $-2\sqrt{3}, \sqrt{3}$ **73.** $\dfrac{-1 \pm \sqrt{3}}{\sqrt{2}} = \dfrac{-\sqrt{2} \pm \sqrt{6}}{2}$ **75.** $-2i, i$ **77.** $-i, 4i$

Problem Set 7.3

1. $D = 16$, two rational **3.** $D = 0$, one rational **5.** $D = 5$, two irrational **7.** $D = 17$, two irrational
9. $D = 36$, two rational **11.** $D = 116$, two irrational **13.** ± 10 **15.** ± 12 **17.** 9 **19.** -16
21. $\pm 2\sqrt{6}$ **23.** $x^2 - 7x + 10 = 0$ **25.** $t^2 - 3t - 18 = 0$ **27.** $y^3 - 4y^2 - 4y + 16 = 0$
29. $2x^2 - 7x + 3 = 0$ **31.** $4t^2 - 9t - 9 = 0$ **33.** $6x^3 - 5x^2 - 54x + 45 = 0$ **35.** $10a^2 - a - 3 = 0$
37. $9x^3 - 9x^2 - 4x + 4 = 0$ **39.** $x^4 - 13x^2 + 36 = 0$ **41.** $(x - 3)(x + 5)^2 = 0$ or $x^3 + 7x^2 - 5x - 75 = 0$
43. $(x - 3)^2(x + 3)^2 = 0$ or $x^4 - 18x^2 + 81 = 0$ **45.** $-3, -2, -1$ **47.** $-3, -4, 2$ **49.** $1 \pm i$
51. $5, 4 \pm 3i$ **53.** $a^{11/2} - a^{9/2}$ **55.** $x^3 - 6x^{3/2} + 9$ **57.** $6x^{1/2} - 5x$ **59.** $5(x - 3)^{1/2}(2x - 9)$
61. $(2x^{1/3} - 3)(x^{1/3} - 4)$ **63.** $-2, 1, i, -i$ **65.** $x = -a$ is a solution of multiplicity 3 **67.** $-1, 5$
69. $1 + \sqrt{2} \approx 2.41, 1 - \sqrt{2} \approx -0.41$ **71.** $\frac{1}{2}, 1, -1$ **73.** $\frac{1}{2}, \sqrt{2} \approx 1.41, -\sqrt{2} \approx -1.41$ **75.** $\frac{2}{3}, 1 + i, 1 - i$

Problem Set 7.4

1. $1, 2$ **3.** $-\frac{5}{2}, -8$ **5.** $\pm 3, \pm i\sqrt{3}$ **7.** $\pm 2i, \pm i\sqrt{5}$ **9.** $\frac{7}{2}, 4$ **11.** $-\frac{9}{8}, \frac{1}{2}$ **13.** $\pm \dfrac{\sqrt{30}}{6}, \pm i$

15. $\pm \dfrac{\sqrt{21}}{3}, \pm \dfrac{i\sqrt{21}}{3}$ **17.** $4, 25$ **19.** Possible solutions 25 and 9; only 25 checks; 25

21. Possible solutions $\frac{25}{9}$ and $\frac{49}{4}$; only $\frac{25}{9}$ checks; $\frac{25}{9}$ **23.** $27, 38$ **25.** $4, 12$ **27.** $t = \dfrac{1 \pm \sqrt{1 + h}}{4}$

29. $t = \dfrac{v \pm \sqrt{v^2 + 1{,}280}}{32}$ **35.** $t = \dfrac{v \pm \sqrt{v^2 + 64h}}{32}$ **37.** $x = \dfrac{-4 \pm 2\sqrt{4 - k}}{k}$ **39.** $x = -y$

41. $3\sqrt{7}$ **43.** $5\sqrt{2}$ **45.** $39x^2 y\sqrt{5x}$ **47.** $-11 + 6\sqrt{5}$ **49.** $x + 4\sqrt{x} + 4$ **51.** $\dfrac{7 + 2\sqrt{7}}{3}$

53. x-intercepts $= 0, 2, -2$; y-intercept $= 0$ **55.** x-intercepts $= -\frac{1}{3}, 3, -3$; y-intercept $= -9$ **57.** $\frac{1}{2}$ and -1

Problem Set 7.5

1.

3.

5.

7.

9.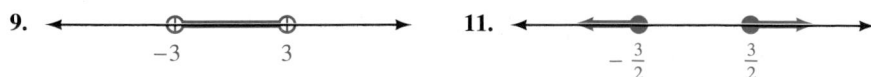

11.

13. ←———○═══○———→
          -1 $\frac{3}{2}$

15. All real numbers   **17.** No solution; $\varnothing$

19. ←———○─○─○→——→
            2 3 4

21. ←—●——●●—→
          -3 -2 -1

23. ←——○━━━━━●——→
          -4 1

25. ←———○━━━━○———→
          -6 $\frac{8}{3}$

27. ←——←○———○→——→
            2   6

29. ←——←○———○─○——→
          -3 2 4

31. ←——○─○─○→——→
            2 3 4

33. ←————●━━━━○——→
              5   6

35. Positive for $x < -5$ or $x > 5$; negative for $-5 < x < 5$

37. Positive for $x < 1$ or $x > 5$; negative for $1 < x < 5$ **39.** Always positive

41. $x \geq 4$; the width is at least 4 inches

43. $5 \leq p \leq 8$; she should charge at least \$5 but no more than \$8 for each radio

45.

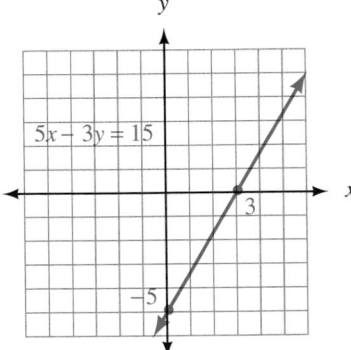

47.

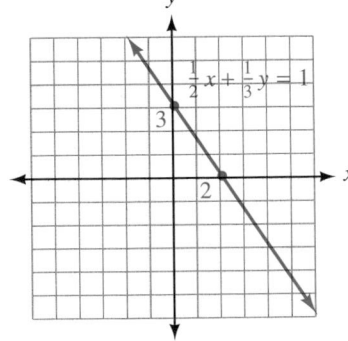

49. $\frac{5}{3}$ **51.** Possible solutions 1 and 6; only 6 checks; 6 **53.**

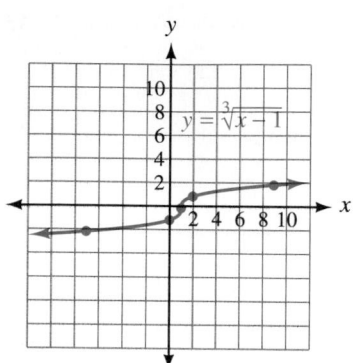

55.
$$1 - \sqrt{2} \qquad 1 + \sqrt{2}$$

57.
$$4 - \sqrt{3} \qquad 4 + \sqrt{3}$$

Problem Set 7.6

1. x-intercepts $= -3, 1$;
vertex $= (-1, -4)$

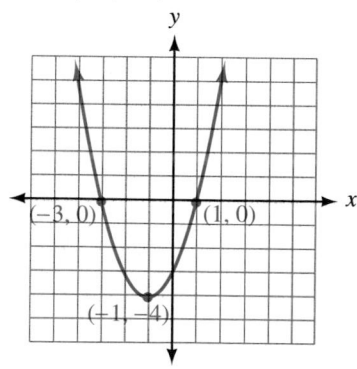

3. x-intercepts $= -5, 1$;
vertex $= (-2, 9)$

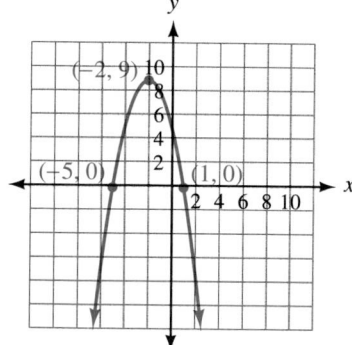

5. x-intercepts $= -1, 1$;
vertex $= (0, -1)$

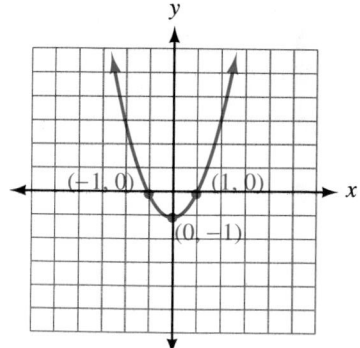

7. x-intercepts $= 3, -3$;
vertex $= (0, 9)$

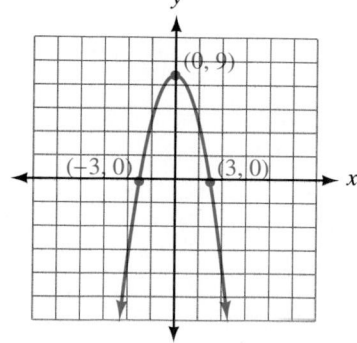

9. x-intercepts $= -1, 3$;
   vertex $= (1, -8)$

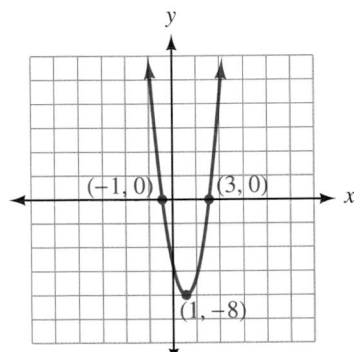

11. x-intercepts $= 1 + \sqrt{5}, 1 - \sqrt{5}$;
   vertex $= (1, -5)$

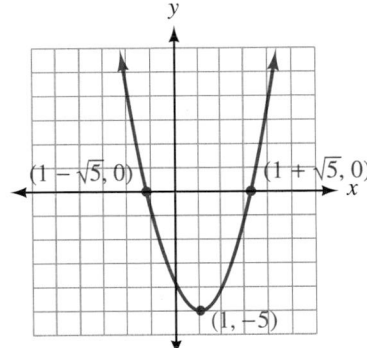

13. Vertex $= (2, -8)$

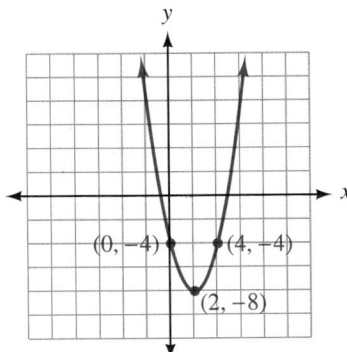

15. Vertex $= (1, -4)$

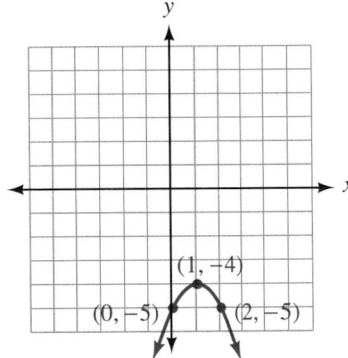

17. Vertex $= (0, 1)$

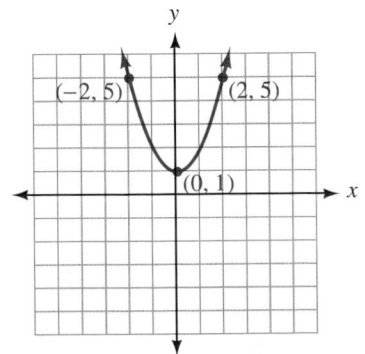

19. Vertex $= (0, -3)$

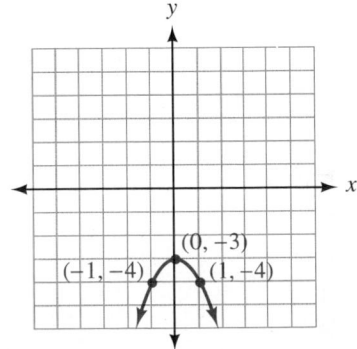

21. Vertex $= (-\frac{2}{3}, -\frac{1}{3})$ **23.** $(3, -4)$, lowest   **25.** $(1, 9)$, highest

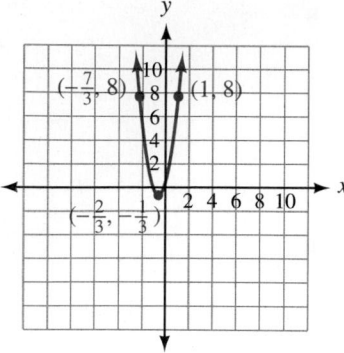

27. $(2, 16)$, highest   **29.** $(-4, 16)$, highest   **31.** 40 items; maximum profit $500
33. 875 patterns; maximum profit $731.25   **35.** 256 feet
37. Maximum $R = \$3,600$, when $p = \$6.00$

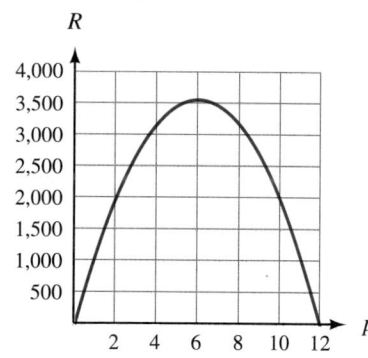

39. Maximum $R = \$7,225$, when $p = \$8.50$ **41.** $1 - i$ **43.** $27 + 5i$ **45.** $\frac{1}{10} + \frac{3}{10}i$

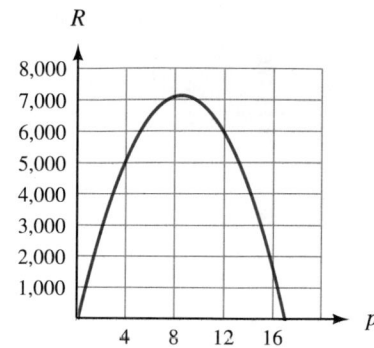

Problem Set 7.7

1.

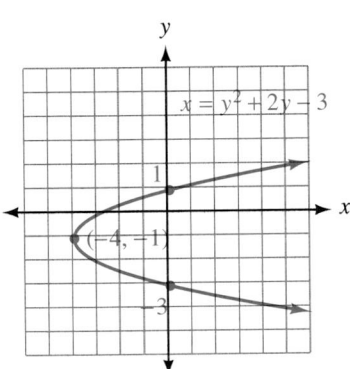

3.

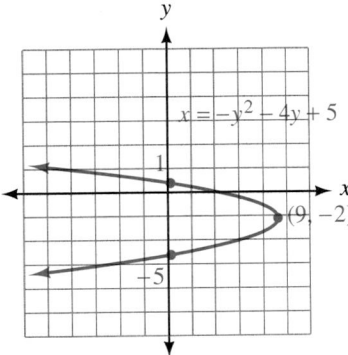

5.

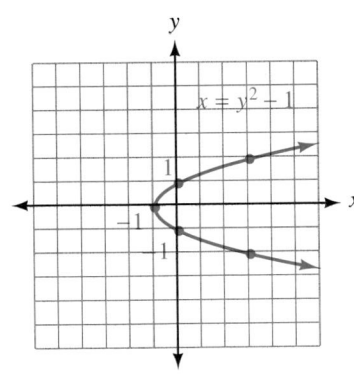

7.

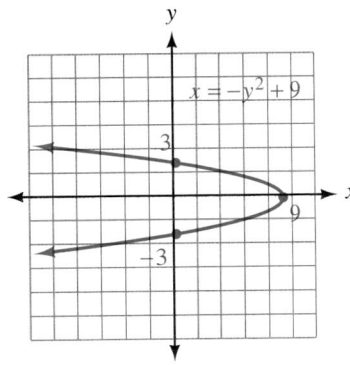

9.

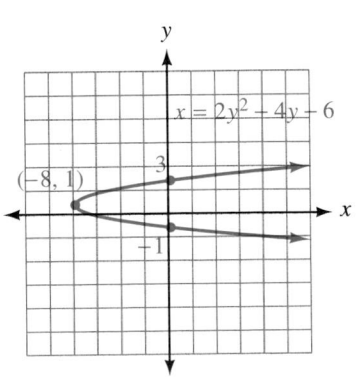

11.

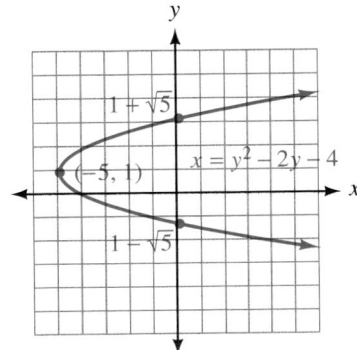

13. $2i\sqrt{5}$ **15.** 1 **17.** $-4 - 3i$

23.

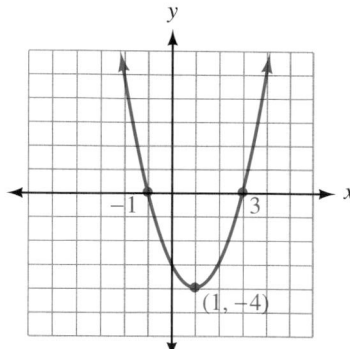

24.

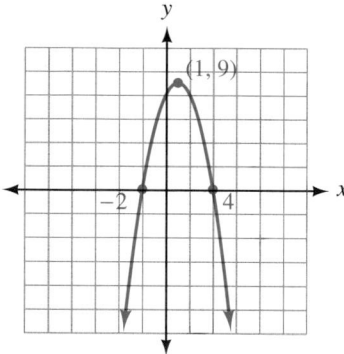

25. Maximum profit = $900 by selling 100 items per week

CHAPTER 8

Problem Set 8.1

1. (4, 3)

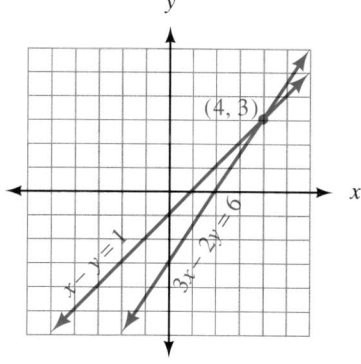

3. (−5, −6)

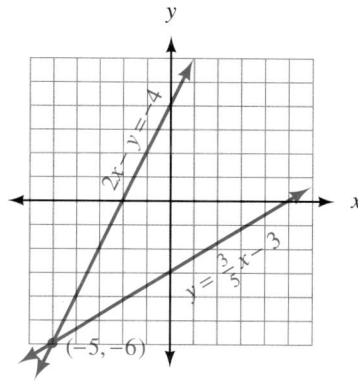

5. (4, 2)

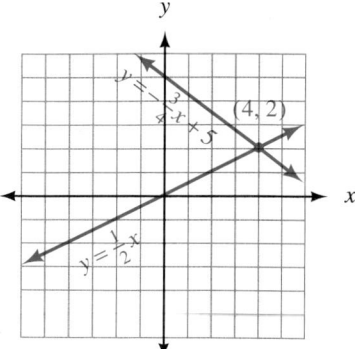

7. Lines are parallel; there is no solution

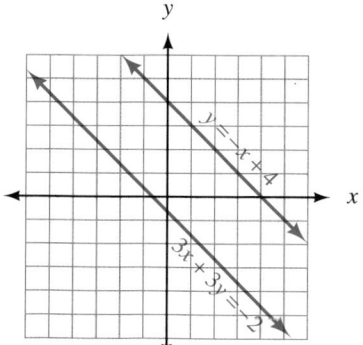

9. Lines coincide; any solution to one of the equations is a solution to the other

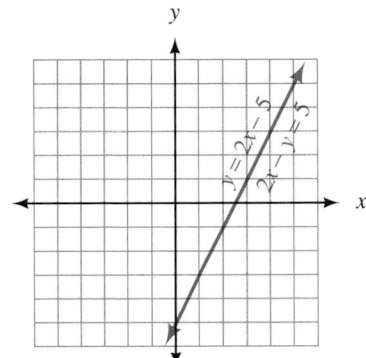

11. $(2, 3)$ **13.** $(1, 1)$ **15.** Lines coincide: $\{(x, y) | 3x - 2y = 6\}$ **17.** $(1, -\frac{1}{2})$ **19.** $(\frac{1}{2}, -3)$
21. $(-\frac{8}{3}, 5)$ **23.** $(2, 2)$ **25.** Lines are parallel; $\varnothing$ **27.** $(12, 30)$ **29.** $(10, 24)$ **31.** $(4, \frac{10}{3})$
33. $(3, -3)$ **35.** $(\frac{4}{3}, -2)$ **37.** $(6, 2)$ **39.** $(2, 4)$ **41.** Lines coincide; $\{(x, y) | 2x - y = 5\}$
43. Lines coincide; $\{(x, y) | x = \frac{3}{2}y\}$ **45.** $(-\frac{15}{43}, -\frac{27}{43})$ **47.** $(\frac{60}{43}, \frac{46}{43})$ **49.** $(\frac{9}{41}, -\frac{11}{41})$ **51.** $(6{,}000, 4{,}000)$
53. 2 **55.** (a) 3.29 (b) $y = 30x + 45$ (c) 2 minutes   **57.** $-2, 3$ **59.** 25 **61.** $5 \pm \sqrt{17}$
63. $1 \pm i$ **65.** $a = 3, b = -2$ **67.** $a = 1, b = -2,$ vertex $= (1, -1)$

Problem Set 8.2

1. $(1, 2, 1)$ **3.** $(2, 1, 3)$ **5.** $(2, 0, 1)$ **7.** $(\frac{1}{2}, \frac{2}{3}, -\frac{1}{2})$ **9.** No solution, inconsistent system
11. $(4, -3, -5)$ **13.** No unique solution   **15.** $(4, -5, -3)$ **17.** No unique solution   **19.** $(\frac{1}{2}, 1, 2)$
21. $(\frac{1}{2}, \frac{1}{3}, \frac{1}{4})$ **23.** $(1, 3, 1)$ **25.** $(-1, 2, -2)$ **27.** 4 amp, 3 amp, 1 amp   **29.** $y = \frac{2}{3}x - 2$

31. $y = \frac{2}{3}x - \frac{5}{3}$ **33.** $\dfrac{-2 \pm \sqrt{10}}{2}$ **35.** $\dfrac{2 \pm i\sqrt{3}}{2}$ **37.** $5, \dfrac{-5 \pm 5i\sqrt{3}}{2}$ **39.** $0, 5, -1$ **41.** $\dfrac{3 \pm \sqrt{29}}{2}$

43. $a = 1, b = -6, c = 5$ **45.** $(1, 2, 3, 4)$

Problem Set 8.3

1. 3 **3.** 5 **5.** -1 **7.** 0 **9.** 10 **11.** 2 **13.** -3 **15.** -2 **17.** $-2, 5$ **19.** 3

21. 0 **23.** 3 **25.** 8 **27.** 6 **29.** -228 **31.** $\begin{vmatrix} y & x \\ m & 1 \end{vmatrix} = y - mx = b; y = mx + b$

33. $D = -7$; two complex   **35.** $x^2 - 2x - 15 = 0$ **37.** $3y^2 - 11y + 6 = 0$ **39.** $2, 3$ (multiplicity 2)
41. 4 **43.** 4

Problem Set 8.4

1. $(3, 1)$ **3.** Lines are parallel; $\varnothing$ **5.** $(-\frac{15}{43}, -\frac{27}{43})$ **7.** $(\frac{60}{43}, \frac{46}{43})$ **9.** $(3, -1, 2)$ **11.** $(\frac{1}{2}, \frac{5}{2}, 1)$
13. No unique solution   **15.** $(-\frac{10}{91}, -\frac{9}{13}, \frac{107}{91})$ **17.** $(\frac{71}{13}, -\frac{12}{13}, \frac{24}{13})$ **19.** $(3, 1, 2)$ **21.** $x = 50$ items

23. $\pm 2, \pm i\sqrt{2}$ **25.** $1, 2$ **27.** $1, \frac{9}{4}$ **29.** $5, 1$ **31.** $\dfrac{-2 \pm \sqrt{4 + k^2}}{k}$

33. $x = -\dfrac{1}{a - b} = \dfrac{1}{b - a}, y = \dfrac{1}{a - b}$ **35.** $x = \dfrac{1}{a^2 + ab + b^2}, y = \dfrac{a + b}{a^2 + ab + b^2}$ **37.** $\begin{aligned} x + 2y &= 1 \\ 3x + 4y &= 0 \end{aligned}$

Problem Set 8.5

1. $(2, 3)$ **3.** $(-1, -2)$ **5.** $(7, 1)$ **7.** $(1, 2, 1)$ **9.** $(2, 0, 1)$ **11.** $(1, 1, 2)$ **13.** $(4, 1, 5)$
15. $(4, \frac{10}{3})$ **21.** $4, 7$ **23.** Boots are \$5; a suit is \$18   **25.** \$18 and \$26

Problem Set 8.6

1. $y = 2x + 3$, $x + y = 18$; the two numbers are 5 and 13   **3.** 10, 16   **5.** 1, 3, 4
7. Let x = the number of adult tickets and y = the number of children's tickets.
   $x + y = \quad 925 \qquad$ 225 adult and 700 children's tickets
   $2x + y = 1,150$
9. Let x = the amount invested at 6% and y = the amount invested at 7%.
   $x + y = 20,000 \qquad$ He has \$12,000 at 6% and \$8,000 at 7%
   $0.06x + 0.07y = 1,280$
11. \$4,000 at 6%, \$8,000 at 7.5%   **13.** \$200 at 6%, \$1,400 at 8%, \$600 at 9%
15. 3 gallons of 50%, 6 gallons of 20%   **17.** 5 gallons of 20%, 10 gallons of 14%
19. Let x = the speed of the boat and y = the speed of the current.
   $3(x - y) = 18 \qquad$ The speed of the boat is 9 mph
   $2(x + y) = 24 \qquad$ The speed of the current is 3 mph
21. 270 mph airplane, 30 mph wind   **23.** 12 nickels, 8 dimes   **25.** 3 of each
27. $x = -200p + 700$, when $p = \$3$, $x = 100$ items   **29.** $C = 0.75x + 21.85$; \$30.85
31. $h = -16t^2 + 64t + 80$, maximum height is 144 feet when $t = 2$ seconds   **33.** $(-2, -23)$, lowest
35. $(\frac{3}{2}, 9)$, highest   **37.** 2 seconds, 64 feet

Chapter 8 Review

1. $(6, -2)$ **2.** $(2, -4)$ **3.** Lines are parallel   **4.** Lines coincide   **5.** $(1, -1)$ **6.** $(0, 1)$
7. Lines coincide   **8.** Lines are parallel   **9.** $(3, 1)$ **10.** $(\frac{3}{2}, 0)$ **11.** $(3, 5)$ **12.** $(\frac{4}{3}, \frac{13}{3})$ **13.** $(4, 8)$
14. $(3, 12)$ **15.** $(3, -2)$ **16.** $(2, 1)$ **17.** $(\frac{3}{2}, \frac{1}{2})$ **18.** $(-\frac{3}{2}, -\frac{7}{2})$ **19.** $(4, 1)$ **20.** $(3, 2)$
21. $(6, -2)$ **22.** $(7, -4)$ **23.** $(-5, 3)$ **24.** $(8, -5)$ **25.** Parallel lines   **26.** Lines coincide
27. $(3, -1, 4)$ **28.** $(-2, 3, 4)$ **29.** $(2, \frac{1}{2}, -3)$ **30.** $(3, 0, \frac{1}{2})$ **31.** $(-1, \frac{1}{2}, \frac{3}{2})$ **32.** $(2, \frac{1}{3}, \frac{2}{3})$
33. Dependent system   **34.** Dependent system   **35.** $(2, -1, 4)$ **36.** $(4, 0, 9)$ **37.** Dependent system
38. $(3, 1, -2)$ **39.** 23   **40.** -3 **41.** -3 **42.** 30   **43.** -16 **44.** 36   **45.** -2
46. -46 **47.** $\frac{4}{7}$ **48.** $\frac{5}{22}$ **49.** $-\frac{3}{2}, \frac{1}{2}$ **50.** $-\frac{3}{2}, \frac{1}{4}$ **51.** $(\frac{7}{29}, -\frac{19}{29})$ **52.** $(\frac{34}{41}, -\frac{18}{41})$
53. Lines coincide   **54.** Lines coincide   **55.** $(\frac{11}{14}, -\frac{1}{14}, \frac{11}{14})$ **56.** $(\frac{1}{7}, -\frac{18}{77}, \frac{45}{77})$ **57.** $(\frac{17}{27}, -\frac{11}{81})$
58. $(-\frac{50}{33}, -\frac{23}{33})$ **59.** $(\frac{113}{46}, \frac{59}{23}, -\frac{7}{23})$ **60.** $(\frac{59}{37}, \frac{92}{37}, \frac{54}{37})$ **61.** Inconsistent   **62.** Inconsistent
63. 2, 9   **64.** 1, 9   **65.** $-6, 3, 5$ **66.** 2, 3, 4   **67.** 47 adults, 80 children
68. 12 dimes, 8 quarters   **69.** \$5,000 at 12%, \$7,000 at 15%   **70.** \$6,000 at 18%, \$3,000 at 13%
71. 7.5 ounces 30%, 7.5 ounces 70%   **72.** 6 gallons 25%, 14 gallons 50%   **73.** 12 mph boat, 2 mph river
74. 8 mph boat, 4 mph river

Chapter 8 Test

1. $(1, 2)$ **2.** $(3, 2)$ **3.** $(15, 12)$ **4.** $(-\frac{54}{13}, -\frac{58}{13})$ **5.** $(1, 2)$ **6.** $(3, -2, 1)$ **7.** -14 **8.** -26
9. $(-\frac{14}{3}, -\frac{19}{3})$ **10.** Lines coincide $\{(x, y) | 2x + 4y = 3\}$ **11.** $(\frac{5}{11}, -\frac{15}{11}, -\frac{1}{11})$ **12.** $(-3, -5)$
13. $(-2, 3, 5)$ **14.** 5, 9   **15.** \$4,000 at 5%, \$8,000 at 6%   **16.** 340 adults, 410 children
17. 4 gallons 30%, 12 gallons 70%   **18.** Boat 8 mph, current 2 mph   **19.** 11 nickels, 3 dimes, 1 quarter
20. 3 ounces of cereal I, 1 ounce of cereal II

CHAPTER 9

Problem Set 9.1

1. $\{x \mid x \geq \frac{1}{2}\}$ **3.** $\{x \mid x \leq \frac{1}{4}\}$ **5.** $\{x \mid x \neq 5\}$ **7.** $\{x \mid x \neq \frac{1}{2}, x \neq -3\}$

9.

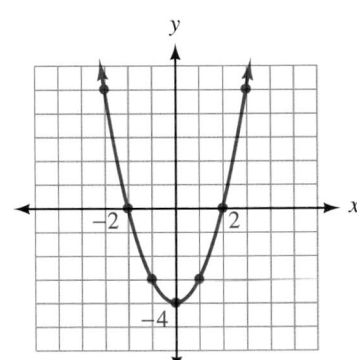

Domain = all reals
range = $\{y \mid y \geq -4\}$
a function

11.

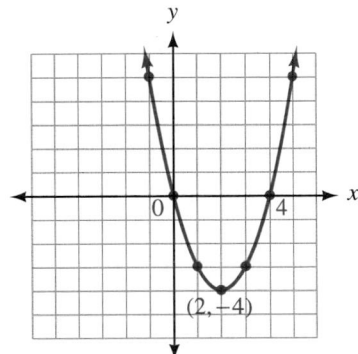

Domain = all reals
range = $\{y \mid y \geq -4\}$
a function

13.

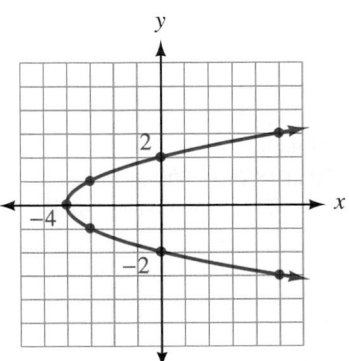

Domain = $\{x \mid x \geq -4\}$
range = all reals
not a function

15.

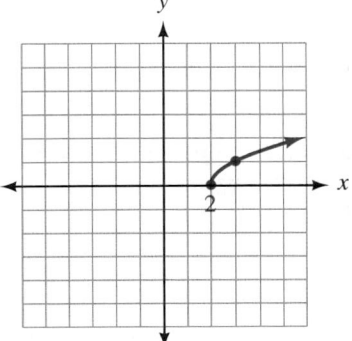

Domain = $\{x \mid x \geq 2\}$
range = $\{y \mid y \geq 0\}$
a function

17.

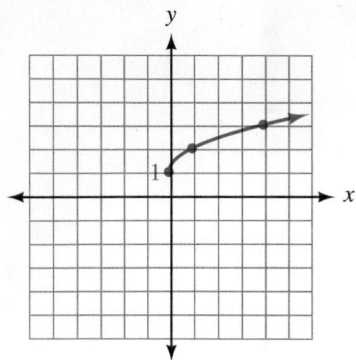

Domain = $\{x \mid x \geq 0\}$
range = $\{y \mid y \geq 1\}$
a function

19.

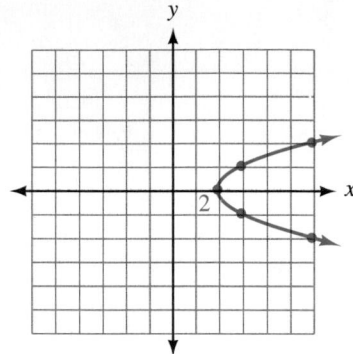

Domain = $\{x \mid x \geq 2\}$
range = all reals
not a function

21.

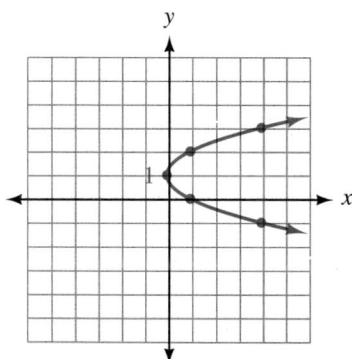

Domain = $\{x \mid x \geq 0\}$
range = all reals
not a function

23. 3   **25.** 4   **27.** $x + a$ **29.** $x + a$ **31.** 2   **33.** -4 **35.** $3(2x + h)$ **37.** $4x - 7$
39. $3x^2 - 10x + 8$ **41.** $-2x + 3$ **43.** $3x^2 - 11x + 10 = h(x)$ **45.** $9x^3 - 48x^2 + 85x - 50$

47. $x - 2 = g(x)$ **49.** $\dfrac{1}{x - 2} = \dfrac{1}{g(x)}$ **51.** $3x^2 - 7x + 3$ **53.** $6x^2 - 22x + 20 = 2h(x)$ **55.** $(1, 2)$

57. $(\frac{5}{2}, -1)$ **59.** $(0, 3)$ **61.** $(3, 4)$ **63.** $y = \dfrac{6}{x - 2}$ **65.** $y = \dfrac{2}{x - 4}$

x	y
1.5	-12
1.7	-20
1.9	-60
2.1	60
2.3	20
2.5	12

x	y
3.9	-20
3.99	-200
4.01	200
4.1	20

Problem Set 9.2

1.

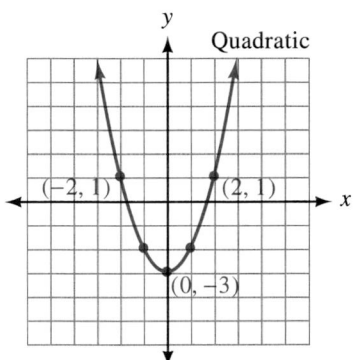

3.

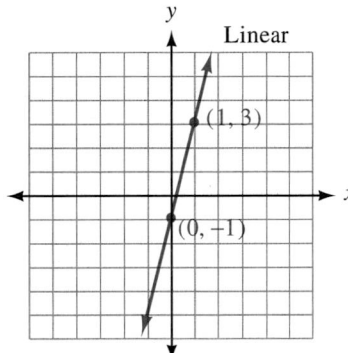

5.

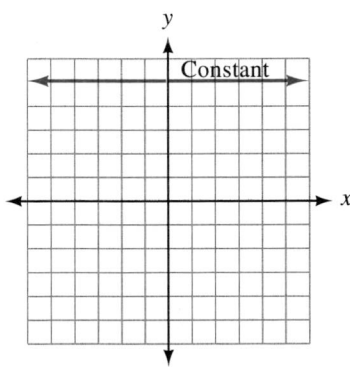

7.

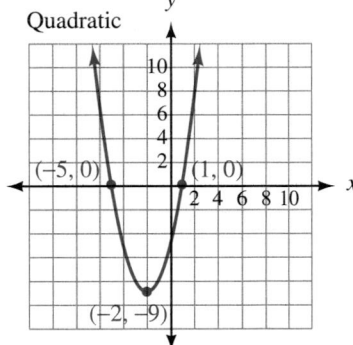

9.

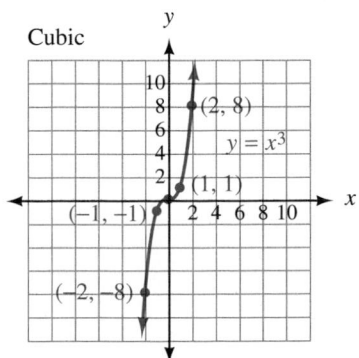

11.

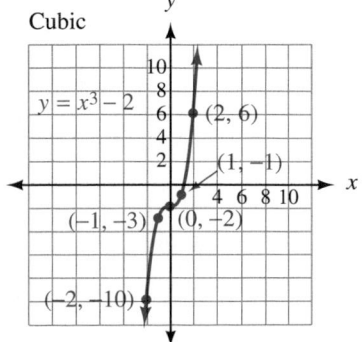

13.

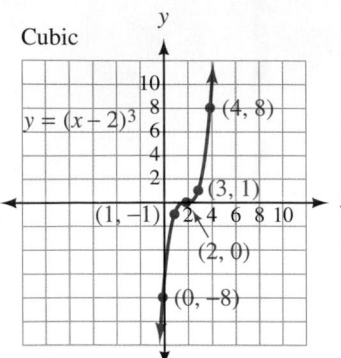

Cubic

$y = (x - 2)^3$

$(4, 8)$

$(3, 1)$

$(1, -1)$

$(2, 0)$

$(0, -8)$

15.

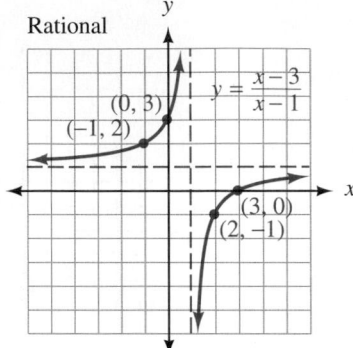

Rational

$y = \dfrac{x - 3}{x - 1}$

$(0, 3)$

$(-1, 2)$

$(3, 0)$

$(2, -1)$

17.

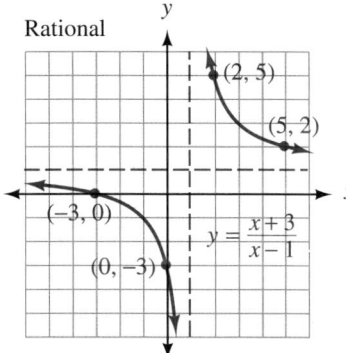

Rational

$(2, 5)$

$(5, 2)$

$(-3, 0)$

$y = \dfrac{x + 3}{x - 1}$

$(0, -3)$

19.

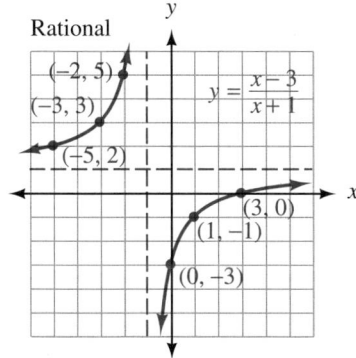

Rational

$(-2, 5)$

$(-3, 3)$

$y = \dfrac{x - 3}{x + 1}$

$(-5, 2)$

$(3, 0)$

$(1, -1)$

$(0, -3)$

21. 3   **23.** $-\frac{3}{2}, \frac{5}{3}$ **25.** $-2 \pm \sqrt{3}$ **27.** $-\frac{1}{2}, 3, -3$

29. At 12:00 noon there is a maximum of 90 people in the store.   **31.** 1   **33.** 2   **35.** $\frac{1}{27}$ **37.** 13

39.

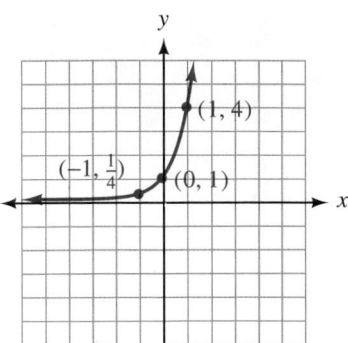

$(1, 4)$

$\left(-1, \frac{1}{4}\right)$

$(0, 1)$

41.

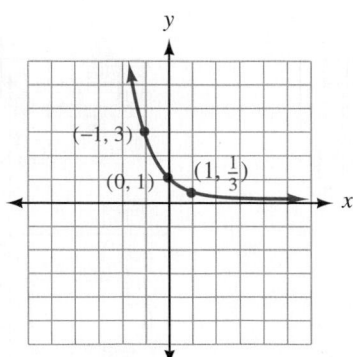

$(-1, 3)$

$(0, 1)$

$\left(1, \frac{1}{3}\right)$

43.

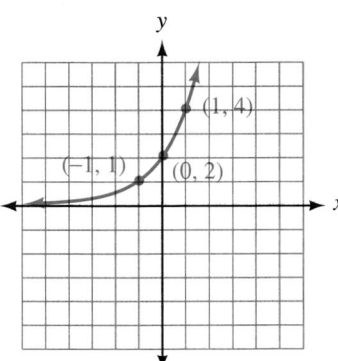

45.

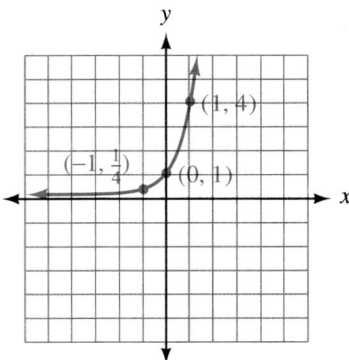

47. 200, 400, 800, 1,600; approximately 10 days   **49.** (1, 2, 3)   **51.** (1, 3, 1)

53.

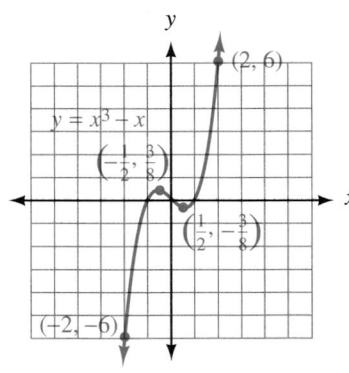

55.

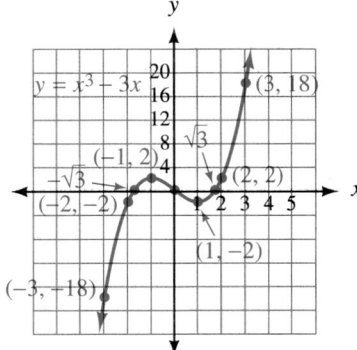

57.

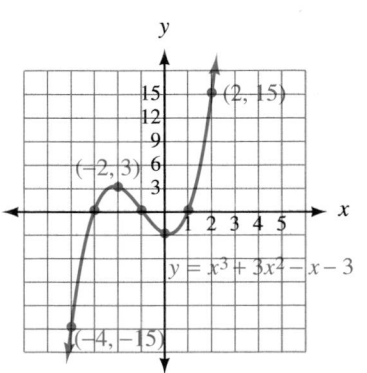

59.

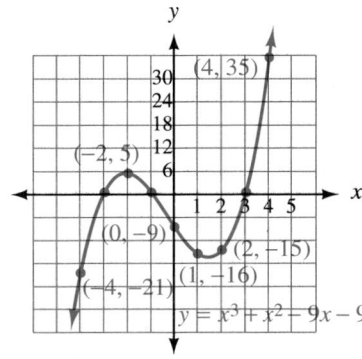

Problem Set 9.3

1. $f^{-1}(x) = \dfrac{x+1}{3}$ **3.** $f^{-1}(x) = \sqrt[3]{x}$ **5.** $f^{-1}(x) = \dfrac{x-3}{x-1}$ **7.** $f^{-1}(x) = 4x + 3$

9. $f^{-1}(x) = 2(x+3) = 2x + 6$ **11.** $f^{-1}(x) = \dfrac{1-x}{3x-2}$

13.

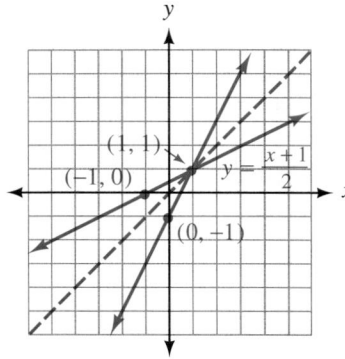

15.

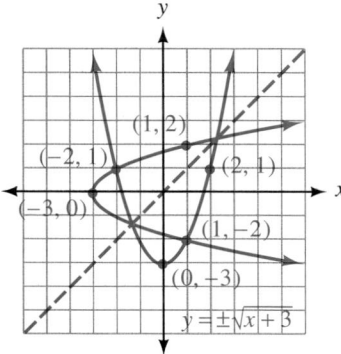

17.

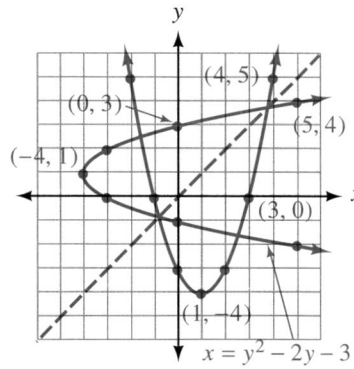

19.

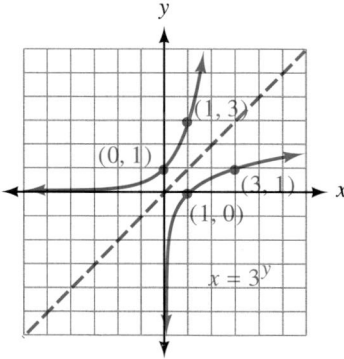

21.

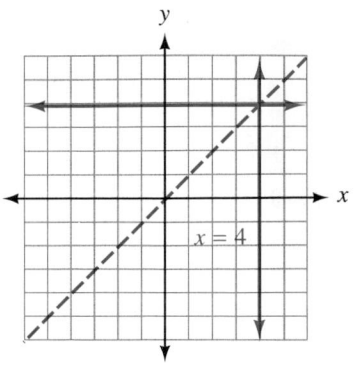

23.

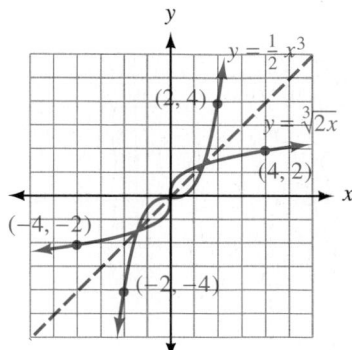

25.   **27.**

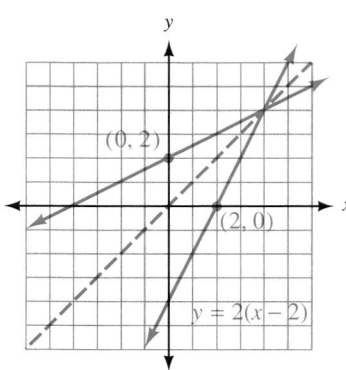   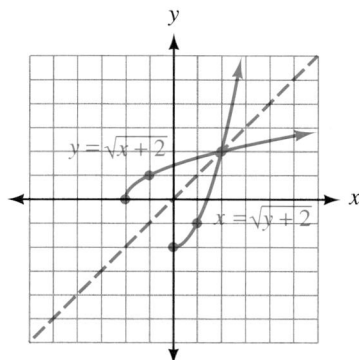

29. (a) 4   (b) $\frac{4}{3}$ (c) 2   (d) 2   **31.** $f^{-1}(x) = \dfrac{1}{x}$ **33.** 60 geese, 48 ducks

35. 150 oranges, 144 apples   **37.** $f^{-1}(x) = \dfrac{x-5}{3}$ **39.** $f^{-1}(x) = \sqrt[3]{x-1}$

41. $f^{-1}(x) = \dfrac{2x-4}{x-1}$

Problem Set 9.4

1. 5   **3.** $\sqrt{106}$ **5.** $\sqrt{61}$ **7.** $\sqrt{130}$ **9.** 3 or -1 **11.** 3   **13.** $(x-2)^2 + (y-3)^2 = 16$
15. $(x-3)^2 + (y+2)^2 = 9$ **17.** $(x+5)^2 + (y+1)^2 = 5$ **19.** $x^2 + (y+5)^2 = 1$
21. $x^2 + y^2 = 4$
23. Center = (0, 0)   **25.** Center = (1, 3)
   Radius = 2   Radius = 5

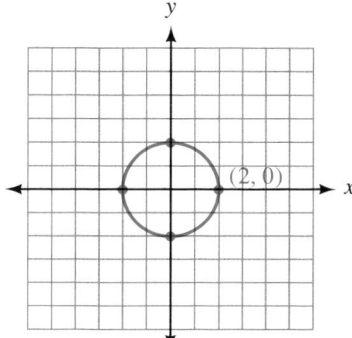   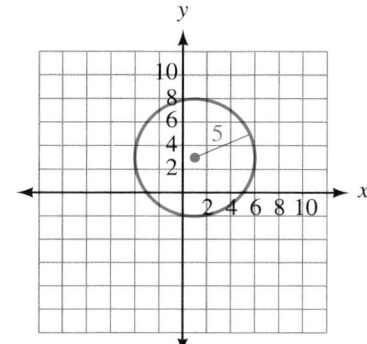

13.

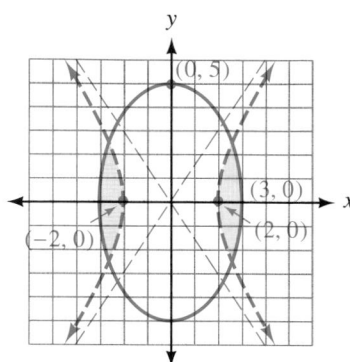

15. No intersection

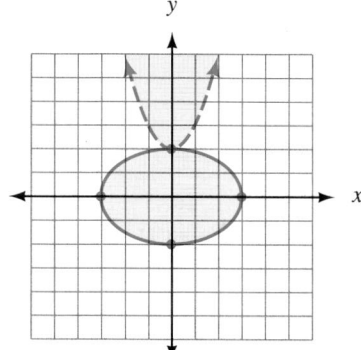

17.

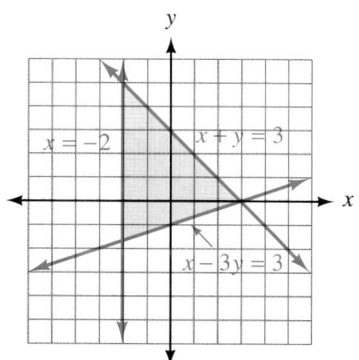

19.

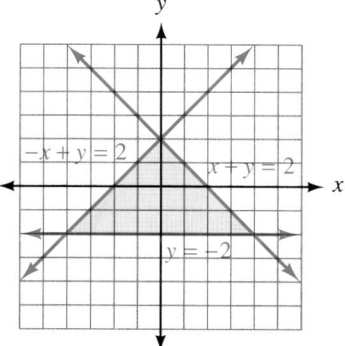

21.

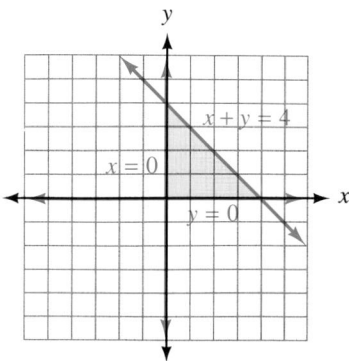

23. $(0, 3)$, $\left(\frac{12}{5}, -\frac{9}{5}\right)$ **25.** $(0, 4)$, $\left(\frac{16}{5}, \frac{12}{5}\right)$

27. $(5, 0), (-5, 0)$ **29.** $(0, -3), (\sqrt{5}, 2), (-\sqrt{5}, 2)$ **31.** $(0, -4), (\sqrt{7}, 3), (-\sqrt{7}, 3)$ **33.** $(-4, 11), (\frac{5}{2}, \frac{5}{4})$
35. $(-4, 5), (1, 0)$ **37.** $(2, -3), (5, 0)$ **39.** $(3, 0), (-3, 0)$ **41.** $(4, 0), (0, -4)$
43. 8, 5 or $-8, -5$ or 8, -5 or $-8, 5$ **45.** 6, 3 or 13, -4 **47.** $(5, -2)$ **49.** $(1, 2, 3)$ **51.** $(1, 3, 1)$

Chapter 9 Review

1. $\{x \mid x \geq -2\}$ **2.** $\{x \mid x \leq \frac{3}{2}\}$ **3.** $\{x \mid x \neq 6\}$ **4.** $\{x \mid x \neq -2, x \neq 4\}$ **5.** 2   **6.** $2x + h$ **7.** 2
8. $x + a$ **9.** $3x - 1$ **10.** $2x^2 + 2x - 12 = h(x)$ **11.** $2x - 4 = g(x)$ **12.** $4x^2 + 4x - 24 = 2h(x)$
13. $4x^2 - 16x + 16$ **14.** $8x - 6$ **15.** 5   **16.** 7   **17.** -12 **18.** 2   **19.** $\frac{5}{2}$ **20.** $-2, 7$
21. $3, -3, -2$ **22.** $2, -2, 3$ **23.** Linear   **24.** Cubic   **25.** Exponential
26. Constant   **27.** Quadratic   **28.** Rational

29.

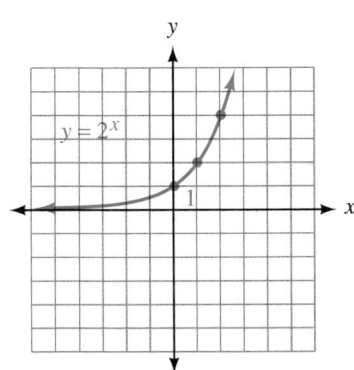

30.

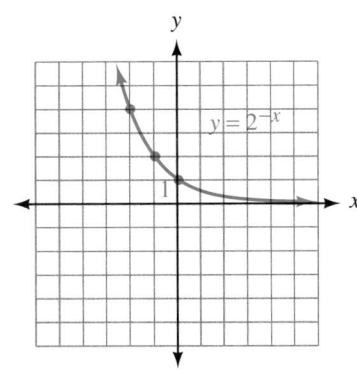

31.

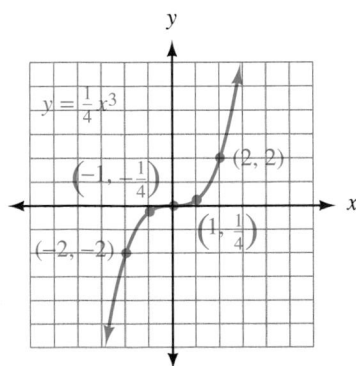

32.

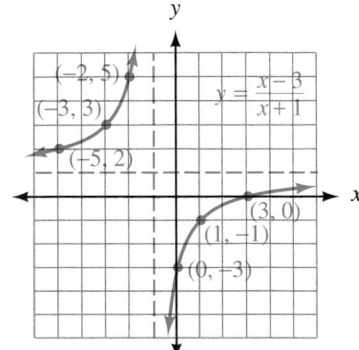

33. 16   **34.** $\frac{1}{2}$ **35.** $\frac{1}{9}$ **36.** -5 **37.** $\frac{5}{6}$ **38.** 7   **39.** $f^{-1}(x) = \dfrac{x - 3}{2}$ **40.** $y = \pm\sqrt{x + 1}$

41. $f^{-1}(x) = 2x - 4$ **42.** $y = \pm\sqrt{\dfrac{4 - x}{2}}$

43.

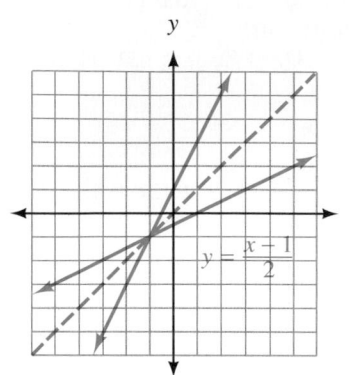

$y = \dfrac{x-1}{2}$

44.

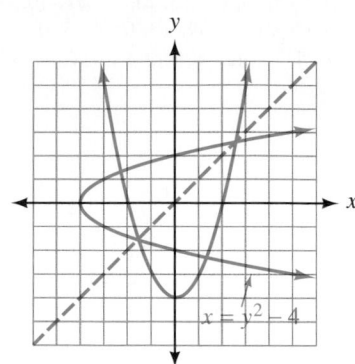

$x = y^2 - 4$

45. $\sqrt{10}$ **46.** $\sqrt{13}$ **47.** 5   **48.** 9   **49.** $-2, 6$ **50.** $-12, 4$ **51.** $(x-3)^2 + (y-1)^2 = 4$
52. $(x-3)^2 + (y+1)^2 = 16$ **53.** $(x+5)^2 + y^2 = 9$ **54.** $(x+3)^2 + (y-4)^2 = 18$ **55.** $x^2 + y^2 = 25$
56. $x^2 + y^2 = 9$ **57.** $(x+2)^2 + (y-3)^2 = 25$ **58.** $(x+6)^2 + (y-8)^2 = 100$
59. $(0, 0); r = 2$ **60.** $(3, -1); r = 4$

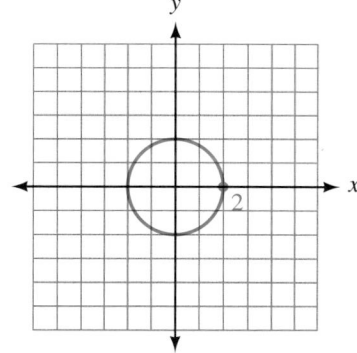

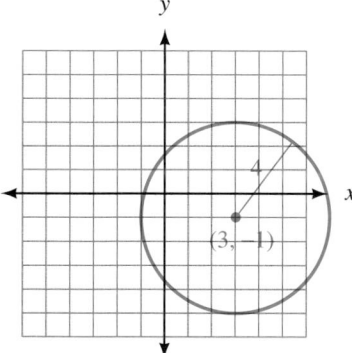

61. $(3, -2); r = 3$ **62.** $(-2, 1); r = 3$

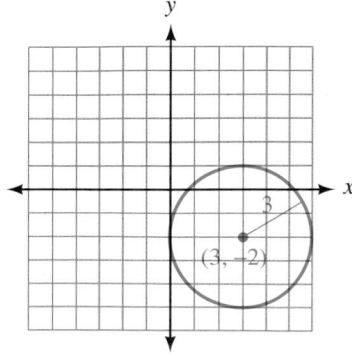

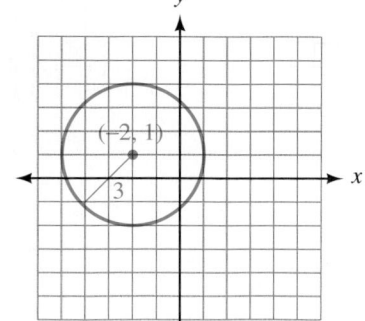

63.

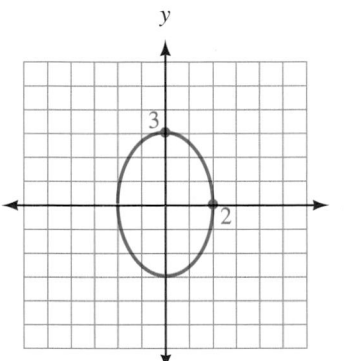

64.

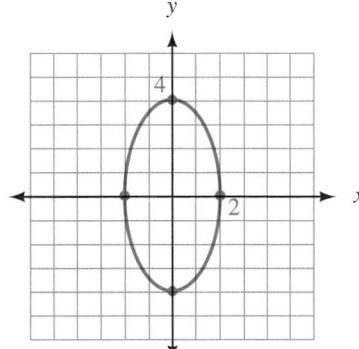

65.

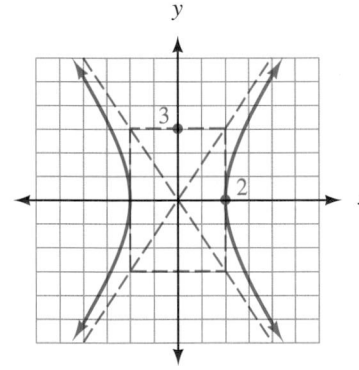

66.

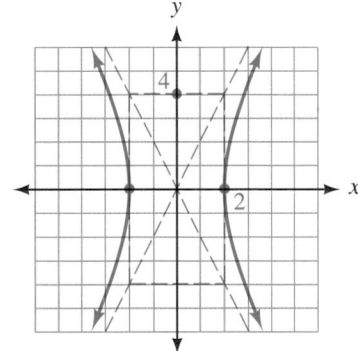

67.

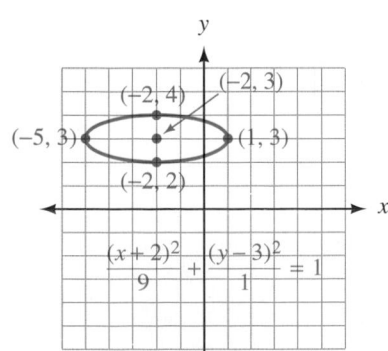

68.

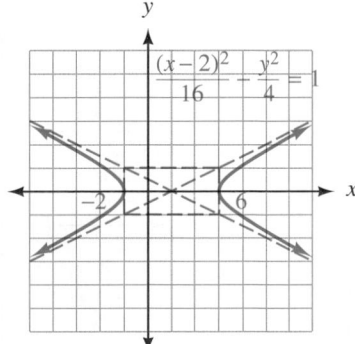

69.

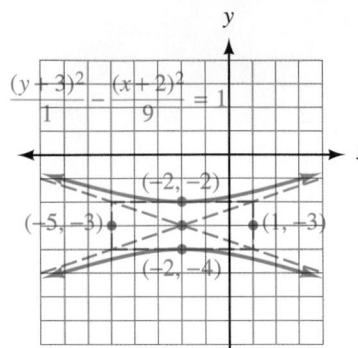

$$\frac{(y+3)^2}{1} - \frac{(x+2)^2}{9} = 1$$

70.

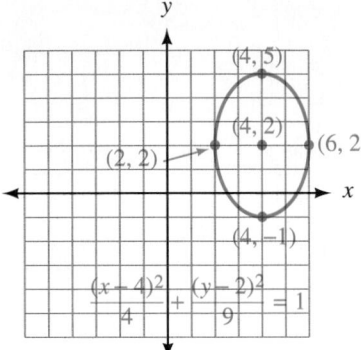

$$\frac{(x-4)^2}{4} + \frac{(y-2)^2}{9} = 1$$

71.

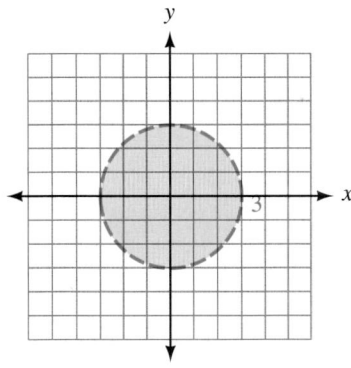

72.

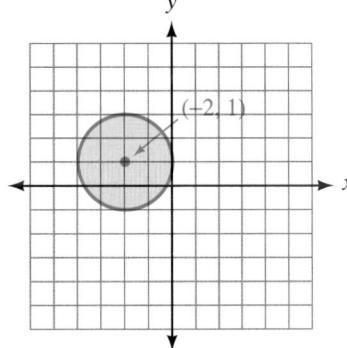

73.

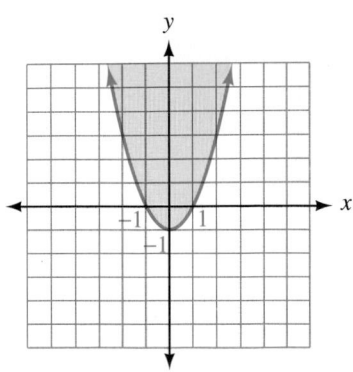

74.

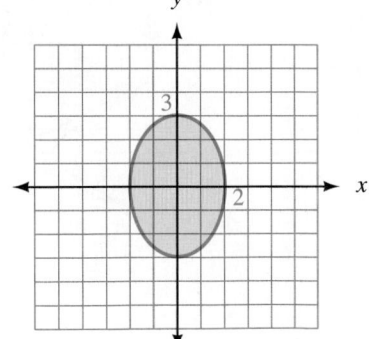

75.

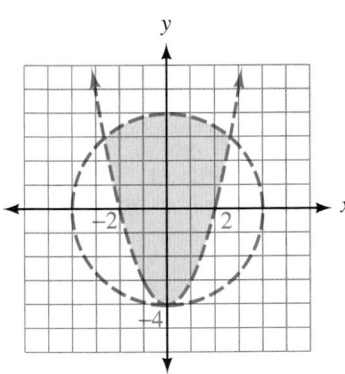

76.

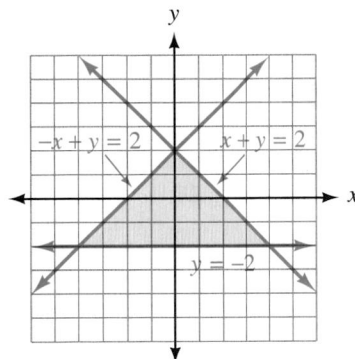

77. $(0, 4)$, $(\frac{16}{5}, -\frac{12}{5})$ **78.** $(0, -2)$, $(\sqrt{3}, 1)$, $(-\sqrt{3}, 1)$ **79.** $(-2, 0)$, $(2, 0)$
80. $(-\sqrt{7}, -\sqrt{6}/2)$, $(-\sqrt{7}, \sqrt{6}/2)$, $(\sqrt{7}, -\sqrt{6}/2)$, $(\sqrt{7}, \sqrt{6}/2)$

Chapter 9 Test

1. $x \geq 4$ **2.** $x \neq 4$, $x \neq -2$ **3.** $3x^2 - 2x - 8 = h(x)$ **4.** $-2x - 6$ **5.** $4x + 2$ **6.** $x - 2 = f(x)$

7.

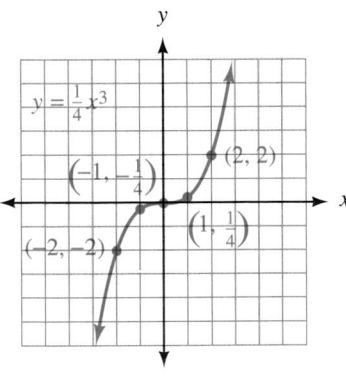

8.

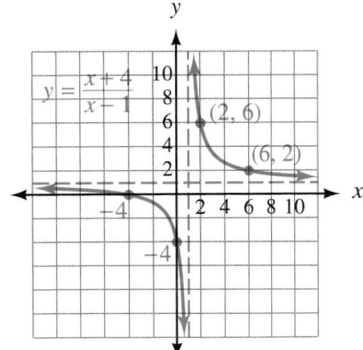

9.

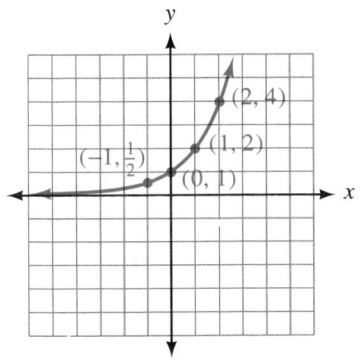

10.

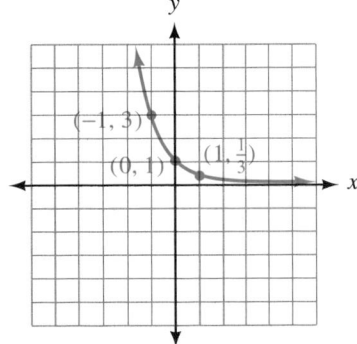

Problem Set 10.2

1. $\log_3 4 + \log_3 x$ **3.** $\log_6 5 - \log_6 x$ **5.** $5 \log_2 y$ **7.** $\frac{1}{3} \log_9 z$ **9.** $2 \log_6 x + 4 \log_6 y$
11. $\frac{1}{2} \log_5 x + 4 \log_5 y$ **13.** $\log_b x + \log_b y - \log_b z$ **15.** $\log_{10} 4 - \log_{10} x - \log_{10} y$
17. $2 \log_{10} x + \log_{10} y - \frac{1}{2} \log_{10} z$ **19.** $3 \log_{10} x + \frac{1}{2} \log_{10} y - 4 \log_{10} z$ **21.** $\frac{2}{3} \log_b x + \frac{1}{3} \log_b y - \frac{4}{3} \log_b z$

23. $\log_b xz$ **25.** $\log_3 \dfrac{x^2}{y^3}$ **27.** $\log_{10} \sqrt{x} \sqrt[3]{y}$ **29.** $\log_2 \dfrac{x^3 \sqrt{y}}{z}$ **31.** $\log_2 \dfrac{\sqrt{x}}{y^3 z^4}$ **33.** $\log_{10} \dfrac{x^{3/2}}{y^{3/4} z^{4/5}}$

35. $\frac{2}{3}$ **37.** 18   **39.** Possible solutions -1 and 3; only 3 checks; 3   **41.** 3
43. Possible solutions -2 and 4; only 4 checks; 4   **45.** Possible solutions -1 and 4; only 4 checks; 4
47. Possible solutions $-\frac{5}{2}$ and $\frac{5}{3}$; only $\frac{5}{3}$ checks; $\frac{5}{3}$ **49.** 2.52

55.

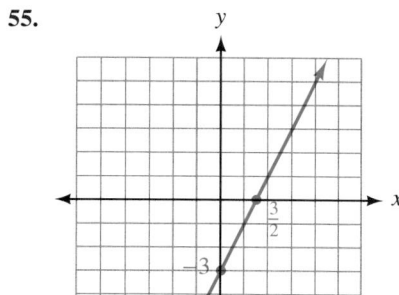

57.

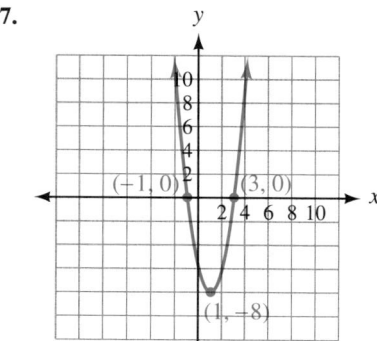

59. $x = \frac{3}{4}$ **61.** $x = -1 \pm \sqrt{2}$ **63.** $f^{-1}(x) = \dfrac{x - 3}{2}$ **65.** $y = \pm\sqrt{x + 4}$ **67.** $f^{-1}(x) = 5x + 3$

Problem Set 10.3

1. 2.5775   **3.** 1.5775   **5.** 3.5775   **7.** -1.4225 **9.** 4.5775   **11.** 2.7782   **13.** 3.3032
15. -2.0128 **17.** -1.5031 **19.** -0.3990 **21.** 759   **23.** 0.00759   **25.** 1,430
27. 0.00000447   **29.** 0.0000000918   **31.** 10^{10} **33.** 10^{-10} **35.** 10^{20} **37.** $\frac{1}{100}$ **39.** 1,000
41. 5   **43.** Approximately 3.19   **45.** 1.78×10^{-5} **47.** 3.16×10^5 **49.** 2.00×10^8
51. 10 times as large   **53.** 12.9%   **55.** 5.3%   **57.** 1   **59.** 5   **61.** x **63.** $\ln 10 + 3t$
65. $\ln A - 2t$ **67.** 2.7080   **69.** -1.0986 **71.** 2.1972   **73.** 2.7724   **75.** -1 or 3
77. Center $= (-3, 2)$, radius $= 4$

79.

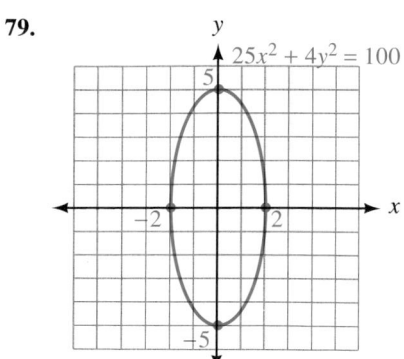

Problem Set 10.4

1. 1.4650   **3.** 0.6826   **5.** −1.5440   **7.** −0.6477   **9.** −0.3333   **11.** 2.0000   **13.** −0.1845
15. 0.1845   **17.** 1.6168   **19.** 2.1131   **21.** $15,529.24   **23.** $438.22   **25.** 11.7 years   **27.** 9.25 years
29. 1.3333   **31.** 0.7500   **33.** 1.3917   **35.** 0.7186   **37.** 2.6356   **39.** 4.1632   **41.** 5.8435
43. −1.0642   **45.** 2.3026   **47.** 10.7144   **49.** 13.9 years later or toward the end of 2007   **51.** 27.5 years

53. $t = \dfrac{1}{r} \ln \dfrac{A}{P}$ **55.** $t = \dfrac{1}{k} \dfrac{\log P - \log A}{\log 2}$ **57.** $t = \dfrac{\log A - \log P}{\log (1 - r)}$

59.

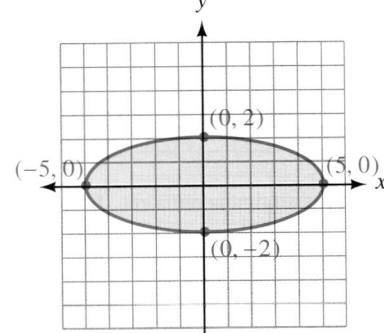

61. $(0, 4)$, $(0, -4)$ **63.** $(0, -2)$, $(\sqrt{3}, 1)$, $(-\sqrt{3}, 1)$

Chapter 10 Review

1. $\log_3 81 = 4$ **2.** $\log_7 49 = 2$ **3.** $\log_{10} 0.01 = -2$ **4.** $\log_2 \frac{1}{8} = -3$ **5.** $2^3 = 8$ **6.** $3^2 = 9$
7. $4^{1/2} = 2$ **8.** $4^1 = 4$ **9.** 25   **10.** 27   **11.** $\frac{3}{4}$ **12.** $\frac{3}{2}$ **13.** 10   **14.** 10
15.   **16.**

17. 2   **18.** 4   **19.** $\frac{2}{3}$ **20.** $\frac{4}{3}$ **21.** 0   **22.** 0   **23.** $\log_2 5 + \log_2 x$ **24.** $\log_3 4 + \log_3 x$
25. $\log_{10} 2 + \log_{10} x - \log_{10} y$ **26.** $2 \log_3 x + 4 \log_3 y$ **27.** $\frac{1}{2} \log_a x + 3 \log_a y - \log_a z$

28. $2 \log_{10} x - 3 \log_{10} y - 4 \log_{10} z$ **29.** $\log_2 xy$ **30.** $\log_2 5x$ **31.** $\log_3 \dfrac{x}{4}$ **32.** $\log_{10} x^2 y^3$ **33.** $\log_a \frac{25}{3}$

34. $\log_2 \dfrac{x^3 y^2}{z^4}$ **35.** 2   **36.** 6   **37.** Possible solutions −1 and 3; only 3 checks; 3

38. Possible solutions −2 and 4; only 4 checks; 4   **39.** 3   **40.** Possible solution $-\frac{9}{2}$, which does not check; ∅
41. Possible solutions −2 and 3; only 3 checks; 3   **42.** Possible solutions −1 and 4; only 4 checks; 4

43. $x^2 + 10x + 25$ **45.** $x^3 + 3x^2y + 3xy^2 + y^3$ **47.** $x^4 + 4x^3y + 6x^2y^2 + 4xy^3 + y^4$

49. 52,428,800   **51.** 199.99924   **53.** $S = 2a$ **55.** $S = \dfrac{a}{b-a}$

Problem Set 11.5

1. $x^4 + 8x^3 + 24x^2 + 32x + 16$ **3.** $x^6 + 6x^5y + 15x^4y^2 + 20x^3y^3 + 15x^2y^4 + 6xy^5 + y^6$
5. $32x^5 + 80x^4 + 80x^3 + 40x^2 + 10x + 1$ **7.** $x^5 - 10x^4y + 40x^3y^2 - 80x^2y^3 + 80xy^4 - 32y^5$
9. $81x^4 - 216x^3 + 216x^2 - 96x + 16$ **11.** $64x^3 - 144x^2y + 108xy^2 - 27y^3$

13. $x^8 + 8x^6 + 24x^4 + 32x^2 + 16$ **15.** $x^6 + 3x^4y^2 + 3x^2y^4 + y^6$ **17.** $\dfrac{x^3}{8} - 3x^2 + 24x - 64$

19. $\dfrac{x^4}{81} + \dfrac{2x^3y}{27} + \dfrac{x^2y^2}{6} + \dfrac{xy^3}{6} + \dfrac{y^4}{16}$ **21.** $x^9 + 18x^8 + 144x^7 + 672x^6$

23. $x^{10} - 10x^9y + 45x^8y^2 - 120x^7y^3$ **25.** $x^{10} + 20x^9y + 180x^8y^2 + 960x^7y^3$ **27.** $x^{15} + 15x^{14} + 105x^{13}$
29. $x^{12} - 12x^{11}y + 66x^{10}y^2$ **31.** $x^{20} + 40x^{19} + 760x^{18}$ **33.** $x^{100} + 200x^{99}$ **35.** $x^{50} + 50x^{49}y$

37. $a_9 = \dfrac{12!}{8!4!}(2x)^4(3y)^8 = 495(16x^4)(6,561y^8) = 51,963,120x^4y^8$

39. $a_5 = \dfrac{10!}{4!6!}x^6(-2)^4 = 210x^6(16) = 3,360x^6$ **41.** $a_4 = \dfrac{9!}{3!6!}x^6(3)^3 = 84x^6(27) = 2,268x^6$

43. $a_{12} = \dfrac{20!}{11!9!}(2x)^9(5y)^{11}$ **45.** $\frac{21}{128}$ **47.** $x = \dfrac{\log 7}{\log 5} \approx 1.21$ **49.** $\frac{1}{6}$ or 0.17   **51.** Approximately 7 years

53. 2.16   **55.** 6.36   **57.** $t = \dfrac{1}{5} \ln \dfrac{A}{10}$ **59.** 56   **61.** 125,970

Chapter 11 Review

1. 7, 9, 11, 13   **2.** 1, 4, 7, 10   **3.** 0, 3, 8, 15   **4.** $\frac{4}{3}, \frac{5}{4}, \frac{6}{5}, \frac{7}{6}$ **5.** 4, 16, 64, 256   **6.** $\frac{1}{4}, \frac{1}{16}, \frac{1}{64}, \frac{1}{256}$

7. $3n - 1$ **8.** $2n - 5$ **9.** n^4 **10.** $n^2 + 1$ **11.** 2^{-n} **12.** $\dfrac{n+1}{n^2}$ **13.** 32   **14.** 25

15. $\frac{14}{5}$ **16.** -5 **17.** 98   **18.** $\frac{127}{30}$ **19.** $\displaystyle\sum_{i=1}^{4} 3i$ **20.** $\displaystyle\sum_{i=1}^{4}(4i-1)$ **21.** $\displaystyle\sum_{i=1}^{5}(2i+3)$ **22.** $\displaystyle\sum_{i=2}^{4} i^2$

23. $\displaystyle\sum_{i=1}^{4}\dfrac{1}{i+2}$ **24.** $\displaystyle\sum_{i=1}^{5}\dfrac{i}{3i}$ **25.** $\displaystyle\sum_{i=1}^{3}(x-2i)$ **26.** $\displaystyle\sum_{i=1}^{4}\dfrac{x}{x+i}$ **27.** Geometric   **28.** Arithmetic

29. Arithmetic   **30.** Neither   **31.** Geometric   **32.** Geometric   **33.** Arithmetic   **34.** Neither
35. $a_n = 3n - 1: 59$ **36.** $a_n = 8 - 3n: -40$ **37.** 34: 160   **38.** 78: 648   **39.** $a_1 = 5, d = 4, a_{10} = 41$
40. $a_1 = 8, d = 3, a_9 = 32, S_9 = 180$ **41.** $a_1 = -4, d = -2, a_{20} = -42, S_{20} = -460$ **42.** 20,100
43. -95 **44.** $a_n = 3(2)^{n-1}, a_{20} = 3(2)^{19}$ **45.** $a_n = 5(-2)^{n-1}, a_{16} = 5(-2)^{15}$
46. $a_n = 4(\frac{1}{2})^{n-1}, a_{10} = 4(\frac{1}{2})^9 = \frac{1}{128}$ **47.** -3 **48.** 8   **49.** $a_1 = 3, r = 2, a_6 = 96$ **50.** $243\sqrt{3}$
51. 28   **52.** 35   **53.** 20   **54.** 36   **55.** 45   **56.** 161,700   **57.** $x^4 - 8x^3 + 24x^2 - 32x + 16$
58. $16x^4 + 96x^3 + 216x^2 + 216x + 81$ **59.** $27x^3 + 54x^2y + 36xy^2 + 8y^3$
60. $x^{10} - 10x^8 + 40x^6 - 80x^4 + 80x^2 - 32$ **61.** $\frac{1}{16}x^4 + \frac{3}{2}x^3 + \frac{27}{2}x^2 + 54x + 81$
62. $\frac{1}{27}x^3 - \frac{1}{6}x^2y + \frac{1}{4}xy^2 - \frac{1}{8}y^3$ **63.** $x^{10} + 30x^9y + 405x^8y^2$ **64.** $x^9 - 27x^8y + 324x^7y^2$
65. $x^{11} + 11x^{10}y + 55x^9y^2$ **66.** $x^{12} - 24x^{11}y + 264x^{10}y^2$ **67.** $x^{16} - 32x^{15}y$ **68.** $x^{32} + 64x^{31}y$

69. $x^{50} - 50x^{49}$ **70.** $x^{150} + 150x^{149}y$ **71.** $\dfrac{10!}{5!5!} x^5(-3)^5 = 252x^5(-243) = -61{,}236x^5$

72. $\dfrac{9!}{3!6!} (2x)^6 = 84(64x^6) = 5{,}376x^6$

Chapter 11 Test

1. $-2, 1, 4, 7, 10$ **2.** $3, 7, 11, 15, 19$ **3.** $2, 5, 10, 17, 26$ **4.** $2, 16, 54, 128, 250$ **5.** $2, \frac{3}{4}, \frac{4}{9}, \frac{5}{16}, \frac{6}{25}$
6. $4, -8, 16, -32, 64$ **7.** $a_n = 4n + 2$ **8.** $a_n = 2^{n-1}$ **9.** $a_n = (\frac{1}{2})^n = 1/2^n$ **10.** $a_n = (-3)^n$
11. (a) 90 (b) 53 (c) 130   **12.** 3   **13.** ± 6 **14.** 320   **15.** 25   **16.** $S_{50} = 3(2^{50} - 1)$
17. $\frac{3}{4}$ **18.** $x^4 - 12x^3 + 54x^2 - 108x + 81$ **19.** $32x^5 - 80x^4 + 80x^3 - 40x^2 + 10x - 1$

20. $x^{20} - 20x^{19} + 190x^{18}$ **21.** $\dfrac{8!}{5!3!} (2x)^3(-3y)^5 = 56(8x^3)(-243y^5) = -108{,}864x^3y^5$

Appendix

1. $x - 7 + \dfrac{20}{x + 2}$ **3.** $3x - 1$ **5.** $x^2 + 4x + 11 + \dfrac{26}{x - 2}$ **7.** $3x^2 + 8x + 26 + \dfrac{83}{x - 3}$

9. $2x^2 + 2x + 3$ **11.** $x^3 - 4x^2 + 18x - 72 + \dfrac{289}{x + 4}$ **13.** $x^4 + x^2 - x - 3 - \dfrac{5}{x - 2}$ **15.** $x + 2 + \dfrac{3}{x - 1}$

17. $x^3 - x^2 + x - 1$ **19.** $x^2 + x + 1$